SOLAR AND STELLAR VARIABILITY:
IMPACT ON EARTH AND PLANETS

IAU SYMPOSIUM No. 264

COVER IMAGE:

Illustration of the impact of the solar wind on Earth's magnetosphere (source: SOHO/NASA), and the dome of a new heliometer developed at the Observatório Nacional at Rio de Janeiro to perform accurate measurements of the solar diameter (see the paper by V. A. d'Ávila et al. in these Proceedings).

INTERNATIONAL ASTRONOMICAL UNION
UNION ASTRONOMIQUE INTERNATIONALE

International Astronomical Union

SOLAR AND STELLAR VARIABILITY: IMPACT ON EARTH AND PLANETS

PROCEEDINGS OF THE 264th SYMPOSIUM OF THE INTERNATIONAL ASTRONOMICAL UNION HELD IN RIO DE JANEIRO, BRASIL AUGUST 3–7, 2009

Edited by

ALEXANDER G. KOSOVICHEV
Stanford University, Stanford, CA, USA

ALEXANDRE H. ANDREI
Observatório Nacional, Rio de Janeiro, Brazil

and

JEAN-PIERRE ROZELOT
OCA-Fizeau Dept, CNRS UMR 6525, & Université de Nice-Sophia-Antipolis, Grasse, France

CAMBRIDGE
UNIVERSITY PRESS

CAMBRIDGE UNIVERSITY PRESS
The Edinburgh Building, Cambridge CB2 8RU, United Kingdom
32 Avenue of the Americas, New York, NY 10013-2473, USA
477 Williamstown Road, Port Melbourne, VIC 3207, Australia
Ruiz de Alarcón 13, 28014 Madrid, Spain
Dock house, The Waterfront, Cape Town 8001, South Africa

First published 2010

Printed in the United Kingdom at the University Press, Cambridge

Typeset in System LaTeX 2ε

A catalogue record for this book is available from the British Library

Library of Congress Cataloguing in Publication data

ISBN 9780521764926 hardback
ISSN 1743–9213

Table of Contents

IV. MAGNETIC ACTIVITY AND DYNAMO MECHANISMS

Chair: V. Martinez-Pilet

V. PHYSICAL MECHANISMS OF SOLAR AND STELLAR VARIABILITY

Chair: H. Shibahashi

X. SUMMARY, OPEN DISCUSSION, AND CLOSING REMARKS

Chair: J.-P. Rozelot

Preface

The Sun is a variable star. Understanding the physical mechanisms of the variability and its effects on the Earth and planets is one of the central and long-standing problems of astronomy and astrophysics. Variability of a similar type has been observed in other stars. It is essential to investigate the similarities and differences between solar and stellar variability in order to understand the physical mechanisms and impacts on planets.

During the past decade multi-wavelength observations from several solar and stellar space missions and ground-based observatories have provided a tremendous amount of new information about the physical processes associated with magnetic activity and variability on the Sun and solar-type stars. Key observational results, new theoretical ideas, and models discussed at Symposium 264 are presented in this volume.

Solar observations have provided new data about the solar cycle variations of the structure and dynamics from the interior to the corona and heliosphere. Intriguing connections exist between the interior dynamics, surface magnetism and coronal phenomena. However, despite the great amount of new data, it is still unclear where and how magnetic fields are generated in the Sun, and why it has a regular 22-year magnetic cycle. Can observations of other stars help us to understand the solar magnetic cycle, which is so important not only for astrophysics but also for our life and society? What are the essential components of the solar variability?

Recent optical, UV and X-ray observations of other stars with significant subsurface convection zones (lower main sequence stars with masses smaller than 1.5 solar masses and cool post-main-sequence stars) reveal starspots and other surface and atmospheric structures that are similar to solar magnetic features. In many cases, the long-term evolution of these features is similar to the solar cycle. Advances in the Doppler imaging technique have enabled mapping of starspot distributions and the tracking of their evolution. This led to the first accurate measurements of stellar differential rotation, a key property of dynamo mechanisms. In addition, UV and X-ray observations detected stellar coronas similar to the solar corona.

The impacts of solar variability on Earth, particularly on the terrestrial atmosphere and climate, have fueled significant debates in recent years. There is no doubt that solar radiation is the major source of terrestrial energy. The solar activity cycle also causes changes in Earth's atmospheric temperature and density, especially in the thermosphere. Energetic particles, which also vary with the solar cycle, cause geomagnetic storms and deposit energy into the magnetosphere and the polar regions of the atmosphere. Variations of the energy input may trigger changes in the climate. However, the complete chain of the impact of the solar variability is not understood.

The unusually long and deep current minimum of solar activity has significantly increased the interest in the mechanisms of solar and stellar variability and the impact of such long periods of low activity on planets. It is well known that a long interval of low solar activity between 1645 and 1715, the Maunder Minimum, coincided with a significant climate change called the "Little Ice Age". Can this happen again? Observations of other Sun-like stars indicate that long periods of low activity are quite common. Various attempts to predict the next solar cycle indicate that it will be significantly weaker than the previous cycle. Some researchers even go so far as to suggest that the Sun is entering a prolonged period of low activity, similar to the Maunder Minimum.

While the first few magnetic sunspot regions of the new solar cycle have appeared on the surface, the solar minimum still continues. The Sun is still mostly spotless. In 2008, 265 days were without sunspots. In the past 160 years of regular solar observations, only 1878, 1901 and 1913 have had more such days. The long solar minimum has had a significant impact on the Earth's space environment and atmosphere. The solar wind pressure is at a 50-year low. This resulted in an increase of the flux of cosmic rays, which are believed to play a significant role in cloud formation. In addition, the total solar irradiance dropped to a 30-year low, the lowest since monitoring from space began. The decrease of UV radiation is particularly significant, about 6%. This caused a reduction of the Earth's atmosphere heating, and decreased its height. Thus, perhaps unexpectedly, investigations of the Sun in a low activity state have become critically important.

Of course, these studies are not only of great practical importance, but are also directly related to some of the most important astronomical problems: what are the basic physical mechanisms of solar and stellar activity, and how does stellar variability affects planetary systems and their habitability?

Perhaps, the most significant highlight of the Symposium, which received coverage in science news, was the presentation of new results showing that newly formed stars spin faster and generate stronger magnetic fields than middle-aged stars like the Sun. These active young stars emit much more intense levels of X rays, ultraviolet light, and charged particles all of which affect the formation and evolution of planetary atmospheres and must have a dramatic effect on the development of emerging life forms.

This highly interdisciplinary Symposium provided an excellent opportunity for astronomers from various fields solar, stellar and planetary physics to establish important connections, discuss mutual ideas and develop partnerships. In particular, this was a unique opportunity to establish collaborations with astronomers from Brazil and other Latin American countries.

The Symposium coincided with the completion of a new heliometer to perform accurate measurements of the solar diameter at the Observatório Nacional in Rio de Janeiro. The LOC organized an excursion to the completion ceremony at the Observatório Nacional. This was one of the most interesting and memorable events of the Symposium.

Alexander G. Kosovichev, Alexandre H. Andrei and Jean-Pierre Rozelot, co-chairs SOC
Rio de Janeiro, August 3, 2009

Participants of IAUS 264 at the completion ceremony of a new heliometer at the Observatório Nacional in Rio de Janeiro. Professor Paolo Benevides-Soares (right) and Dr. Christian Delmas (left) receive commemorative medals from Dr. Jucira Penna, Chair of the LOC.

The heliometer of the Observatório Nacional and its creators,
Drs Victor d'Ávila (right) and Eugenio Reis (left).

The LOC members (from left to right): Drs. Eugenio Reis, Jucira Penna and Victor d'Ávila

A group of participants after a discussion (left to right): Victor Matias (sitting), Ricardo Dunna (standing), Carlos Eduardo Portela, Eugenio Reis, Todd Hoeksema, Maria Cristina Rabello-Soares, Jucira Penna, Sergio Lomonaco and Victor d'Ávila.

THE ORGANIZING COMMITTEE

Scientific

Alexandre H. Andrei (co-chair, Brasil)
Jean-Pierre Rozelot (co-chair, France)
Maria Pia Di Mauro (Italy)
Eduardo Janot Pacheco (Brasil)
Cristina H. Mandrini (Argentina)
Mudumba Parthasarathy (India)
Franck Selsis (France)
Jingxiu Wang (China)
Alexander Kosovichev (co-chair, USA)
Annie Baglin (France)
Julio A. Fernandez (Uruguay)
John D. Landstreet (Canada)
Hiroko Miyahara (Japan)
Ignasi Ribas (Spain)
Jill C. Tarter (USA)
Lev M. Zeleny (Russia)

Local

Jucira Lousada Penna (Chair)
Victor Amorim d'Ávila
Alexandre Humberto Andrei
Ramachrisna Teixeira
Gustavo Porto de Mello

Acknowledgements

The symposium is sponsored and supported by the IAU Divisions II (Sun & Heliosphere), III (Planetary Systems Sciences) and IV (Stars).

The Local Organizing Committee operated under the auspices of the
Observatório Nacional of Brazil

Funding by the
International Astronomical Union,
and
Observatório Nacional, Brazil
is gratefully acknowledged.

Participants

Hebe Cremades, UTN-FRM/CONICET, Argentina	hebe.cremades@hotmail.com
Eduardo Fernandez Lajus, Fac. Ciencias Astronómicas y Geofísicas - UNLP, Argentina	eflajus@fcaglp.unlp.edu.ar
Luciano Garcia, Observatorio Astronomico de Cordoba, Argentina	lucianog@oac.uncor.edu
Ana G. Elias, UNT-CONICET, Argentina	aelias@herrera.unt.edu.ar
Anahi Granada, Universidad nacional de La Plata, Argentina	granada@fcaglp.unlp.edu.ar
Cristina H. Mandrini, Instituto de Astronomia y Fisica del Espacio, Argentina	mandrini@iafe.uba.ar
Olga Pintado, INSUGEO-CONICET, Argentina	olga.pintado@gmail.com
Alejandra Daniela Romero, Instituto de astrofisica La Plata, Argentina	aromero@fcaglp.unlp.edu.ar
Emanuel Sainz, Observatorio Astronomico de Cordoba, Argentina	sainz@mail.oac.uncor.edu
Pablo Daniel Sisterna, Universidad Nacional de Mar del Plata, Argentina	sisterna@mdp.edu.ar
Michael Ashley, University of New South Wales, Australia	m.ashley@unsw.edu.au
Paul Cally, Monash University, Australia	paul.cally@sci.monash.edu.au
Allan Ernest, Charles Sturt University, Australia	aernest@csu.edu.au
David Jauncey, Australia Telescope National Facility, CSIRO, Australia	david.jauncey@csiro.au
Stephen Marsden, Anglo-Australian Observatory, Australia	scm@aao.gov.au
Tiago M. D. Pereira, Australian National University, Australia	tiago@mso.anu.edu.au
David Warren, Astronomy Australia Ltd., Australia	david.warren@astronomyaustralia.
Günter Houdek, University of Vienna, Austria	guenter.houdek@univie.ac.at
Werner Weiss, University Vienna, Austria	weiss@astro.univie.ac.at
Shaig Nabiyev, Shamakhy Astrophysical Observatory, Azerbaijan	snebiyev@qafqaz.edu.az
Emile Biemont, University of Liège, Belgium	e.biemont@ulg.ac.be
Anne Lemaitre, University of namur (FUNDP), Belgium	anne.lemaitre@fundp.ac.be
Alex Lobel, Royal Observatory of Belgium, Belgium	alobel@sdf.lonestar.org
Marcos Antonio Albarracian Manrique, Universidade Federal do ABC, Brazil	marcos.manrique@ufabc.edu.br
Silvia Alencar, UFMG, Brazil	silvia@fisica.ufmg.br
Victor Alves Alencar, Universidade Federal do Ceara, Brazil	victoralencar@fisica.ufc.br
Leonardo Almeida, Instituto Nacional de pesquisas Espaciais, Brazil	leonardo@das.inpe.br
Joao Antonio Amarante, UFRJ, Brazil	amarante@astro.ufrj.br
Diana Andrade-Pilling, Pontificia Universidade Católica do Rio de Janeiro, Brazil	dianaufrj@gmail.com
Alexandre Humberto Andrei, Observatório Nacional/MCT, Brazil	oat1@on.br
Silvia Calbo Aroca, Instituto de Fisica de São Carlos, Brazil	silviaaroca@uol.com.br
Victor Avila, Observatório Nacional, Brazil	victor@on.br
Julio Daniel Blanco Zarate, Observatório Nacional, Brazil	jblanco@on.br
Heloisa Maria Boechat-Roberty, Universidade Federal do Rio de Janeiro, Brazil	heloisa@astro.ufrj.br
Bernardo Borges, IAG/USP, Brazil	bernardo@astro.iag.usp.br
Sergio Boscardin, Observatório Nacional, Brazil	sergio.boscardin@on.br
Albert Bruch, Laboratorio Nacional de Astrofisica, Brazil	albert@lna.br
Leonardo Wencioneck Cacholi, Universidade Federal de Viçosa, Brazil	klamanter@hotmail.com
Denise Castro, Observatório Nacional, Brazil	denise@on.br
Joao Paulo Nogueira Cavalcante, Observatório Nacional, Brazil	jpncavalcante@yahoo.com.br
Abraham Chian, National Institute for Space Research, Brazil	achian@dge.inpe.br
Gabriela Conde Saavedra, Observatório do Valongo / UFRJ, Brazil	gconde@astro.ufrj.br
Wagner J B Corradi, Universidade Federal de Minas Gerais, Brazil	wag@fisica.ufmg.br
Jonathan Cristiano Costa, UEPG, Brazil	jonathanuepg@yahoo.com.br
Flavio D'Amico, INPE, Brazil	damico@das.inpe.br
Elisabete de Gouveia Dal Pino, IAG Universidade de São Paulo, Brazil	dalpino@astro.iag.usp.br
Gabriel Dalmaso, Universidade Federal do Rio de Janeiro, Brazil	gdalmaso@gmail.com
Marcos Diaz, IAG, Brazil	marcos@astro.iag.usp.br
Isadora Chaves Bicalho Domingos, UNB, Brazil	isadora.bicalho@gmail.com
Alessandro Saldanha Chantre Dutra, Observatório Nacional, Brazil	ascdutra@on.br
Marcelo Emilio, Universidade Federal de Ponta Grossa, Brazil	marcelo_emilio@yahoo.com
Francisco Fernandes, UNIVAP, Brazil	guga@univap.br
Marcelo Leal Ferreira, UFRJ-Observatorio de Valongo, Brazil	mllferreira@gmail.com
Leticia Dutra Ferreira, Observatorio de Valongo-UFRJ, Brazil	leticia@astro.ufrj.br
Vanessa Ferreira, Universidade Federal de Itajuba, Brazil	vanessafisi@gmail.com
Ana Monica Ferreira-Rodrigues, Catholic University of Rio de Janeiro, Brazil	anamonica.rodrigues@gmail.com
Douglas Galante, Universidade de São Paulo, Brazil	douglas@astro.iag.usp.br
Luiz Paulo Carneiro Gama, Observatório Nacional, Brazil	luizpaulo@gmail.com
Luiz Garcia de Andrade, UERJ, Brazil	-
Luan Ghezzi, Observatório Nacional, Brazil	luan@on.br
Altair Ramos Gomes Junior,, Observatorio de Valongo, Brazil	altairastronomia@hotmail.com
Jorge Gonzales, Observatório Nacional, Brazil	gonzales@on.br
Guilherme Grams, Southern Regional Space Research Center, Brazil	ggrams@lacesm.ufsm.br
Marcia Regina Guimarães Guedes, UNIFEI, Brazil	mrgguedes@gmail.com
Marcelo Guimaraes, UFMG, Brazil	mmg@fisica.ufmg.br
Pedro Henrique A Hasselmann, Observatório Nacional, Brazil	pedrohasselmann@gmail.com
Luiz Jafelice, UFRN, Brazil	jafelice@gmail.com
Eduardo Janot-Pacheco, Instituto de Astronomia, Geofisica e Ciencias Atmosfericas, Brazil	janot@astro.iag.usp.br
Flavia Luzia Jasmim, Observatório Nacional, Brazil	flavialuzia@on.br
Vera Jatenco-Pereira, USP, Brazil	jatenco@astro.iag.usp.br
Tiago Jota, UFMG, Brazil	jotaatiago@ufmg.br
Natalia Landin, Federal University of Minas Gerais, Brazil	nlandin@fisica.ufmg.br
Nathalia Lia, UFRJ, Brazil	nlia@astro.ufrj.br
Dennis Lima, Observatorio Astronomico Christus, Brazil	dwastronomia@yahoo.com.br
Dalton Lopes, Observatório Nacional, Brazil	dalton@on.br
Victor de Souza Magalhaes, Universidade de Brasilia, Brazil	victor.magalhaes.2802@gmail.com
Manuel Malheiro, InstitutoTecnologico de Aeronautica, Brazil	malheiro@ita.br
Rubens de Melo Marinho Jr., Instituto Tecnologico de Aeronautica, Brazil	marinho@ita.br
Eder Martioli, Instituto Nacional de Pesquisas Espaciais, Brazil	edermartioli@gmail.com
Thiago Matheus, Inst. de Astronomia,Geofísica e Ciências Atmosféricas da USP, Brazil	thiago@astro.iag.usp.br
Pauline McGinnis, Universidade Federal de Minas Gerais, Brazil	pauline@fisica.ufmg.br
Gustavo Frederico Porto de Mello, Universidade Federal do Rio de Janeiro, Brazil	gustavo@astro.ufrj.br
Vinicius Melo, UFRJ, Brazil	viniciusvbm@hotmail.com
Carlo Miceli, IFGW, Unicamp, Brazil	carlomiceli@gmail.com
Manoel Moraes, IAG - USP, Brazil	mcvmjr@astro.iag.usp.br
Decio Mourao, UNESP, Brazil	dcm134@uol.com.br

Raquel Santiago Nascimento, Universidade federal de Santa Cruz, Brazil	quelg@yahoo.com.br
Natalia Amarinho Nunes, Universidade Federal de Itajubá - UNIFEI, Brazil	natalia_astro@unifei.edu.br
Diego Lorenzo de Oliveira, UFRJ, Brazil	diegolorenzo83@gmail.com
Alex Dias Oliveira, Universidade Federal do Rio de Janeiro, Brazil	aoliveira@astro.ufrj.br
Alessandra Pacini, Instituto Nacional de Pesquisas Espaciais, Brazil	pacini@dge.inpe.br
Andres Reinaldo Rodriguez Papa, Observatório Nacional, Brazil	papa@on.br
Jucira L. Penna, Observatório Nacional, Brazil	jucira@on.br
Eduardo Penteado, Universidade Federal do Rio de Janeiro, Brazil	monfpent@astro.ufrj.br
Fabio Pereira Santos, ICEx/UFMG, Brazil	fbpsantos@gmail.com
Andres Felipe Perez Sanchez, Observatorio do Valongo - UFRJ, Brazil	aperez@astro.ufrj.br
Tereza Satiko Nishida Pinto, INPE, Brazil	tereza.satiko@gmail.com
Sarah Carvalho Pinto, Observatório Nacional, Brazil	sarah@on.br
Raul Eduardo Puebla Puebla, IAG/Universidade de São Paulo, Brazil	raul@astro.iag.usp.br
Jean-Pierre Raulin, Universidade Presbiteriana Mackenzie, Brazil	raulin@craam.mackenzie.br
Mauricio Redaelli, UFRGS, Brazil	maukeyboard@gmail.com
Eugenio Reis Neto, Observatório Nacional, Brazil	eugenio@on.br
Leandro Rimulo, Universidade Federal de Minas Gerais, Brazil	lrrimulo@fisica.ufmg.br
Artur Justiniano Roberto Junior, Univ. Federal de Alfenas, Brazil	arturjustiniano@yahoo.com.br
Helio Jaques Rocha-Pinto, Observatório de Valongo, Brazil	helio@astro.ufrj.br
Luiz Teixeira Rodrigues, OAB/RJ, Brazil	luizadvogado@hotmail.com
Thaise da Silva Rodrigues, UFRJ, Brazil	tsrodrigues@astro.ufrj.br
Gustavo Rojas, Universidade Federal de São Carlos, Brazil	grojas@ufscar.br
Fabiano Rollo, Observatório Nacional, Brazil	fabiano.gr@hotmail.com
Reinaldo Roberto Rosa, INPE, Brazil	reinaldo@lac.inpe.br
Marcelo Rubinho, IAG, Brazil	esteemeuemail@gmail.com
Maria de Fátima Oliveira Saraiva, Univ. Federal do Rio Grande do Sul, Brazil	fatima@if.ufrgs.br
Hanumant Sawant, INPE, Brazil	sawant@das.inpe.br
Nelson Jorge Schuch, INPE, Brazil	njschuch@lacesm.ufsm.br
Caius Lucius Selhorst, CRAAM - Univ. Presbiteriana Mackenzie, Brazil	caiuslucius@gmail.com
Felipe Serro, -, Brazil	ffserro@hotmail.com
Joao Victor Silva, Observatório Nacional, Brazil	joaovictor@on.br
Loloano Silva, UFRJ, Brazil	loloano@if.ufrj.br
Wallace Silva, Universidade Federal do Rio de Janeiro, Brazil	wllc.silva@yahoo.com.br
Telma C. Couto da Silva, Universidade federal de mato Grosso, Brazil	tccs@ufmt.br
Alana Paixão Sousa, UFMG, Brazil	alanasousa@ig.com.br
Elton Rodrigues de Souza, Observatorio do Valongo, Brazil	elton@astro.ufrj.br
Tardelli Ronan Coelho Stekel, Universidade Federal de Santa maria, Brazil	tardelli@lacesm.ufsm.br
Carlos Alberto Torres, Laboratório Nacional de Astrofísica/MCT, Brazil	beto@lna.br
Flavia dos Prazeres Trindade, UFRJ, Brazil	flaviafp07@astro.ufrj.br
Adriana Valio, Univ. Presbiteriana Mackenzi, Brazil	adrivalio@gmail.com
Luiz Paulo Ribeiro Vaz, Universidade Federal de Minas Gerais, Brazil	lpv@fisica.ufmg.br
Aline Vidotto, University of São Paulo, Brazil	aline@astro.iag.usp.br
Sergio Murilo Rezende Vieira, UFTJ, Brazil	murilo_rezende9@yahoo.com.br
Lucas Vieira, Universidade Federal de Santa Maria, Brazil	lucasfisio@gmail.com
Jaime Fernando Villas da Rocha, UNIRIO, Brazil	jfvroch@pq.cnpq.br
Silvio de Andrade, Universidade Federal do Rio de Janeiro, Brazil	silviodeandrade@gmail.com
Juarez Barbosa de Carvalho, UNIFEI, Brazil	juarezbcarvalho@gmail.com
Marcelo Vargas dos Santos, Universidade Federal do Rio de Janeiro, Brazil	vargas@if.ufrj.br
Yavor Chapanov, Bulgarian Academy of Sciences, Bulgary	astro@bas.bg
Matthew Browning, Canadian Inst. for Theoretical Astrophysics, Canada	browning@cita.utoronto.ca
Eugene F. Milone, University of Calgary, Canada	milone@ucalgary.ca
Richard Querel, University of Lethbridge, Canada	richard.querel@uleth.ca
Slavek Rucinski, University of Toronto, Canada	rucinski@astro.utoronto.ca
James Silvester, RMC & Queen's University, Canada	james.silvester@rmc.ca
Gregg Wade, Royal Military College, Canada	wade-g@rmc.ca
Taner Akgun, Pontificia Universidad Católica de Chile, Chile	akgun@astro.cornell.edu
Melina Bersten, Universidad de Chile, Chile	melina@das.uchile.cl
Claudio Caceres, European Southern Observatory, Chile	ccaceres@eso.org
Carlos Contreras, Pontificia Universidad Catolica de Chile, Chile	ccontrer@astro.puc.cl
Alexandre Gallenne, ESO, Chile	agallenn@eso.org
Karen Kinemuchi, Universidad de Concepcion / Univ. of Florida, Chile	kkinemuchi@astro-udec.cl
Cristian Lopez, Universidad de Chile, Chile	clopez@das.uchile.cl
Gautier Mathys, European Southern Observatory, Chile	gmathys@eso.org
Maria Jose Maureira, Universidad de Chile, Chile	mmaureir@das.uchile.cl
David Orellana, Cerro Tololo Inter-American Observatory, Chile	dorellana@ctio.noao.edu
Lara Rodrigues, Depto Astronomía/Astrofisica. PUC, Chile, Chile	lara@astro.puc.cl
Klaus Simon Rubke Zuniga, Universidad de Chile, Chile	krubke@hotmail.com
Maria Teresa Ruiz, Universidad de Chile, Chile	mtruiz@das.uchile.cl
Matthias R Schreiber, Universidad de Valparaiso, Chile	matthias@dfa.uv.cl
Simon Silva, Universidad de Chile, Chile	ssilva@das.uchile.cl
Eer Afrad, Yunnan Astronomical Observatory, China	xjq_wuhan@yahoo.com.br
Kwing L Chan, Hong Kong U. of Science and Technology, China	maklchan@ust.hk
Cheng Fang, Nanjing University, China	chengfang38@yahoo.com.cn
Weiqun Gan, Purple Mountain Observatory, China	wqgan@pmo.ac.cn
Cheng-Li Huang, Shangai Astronomical Observatory, China	clhuang@shao.ac.cn
Guangli Huang, Purple Mountain Observatory, China	glhuang@pmo.ac.cn
Cheng Huang, Shangai Astronomical Observatory, China	hc@shao.ac.cn
Lifang Li, National Astronomy Observatories, Yunnan Observatory, China	llf@ynao.ac.cn
Ganghua Lin, National Astronomical Observatoires, China	lgh@bao.ac.cn
Yu Liu, Yunnan Astronomical Observatory, China	lyu@ynao.ac.cn
Xiaoming Lu, Beijing Academy of Science and Technology, China	angelinxiao@gmail.com
Yuehua Ma, Purple Mountain Observatory, China	yhma@pmo.ac.cn
Yuan Ma, Yunnan Astronomical Observatory, China	mayuan@ynao.ac.cn
Zongjun Ning, Purple Mountain Observatory, China	ningzongjun@pmo.ac.cn
Shengbang Qian, National Astron. Observatories Yunnan Observatory, China	qsb@ynao.ac.cn
Yuhua Tang, Nanjing University, China	yhtang@nju.edu.cn
Jingxiu Wang, National Astronomical Observatories, China	wangjx@bao.ac.cn
Yongqi Xia, Beijing Academy of Science and Technology, China	angellinxiao@hotmail.com
Yihua Yan, National Astronomical Observatories, China	yyh@bao.ac.cn
Shuhong Yang, National Astronomical Observatories, China	shuhongyang@ourstar.bao.ac.cn
Liheng Yang, Yunnan Observatory, China	yangliheng@ynao.ac.cn
Hong-Xin Yin, Shandong University at Weihai, China	yinhx@sdu.edu.cn
Xizhen Zhang, National Astronomical Observatories, CAS, China	zxz@bao.ac.cn

Jin Zhu, Beijing Planetarium, China jinzhu@bjp.org.cn
Liying Zhu, National Astronomical Observatories Yunnan Observatory, China zhuly@ynao.ac.cn
Erquinio Alberto Taborda Martinez, Secretaria de Educ. del Depart. del Atlantico, Colombia erquiniot_74@yahoo.com
Jorge Zuluaga, Universidad de Antioquia, Colombia zuluagajorge@gmail.com
Guy Kabongo Leba, Université Pedagogique Nationale, Congo geekale@gmail.com
Andrea Kotrlova, Silesian University in Opava, Czech Republic andrea.kotrlova@seznam.cz
Michal Krizek, Institute of Mathematics, Czech Republic krizek@cesent.cz
Cyril Ron, Astronomical Institute, Academy of Sciences, Czech Republic ron@ig.cas.cz
Marek Vandas, Astronomical Institute, Academy of Sciences, Czech Republic vandas@ig.cas.cz
Jan Vondrak, Astronomical Institute,Acd. Sci., Czech Republic vondrak@ig.cas.cz
Åke Nordlund, University of Copenhagen, Denmark aake@nbi.dk
Laurits Leedjarv, Tartu Observatory, Estonia leed@aai.ee
Jaan Pelt, Tartu Observatory, Estonia pelt@aai.ee
Marjaana Lindborg, University of Helsinki, Finland marjaana.lindgorg@helsinki.fi
Dagmara Anna Oszkiewicz, Observatory, University of Helsinki, Finland dagmara.oszkiewicz@helsinki.fi
Ilkka Tuominen, University of Helsinki, Finland ilkka.tuominen@helsinki.fi
France Allard, Centre de Recherche Astrophysique de Lyon, France fallard@ens-lyon.fr
Philippe Bendjoya, Laboratoire H. Fizeau UNS-OCA-CNRS, France bendjoya@unice.fr
Marcelo Borges Fernandes, Observatoire de la Cote d'Azur, France marcelo.borges@obs-azur.fr
Jean-Louis Bougeret, Observatoire de Paris-CNRS, France jean-louis.bougeret@obspm.fr
Carine Briand, Observatoire de Paris, France carine.briand@obspm.fr
Allan Sacha Brun, CEA, Saclay, France sacha.brun@cea.fr
Armando Domiciano, Univsersité de Nice, France armando.domiciano@unice.fr
Anne Dutrey, Laboratoire d'Astrophysique de Bordeaux, France anne.dutrey@obs.u-bordeaux1.fr
Rubens Freire Ferrero, Observatoire Astronomique de Strasbourg, France freire@astro.u-strasbg.fr
Rafael A Garcia, Service d'Astrophysique, CEA, France rgarcia@cea.fr
Dominique Gilles, CEA, France dominique.gilles@cea.fr
Guillaume Hebrard, CNRS, France hebrard@iap.fr
Sandrine Lefebvre, CNRS, France sandrine.lefebvre@latmos.ipsl.fr
Stephane Mathis, CEA/DSM/IRFU, France stephane.mathis@cea.fr
Thierry Montmerle, Laboratoire d'Astrophysique de Grenoble, France montmerle@obs.ujf-grenoble.fr
Coralie Neiner, Observatoire de Paris-Meudon, France coralie.neiner@obspm.fr
Jean-Claude Pecker, College de France, France j.c.pecker@wanadoo.fr
Bernard Rouge, CESBIO, France rougebe@free.fr
Jean-Pierre Rozelot, Universite de Nice Sophia-Antipolis, France rozelot@obs-azur.fr
Alain Sarkissian, LATMOS, France alain.sarkissian@latmos.ipsl.fr
Sylvaine Turck-Chieze, CEA, France cturck@cea.fr
Jean-Claude Vial, Institut d'Astrophysique Spatiale, France jean-claude.vial@ias.u-psd.fr
Maria Bergemann, Max-Planck Institute for Astrophysics, Germany mbergema@mpa-garching.mpg.de
Gregory Herczeg, Max-Planck-Institut für extraterrestriche Physik, Germany gregoryh@mpe.mpg.de
Heidi Korhonen, ESO, Germany hkorhone@eso.org
Joachim Krautter, Landesssternwarte - Zent. für Astron. Univ. Heidelberg, Germany j.krautter@lsw.uni-heidelberg.de
Markus Mugrauer, AIU Jena, Germany markus@astro.uni-jena.de
Theodor Pribulla, Friedrich-Schiller - University Jena, Germany pribulla @astro.uni-jena.de
Andreas Quirrenbach, Landessternwarte, Germany a.quirrenbach@lsw.uni-heidelberg
Ulf Seemann, ESO, Germany useemann@eso.org
Michael Weber, Astrophysikalisches Institut Potsdam, Germany mweber@aip.de
Kristof Petrovay, Eotvos University, Hungary k.petrovay@astro.elte.hu
Sirajul Hasan, Indian Institute of Astrophysics, India hasan@iiap.res.in
Parag Mahajani, Milkyway Citizens, India milkywaycitizen@yahoo.co.in
Ehsan Moravvji, Institute for Advanced Studies in Basic Sciences, Iran moravveji@iasbs.ac.ir
Giuseppe Galletta, Universita de Padova, Italy giuseppe.galletta@unipd.it
Giuseppina Micela, INAF, Italy giusi@astropa.inaf.it
Giovanni Peres, Universita di Palermo, Italy peres@astropa.unipa.it
Salvatore Sciortino, INAF, Italy sciorti@astropa.inaf.it
Giuseppe Severino, INAF, Italy severino@na.astro.it
Alessandro Sozzetti, INAF, Italy sozzetti@oato.inaf.it
Beate Stelzer, INAF - Observatorio Astronomico di Palermo, Italy stelzer@astropa.unipa.it
Hitoshi Fujiwara, Tohoku University, Japan fujiwara@pat.geophys.tokohu.ac.j
Kanya Kusano, JAMSTEC, Japan kusano@jamstec.go.jp
Hiroko Miyahara, The University of Tokyo, Japan hmiya@icrr.u-tokyo.ac.jp
Yuko Motizuki, RIKEN Nishina Center, Japan motizuki@riken.jp
Hiromoto Shibahashi, University of Tokyo, Japan shibahashi@astron.s.u-tokyo.ac.j
Hitoshi Yamaoka, Kyushu University, Japan yamaoka@phys.kyushu.ac.jp
Damdin Batmunkh, Research Centre of Astronomy and Geophysics, Mongolia btmnh_d@yahoo.com
Johnson Urama, University of Nigeria, Nsukka, Níger johnson@hartrao.ac.za
Gyan Bahdur Shrestha, Tribhuvan University, Nepal gbstha@hotmail.com
Willem Brouw, Kapteyn Institute, Netherlands w.n.brouw@rug.nl
Matteo Cantiello, Institute for Astronomy Utrecht, Netherlands m.cantiello@uu.nl
Jean-Mathias Griessmeier, ASTRON, Netherlands griessmeier@astron.nl
Isa Oliveira, Leiden University, Netherlands oliveira@strw.leidenuniv.nl
Daniel Risquez, Leiden Observatory, Netherlands risquez@strw.leidenuniv.nl
Mats Carlsson, University of Oslo, Norway mats.carlsson@astro.uio.no
Oyvind Sorensen, The Norwegian Academy of Sciences and Letters, Norway oyvind.sorensen@dnva.no
Sven Wedemeyer-Bohm, University of Oslo, Norway svenwe@astro.uio.no
Jesus Antonio Dalmau Cam, Instituto Geofisico del Peru, Peru antoniodalmau@gmail.com
Walter Guevara Day, Universidad Nacional Mayor de San, Peru walter@conida.gob.pe
Jose Luis Ricra Mayorca, Universidad Nacional de Ingeniería, Peru jricram@uni.pe
Jadwiga Daszynska-Daszkiewicz, Uniwersytet Wroclawski, Poland daszynska@astro.uni.wroc.pl
Jerzey M. Kreiner, Pedagogical University, Poland sfkreine@cyf-kr.edu.pl
Bogumil Pilecki, Warsaw University Observatory, Poland pilecki@astrouw.edu.pl
Milena Ratajczak, Polish Academy of Sciences, Poland milena@ncac.torun.pl
Pawel Rudawy, Uniwersytet Wroclawski, Poland rudawy@astro.uni.wroc.pl
Piotr Sybilski, Polish Academy of Sceinces, Poland sybilski@ncac.torun.pl
Dorota Szczygiel, Warsaw University Observatory, Poland dszczyg@astrouw.edu.pl
Krzysztof Ulaczyk, Warsow University Observatory, Poland kulaczyk@astrouw.edu.pl
Edwin Wnuk, Adam Mickiewicz University, Poland wnuk@amu.edu.pl
Jorge Melendez, Centro de Astrofisica da Universidade de Porto, Portugal jorge@astro.up.pt
Joana Sousa, CAUP-Centro de Astrofisica da Universidade do Porto, Portugal jsousa@astro.up.pt
William Bruckman, University of Puerto Rico at Humacao, Puerto Rico miguelwillia.bruckman@upr.edu
Young-Woon Kang, Sejong University, Republic of Korea kangyw@sejong.ac.kr

Georgeta Maris, Institute of Geodynamics, Romania gmaris@geodin.ro
Petre Popescu, Astronomical Institute of Romanian Academy, Romania petre@aira.astro.ro
Gennadij Bisnovatyi-Kogan, Space Research Institute RAS, Russian Federation gkogan@iki.rssi.ru
Elena Gavryuseva, Institute for Nuclear Research, Russian Federation elena.gavryuseva@gmail.com
Elena Glushkova, Sternberg Astronomical Institute, Russian Federation elena@sai.msu.ru
Alexander Kholtygin, Astron. Inst of Saint-Petersburg State Univ., Russian Federation afkholtygin@gmail.com
Vladimir Obridko, Izmiran, Russian Federation obridko@izmiran.ru
Elena Pitjeva, Inst. Appl. Astron., Russian Academy of Sciences, Russian Federation evp@ipa.nw.ru
Maxim Pshirkov, PRAO ASC LPI, Russian Federation pshirkov@gmail.com
Tatiana Ryabchikova, Institute of Astronomy RAS, Russian Federation ryabchik@inasan.ru
Sergey Samsonov, Yu. G. Shafer Inst. Cosmophys. & Aeronomy, Russian Federation s_samsonov@ikfia.ysn.ru
Dmitry Sokoloff, Moscow State University, Russian Federation sokoloff@dds.srcc.msu.su
Vladislav Timofeev, Yu. G. Shafer Inst. Cosmophys. & Aeronomy, Russian Federation vetimofeev@ikfia.ysn.ru
Andrey Tlatov, The Central Astron. Observatory of RAS at Pulkovo, Russian Federation tlatov@mail.ru
Valeria Vasilyeva, The Central Astron. Observatory of RAS at Pulkovo, Russian Federation xyzlera@rambler.ru
Irina Voloshina, Stenberg Astronomical Institute, Russian Federation vib@sai.msu.ru
Jan Budaj, Astronomical Institute, Slovak Academy of Sciences, Slovakia budaj@ta3.sk
Peter Martinez, South African Astronomical Observatory, South Africa peter@saao.ac.za
Daniel Christopher Opolot, University of the Western Cape, South Africa opodanchris@yahoo.ca
Brian Warner, University of Cape Town, South Africa warner@physci.uct.ac.za
Juan Fabregat, Universidad de Valencia, Spain juan.fabregat@uv.es
Alvaro Gimenez, Centro de Astrobiologia (INTA-CSIC), Spain gimeneza@inta.es
Alexis Klutsch, UCM, Spain klutsch@astrax.fis.ucm.es
Valentin Martinez Pillet, Instituto de Astrofisica de Canarias, Spain vmp@iac.es
David Montes, UCM, Universidad Complutense de Madrid, Spain dmg@astrax.fis.ucm.es
Ignasi Ribas, Institut de Ciencies de l'Espai (CSIC-IEEC), Spain ribas@ice.csis.es
Jorge Sanz-Forcada, Center of Astrobiology-INTA/CSAI, Spain jsanz@laeff.inta.es
Pilar de Teodoro, ESAC, Spain pilar.teodoro@esa.int
Dainis Dravins, Lund Observatory, Sweden dainis@astro.lu.se
Bengt Gustafsson, Uppsala University, Sweden bg@astro.uu.se
Oleg Kochukhov, Uppsala University, Sweden oleg.kochukhov@fysast.uu.se
Andres Carmona Gonzalez, Geneva University, Switzerland andres.carmona@unige.ch
Bruno Chazelas, Observatoire de Genéve, Geneva University, Switzerland bruno.chazelas@unige.ch
Laurent Eyer, University de Geneve, Switzerland laurent.eyer@unige.ch
Claudia Greco, Observatoire de Genéve, Switzerland claudia.greco@unige.ch
Manuel Guedel, ETH Zurich, Switzerland guedel@astro.phys.ethz.ch
Jan Stenflo, ETH Zurich, Switzerland stenflo@astro.phys.ethz.ch
Subhon Ibadov, Institute of Astrophysics, Tajik Academy of Sciences, Tajikistan ibadovsu@yandex.ru
Boonrucksar Soonthornthum, National Astronomical Research Institute of Thailand, Thailand boonrucksar@narit.or.th
Solen Balman, Middle East Technical University, Turkey solen@metu.edu.tr
Atila Ozguc, Bogazici University, Turkey ozguc@boun.edu.tr
Alex Golovin, Main Astron. Obs. National Academy of Sciences of Ukraine, Ukraine golovin.alex@gmail.com
Nataliya Kovalenko, KYIV Planetarium, Ukraine kievplanet@ukr.net
Yuri Tsap, Crimean Astrophysical Observatory, Ukraine yur_crao@mail.ru
Nadine Afram, Imperial College London, UK n.afram@ic.ac.uk
William Chaplin, University of Birmingham, UK w.j.chaplin@bham.ac.uk
Giulio Del Zanna, University of Cambridge, UK g.del-zanna@damtp.cam.ac.uk
Christoffer Karoff, School of Physics and Astronomy, UK karoff@bison.ph.bham.ac.uk
John Landstreet, Armagh Observatory, UK jlandstr@astro.uwo.ca
Urmila Mitra-Kraev, University of Cambridge, UK u.mitra-kraev@damtp.cam.ac.uk
Danail Obreschkow, University of Oxford, UK dobresch@astro.ox.ac.uk
Alan Pickwick, EAAE & MGS, UK alan_c_pickwick@btinternet.com
Christophe Pinte, University of Exeter, UK pinte@astro.ex.ac.uk
Vinothini Sangaralingam, University of Birmingham, UK vs@star.sr.bham.ac.uk
Helen Walker, STFC Rutherford Appleton Laboratory, UK helen.walker@stfc.ac.uk
Lidia van Driel-Gesztelyi, University College London, UK lidia.vandriel@obspm.fr
Douglas Arion, Carthage College, USA darion@carthage.edu
Thomas Ayres, University of Colorado, USA thomas.ayres@colorado.edu
Peter Beiersdorfer, Lawrence Livermore National Laboratory, USA beiersdorfer@llnl.gov
Alexander Brown, University of Colorado, USA alexander.brown@colorado.edu
Joseph Burns, Cornell University, USA jab16@cornell.edu
Manfred Cuntz, University of Texas at Arlington, USA cuntz@uta.edu
Edward Devinney, Villanova University, USA edward.devinney@villanova.edu
Pamela L Gay, Southern Illinois University Edwardsville, USA pgay@siue.edu
Ciriaco Goddi, Harvard-Smithsonian Center for Astrophysics, USA cgoddi@cfa.harvard.edu
Richard Gray, Appalachian State University, USA grayro@appstate.edu
Daniel Green, Harvard-Smithsonian Center for Astrophysics, USA green@cfa.harvard.edu
Erika Grundstrom, Vanderbilt University, USA erika.grundstrom@vanderbilt.edu
Javiera Guedes, University of California, Santa Cruz, USA javiera@ucolick.org
Edward Guinan, Villanova University, USA edward.guinan@villanova.edu
Margit Haberreiter, University of Colorado, USA haberreiter@lasp.colorado.edu
Thomas Hartlep, Stanford University, USA thartlep@stanford.edu
Donald Hassler, Southwest Research Institute, USA hassler@boulder.swri.edu
Frank Hill, National Solar Observatory, USA fhill@noao.edu
Kenneth Hinkle, NOAO, USA hinkle@noao.edu
Jon Todd Hoeksema, Stanford University, USA jthoeksema@spd.aas.org
Christopher Johns-Krull, Rice University, USA cmj@rice.edu
George Kaplan, -, USA gkaplan@zoominternet.net
Irina Kitiashvili, Stanford University, USA irinasun@stanford.edu
Alexander Kosovichev, Stanford University, USA sasha@sun.stanford.edu
Chryssa Kouveliotou, NASA, USA chryssa.kouveliotou@nasa.gov
John Leibacher, National Solar Observatory, USA jleibacher@nso.edu
Steven Majewski, University of Virginia, USA srm4n@virginia.edu
Rocco Mancinelli, SETI Institute, USA rmancinelli@seti.org
Bruce McCollum, California Institute of Technology, USA Mccollum@ipac.caltech.edu
Alexei Pevtsov, National Solar Observatory, USA apevtsov@nso.edu
Peter Plavchan, Caltech / IPAC/ NExSci, USA plavchan@ipac.caltech.edu
Louise Riofrio, Oceaneering Space Systems, USA l.m.riofrio@nasa.gov
Rachel Roettenbacher, Lehigh University, USA rmr207@lehigh.edu
Ronald G. Samec, Bob Jones University, USA rsamec@bju.edu
Ashit Sanyal, Capitol College, USA asanyal@comcast.net

Ethan J. Schreier, Associated Universities Inc., USA ejs@aui.edu
Ryan Shannon, Cornell University, USA rms72@cornell.edu
David Soderblom, Space Telescope Science Institute, USA drs@stsci.edu
John Stauffer, California Institute of Technology, USA stauffer@ipac.caltech.edu
David Tholen, University of Hawaii, USA tholen@ifa.hawaii.edu
Guillermo Torres, Harvard-Smithsonian Center for Astrophysics, USA gtorres@cfa.harvard.edu
Wesley Traub, Jet Propulsion Laboratory, USA wtraub@jpl.nasa.gov
Jeff Valenti, Space Telescope Science Institute, USA valenti@stsci.edu
Steven Vogt, UCO/Lick Observatory, USA vogt@ucolick.org
Lucianne Walkowicz, University of California Berkeley, USA walkowicz@berkeley.edu
David Webb, Boston College, USA david.webb@hanscom.af.mil
Junwei Zhao, Stanford University, USA junwei@sun.stanford.edu
Xuepu Zhao, Stanford University, USA xuepu@sun.stanford.edu
Tabare Gallardo, Faculdad de Ciencias, Uruguay gallardo@fisica.edu.uy
Andrea Maciel, Facultad de Ciencias, Uruguay amaciel@fisica.edu.uy
Andrea Sanchez Saldias, Faculdad de Ciencias, Uruguay andrea@fisica.edu.uy
Gonzalo Tancredi, Facultad de Ciencias, Uruguay gonzalo@fisica.edu.uy
Alisher S Hojaev, Nat. Univ. Uzbekistan & UBAI/Uzbek Academy of Sciences, Uzbekistan ash@astrin.uzsci.net
Nguyen Lan, Hanoi National University of Education, Viet Nam nquynhlan@hnue.edu.vn
Prospery C Simpemba, Copperbelt University, Zambia pcsimpemba@yahoo.com

Session I

Introduction

Chair: C. Fang

Solar and Stellar Variability: Impact on Earth and Planets
Proceedings IAU Symposium No. 264, 2009
A. G. Kosovichev, A. H. Andrei & J.-P. Rozelot, eds.

doi:10.1017/S1743921309992298

The Sun and stars as the primary energy input in planetary atmospheres

Ignasi Ribas

Institut de Ciències de l'Espai (CSIC-IEEC), Facultat de Ciències, Torre C5, parell, 2a pl, Campus UAB, 08193 Bellaterra, Spain
email: iribas@ice.csic.es

Abstract. Proper characterization of the host star to a planet is a key element to the understanding of its overall properties. The star has a direct impact through the modification of the structure and evolution of the planet atmosphere by being the overwhelmingly larger source of energy. The star plays a central role in shaping the structure, evolution, and even determining the mere existence of planetary atmospheres. The vast majority of the stellar flux is well understood thanks to the impressive progress made in the modeling of stellar atmospheres. At short wavelengths (X-rays to UV), however, the information is scarcer since the stellar emission does not originate in the photosphere but in the chromospheric and coronal regions, which are much less understood. The same can be said about particle emissions, with a strong impact on planetary atmospheres, because a detailed description of the time-evolution of stellar wind is still lacking. Here we review our current understanding of the flux and particle emissions of the Sun and low mass stars and briefly address their impact in the context of planetary atmospheres.

Keywords. Sun: activity, Sun: particle emission, stars: activity, planetary systems, ultraviolet: stars, X-rays: stars, planets and satellites: general

1. Introduction

Our knowledge of planets, including our Earth, is directly driven by our knowledge of the parent star. Indeed, the host star to a planet is the overwhelmingly larger source of energy and its emissions critically affect the composition, thermal properties and the mere existence of planetary atmospheres. Our current description of the internal structure, evolution and radiative flux of stars has reached a relatively mature level and theoretical models can now predict stellar attributes to fairly good accuracy. While this refers to the bulk properties of stars, some other components related to stellar emissions still defy detailed understanding. This is the case of the emissions related to magnetic activity, a common feature of all low-mass stars. High-energy emissions of solar-type stars and their variations over different timescales are now quite well characterized but not so active stars in general. The situation for particle winds is even more frustrating, with a lack of direct detections and large uncertainties on the predicted wind of active stars from indirect measurements. However, all evidence collected so far supports the fact that low-mass stars undergo an early phase of strong magnetic activity (lasting from tens of Myrs to Gyrs) in which their high-energy and particle emissions are enhanced by orders of magnitude with respect to our current Sun. If the effects of the active Sun are even noticeable today on our Earth, it is undeniable that such effects *must* have been even stronger for our planet and all Solar System planets in the past. In general, all planets orbiting low-mass stars are affected by the magnetic evolution of their parent star.

In the following sections we review the current state of knowledge of the properties of stars from the point of view of planet hosts, including their bulk emissions and those related with magnetic activity. Further, we discuss the possible impact that long-term

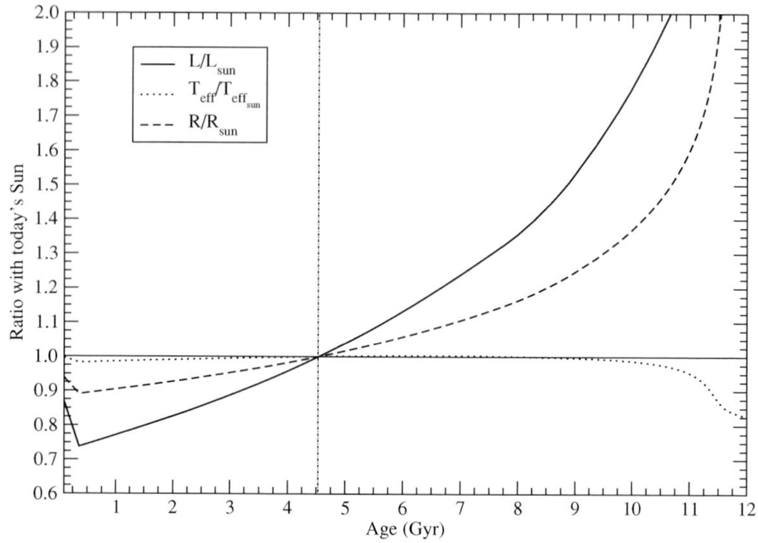

Figure 1. The evolution of the effective temperature, radius, and luminosity of the Sun from the zero-age main sequence to the start of its red giant phase. The vertical line marks the current age of the Sun. Based on the Yonsei-Yale evolutionary sequences (see text for references).

variability of high-energy radiations and particle fluxes has on the Solar System planets and exoplanets around solar-type and lower-mass stars.

2. The structure and evolution of the Sun and stars

Much progress has been made in understanding the internal structure and evolution of stars. Stellar models are now able to predict with relatively good accuracy the fundamental properties of stars over a wide range of masses and a large fraction of the Hertzsprung-Russell diagram. This is especially true for the better-behaved evolutionary stages (i.e., main sequence and giants) thanks to great advances in our knowledge of the physical processes that govern the interiors of stars, such as opacities, nuclear reaction rates, energy transport, equation of state, etc (see, e.g., Lebreton 2000; Chabrier & Baraffe 2000). Grids of stellar evolution models have been proposed by a number of different groups and they mostly manage to reproduce the observable properties of main sequence and giant stars satisfactorily. In spite of the long-lasting major concern that our understanding of the interior of the Sun (and, by extension, of stars) was incomplete, the resolution of the so-called "solar neutrino problem" (Ahmad *et al.* 2001; Bahcall 2004) ultimately vetted the calculations and lent confidence on the ability to model stellar interiors.

The relative good performance of stellar models has permitted the calculation of the past and future evolution of the Sun, and thus the variation of its main properties over time. An illustration of this is given in Fig. 1, where the normalized luminosity, effective temperature, and radius of the Sun are plotted as a function of age as calculated from the Yonsei-Yale theoretical models (Kim *et al.* 2002; Yi *et al.* 2003). The plot shows how the luminosity of the Sun is found to be some 75% of today's when life supposedly arose on Earth, thus giving rise to the so-called "Faint Young Sun" paradox (e.g., Sagan & Mullen 1972).

The optimistic picture of the theory of stellar evolution is not without its blemishes. There are still nagging deficiencies in the models linked to our poor understanding of

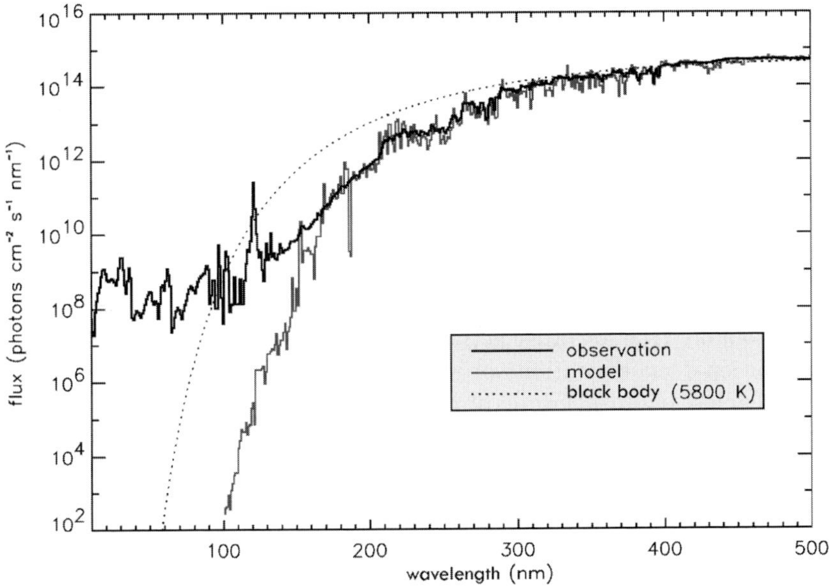

Figure 2. Comparison of the observed solar energy distribution with the prediction of a matching photospheric model and a black body. Note the divergence below 170 nm when the contribution from the chromosphere dominates. Adapted from Selsis (2000).

phenomena such as convective energy transport or mass loss (e.g., Lafon & Berruyer 1991; Cassisi 2009). While these have a stronger impact on high-mass stars, the effects of mass loss have also been considered in low-mass stars such as the Sun (Sackmann & Boothroyd 2003). Also in the in the case of the Sun, an apparent victory of theory was the almost perfect agreement with the results of helioseismology, but the idyllic situation broke down, so far irreversibly, with the revision of the solar abundances (see Christensen-Dalsgaard *et al.* 2009; Asplund *et al.* 2009). In the particular case of the cool end of the main sequence, the chief open issues are related to the inability of theory to correctly predict the radii and effective temperature of stars, presumably because of the complete absence of the effects of magnetic activity in the model calculations (e.g., Ribas *et al.* 2008).

Besides the integrated radiative flux of stars, models have also come a long way in trying to reproduce the spectral energy distribution of the emissions. Theoretical atmosphere models provide today a fairly realistic picture of the bulk energy output of stars thanks to intensive efforts in compiling large databases of atomic and molecular transitions and in including fine details in the calculations of radiative transfer. Direct comparison with the spectral energy distribution of the Sun (arguably the star for which the best data are available) shows that models are able to predict the spectral distribution of flux near the peak of the emission within 5%, as shown by Edvardsson (2008) in the case of the MARCS models and Bohlin (2007) in the case of the Kurucz models. Such models consider the radiation from the photosphere, which dominates over the visible and near-IR spectrum. However, at shorter wavelengths (X-rays to UV) the predictions start to deviate from the calculations. This is nicely illustrated in Fig. 2, where the synthetic model and real data diverge below 170 nm. The reason for the divergence is the contribution from the hot atmosphere layers of the Sun, which are associated to magnetic activity. This is addressed in the following section.

3. The magnetic Sun and stars

Activity in late-type stars has been the subject of attention for many years. Early studies already pointed out a strong correlation between the rotation rate of a star and its activity level (Wilson 1966; Kraft 1967). This evidence, together with the growing body of information on our nearest active star, the Sun, has shaped up the currently accepted model of stellar activity in cool stars as being a consequence of the operation of a magnetic dynamo (Parker 1970). In this model, the interplay between the stellar rotation and the gas motions in the convective layer of the star generates magnetic fields that give rise to the observed phenomena. It is well established for the Sun that active regions arise from magnetic field lines emerging from (and plunging into) the surface and the same is probably true for other active stars.

The observational manifestations of stellar activity are plentiful and include modulations of the stellar photospheric light due to stellar spots, high-energy emissions, and flares. The magnetic dynamo generates energy that heats up the upper stellar atmosphere creating a vertical temperature profile with distinct regions such as the chromosphere and the corona. These layers, with estimated temperatures of $10^4 - 10^7$ K, are the source of the high-energy emissions (from X rays to the UV) observed in active cool stars. The stratified nature of the solar atmosphere makes that observations at different wavelengths are typically associated with the various layers (X-rays & EUV – corona, FUV – transition region; UV – chromosphere) and thus serve as probes for their physical conditions. A consequence of a hot upper layer in the Sun is the existence of a particle wind, for which there is ample evidence and has been subject to intensive scrutiny (e.g., Zurbuchen 2007).

A component intimately associated with stellar activity is its pronounced variability over time. Such variations cover nearly all timescales, including hours (flares), days (rotational modulation; Fröhlich & Lean 2004), years (the 11-yr sunspot cycle; Fröhlich & Lean 2004), centuries (Maunder minima and the likes; e.g., Soon & Yaskell 2004), and up to billions of years. The latter is of the same order as the nuclear evolution timescale and it is related with the rotational spin down from magnetic braking. This is discussed in more detail in the following section.

4. Long-term evolution of solar/stellar activity

4.1. *High-energy emissions*

Compelling observational evidence (Güdel *et al.* 1997) shows that zero-age main sequence (ZAMS) solar-type stars rotate over 10 times faster than today's Sun. As a consequence of this, young solar-type stars, including the young Sun, have vigorous magnetic dynamos and correspondingly strong high-energy emissions. From the study of solar type stars with different ages, Skumanich (1972), Simon *et al.* (1985), and others showed that the Sun loses angular momentum with time via magnetized winds (magnetic braking) thus leading to a secular increase of its rotation period (Durney 1972). This rotation slow-down is well fitted by a power law roughly proportional to $t^{-1/2}$ (e.g., Skumanich 1972; Soderblom 1982; Ayres 1997; Barnes 2007; Mamajek & Hillenbrand 2008). This is illustrated in Fig. 3. Note that the age–rotation period relationship is tighter for intermediate/old stars, while young stars (a few 10^8 yr) show a larger spread in rotation periods. In response to slower rotation, the solar dynamo strength diminishes with time causing the Sun's high-energy emissions also to undergo significant decreases. Comprehensive studies on this subject were published by Zahnle & Walker (1982) and Ayres (1997).

The *Sun in Time* program (Dorren & Guinan 1994; Ribas *et al.* 2005) was established to study the magnetic evolution of the Sun using a homogeneous sample of single, nearby

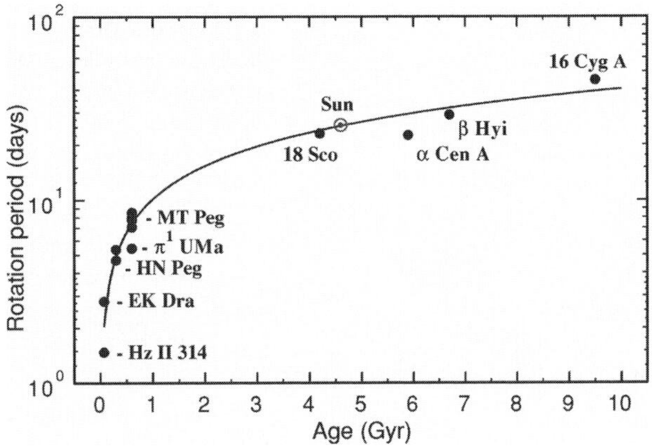

Figure 3. Rotation period as a function of age for a sample of solar-type stars. The solid line represents a power-law fit.

G0-5 main sequence stars, which have known rotation periods and well-determined physical properties, including temperatures, luminosities, metal abundances and ages. Such stars could be used as proxies for the Sun at evolutionary stages covering from 0.1 up to about 7 Gyr. Detailed analyses indicated that all proxies have masses within 10% of that of the Sun.

One of the primary goals of the *Sun in Time* program was to reconstruct the spectral irradiance evolution of the Sun. To this end, a large amount of multiwavelength (X-ray, EUV, FUV, UV, optical) data was collected using various space missions (ASCA, ROSAT, EUVE, FUSE, IUE, HST) covering a range between 0.1 and 330 nm†. Details of the data sets and the flux calibration procedure employed are provided in Ribas *et al.* (2005). Full spectral characterization was completed for five of the stars in the sample, which cover key stages in the evolution of the Sun (and solar-type stars): EK Dra – 0.1 Gyr, π^1 UMa – 0.3 Gyr, κ^1 Cet – 0.6 Gyr, β Com – 1.6 Gyr, and β Hyi – 6.7 Gyr. The reconstructed irradiances of these solar proxies indicate that the X-ray and EUV emissions of the young main-sequence Sun were about 100 to 1000 times stronger than those of the present Sun. Similarly, the FUV and UV emissions of the young Sun were 10 to 100 and 5 to 10 times stronger, respectively, than today. The spectral energy distribution of the Sun and its time evolution is illustrated in Fig. 4 (top).

But, additionally, the results suggest a very good correlation of the emitted flux with age that can be modeled accurately by power laws (see Fig. 4 bottom). A fit to the overall XUV flux (integrated between 0.1 and 120 nm) as a function of age (τ) and normalized to a distance of 1 AU yields the following expression:

$$F_{\rm XUV} = 29.7\,[\tau({\rm Gyr})]^{-1.23}\ \ {\rm erg\ s^{-1}\ cm^{-2}} \tag{4.1}$$

The equation reveals that the XUV emissions of the Sun were about 3 times higher 2.5 Gyr ago and 6 times higher 3.5 Gyr ago, roughly when life appeared on Earth. Also, the young 0.1-Gyr old Sun could have had up to 100 times stronger XUV radiation (albeit for a short period of time). At longer wavelengths, the H Lyman-α emission feature can contribute to a significant fraction of the XUV flux (over 50% in the case of the Sun; Woods *et al.* 1998). High-resolution Hubble Space Telescope spectroscopic

† Note the existence of a gap between 36 and 92 nm that is unobservable because of the very strong interstellar absorption

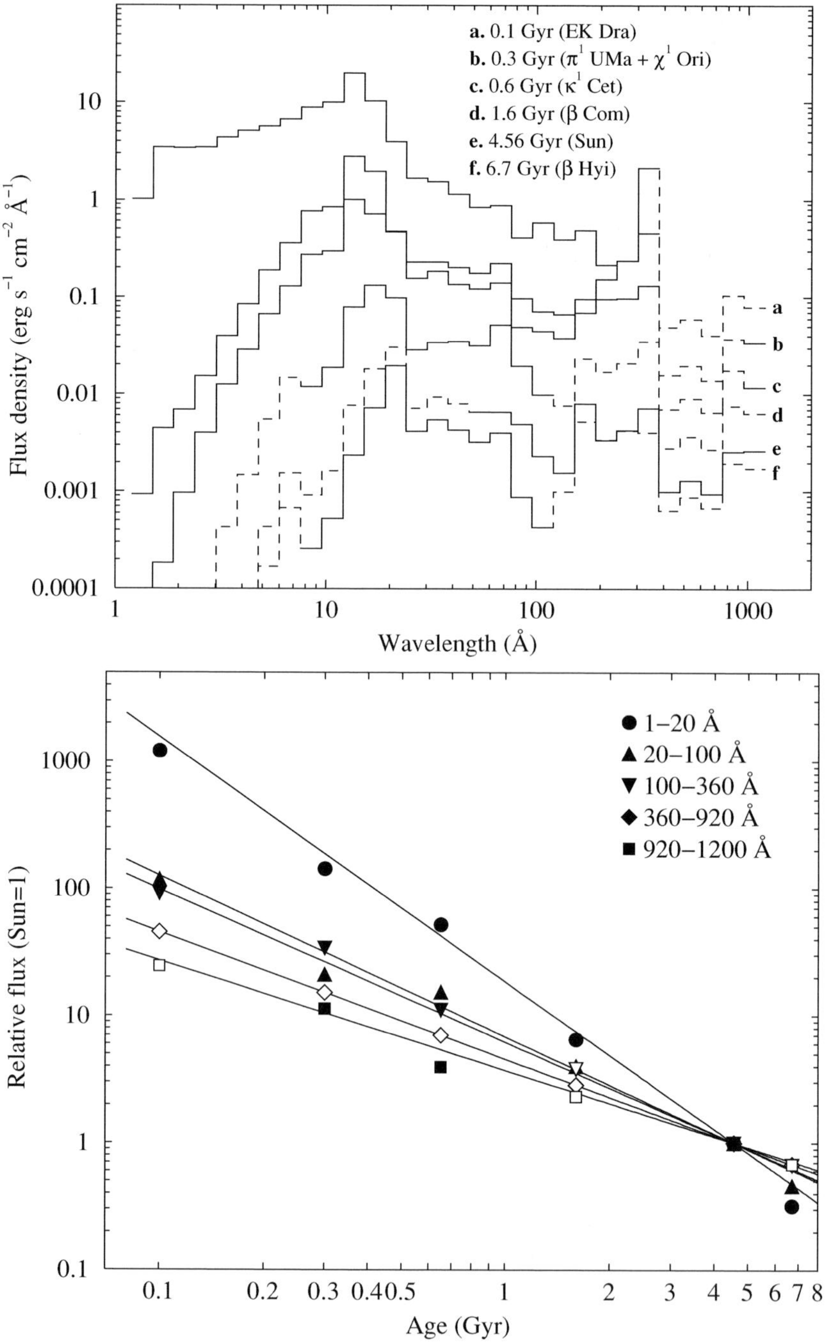

Figure 4. *Top:* Full spectral energy distribution of the solar-type stars at different stages of the main sequence evolution. The solid lines represent measured fluxes while the dotted lines are fluxes calculated by interpolation using a power-law relationship. *Bottom:* Solar-normalized fluxes vs. age for different stages of the evolution of solar-type stars. Plotted here are the measurements for different wavelength intervals (filled symbols) and the corresponding fits using power-law relationships. Represented with empty symbols are the inferred fluxes for those intervals with no available observations. Adapted from Ribas *et al.* (2005).

observations were used to estimate the net stellar flux. These measurements, together with the observed solar Lyman-α define the following power-law relationship with high correlation:

$$F_{\mathrm{Ly}\alpha} = 19.2\,[\tau(\mathrm{Gyr})]^{-0.72} \ \mathrm{erg\ s^{-1}\ cm^{-2}} \tag{4.2}$$

While the long-term variation of the high-energy emissions of the Sun and solar-type stars is now well characterized, the same is not true for stars of lower mass. The situation is further aggravated by the fact that K- and M-type stars are known to be more active than their solar counterparts. We are currently generalizing the Sun in Time study to all stars of GKM spectral types (Garcés *et al.* 2010). For an initial estimate of the evolution of XUV irradiances we have used a proxy indicator, which is the ratio of the X-ray luminosity to the bolometric luminosity ($\log\,[L_{\mathrm{X}}/L_{\mathrm{bol}}]$). This ratio is highest for the more active stars (i.e. fastest rotation period) and decreases monotonically with decreasing level of chromospheric activity (e.g., Stelzer & Neuhäuser 2001; Pizzolato *et al.* 2003). From the analysis of open cluster stars it is now well established that all single GKM stars spin down as they age, their activity decreases with time, and so does he ratio $\log(L_{\mathrm{X}}/L_{\mathrm{bol}})$. It is also a well-known effect that $\log(L_{\mathrm{X}}/L_{\mathrm{bol}})$ cannot reach values arbitrarily close to unity (in quiescence) for very active stars. A "saturation" phenomenon occurs at $\log(L_{\mathrm{X}}/L_{\mathrm{bol}}) \approx -3$ (e.g., Vilhu & Walter 1987; Stauffer *et al.* 1994). Qualitatively, the evolution of $\log(L_{\mathrm{X}}/L_{\mathrm{bol}})$ for a late-type star has a flat plateau from its arrival on the main sequence up to a certain age (end of saturation phase) and then decreases monotonically with age.

Assuming that all active stars share the same underlying physical mechanism responsible for the emission and a supposedly similar spectral energy distribution, one can infer that stars with similar values of $\log(L_{\mathrm{X}}/L_{\mathrm{bol}})$ will also have similar $\log(L_{\mathrm{XUV}}/L_{\mathrm{bol}})$. From this assumption and the available results for solar analogs, estimates of stellar ages for late G-, K-, and M-type stars yield the following preliminary expressions, generalizing Eq. 4.1:

$$\begin{aligned} F_{\mathrm{XUV}} &= 4.04 \cdot 10^{-24}\, {L_{\mathrm{bol}}}^{0.79} && \text{for } 0.1\ \mathrm{Gyr} < \tau < \tau_i \\ F_{\mathrm{XUV}} &= 29.7\, \tau^{-1.23} && \text{for } \tau < \tau_i \\ \tau_i &= 1.66 \cdot 10^{20}\, {L_{\mathrm{bol}}}^{-0.64} \end{aligned} \tag{4.3}$$

with F_{XUV} (0.1–120 nm) in erg s^{-1} cm^{-2} at 1 AU, τ in Gyr and L_{bol} in erg s^{-1}. The expression indicates that, solar-type stars stars stay at saturated emission levels until ages of $\sim$0.1 Gyr and then their XUV flux rapidly decreases following a power law relationship as a function of age. Early K-type stars stay at saturated emission levels for a little longer and then also decrease following a power law relationship of very similar slope. Finally, early M-type stars have saturated emission levels for up to 1 Gyr (and possibly longer) and then decrease in an analogous way to G- and K-type stars. More quantitatively, our preliminary results (to be further discussed in Garcés *et al.* 2010) indicate that early K-type stars and early M-type stars may have XUV fluxes that are 3–4 times higher and 10–100 times higher, respectively, than solar-type stars of the same age.

4.2. *Particle winds*

High energy fluxes are not the only trait of active stars. As mentioned above, particle emissions (winds) are also associated with the hot coronal plasma. If the Sun was more active in the past one may intuitively assume that its particle wind was also more intense. This agrees with evidence from lunar rocks (Newkirk 1980) and from the relative nitrogen isotopic abundances in the atmosphere of Titan (Lammer *et al.* 2000). Unfortunately, the direct detection of the stellar counterparts to the solar wind has, so far, remained elusive even for the nearest and more active stars (Gaidos *et al.* 2000; Ayres *et al.* 2001). However, indirect evidence for the existence and strength of stellar winds was obtained by the pioneering work of Wood *et al.* (2002). The authors devised a technique based on the detailed modeling of the H Lyman-α emission feature that reveals an absorption component associated with the interaction between the stellar wind and the surrounding interstellar medium. Quantitative analysis yielded an estimate of the mass loss rate of the stars in the most favorable cases (appropriate geometry, clean line of sight). Wood *et al.* (2002) proposed the existence of a correlation between the X-ray surface flux (as activity indicator) and the mass loss rate, which was found to be proportional to τ^{-2}. This would imply a much stronger (up to 1000 times) solar wind in the past. Recent results from the same team (Wood et al. 2005) have casted doubt on the validity of the results for the younger stars, with a break down of the relation for ages below $\approx$0.7 Gyr. In addition, Holzwarth & Jardine (2007) ague that K- and M-type stars should not be considered jointly and propose a new, much shallower power law.

The question of the strength of the wind of single active stars remains unsolved. Current best estimates for the young Sun suggest a stronger wind by a very uncertain factor between 10–1000. All these results are based on a few observations taken before the demise of the STIS spectrograph on board HST. The recent recovery of the instrument should permit the continuation of the observations and, hopefully, provide the needed proof to constrain the time-evolution of the mass loss of the Sun and active stars in general. On the down side, other important parameters will remain unknown for the active stellar phases, such as the geometry of the wind or the intensity of the interplanetary magnetic field. For example, active stars are known to present active regions at high latitudes (e.g., Strassmeier 2009) and the simple scaling of the geometry of the current solar wind may just be incorrect.

5. Short-term variability

Magnetic activity phenomena have characteristic variations over a very wide range of timescales. In mid-range timescales (years and decades), the variations are associated with activity cycles, which in the case of the Sun is of the other of 11 years. Over this time period, the amplitude of the variations is of about 20–50% in the UV range up to a factor 10–20 in X-rays (Rottman 1988; Lean 1997). Most of what we know about activity cycles in field stars is due to the Mount Wilson, based on the monitoring of the Ca II H&K emission lines (e.g., Donahue *et al.* 1996). The data collected has shown the existence of activity cycles in many stars which, interestingly, have periods similar to that of the Sun (Baliunas & Vaughan 1985). Even in the case of very active stars, such as EK Dra, the activity cycle seems to be around 10 years (Güdel *et al.* 2003; Järvinen *et al.* 2005). But the most relevant factor to the potential impact on planets is the amplitude of the variations. Such amplitude is seen to decrease with increasing activity. Thus, a young solar analog has peak-to-peak variations of a factor 2–2.5 in X-rays (some 4 times smaller than those of the Sun; Micela & Marino 2003). This apparent contradiction is probably explained by the lower contrast between the epochs of maximum and minimum activity for stars that are already strongly active in quiescence.

At short timescales, variability is driven by spot modulation and, more interestingly, by flare events. Thanks to our privileged view of the Sun, links have been established between these flares, high-energy emissions and particle emissions (coronal mass ejections) (e.g., Aschwanden *et al.* 2001). Flares can be very energetic phenomena that trigger rapid increases over the quiescent high-energy and particle emissions. For example, the increase in the flux for strong solar flares can be of 20–50% in the FUV–UV and 2-10 fold in X-rays.

A relevant question is whether flare rates change over the course of the stellar lifetime. Indeed, observations by the EXOSAT X-ray satellite revealed a strong flare (10-fold increase in flux) of the 0.3-Gyr-old solar analog π^1 UMa (Landini *et al.* 1986), thus suggesting an enhanced flare activity in young stars. This was corroborated by the thorough analysis carried out by Audard *et al.* (2000) using EUVE data, who found a well-defined correlation between the rate of flare occurrence and magnetic activity (using L_X as proxy). The expressions in Audard *et al.* (2000) indicate that the cumulative flare distribution follows the power law:

$$N(>E) \approx 0.08\, {L_X}^{0.95} E^{-0.8}\ \ \mathrm{day}^{-1} \qquad (5.1)$$

with E being the flare energy. The expression indicated that a young, active solar-type star ($L_X \approx 10^{30}$ erg s^{-1}) can undergo several tens of large ($E > 10^{32}$ erg) flares per day, while the current Sun at its peak of activity experiences such strong flares once every two weeks on average.

Flares are an important addition to the high-energy fluxes and particle emissions in quiescence. The high flare rates of active stars indicate that this is a factor that needs to be taken into account for a complete picture of stellar activity and its potential impact on planets.

6. Effects of solar/stellar activity on planets

The evidence discussed above unambiguously draws a picture of the young Sun (and young, active stars in general) of enhanced high-energy and particle emissions, with frequent energetic flares. To put these results in context it is useful to consider that high-energy emissions of the Sun today account for about 3 millionths of the total radiative flux that the Earth received. Even in the most active phase of the young Sun, this high-energy emissions just amount to 5 parts in ten thousand, a very small fraction of the total flux. This implies, for example, that the higher flux cannot explain the Faint Young Sun paradox using arguments related to the radiative budget. However, an important feature of high-energy emissions is that they are absorbed at very high altitude in the planetary atmosphere (because of the large cross section), in a region, the exosphere, where the density is low. Even the relatively small contribution of energetic photons can increase the temperature of this atmospheric layer dramatically, leading to the onset of escape processes. This illustrates the often non-linear behavior of Nature, in which it can be the process instead of sheer power what explains physical phenomena. Besides, the enhanced UV radiation can trigger photochemical reactions thus altering the chemical composition (see Hunten *et al.* 1991 for a review). Finally, particle emissions also play their role by contributing to the erosion of planetary atmospheres via non-thermal processes, mostly through so-called sputtering and ion pick-up. Only the strength of the planetary magnetic field can offer some protection by diverting the incoming charged particles. All the loss processes, both thermal and non-thermal, are thoroughly covered in the monograph by Bauer & Lammer (2004).

All the evidence discussed so far compellingly demonstrates that radiation and particle emissions from active stars (including the young Sun) can be orders of magnitude stronger than those of the current Sun and thus potentially have an impact on the properties of planetary atmospheres. In the following sections we review some of these effects on both Solar System planets and exoplanets.

6.1. *Mars and Venus*

An immediately obvious application of the time-evolution of solar magnetic emissions is the study of the volatile inventory of planet Mars. The small mass of the planet and the lack of a protecting magnetic field make it specially vulnerable to the charges from the active Sun. Relevant processes include the loss of atmospheric constituents to space (basically molecules and H, He, C, N and O ions) and the balance with the incorporation in the Martian soil via weathering processes. Recent sophisticated studies based on 2D and 3D modeling (Lammer *et al.* 2003a; Donahue 2004; Terada *et al.* 2009) suggest that the loss of volatiles from Mars could amount to a global water ocean with a depth of several tens of meters. This is in agreement with the inference from the water-related features visible on the Martian surface as shown by the images of the Mars Global Surveyor (Carr & Head 2003). The loss of water was driven by photolysis in the atmosphere and subsequent loss of oxygen and hydrogen ions by thermal escape. Instead of the expected 2:1 (H:O) loss ratio, the observed and modeled value is 20:1 because of hydrogen being lighter and escaping more easily. The remaining O is incorporated in the Martian surface giving it the characteristic rusty aspect. The thickness of the oxidized layer is relevant to future searches for organic matter and the calculations indicate that it could be at least of 2–5 m.

A related aspect is the investigation of the conditions of the Martian atmosphere that allowed the presence of liquid water on the surface 3.8 Gyr ago, as indicated by the observations (Baker 2001). The studies that account for the effects of the solar high-energy emissions show that a greenhouse effect driven by CO_2 could have provided the necessary conditions (Kulikov *et al.* 2007), although there are still numerous uncertainties. At the same time, Mars should have been protected by a magnetic field to avoid massive erosion of its atmosphere from non-thermal processes. The subsequent evolution of Mars can just be guessed. Its core possibly ceased to generate a magnetic dynamo because of several possible reasons (Stevenson 2009) and the loss of the magnetic field exposed the atmosphere of the planet to intense erosion. The atmosphere started to evaporate (including the water and volatile inventory) and the planet eventually cooled down because of the lack of a greenhouse effect. During this cooling process, which could have been quite rapid, part of the water could have been incorporated into the Martian soil as water ice and explain the observations from several probes (Murray *et al.* 2005).

The other interesting terrestrial planet of the Solar System is Venus, with a mass similar to the Earth's but at the same time with a very different atmosphere. The question is whether Venus was once a wet planet and, if so, why it is now so dry. The past of Venus is still quite mysterious but there is strong evidence that Venus had a significant water inventory, both from the current understanding of the formation of the Solar System planets and from certain isotope ratios. Estimates using D/H ratios suggest that the water content of Venus was at least 0.3% of an Earth ocean (Hartle *et al.* 1996), and possibly closer to the total Earth water inventory. The combination of a runaway greenhouse effect caused by the increasing solar flux with the high-energy emissions heating up the upper atmosphere and producing photolysis led to the loss of water to space. Depending on different assumptions, the loss of water could easily amount to over an Earth ocean in just a few tens or hundreds of Myr (Kulikov *et al.* 2006). This was probably further exacerbated by non-thermal losses caused by the lack of a protecting magnetic field in Venus. The current CO_2 rich atmosphere could come from out-gassing processes resulting from volcanism.

Similar investigations can be carried out on the satellites of the Solar System with an atmosphere, such as Titan. An excellent review on the atmospheric escape processes in terrestrial planets (Venus, Mars and Earth) and satellites (Titan) was published recently by Lammer *et al.* (2008).

6.2. *Earth*

The question of the influence of variations in the Sun on Earth's climate has been a recurrent one. Correlations between climate variations (temperature, rainfall) and solar activity have been proposed even on short timescales, such as those related to the 11-yr solar cycle (see, e.g., Hoyt & Schatten 1997). The overall solar irradiance variations over the sunspot cycle are of about 0.1–0.2% (Fröhlich & Lean 2004). This translates into a cyclic temperature variation of the Earth of about 0.1°C when computed from simple black body considerations, which does not seem sufficient by itself to produce observable climate changes.

Even though the observed global light variations of the Sun are very small over its activity cycle, these are much larger at shorter wavelengths (e.g., Lean 1997). For example, the typical variations of solar X-ray and EUV emissions from the minimum to the maximum of the ~11 year activity cycle are over a factor of 5, while at FUV and UV they range from 10 to 50 percent. Also the frequencies and intensities of flaring events and coronal mass ejections (CME) are strongly correlated with the Sun's activity cycle. For example, the rate of CME occurrences is larger during the sunspot maxima than during the times sunspot minima (Webb & Howard 1994). An interesting – and perhaps definitive – proof of forcing from solar activity cycle on Earth's climate was presented recently by Meehl et al. (2009). The authors link rainfall and temperature anomalies in the Pacific to the response of the stratosphere to ozone fluctuations caused by solar high-energy emissions. This mechanism illustrates how a small change in the energy budget can amplify via feedback mechanisms to measurable variations in the Earth climate.

One of the most tantalizing hypotheses on possible climate–solar activity connection is the coincidence of the extended low solar activity level that occurred during ~1450–1850 with an interval of cooler/stormier weather known as the "The Little Ice Age". The early observations of sunspot number have been reconstructed from historical sources starting from the telescopic observations of sunspots by Galileo in 1610/11. There is a paucity of sunspots during 1645–1715 and, to a lesser extent, again during 1800–1840. These intervals of low sunspot numbers are known as the Maunder Sunspot Minimum and the Dalton Sunspot Minimum, respectively. Attempts have been made to compute the solar total irradiance during the Maunder Minimum from sunspot activity (see Lean et al. 1995). The reconstruction of the solar irradiance back to 1600 suggests an estimated decrease of 0.24% from the present to the time of the Maunder Minimum in the late 17th century. Using simple radiative laws, this irradiance change should produce a global temperature decrease of about 0.2°C. However, proxies of surface temperature indicate a temperature decrease of about 0.7°C during the Maunder Minimum (Bradley & Jones 1993). The connection of stellar activity and the "Little Ice Age" remains unsolved, but, as before, it could be related to non-linear feedback mechanism in the climate system. There is even controversy over how widespread was the cool period and it has been suggested that the temperature anomaly could just have affected the Northern Hemisphere (e.g., Jones *et al.* 1998), thus lessening its global impact.

Over long timescales, the Earth has suffered the same processes as explained above for Venus and Mars (e.g., Lundin *et al.* 2007), but it is obvious that Earth managed to keep its volatiles, probably thanks to the large mass and protecting magnetic field. In the case of the Earth it is very interesting to study the effects of high-energy radiation on the photochemistry of the early atmosphere. A particularly noteworthy period was the time of appearance of life on Earth, about 3.9 Gyr ago. Remarkably, one of the proxies

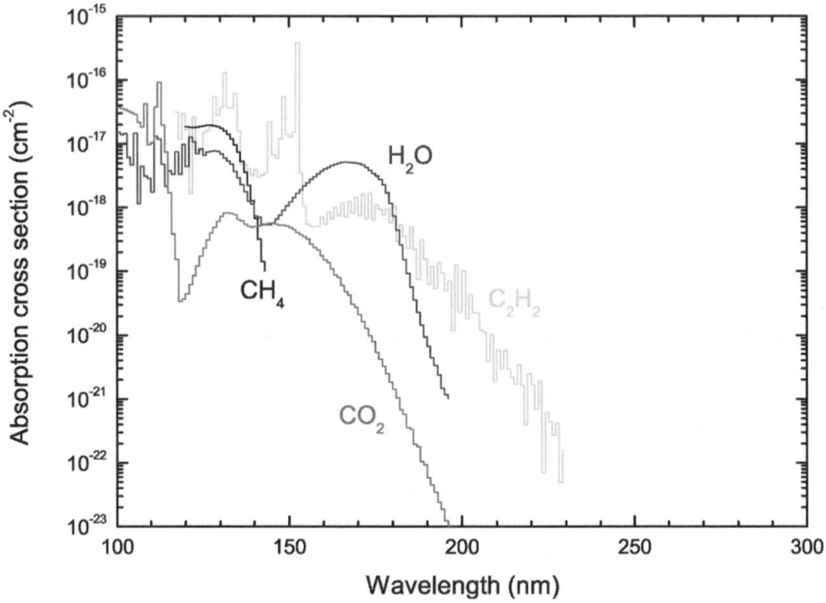

Figure 5. Photoabsorption cross sections of some molecules suspected to have been present in early Earth's atmosphere according to the model by Pavlov *et al.* (2001). Note the sensitivity of these molecules to photodissociation by UV radiation, which is found to be significantly enhanced in active stars, including the young Sun.

used in the Sun in Time program, κ^1 Cet, happens to be an excellent match for the Sun at at key time in Earth's past. As discussed Ribas *et al.* (2010), κ^1 Cet's UV flux is some 35% lower than the current Sun's between 210 and 300 nm, it matches the Sun's at 170 nm and increases to at least 2–7 times higher than the Sun's between 110 and 140 nm. This wavelength regime is important to the photodissociation of some important molecules expected in the atmosphere of the Earth, such as CO_2, CH_4, H_2O, and C_2H_2 (Pavlov *et al.* 2001), as shown in Fig. 5. The use of the correct UV fluxes indicate that photodissociation rates should have been several times higher than those resulting from the simplistic "theoretical" solar spectrum, as used in most of the previous calculations of the young Earth's atmosphere. This is likely to trigger a very peculiar atmospheric chemistry that will be investigated by using a photochemical model of the primitive Earth's atmosphere. The enhanced H_2O photolysis would also result in a higher escape rate of hydrogen to space, particularly important for the early evolution of Mars and Venus.

Our calculations vividly show that self-consistent planetary atmosphere calculations must account for the much stronger photodissociating radiation of the young Sun. The resulting chemistry could be markedly different from that assumed in most investigations. This is obviously relevant to a very significant point in the Solar System evolution, when life was developing on Earth.

6.3. *Hot Jupiters*

Thermal escape in Solar System planets is today almost negligible, but this is not the case of some of the exoplanets discovered. The most suggestive case is that of "Hot Jupiters", giant planets orbiting at very close distances (a few hundredths of an AU) of their host star. The influence of the stellar radiations can be very important for this class of planets. High-energy emissions heat up the exosphere of the planet and lead to intense evaporation of hydrogen and other atmospheric constituents through an outflow (Lammer *et al.* 2003b). Initial estimates suggested that the escape rate could be high enough to fully evaporate giant planets over a timescale of 1 Gyr (Baraffe *et al.* 2004), leaving only a naked terrestrial core. Further more refined simulations (Yelle *et al.* 2008, and references therein) indicate lower evaporation rates that, when integrated over time, would not imply significant loss of mass for a planet during the course of its lifetime. Certain processes in the upper atmosphere would provide the necessary cooling to counteract the high-energy flux, such as the creation of H_3^+ ions. However, the expected strong emission associated to H_3^+ has not been detected yet (Shkolnik *et al.* 2006).

Besides thermal losses, evaporation driven by particle winds can also be very relevant. The key element here is the existence of a protecting magnetic field. The uncertainties in this case are difficult to overcome. For example, Grießmeier *et al.* (2004) show that the evaporation rate will depend on the magnetic moment of the planet (which is tidally locked) and different assumptions lead to very different conclusions. Direct observations of mass loss from Hot Jupiters should hold the key to understanding the processes that drive the evolution of these planets. Unfortunately, the only observation that could relate to such evaporation process at play (Vidal-Madjar *et al.* 2003, 2008) has raised significant controversy (Holmström *et al.* 2008; Ben-Jaffel 2008).

6.4. *Habitability of terrestrial exoplanets*

The milestone of identifying life outside of our planet and, eventually, beyond our own Solar System is of the utmost impact to science and to humanity as a whole. A key component in this direction is to define the main requirements to habitability. There is substantial agreement in pointing out liquid water as a requisite for the existence of life (or, as least, life that could be recognizable to us). Kasting *et al.* (1993) were pioneers in employing a simple climate model to predict the range of distances around a star that could sustain liquid water bodies on the surface of a planet. This led to the definition of the so-called Habitable Zone. However, even today, there are significant uncertainties as to the extension of the habitable zone because processes like CO_2 cloud formation are poorly known and so are their thermal properties. An example is the case of Mars, which is well within the habitable zone according to the calculations yet it does not have liquid water on its surface (at least in a stable condition). Its mass is not sufficient to keep the volatiles neither to possess active plate tectonics providing the necessary climate stabilization feedback. All this is clearly discussed in the review by Kasting & Catling (2003).

It has become evident over time that the simple definition of the habitable zone can be very incomplete. Habitability is influenced by many factors in addition to the incoming stellar radiation, such as the presence of a magnetic field, the mass of the planet, the existence of plate tectonics, the composition of the atmosphere, and even some second order effects like the frequency of catastrophic impacts or the stability of the obliquity. All these further ingredients need to be put in the models to assess the presence and properties of a planet atmosphere and, ultimately, whether the planet could habitable. A thorough discussion on the wide concept of habitability and its implications can be found in Lammer *et al.* (2009).

A particular case that has raised much attention lately is that of planets around very low mass stars (M-type). Such stars, because of their relative abundance and good prospects for direct study of habitable planets, have become very appealing. There are two main issues that planets around M-type stars need to face. Firstly, according to tidal theory, they would have their rotation captured and thus spin synchronously with the orbital translation motion (implying diurnal and nocturnal hemispheres). Secondly, as explained above, low-mass stars could have extended periods of strong magnetic activity with accompanying intense high-energy and particle emissions (plus frequent flares with associated coronal mass ejections). Both effects have raised concern whether a terrestrial planet around an M-type star could have a stable atmosphere with a climate allowing for the existence of liquid water. Calculations by Joshi *et al.* (1997) showed that a dense atmosphere with efficient circulation could avoid condensation on the night-side of the planet. Regarding evaporation, several studies have evaluated the different conditions (Scalo *et al.* 2007; Lammer *et al.* 2007; Khodachenko *et al.* 2007). Models seem to suggest that a CO_2 rich and dense atmosphere could survive the strong incoming high-energy radiation by efficient IR cooling, but the degree of uncertainty is still large. Photochemical reactions may also play a key role in the atmosphere of terrestrial planets around M-type stars because of the strong UV radiations (Segura *et al.* 2005). A practical example of habitability criteria applied to the Gl 581 system was presented by Selsis *et al.* (2007).

7. Conclusions

Our planet Earth orbits around a dependable star that has provided a suitable environment for life to thrive. But the apparent stability of the Sun is just an illusion. Thanks to our vantage point we are able to scrutinize the solar properties with exquisite detail. Centuries of observation have shown a variety of effects related to the magnetic activity that reveal our Sun as a variable star. Sunspots, faculae, coronal mass ejections, are all different facets of the active Sun. Firsthand experience on Earth has demonstrated that such phenomena have a powerful influence on our planet's upper atmosphere and, perhaps, even on our climate. But consider the current Sun with an increase of its high-energy and particle emissions by two or three orders of magnitude. This would have been the infant Sun. Our planet and the rest of its Solar System siblings had to endure an epoch of heavy irradiations early in their history. The atmosphere of Mars, once it lost its magnetic shielding, succumbed to the intense erosion. The strong emissions also left their imprint in the isotopic ratios of the atmospheres of Venus and Titan, for example. Fortunately, the mass of the Earth and its magnetic protection helped our planet to keep its volatile inventory and eventually allowed for the development of complex life. While our Sun's wild ages lasted for a few hundred million years, other less massive stars may have kept the strong emissions for much longer. Future research will tell us what could happen to a terrestrial planet orbiting a red dwarf star, which stays active for billions of years, and whether habitability can exist in such hostile environment.

References

Ahmad, Q. R., *et al.* 2001, *Physical Review Letters*, 87, 071301
Aschwanden, M. J., Poland, A. I., & Rabin, D. M. 2001, *ARA&A*, 39, 175
Asplund, M., Grevesse, N., Sauval, A. J., & Scott, P. 2009, *ARA&A*, 47, 481
Audard, M., Güdel, M., Drake, J. J., & Kashyap, V. L. 2000, *ApJ*, 541, 396
Ayres, T. R. 1997, *JGR*, 102, 1641
Ayres, T. R., Marstad, N. C., & Linsky, J. L. 1981, *ApJ*, 247, 545
Bahcall, J. 2004, *arXiv:physics/0406040*
Baker, V. R. 2001, *Nature*, 412, 228
Baliunas, S. L. & Vaughan, A. H. 1985, *ARA&A*, 23, 379
Baraffe, I., Selsis, F., Chabrier, G., Barman, T. S., Allard, F., Hauschildt, P. H., & Lammer, H. 2004, *A&A*, 419, L13
Barnes, S. A. 2007, *ApJ*, 669, 1167
Bauer, S. J. & Lammer, H. 2004, Planetary Aeronomy, Springer, Berlin

Ben-Jaffel, L. 2008, *ApJ*, 688, 1352

Bohlin, R. C. 2007, in The Future of Photometric, Spectrophotometric and Polarimetric Standardization, *ASP Conference Series*, 364, 315

Bradley, R. S. & Jones, P. D. 1993, *The Holocene*, 3, 367

Carr, M. H. & Head, J. W. 2003, *JGR (Planets)*, 108, 5042

Cassisi, S. 2009, *American Institute of Physics Conference Series*, 1111, 55

Chabrier, G. & Baraffe, I. 2000, *ARA&A*, 38, 337

Christensen-Dalsgaard, J., di Mauro, M. P., Houdek, G., & Pijpers, F. 2009, *A&A*, 494, 205

Donahue, T. M. 2004, *Icarus*, 167, 225

Donahue, R. A., Saar, S. H., & Baliunas, S. L. 1996, *ApJ*, 466, 384

Dorren, J. D. & Guinan, E. F. 1994, in IAU Coll. 143, The Sun as a Variable Star, eds. J. M. Pap, C. Fröhlich, H. S. Hudson, & S. K. Solanki (Cambridge: CUP), p. 206

Durney, B. 1972, in Solar Wind, eds. C. P. Sonett, P. J. Coleman, & J. M. Wilcox (Washington: NASA), p. 282

Edvardsson, B. 2008, *Physica Scripta Volume T*, 133, 014011

Fröhlich, C. & Lean, J. 2004, *A&A Rev*, 12, 273

Gaidos, E. J., Güdel, M., & Blake, G. A. 2000, *Geophys. Res. Lett.*, 27, 501

Garcés, A., Catalán, S., & Ribas, I. 2010, in preparation

Grießmeier, J.-M., Stadelmann, A., Penz, T., *et al.* 2004, *A&A*, 425, 753

Güdel, M., Guinan, E. F., & Skinner, S. L. 1997, *ApJ*, 483, 947

Güdel, M., Audard, M., Smith, K. W., *et al.* 2003, in Future of Cool-Star Astrophysics: The Twelfth Cambridge Workshop on Cool Stars, Stellar Systems and the Sun, eds. A. Brown, G. M. Harper, & T. R. Ayres (Boulder: Univ. of Colorado), p. 303

Hartle, R. E., Donahue, T. M., Grebowsky, J. M., & Mayr, H. C. 1996, *JGR*, 101, 4525

Holmström, M., Ekenbäck, A., Selsis, F., Penz, T., Lammer, H., & Wurz, P. 2008, *Nature*, 451, 970

Holzwarth, V. & Jardine, M. 2007, *A&A*, 463, 11

Hoyt, D. V. & Schatten, K. H. 1997, The role of the sun in climate change, New York : Oxford University Press

Hunten, D. M., Gérard, J.-C., & François, L. M. 1991, in The Sun in Time, eds. C. P. Sonett, M. S. Giampapa, & M. S. Matthews, (Tucson: The University of Arizona Press), p. 463

Järvinen, S. P., Berdyugina, S. V., & Strassmeier, K. G. 2005, *A&A*, 440, 735

Jones, P. D., Briffa, K. R., Barnett, T. P., & Tett, S. F. B. 1998, *The Holocene*, 8, 455

Joshi, M. M., Haberle, R. M., & Reynolds, R. T. 1997, *Icarus*, 129, 450

Kasting, J. F. & Catling, D. 2003, *ARA&A*, 41, 429

Kasting, J. F., Whitmire, D. P., & Reynolds, R. T. 1993, *Icarus*, 101, 108

Khodachenko, M. L., *et al.* 2007, *Astrobiology*, 7, 167

Kim, Y. C., Demarque, P. & Yi, S. K. 2002, *ApJS*, 143, 499

Kraft, R. P. 1967, *ApJ*, 150, 551

Kulikov, Y. N., *et al.* 2006, *P&SS*, 54, 1425

Kulikov, Y. N., Lammer, H., Lichtenegger, H. I. M., Penz, T., Breuer, D., Spohn, T., Lundin, R., & Biernat, H. K. 2007, *Space Science Reviews*, 129, 207

Lafon, J.-P. J. & Berruyer, N. 1991, *A&A Rev*, 2, 249

Lammer, H., Stumptner, W., Molina-Cuberos, G. J., Bauer, S. J., & Owen, T. 2000, *P&SS*, 48, 529

Lammer, H., Lichtenegger, H., Kolb, C., Ribas, I., Guinan, E. F., & Bauer, S. J. 2003a, *Icarus*, 165, 9

Lammer, H., Selsis, F., Ribas, I., Guinan, E. F., Bauer, S. J., & Weiss, W. W. 2003b, *ApJ*, 598, L121

Lammer, H., *et al.* 2007, *Astrobiology*, 7, 185

Lammer, H., Kasting, J. F., Chassefière, E., Johnson, R. E., Kulikov, Y. N., & Tian, F. 2008, *Space Science Reviews*, 139, 399

Lammer, H., *et al.* 2009, *A&A Rev*, 17, 181

Landini, M., Monsignori Fossi, B. C., Pallavicini, R., & Piro, L. 1986, *A&A*, 157, 217

Lean, J. 1997, *ARA&A*, 35, 33

Lean, J., Beer, J., & Bradley, R. 1995, *GeoRL*, 22, 3195
Lebreton, Y. 2000, *ARA&A*, 38, 35
Lundin, R., Lammer, H., & Ribas, I. 2007, *Space Science Reviews*, 129, 245
Mamajek, E. E. & Hillenbrand, L. A. 2008, *ApJ*, 687, 1264
Meehl, G. A., Arblaster, J. M., Matthes, K., Sassi, F., & van Loon, H. 2009, *Science*, 325, 1114
Micela, G. & Marino, A. 2003, *A&A*, 404, 637
Murray, J. B., Muller, J.-P., Neukum, G., *et al.* 2005, *Nature*, 434, 352
Newkirk, G., Jr. 1980, in The ancient Sun: Fossil record in the Earth, Moon and meteorites, eds. R. O. Pepin, J. A. Eddy, & R. B. Merrill (New York: Pergamon Press), p. 293
Parker, E. N. 1970, *ARA&A*, 8, 1
Pavlov, A. A., Brown, L. L., & Kasting, J. F. 2001, *JGR*, 106, 23267
Pizzolato, N., Maggio, A., Micela, G., Sciortino, S., & Ventura, P. 2003, *A&A*, 397, 147
Ribas, I., Guinan, E. F., Güdel, M., & Audard, M. 2005, *ApJ*, 622, 680
Ribas, I., Morales, J. C., Jordi, C., Baraffe, I., Chabrier, G., & Gallardo, J. 2008, *Memorie della Societa Astronomica Italiana*, 79, 562
Ribas, I., Porto de Mello, G. F., Ferreira, L. D., Selsis, F., Hébrard, E., Catalán, S., Garcés, A., Nascimento, J. D. Jr, & de Medeiros, J. R. 2010, in preparation
Rottman, G. J. 1988, in Solar Radiative Output Variation, ed. P. Foukal (Boulder: NCAR), p. 71
Sackmann, I.-J. & Boothroyd, A. I. 2003, *ApJ*, 583, 1024
Sagan, C. & Mullen, G. 1972, *Science*, 177, 52
Scalo, J., *et al.* 2007, *Astrobiology*, 7, 85
Segura, A., Kasting, J. F., Meadows, V., Cohen, M., Scalo, J., Crisp, D., Butler, R. A. H., & Tinetti, G. 2005, *Astrobiology*, 5, 706
Selsis, F. 2000, *PhD Thesis*, Univ. Bordeaux
Selsis, F., Kasting, J. F., Levrard, B., Paillet, J., Ribas, I., & Delfosse, X. 2007, *A&A*, 476, 1373
Shkolnik, E., Gaidos, E., & Moskovitz, N. 2006, *AJ*, 132, 1267
Simon, T., Boesgaard, A. M., & Herbig, G. 1985, *ApJ*, 293, 551
Skumanich, A. 1972, *ApJ*, 171, 565
Soderblom, D. R. 1982, *ApJ*, 263, 239
Soon, W. W.-H. & Yaskell, S. H. 2004, The Maunder minimum and the variable Sun-Earth connection, World Scientific Publishing
Stauffer, J. R., Caillault, J.-P., Gagne, M., Prosser, C. F., & Hartmann, L. W. 1994, *ApJS*, 91, 625
Stelzer, B. & Neuhäuser, R. 2001, *A&A*, 377, 538
Stevenson, D. J. 2009, *Space Science Reviews*, in press
Strassmeier, K. G. 2009, *A&A Rev*, 17, 251
Terada, N., Kulikov, Y. N., Lammer, H., Lichtenegger, H. I. M., Tanaka, T., Shinagawa, H., & Zhang, T. 2009, *Astrobiology*, 9, 55
Vidal-Madjar, A., Lecavelier des Etangs, A., Désert, J.-M., Ballester, G. E., Ferlet, R., Hébrard, G., & Mayor, M. 2003, *Nature*, 422, 143
Vidal-Madjar, A., Lecavelier des Etangs, A., Désert, J.-M., Ballester, G. E., Ferlet, R., Hébrard, G., & Mayor, M. 2008, *ApJ*, 676, L57
Vilhu, O. & Walter, F. M. 1987, *ApJ*, 321, 958
Webb, D. F. & Howard, R. A. 1994, *JGR*, 99, 4201
Wilson, O. C. 1966, *ApJ*, 144, 695
Wood, B. E., Müller, H.-R., Zank, G., & Linsky, J. L. 2002, *ApJ*, 574, 412
Wood, B. E., Müller, H.-R., Zank, G. P., Linsky, J. L., & Redfield, S. 2005, *ApJ*, 628, L143
Woods, T. N., Rottman, G. J., Bailey, S. M., Solomon, S. C., & Worden, J. R. 1998, *Solar Phys.*, 177, 133
Yelle, R., Lammer, H., & Ip, W.-H. 2008, *Space Science Reviews*, 139, 437
Yi, S. K., Kim, Y. C., & Demarque, P. 2003, *ApJS*, 144, 259
Zahnle, K. J. & Walker, J. C. G. 1982, *Rev. Geophys. Space Phys.*, 20, 280
Zurbuchen, T. H. 2007, *ARA&A*, 45, 297

Session II

Observations of solar and stellar variability

Chair: A. G. Kosovichev

Solar and Stellar Variability: Impact on Earth and Planets
Proceedings IAU Symposium No. 264, 2009
A. G. Kosovichev, A. H. Andrei & J.-P. Rozelot, eds.

doi:10.1017/S1743921309992304

One solar cycle of solar astrometry with MDI/SOHO

Marcelo Emilio[1], Jeff R. Kuhn[2] and Rock I. Bush[3]

[1]Observatório Astronômico-Departamento de Geociências Universidade Estadual de Ponta Grossa, Paraná, Brazil
email:memilio@uepg.br

[2]Institute for Astronomy, University of Hawaii, Woodlawn Dr. 96822, HI, USA
email:kuhn@ifa.hawaii.edu

[3]Solar Physics Group, Stanford University, Stanford, CA, 94305, USA
email:rbush@solar.stanford.edu

Abstract. In this work we describe the method and results of precise solar astrometry made with the Michelson Doppler Imager (MDI), on board the Solar and Heliospheric Observatory (SOHO), during one complete solar cycle. We measured an upper limit to the solar radius variation, the absolute solar radius value and the solar shape. Our results are 22 mas peak-to-peak upper limit for the solar radius variation over the solar cycle, the absolute radius was measured as 959.28 ± 0.15 arcsec at 1 AU and the difference between polar and equatorial solar radii in 1997 was 5 km and about three times larger in 2001.

Keywords. Sun: fundamental parameters, (Sun:) solar-terrestrial relations, Sun: oscillations, Sun: photosphere

1. Introduction

The Michelson Doppler Imager (MDI) on board the Solar and Heliospheric Observatory (SOHO) satellite measured the variation and the absolute value of the solar radius and the solar shape (Kuhn *et al.* 1998, Emilio *et al.* 2000, Kuhn *et al.* 2004, Emilio *et al.* 2007, Kuhn *et al.* 2009) over a complete solar cycle now.

Several ground-based experiments have been proposed in order to describe the phenomenon of solar radius variations that led to contradictory results. Gilliland (1980) and Laclare *et al.* (1996) claim to have found a negative correlation when the variation in solar diameter was compared with the solar cycle, while other groups (Ulrich & Bertello 1995 and Noël 1997) found a direct correlation or no correlation (Emilio & Leister 2005, Wittman & Bianda 2000, Brown & Chistensen-Dalsgaard 1998) at all. By eliminating the effects due the terrestrial atmosphere, the precision for relative variations was found to be better than any previously-reported ground-based measurements.

The problem of determining the solar radius from CCD imagery is, in principle, well defined. In general its solution involves: 1) detecting a fiducial reference height in the solar atmosphere, that for this work is defined from the inflection point in the radial limb darkening function (LDF), and 2) deriving the physical image scale of the optical system. Solar radius measurements are obviously limited when changes in the real or apparent limb profile obscure the fiducial height in the atmosphere. This occurs, for example, if the vertical temperature structure in the solar atmosphere changes or, more importantly, if the point spread function of the optical system varies significantly. The image scale is also affected by perturbations in the optical instrument. The precision of the MDI radius measurements is of order 1 milliarcsec and at this scale several instrumental apparent

0.01 pixel. Errors in the early FB=1 data were as large as 0.01 pix and we discard these in the following analysis.

Periodically we empirically determined the best MDI focus. This was derived from solar images obtained at each focus block setting. The intensity contrast from each image was determined and averaged over a region near the image center. The best fractional focus block setting was then obtained by interpolating the integer domain empirical dependence of contrast on focus block setting to find the maximum contrast and best focus condition.

Each of these thermally induced perturbations affect the final image scale and the best focus of the instrument. We compute the change in the apparent solar radius from our ZEMAX ray trace model by computing how the middle of the blur circle of a field point at an angle of 0.267° (approximately one solar radius) is shifted. The best focus is obtained by computing the rms blur radius at each focus block (FB) optical configuration and then interpolating to the fractional FB value which minimizes the blur diameter. The model optical configuration is symmetric about the optic axis so that this calculation can only depend on the radial field angle.

The accuracy of these corrections is limited by our knowledge of the precise form of the temperature perturbations. For example, we do not know the true temperature gradients in the MDI OSS, nor do we know the actual temperature distribution across the front window. Based on our knowledge of the model parameters and experience with the perturbed optical system we believe the OSS expansion term is accurate to about 50%, while the primary lens perturbation is accurate to 10%. Thus, the window gradient term is particularly sensitive to the absolute instrument focus condition, which is relatively weakly determined. The gradient sensitivity we calculate here is consistent with the MDI optics model but could change by a factor of two within our parametric model uncertainties. Here we refer to the "accuracy" of a perturbation as the fractional uncertainty in derived focus and radius changes caused by a given temperature change.

Figure 4 indicates that an accurate solar radius determination must rely on understanding and correcting for systematic radius noise sources and that the instrument temperature perturbations clearly dominate these systematics. Figure 1 showed that the window temperature, primary lens, and secondary temperatures vary in phase with the

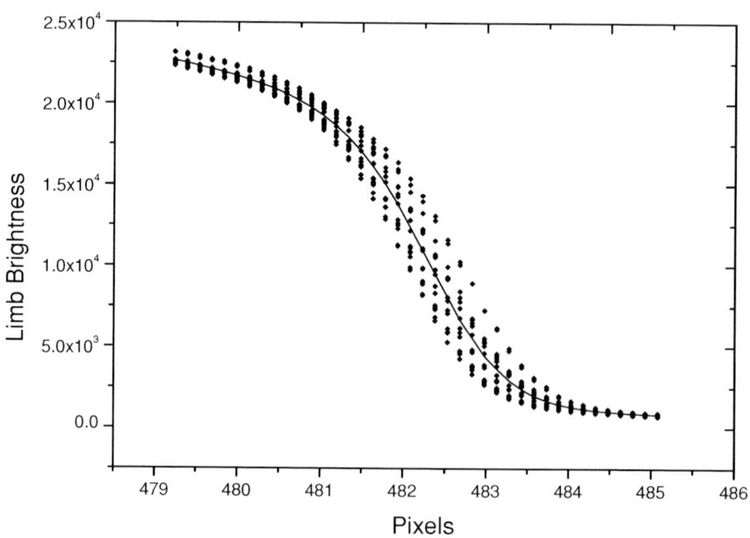

Figure 2. This figure shows a family of limb darkening functions (LDF's), $I(r, \theta_i)$, for a typical image. Each 22°.5 sector yelds a LDF for that angular bin within about 6" of the limb.

yearly apparent radius change of 0.2 arcsec-AU in Figure 4. In our discussion below we use the secondary lens measurements as a proxy for the average OSS temperature.

Important features of the MDI temperature record are: 1) the yearly variation due to the changing incident solar heat flux, 2) the front window opening transient response near the end of 1998 and, 3) the deliberate change in MDI heater power levels in 1998 and 2002. These temperature perturbations can be directly related to the apparent radius and focus changes using the optical perturbation model.

An important check of the thermal response model comes from a calculation of the *a priori* apparent solar radius changes caused by the yearly fluctuations in the window temperature gradient, the primary lens temperature, and OSS structure temperature. Thus, using empirical yearly temperature changes and the apparent radius sensitivities computed, we calculate a radius change of 0.1 pixel and focus change of -0.1 FB that are synchronous with the annual variation in solar heating. Given the systematic errors in the three components which add and subtract to yield it, it may be fortuitous that this expected change of 0.1 pixel is in such good agreement with the observed 0.1 pixel (0.2 arcsec-AU) annual variation. Evidently our simplified thermal model is consistent with the observed annual MDI fluctuations.

The transient response of MDI, after opening the front window and allowing solar radiation to enter the instrument in late 1998, is expanded in Figure 5. From the thermal model of the front window we've also plotted the edge-to-center temperature difference versus time. Figure 5 also shows the derived best focus position and the change in the observed solar radius (at focus block 4). The data obtained from MDI as it responded to this relatively large thermal transient allow us to quantitatively test the different thermal contributions to the optical variability of the best focus and apparent radius.

The front window eventually heats by 28 °C, and after 95 minutes has risen in temperature to 80% of this maximum. According to the finite element thermal model, the

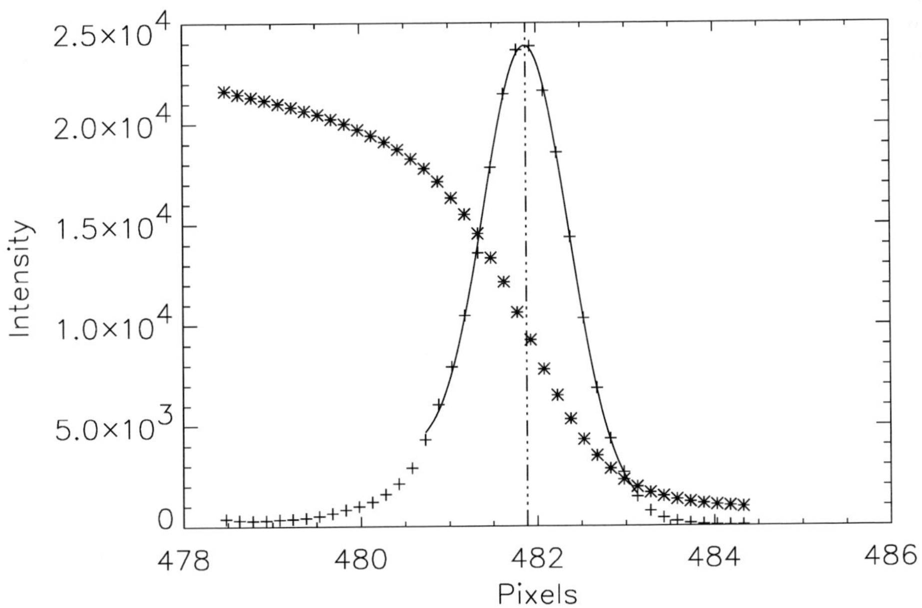

Figure 3. Typical limb-darkening function data from a single observation. The limb intensity is plotted with star symbols. The squared radial derivative is shown with plus sysmbols, and the curve shows the derived fit whose maximum value determines the solar radius.

window temperature gradient reaches a maximum of about 4 °C after 40 minutes, while after only 15 minutes it is at 80% of this maximum. The primary lens heats by 13 °C but achieves 40% of this after 95 minutes. The secondary lens has an even smaller and slower response, increasing by 8 °C and only 25% of this maximum after 95 minutes.

The best focus setting reaches a minimum after only 40 minutes and declines by 3.7 focus blocks. It then increases by 0.8 FB after 95 minutes. The apparent radius increases by 0.3 pixel over this short period, then decreases by 0.05 pixel between 40 and 95 minutes after opening the front window aperture cover.

Qualitatively these results are consistent with the temperature perturbations of our model. The window gradient dominates the focus and radius changes during the first 45 minutes. According to the model a gradient of 4 °C should decrease focus by 3 FB but increase the apparent radius by 0.3 pixel as it does. This result is in, perhaps fortuitous, agreement with the observed radius change. Nevertheless the window gradient is the only effect which accounts for the large focus shift and apparent radius increase over this short duration.

As part of monthly calibration observations, MDI obtains solar images over a range of focus blocks in order to determine the instrument best focus. Figure 6 plots the best focus over the MDI history. While these data are noisy, a hint of a yearly focus variation around the trend $F(t)$ is apparent and described by a quadratic function.

Encouraged by our success in describing the apparent radius changes caused by variable thermal effects we attempt a "non-statistical" (*a priori*) direct correction for the radius timeseries by adjusting, and scaling the known temperature perturbations using the model temperature sensitivities. Because the temporal linear increasing trend in the front window temperature does not directly imply an increasing window radial gradient we fit and remove this from the window timeseries data. We already showed that it has an insignificant effect on the OSS expansion and primary lens focal length changes so, for simplicity, we also remove this trend term from the other temperature records.

As we argued above, the internal optics heater changes do not affect the window gradient so we also remove this jump from the window temperature record. This corrected

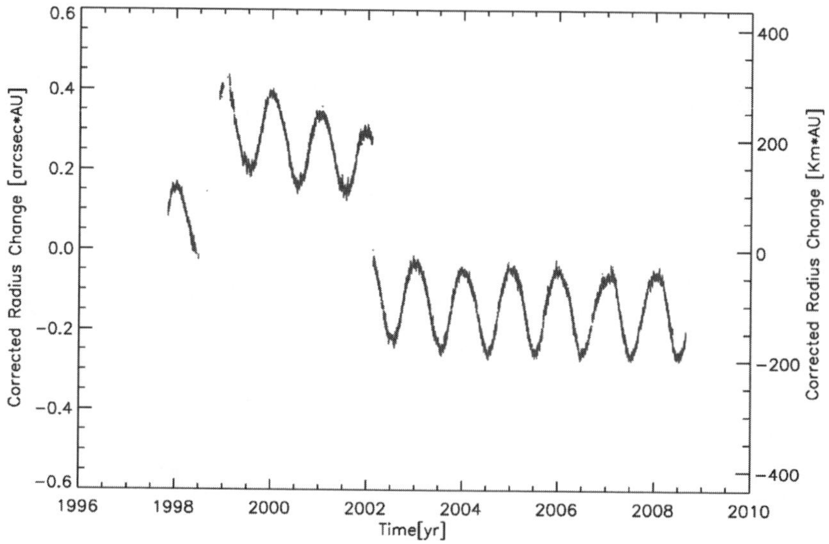

Figure 4. Residual apparent solar radius changes are plotted here after correcting for focus block changes in the instrument to an effective observing distance of 1 AU.

window temperature, $W(t)$, is then a proxy for apparent radius changes due to the annual window gradient variations when scaled by 0.01 pixel/°C. Our previous results (Kuhn *et al.* 2004) for the *a priori* model we scaled the primary lens temperature, $P(t)$ by -0.009 pixel/°C and the gradient-adjusted secondary lens (OSS) temperature, $S(t)$, scales by 0.038 pixel/°C. The final gradient correction comes from the linear trend in best focus from Figure 6. As we argued above, this is due to the decreasing radial window gradient from the increasing front window opacity.

This final correction to the apparent radius timeseries $\delta r(t)$, comes from scaling the linear trend in focus, $F(t)$, by -0.097 pixel/FB.

Recognizing that our *a priori* temperature sensitivities are based on an imperfect model we have also allowed the scaling coefficients to vary in a least-squares fitting procedure which minimizes the deviation between the radius variation data and the sum of the scaled temperature proxy functions. Thus we seek a solution for coefficients w, p, s, and f that minimizes $\sum_t (dr(t) - wWA(t) - pP(t) - sS(t) - fF(t))^2$. Some of these functions are strongly correlated and in order to constrain the solution and decrease the number of free parameters we fix $w = 0.01$ pixel/°C to the *a priori* value and allow only three fitting variables.

We obtain values for sensitivity parameters of $p = -0.011 \pm 0.002$, $s = 0.041 \pm 0.002$ pixel/°C and $f = -0.079 \pm 0.005$ pixel/FB. Those values are different from our previous statistical model in Kuhn et al (2004). Here we obtained the coefficients from the full solar cycle dataset. The coefficients are consistent with our physical model and their expected uncertainties. The residual radius variation obtained after subtracting the best-fit function is plotted in Figure 7.

The data with fitted corrections in Figure 7 are insensitive to any real secular solar radius variability because they are degenerate with the fitted secular trend in the window

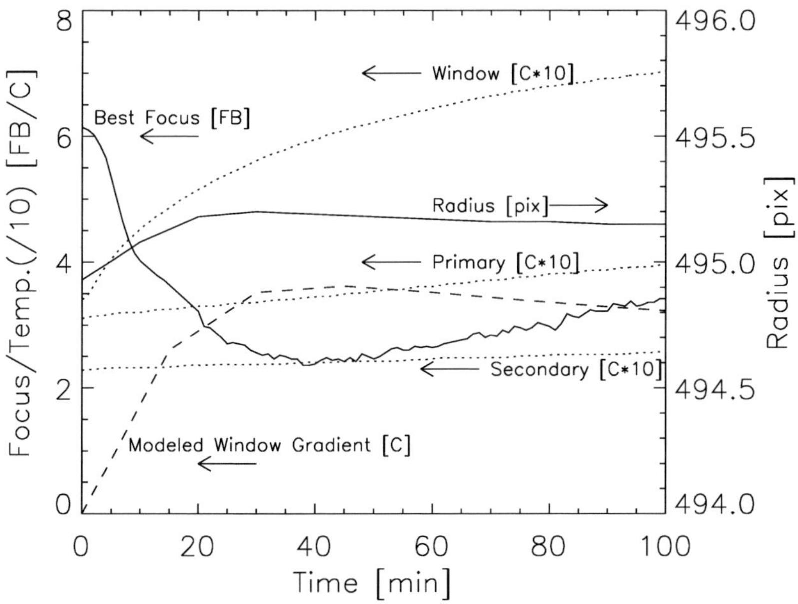

Figure 5. The transient thermal and optical response of MDI after the front window was opened 1998 November 20. Time is measured along the horizontal axis in minutes after the instrument is exposed to sunlight. The best-focus setting in focus block units, the modeled window gradient (in °C), the temperatures of the window, primary and secondary lens (in 10's of °C) are indicated on the left. The apparent solar radius (in pixels) is shown on the right vertical scale.

gradient correction. While we believe the trend in the radius variation data is only caused by the imperfect thermal model, we use these residuals to derive a conservative upper limit to any secular solar radius change. The best-fit line to these data suggests that any linear trend in the solar radius must be smaller than 0.9 ± 1.6 milliarcsec/year. The uncertainty were calculated using the bootstrap procedure. This is not a statistically significant trend and implies a 2 σ upper limit of 3.2 mas/yr or 0.3 arcsec/century. In comparison Shapiro (1980) obtained the same formal upper limit from about 200 years of Mercury transit data.

After scaling the sunspot number solar cycle variation to unity range we have fit this function to the residual radius data using the bootstrap procure. The derived solar cycle radius variation in the statistical corrected data is -17 ± 11 mas peak-to-peak over a solar cycle. This result, which is not statistical significant implies a 2σ limit on the amplitude of the solar cycle of 22 mas.

4. Absolute Solar Radius Determination

The image scale has been determined from the May 7 2003 Mercury transit data separately for each focus between FB 3 and 6 (Kuhn *et al.* 2004). Each of these absolute image scale measurements has a statistical uncertainty of less than 0.0001 arcsec/pixel. In contrast the ray trace model and solar limb data provide accurate measurements of the *differential* scale change between focus settings. For the solar limb data obtained at focus 3 and 4 (essentially all of the useful radius data) these changes imply an uncertainty in the differential image scale of about 2×10^{-5} arcsec/pixel, which is somewhat better than the fractional absolute scale determination uncertainty.

We compute the solar radius from the limb average of the MDI image radius. These data must be corrected for the average radial image distortion which causes the Sun to appear larger than it is. This amounts to 0.92 ± 0.05 pixel where 0.32 pixel results

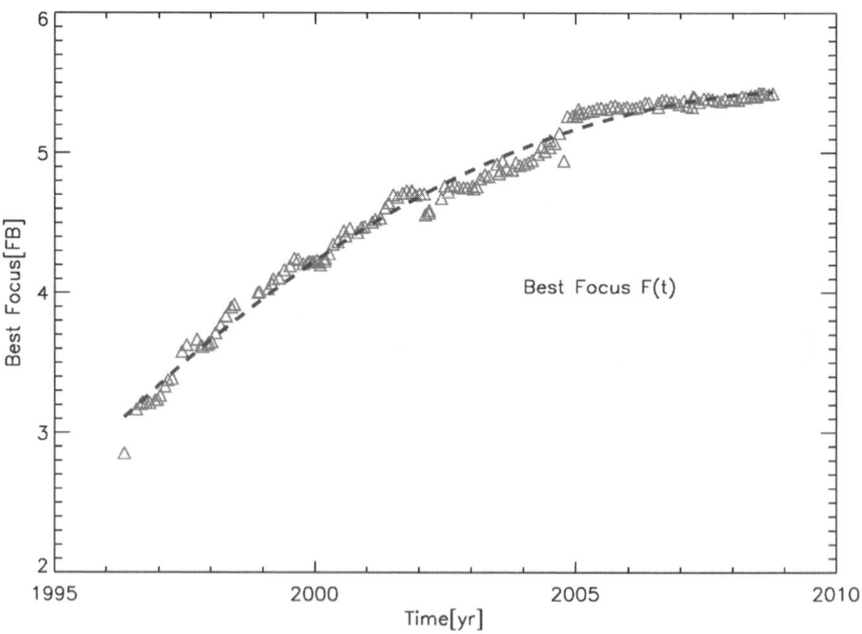

Figure 6. Change in the best-focus in units of focus blocks. The quadratic temporal trend is plotted as the dashed line.

from image tilt and 0.6 pixel comes from the symmetric image distortion. The distortion uncertainty quoted here comes from the 70% systematic uncertainty limits in the transit data fit to the model distortion function.

Immediately before and after 2003 May 7, MDI was operating with FB 4. The mean solar image radius near this time was 484.371 ± 0.0015 pixel. Correcting for distortion we obtain 483.43 ± 0.05 pixel. With FB 4 calibration of 1.9870 ± 0.0002 arcsec/pixel we derive a solar radius of 960.58 ± 0.15 arcsec (as seen from SOHO) where our quoted error includes the statistical, distortion, and calibration uncertainties. During the transit SOHO was 0.99865 AU from the Sun, so that the apparent solar radius at 1 AU is 959.28 ± 0.15 arcsec and the error reflects all known statistical and systematic uncertainties.

5. Solar Shape

Small changes in the form of the local LDF, $L(r, \theta)$, at solar colatitude θ, contain solar shape and radius information. We measure θ in the detector array coordinates, but SOHO/MDI keeps the projected rotation axis of the sun accurately lined with the vertical axis of the array. Distinguishing the raw shape signal, $\beta(\theta)$ – the radially displaced mean LDF defined by $L(r - \beta(\theta))$ – from other solar and instrumental perturbations to $L(r, \theta)$ is non-trivial. This is because the solar shape is typically 2 to 3 orders of magnitude smaller than the LDF measurement pixel size (approximately 2 arcsec in MDI) and about the same factor smaller than the diffraction limit of the telescope. At this level, small latitudinal variations in the solar atmosphere and instrumental effects can contribute to the observed LDF and the apparent shape signal.

Full-disk images were obtained in 2001 as they were in 1997 by stepping SOHO by 30 degree increments through 360 degrees in roll angle. At each position approximately 5 MDI continuum filtergram images were obtained with the MDI fast tip-tilt steering mirror engaged. These images were reduced to measurements of the local LDF in the

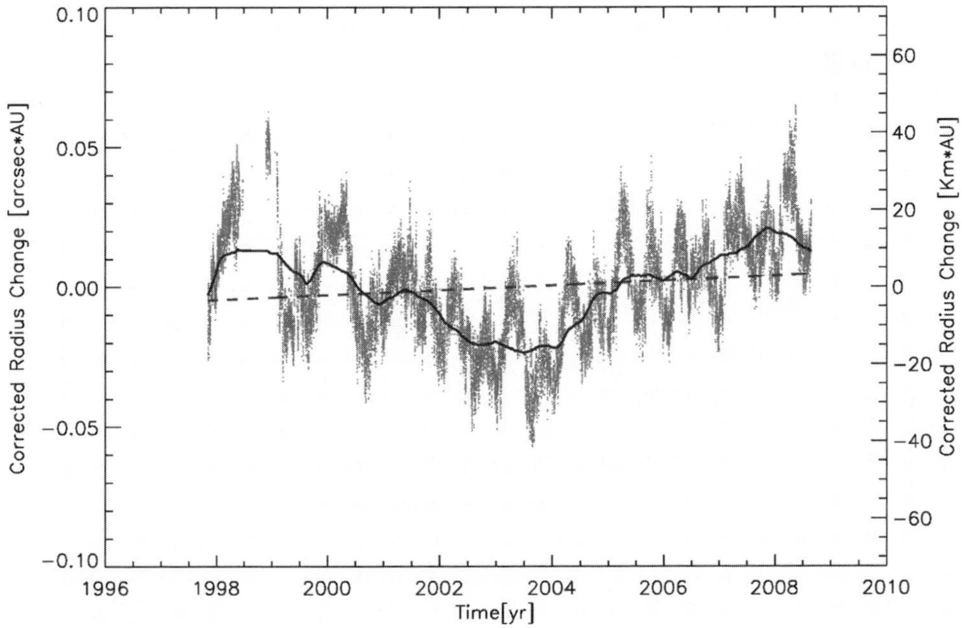

Figure 7. Residual variation of the statistical least-square model.

same way 1997 observations were analyzed except that the data were binned into 360, 1 degree angular bins.

In 1997 we also suppressed instrumental effects by rotating the spacecraft. The derived shape signal $\beta = \beta_s + \beta_i$ consists of slow spatially varying instrumental (β_i) and solar (β_s) contributions. The stability of the instrument during the satellite roll procedure allows us to recover β_s even when it is much smaller than β_i. We modify our 1997 analysis to account for localized shape function contamination from active regions near the limb. We also improved the determination of the mean LDF used as the fitting functions in our least-squares fit for α, β and γ parameters. The previous analysis used the first order LDF which was smeared by the instrumentally induced limb shape distortion. In this work we corrected this effect by measuring the r shift of the smeared $L(r)$ with relation to $L(r, \theta)$. These shifts, $\beta(\theta)$, are then used to shift $L(r - \beta(\theta), \theta)$ and then average over θ to compute an unsmeared LDF.

Our model for the apparent LDF is

$$L(r, \theta) = (1 - \alpha(\theta))L(r - \beta_s(\theta) - \beta_i(\theta) - F(\theta)) + \gamma(\theta)L''(r)$$

In practice we made a least square fit using our empirical (unsmeared) LDF function $L(r)$ and its first and second derivatives as fitting functions. The corresponding coefficients $1 - \alpha(\theta)$, $\beta(\theta)$ and $\gamma(\theta)$ are found for each roll angle. Initially $F(\theta)$ is taken to be zero, but we redo this fit using the $\alpha(\theta)$ obtained from the previous iteration with $F = f\alpha$ (the constant f is described below). The solar limb shape function, β_s is recovered, as it was in 1997, from the sequence of 12 rotated LDF datasets using a phase modulation

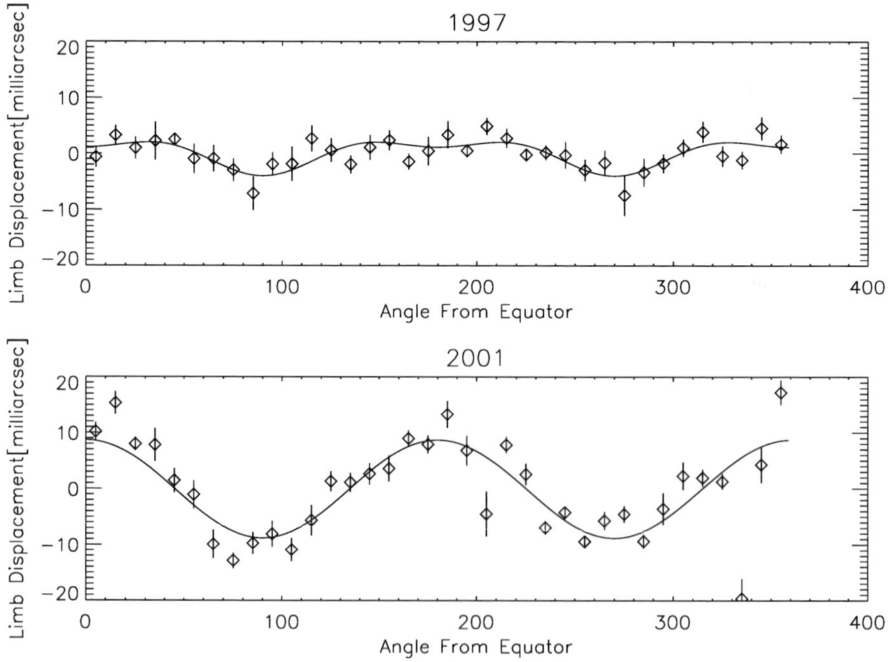

Figure 8. The limb shape from 1997 and 2001 is plotted here. The equator corresponds to position angle 0. Data have been averaged into 10 degree wide bins and the best fit $l = 2$ (oblateness) and $l = 4$ (hexadecapole) Legendre Polynomial have been plotted. The shape distortion after correcting for bright contamination is nearly a pure oblateness term in 2001, while 1997 has a significant hexadecapolar shape contribution.

approach (Kuhn *et al.* 1998) This method averages azimuthal fourier transforms of the $\beta(\theta)$ transforms from each roll angle to separate β_i from β_s.

Figure 8 shows how the faculae-corrected limb shape changed from 1997 to 2001 (Emilio *et al.* 2007). The solid curve in each graph is a fit to a quadrupole and hexadecapole shape distortion. Our previous analysis of the 1997 data (Kuhn *et al.* 1998) yielded marginally different (at the 2-σ level) results. That analysis did not use all of the image stabilized limb observations from the spacecraft and it did not include fitting terms for the known tilted-CCD instrumental effect in the MDI detector. The binned data in figure 2 shows a difference between the equatorial and polar radii in 1997 of 8.7 ± 2.8 milliarcsec and 18.9 ± 1.9 milliarcsec in 2001. Least squares fits of a second and fourth order legendre polynomial give a solar shape that is nearly purely oblate near solar maximum, but has a significant hexadecapole shape near minimum. The oblateness shape term changed from -3.1 ± 0.7 to -12 ± 2 milliarcsec while the hexadecapole went from -1.7 ± 0.9 to 0.3 ± 2.5 milliarcsec from 1997 to 2001.

6. Conclusion

The SOHO MDI experiment has yielded accurate measurements of the solar radius and radius variability from above the Earth's atmosphere. We have shown that it is possible to determine the temperature sensitivity and optical characteristics of MDI with sufficient accuracy so that a statistical calibration of instrumental systematics is not necessary. The limit MDI achieves on possible solar cycle radius variations is 22 milliarcsec.

This is the first accurate optical solar radius measurement from above the atmosphere. The derived radius is precisely defined with respect to the inflection point in the unblurred solar limb darkening profile. Adopting the value 1 AU $= 1.495979 \times 10^8$ km we find the Sun's radius to be $R_\odot = 6.9574 \pm 0.0011 \times 10^5$ km. This is slightly smaller than recent ground-based measurements (Brown and Christensen-Dalsgaard 1998) but is consistent with the highly precise helioseismic determination (Schou *et al.* 1997) of $6.9568 \pm 0.0003 \times 10^5$ km. The fundamental limitation to our measurement uncertainty comes from incomplete knowledge of the MDI optical distortion. Our limits on solar radius variability are consistent with earlier precise MDI and helioseismic measurements (Emilio *et al.* 2000 and Dziembowski et al. 2001) but are established here from image data without assumptions about systematic temporal instrument variations.

A significant discrepancy between ground-based observations and these results is clear. The ensemble of solar radius and variability data from the ground instruments suggests that they may be seeing effects caused by the terrestrial atmosphere and potentially solar cycle induced variability in the Earth's atmosphere. This may explain the apparent solar radius variations with instrument latitude and phase during the solar cycle, although this remains an important issue to be resolved.

The difference between polar and equatorial solar radii in 1997 was 5 km and it was about 3 times larger than this in 2001. While these are highly statistically significant values they are tiny in comparison to physically interesting length scales in the photosphere. For example the density scale height is 150 km and the smallest (granulation) convection scale near the photosphere is 700 km.

Acknowledgements

This research was supported by the NASA SOHO/MDI grant NNX07AK36G to Stanford University and the NASA SDO/Helioseismic and Magnetic Imager contract NAS5-02139 to Stanford University and subcontract to the University of Hawaii.

References

Brown, T. M. & Christensen-Dalsgarrd, J. 1998, *ApJ*, 500, L195

Dziembowski, W. A., Goode, P. R., Kosovichev, A. J., & Schou, J. 2001, *ApJ*, 553, 897

Emilio, M., Kuhn, J. R., Bush, R. I., & Scherrer, P. 2000, *ApJ*, 543, 1007

Emilio, M. & Leister, N. V. 2005, *MNRAS*, 361, 1005

Emilio, M., Bush, R. I., Kuhn, J. R., & Scherrer, P. 2007, *ApJ*, 660, L161

Gilliland, R. L. 1980, *ApJ*, 248, 1144

Kuhn, J. R., Bush, R. I., Scherrer, P., & Scheick, X. 1998, *Nature*, 392, 155

Kuhn, J. R., Bush, R. I., Emilio, M., & Scherrer, P. H. 2004, *ApJ*, 613, 1241

Kuhn, J. R., Emilio, M., & Bush, R. I. 2009, *Science*, 324, 1143-b

Laclare, F., Delmas, C., Coin, J. P., & Irbah A. 1996, *Solar Phys.*, 166, 211

Wittmann, A. D. & Bianda, M. 2000, In: Proceedings of the 1st Solar and Space Weather Euroconference, ESA SP 463, 113

Scherrer, P. H., Bogart, R. S., Bush, R. I., Hoeksema, J. T., Kosovichev, A. G., Schou, J., Rosenberg, W., Springer, L., Tarbell, T. D., Title, A., Wolfson, C. J., Zayer, I., & MDI Engineering Team 1995, *Solar Phys.*, 162, 129

Schou, J., Kosovichev, A. G., Goode, P. R., & Dziembowski, W. A. 1997, *ApJ*, 489, 197

Shapiro, I. I. 1980, *Science*, 208, 51.

Ulrich, R. K. & Bertello, L. 1995, *Nature*, 377, 214

Solar and Stellar Variability: Impact on Earth and Planets
Proceedings IAU Symposium No. 264, 2009
A. G. Kosovichev, A. H. Andrei & J.-P. Rozelot, eds.

doi:10.1017/S1743921309992316

Helioseismic inferences

Hiromoto Shibahashi

Department of Astronomy, University of Tokyo, Tokyo 113-0033, Japan
email: shibahashi@astron.s.u-tokyo.ac.jp

Abstract. The brilliant outcome of some 30 years of helioseismology spreads over a wide range of topics. Some highlights relevant to the cause of the solar activity cycle are listed up. The rotation profile in the solar convective zone is discussed as an important source of the dynamo mechanism. The kinematic dynamo model is described in the linear approximation, and the condition for the solar type dynamo is derived. It is shown that comparison of this condition with the rotation profile determined from helioseismology is useful to identify the possible seats of the dynamo.

Keywords. magneothydrodynamics: MHD, Sun: activity, Sun: helioseismology, Sun: magnetic fields

---

1. Introduction

It is almost half a century ago that the solar surface was found by Leighton *et al.* (1962) to be almost entirely covered with vertically oscillating elements. As observations were accumulated and they were compared in detail with various theoretically proposed interpretations, it became clear by the late 1970s that these oscillations are superposition of nonradial p-modes of the sun with degrees $l \simeq$ 200-1000. Observations of the sun using the integrated sunlight over the whole disk, which were first performed by the Birmingham group also in the late 1970s, clarified that the sun is also oscillating with p-modes with low degrees $l \simeq$ 0-4. The gap between low l and high l p-modes in detection of solar oscillations was later filled, and it became eventually clear that the sun is pulsating with millions of its eigenmodes. The detection of the eigenmodes of the sun has opened a new field of research called helioseismology, providing a new tool to see the invisible interior of the sun. Ground-based network observations (BiSON, GONG, IRIS, and TON) and uniterrupted observations from space (SOHO) developed significantly helioseismolgy, and various helioseismic methods are now providing new insight into understanding of the solar acitivity, which is one of the most important issues in solar physics.

2. Outcome of some 30 years of helioseismology

The outcome of some 30 years of helioseismology is brilliant and spreads over a wide range of topics. Listed below are only some highlights.

• Finding of the fact that the sun is oscillating with a lot of its eigenmodes, —that is, the sun is now regarded as a pulsating star while it had never been thought so before.

• The recent observations of solar oscillations provide us with accurate frequencies of a large number of solar eigenmodes. The relative errors in some of the measured frequencies are as small as 10^{-6}.

• Determination of the sound-speed and density profiles in the sun: By the inversion of these frequencies, we can determine functional forms of the acoustic quantities, the sound speed profile and the density profile, as a function of radius respectively. We now know the sound speed profile within the errors of a few tenths of a percent, and the density profile with the order of 1% errors, in the range of $0.2 \lesssim r/R_\odot \lesssim 0.95$.

• Calibration of the depth of the convection zone: Convection is still one of the poorly understood physics in making theoretical models of stars. However, the extent of the convection zone in the sun is now accurately determined from the sound-speed profile determined helioseismically .

• Finding of a new mechanism of driving eigenoscillations, —that is, turbulent convection in a star generates acoustic noise propagating inside the star, and the waves resonating in the star, which is regarded as a wave cavity, survive and are maintained at some level.

• Based on the discovery of very small amplitude pulsations found from observations of the sun as a star, we now expect that pulsations, at least with small amplitudes, will be ubiquitous among stars in general. This opens the possibility of a seismological approach to stars, which is now called "asteroseismology."

• Calibration of "standard evolutionary models": The sound-speed profile determined from helioseismology matches fairly well up to that of solar models constructed with standard evolution theory of stars, though such beautiful matching might be broken if the recently determined chemical abundances of the sun.

• Inference of helium abundance: A lack of appropriate spectroscopic lines of helium makes it hard to determine spectroscopically abundance of helium in the solar atmosphere. Helioseismology provides new methods of determination of helium abundance.

• Proof of importance of chemical diffusion and gravitational settling: Theoretical models of the sun constructed taking account of chemical diffusion and gravitational settling provides much better agreement with the frequencies of various modes, and then these processes are now regarded as being important even in the chemically normal stars.

• Inference of the solar internal rotation: If the sun would not be rotating, the eigenfrequencies of the modes with the same degree l and the radial order n would show the $(2l+1)$-fold degeneracy. The degeneracy is lifted by rotation in reality. Analyses of this rotational splitting make it possible to determine the rotation profile in the sun as a function of the distance from the center and of the latitude. The rotation profile deduced from helioseismology is described more in detail in the next section.

• Finding of the "tachocline" layer: It was found that there is a shear layer just beneath the base of the convection zone, which is named the "tachocline" layer. A mixing process related with convective overshooting or rotational shear is likely to be induced. If the small scale turbulence in the thin layer is induced by shear instability, it will help to maintain the rigid rotation in the deeper zone.

• Finding of the "torsional oscillation" in the convection zone: Helioseismic investigation using p-modes as well as f-modes makes it possible to see the alternating zonal bands of slightly faster and slower rotation are superposed on the relatively smooth latitudinal variation in the rotational profile. These bands are so far found to reach to the depth of about 5% of the solar radius.

• Finding of the meridional motion in the convection zone: The latitudinal differential rotation would not be maintained without the transport of angular momentum from higher latitudes to lower latitudes or the other way around. Thus the presence of a meridional flow has long been expected. While the observational results about the meridional flow so far available are still limited to the horizontal motion in the meridional, the meridional flow thus obtained is poleward with a mean amplitude of $10\,\mathrm{m\,s^{-1}}$ at latitude 20°, and shows the peak-to-peak variation of $7\,\mathrm{m\,s^{-1}}$.

• Detection of solar-cycle variation in the frequencies and amplitudes of eigenmodes: A frequency change over the solar cycle has indeed been detected. However, it has not

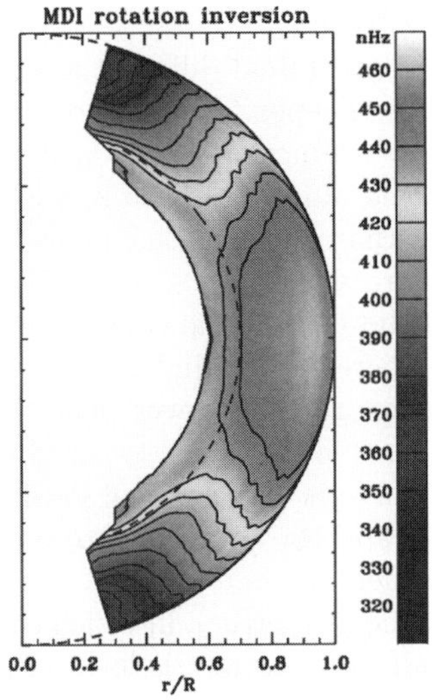

Figure 1. The rotation profile of $\Omega(r, \theta)$ in the solar envelope, inferred from helioseismic observations (Schou *et al.* 1998). The vertical axis is the rotation axis, and the dashed circle indicates the base of the convection zone. Contour labels are in nHz. The white region indicates the region where reliable inference was not made at that time.

yet clarified the change in the thermal structure near the base of the convection zone which is believed to be the most plausible seat of the solar dynamo.

• Detection of solar-cycle variation in the solar internal dynamics: The bands of slower- and faster-than-average rotation migrating in latitude over the solar cycle in a manner similar to the zones of solar activity. The solar-cycle dependence of the meridional flow has also been investigated.

• Making it possible to look at the far-side of the sun and beneath the surface: Various techniques of local helioseismology have made it possible to infer thermal inhomogeneity and flows beneath the surface. Also active regions on the invisible far-side are now well inferred from helioseismic analyses of the visible side of the sun.

3. Solar internal rotation

The differential rotation is believed to be essential, with convective motion in the convection zone, for the solar dynamo mechanism, which must be responsible for the well-known 22-year solar activity cycle. The solar internal rotation is now observationally measurable from the m-splittings of the p-mode frequencies. Since the modes which penetrate into the very deep interior are limited to those of low degrees, and their splitting is smaller (being proportional to $2l + 1$ with the decrease of the degree l), it is hard to determine the internal rotation rate in the very deep interior. The same is true for the polar regions, since the modes which are sensitive to the polar regions are limited to low azimuthal order modes ($m \sim 0$). Furthermore, since the sound speed becomes substantially higher with depth, p-modes are less sensitive to the deeper layers. This fact makes it harder to determine the internal rotation rate in the very deep interior. Hence, in figure 1, which is based on the analyses carried out about a decade ago, only the rotation profile of the accurately determinable region is shown.

Conspicuous features are as follows:

• In the convection zone, the latitudinal differential rotation seen at the solar surface persists, more-or-less, even in the deeper layers. The rotational angular velocity in the convection zone is approximately a function of the latitude.

• The radiative core is rotating almost uniformly with a speed corresponding to the surface at mid-latitude – slower than the surface equatorial zone. According to the more recent observational results, the rotation rate seems to be approximately constant in the region between $r \simeq 0.2R_\odot$ and the base of the convection zone.

• In the low-latitude zone, there is a radial gradient in the rotation profile in the convection zone, and the rotation rate is slower near the surface than in the slightly deeper layer .

• In the high-latitude zone, there is also a radial gradient in the rotation profile near the base of the convection zone, and the rotation rate becomes higher with approaching the bottom of the convection zone.

• The transition layer (the tachocline layer), in which the almost-latitudinal differential rotation turns to the almost-uniform rotation, is located just beneath the bottom of the convection zone, and it is very thin.

4. Kinematic dynamo

Let us consider the induction equation describing temporal variation of magnetic fields $\boldsymbol{B}$ in the MHD approximation:

$$\frac{\partial \boldsymbol{B}}{\partial t} = \nabla \times (\boldsymbol{V} \times \boldsymbol{B}) + \eta \nabla^2 \boldsymbol{B}, \tag{4.1}$$

where $\boldsymbol{V}$ denotes the fluid velocity and η is magnetic diffusivity. We consider the fluid motion in the solar convection zone and the magnetic fields there. At any given point and time, the physical variables of a system may be expressed in terms of mean values and fluctuating values. For such a decomposition to make sense, we assume that a suitable averaging spatial size is found so that the mean values are substantially independent of this averaging size. The fluid motion is then divided into the two parts, —the velocity field due to rotation of the sun, $\overline{\boldsymbol{V}}$, and that due to turbulent convection, $\boldsymbol{v}$; $\boldsymbol{V} = \overline{\boldsymbol{V}} + \boldsymbol{v}$. Similarly, we divide the magnetic field into two components, —the mean field, $\overline{\boldsymbol{B}}$, and fluctuation, $\boldsymbol{b}$; $\boldsymbol{B} = \overline{\boldsymbol{B}} + \boldsymbol{b}$.

We assume the influence of the magnetic field upon the mean motion is negligibly small, while the mean magnetic filed is generated and amplified by the presence of motion. This picture is called a kinematical dynamo. Such a model may be good enough to see where the magnetic field can be generated in the solar convection zone. With some appropriate assumptions (Parker 1955), the induction equation for the mean magnetic field is reduced to

$$\frac{\partial \overline{\boldsymbol{B}}}{\partial t} = \nabla \times (\overline{\boldsymbol{V}} \times \overline{\boldsymbol{B}} + \alpha \overline{\boldsymbol{B}}) + (\eta + \eta_t) \nabla^2 \overline{\boldsymbol{B}}, \tag{4.2}$$

where α and η_t are approximately given by

$$\alpha \simeq -\frac{\tau}{3} \overline{\boldsymbol{v} \cdot \nabla \times \boldsymbol{v}} \tag{4.3}$$

and

$$\eta_t \simeq \frac{\tau}{3} \overline{\boldsymbol{v}^2}, \tag{4.4}$$

respectively, and τ is a representative lifetime of fluctuation. Note that the quantity

$\overline{\boldsymbol{v} \cdot \nabla \times \boldsymbol{v}}$ is a measure of twist in convective motion and that it is positive in the case of clockwise twist with moving ahead.

Let us divide the mean magnetic field into the poloidal component $\overline{\boldsymbol{B}}_{\rm p}$ and the toroidal component $\overline{B}_\phi$, and consider these two components of equation (4.2). Also we hereafter ignore "bar" to express the mean field, as far as no confusion is expected. We further introduce the vector potential A_ϕ for the poloidal component:

$$\boldsymbol{B}_{\rm p} = \nabla \times (A_\phi \mathbf{e}_\phi). \tag{4.5}$$

Then, equation (4.2) is reduced to the following two equations:

$$\frac{\partial A_\phi}{\partial t} = -\frac{\boldsymbol{V}_{\rm p}}{r} \cdot \nabla(rA_\phi) + \eta \left(\nabla^2 - \frac{1}{r^2}\right) A_\phi + \alpha B_\phi \tag{4.6}$$

and

$$\frac{\partial B_\phi}{\partial t} = r(\boldsymbol{B}_{\rm p} \cdot \nabla)\frac{V_\phi}{r} - r(\boldsymbol{V}_{\rm p} \cdot \nabla)\frac{B_\phi}{r} + \eta \left(\nabla^2 - \frac{1}{r^2}\right) B_\phi. \tag{4.7}$$

The last term in the RHS of equation (4.6) expresses the so-called α-effect, by which the poloidal component of the magnetic field is generated by the toroidal component. The first term in the RHS of equation (4.7) means that the toroidal component is generated from the poloidal component with the help of rotation.

5. Where does the dynamo work?

It is instructive to consider the induction equations (4.6) and (4.7) for the mean magnetic field with the Cartesian coordinates and to treat them locally (Yoshimura 1975). Let us take the z-axis to point the vertical direction and let the (x, y)-plane be the horizontal plane at a given point. We take the y-direction eastward and the x-direction southward, respectively. Then, equations (4.6) and (4.7) are reduced to

$$\left[\frac{\partial}{\partial t} - \eta\nabla^2\right] A_y = \alpha B_y \tag{5.1}$$

and

$$\left[\frac{\partial}{\partial t} - \eta\nabla^2\right] B_y = \frac{dV_y}{dz}\frac{\partial A_y}{\partial x}, \tag{5.2}$$

respectively.

In the local treatment, dV_y/dz is treated as a constant, and A_y and B_y are assumed to be proportional to $\exp[i\omega t + i(k_x x + k_z z)]$. We take here $k_x > 0$ without losing generality. Then, from equations (5.1) and (5.2), the following dispersion relation is derived:

$$i\omega = \eta k^2 \left[-1 + (1 \pm i)\sqrt{|N_{\rm D}|}\right], \tag{5.3}$$

where

$$N_{\rm D} \equiv \frac{\alpha k_x}{2\eta^2 k^4}\frac{dV_y}{dz}. \tag{5.4}$$

In the RHS of equation (5.3), the plus sign should be taken if $N_{\rm D} > 0$ while the minus sign should be taken otherwise.

We know from the observed butterfly diagram, which shows the long-term variation in the latitude of the solar active regions, that the solar active region migrates equatorward (toward the direction of increase in x, in the northern hemisphere). Such a wave is

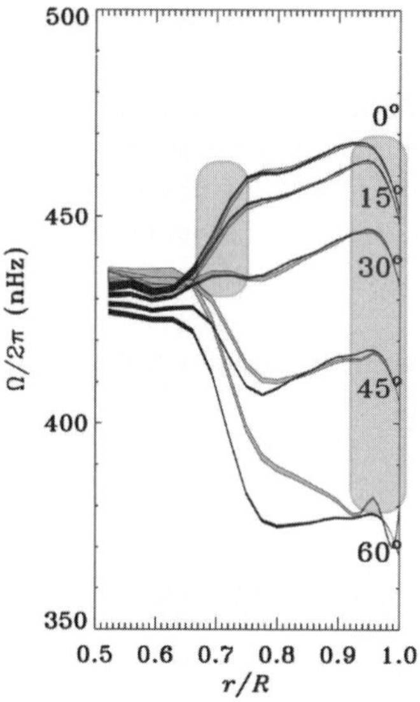

Figure 2. The rotation profiles of $\Omega(r,\theta)$ at various latitudes ($0^\circ, 15^\circ, 30^\circ, 45^\circ, 60^\circ$) in the solar envelope, inferred from helioseismic observations. The hatched parts of the profiles satisfy the condition for the dynamo action.

expressed in terms of the dispersion relation with the minus sign in equation (5.3):

$$i\omega = \eta k^2 \left[-1 + (1-i)\sqrt{|N_{\rm D}|}\right]. \tag{5.5}$$

Hence $N_{\rm D}$ should be negative:

$$N_{\rm D} \equiv \frac{\alpha k_x}{2\eta^2 k^4}\frac{dV_y}{dz} < 0. \tag{5.6}$$

Note that the quantity α is related to kinetic helicity (α has the opposite sign to kinetic helicity, $\overline{\boldsymbol{v}\cdot\nabla\times\boldsymbol{v}}$). Convective motion may be regarded as a plume moving upward from the bottom of the convection zone, where a fraction of gas elements converges horizontally to generate the plume, to the top layer, where the motion diverges horizontally. Coriolis force tends to twist counterclockwise the horizontally diverging flow near the top of the convection zone in the northern hemisphere while it twists clockwise the horizontally converging flow near the bottom of the convection zone. Hence α in the convection zone in the northern hemisphere is expected to be positive in the upper layer and negative in the lower layer. In this situation, the condition (5.6) requires $dV_y/dz < 0$, which corresponds to $\partial\Omega/\partial r < 0$ in the spherical coordinates, in the upper layer in the convective zone and $dV_y/dz > 0$, which corresponds to $\partial\Omega/\partial r > 0$, in the lower layer.

From this consideration, it is concluded that the dynamo action can work only near the upper layer of the convection zone at low and middle latitudes and/or near the low layer of the convective zone at low latitudes, as indicated in figure 2.

References

Leighton, R. B., Noyes, R. W., & Simon, G. W. 1962, *ApJ*, 135, 474
Parker, E. N. 1955, *ApJ*, 122, 293
Schou, J. *et al.* 1998, *ApJ*, 505, 390
Yoshimura, H. 1975, *ApJ*, 201, 740

Solar and Stellar Variability: Impact on Earth and Planets
Proceedings IAU Symposium No. 264, 2009
A. G. Kosovichev, A. H. Andrei & J.-P. Rozelot, eds.

doi:10.1017/S1743921309992328

Variability of the solar spectral irradiance and energetic particles

Adriana Silva-Válio[1]

[1]CRAAM, Mackenzie University, São Paulo, Brazil
email: avalio@craam.mackenzie.br

Abstract. The total spectral irradiance of the Sun is seen to vary on many time scales. Three timescales are more prominent: (1) the longest one of about 11 years; (2) an intermediate timescale of the order of a few weeks; and (3) the shortest variation from hours to seconds. Every 11 years, the total solar irradiance periodically shows intervals of great activity and periods of almost no activity. The peak to peak variability, however, is less than 0.1%. This periodic variation of 11 years is called the solar cycle, which main tracer are sunspots. During times of maximum activity, there are many sunspots on the surface of the Sun, whereas during minimum there may be none. Sunspots are dark, and therefore cool, regions of enhanced magnetic fields of about a few hundred Gauss, that usually appear in groups on the solar photosphere. Basically, the solar cycle is regulated by the magnetic dynamo acting below the solar surface. Right now, the Sun is going through a time of minimum activity. The sunspot lifetime is of the order of one to two weeks, and are thus responsible for the intermediate variability timescale. The magnetic loops seen in ultraviolet and X-ray images have their footpoints anchored on sunspots. The most energetic phenomena of solar activity are flares and coronal mass ejections. Flares are large explosions that occur on the solar atmosphere and may last from a few seconds to hours. A solar flare is caused by a sudden, and yet unpredicted, energy release high above the magnetic loops. This magnetic energy is then used into particle acceleration and heating of the surrounding atmosphere. Both the energetic particles and the hot gas produce emission throughout the whole electromagnetic spectrum, from the very energetic gamma-rays all the way to long radio waves. From the observation of the emission produced during flares it is possible to infer the energetic particles spectra and thus have a clue on the acceleration mechanism that produced these particles. The recent findings of flare observations at gamma-rays by the RHESSI satellite and at high radio frequencies by the Solar Submillimeter Telescope are presented and discussed.

Keywords. Sun: activity, Sun: flares, Sun: particle emission

1. Solar activity

The solar variability can occur on three timescales: (1) Solar dynamo: is the longest timescale of about 11 years; (2) Solar rotation and sunspot lifetime: an intermediate timescale of the order of a few weeks; (3) Flares and coronal mass ejections: the shortest timing, varying from hours to seconds.

Sunspots are dark regions seen in solar images in visible light, where high concentration of magnetic fields are located (Figure 1). Because magnetic fields hamper the energy transfer to the photosphere from the convective zone below, these regions end up being cooler than the surrounding photosphere (T_{eff} = 5780 K), with temperatures of about 4000 - 5000 K.

Since 165 B.C. there have been reports of sunspots sighting (Wittmann & Xu 1987) by the Chinese with the naked eye during sunsets and sunrises when the Sun was close to the horizon, and its brightness was not so intense. Nevertheless, it was only with the first

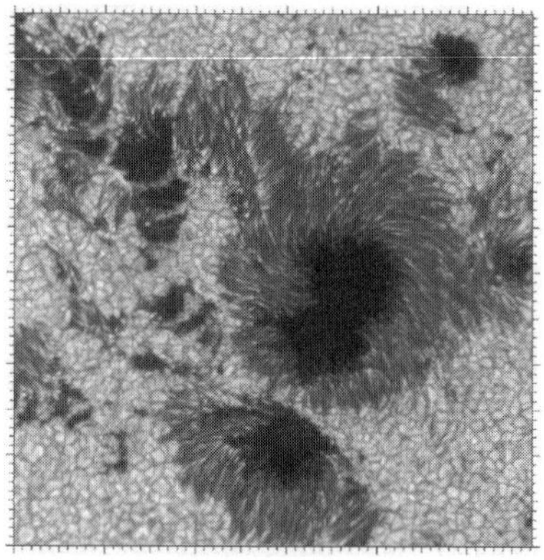

Figure 1. Sunspot observed with the Swedish Solar Telescope (SST). Credit: The SST is operated by the Royal Swedish Academy of Sciences,

observation through a telescope by Galileo in 1610 that sunspots have been monitored regularly. In 1843, Schwabe noticed that the number of sunspots varied periodically every 11 years (Schwabe 1844). This can be seen in the left panel of Figure 2, where the number of sunspots is plotted over the last four centuries.

The solar activity cycle lasts for approximately 11 years, when during periods of maximum activity the Sun is covered with spots. Not only does the sunspot number vary with a period of 11 years, but also the soft X-ray brightness of the Sun, as can be seen from the composite of solar images by the Yohkoh satellite taken every 6 months (right panel of Figure 2). Presently, in 2009, the Sun is still displaying minimum activity.

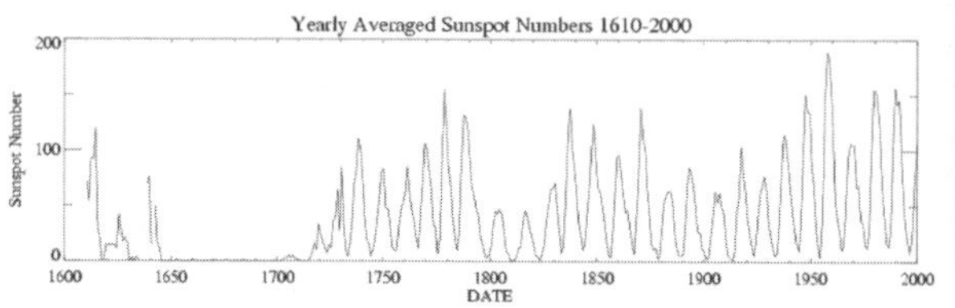

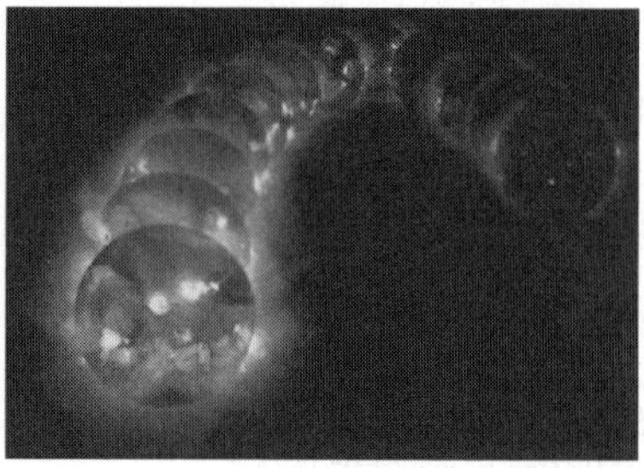

Figure 2. Left: The 11-year periodicity in the Sun activity is easily observed in the sunspot number (http://www.physics.unlv.edu/ jeffery/astro/sun/sun.html) Right: Solar activity cycle seen in the global soft X-ray emission from the Yohkoh satellite.

The solar flux above the Earth's atmosphere, also known as the *solar constant*, was measured by several satellites. The result of the observations after cross calibration is plotted in Figure 3. As can be seen from the figure, the solar constant is not so constant, but varies by about 0.1 % with an 11 year period, the variation of the total solar output being in phase with the solar cycle. This is at first sight contradictory since during periods of maximum activity there are more sunspots, or dark regions, on the solar surface. However, the plages that are bright regions around sunspots are also more abundant during maximum activity periods and their brightness more than counterbalances the dimming caused by the spots.

Other phenomena also follow the 11 year cycle, such as solar flares, filament eruption, and coronal mass ejections. Flares and mass ejections are the most energetic events in

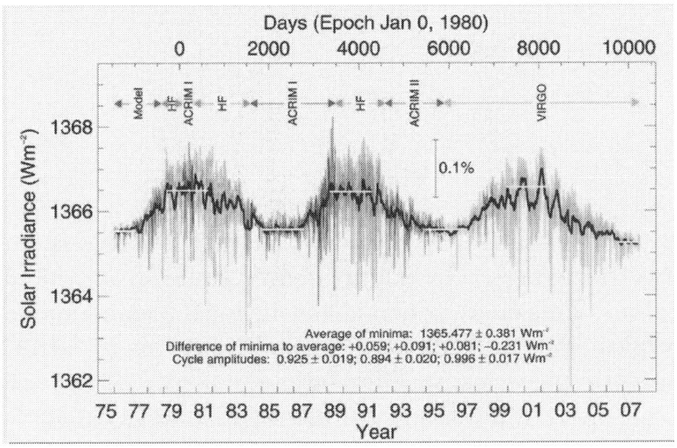

Figure 3. The solar irradiance, or flux at the top of Earth atmosphere, for the past three decades. The data were collected from different satellites and cross calibrated.

the Solar System, and are powered with the magnetic energy contained in the magnetic field loops of the solar atmosphere.

Coronal mass ejections are huge bubbles of solar plasma and magnetic fields that are ejected toward the interplanetary medium during the course of many hours. Some of these mass ejections are directed at Earth and may cause many disturbances once the particles hit the Earth's magnetosphere and atmosphere generating geomagnetic storms, power outages, satellite damage, and auroras, among others. Figure 4 illustrates the mass being ejected from the Sun (left panel) and reaching the Earth magnetosphere (right panel).

Figure 4. Left: Image of a coronal mass ejection from the Sun observed by the coronograph LASCO on board the SOHO satellite. The occulted solar disc is depicted by the white circle. Right: Illustration of a mass ejection from the Sun and its interaction with the Earth magnetosphere (distances not to scale).

During a solar flare, up to 10^{25} J of energy is released in only 100-1000 s. Flares always occur within active regions. Filaments (or prominences when seen on the limb) are cold material (5000-10000 K) suspended by magnetic fields. In response to a magnetic instability, they can be suddenly thrown upward. A coronal mass ejection (CME) may throw up to 5×10^{13} kg of plasma into interplanetary medium at velocities of 200-2000 km/s, totaling $10^{23} - 10^{24}$ J of energy. What all these phenomena have in common is that their energy source originates from the reconnection of magnetic field lines. Flares and CMEs also show a periodicity of 11 years, occurring more frequently and with greater intensity during periods of solar maximum.

2. Solar flares

A solar flare may be thought of as a sudden release of large quantities of energy in timescales of seconds to minutes. The energy source comes from the magnetic field loops located in active regions. This energy then heats the local plasma and accelerates particles up to very high energies (electrons up to $\sim$100 MeV and ions to $\sim$10 GeV). Both the hot plasma and the energetic particles produce large quantities of radiation throughout the whole electromagnetic spectrum (radio, visible, ultraviolet, X-rays)

An idealized picture of a solar flare has emerged from the observations at all wavelengths of the electromagnetic spectrum. The idealized flare scenario is that magnetic reconnection occurs above the loop top accelerating the particles (see Figure 5). A fraction of these energetic particles escapes toward the interplanetary medium through open field lines, whereas the remaining particles are injected into the loop below. Some of them may become trapped, spiraling around the magnetic lines and thus producing high frequency radio emission by gyrosynchrotron. The ones that are not trapped, or when these trapped particles finally overcome the magnetic bottle through collisions, reach the denser chromosphere below and interact with the ions creating hard X- and gamma-rays. The energy deposited close to the loop footpoints then heats the chromospheric plasma that expands filling the magnetic loop and emitting copious amounts of soft X-ray.

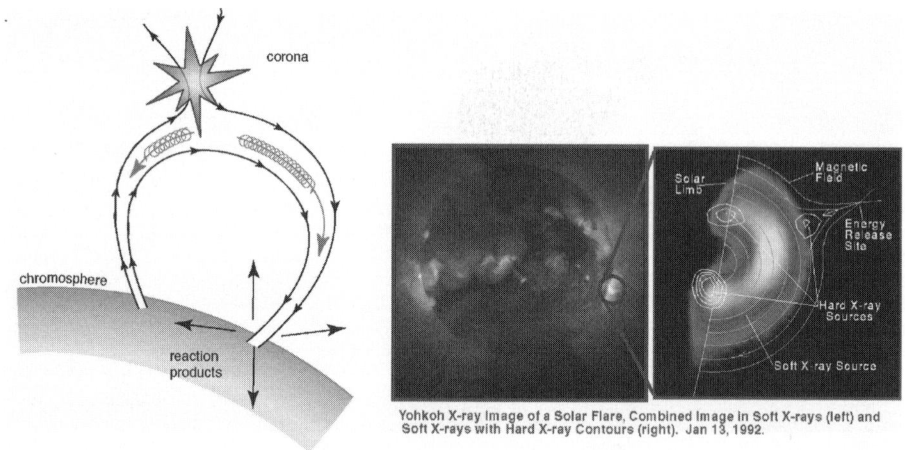

Figure 5. Left: Idealized magnetic loop where a flare occurs. Right: A hard X-ray source detected by the Yohkoh satellite above the loop top for a limb event.

An observational evidence of such model was obtained from the detection of a hard X-ray source above the magnetic field loops for a limb flare (Masuda *et al.* 1994). The magnetic field loop is that outlined from the low energy soft X-rays, whereas the footpoints are seen as hard X-ray sources (white countours in the left panel of Figure 5). The novelty of this observation is the detection of a hard X-ray source above the loop top at the location where the energetic particles, accelerated at the magnetic reconnection site, were injected downward into the loop and interacted with the plasma within the looptop yielding the hard X-ray source by bremsstrahlung. The response of the chromosphere to a flare can also be seen in H-α images (Figure 6, left panel).

It is clear from the model depicted above that flares generate emission at all wavelengths of the electromagnetic spectrum. Nevertheless, the temporal evolution is different for each wavelength as shown in the right panel of Figure 6. The time profile of flare emission can be divided into an impulsive phase due to the accelerated particles and a gradual one created from the thermal plasma. The impulsive peak is clearly seen at high frequency (>1 GHz) radio emission and hard X-rays, whereas the gradual phase is best seen in soft X-rays. The next subsections depict the main discoveries obtained at hard X-ray and submillimeter wavelengths that are thought to be produced by the same population of accelerated particles.

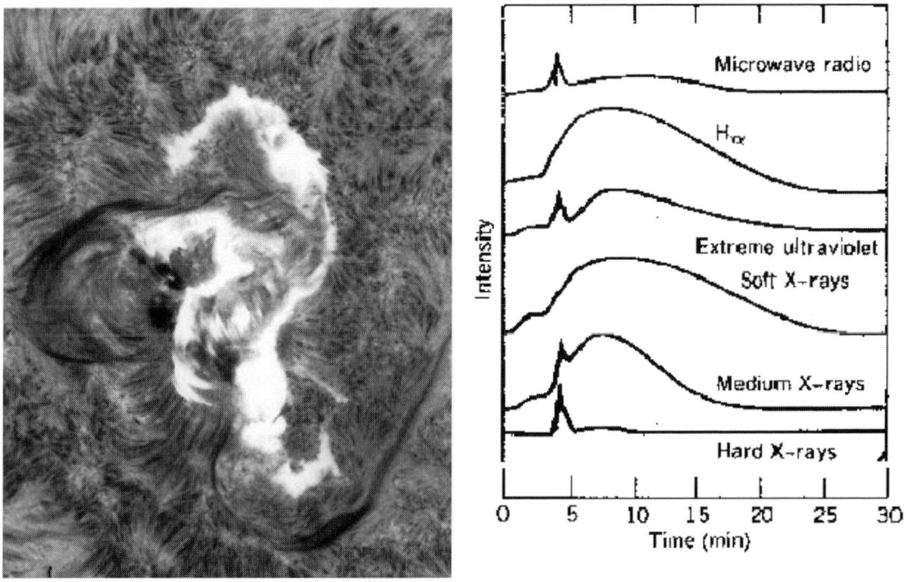

Figure 6. Left: Image of a solar flare through a H-α filter centered at a wavelength of 6563Å from the Bib Bear Solar Observatory. Right: Time profiles of solar flare emission at several wavelengths showing the initial impulsive phase and later on the more gradual emission.

2.1. *Main discoveries of flares at hard X-rays*

Many new results were obtained with the Reuven Ramaty High Energy Solar Spectroscopic Imager (RHESSI) (Lin *et al.* 2002). The goal of this solar dedicated satellite, launched on February 2nd, 2002, was to explore the physics of particle acceleration mechanism and energy release in solar flares. Hard X- and gamma-ray images with high spatial (>2"), spectral (1-5 keV), and temporal (10 ms) resolution can be obtained. The spectral range of the observations varies from 3 keV to 15 MeV. Among the main discoveries are the first image of gamma-rays from a solar flare and the resolved gamma-ray lines of Fe, Mg, Si, Ne, C, and O. A complete issue of the *Astrophysical Journal Letters* (volume 595) was dedicated to the results obtained from the observations of the flare on July 23, 2003 (Lin *et al.* 2003).

Gamma-ray lines are produced by nuclear collisions of energetic ions with the solar atmosphere. This is in agreement with the simple flare loop model described above. The detected line centered at an energy of 2.223 MeV is due to Deuterium formation by neutron capture. The left panel of Figure 7 shows the centroid of the 2.223 MeV emission detected from the July 23, 2003 flare (Hurford *et al.* 2003). This flare emission from an ion was found to be shifted by 20" from the lower energy sources at 0.05-1.4 MeV, that are due to electron bremsstrahlung. This implies a different acceleration and/or propagation mechanisms for electrons compared to that of ions in solar flare.

Other gamma-ray lines produced by the de-excitation of ions (C, O, Ne, Mg, Si, Fe) were resolved for the first time and are shown in the right panel of Figure 7 (Smith *et al.* 2003). The measured lines display a redshift of 0.1%-0.8%, and this redshift is seen to decrease with the ion mass. This redshift is larger than that expected for an isotropic distribution of ions in a radial magnetic field. The explanation for the measured redshift is that the magnetic field lines are inclined by 40° with respect to the solar surface.

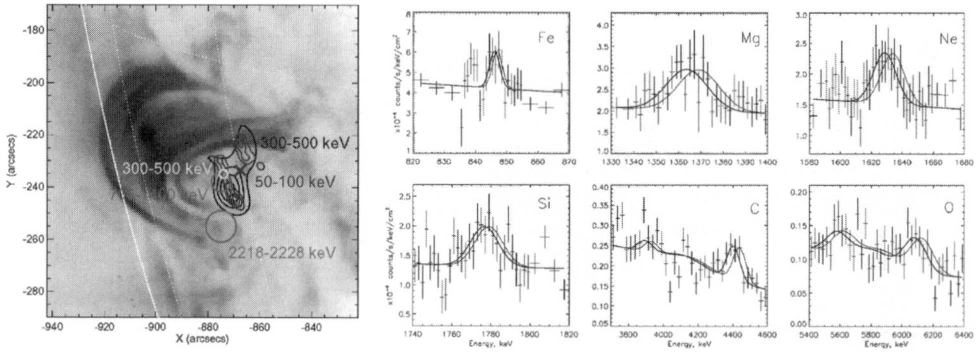

Figure 7. Left: First image at gamma-rays showing that the gamma-ray source at 2.223 MeV, due to ions, occurs at a different location than the lower energy sources (50 - 300 keV) produced by electrons. Right: Resolved gamma-ray lines for six different species of ions.

2.2. *Flare at submillimeter wavelengths*

On the other end of the spectrum is the emission at submillimeter wavelengths. Nevertheless, this emission is thought to be produced by > MeV electrons (White & Kundu 1992), the same ones that generate the 100-1000 keV hard X-rays. The results described here were obtained with the Solar Submillimeter Telescope (Figure 8), a 1.5 m dish installed at the CASLEO Observatory in the Argentinean Andes at an altitude of 2500 m (Kaufmann *et al.* 2001). It is a solar dedicated instrument operating with 4 receivers at 212 GHz and 2 receivers at 405 GHz, with sub-second temporal resolution.

The SST observed the greatest explosion ever recorded, a GOES class X28, on November 4, 2003 (Kaufmann *et al.* 2004). This limb event is shown in the left panel of Figure 9, where the white crosses represent the beams of the telescope. The data showed a striking new result, that the flux at 405 GHz was higher than that at 212 GHz. This can be seen in the temporal evolution of the submm emission plotted in the middle panel of the figure, where 4 main peaks can be seen. Radio spectra taken at the time of the first (P1) and fourth (P4) peaks were obtained by combining the SST data with microwave data from the Owens Valley Solar Array (OVSA). The spectra plotted in the right panel of Figure 9 clearly show that the submm data points at 212 and 405 GHz do not follow the microwave spectral decrease but rather increase toward higher frequencies. This new spectral component has been named the "THz component" because the peak of the spectrum is expected to be located somewhere in the THz domain.

The November 4, 2003, flare was one of the so called Halloween events and was not the only flare observed to display an increasing spectral component at submm wavelengths. Three other flares showed the same increasing spectra at frequencies >200 GHz: the

Figure 8. The Solar Submillimeter Telescope, a solar dedicated radio antenna installed in Argentina operating at 212 and 405 GHz.

GOES X2.0 flare on April 12, 2001 (Lüthi, Magun, & Miller 2004b) and the GOES X17.2 flare on October 28, 2003 (Lüthi, Lüdi, & Magun 2004a) were both observed by the KOSMA submillimeter telescope at 230 and 345 GHz. The other flare, a GOES X8.7 on November 2, 2003, occurred on the same active region as the one on November 4th, was the first flare detected simultaneously by SST and RHESSI (Silva *et al.* 2007).

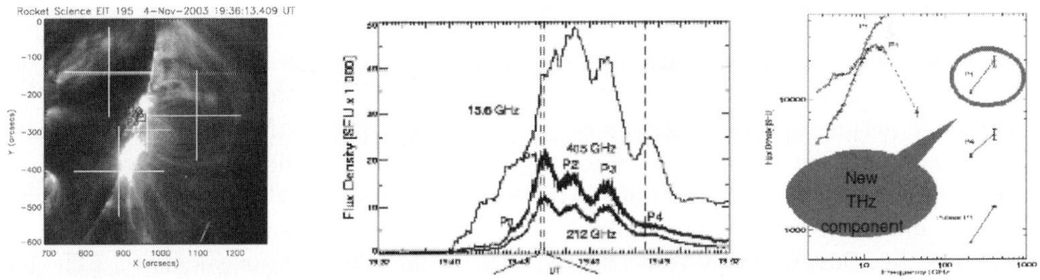

Figure 9. The November 4, 2003. Left: The gray scale shows the ultraviolet emission from the flare, whereas the symbols represent the centroid of 212 GHz emission at different times. The white crosses correspond to the size of SST beams. Middle: Temporal evolution of the November 4, 2003, flare at 15 GHz and submm wavelengths. Right: Radio spectra of the flare combining the microwaves from OVSA and the submm data from SST. A new spectral component peaking somewhere in the THz is depicted.

The flare emission was located mainly in the magnetic loop outlined in ultraviolet light shown in Figure 10 (left panel), where the contour levels represent the hard X-rays at several energies and the centroid of the submm emission is depicted by the small black points. As expected, the high energy hard X-rays are located close to the loop footpoints. The temporal evolution of the flare emission at several wavelengths are plotted in the middle panel of Figure 10.

Two impulsive peaks are seen in the microwaves and <500 keV emission. The submm flux also displays the second impulsive peak (the first peak occurred during a calibration scan of the telescope) and a long duration emission that lasts for several minutes. This gradual emission was also detected at greater than 2 MeV hard X-rays. This good correlation implies that the submm emission is in fact associated to the high energy electrons (>5 MeV) that also produce the high energy hard X-rays.

The radio spectra of this flare, composed of microwave data from OVSA and submm emission observed by the SST, are plotted in Figure 10 (right panel) for nine time intervals. Again the THz spectral component is evident with the flux at 405 GHz even higher

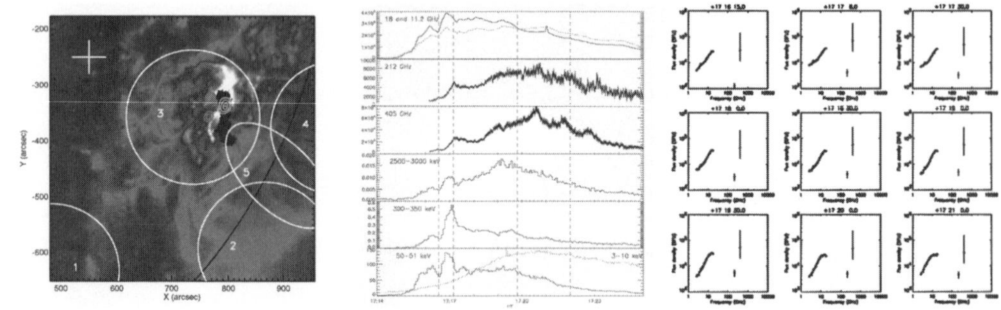

Figure 10. The November 2, 2003, flare. Left: The gray scale corresponds to the ultraviolet emission of this event, whereas the superimposed contours represent the hard X-ray emission (12 - 25 keV in gray and 100 - 300 keV in red. The black points are the centroid position of the 212 GHz emission, where the cross mark the absolute uncertainty in position. The white circles show the SST beams (1, 2, 3, and 4 at 212 GHz, and 5 is the 405 GHz beam). Middle: Time profiles of the flare emission at microwaves from OVSA, submm waves from SST, and hard X-rays from RHESSI. Right: Radio spectra composed by microwave and submm data of the flare at nine time intervals.

than that at microwaves. Silva *et al.* (2007) analyzed three possible emission mechanisms in order to unravel the nature of this novel spectral component. The authors propose three possibilities for the submm source: (1) Free-free emission from thermal electrons; gyrosynchrotron radiation from (2) positrons or from (3) non-thermal electrons.

If the increasing spectral component at frequencies larger than 200 GHz is due to bremsstrahlung from thermal electrons, then the submm emission could have been produced by a 20" source of temperature $\sim 10^7$ K and densities of $10^{12} - 10^{14}$ cm^{-3}, shown as a dashed line in the left panel of Figure 11 (for a full discussion, see Silva *et al.* 2007). The soft X-ray flux that would have been produced by such source was calculated (Thomas, Crannel, & Starr 1985) and is about a 100 times larger than the soft X-rays observed by the GOES satellite.

The second possibility is that the submm emission originates from positrons as they gyrate around the magnetic field lines. Positrons are believed to be produced by pion or radioactive decay after nuclear interactions during a large flare. RHESSI detected the 511-keV line due to the electron-positron annihilation (Share *et al.* 2004). The number of positrons estimated from the annihilation-line fluence detect by RHESSI, however, is 3000 to 2×10^5 times smaller than that need to account for the submm flux. Thus this explanation can also be ruled out.

When considering gyrosynchrotron emission from non-thermal electrons it is necessary to match simultaneously the submm flux and the hard X-rays since both types of emission are believed to be produced by the same population of energetic electrons. The hard X-ray spectra from RHESSI yield 5×10^{35} electrons between 50 keV and 20 MeV and an electron spectral distribution of index $\delta = 2.5$ above ~ 3 MeV. Using this same electron spectra to fit the submm emission, Silva *et al.* (2007) found that this required a very compact source of 0.5" radius, with high density (10^{12} cm^{-3}) in a region of very strong magnetic field (4500 G). This same electron spectrum can also model the microwave spectra (1-15 GHz) as produced by a more extended source of 50" radius in a 480 G magnetic field region. An example of the gyrosynchrotron fits (Ramaty 1969, Ramaty *et al.* 1994) to the radio spectra at the time of maximum submm flux are shown as solid lines in Figure 11 (left panel).

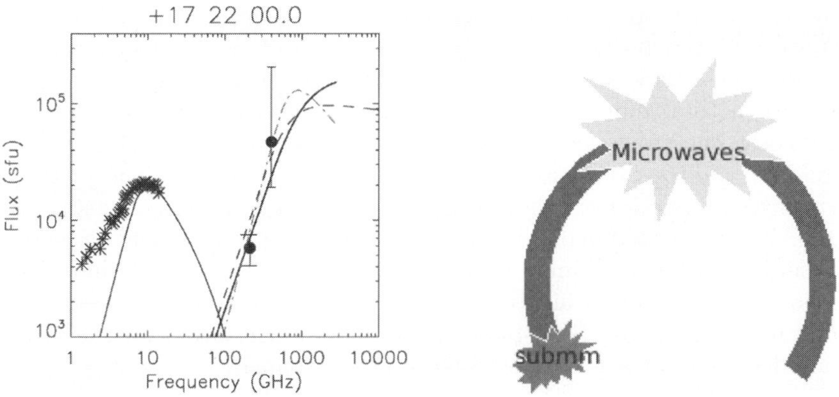

Figure 11. Left: Radio spectra at the time of the 405 GHz peak emission with the fitting results for different emission mechanisms: bremsstrahlung from thermal electrons (dashed line), gyrosynchrotron from positrons (dot-dashed curve) and non-thermal electrons (solid line). Right: Cartoon depicting the microwave loop top source and the submm source at the footpoint.

3. Conclusions

The Sun despite being considered a quiet star when compared to others is still the source of very energetic events such as solar flares and coronal mass ejections. Since the XIX century mainly from the monitoring of sunspots on its surface, it is known that the Sun follows an 11 year cycle. Actually, the true solar cycle due to the magnetic dynamo has a duration of 22 years, when the global magnetic field returns to its original polarity. Sunspots are the main indicators of solar activity, but the occurrence of flares and CMEs is also observed to have an 11 year periodicity.

The analysis of the flare emission at several wavelengths is crucial for an understanding of the global picture of a solar flare that follows. A solar flare is powered by the magnetic energy released in magnetic field reconnection above looptops. This energy then goes into particle acceleration, both electrons and ions, and plasma heating. A fraction of the energetic particles is injected into the magnetic loop below. Some of them may become trapped and emit radio waves by gyrosynchrotron, while the ones that escape the magnetic trap reach the loop footpoints below and produce hard X- and gamma-rays as a result of the interaction with the denser material of the chromosphere. All this input of energy heats the plasma that then expands filling the loop, this is known as "chromospheric evaporation" and the whole heated loop at temperatures of tens of millions of K radiates in soft X-rays and radio wavelengths.

With the launch of the RHESSI satellite in 2002 and the installation of the SST in 1999, many interesting results have been found about solar flares. A very well studied flare that occurred on July 23, 2003, provided new insight into the physics of gamma-ray production. For the first time, an image of a source of ion emission was obtained and showed that possibly a different acceleration mechanism or distinct propagation paths from that of the accelerated electrons is at work. Gamma-ray lines displayed redshifts that corroborated the picture of down streaming ions into the chromosphere during a flare and determined the inclination of magnetic loops with respect to the solar surface.

Observations of solar flares at high radio frequencies have uncovered a new spectral component that is likely to peak in the THz domain. Four flares observed with two different radio telescopes (SST and KOSMA) have shown an increasing emission toward higher frequencies ($>$ 200 GHz). Here I have highlighted the main results of the flares that occurred on November 4th (Kaufmann *et al.* 2004) and 2nd (Silva *et al.* 2007).

Simultaneous observations of the November 2, 2003, flare in hard X-rays and radio waves allowed the conclusion for the best explanation of this new spectral component. The data were best fit as produced by emission of a single population of non-thermal electrons with a total number of 5×10^{35} and a power-law index of $\delta = 2.5$ above ~3 MeV. These energetic electrons are responsible for producing the hard X-rays by bremsstrahlung and the radio waves by gyrosynchrotron in two distinct emitting regions. The microwaves originate from the same electrons trapped in a low density (10^9 cm^{-3}) region, 100" in size, located higher in the loop where the magnetic field strength is 480 G. On the other hand the submillimetric source was very small (size of 1") and located at the loop footpoints in a region of high density (accelerated electron densities of 10^{12} cm^{-3}) and strong magnetic fields (B = 4500 G). Both the microwave and the submillimeter sources are depicted in the cartoon of Figure 11 (left panel).

References

Hurford, G. J., Schwartz, R. A., Krucker, S., Lin, R. P., Smith, D. M., & Vilmer, N. 2003, *ApJ* (Letters), 595, 77

Kaufmann, P., Costa, J. E. R., Giménez de Castro, C. G., Hadano, Y. R., Kingsley, J. S., Kingsley, R. K., *et al.* 2001, In: J. J. Pinho, G. P. S. Cavalcante, & A. H. Oliveira (eds.), *Proc. 2001 SBMO/IEEE MTT-S International Microwave and Optoeletronics Conference*, p. 439

Kaufmann, P., Raulin, J.-P., de Castro, C. G. G., Levato, H. Gary, D. E., Costa, J. E. R., *et al.* 2004, *ApJ* (Letters), 603, 121

Lin, R. P., Dennis, B. R., Hurford, G. J., Smith, D. M., Zehnder, A., Harvey, P. R., *et al.* 2002, *Solar Phys.*, 210, 3

Lin, R. P., Krucker, S., Hurford, G. J., Smith, D. M., Hudson, H. S., Holman, G. D., Schwartz, R. A., Dennis, B. R., Share, G. H., Murphy, R. J. *et al.* 2003, *ApJ* (Letters), 595, 69

Lüthi, T., Lüdi, A., & Magun, A. 2004a, *A&A*, 420, 361

Lüthi, T., Magun, A., & Miller, M. 2004b, *A&A*, 415, 1123.

Masuda, S., Kosugi, T., Hara, H., Tsuneta, S., & Ogawara, Y. 1994, *Nature*, 371, 495

Ramaty, R. 1969, *ApJ*, 158, 753

Ramaty, R., Schwartz, R. A., Enome, S., & Nakajima, H. 1994, *ApJ*, 436, 941

Schwabe, M. 1844, *Astronomische Nachrichten*, 21, 223

Share, G. H., Murphy, R. J., Smith, D. M., Schwartz, R. A., & Lin, R. P. 2004, *ApJ* (Letters), 615, 169

Silva, A. V. R., Share, G. H., Murphy, R. J., Costa, J. E. R., de Castro, C. G. Gimnez, Raulin, J.-P., & Kaufmann, P. 2007, *Solar Phys.*, 245, 311

Smith, D. M., Share, G. H., Murphy, R. J., Schwartz, R. A., Shih, A. Y., & Lin, R. P. 2003 *ApJ* (Letters), 595, 81

Thomas, R. J., Crannell, C. J., & Starr, R. 1985, *Solar Phys.*, 95, 323

White, S. M. & Kundu, M. R. 1992, *Solar Phys.*, 141, 347

Wittmann, A. D. & Xu, Z. T. 1987, *A&A Suppl. Ser.*, 70, 83

Solar and Stellar Variability: Impact on Earth and Planets
Proceedings IAU Symposium No. 264, 2009
A. G. Kosovichev, A. H. Andrei & J.-P. Rozelot, eds.

doi:10.1017/S174392130999233X

A solar cycle lengthwise series of solar diameter measurements

J. L. Penna[1], A. H. Andrei[1,2,*], S. C. Boscardin[1], E. Reis Neto[1] and V. A. d'Ávila[1,3]

[1] Observatório Nacional/MCT, R. Gal. Jose Cristino 77, Rio de Janeiro, Brazil
email(*): oat1@on.br

[2]Obs. do Valongo/UFRJ-BR, Rio de Janeiro, Brazil
[3]UERJ-BR, Rio de Janeiro, Brazil

Abstract. The measurements of the solar photospheric diameter rank among the most difficult astronomic observations. Reasons for this are the fuzzy definition of the limb, the SNR excess, and the adverse daytime seeing condition. As a consequence there are very few lengthy and consistent time series of such measurements. Using modern techniques, just the series from the IAG/USP and from Calern/OCA span more than one solar cycle. The Rio de Janeiro Group observations started in 1997, and therefore in 2008 one complete solar cycle time span can be analyzed. The series shares common principles of observation and analysis with the ones afore mentioned, and it is complementary on time to them. The distinctive features are the larger number of individual points and the improved precision. The series contains about 25,000 single observations, evenly distributed on a day-by-day basis. The typical error of a single observation is half an arc-second, enabling us to investigate variations at the expected level of tens of arc- second on a weekly basis. These features prompted to develop a new methodology for the investigation of the heliophysical scenarios leading to the observed variations, both on time and on heliolatitude. The algorithms rely on running averages and time shifts to derive the correlation and statistical incertitude for the comparison of the long term and major episodes variations of the solar diameter against activity markers. The results bring support to the correlation between the diameter variation and the solar activity, but evidentiating two different regimens for the long term trend and the major solar events.

Keywords. Sun: photosphere, fundamental parameters, activity; instrumentation: photometers; methods: statistical

1. Introduction

The bulk of solar light and heat received by the Earth is released from the solar photosphere. Though virtually all the energy is produced in the solar core through the proton-proton fusion chain, it undergoes a lengthy pathway, across the radiation zone and up to the convective zone, during a million years process, to finally reach what is purposely called the solar photosphere (Mursula, 2005).

The solar constant (**W**) expresses the total irradiance, i.e., the total power of solar radiation over the full frequency spectrum per perpendicular area at the mean distance of the Earth. The solar constant is hardly constant though. It varies by a factor of

(*a*) 10^{-6} over minutes. (These are related to pressure variations called solar oscillations),

(*b*) 10^{-3} over several days. (These are related to sunspots and other active regions),

(*c*) 10^{-3} over solar cycle (exact value uncertain).

The solar energy is one of the major driving inputs for terrestrial climate. Some evidences of correlations exist between surface temperature changes and solar activity. It is then important to know on what time scale the solar irradiance and other fundamental solar parameters, like the diameter, vary in order to better understand and assess the origin and mechanisms of the terrestrial climate changes.

Global effects, such as diameter changes, large convective cells, the differential rotation of the Sun's interior and the solar dynamo at the base of the convective zone, can probably produce variations in the total irradiance. Or, at least, be correlated to these variations in association with the changing emission of bright faculae and the magnetic network (Damé *et al.*, 2000).

Since 1981 the Observatório Nacional/MCT (ON) group makes daily solar observations, in particular, since 1997, aiming to record and study the diameter variations (Jilinski *et al.*, 1998). In 2002 an international network (Réseau de Suivi au Sol du Rayon Solaire – R2S3) was formed to combine the similar work pursued in five countries worldwide (Andrei *et al.*, 2002). In 2006 the ON associated to the SCOSTEP/CAWSES project for the investigation of Space Weather and the Earth-Sun System issues (Andrei *et al.*, 2006). The goal of the work done at ON is therefore multiple: to record lengthy, coherent series of digitized solar diameter observations; to use this series and those from the R2S3 and other groups to investigate the physics of the photosphere and convection zone; to contribute to the understanding of the Earth-Sun relationship as driven by the heliosphere, with emphasis on short term, geophysics local effects; and finally to upgrade and develop data acquisition and treatment methods and apparatus.

2. Solar diameter variations

ON and R2S3 measures have been consistently obtaining variations of the observed diameter on the order of 100 mas/solar cycle. Several past and current independent experiments from different groups find (at least) likewise. However, from gravitational energy (**E**) variation considerations only, a radius (**R**) variation gives rise to

$$\Delta E = \Delta R \times (3GM^2/5R^2), \tag{2.1}$$

where **G** is the gravitation constant and **M** is the solar mass.

Therefore, to vary the Sun stellar radius by 1exp(-4) along the 11y solar cycle, the required energy variation is about 150 times the total solar output (Durney, 1972). Yet, if just the irradiating surface is considered, and assuming that the flux per unit area is kept, a radius variation gives rise to

$$\Delta I = \Delta R \times (2/R). \tag{2.2}$$

That is, the observed 0.1% radius variation would explain half of the irradiance **I** variation along the solar cycle. Now, if the radius variation is assumed confined to the convective zone (for which in turn an adiabatic regimen is granted), the same 0.1% variation becomes inexpensive in gravitational energy terms (0.01% in the solar cycle). At the same time, a variation on the convective zone height scale would be quite effective to produce a change of the radius defined by hydrostatic equilibrium, at no expense of gravitational energy and very little change of irradiance.

$$dI \propto d\rho = dh_0 \times \rho(exp[h/{h_0}^2]), \tag{2.3}$$

where ρ is the density and $\mathbf{h_0}$ is the height scale.

Heliosismology holds that "Results from f-mode oscillations that the solar radius only changes by about 1 km/year do not preclude the less-sensitive efforts to measure variations of the solar radius at the photosphere by limb observations since the latter are likely to be much larger than the former." (Sofia *et al.*, 2006). So, prevented gravitational or irradiance catastrophes by keeping the physics confined to the convection zone, or better to the photosphere, the massive vault of observations reporting radius variations can be exploited just as the experimental evidence they are.

Some hypothesis to explain the diameter variations have been put forward: localized variations depending of heliolatitude (in special by the equatorial bulge, by the boundary of the royal zone, and at 60° heliolatitude) – following the magnetic network evolution along the cycle; shear in the differential rotation in the convective zone; self-refraction in the solar atmosphere.

On the other hand, disturbances of the troposphere were for quite a while used as an alternative explanation for the observed variations. Quantitative results were obtained for non-linear quasi-zenith distance dependencies in the case of non-varying prism astrolabes, while an analytic description relates the coherency window (Fried's parameter) and the placement of the inflection point defining the solar limb effective stop. The episode analysis and the compound R2S3 series, bringing together multi-site results from widely separate stations, enforce limits on the effectiveness of those mechanisms (Andrei *et al.*, 2004). Either way, they before all produce strong yearly signals that are not verified, and if prevalent should smear the multi-site signal.

More recently, attention was drawn to the stratosphere modulation with the solar cycle. It would imprint both a mark on multi-site observations (in special from alike instruments, leading to alike biases), due to the effective horizontal mixing, and at the same time as a solar activity related signature. However, it should be considered that the QBO and the yearly stratosphere variations are stronger than the 11 years one, respectively by factors of 2 and 10. On the contrary, for the R2S3 series the 11 year signal is much more evident than one or two year periods. In support to the amplitudes mismatch, it can be also remarked that the 11 year stratospheric modulation befalls on the upward structure, while the solar astrolabe observations are, at least to first order, insensitive to small amplitude refraction and focusing variations.

Combining the self-refraction and height scale variation principles, another mechanism can be postulated. The observed solar radius is just the level beyond which the optical depth reaches unity ($\tau = 1$). The optical depth can be expressed by

$$\tau = \kappa\rho \times (KT/mg), \tag{2.4}$$

where κ is the opacity, **K** is the Boltzmann's constant, **T** is the temperature, **m** is the molecular mass, and **g** is the gravitational acceleration.

Thus, the optical depth can vary and in so making vary the observed radius in different ways. Either of them will leave a signature on the appearance of the photosphere.

Though granulation forms the very face of sun's photosphere, there are no long term registers of it. Observational and computational hardships to define and follow such highly variable face have so far prevented the realization of those registers (but see Roudier & Reardon, 1998) even much they are useful for solar physicists. However, in recent years a large, coherent body of white light images became available (Hirzberger *et al.*, 1997). The resolution of the images is at the level of 1 arcsec, thus enabling to perceive granulation structures and individual grains. We retrieved the full solar disk white light images from the Big Bear Solar Observatory (BBSO), at California/USA, covering the raising, peaking, and pos-peak periods of the solar cycle 23.

Boscardin *et al.* (2009) present the results obtained. In conclusion, it is found that the grains sizes are the largest by the solar maximum, in excellent agreement with the maximum of the measured diameter. The grains brightness, on the contrary, is minimum at the solar maximum, and again an excellent agreement is verified with the maximum of the measured diameter. Accordingly, the granulation variation would change the photosphere density, mixing, and eventually its height scale, therefore providing a mechanism to vary the optical depth, and hence the measured diameter.

At the ON ($\Phi = -22°\ 53'\ 42"$, $\Lambda = +2h\ 52m\ 53s.5$, h = 33m) the series of solar semi-diameter measurements started in 1997. The original Danjon Astrolabe was specially redesigned for this type of observations. The most important features were the installation of a variable angle front prism enabling the continuous observations between the zenith distance of 26° and 56°, the concurrent installation of a moving density filter, and the installation of a CCD camera, which allowed the observations to become fully freed of personal equations (Jilinski *et al.*, 1999).

The solar semi-diameter series treated here was observed with the CCD Astrolabe from March 2nd 1998 to November 27th 2003. It comprises more than 18,000 observations, with mean internal error of 0".20 and standard deviation of 0."59. The observations are made daily, to an average of 20 observations (actually observing days considered), and even distribution of the measurements all year around. A major gap is verified on the series from Sept. 21th to Dec. 19th, 2001, due to apparatus problems. The observations are taken on before and after meridian sessions. As a rule, there is no significant difference between the measurements from the two sessions. The heliolatitude coverage covers the whole solar figure in a semi-annual cycle (Boscardin, 2004).

From an analysis on heliolatitude bins, the solar oblateness can be derived. A conservative approach was chosen, and an ellipsoid of revolution was directly adjusted to the data, in view of the errors involved in the measurements as well as possible systematic effects due to brightness variations on the observed limb. All data considered, we obtain an equatorial radius (**Re**) equal to 959.113± 0.007 and a polar radius (**Rp**) equal to 959.100 ± 0.011. In order to improve on the precision of the determination, and taking advantage of the large number of independent measurements and their normal distribution, the series was sampled by progressively removing the more discrepant points (relatively to an ellipsoid adjustment). We took steps of 0.1 units of standard deviation, removing from 2σ up to 3σ. In the first case, the strictest, 7556 measurements are used, while in the more relaxed case 9047 measurements are used. By this method the robustness of the solution was tested by bootstrapping. The final equatorial and polar radii are given by the average of the values at each step. The final errors are calculated by the root mean square of the squared error at each step plus the covariance of adjacent steps. The final value obtained for the oblateness is 13 ± 4 mas, corresponding to slightly smaller radii $\mathbf{Re} = 959.110 \pm 0.002$ and $\mathbf{Rp} = 959.097 \pm 0.003$. Adopting the component of the oblateness due to surface rotation alone as $\delta \mathbf{r} = 7.8$ mas, the gravitational quadrupole moment of the Sun is obtained as $\mathbf{J_2} = 3.61 \pm 2.90 \times 10^{-6}$ (Reis Neto *et al.*, 2003).

The data file here concerned comprises the rise, crest and beginning of the subduing of the solar cycle 23. Combined to the large quantity and even spreading of the observations, this offers the conditions to statistically compare the observed solar semi-diameter variations against estimators of the solar activity, which could also be expressed as continuous daily series. The estimators were the sunspot count number and its proxy, the 10.7cm radio flux, both sensing the photosphere state; the total solar irradiance and the strength of the integrated solar magnetic field, sensing directly the solar cycle age, and finally the flare index to assess the major solar outbursts. All estimator data were retrieved from the National Geophysical Data Center – NGDC. The hypothesis that the variation of the

Table 1. Maximum of the Pearson Linear Correlation between the Time Variations of the Semi-Di ameter and Estimators of the Solar Activity (1998-2003).

Pair	SD-SS[1]	SD-FL[2]	SD-RF[3]	SD-IR[4]	SD-MF[5]
Correlation	0.80	0.66	0.88	0.78	0.62
Delay (day)	90	100	90	180	120

Notes:
[1]Measured Semi-diameter against sunspot count variations.
[2]Measured Semi-diameter against flare index.
[3]Measured Semi-diameter against 10.7cm emission variations.
[4]Measured Semi-diameter against total irradiance variations.
[5]Measured Semi-diameter against compound magnetic field variations.

solar semi-diameter could be linked to the solar activity was checked by calculating the correlations between the solar semi-diameter series against each one of those estimators. Afterwards, the correlations were re-calculated, taking pairs of correlated series and allowing variable time delays between them (Boscardin *et al.*, 2009). This may points to interconnected phenomena, either with some time delay between them, or even a causal relationship.

In particular, when the periods where had occurred peaks of solar activities are removed from the solar semi-diameter and from the total solar irradiance series, the time delay for the largest correlations between them changes prominently. Taking the complete, no-removal series, the semi-diameter versus irradiance correlation shows two maxima, one close to zero and one close to -380 days. The periods of most intense solar activity are between the Julian modified dates 1709.0 and 1919.0 and between 2213.0 and 2378.0. They are removed from the series. In this case the maximum close to zero disappears. This suggests two modes of response of the semi diameter relatively to the solar irradiance. Along the solar activity cycle the semi-diameter variation trails behind the cycle. However when peaks of activity occur, it also occurs a rapid variation on the measured semi diameter. In these cases the semi diameter variation actually acts as a predictor of intense solar activity.

3. Conclusions and Perspectives

The physics of the photosphere is important for the understanding of the solar flux, from the convection zone to the violent outbursts that emerge from the solar outer layers. It is also all relevant to understand the space weather and relationships to the Earth's magnetosphere, ionosphere, and atmosphere.

The ON series enables to obtain a broader description of the photosphere physical processes, including the relationships between the variations of the diameter and of the main estimators of the solar activity. The correlations between the variations along the solar cycle attain 0.78 for the pair of diameter and total solar irradiance, 0.80 for the pair of diameter and sunspot number, and 0.88 for the pair of diameter and 10.7cm radio flux.

Also, from the analysis grouping the results in heliolatitude bins, the solar oblateness is derived to a precision of -3 dex.

Acknowledgements

The authors acknowledge partial financial support from the Brazilian Science Funding Agencies FAPERJ, CNPq, and FINEP. We also acknowledge the use of solar data from the BBSO and from the NGDC. We register the fruitful relationship between the national institutions belonging in the R2S3 solar network and in the SCOSTEP/CAWSES project.

References

Andrei, A. H., Penna, J. L., Reis Neto, E., D'Ávila, V. A., de Almeida, W. G., & Jilinski, E. G. 2002, What are the astrolabe observations of the Sun measuring?, First Annual Meeting of the R2S3, Calern, France.

Andrei, A. H., Boscardin, S. C., Chollet, F., Delmas, C., Golbasi, O., Jilinski, E. G., Kili, H., Laclare, F., Morand, F., Penna, J. L., & Reis Neto, E. 2004, Comparison of CCD astrolabe multi-site solar diameter observations, Astronomy & Astrophysics, 427, 717–723.

Andrei, A. H., Boscardin, S. C., Reis Neto, E., Penna, J. L., & D'Ávila, V. A. 2006, Variations of the solar radius in solar cycle 23, SCOSTEP-CAWSES Meeting, 1, 43–44.

Boscardin, S. C. 2004, Diameter observations analysis in the solar activity context, MSc Thesis, UFRJ, Observatorio do Valongo.

Boscardin, S. C., Reis Neto, E., Penna, J. L., Papa, A. R. R., Andrei, A. H., & D'Ávila, V. A. 2009, Observed variations of the solar photospheric diameter, 2009, IAU Symposium 264, Solar and Stellar Variability – Impact on Earth and Planets.

Damé, L., Brun, J.-F., Cugnet, D., Derrien, M., Leroy, C., Meftah, M., Meissonnier, M., & Porteneuve, J. 2000, A solar diameter metrology measurement: the PICCARD microsatellite program, ICSO 2000.

Durney, B. R. & Roxburgh, I. W. 1971, Inhomogeneous convection and the equatorial acceleration of the Sun, Solar Physics, 16, 3–11.

Hirzberger, J., Vazquez, M., Bonet, J. A., Hansmeier, A., & Sobotka, M. 1997, The series of solar granulation inages. I. Differences between small and large granules in quiet regions, The Astrophysical Journal, 480, 406–419.

Jilinski, E. G., Puliaev, S. P., Penna, J. K., Andrei, A. H., Sinceac, V., Chollet, F., & Delmas, C. 1998, Solar diameter observations with the astrolabe at Observatório Nacional Rio de Janeiro, Astronomy & Astrophysics Supplement Series, 130, 317–321.

Mursula, K. 2005, Sun and Heliosphere, University of Oulu.

Reis Neto, E., Andrei, A. H., Pennal, J. L., Jilinski, E. G., & Puliaev, S. P. 2003, Observed variations of the solar diameter in 1998–2000, Solar Physics, 212, 7–16.

Roudier, Th. & Reardon, K. 1998, Variations of the solar granulation over the cycle: previous results and future observations, Synoptic Solar Physics, 140, 455–466.

Sofia, S., Basu, S., Demarque, P., Linghuai, L., & Thuiller, G. 2006, The nonhomologous nature of solar diameter variations, Astrophysical Journal, 632, 439.

Solar and Stellar Variability: Impact on Earth and Planets
Proceedings IAU Symposium No. 264, 2009
A. G. Kosovichev, A. H. Andrei & J.-P. Rozelot, eds.

doi:10.1017/S1743921309992341

Solar-like stars seismology

Dumitru Pricopi and Marian Doru Suran

Astronomical Institute of the Romanian Academy, Str. Cutitul de Argint 5, RO-040557 Bucharest, Romania
emails: dpricopi@aira.astro.ro, suran@aira.astro.ro

Abstract. Many solar like stars were observed by CoRoT and MOST missions. We try to investigate some of these stars using asteroseismic inversion method using both frequencies fitting and separation fitting. Our results are compared with semiempirical method based on direct power spectrum analysis.

Keywords. stars: oscillations, methods: data analysis

1. Introduction

In this paper we are focused on the asteroseismic interpretation of the F8V star HD203608. Basic parameters of this star are found using the asteroseismic inversion method i.e. fitting the observed and calculated frequency spectrum. In our study we are using the 21 observed frequencies of this star (Mosser *et al.* 2008) in a frequency range [1.5;3.0mHz]. HD203608 is a low-metallicity star of the old galactic disk. The surface metallicity [Z/X] = − 0.5 ± 0.10 with respect to the solar metallicity (Grevesse & Noels (1993)). Because this star belongs to the old disk, we shall adopt for the calibration an initial abundance Y = 0.25. Using photometric data, Thevenin (1998) and more recently Jonsell *et al.* (2005) and del Peloso *et al.* (2005) adopted a value of effective temperature: Teff = 6070K with an uncertainty of ±100 K. The luminosity were estimeted by Mosser *et al.* (2008): $L/L_\odot = 1.39 \pm 0.13$. This star has a low projected rotational velocity: vsini = 2.4 ± 0.5 km/s. The age of the star is now determined in the range 7.25 ± 0.07Gyr.

2. Results and conclusion

We computed evolutionary models using CESAM2K stellar evolutionary code for stellar masses between 0.80M8 and 1.10M8, with a step of 0.02M8, for chemical composition X = 0.71, X = 0.72, Z = 0.006, a mixing-length parameter of 1.65 times the pressure scale height and no core overshooting. Subsequently, we computed the frequencies of modes having l = 0,1,2,3 of all the models, using LNAWENR non-adiabatic code. As is well-known, a necessary condition for a model frequency to be detected at the real star, is that the model frequency to be stable. We tested the stability of model's oscillations using the usual condition $\eta > 0$, where $\eta = W/\int_0^1 |dW/dx|dx \propto Im(\omega)/Re(\omega)$, W is the total mechanical work done by pulsation, ω is the dimensionless frequency. The fundamental stellar parameters of the best fitting model for HD203608 are: $M = 0.86M_\odot$, $Teff = 6100K$, $L = 1.34L_\odot$, $R = 1.05R_\odot$ and $age = 9.35Gyr$. These values are very similar to those obtained by Mosser *et al.* (2008). The best fitting model was able to match 5 frequencies in 1σ, 10 in 2σ and 13 in 3σ. All the calculated frequencies are found to be unstable. Our best fitting model failed to represent the radial modes and non-radial modes with l = 3.

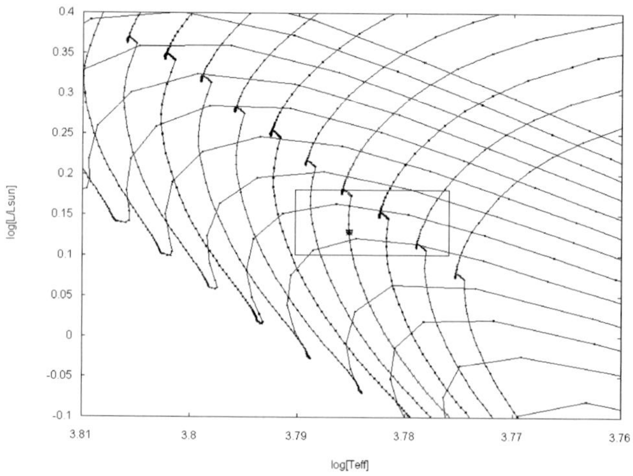

Figure 1. The location of the best fitting model in the grid corresponding to X = 0.72. This location correspond to a stellar mass of $0.86 M_{\odot}$ and age of 9.35Gyr. The rectangle denotes the error box of observed values of effective temperature and luminosity.

References

del Peloso, E. F., Cunha, K., da Silva, L., & Porto de Mello, G. F. 2005, *A&A*, 441, 1149

Grevesse, N. & Noels, A. 1993, *Cosmic abundance of the elements*, in *Origin and evolution of the elements*, edited by N. Prantzos, E. Vangioni Flam and M. Casse, Cambridge University, England, pp. 15–25

Jonsell, K., Edvardsson, B., Gustafsson, B. *et al.* 2005, *A&A*, 440, 231

Mosser, B., Deheuvels, S., Michel, E. *et al.* 2008, *A&A* (to appear)

Thevenin, F. 1998 VisieR Online Data Catalog, 3193, 0

Solar and Stellar Variability: Impact on Earth and Planets
Proceedings IAU Symposium No. 264, 2009
A. G. Kosovichev, A. H. Andrei & J.-P. Rozelot, eds.

doi:10.1017/S1743921309992353

Sounding stellar cycles with Kepler – preliminary results from ground-based chromospheric activity measurements†

C. Karoff[1], T. S. Metcalfe[2], W. J. Chaplin[1], S. Frandsen[3], F. Grundahl[3], H. Kjeldsen[3], D. Buzasi[4], T. Arentoft[3] and J. Christensen-Dalsgaard[3],

[1]School of Physics and Astronomy, University of Birmingham, Edgbaston, Birmingham B15 2TT, UK
[2]High Altitude Observatory and Scientific Computing Division, NCAR, PO Box 3000, Boulder, CO 80307, USA
[3]Department of Physics and Astronomy, Aarhus University, DK-8000 Aarhus C, Denmark
[4]Eureka Scientific, Inc., 2452 Delmer Street Suite 100, Oakland, CA 94602, USA
email: `karoff@bison.ph.bham.ac.uk`

Abstract. Due to its unique long-term coverage and high photometric precision, observations from the Kepler asteroseismic investigation will provide us with the possibility to sound stellar cycles in a number of solar-type stars with asteroseismology. By comparing these measurements with conventional ground-based chromospheric activity measurements we might be able to increase our understanding of the relation between the chromospheric changes and the changes in the eigenmodes.

In parallel with the Kepler observations we have therefore started a programme at the Nordic Optical Telescope to observe and monitor chromospheric activity in the stars that are most likely to be selected for observations for the whole satellite mission. The ground-based observations presented here can be used both to guide the selection of the special Kepler targets and as the first step in a monitoring programme for stellar cycles. Also, the chromospheric activity measurements obtained from the ground-based observations can be compared with stellar parameters such as ages and rotation in order to improve stellar evolution models.

Keywords. Sun: activity, Sun: oscillations, stars: activity, stars: oscillations

1. Introduction

During recent years there has been a growing interest in understanding stellar cycles in general and the solar cycle in particular. The reason for this is partly the indications that the solar activity cycle could cause climate change (see e.g. Svensmark & Friis-Christensen 1997) and partly that solar cycle 24 now seems to be around two years delayed compared to theoretical predictions (Zimmerman 2009).

The "Sounding stellar cycles with Kepler" programme will monitor stellar cycles from the amplitude and frequency shifts of the oscillation modes observed in these stars. By comparing these measurements with conventional ground-based chromospheric activity measurements we might be able to increase our understanding of the relation between the chromospheric changes and the changes in the eigenmodes. Asteroseismic analysis of the Kepler observations also allows us to test the hypothesis put forward by Böhm-Vitense (2007) – that stellar cycles in active stars are generated by the gradient in the rotation

† Based on observations made with the Nordic Optical Telescope, operated on the island of La Palma jointly by Denmark, Finland, Iceland, Norway, and Sweden, in the Spanish Observatorio del Roque de los Muchachos of the Instituto de Astrofísica de Canarias.

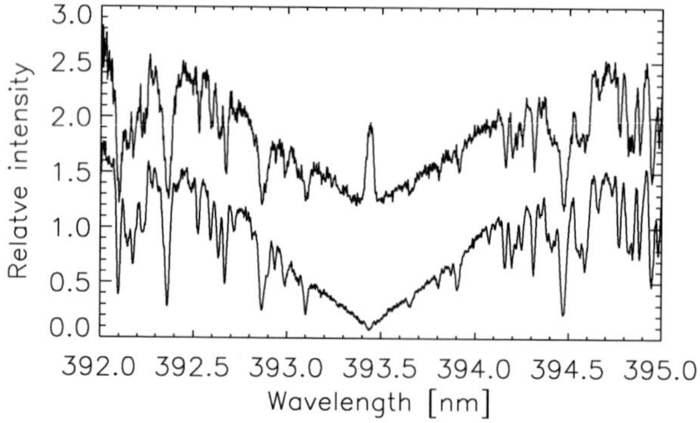

Figure 1. Two spectra illustrating stars with and without chromospheric emission. The wavelength region covers the region around the Ca II K line. The upper star (SAO 48081) clearly shows emission in the middle of the Ca II K absorption line which proves that this star contains a high level of chromospheric activity. The lower star (SAO 31656) shows no emission which indicates that this star is inactive. The relative intensities have been normalized to the mean intensity in the echelle order and the spectrum for SAO 48081 has been shifted one unit upwards. No radial velocities have been calculated for the stars, but the spectra have been arbitrary aligned.

rate close to the surface, whereas the stellar cycles in inactive stars are generated by the gradient in the rotation rate at the base of the convection zone. This can be done as the asteroseismic analysis of the Kepler observations will allow us to measure the depth of the convection zone and give estimates of the internal rotation profile. The details of the "Sounding stellar cycles with Kepler" programme are described by Karoff *et al.* (2009).

The ground-based chromospheric activity measurements can be used not only to compare the changes on the stellar surfaces to those in the interior over the stellar cycles and to test stellar dynamo models, but also to calibrate rotation-activity-age relations. This can be done as asteroseismology offers a unique possibility to measure ages of solar-type field stars to a precision of 5–10% (Christensen-Dalsgaard *et al.* 2007) and because asteroseismology provides the possibility to measure the inclination of the stellar rotation axis so that the spectroscopic measured $v\sin i$ can be converted into equatorial surface rotation rates (Gizon & Solanki 2003). Also, the first ground-based chromospheric activity measurements that we present here on the potential candidates for observations for the whole mission can be used to rate the different candidates in order to ensure that both active and inactive stars are observed.

2. Observations

During the summer of 2009 we have measured chromospheric activity for 40 of the stars that have also been observed for one month in the first part of the Kepler survey phase and are therefore candidates for being selected for observations during the whole mission. These stars were selected as described by Karoff *et al.* (2009) – i.e. as the brightest stars on the cool end of the main sequence. The observations were performed with the high-resolution FIbre-fed Echelle Spectrograph mounted on the 2.6-m Nordic Optical Telescope (Frandsen & Lindberg 2000). One echelle spectrum with a resolution of 25000 was obtained for each star with an integration time of 420 sec. The raw spectra were reduced and wavelength calibrated using the software package *FIEStool*. Examples of reduced spectra for an active and an inactive star are shown in Fig. 1.

In order to measure the chromospheric activity we first measure the dimensionless S index as described by Hall *et al.* (2007) by measuring the flux in four 1-Å-wide

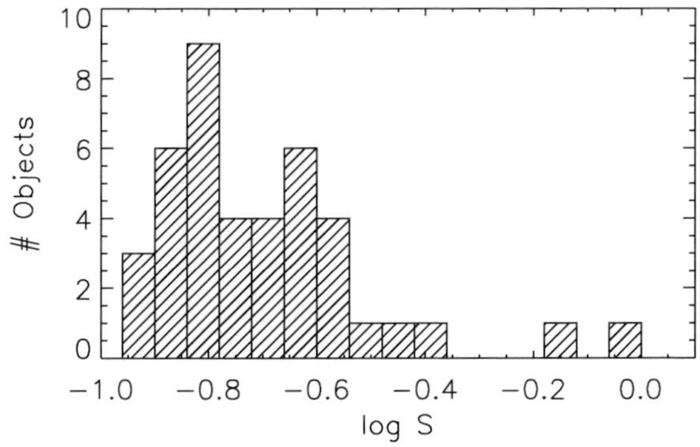

Figure 2. Histograms of the S indices for the 40 Kepler targets observed in the summer of 2009 with FIES at the NOT as part of the "Sounding stellar cycles with Kepler" program. The figure nicely shows the separation of the observed stars into active stars with log S around −0.6 and inactive stars with log S around −0.8.

bandpasses – two in the Ca II H & K line cores and two in the continuum away from the Ca II H & K lines. During the observations we started each night with observing the star HD 157214 which is part of the HK project at Lowell Observatory (Hall *et al.* 2007). It has therefore been possible to calibrate our measured S indices to the S indices measured at Lowell Observatory using the spectrum and S index of HD 157214 measured at Lowell Observatory around the same time as our observations and kindly provided to us by J. C. Hall. We also plan to transform our measured S indices into chromospheric emission fractions R'_{HK} and absolute flux $\mathcal{F}_{\mathrm{HK}}$ (see Hall *et al.* 2007 for details), but this will only be done when the Kepler observations have provided us with precise asteroseismic measurements of the stellar effective temperatures.

Figure 2 shows a histogram of the measured S indices. The stars separate into a nice bifurcated distribution known as the Vaughan–Preston gap (Vaughan & Preston 1980) and the goal for the selection of stars for observations during the whole mission is that these stars will also reflect this distribution.

References

Böhm-Vitense, E. 2007, ApJ, 657, 486

Christensen-Dalsgaard, J., Arentoft, T., Brown, T. M., Gilliland, R. L., Kjeldsen, H., Borucki, W. J., & Koch, D. 2007, Communications in Asteroseismology, 150, 350

Frandsen, S. & Lindberg, B. 2000, The Third MONS Workshop: Science Preparation and Target Selection, 163

Gizon, L. & Solanki, S. K. 2003, ApJ, 589, 1009

Hall, J. C., Lockwood, G. W., & Skiff, B. A. 2007, AJ, 133, 862

Karoff, C., Metcalfe, T. S., Chaplin, W. J., Elsworth, Y., Kjeldsen, H., Arentoft, T., & Buzasi, D. 2009, MNRAS, 1226

Svensmark, H. & Friis-Christensen, E. 1997, Journal of Atmospheric and Terrestrial Physics, 59, 1225

Vaughan, A. H. & Preston, G. W. 1980, PASP, 92, 385

Zimmerman, R. 2009, ScienceNOW Daily News, 8 May

Solar and Stellar Variability: Impact on Earth and Planets
Proceedings IAU Symposium No. 264, 2009
A. G. Kosovichev, A. H. Andrei & J.-P. Rozelot, eds.

doi:10.1017/S1743921309992365

Differential rotation on active late-type stars observed with *Corot*

P. Gondoin[1], M. Fridlund[1], D. Gandolfi[2] and E. Güenter[2]

[1]European Space Agency, ESTEC - Postbus 299, 2200 AG Noordwijk, The Netherlands

[2]Thüringer Landessternwarte Tautenburg, Sternwarte 5, 07778 Tautenburg, Germany

Abstract. One major topic in studying stellar activity is to explain how phenomena seen on the Sun and stars, and specially magnetic phenomena, depend on stellar properties such as rotation and age. Differential rotation is an important physical process in theories of stellar magnetic field generation. The solar surface differential rotation was initially discovered via the simple method of tracking the rotation rates of individual starspots at different latitudes. Adopting a similar principle, high accuracy light curves of active stars observed with the *CoRoT* satellite are analyzed using a model based on the rotational modulation of the visibility of active regions.

Keywords. stars: activity, stars: late-type, stars: rotation, stars: spots

1. Introduction

Differential rotation measurements have been performed on F and early G dwarfs (e.g. Reiners & Schmitt 2003; Reiners 2006) by studying the rotational broadening of spectral lines with a Fourier transform method. Differential rotation has also been measured using Doppler imaging techniques on a few young rapidly rotating G and K stars (e.g. Donati *et al.* 2000; Petit *et al.* 2002; Collier-Cameron & Donati 2002; Marsden *et al.* 2004; Barnes *et al.* 2005).

One magnetic-field diagnostic for cool stars is the presence of starspots. Starspots cooler than the unperturbed photosphere can account for the rotation modulated attenuation of stellar fluxes in broad optical pass-bands. The photometric periods of active stars have been measured by studying the rotation modulation of long sequences of broad-band photometric measurements (e.g. Hall 1991; Henry *et al.* 1995; Messina & Guinan 2003). We have analyzed high accuracy light curves of active F, G and K stars, that were obtained with the *CoRoT* satellite during its initial run.

2. Lightcurves analyses

The *CoRoT* satellite (Baglin *et al.* 2006; Auvergne *et al.* 2009) simultaneously registers the visible lightcurves of a large number of stars for durations of several months with the aim to detect extrasolar planets by the transit method. *CoRoT* observations started with a pointing direction close to the anticenter of the Galaxy. This initial run lasted from February 6th to April 2nd 2007 and supplied high accuracy lightcurves. These time series are nearly continuous over 57 days with only a small number of gaps that mainly result from the crossing of the South-Atlantic Anomaly. A fraction of these light curves have a periodic modulation indicative of starspots.

The rotation modulation of stellar photometric lightcurves by stellar active regions can be modelled using two numerical approaches: the surface integration methods and the analytical method (Ribarik *et al.* 2003). The former assigns a temperature to each pixel of the spherical integration net and then varies each value until an optimal fit to the data is achieved. In this study, we use the analytical approach described by Lanza *et al.* (2006) and based on a model used to fit the time variations of the solar bolometric and spectral irradiance.

The relative flux variations of active stars are fitted with a two or three spots model. Best fit models to the lightcurves are obtained by minimizing the sum of squared residuals. The fixed parameters in the simulations included stellar parameters determined spectroscopically (Gandolfi *et al.* 2009) and active region parameters. The surfaces, longitudes, latitudes and rotation periods of the active regions are left as free parameters. The analyses of the lightcurves are performed as a grid search using a two or three spots model and iterating on 8 or 12 variable parameters. Series of fitting processes varying the inclination of the star rotation axis are conducted in order to identify the inclination angle that minimizes the overall χ^2 and the fitting parameters for which the difference in χ^2 becomes significant at a more than 80 % confidence levels.

3. Preliminary results

Preliminary analysis were conducted on a sample of active dwarfs that were arbitrarily selected. One first result of this analysis is that all high accuracy lightcurves with evidence for rotation modulation cannot be parameterized with a simple two or three spots model. This may be due to complex topologies, time evolution or latitude migration of the active centers during the observing run. Among an initial sample of six dwarfs, the lightcurve of only one star could be fitted with a two spots model on the entire 57 days observation duration (see Fig. 1 top). The lightcurves of two other sample stars (see Fig. 1 middle and bottom) could be fitted on a large fraction (about 45 days) of the observing run.

Nevertheless, within the frame of the analysis method, differential rotation has been detected at a significant confidence level on these three stars. The differential rotations ΔP between the high and low latitude spots are 0.03, 0.42 and 0.5 days for average rotation periods of 3.05, 4.24 and 15.23 days, respectively. Although these first results may be underestimates of the latitudinal differential rotation shear, comparisons with measurements obtained with different analysis methods support the trend of increasing differential rotation rate $\Delta P/P$ with increasing rotation period.

References

Auvergne, M., Bodin, P., Boisnard, L. *et al.* 2009, *A&A*, in press

Baglin, A., Auvergne, M., Boisnard, L., *et al.* 2006, *36th COSPAR Scientific Assembly*, 36, 3749

Barnes, J. R., Collier Cameron, A., Donati, J.-F., *et al.* 2005, *MNRAS*, 357, L1

Collier Cameron, A. & Donati, J.-F. 2002, *MNRAS*, 329, L23

Donati, J.-F., Mengel, M., Carter, B., *et al.* 2000, *MNRAS*, 316, 699

Gandolfi, D. *et al.* 2009, in preparation.

Hall, D. S. 1991, in Tuominen I., Moss D., Rüdiger G., (eds), *Proc. IAU Colloq. 130, The Sun and Cool Stars*, Springer-Verlag, p. 353

Henry, G. W., Eaton, J. A., Hamer, J., & Hall, D. S., 1995, *ApJS*, 97, 513

Marsden, S. C., Waite, I. A., Carter, B. D., & Donati, J.-F. 2004, *AN*, 325, 246

Lanza, A. F., Messina, S., Pagano, I., *et al.* 2006, *AN*, 327, 28

Petit, P., Donati, J.-F., Collier Cameron, A. 2002, *MNRAS*, 334, 374

Reiners, A. & Schmitt, J. H. M. M. 2003, *A&A*, 398, 647

Reiners, A. 2006, *A&A*, 446, 267

Ribarik, G., Olah, K., & Strassmeier, K. G. 2003, *AN*, 324, 202

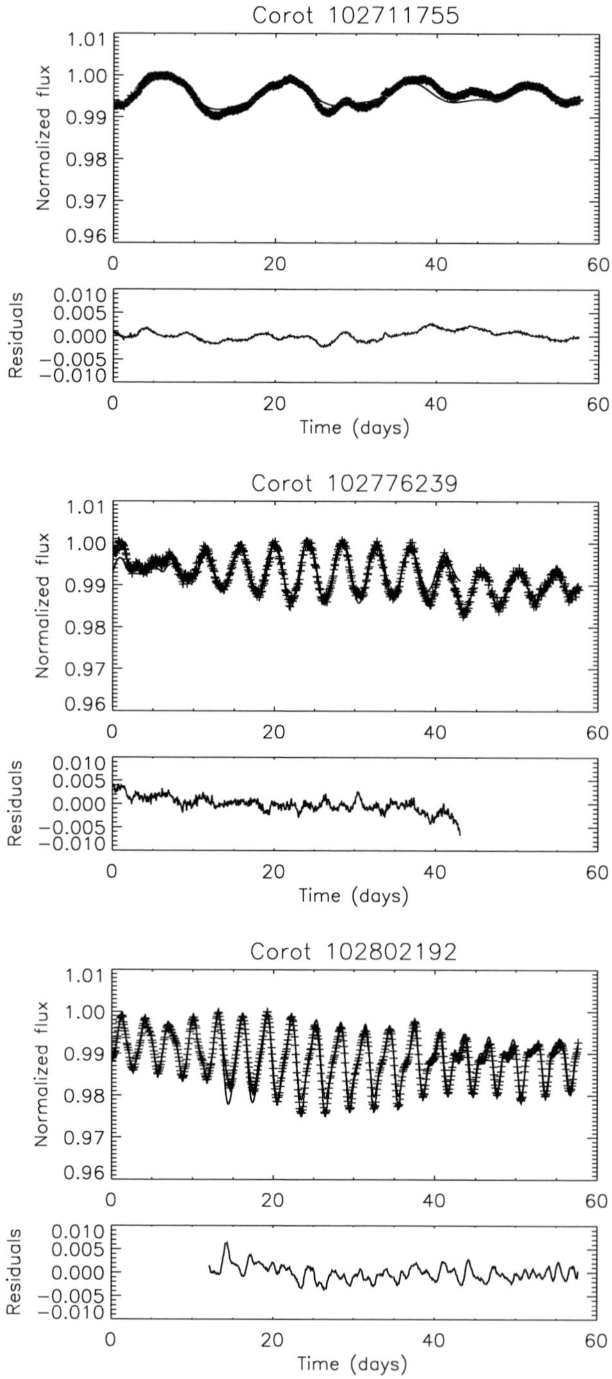

Figure 1. Best fit models and residuals to the light curve of a sample of active dwarfs observed with *Corot*.

Solar and Stellar Variability: Impact on Earth and Planets
Proceedings IAU Symposium No. 264, 2009
A. G. Kosovichev, A. H. Andrei & J.-P. Rozelot, eds.

doi:10.1017/S1743921309992377

A comparison of measured and simulated solar network contrast

N. Afram[1], Y. C. Unruh[1], S. K. Solanki[2], M. Schüssler[2] and S. K. Mathew[3]

[1]Imperial College, London, UK
[2]Max-Planck-Institut für Sonnensystemforschung, D-37191 Katlenburg-Lindau, Germany
[3]Udaipur Solar Observatory (USO), Rajasthan, India

Abstract. Long-term trends in the solar spectral irradiance are important to determine the impact on Earth's climate. These long-term changes are thought to be caused mainly by changes in the surface area covered by small-scale magnetic elements. The direct measurement of the contrast to determine the impact of these small-scale magnetic elements is, however, limited to a few wavelengths, and is, even for space instruments, affected by scattered light and instrument defocus. In this work we calculate emergent intensities from 3-D simulations of solar magneto-convection and validate the outcome by comparing with observations from Hinode/SOT. In this manner we aim to construct the contrast at wavelengths ranging from the NUV to the FIR.

Keywords. Sun: general, solar-terrestrial relations, Sun: magnetic fields, Sun: granulation

1. Introduction

Variations in the long-term spectral solar irradiance, i.e. the changes in the Sun's brightness at a certain wavelength, are significant for their effects on Earth's atmospheric temperature and composition. Magnetic fields concentrated in small structures influence these long-term changes, with increasing relevance towards the solar limb. Here, we compare observed and simulated intensity contrasts and investigate their behaviour from disk centre to the limb in 3D MHD simulations with different average vertical magnetic fields.

2. Simulations and observations

In this work, we use 3D MHD simulations that have been run with non-grey radiative transfer using the MURaM code (Vögler *et al.* 2005), which enables realistic simulations of solar magneto-convection in the photosphere and the upper layers of the convection zone. The simulations considered here had an average vertical field of 0G, 50G, 200G. The simulations run on a 288 × 288 surface grid with 100 depth points, corresponding to a solar surface area of 6 × 6 Mm and a depth of 1.4 Mm. The intensities are calculated using the spectral synthesis code ATLAS9 (Kurucz 1993) on the simulated model atmospheres. We determine the root-mean-square (rms) contrast (standard deviation divided by the mean value) to characterise intensity variations between dark and bright features.

The data used in this work were recorded with the broad-band filter imager mounted on the Solar Optical Telescope (SOT, Tsuneta *et al.* 2008, Ichimoto *et al.* 2008, Suematsu *et al.* 2008, Shimizu *et al.* 2008) onboard Hinode on 8th Nov. 2006 (during the Mercury transit), and were also used to obtain the point-spread-function (PSF) (see Mathew *et al.* 2009). The pixel resolution is 0.05 arsec. The dark correction and flat fielding were done with SSW-IDL routines. We have calculated intensity (rms) contrasts for the Hinode

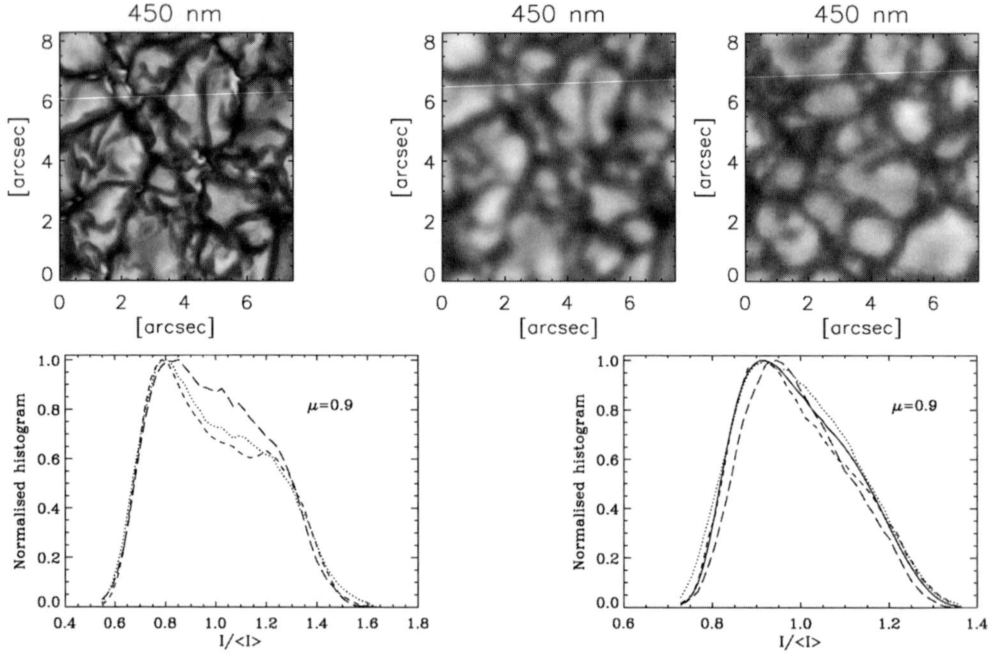

Figure 1. Simulations and observations in the blue continuum filtergram. *Top left:* Original simulation snapshot with an average vertical magnetic field of 50G. *Top middle:* Convolved simulation snapshot with an average vertical magnetic field of 50G. *Top right:* Extract from observed image. *Bottom left:* Normalised histograms for the intensity distribution in the original simulations: 0G (dotted), 50G (dashed), 200G (long dashes). *Bottom right:* Normalised histograms for the intensity distribution in convolved simulations and in observations: 0G (dotted), 50G (dashed), 200G (long dashes), observation (thick solid).

images at different wavelengths (388nm (CN), 430.5nm (G-band), 450nm (blue continuum), 555nm, 668nm) and compared them with contrasts for the MURaM/Atlas images (which were convolved with the corresponding PSF). Figure 1 shows the comparison of simulated (original and convolved) and observed images and their contrast histograms for the blue continuum, where we obtain an rms contrast of 12% for the convolved MURaM simulation as well as for the HINODE observations, while the rms contrast for the original simulation is 21.6%. Similar results are obtained for the other wavelengths.

3. Centre to limb variation in simulations

A spatial element of the solar atmosphere radiates differently if it is located at disk centre or nearer to the solar limb, and this spatial evolution of the irradiance behaves differently for different wavelengths. In the UV magnetic features are comparatively brighter. These changes become even stronger near the limb and are therefore crucial for spectral solar irradiance variations. In Figure 2, we study the simulated contrasts (i.e. the mean intensity difference between the magnetic and quiet Sun simulation compared to the quiet Sun) as a function of heliocentric position in five different wavelengths (corresponding to the Hinode/SOT filters). We find that the contrast increases towards the limb and is higher for shorter wavelengths. For the simulation with the higher average vertical field of 200G, the contrasts are significantly larger. The flattening and turnover (e.g. for 388 nm and 430 nm) agree better with measurements of facular contrasts than 1D calculations (e.g. Unruh et al 1999, Ortiz *et al.* 2002).

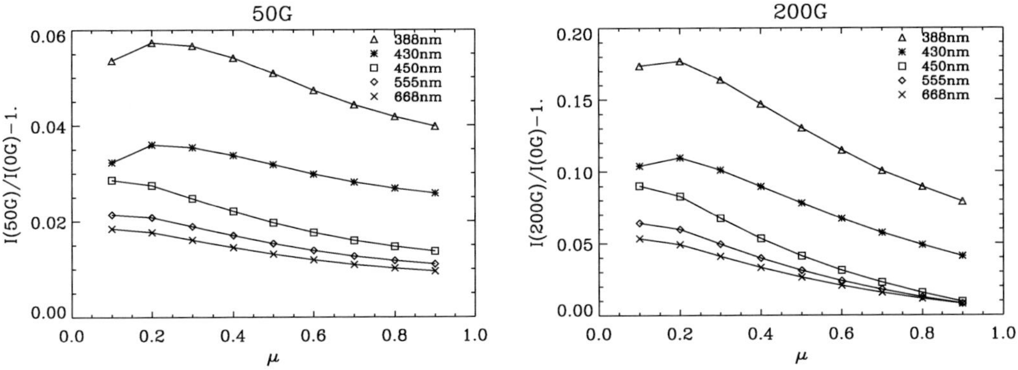

Figure 2. Contrast as a function of limb angle ($\mu = cos\theta$) for the 5 different Hinode/SOT wavelengths for an average vertical magnetic field of 50G and 200G, respectively. The contrast is given as the mean intensity difference between the magnetic and quiet Sun simulation compared to the quiet Sun.

4. Conclusions

This analysis will allow us to determine the network and facular contrast as a function of magnetic flux, limb angle, and wavelength. We will then be able to remove the single free parameter in the SATIRE model (e.g. Krivova *et al.* 2003) and help improve irradiance reconstructions.

References

Ichimoto, K., Katsukawa, Y., Tarbell, T., Shine, R. A., Hoffmann, C., Berger, T., Cruz, T., Suematsu, Y., Tsuneta, S., Shimizu, T., & Lites, B. W. 2008, *in First Results From Hinode, ed. S. A. Matthews, J. M. Davis, & L. K. Harra, ASP-CS*, 397, 51

Krivova, N. A., Solanki, S. K., Fligge, M., & Unruh, Y. C. 2003, *A&A*, 399, L1

Kurucz, R. L. 1993, *ATLAS9 Stellar Atmosphere Programs and 2km/s grid (Kurucz CD-ROM 13)*

Mathew, S. K., Zakharov, V., & Solanki, S. K. 2009, *A&A*, 501, L19

Ortiz, A., Solanki, S. K., Domingo, V., Fligge, M., & Sanahuja, B. 2002, *A&A*, 388, 1036

Shimizu, T., Nagata, S., Tsuneta, S., Tarbell, T., Edwards, C., Shine, R., Hoffmann, C., Thomas, E., Sour, S., Rehse, R., Ito, O., Kashiwagi, Y., Tabata, M., Kodeki, K., Nagase, M., Matsuzaki, K., Kobayashi, K., Ichimoto, K., & Suematsu, Y. 2008, *Solar Phys.*, 249, 221

Suematsu, Y., Tsuneta, S., Ichimoto, K., Shimizu, T., Otsubo, M., Katsukawa, Y., Nakagiri, M., Noguchi, M., Tamura, T., Kato, Y., Hara, H., Kubo, M., Mikami, I., Saito, H., Matsushita, T., Kawaguchi, N., Nakaoji, T., Nagae, K., Shimada, S., Takeyama, N., & Yamamuro, T. 2008, *Solar Phys.*, 249, 197

Tsuneta, S., Ichimoto, K., Katsukawa, Y., Nagata, S., Otsubo, M., Shimizu, T., Suematsu, Y., Nakagiri, M., Noguchi, M., Tarbell, T., Title, A., Shine, R., Rosenberg, W., Hoffmann, C., Jurcevich, B., Kushner, G., Levay, M., Lites, B., Elmore, D., Matsushita, T., Kawaguchi, N., Saito, H., Mikami, I., Hill, L. D., & Owens, J. K. 2008, *Solar Phys.*, 249, 167

Unruh, Y. C., Solanki, S. K., & Fligge, M. 1999, *A&A*, 345, 635

Vögler, A., Shelyag, S., Schüssler, M., Cattaneo, F., Emonet, T., & Linde, T. 2005, *A&A*, 429, 335

Solar and Stellar Variability: Impact on Earth and Planets
Proceedings IAU Symposium No. 264, 2009
A. G. Kosovichev, A. H. Andrei & J.-P. Rozelot, eds.

doi:10.1017/S1743921309992389

Multifractal spectrum of solar active region NOAA 10960 in the H_α spectral line

Batmunkh D.

Research centre for Astronomy and Geophysics, Mongolian Academy of Sciences
Box-152, Ulaanbaatar-51, Mongolia
email: batmunkh@rcag.ac.mn

Abstract. We obtained the H_α images of some solar active regions and prominences. Our astronomical observatory has the telescope-coronagraph which was equipped with the birefringent Halle H_α filter and CCD camera Apogee U4 (2048x2048 pixels). This paper presents multifractal spectra of images of some solar active regions in H_α line obtained with the coronagraph. The Pointwise Hoelder exponents (α) and Hausdorff spectrum $f_n(\alpha)$ for a part of chromosphere with active region and without it, have been obtained. It is visible, that curves $f_n(\alpha)$ for quiet and active regions of chromosphere differ very strongly. In particular, $f_n(\alpha)$ for region with a sunspot and flare has very complicated form, and for region with filament the curves do not considerably differ from quiet chromosphere. The multifractal spectrum of quiet chromosphere shows that the quiet chromosphere is very well described by fractals with different dimensions. We analyze this result.

Keywords. Sun: atmosphere, activity, flares, multifractal spectrum

1. Introduction

Some publications were devoted to research about existence of scale invariant solar magnetic field at a level of photosphere (Abramenko *et al.* (2002), Lawrence *et al.* (1993), Salakhutdinova and Golovko (2004), Stark *et al.* (1997)). They emphasized the following. The quiet-Sun photospheric fields are generated by local dynamo action based on random convective motions at high magnetic Reynolds number. The properties of active region images are determined by the presence of fields generated by the global, mean field dynamo (Lawrence *et al.* (1993)). Fractal and multifractal techniques have been applied to various types of solar data to study the fractal properties of sunspots as well as the distribution of photospheric magnetic fields and the role of random motions on the solar surface in this distribution. Other research includes the investigation of changes in the fractal dimension as an indicator for solar flares (Stark *et al.* (1997)). The B_z field in an active region is an intermittent (multifractal) structure and that the degree of intermittency increases when the flaring activity is rising (Abramenko *et al.* (2002)). The medium is in the state of intermittent turbulence, strongly different from the classic Kolmogorov's turbulence. The difference is observed not only for larger moments of structure functions, but also for smaller ones (Salakhutdinova and Golovko (2004)). Magnetic fields of the solar chromosphere are frozen into plasma. Therefore the H_α filtergram exhibits a properties of magnetic field of active regions of the Sun. In this work we have considered behaviour of multifractal spectrum of some active phenomena observed in the H_α line at the minimum of solar activity.

2. Observational Data

The Ulaanbaatar astronomical observatory of the RCAG (Research Centre for Astronomy and Geophysics of Mongolian Academy of Sciences) carries out observations of solar active phenomena in the H_α line with the coronagraph telescope equipped with the Halle birefringent filter with the $0.5\AA$ passband. The telescope was equipped with the Apogee CCD camera with 2048x2048 pixel.
This paper is based on the observations of two active regions: NOAA 10898 (30 June 2006)(Figure 1, left panel), and the activity complex of NOAA 10960 (4 June 2007) (Figure 1, right panel). All digital images were preprocessed using a fast Fourier transform filters (using CCD imaging software MaxIm DL). The flare develops from photosphere to corona (Figure 1, right panel and figure 2). The data used in our investigation, are obtained with this telescope.

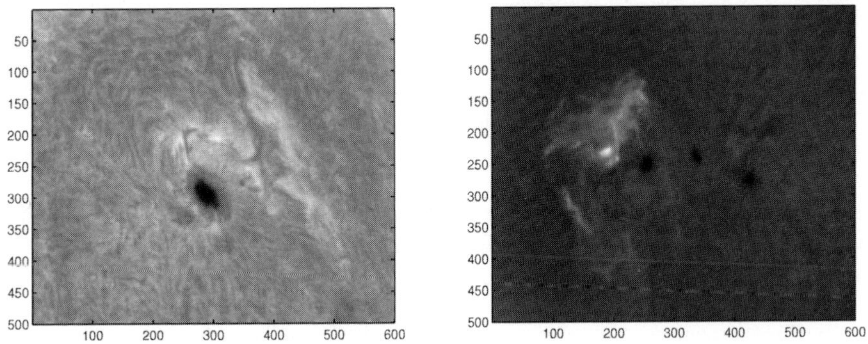

Figure 1. H_α filtergrams for the active regions: left panel- NOAA 10898(30 June 2006) and right panel- NOAA 10960 (04 June 2007). In the images the spatial pixel scale is 0.51arcsec. Observations were made in Ulaanbaatar Astronomical Observatory.

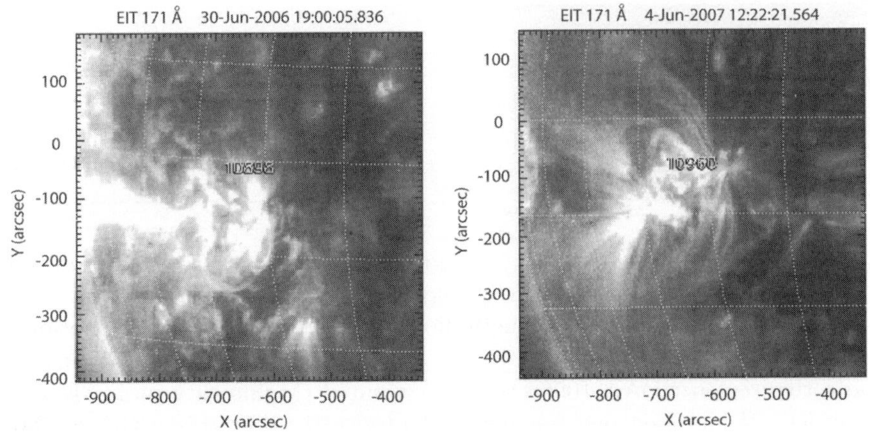

Figure 2. Same active regions(as figure 1) in EIT $171\AA$ from SOHO station(see soho): left panel- NOAA 10898 and right panel- NOAA 10960.

3. Results and discussion

We have chosen images of active regions with the size 300×200 $pixel^2$ ($1pixel = 0.51arcsec$ in the solar disk) (Fig. 3–12, left panel). We have estimated multifractal spectra of active regions images in H_α line using "FracLab" software (see FracLab). For quiet chromosphere areas (Fig. 3, 8; left) multifractal spectrum is stable and a variations (fluctuations) of the singularity are weak (Fig. 3, 8; right).

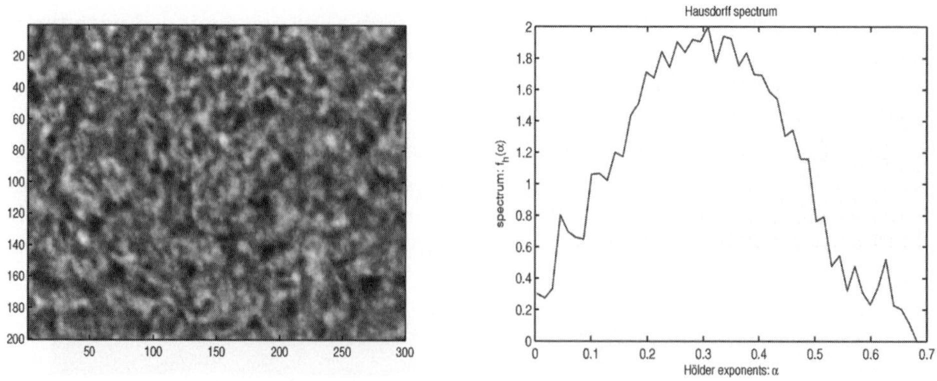

Figure 3. Left panel- H_α filtergram for a some part of the quiet solar chromosphere and right panel- multifractal spectrum for this part.

In figures 4 and 9, the solar area with the filament (near NOAA 10898) and their multifractal spectrum are presented. An effect of the filament to multifractal spectra shows itself in wing, but not considerably.

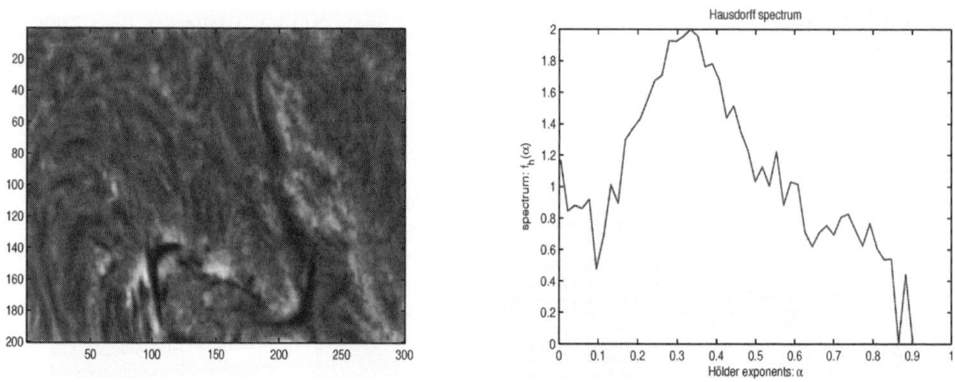

Figure 4. Left panel- H_α filtergram for the filament of the active region NOAA10898 and right panel- multifractal spectrum for this filament.

In the active region NOAA 10960 the flare occured (McIntosh type is Fkc, from (see solarmonitor)). We carried out the series of observations of this active region. Some observations are shown in Figures 5–7 and 10–12. In the Figure 5 multifractal spectrum at pre-flare stage is shown and a fluctuations of the singularity are very strong in comparison to the spectrum of quiet chromosphere (Fig. 3). In the Figure 6 multifractal spectra at flare stage and the right wing with the fluctuations of the singularity of this spectrum are shown. In other words, the width of the spectrum expands strongly and is very different from Fig. 5 (right panel). Figure 7 shows H_α filtergram of image after the flare and the corresponding multifractal spectrum. Here it can be seen,that the process of event goes to the initial state. Thus, primary results show,that multifractal spectra are very sensitive

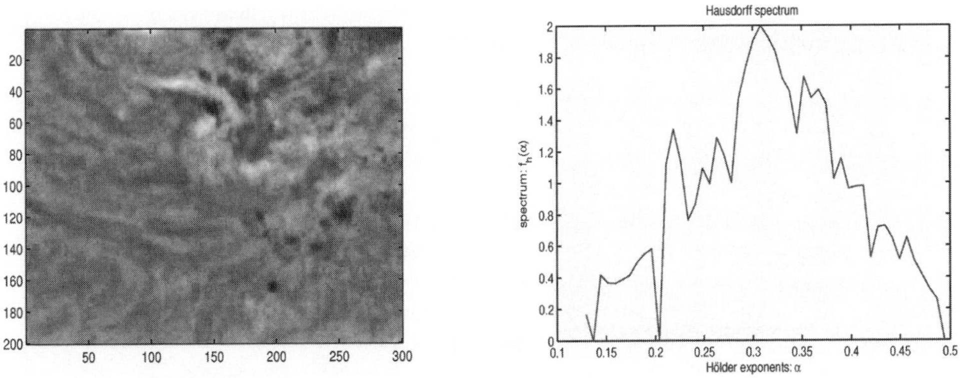

Figure 5. Left panel- H_α filtergram at pre-flare stage of the active region NOAA10960 and right panel- multifractal spectrum for this image. Time: 03:17:41UT.

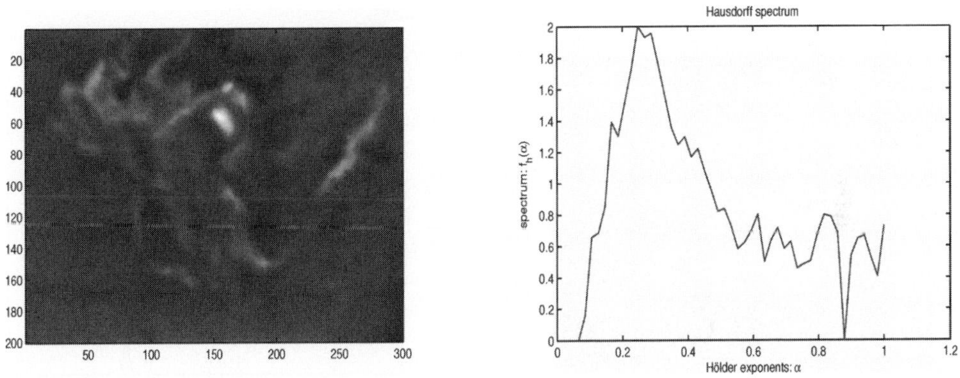

Figure 6. Left panel- H_α filtergram at the flare of the active region NOAA10960 and right panel- multifractal spectrum for this image. Time: 04:17:21UT.

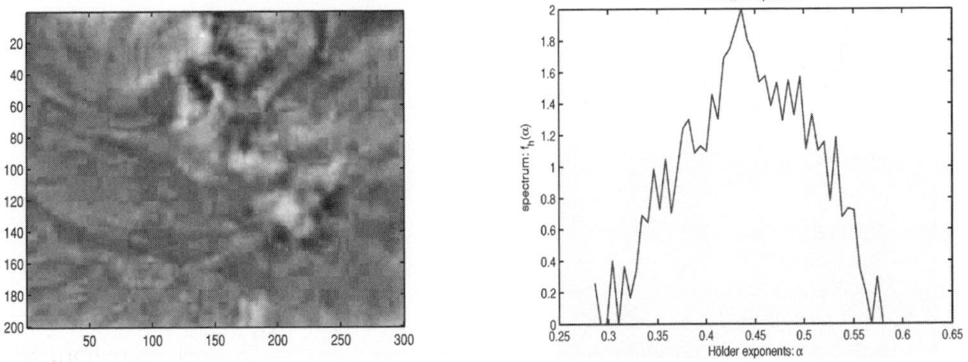

Figure 7. Left panel- H_α filtergram of image after flash of the active region NOAA10960 and right panel- multifractal spectrum for this image. Time: 05:37:16UT.

to reconstruction of structures in the active region. It can be seen, that the changes in the multifractal spectrum can be regarded as an indicator for solar flares.

Also, in the Figures 8–12 were shown Hoelder images and their multifractal spectrum. They yield too the results as previous.

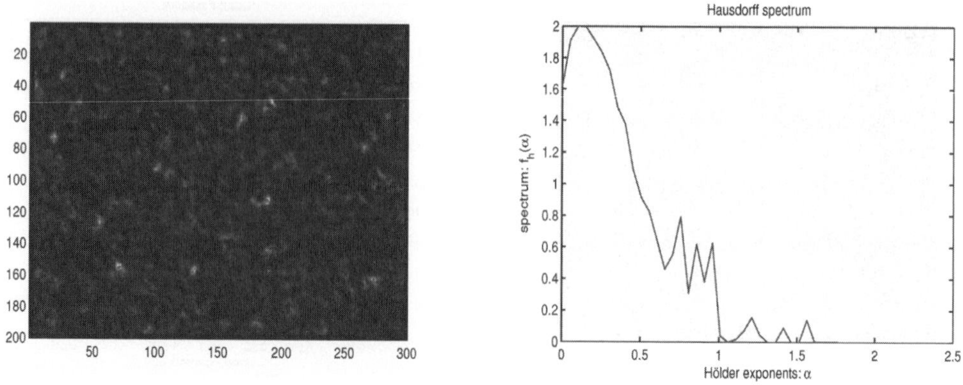

Figure 8. Left panel- Hoelder image for a part of the quiet solar chromosphere and right panel- multifractal spectrum for this part.

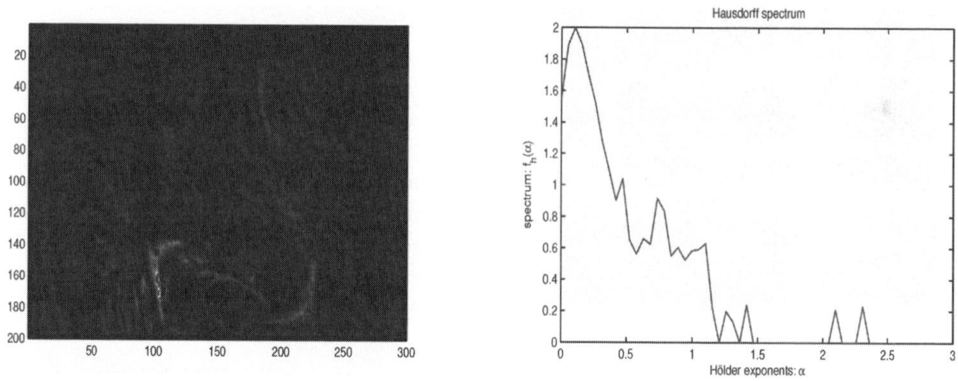

Figure 9. Left panel- Hoelder image for the filament of the active region NOAA10898 and right panel- multifractal spectrum for this image.

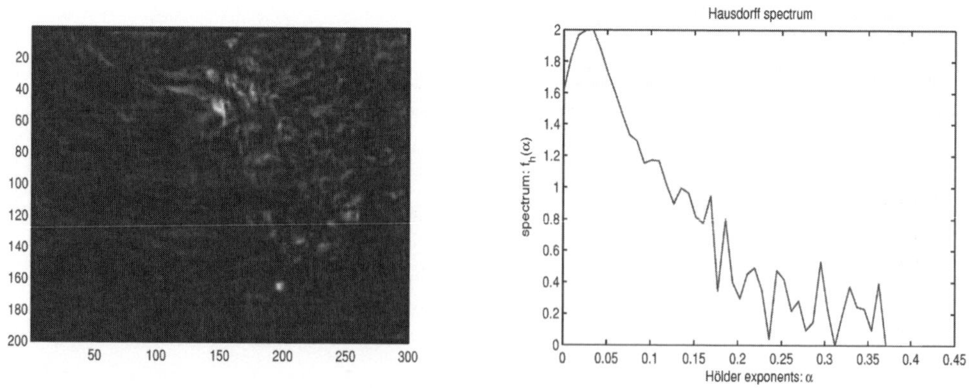

Figure 10. Left panel- Hoelder image at pre-flare stage of the active region NOAA10960 and right panel- multifractal spectrum for this image.

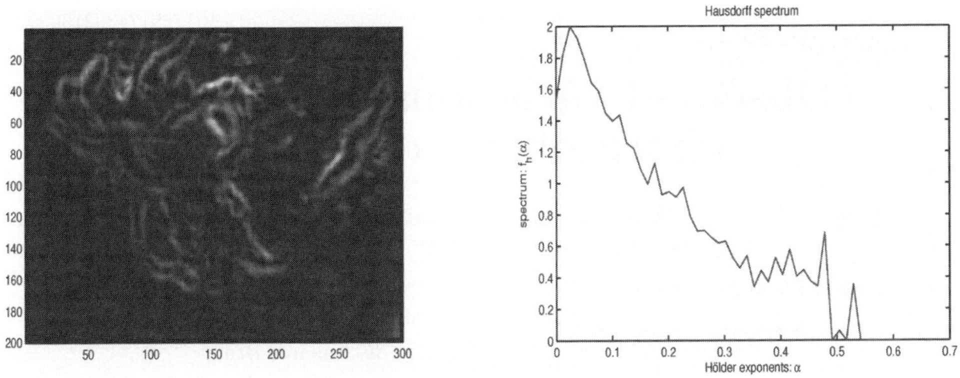

Figure 11. Left panel- Hoelder image at the flare of the active region NOAA10960 and right panel- multifractal spectrum for this image.

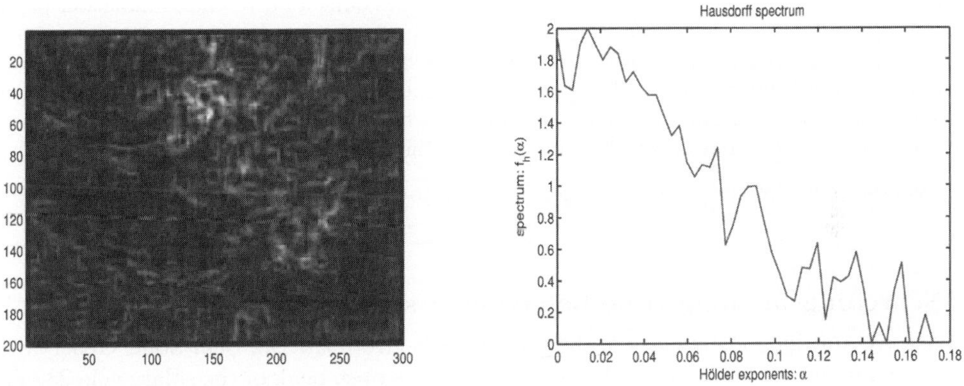

Figure 12. Left panel- Hoelder image after flash of the active region NOAA10960 and right panel- multifractal spectrum for this image.

4. Concluding Remarks

Based on the observation data in H_α line and the software "FracLab" we have obtained Hausdorff spectrum $f_n(\alpha)$ of the active regions at the minimum of solar activity. Multifractal spectra are very variable at different stages of solar activity. In particular, fluctuation of the singularity in the spectrum at different stages of development of the flare noticeably differs each from other (Fig. 5–7, Fig. 10–12; right panel). Also at a phase of flare the width of a spectrum extends. Therefore primary results show that multifractal spectra can be one of indicators of the stage of development of solar active regions.

References

Abramenko, V. I., Yurchyshyn, V. B., *et al.* 2002, *ApJ*, 577, 487
Lawrence, J. K., Ruzmaikin, A. A., & Cadavid, A. C. 1993, *ApJ*, 417, 805
Salakhutdinova, I. I. & Golovko, A. A. 2004, *Solar Phys.*, 225, 59
Stark, B., Adams, M., *et al.* 1997, *Solar Phys.*, 174, 297
FracLab(http://apis.saclay.inria.fr/FracLab/)
soho(http://sohowww.nascom.nasa.gov/)
solarmonitor(http://www.solarmonitor.org/)

Solar and Stellar Variability: Impact on Earth and Planets
Proceedings IAU Symposium No. 264, 2009
A. G. Kosovichev, A. H. Andrei & J.-P. Rozelot, eds.

doi:10.1017/S1743921309992390

Observed variations of the solar photospheric diameter

S. C. Boscardin[1], E. Reis Neto[1], J. Penna[1], A. R. Rodriguez Papa[1,3], A. H. Andrei[1,2] and V. A. d'Ávila[1,3]

[1]Observatório Nacional/MCT, R. Gal. Jose Cristino 77, Rio de Janeiro, Brazil
email: sergio.boscardin@on.br

[2]Obs. do Valongo/UFRJ-BR, Rio de Janeiro, Brazil
[3]UERJ-BR, Rio de Janeiro, Brazil

Abstract. Here we derive a formulation connecting the observed variations of the solar diameter to the heliophysics of the photosphere, in particular in connection to the granulation pattern and morphology. The results from the measurements are next used to correlate the variations of the semi-diameter and of estimators of the solar activity along the solar cycle 23. The values obtained strongly support a broader physical description of the photosphere, intertwining the diameter variations with the irradiance, the sunspots, the 10.7 cm radio emission, and to a lesser degree with the integrated magnetic field and with the flares count.

Keywords. Sun: fundamental parameters, granulation, flares, activity, radio radiation

---

1. Searching for long time behavior of solar granulation

We have obtained the images of the full solar disc from the Big Bear Solar Observatory. These images cover the beginning, the peak, and the post-peak of the solar cycle 23. They are densely distributed between 2000 to 2006. The images are all of 1364×1035 pxs. The same telescope, the same camera and the same orientation were used. The exposition time varies from 50 to 80 ms. We have examined two kinds of images: **Fl** images - subtracted from dark and from flat field (1261 images), and **Fr** images - subtracted from limb darkening (1341 images)

To avoid limb darkening influences, we took only the central part of the images. Thus, we considered only the central part from each original image up to 0, 35 of the solar radius. In this way only 2% of intensity variation remains even for **Fl** images. The central portion of the Sun was next divided in 10×10 collateral sectors. Each sector has 30×30 pxs. In each sector the statistics described below have been independently applied, and sectors deviating from the average by 3σ or more were removed from all the final analyses. This strategy seeks to discard the presence of sunspots from the solar granulation description.

Three estimators are here used as first assessment to the granulation state. For each estimator the average of the sectors is calculated (after removing the deviating ones).

(*a*) standard deviation of counts (**S**) as probe of the grains' mean size.

(*b*) difference between the uppermost and lowest counts tenths (**Q**) as probe of the grains' brightness.

(*c*) degree of the better adjusted polynomial along lines and columns (**N**) as probe of the grains' number.

The model then assumes a patch of grains formed by bright centers and dark intergrain contours. Through large number statistics the balance between the two structures along the solar cycle is assessed. After all **Fl** and **Fr** images have been treated, and the three statistics obtained for each one, a final filter was applied removing within each year the images for which all three statistics mean value were afar more than 3σ from the yearly average.

The number of used images is 1104 **Fl** and 1245 **Fr**. The average number of subregions used (that is, not suspected of containing spots or faculae) per image is very similar, being 94.1 (σ 2.6) for **Fl** and 94.6 (σ 2.4) for **Fl**.

The autocorrelation **S**, **Q**, and **N** statistics, for both the **Fl** and **Fr** images, extends up to about two months. The exception is **N**(**Fr**), for which it extends beyond half a year (this feature is better understood when of the analysis of the statistics itself). In agreement to the autocorrelation distributions, the Runs test indicates statistically significant presence of signal for all the statistics.

The time series for the **S** and **Q** statistics, for both types of images, shows anti-phase correlation larger than 0.99 to the solar cycle evolution. The position of the minimum of the time series fitted to the **S** and **Q** statistics always. They are: **S**(**Fr**)**min** = 2003.15; **Q**(**Fr**)**min** = 2003.29; **Q**(**Fr**)**min** = 2002.83; **Q**(**Fl**)**min** = 2002.89

On the other hand, there is no significantly variation for the **N** statistics. That is, according to the **N** statistics, there is no significant variation in the number of grains.

In conclusion, the grains sizes are the largest by the solar maximum, in excellent agreement with the maximum of the measured diameter. The grains brightness, on the contrary, is minimum at the solar maximum, and again an excellent agreement is verified with the maximum of the measured diameter. Accordingly, the granulation variation varies the photosphere density, mixing, and eventually its height scale, therefore providing a mechanism to vary the optical depth, and hence the measured diameter.

2. Correlations between solar diameter and other solar indexes

At the Observatório Nacional/MCT in Rio de Janeiro ($\phi = -$ 22° 53' 42", $\lambda = +$ 2h 52m 53s.5, h = 33m) the series of solar semi-diameter measurements started in 1997, in some way provoked by systematic biases found during several preceding years of astrometric observations of the Sun. For the ensuing semi-diameter long term campaign the instrument underwent several modifications. The most important were the installation of a variable angle front prism enabling the continuous observations between the zenith distance of 26° and 56°, the concurrent installation of a moving density filter, and the installation of a CCD camera, which allowed the observations to become fully freed of personal equations. Here we analyze the period from 1998 up to 2003.

The raw data were corrected from effects related to the observation conditions: the air temperature, its first derivative, the Fried factor and the standard deviation of the adjusted parable to the directly observed solar edge. The instrumental conditions were inspected in order to detect effects caused by any instability of the objective prism and from the lacking of leveling of the astrolabe that could cause errors as function of the observed azimuth. The standard deviation of the data fell to $0''.56$, which showed that all corrections applied were small and did not introduce any spurious long-term modulation upon the series. The very small gain on the standard deviation after the corrections shows that seasonal or annual effects on the raw measurements are very small, coherent with the standard refraction theory.

The final series of solar semi-diameter values was correlated against the series of solar activity parameters in the common period. Although the length of all the series is the same, clearly the number of points is different. It is always much larger for the semi-diameter series, even though it presents a less continuous distribution. In addition, one could hardly expect a physically meaningful variation of the solar diameter, at the level of several milli arcsec within the day, and except for truly exceptional events, within the week. So, in order to deal with significant variations and in order to homogenize the number of points for all the series, those were split into equal sampling intervals.

The series are divided the from 6 up to 72 equal sampling intervals. That is, roughly corresponding to aliasing periods from one month to one year. In this way the correlations cover a range from the more detailed to the broader features. The correlation coefficient is calculated independently for each aliasing period. But, in order to not privilege particular intervals, the beginning date of the sampling intervals is displaced in steps of 10 days. Thus for each aliasing period the doubles of points taken to calculate the correlation follow similarly to a running average with step of 10 days.

All the correlation coefficients have larger values when we consider longer periods. This indicate common trends, suggesting a common set of precursors for the variations of the semi-diameter as well as for the solar activity estimators. The declining of the correlation values towards the smaller aliasing periods, i.e. the recognizing of detailed features, is however quite smooth. This endorses the choice of 1 month as the shortest aliasing period, because it is not dominated by random errors. As it is, the smooth decline is suggestive of a complex interplay between the studied pairs of semi diameter and solar activity estimators variations, and a complex derivation from the set of precursors to the estimators' response.

In general, the allowance of a time delay did not increase the correlation of any given pair beyond one tenth. The noteworthy exception relates the pair of the flare index and the semi-diameter. This can be accounted for by the different photospheric and chromospheric regimens. On the other hand, in many instances statistically significant second peaks do show up.

Also the correlations between the pairs of the estimators of the solar activity were calculated. In the overall, the set of correlation coefficients for pairs including the semi-diameter measured variations fares as statistically significant as the pairs including only the independent estimators of the solar activity. All values though are chiefly representative of the most active half of the solar cycle 23. Their extension to other solar cycles and for the calmer parts of the cycles must made with caution. This is also particularly true for satellite borne experiments that span at most a third of one solar cycle. It emphasizes the need for the support deserved to long series of measurements. Nevertheless the time series covered here also spans some years of calm or moderate solar activity.

Finally, the great number of individual observations from 1998 to 2004 allows us to draw the shape of the Sun and its variations along the solar activity cycle. It transpires that the solar shape has complex features, with distinct features on different heliolatitude zones . They change on the time, and even the solar oblateness results changing according to the values averaged along the solar cycle.

References

Boscardin, S. C. 2005, Análise de Observações do Diâmetro no Contexto da Atividade Solar (Analysis of Observations of the Diameter in the context of the Solar Activity), MsC discertation held at Observatório do Valongo/Universidade Federal do Rio de Janeiro.

Boscardin, S. C., Andrei, A. H., Penna, J. L., Reis Neto, E., & d'Ávila,V. 2009, Solar diameter observed variation: the correlation to activity and the results from different stations, in preparation.

Jilinski, E. G., Puliaev, S. P., Penna, J. L., Andrei, A. H., & Laclare, F. 1998, Solar diameter observations with the astrolabe at Observatório Nacional - Rio de Janeiro, Astronomy and Astrophysics Supplement Series, 130, 317.

Solar and Stellar Variability: Impact on Earth and Planets
Proceedings IAU Symposium No. 264, 2009
A. G. Kosovichev, A. H. Andrei & J.-P. Rozelot, eds.

doi:10.1017/S1743921309992407

The asymmetric light curves of the GSC 2764 1417 (And), GSC 3355 0394 (Per) and GSC 2537 0520 (Psc)

Ronald G. Samec[1], Evan R. Figg[1], Reid Melton[1], Christa M. Labadorf[1], Jess Miller[1], Robert McKenzie[1],Danny R. Faulkner[2] and Walter Van Hamme[3]

[1]Astronomy program, Physics Department, Bob Jones University, Greenville, SC, USA
email: rsamec@bju.edu
[2]University of South Carolina, Lancaster, USA
email: drfaulkn@gwm.sc.edu
[3]Florida International University, Florida, USA
email: vanhamme@fiu.edu

Abstract. Solar-type variability is enhanced in short period close binaries with increased dynamo driven activity. This activity is studied in our analysis of recent light curves taken of the newly discovered eclipsing binaries GSC 2764 1417 (And), GSC 3355 0394 (Per) and GSC 2537 0775 (CVn).

Keywords. binaries: close, stars: activity, spots, mass loss, winds, outflows; techniques: photometric

1. Observations and Analyses

Our recent CCD observations of these variables were taken with the Lowell 0.81-m though our association with NURO (the National Undergraduate Research Observatory) and the SARA (Southeastern Association for Research in Astronomy) 0.9-m telescope. Times of minima were calculated, period studies were carried out and orbital behaviors were determined and used to phase present light curves. B,V,R,I light curves were hand premodeled with Binary Maker 3.0 (Bradstreet 2002). Average parameters were used as initial input values for a simultaneous 4 color synthetic light curve calculation with the 2004 Wilson Differential Corrections Code (Wilson & Devinney 1971, Wilson, 1990, Wilson 1994, Van Hamme & Wilson 1998).

2. Results

GSC 2764 1417 is a contact binary probably in a transition from a semidetached configuration to contact. The primary, more massive component is filling its respective Roche Lobe while the secondary component is under filling. This is probably a V1010 Oph binary, that is, it is making its first transition into contact. The driving mechanism behind such a transition is physically provided by torques via stellar winds leaving the star along stiff rotating magnetic field lines. Rapid mass transfer is indicated by a hot spot on the secondary component near the L1 point slightly shifted to the following side. A large dark spot region is also apparent.

GSC 3355 0394 has an EB type light curve which is dominated by dark spot activity. It displays night to night variations in light curve shapes. The period study

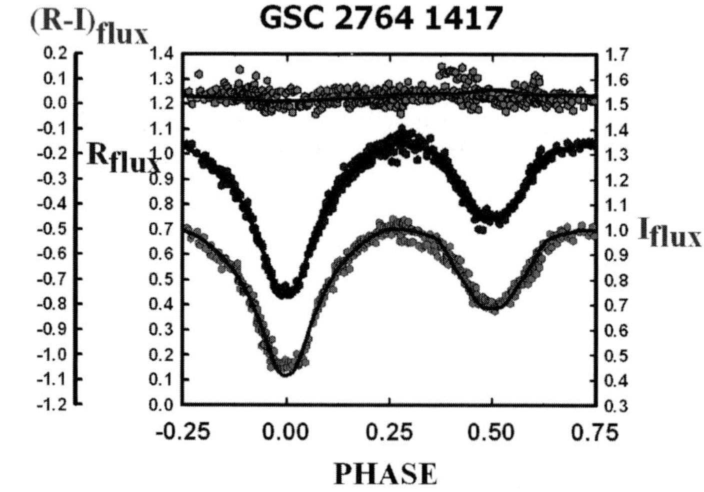

Figure 1. R and I light curves of GSC2764 1417 overlaid with Wilson code solution.

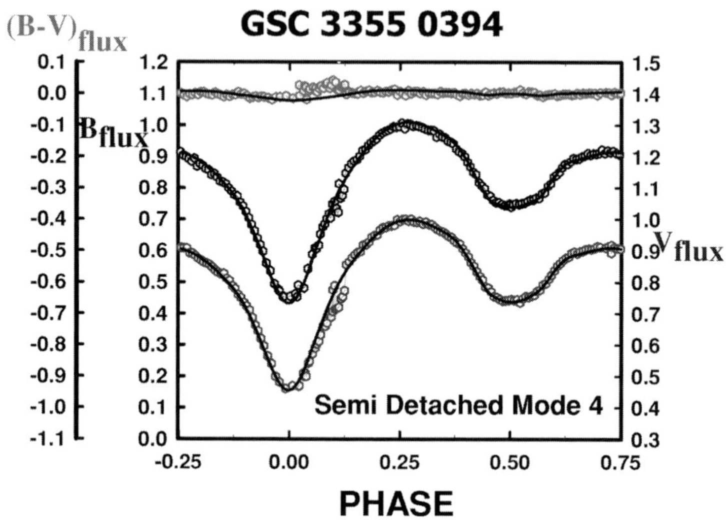

Figure 2. R and I light curves of GSC3355 0394 overlaid with Wilson code solution.

yields 6 new time of minimum light, and the first precision ephemeris: JD Hel Min I = 2454427.903($\pm$0.0016)+0.4621606($\pm$0.0000007). BVRI Wilson Synthetic light curve solutions are presented for both a mode 4 (V1010 Oph) configuration and a mode 3, critical contact configuration. It is the lowest residual solution. Four major spot regions are needed to model this binary, and one is evidently a stream spot.

GSC 2537 0520 is a rare EB-type extreme mass ratio solar type contact binary. The component temperature difference is an amazing 675K. Most other members of this group have nearly identical temperatures, $\Delta T = 168 \pm 146$. Its period is apparently decreasing as it is steadily losing angular momentum. The several spot regions are indicative of strong dynamo-driven magnetic activity.

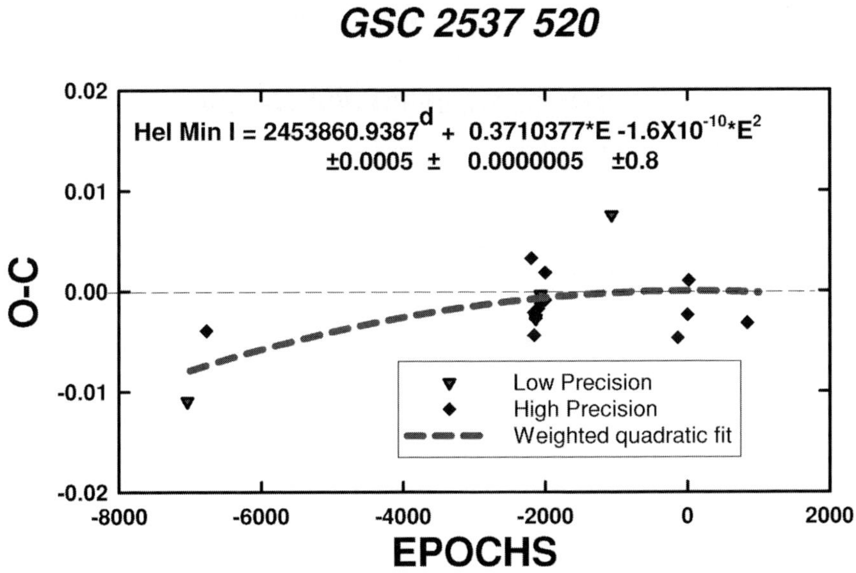

Figure 3. O-C residual plot showing possible decrease in GSC 2537 0520 due to AML.

Acknowledgements

We gratefully acknowledge the support received through the Small Research Grant program and the International Travel Grant Program of the American Astronomical Society.

References

Bradstreet, D. H. 2005, *BAAS*, 201, 201.7502B
Wilson, R. E. & Devinney, E. J. 1971, *ApJ*, 166, 606
Wilson, R. E. 1990, *ApJ*, 356, 613
Wilson, R. E. 1994, *PASP*, 106, 921
Wilson, R. E. & Van Hamme, W. V. 1998, *BAAS*, 19310203V

Solar and Stellar Variability: Impact on Earth and Planets
Proceedings IAU Symposium No. 264, 2009
A. G. Kosovichev, A. H. Andrei & J.-P. Rozelot, eds.

doi:10.1017/S1743921309992419

The EUV spectral irradiance of the Sun from 1997 to date

Giulio Del Zanna[1] **and Vincenzo Andretta**[2]

[1]Department of Applied Mathematics and Theoretical Physics, University of Cambridge, Wilberforce Road, Cambridge CB3 0WA, UK
email: g.del-zanna@damtp.cam.ac.uk

[2]INAF – Osservatorio Astronomico di Capodimonte, Salita Moiariello, 16, I-80131, Napoli, Italy
email: andretta@oacn.inaf.it

Abstract. We present measurements of the EUV spectral irradiance we have obtained from radiance measurements with the SOHO Coronal Diagnostic Spectrometer, from 1997 to date. We discuss the contribution of the various regions of the Sun to the total EUV irradiance, and how they varied dramatically between the last two solar minima. These observations allow us to interpret spectral irradiance measurements in the EUV which have been available since 2002 with the TIMED mission. We also briefly discuss how changes in various activity indices compare with the EUV spectral variability, and the limitations of spectral modelling.

Keywords. Sun: UV radiation, Sun: activity, solar-terrestrial relations

1. Introduction

The importance of studying the EUV solar irradiance is twofold: the study of the Sun-as-a-star, as a key to interpret EUV emission from similar stars, and the effect of the solar EUV radiation on the Earth system. The study of the EUV solar irradiance is in fact the study of the emission from a star for which we also have detailed information about the distribution of the sources on its surface. Thus far, studies in this direction have been carried out only in the X-ray range using broad-band filter observations (e.g. Orlando *et al.* (2004)). Some effects of the EUV radiation on Earth's atmosphere – particularly on the thermosphere – are well known. More subtle effects on the Earth system, perhaps on the climate, are currently debated. Some recent results (Meehl *et al.* (2009)) seem to hint that even the very small amount of variation along the activity cycle of the solar energy output (of the order of 0.1%) can produce significant effects on some local climate patterns. The FUV and EUV radiation, although only a small fraction of the total solar output, display variations of much larger amplitude as the magnetic activity of the Sun varies across its 11-year cycle. Whether and how such large variations can produce some effect – however small – on Earth's climate is still unclear (Lean *et al.* (1995), Haigh (1996), Priest *et al.* (2007), Foukal *et al.* (2009)).

2. Measurements of the EUV spectral irradiance

Despite the importance of the EUV solar irradiance, few measurements are actually available, and those few either are taken from sounding rockets, and thus give only a snapshot of the solar irradiance, or cover less than one solar cycle. Data from the NASA TIMED SEE experiment (Woods *et al.* (2005)), for instance, cover the solar cycle only since 2002. Therefore, all studies on the solar irradiance and its effects, for example, on the Earth's atmosphere are based on proxies. Such an approach, of course, adds an

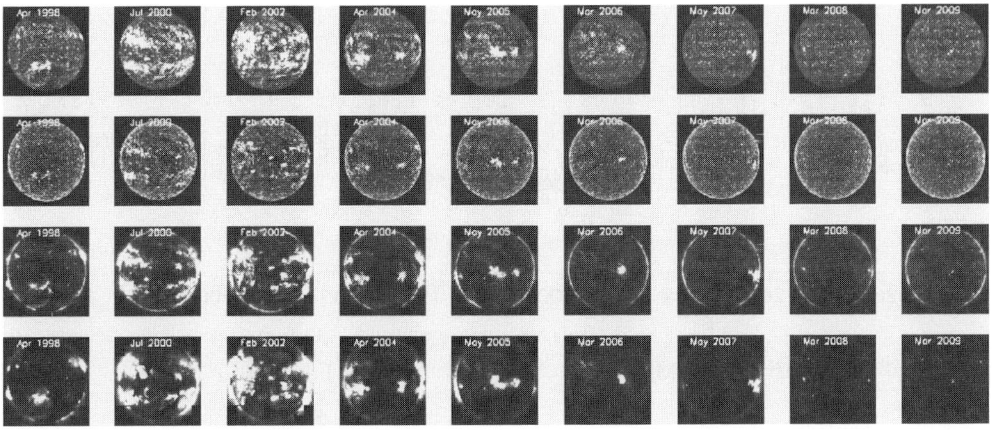

Figure 1. CDS radiances from (top from bottom) He I, O V (~ 0.25 MK), Mg X (~ 1 MK), Fe XVI (~ 2.5 MK), from the beginning of cycle 23, to the current extended minimum.

additional layer of complexity and uncertainty to those studies. Hence the need for direct, reliable measurement of the solar EUV, spectrally resolved irradiance.

SOHO has been observing the Sun since 1996. Apart from the broad-band measurements of CELIAS/SEM (Hovestadt *et al.* (1995)), the Coronal Diagnostics Spectrometer (CDS; Harrison *et al.* (1995)), has been observing routinely the full solar disc in the ranges 31–38 and 51.5–63 nm with its *Normal Incidence Spectrograph* (NIS). We have thoroughly re-examined the performance and radiometric calibration of the NIS instrument, and found that, in fact, it has performed exceptionally well, with a sensitivity loss by only a factor of 2 across all wavelengths in 13 years of operations (Del Zanna *et al.* (2009)). In particular, we checked our results against those of the prototype of the NASA SDO EUV irradiance monitor (EVE), launched in April 2008 (Woods *et al.* (2009), Chamberlin *et al.* (2009)), and found agreement in the radiometric calibration within 30%. Discrepancies were instead found with the TIMED/SEE irradiances (version #9).

The CDS/NIS irradiances we computed were derived from mosaics of spectrally-resolved radiance measurements taken in average once a month with the CDS study `USUN`, comprising $\sim$1000 exposures (spectra) sampling the whole Sun in about 13 hours (Fig. 1). The strongest, unblended lines in the spectra were extracted, and their integrated radiance computed. Estimates of the contribution of off-limb radiance were also made. The resulting irradiances in a selection of EUV lines are shown in Fig. 2, with the values obtained from the EVE prototype (diamonds) and those from the CDS/ rocket EGS measurements of May 1997 by Brekke *et al.* (2000) [asterisks].

3. Conclusions and future work

From the analysis of the CDS irradiances (and of historical measurements from the 60's), we found that lines formed in the low transition region show relatively modest changes (up to 50%). Lines forming around 1 MK already change during the cycle by a factor $\sim$ 5; for hotter lines (2.5 MK) the variability reaches factors of the order of 40. The presence of active regions during the solar cycle has a strong influence even in regions of the more "quiet corona" which change considerably along the cycle. Furthermore, we started investigating the reliability of various proxies of solar EUV irradiance. As expected, we are finding that it is virtually impossible to predicy solar EUV irradiances

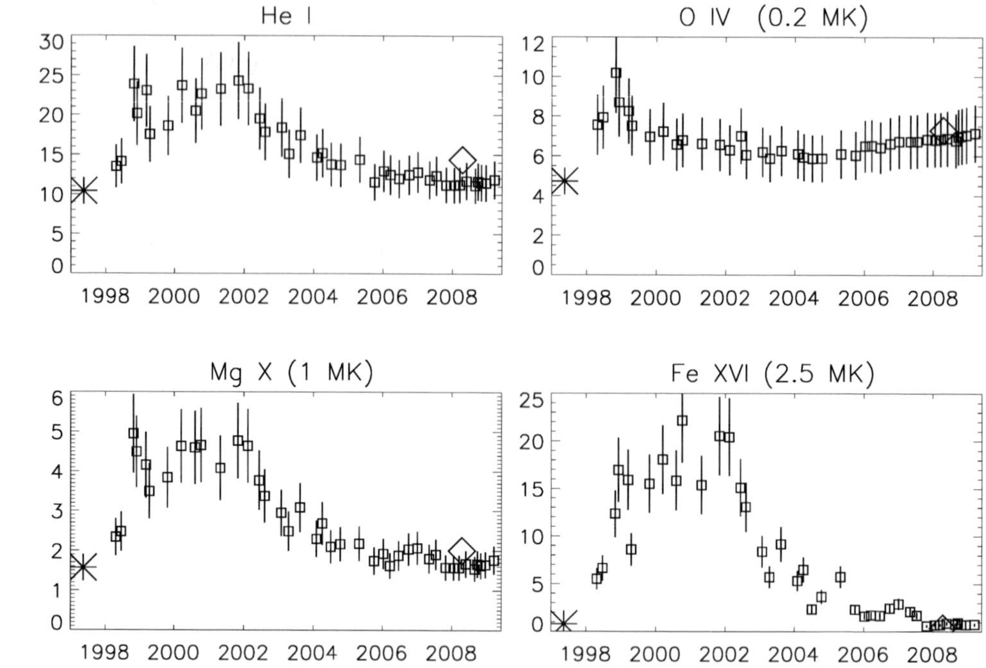

Figure 2. CDS spectral irradiances (boxes) compared with SDO/EVE prototype spectrum.

based on any proxy variation. Moreover, even if a proxy or a set of proxies were found to be sufficiently reliable to reproduce the solar irradiance at some EUV wavelength or band, translating such a result to the analysis of the EUV emission of other stars is very dangerous: for example the spot-to-plage ratio likely depends on the stellar activity level. Future work will focus on improving models of the solar EUV irradiance, with the goal of providing reliable input to climate models and to models of the effects of the magnetic activity of young stars (including the young Sun) on planets.

Acknowledgements

GDZ acknowledges support from STFC (UK) and a travel grant from the IAU. SOHO is a project of international collaboration between ESA and NASA.

References

Brekke, P., Thompson, W. T., Woods, T. N., & Eparvier, F. G. 2000, *ApJ*, 536, 959
Chamberlin, P. C., Woods, T. N., Crotser, D. A., *et al.* 2009, *GRL*, 36, 5102
Del Zanna, G., Andretta, V., Chamberlin, P. C., *et al.* 2009, *A&A*, accepted
Foukal, P., Bertello, L., Livingston, W. C., *et al.* 2009, *Solar Phys.*, 255, 229
Haigh, J. D. 1996, *Science*, 272, 981
Harrison, R. A., Sawyer, E. C., Carter, M. K., *et al.* 1995, *Solar Phys.*, 162, 233
Hovestadt, D., Hilchenbach, M., Bürgi, A., *et al.* 1995, *Solar Phys.*, 162, 441
Lean, J., Beer, J., & Bradley, R. 1995, *GRL*, 22, 3195
Meehl, G. A., Arblaster, J. M., Matthes, K., Sassi, F., & van Loon, H. 2009, *Science*, 325, 1114
Orlando, S., Peres, G., & Reale, F. 2004, *A&A*, 424, 677
Priest, E., Lockwood, M., Solanki, S., Wolfendale, A., & Coustenis, A. 2007, *A&G*, 48, 03
Woods, T. N., Chamberlin, P. C., Harder, J. W., *et al.* 2009, *GRL*, 36, 1101
Woods, T. N., Eparvier, F. G., Bailey, S. M., *et al.* 2005, *JGR* (Space Physics), 110, 1312

Solar and Stellar Variability: Impact on Earth and Planets
Proceedings IAU Symposium No. 264, 2009
A. G. Kosovichev, A. H. Andrei & J.-P. Rozelot, eds.

doi:10.1017/S1743921309992420

Semi-harmonic and intermittent solar decimetric spikes

Francisco C. R. Fernandes[1], Maurício J. A. Bolzan[1], Reinaldo R. Rosa[2], José A. S. S. Dutra[1], José R. Cecatto[2], Hana Mészárosová[3] and Hanumant S. Sawant[2]

[1]Instituto de Pesquisa e Desenvolvimento, Universidade do Vale do Paraiba - UNIVAP, São José dos Campos, SP, Brazil
email: guga@univap.br

[2]Divisão de Astrofisica, Instituto Nacional de Pesquisas Espaciais - INPE, São José dos Campos, SP, Brazil
email: reinaldo@lac.inpe.br

[3]Ondrejov Observatory, Astronomical Institute Academy of Sciences of the Czech Republic Ondrejov, Czech Republic
email: hana@asu.cas.cz

Abstract. We present the analysis of the narrowband solar radio spike emissions observed by the Brazilian Solar Spectroscope (BSS) on 24 June, 1999, at 16:53:55–16:56:48 UT. The data were recorded with temporal resolution of 50 ms and frequency resolution of 5 MHz. The dynamic spectra of clusters of spikes were recorded in 100 channels in the 1.2-1.7 GHz frequency band. The groups of radio spikes observed clearly show spectral semi-harmonic structures and an intermittent temporal patter. The observational parameters were determined from the dynamic spectra analysis. The wavelet analysis technique able us to determine the temporal cadence for the clusters. The frequency ratio between two clusters of spikes recorded at same time were inferred. The preliminary results show average cadence of clusters of spikes occurrence of about 32 seconds and a semi-harmonic average frequency ratio of the order of 1:1.2.

Keywords. Sun: activity, radio radiation; decimetric spikes; instrumentation: spectrographs

1. Introduction

Observations and analysis of clusters of millisecond spikes exhibiting harmonic bands have been reported by Benz & Güdel (1987), Güdel (1990), and Krucker & Benz (1994). Benz & Güdel (1987) found a semi-harmonic frequency ratio between two spectral peaks of about 1:1.39. Schwarz *et al.* (1993) found frequency ratio raging between 1:1.03 and 1.3. Güdel (1990) reported a large sample of events showing only integer ratios of 2, 3, 4 and 5. The identifications of harmonic structures and their characteristics can contribute to improve the understanding of the emission mechanisms. According to Wu *et al.* (1985), such simultaneous emission at multiple harmonic with comparable intensity can be hardly explained by loss-cone distribution, but can be generated by mildly relativistic beam or high-relativistic synchrotron maser (Louarn *et al.*, 1987). The investigations of the semi-harmonic structures in spike clusters recorded by the Brazilian Solar Spectroscope (BSS) is on going, including the application of wavelet techniques.

2. Spikes observations

Frequently, the spikes form clusters in frequency and time domains, identified at radio dynamic spectra. However, usually, the clusters do not show related behavior

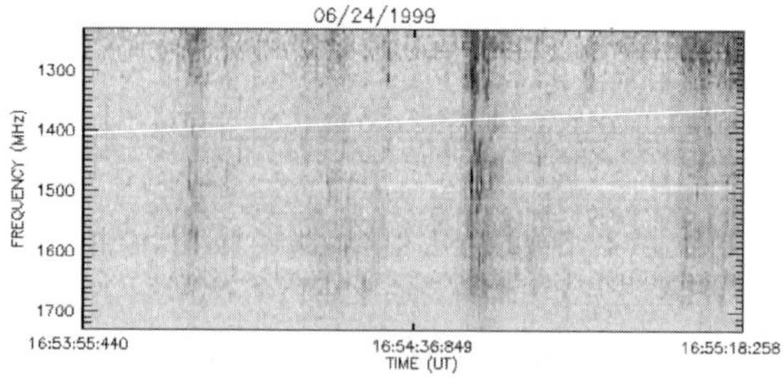

Figure 1. Dynamic spectra of decimetric spikes recorded in 24 June, 1999 (~16:53:55 UT).

between each other or do not occur simultaneously. Some clusters of decimetric spikes were recorded by BSS in 24 June, 1999, in the time interval of 16:53:55-16:56:48 UT. The data were recorded in 100 frequency channels, equally spaced in the frequency range of 1230-1730 MHz, with time resolution of 50 milliseconds and frequency resolution of 5 MHz. The dynamic spectra of radio spikes clusters, shown in the Figure 1, indicated splitting in frequency and intermittent behavior.

3. Wavelet Analysis

According to Morettin (1999), wavelet transform breaks a signal into components formed by dilated and shifted versions of an original wavelet, called mother wavelet. Given a mother wavelet $\psi(\mathrm{t})$, the dilated and translated daughter wavelets are given by

$$\psi_{a,b}(t) = \frac{1}{\sqrt{a}}\,\psi\left(\frac{t-b}{a}\right), \tag{3.1}$$

where $a > 0$ is the scaling factor controlling the dilation of the mother wavelet and b is the translation parameter determining the shift of the mother wavelet. The scaling factor is directly related to the frequency, or period. The coefficients for the continuous wavelet transform are given by

$$W(a,b) = \frac{1}{\sqrt{a}} \int_{-\infty}^{\infty} H(t)\,\psi\left(\frac{t-b}{a}\right) dt. \tag{3.2}$$

In the wavelet analysis, the mother wavelet was chosen as Gaussian Morlet given by

$$\psi(t) = \pi^{-1/4}\, e^{i\omega_o t}\, e^{-t^2/2}, \tag{3.3}$$

where ω_o is the frequency parameter that allows one to shift the frequency range for investigations.

4. Results

We used this wavelet transform to decompose each time-series of spikes clusters (100 frequency channels) into the time and frequency domains. The Figure 2 shows the results of the wavelet analysis for the frequency channel of 1245 MHz. The figure shows the average cadence of clusters of spikes is of 32 (± 2) seconds and its harmonics of 16 and 64 seconds. Some clusters of spikes indicated splitting in frequency. The frequency ratio between the mean frequencies of clusters in two simultaneous bands (upper and lower), was estimated for many distinct times. The average frequency ratio is 1:1.2 (± 0.3).

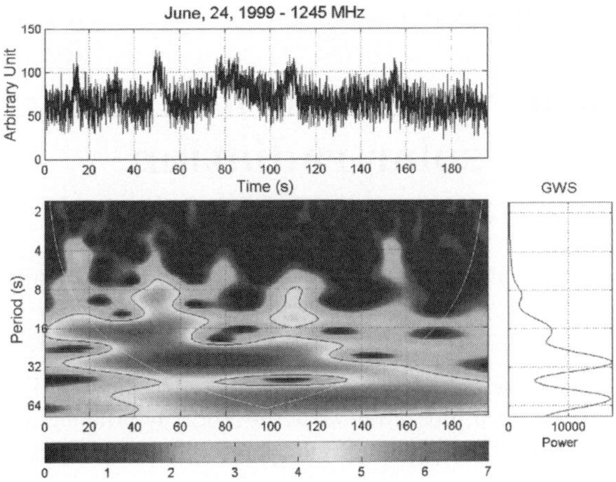

Figure 2. The top panel shows the spikes time evolution for 1245 MHz channel. The wavelet analysis output for this channel is given in the bottom panel.

5. Concluding Remarks

Many emission mechanisms have been proposed for the decimetric millisecond spikes bursts: electron-cyclotron maser emission, synchrotron maser emission, plasma emission excited by electrostatic and electromagnetic waves or current-driven direct radiation, according to Aschwanden & Güdel (1990) (see reviews by Wu *et al.*, 2007; Benz, 1986 and Fleishman & Melnikov, 1998).

The results obtained to the analysis of the clusters of spikes can support the investigations about their emission mechanisms. In case of the semi-harmonic spikes, according to Benz & Güdel (1987), their observations cannot be explained considering two integer multiples of the gyrofrequency. However, the maser emission occurs at frequencies slightly different of the harmonics of the gyrofrequency and the radio emission at semi-harmonic frequencies can be generated. Also, the intermittent behavior of the clusters, with of about 32 seconds could be interpreted in terms of magnetic field modulation by MHD waves (Zhao *et al.*, 1990). The detailed analysis of non-harmonic and intermittent decimetric spikes and the implication of their detection to the emission mechanisms, briefly mentioned above, is given by Fernandes *et al.* (2010).

Acknowledgements

The authors acknowledge the Brazilian agencies CNPq (480045/2008) and FAPESP (06/55883-0).

References

Aschwanden, M. J. & Güdel, M. 1990, *Solar Physics*, 30, 490
Benz, A. O. 1986, *Solar Physics*, 104, 99
Benz, A. O. & Güdel, M. 1987, *Solar Physics*, 111, 175
Fleishman, G. D. & Melnikov, V. F. 1998, *Physics-Uspekhi*, 41(12), 1157
Fernandes, F. C. R., Bolzan, M. J. A., Rosa, R. R., Dutra, J. A. S. S., Mészárosová, H., & Sawant, H. S. 2010, *A&A* (in preparation)
Güdel, M. 1987, *A&A*, 245, L1
Krucker, S. & Benz, A. O. 1994, *A&A*, 285, 1038
Louarn, P., Le Queau, D., & Roux, A. 1987, *Solar Physics*, 111, 201
Morettin, P. A. 1999, *EdUSP, São Paulo*, 1, 42
Schwazs, U., Benz, A. O., Kurths, J., & Witt, A. 1993, *A&A*, 277, 215
Wu, D. J., Huang, J., Tang, J. F., & Yan, Y. H. 2007, *ApJ*, 665, L171
Zhao, R.-Y., Jin, S.-Z., Fu, Q.-J., & Li, X.-C. 1990, *Solar Physics* 130, 151

Solar and Stellar Variability: Impact on Earth and Planets
Proceedings IAU Symposium No. 264, 2009
A. G. Kosovichev, A. H. Andrei & J.-P. Rozelot, eds.

doi:10.1017/S1743921309992432

The periodic variation of 6.7 days for total solar radiation

W. Q. Gan and Y. P. Li

Purple Mountain Observatory, Chinese Academy of Sciences, Nanjing 210008, China
email: wqgan@pmo.ac.cn

Abstract. Analyzing the power spectrum of Total Solar Irradiance (TSI) for the period from 2003 February 25 to 2009 July 6, observed with the Total Irradiance Monitor (TIM) onboard SOlar Radiation and Climate Experiment (SORCE), we found that there are quite a number of periodic variations. The outstanding shortest one is the period of 6.7 days, about one fourth of the period of solar rotation. Checking the solar magnetic field for the same period of time observed with MDI onboard SOHO, we found that there is about 90 degree difference in longitude for the distribution of solar magnetic field. We therefore conclude that both the 90 degree difference in longitude for the distribution of solar magnetic field and the solar rotation are the reason resulting in the periodic variation of 6.7 days for the total solar radiation.

Keywords. Sun: activity, Sun: magnetic fields, Sun: solar-terrestrial relations

1. Introduction

The space observation for the Total Solar Irradiance (TSI) with high accuracy began at 1978 (see Fröhlich 2006). It was discovered in the early time that the variation of TSI is related to the disc sunspots and faculae (e.g., Willson *et al.* 1981). Further studies (e.g., Solanki & Krivova 2006) show that faculae and bright network made an increase of TSI, while the sunspots let TSI be decreased. The relationships of TSI with some parameters describing the solar activity were studied by many authors (e.g., Minarovjech *et al.* 2007; Krivova *et al.* 2007; Tapping *et al.* 2007) and the TSI was extrapolated backward to the time interval when the observation was unavailable. The connection of TSI with solar magnetic field has also been studied (e.g., Withbroe 2009). In this paper, we use the latest advanced data to study the variation of TSI and try to explain it with a combined effect of solar magnetic field and solar rotation.

2. Data and Analysis

The data were taken from the observations with the Total Irradiance Monitor (TIM) instrument onboard SOlar Radiation and Climate Experiment (SORCE) (Kopp *et al.* 2005). The TSI measurements were provided with a relative standard uncertainty (absolute accuracy) of approximately 0.01% (100 parts per million, ppm) based on SI units and with a long-term precision (relative accuracy) of 0.001% yr^{-1} (10 ppm). The temporal resolution of the data we used is 6 hours. Figure 1 shows the TSI for the period from 2002 February 25 to 2009 July 6, observed at about 1 AU with TIM. We see from the figure that there is a dip drop ($\sim$ 0.33%) on 2003 October 29, when there was a largest solar flare and correspondent big group of sunspots. On average, the TSI is larger at the peak years (when the fluctuation of TSI is also bigger) than at the valley years.

The power spectrum of TSI is presented in Figure 2. Although there looks quite a number of periods, the obviously shortest one is the period of 6.7 days. This period seems to be related to the discovery of the cutoff of 7 days, based on the data analysis for the initial 153 days, observed with ACRIM onboard SMM (Willson *et al.* 1981).

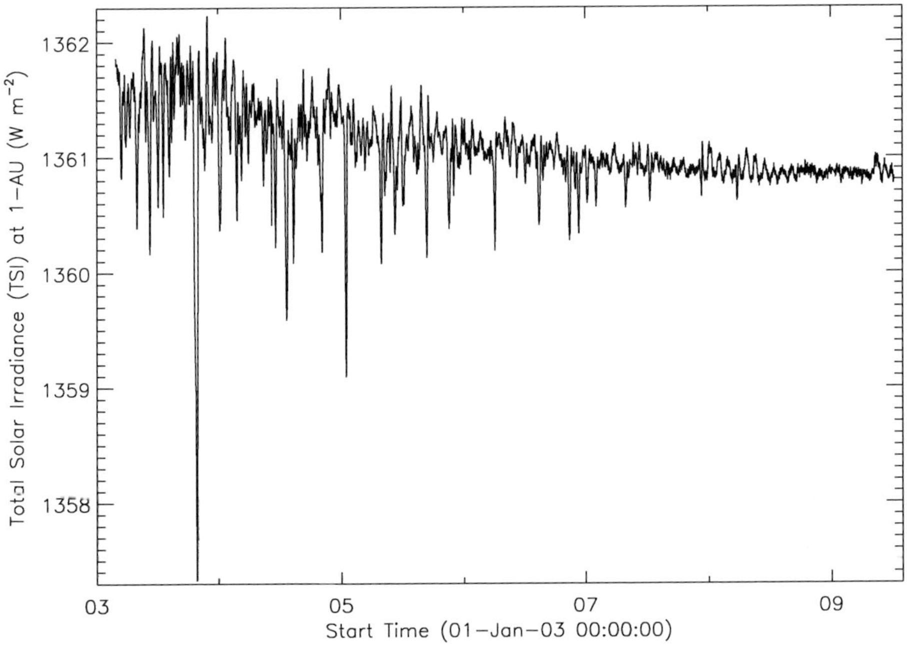

Figure 1. The TSI for the period from 2003 February 25 to 2009 July 6, observed with TIM onboard SORCE.

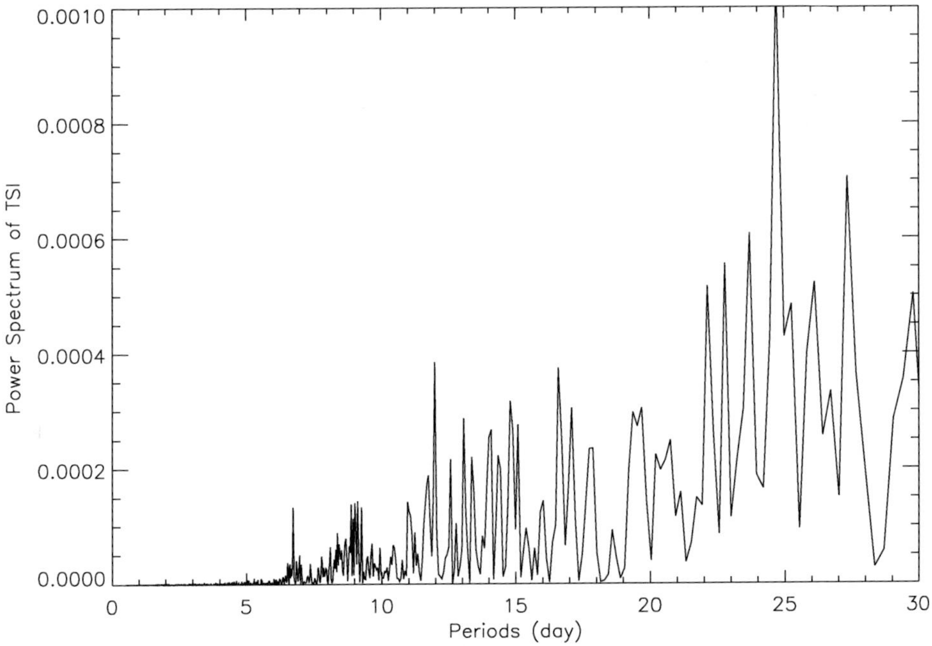

Figure 2. The power spectrum of Figure 1.

It is so evident that 6.7 days take about one fourth of solar rotational period. Motivated by the relevance of the TSI to the solar magnetic field (e.g., Withbroe 2009), we examined the observations with MDI onboard SOHO. It is indeed found that for some period of time, solar magnetic pattern presents an obvious concentration in the longitude difference of 90°. As an example, Figure 3 shows the magnetic field synoptic chart for the period from 2004 July 22 to 2004 September 14, together with the corresponding TSI. Within two Carrington rotations, the strong magnetic field appears periodically every 90° in longitude difference. Solar rotation, plus this 90° distribution of strong magnetic field, provides the most natural explanation on the TSI period of 6.7 days.

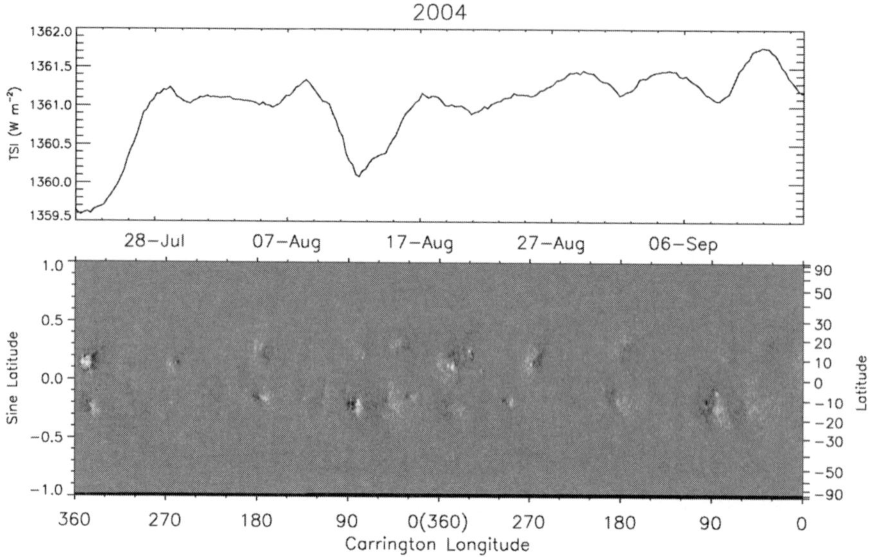

Figure 3. Top panel: TSI for the period from 2004 July 22 to 2004 September 14. Bottom panel: MDI magnetic field synoptic charts.

3. Summary and Conclusions

The analysis on TSI shows an obvious period of 6.7 days. We connect this variation to the solar magnetic field. It is concluded that both solar rotation and 90° longitude difference for the strong magnetic configuration explain naturally the TSI period of 6.7 days.

Acknowledgements

This work is supported via grants 10773031, 10833007, and 2006CB806302.

References

Fröhlich, C. 2006, *Space Sci. Revs*, 125, 53
Kopp, G. & Lawrence, G. 2005, *Solar Phys.*, 230, 91
Krivova, N. A., Balmaceda, L., & Solanki, S. K. 2007, *A&A*, 467, 335
Minarovjech, M., Rušin, V., & Saniga, M. 2007, *Solar Phys.*, 241, 269
Solanki, S. K. & Krivova, N. A. 2006, *Space Sci. Revs*, 125, 25
Tapping, K. F., Boteler, D., Charbonneau, P., *et al.* 2007, *Solar Phys.*, 246, 309
Willson, R. C., Gulkis, S., Janssen, M., *et al.* 1981, *Science*, 211, 700
Withbroe, G. L. 2009, *Solar Phys.*, 257, 71

Solar and Stellar Variability: Impact on Earth and Planets
Proceedings IAU Symposium No. 264, 2009
A. G. Kosovichev, A. H. Andrei & J.-P. Rozelot, eds.

doi:10.1017/S1743921309992444

Looking for variable stars in galactic open clusters

Claudia Greco[1], Nami Mowlavi[1], Laurent Eyer[1], Maxime Spano[1], Mihaly Varadi[1] and Gilbert Burki[1]

[1]Geneva Observatory, University of Geneva, ch. des Maillettes 51, CH-1290, Sauverny, Switzerland
email: claudia.greco@unige.ch

Abstract. A long-term project, aiming at systematic search for variable stars in Galactic Open Clusters (OCs), was started at the Geneva Observatory in 2002. We have been observing regularly a sample of twenty-seven Galactic Open Clusters in the U, B, V Geneva filters (hereafter U, B, V). The goal is to identify and to study their variable stars, as well as the connection between the variable stars in a cluster and the cluster properties. We present the status of this work in progress, and show preliminary results for one of these clusters, IC 4651.

Keywords. open clusters and associations: general, individual (IC 4651); stars: variable

1. Introduction

Long-term observations of a sample of twenty-seven Galactic OCs have been carried out in U, B and V (see Cherix *et al.*, 2006). Twelve OCs in the Southern Hemisphere have been observed with Euler-Cam, mounted on the 1.2m Euler Swiss Telescope at La Silla, Chile and fifteen OCs in the Northen Hemisphere OCs with Merope-Cam installed on the 1.2m Mercator Belgian telescope, in La Palma, Canary Islands, Spain. The field of view of the two cameras (Euler: $11.5' \times 11.5'$; Mercator: $6.5' \times 6.5'$) is centered on the cluster and contains cluster stars as well as Milky Way field stars.Data acquisition and data reduction are still in progress. For the Southern clusters we already collected more than 200 U, 200 B and 2000 V images on a time baseline of 7 years. In Table 1, we summarize some properties for the OCs included in this project, includeing coordinates, estimated age and metallicities. Times series for Northern OCs include so far almost 1600 V image over a 5 year baseline.

The goal of the project is to map the whole variable star population in each cluster in our database. Stellar variability provides independent measurements for the physical parameters of OCs.

2. Data Reduction

We developed a semi-automatic pipeline to prereduce and reduce all the images. We perform PSF photometry with the DaophotII/Allstar (Stetson 1998, 1999) packages. For each image, we build a proper PSF, using at least one hundred selected stars chosen among the brightest and most isolated ones. All the reduced images are then aligned to a common reference list of stars in order to build the Color Magnitude Diagram and the time series for each star. Time series are then searched for objects showing variability. The time series of the candidate variables have been studied using two different codes. We used GrATis (Graphical Analyzer of Time Series; Di Fabrizio 1999; Clementini 2000) and Period04 (Lenz & Breger 2005).

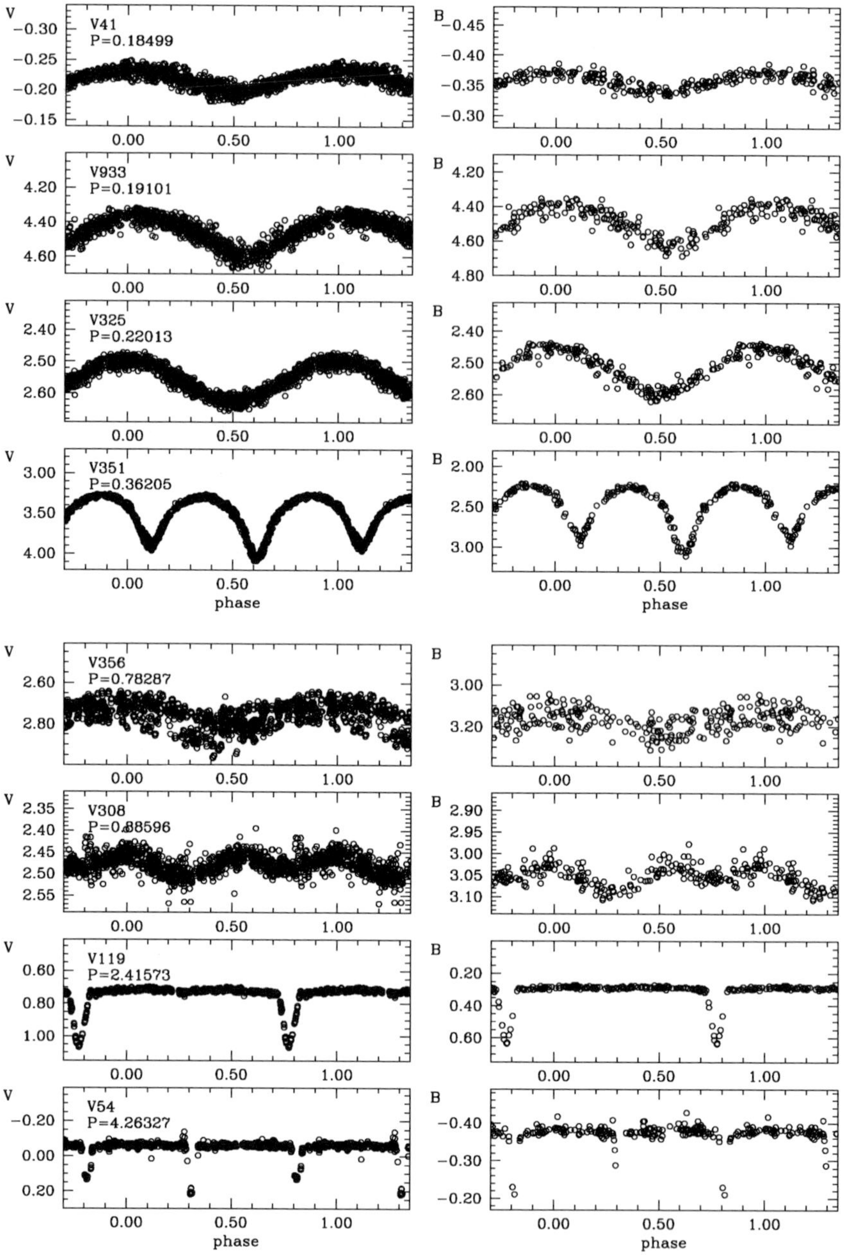

Figure 1. Differential magnitude folded light curves for the monoperiodic variable stars we identified in IC 4651. Periods are in days. Period of the variability increase from top to bottom.

The first open cluster to be analyzed has been IC 4651. IC 4651 is a quite old and metal rich open cluster. We found variable stars with periods spanning from 0.18 to more than 53 days, including *bona fide* δ Scuti and binary stars. Several candidates show multi-periodic variability and variation of the period along the time series. In Fig. 1 we show differential magnitude folded light curves for the monoperiodic variable stars we identified.

Table 1. Basic information about the twenty-seven OCs in the Geneva database. In bold, the eight clusters we have already reduced. References to previous works from the same database are in footnotes. Literature data are taken from WEBDA.

Open Cluster	RA	DEC	log(age)	⟨[Fe/H]⟩	Telescope
	(J2000)	(J2000)	(years)	(dex)	
NGC 3766	11 : 36 : 14	−61 : 36 : 30	7.160	–	Euler
NGC 4103	12 : 06 : 40	−61 : 15 : 00	7.393	–	Euler
NGC 6755	19 : 07 : 49	+04 : 16 : 00	7.719	–	Mercator
NGC 7654	23 : 24 : 48	+61 : 35 : 36	7.764	–	Mercator
NGC 7039	21 : 10 : 48	+45 : 37 : 00	7.280	–	Mercator
NCG 5617[1]	14 : 29 : 44	−60 : 42 : 42	7.915	–	Euler
NGC 6996	20 : 56 : 30	−54 : 13 : 06	8.000	–	Euler
NGC6067	16 : 13 : 11	−54.13 : 06	8.076	0.13	Euler
NGC3247	10 : 25 : 51	−57 : 55 : 24	8.083	–	Euler
NGC 2323	07 : 02 : 42	−08 : 23 : 00	8.096	0.02	Euler
NGC 1513	04 : 09 : 57	+49 : 30 : 54	8.110	–	Mercator
NGC 6705	18 : 51 : 05	−06 : 16 : 12	8.302	0.13	Mercator
NGC2437	07 : 41 : 46	−14 : 48 : 36	8.390	0.05	Euler
NGC3532	11 : 05 : 39	−58 : 45 : 12	8.492	−0.02	Euler
NGC 2194	06 : 13 : 45	+12 : 48 : 24	8.515	–	Mercator
NGC 1907	05 : 28 : 05	+35 : 19 : 30	8.567	–	Mercator
NGC2447	07 : 44 : 30	−23 : 51 : 24	8.588	0.03	Euler
NGC 2324	07 : 04 : 07	+01 : 02 : 42	8.630	−0.52	Mercator
NGC 1245	03 : 14 : 41	+47 : 14 : 12	8.704	0.10	Mercator
NGC 6811	19 : 37 : 17	+46 : 23 : 18	8.799	–	Mercator
NGC 1901[2]	05 : 18 : 15	−68 : 26 : 12	8.920	–	Euler
NGC 6134	16 : 27 : 46	−49 : 09 : 06	8.968	0.18	Euler
NGC 2420	07 : 38 : 23	+21 : 34 : 24	9.048	−0.26	Mercator
IC 4651	17 : 24 : 49	−49 : 56 : 00	9.057	0.10	Euler
NGC 7789	23 : 57 : 24	+56 : 42 : 30	9.235	−0.08	Mercator
NGC 6939	20 : 31 : 30	+60 : 39 : 42	9.346	0.02	Mercator
NGC 188	00 : 47 : 28	+85 : 15 : 18	9.632	−0.02	Mercator

Notes:
[1] *see Carrier 2009*
[2] *see Cherix 2006*

3. Conclusions

An overall accuracy of a few mmag, plus the long time baseline, allows the identification of various kinds of variable stars, including small period and small amplitude ones. The analysis of the variable stars found in IC 4651 indeed confirms the potentialities of this project to identify variable stars and to use them to characterize the properties of OCs.

References

Carrier, F. *et al.* 2009, *ArXiv e-prints*, **0901.0627**.

Cherix, M., Carrier, F., Burki, G., & Blecha, A. 2006, *Memorie della Societa Astronomica Italiana*, 77, 328–+

Clementini, G., *et al.* 2000, *AJ*, 120, 2054–2064

Dias, W. S., Alessi, B. S., Moitinho, A., & Lépine, J. R. D. 2002, *A&A*, 389, 871–873

Di Fabrizio, L. 1999, *Master Thesis*, Universitá di Bologna, Dipartimento di Astronomia, via Ranzani 1

Lenz, P. & Breger, M. 2005, *Communications in Asteroseismology* 146, 53–136

Lynga, G. 1987, *Catalogue of Open Cluster Data (5th Ed.)*, Lund Observ., Solvegatan 27, Lund

Stetson, P. B. 1996, *PASP*, 108, 851

Stetson, P. B., Hesser, J. E., & Smecker-Hane, T. A. 1998, *PASP*, 110, 533–552

Solar and Stellar Variability: Impact on Earth and Planets
Proceedings IAU Symposium No. 264, 2009
A. G. Kosovichev, A. H. Andrei & J.-P. Rozelot, eds.

doi:10.1017/S1743921309992456

Doppler imaging of the active star PW And

Sheng-hong Gu[1], Andrew Collier Cameron[2] and Kang Min Kim[3]

[1]National Astronomical Observatories/Yunnan Observatory, Chinese Academy of Sciences, P.O. Box 110, Kunming 650011, P. R. China
email: shenghonggu@ynao.ac.cn
[2]SUPA, School of Physics and Astronomy, University of St Andrews, North Haugh, St Andrews, Fife KY16 9SS, UK
[3]Korea Astronomy and Space Science Institute, 61-1 Whaam-Dong Yuseong-Gu, Daejeon 305-348, Korea

Abstract. We present spot activity of the K2V pre-main-sequence star PW And based on the high-resolution spectroscopic data obtained at Xinglong station and BOAO in 2005 November and December. Using the Least-Squares Deconvolution method, we have derived time series of composite profiles of PW And with high signal to noise ratio. These have been used to reconstruct its starspot pattern via the Doppler imaging method. The result shows that intermediate to high-latitude spots are the dominant features and weak low-latitude spots also appear. Comparing Doppler images from two data subsets, it can be found that almost no difference between two images exists except the small position evolution of weak low-latitude spots, which suggests that the intermediate to high-latitude spots have longer lifetimes than one month, and the low-latitude spots have shorter lifetimes.

Keywords. stars: late-type, stars: activity, stars: spots

1. Introduction

PW And (HD1405) is one member of nearby AB Dor moving group (Montes *et al.* 2001b, Zuckerman & Song 2004). It is a young pre-main-sequence K2V dwarf with a large rotation velocity vsini=22-24km/s (Lopez-Santiago *et al.* 2003, Strassmeier & Rice 2006). Previous results show that PW And is an active star with starspots and chromospheric emission (Bidelman 1985, Hooten & Hall 1990, Montes *et al.* 2001a, Lopez-Santiago *et al.* 2003, Strassmeier & Rice 2006).

2. Observations and data reduction

PW And was observed at Xinglong station of National Astronomical Observatories, China (NAOC) on 2005 November 18-24 and December 17-22 using the 2.16m telescope and Coudé echelle spectrograph with a 1k×1k CCD detector (Zhao & Li 2001), and Bohyunsan Optical Astronomy Observatory, Korea (BOAO) on 2005 December 08-11 using the 1.8m telescope and BOES spectrograph with a 2k×4k CCD detector (Kim *et al.* 2002, Kim *et al.* 2007). The spectral resolution is about 36000 for Xinglong observations and about 48000 for BOAO one.

The observed data are reduced using IRAF package following standard procedure, including bias subtraction, flat fielding, background subtraction, cosmic ray removal, optimal extraction, wavelength calibration and continuum fitting.

3. Image reconstruction

In order to check whether the spot pattern evolves during our observing period spanned just more than one month, we separate the all data to two data subsets to do Doppler imaging respectively. Data subset 1 is corresponding to the observations made on Nov.18-24 (Xinglong) and Dec. 8-11 (BOAO), 2005, data subset 2 to the observations on Dec.8-11 (BOAO) and Dec.17-22 (Xinglong), 2005. To obtain time series of line profiles with the highest signal to noise ratios from the data for Doppler imaging purpose, we employ the Least-Squares Deconvolution (LSD) method (Donati *et al.* 1997) to combine the profile information from all the observed photospheric lines in each observed spectrum.

We use the single star mode of our Doppler imaging code DoTS (Collier Cameron 1997) to reconstruct the spot pattern of PW And based on the above LSD profiles. The resulting Doppler images based on two data subsets are shown in Fig. 1 where the mean spot filling vs. latitude is also displayed.

4. Discussion and conclusions

In Fig. 1, it can be seen that the predominant features of Doppler images of PW And in our observing season are intermediate to high-latitude spots, which are obviously different from the results derived by Strassmeier & Rice (2006), which show no spots appeared above the latitude 40°.

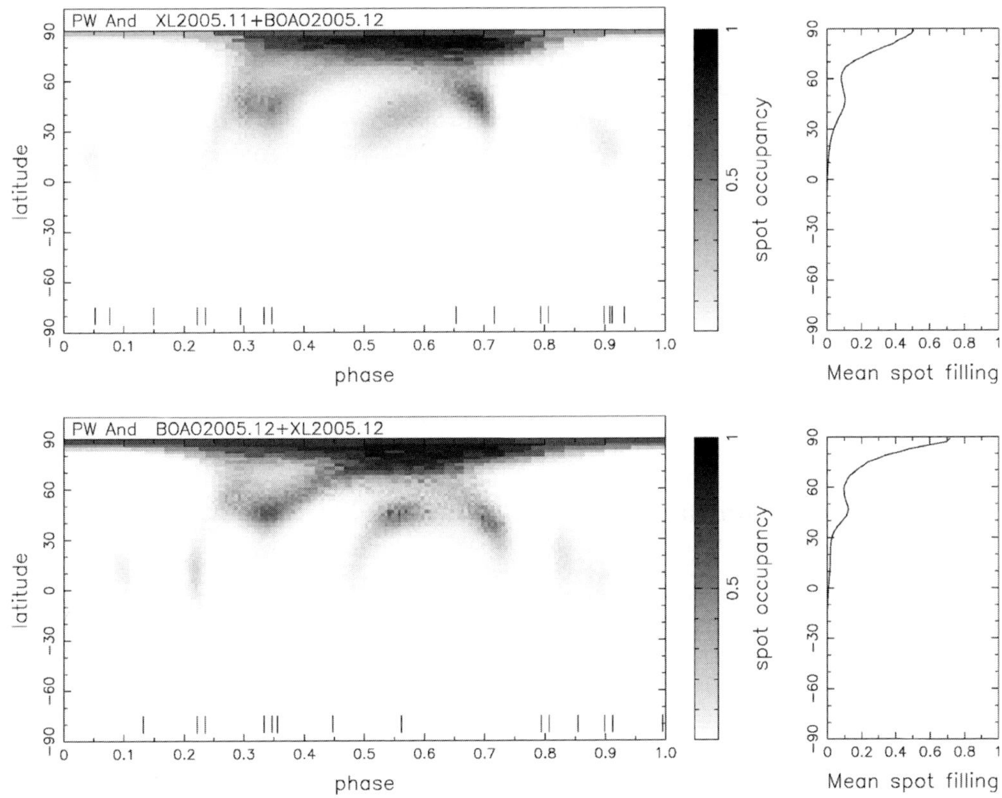

Figure 1. Doppler images of PW And for two data subsets. The upper image is for the data subset 1 and the lower one for the data subset 2.

From the Doppler images in Fig. 1, it is clear that two spot patterns are almost the same as each other except small low-latitude spots. This indicates that the main structure of spot pattern is stable during one month time scale. However, the low-latitude spots have some changes either in position or in intensity. Such a result implies that the low-latitude spots seem to have shorter lifetimes than the intermediate to high-latitude spots, like most of other magnetic active stars.

Hooten & Hall (1990) obtained four groups of photometric data, and found the second dataset spanned more than 5 months gave a large scatter when searching the period although it contains the largest number of data points, whereas the fourth dataset spanned less than one month gave a small scatter for the period obtained. This suggests that the spot pattern is stable for short duration, and changes for long time interval.

From the observations collected in Aug. 2004, Strassmeier & Rice (2006) derived a spot pattern similar to our Sun, in which the six main cool spots concentrate below latitude 40°. From new observations made in Nov. and Dec. 2005, we get a obviously different spot pattern with the main spots distributed from latitude 41° to 83°. Such a large difference suggests that the lifetime of main spot structure cannot last more than one year.

These information suggests that the spot pattern of PW And should change largely during its one activity cycle, like another active single star LQ Hya with the same spectral type (K2V) and similar rotational velocity, whose Doppler imaging results show that sometimes the low-latitude spots were predominant, and the obvious polar spots appeared in other times (Donati 1999, Donati *et al.* 2003, kovari *et al.* 2004).

Acknowledgments

Gu and Cameron acknowledge the support of Chinese Academy of Sciences and the Royal Society through an International Joint Project award that provided travel funds and living expenses during Gu's two visits to the University of St Andrews, and Cameron's visit to Yunnan Observatory in the course of this work. This work is also supported by NSFC under grants No. 10373023 and No. 10773027.

References

Bidelman, W. P. 1985, *AJ*, 90, 341
Collier Cameron, A. 1997, *MNRAS*, 287, 556
Donati, J.-F. 1999, *MNRAS*, 302, 457
Donati, J.-F., Collier Cameron, A., Semel, M. *et al.* 2003, *MNRAS*, 345, 1145
Donati, J.-F., Semel, M., Carter, B. D. *et al.* 1997, *MNRAS*, 291, 658
Hooten, J. T. & Hall, D. S. 1990, *ApJS*, 74, 225
Kim, K. M., Han, I., Valyavin, G. *et al.* 2007, *PASP*, 119, 1052
Kim, Kang-Min, Jang, Be-Ho, Han, Inwoo *et al.* 2002, *JKAS*, 35, 221
Kovari, Zs., Strassmeier, K. G., Granzer, T. *et al.* 2004, *A&A*, 417, 1047
Lopez-Santiago, J., Montes, D., Fernandez-Figueroa, M. J. *et al.* 2003, *A&A*, 411, 489
Montes, D., Lopez-Santiago, J., Fernandez-Figueroa, M. J. *et al.* 2001a, *A&A*, 379, 976
Montes, D., Lopez-Santiago, J., Galvez, M. C. *et al.* 2001b, *MNRAS*, 328, 45
Strassmeier, K. G. & Rice, J. B. 2006, *A&A*, 460, 751
Zhao, G. & Li, H.-B. 2001, *ChJAA*, 1, 555
Zuckerman, B. & Song, I. 2004, *ApJ*, 613, L65

Solar and Stellar Variability: Impact on Earth and Planets
Proceedings IAU Symposium No. 264, 2009
A. G. Kosovichev, A. H. Andrei & J.-P. Rozelot, eds.

doi:10.1017/S1743921309992468

New improvements of HASTA for the analysis of chomospheric solar events

L. Leuzzi[1,2], C. Francile[1,2], M. L. Luoni[3], M. Rovira[3] and J. I. Castro[1,2]

[1]Facultad de Ciencias Exactas, Físicas y Naturales (FCEFyN), Universidad Nacional de San Juan (UNSJ), San Juan, Argentina
email: luisleuzzi@yahoo.com.ar
[2]Félix Aguilar Astronomical Observatory (OAFA), UNSJ, San Juan, Argentina
email: cfrancile@unsj-cuim.edu.ar
[3]Instituto de Astronomía y Física del Espacio (IAFE) , Argentina
email: mluoni@iafe.uba.ar

Abstract. It is well known that chromospheric observations in the hydrogen alpha line give relevant information about solar flares, plages and protuberances, among other typical features of the Sun. From 1998 to 2006, the **HA**lpha **S**olar **T**elescope of **A**rgentina (**HASTA**) has provided solar images to the scientific community with the technological resources available at that time. Starting in 2007, major improvements have been incorporated, like a new CCD camera with enhanced spatial and temporal resolution, filter replacement, the automatic focusing system, and a new flat-fielding procedure. The hardware changes also called for software improvements, and a new solar-flare classification routine was implemented. At present, the Félix Aguilar Observatory (**OAFA**) of the University of San Juan (**UNSJ**) has a permanent staff of observers which now permits continuous solar monitoring. We expect that all these advances will allow to analyze chromospheric solar activity, especially solar flares, in more detail.

Keywords. Instrumentation: adaptive optics - Sun: chromosphere - Sun: flares - Sun: filaments - Sun: activity - waves - telescopes

1. Introduction

The **HASTA** telescope, located 50km eastwards from the Andes mountains at an altitude of 2370 m, has been operating continuously since 1998 at the Carlos U. Cesco Station (**EACUC**), El Leoncito, San Juan, Argentina as part of an agreement for scientific cooperation involving three institutions: Max Planck Institut für Extraterrestrische Physik (MPE) of Germany, and Instituto de Astronomía y Física del Espacio (**IAFE**) and "Félix Aguilar" Observatory (**OAFA**) the last two from Argentina, to observe the Solar Chromosphere (Bagalá *et al.* 1999). During the past two years new instrumental and software developments have been implemented to the telescope carried out by **OAFA** (Francile *et al.* 2008), to improve and extend its observational capabilities of the dynamics of solar atmosphere and solar activity.

2. Characteristics of the instrument and its improvements

The place where the telescope is located is extremely low air humidity and without human activity that might cause atmospheric pollution.
HASTA is a solar refractor telescope that provides full sun CCD images in the Hydrogen Hα emission line at 656.3 nm. Table 1 shows the main features of the telescope.

Telescope	Diameter Focal Distance	11 cm 165 cm
Location	Longitude Latitude Coverage of Time	69.3 W 31.8 S 11:30 UT to 20:30 UT
Filter	Type Central Wavelengh Bandwidth Tuning Range	Tunable Lyot - Öhmann 656.27 nm 0.03 nm ± 0.11 nm
Images	Normal Size Integration Time Type Resolution	(1280 x 1024) pixels - Full Sun Disk 50 - 100 msec Center Line & Line Wings, sequential ≈ 2 arc sec
Operational Modes	Patrol Mode Flare Mode	Images every ≈ 3 min Images every 0.5 sec

Table 1. Main characteristics of the **HASTA**

The solar activity is analyzed in real time and, if a fast event is detected, the camera switches into high-speed mode or Flare mode. In this mode it can take and store full-frame images up to 2 per second during the Impulsive Phase of a Flare.

The New CCD Camera. The images are registered with a PCO-1600 CCD Camera with 512 Mbytes internal memory allowing high speed and resolution recording. The typical exposure times are 50-100 milliseconds (see the Table 1).

The New Flat Fielding Procedure. A new procedure to obtain a flat-field image was developed using a diffuser of the type holographic 60°. Some flat field images are generated daily by means of the new diffuser placed at the entrance of the telescope. These images are pre-processed and median averaged to reduce the statistic noise to get a better image for the flat field correction.

Automatic Focusing System. The continuous focus misalignments of the telescope due to the daily thermal variations that modifies its mechanical properties and the physical properties of the air inside it, are automatically corrected by the **HASTA** control software.
This software analyzes the spectral components of the measured intensities on the sun disk and limb, and obtains a focusing factor. As this factor is affected by the local conditions at the telescope place (clouds, seeing, weather) and also by the solar activity conditions (flares, plages, filaments), a feedback loop is needed to maintain the focus factor as optimum as can be possible for these conditions during the whole day. This action is performed moving a focusing stepper motor recently installed. In this way the quality of the sun observations are improved considerably.

The New Controlled Tuning Device for the "Lyot-Öhman" Hα Filter. **HASTA** takes full-disk images as Hα line center as wings. By means of a software the scientist can define automatic wavelength-shift sequences for the tunable Hα filter. It allows to control the motors that rotate the input and output filter polarizers to achieve the desired frequency shift.

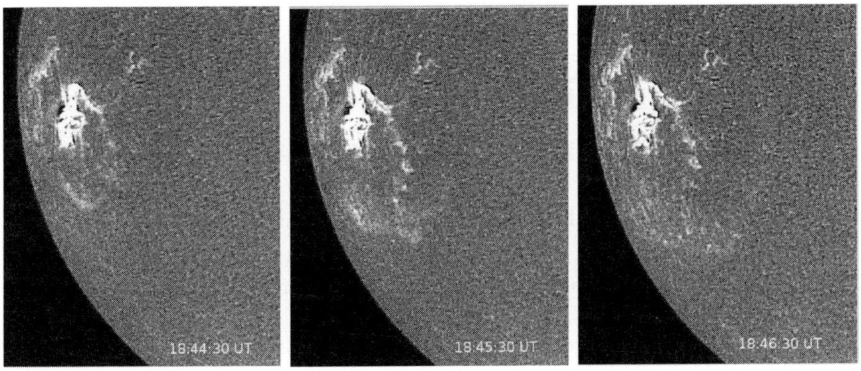

Figure 1. Sequence of Hα images shows a tsunami-like shock wave spreading from a class X6.5 flare erupting from AR 10930 on 2006 December 06. The contrast and brightness of the images have been enhanced to bring up details of the shock wave incoming.

3. Example of observation of a Moreton Wave

The Moreton waves are shock wave, they are large arch-shaped brightenings propagating through the Sun observed in Hα. They are observed immediately following large solar flares (Athay & Moreton 1960). By other side, although such flares do occur there are not reports of Moreton waves. Some recent studies (Cliver *et al.* 1999) suggest that Moreton waves may originate from fast coronal mass ejections.
These images are a solar full-disk fraction and were recorded on 2003 December 06 at Hα line center (656.27 nm) with a cadence of 0.5 sec (pixel size = 2"). We note that the observations show the details of the expanding Moreton wave as an intensity enhancement.

4. Conclusions

HASTA represents an important tool, both for solar physics and space weather research, which will certainly increase our knowledge about the physical processes behind the various chromospheric phenomena and their interrelations.
In this part of Southern hemisphere this solar telescope is the only one with such high spatial and temporal resolution. All the improvements taken part in the **HASTA** enable to follow the evolution of rapid changing phenomena like flares and propagating waves as well as filaments, in the solar chromosphere.

References

Athay, R. & Moreton, G. 1961, *ApJ*, 133, 935

Bagalá, G., Bauer, O., Fernández Borda, R., Francile, C., Haerendel, G., & Rovira, M. 1999, *Proceedings of the 9th European Meeting on Solar Physics, ESA-SP*, 448, 469

Cliver, E., Webb, D., & Howard, R. 1991, *Sol. Phys.*, 187, 89

Francile, C., Castro, J. I., Leuzzi, L., Luoni, M. L., Rovira, M. G., Cornudella, A., Gomez, W., & Sarmiento, R. 2008, *Bol. Asoc. Argen. Astr.*, 51, 339

Solar and Stellar Variability: Impact on Earth and Planets
Proceedings IAU Symposium No. 264, 2009
A. G. Kosovichev, A. H. Andrei & J.-P. Rozelot, eds.

doi:10.1017/S174392130999247X

Coronal magnetic fields from the inversion of linear polarization measurements

Yu Liu[1,2], Haosheng Lin[2] and Jeff Kuhn[2]

[1]National Astronomical Observatories,
Yunnan Astronomical Observatory, Kunming 650011, China
email: lyu@ynao.ac.cn

[2]Institute for Astronomy, University of Hawaii, Ohia Ku Street, Pukalani, HI 96768, USA

Abstract. Real 3-D coronal magnetic field reconstruction is expected to be made based on the technologies of IR spectrometry and tomography, in which the data from other wavelengths can be used as critical reference. Our recent studies focused on this issue are briefly reviewed in this paper. Liu & Lin (2008) first evaluated the validity of potential field source surface model applied to one of five limb regions in the corona by comparing the theoretical polarization maps with SOLARC observations in the IR Fe XIII 10747 Å forbidden coronal emission line (CEL). The five limb coronal regions were then studied together in order to study the spatial relation between the bright EUV features on the solar disk and the inferred IR emission sources, which were obtained from the inversion of the SOLARC linear polarization (LP) measurements (Liu 2009). The inversion for each fiber data in the field of view was made by finding the best location where the difference between the synthesized and the observed polarizations reaches the minimum in the integration path along the line of sight. We found a close relationship between the inferred IR emission source locations and the EUV strong emission positions.

Keywords. Sun: corona; Sun: magnetic fields; Sun: infrared

1. Introduction

The solar coronal magnetic fields play a dominant role in the coronal plasmas distribution and dynamics. During the last decades, a few useful methods have been developed based on model assumptions for radio observations to derive coronal magnetic field strength; or based on the scattering polarization of CELs to map coronal magnetic field direction. Some valuable progresses have been made recently which use the strong infrared Fe XIII 1075 nm CEL for Zeeman splitting observations (Kuhn 1995; Lin *et al.* 2000, 2004). Penn *et al.* (2004) studied the noise sources in coronagraphic magnetic field strength measurements. Tomczyk and colleagues found the direct evidence of Alfven waves in the corona with their new instrument CoMP(Tomczyk *et al.* 2008).

The imaging coronal spectropolarimeter SOLARC (Solar Observatory for Limb Active Regions and Coronae) was installed on Haleakala. A line-of-sight (LOS) coronal magnetogram was shown by Lin *et al.* (2004) as one of the first results taken by SOLARC. Liu & Lin (2008) studied in detail a limb coronal region by comparing the observed circular and linear polarizations with the synthesized maps based on potential field model. The best consistence for both linear and circular polarizations was found at the layer in the corona above a sunspot near the solar limb, indicating that local coronal magnetic fields may be revealed from the IR polarization data with the proper coronal magnetic field inversion method. As a next step, only the LP data were used to study the five limb coronal regions together. The single points were solved along each LOS using the same method, by minimizing differences between observed and computed LP data which were based on the extrapolation code by Liu & Zhang (2002). The derived points from the IR data were compared with the EUV images of Fe XII 195 Å (Liu 2009).

2. Results

The five west limb coronal regions had been shown in Fig. 2 of Lin *et al.* (2004) with PAH (i.e., the Position Angle in the Heliocentric coordinate system) of 296, 291, 286, 281, and 276 degrees, separately. The coordinate system and the parameters used in the paper

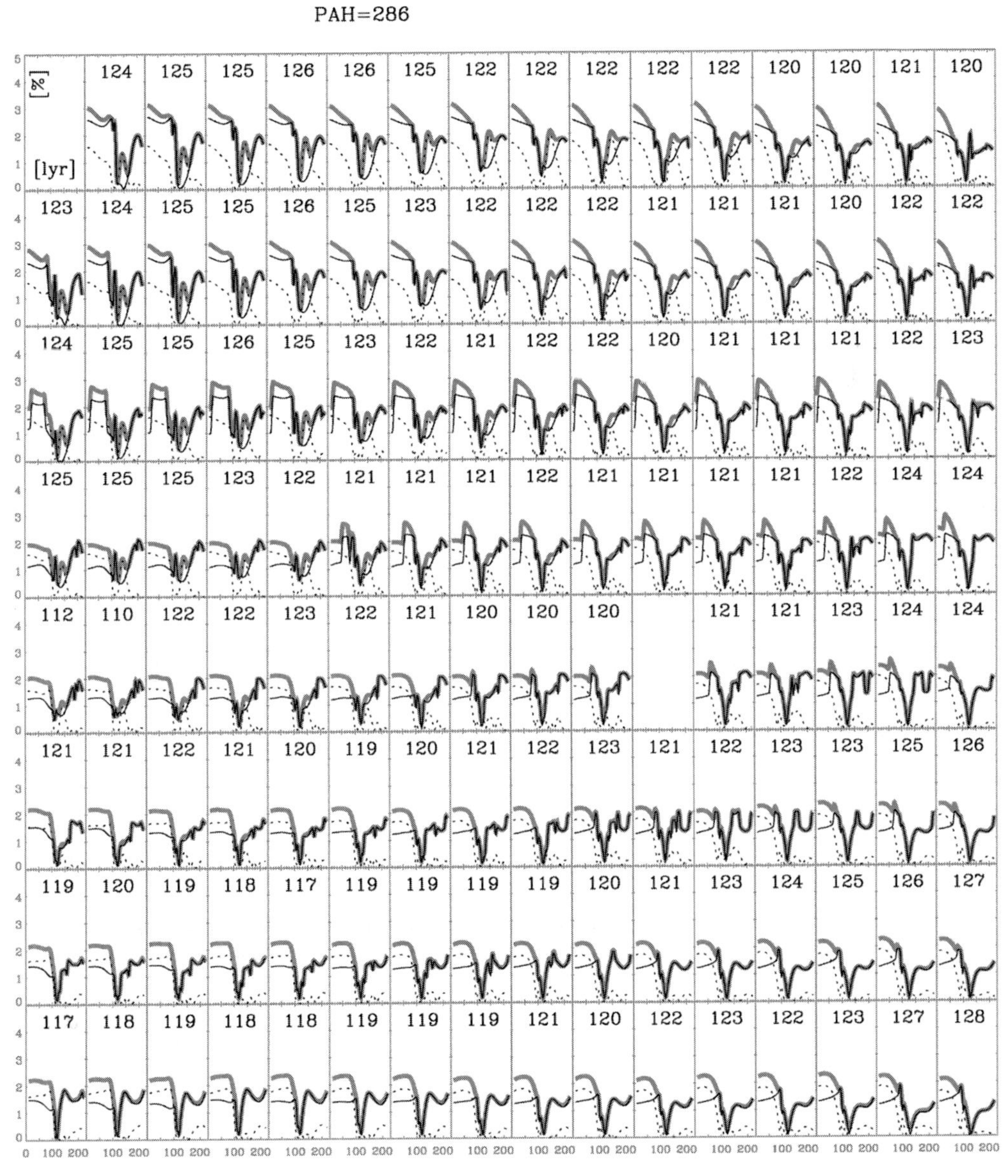

Figure 1. Error profiles for the coronal region at PAH 286 degrees. All the 16 × 8 fibers are shown for each coronal position. The location (z or layer$_{best}$) of the minimum error in a σ_{LP} profile (thick gray line) is shown in each frame. Most σ_{LP} profiles show a clear minimum value around the layer 120 above a sunspot located at the solar west limb. Note, 1 lyr = 4.5 Mm.

of Liu (2009) are the same as in Liu & Lin (2008). Similar to results of the region PAH 286 (Liu & Lin 2008), the profiles of LP *vs.* z values (in unit of layer, i.e., lyr; Layerbest) in the other four regions also show a clear dip (Liu 2009). The identical peak/valley locations were noticed for different profiles. Such a common feature for different coronal regions is worthy of study. As a clear demonstration, Fig. 1 shows the LP profiles for each position in the coronal region with PAH of 286 degrees.

It can be seen that most derived IR emission sources were mainly distributed at around $z = 120$ where the sunspot region was located near the solar limb on the photosphere (Liu & Lin 2008). If we presume that the basic coronal structures in EUV wavelengths be kept stable and the same during the days before the solar limb SOLARC observation, it would be interesting to compare the inferred IR emission sources with the EUV bright features (Liu 2009). The EUV data have been pre-processed with the standard solar software procedures. For the coronal regions close to strong photospheric fields, their IR emission sources were mainly distributed in a cluster which discloses the root parts for bright EUV loops; while for a coronal region on the top of a bundle of transequatorial loops with weaker EUV brightness, their IR emission sources appeared to be distributed rather diffusely.

3. Conclusion

We applied the optimized method built in the paper of Liu & Lin (2008) for locating observed coronal magnetic fields from IR linear polarization data to all five coronal regions observed on 7 April 2004. The idea of this method is different from the traditional ways which take the corona emission as optically thin for IR observations along sight line (e.g., Judge *et al.* 2006). Practically, we usually operate SOLARC to observe the limb coronal regions with strong EUV 195 Å emissions for routine IR polarization measurements. Our recent results have confirmed the advantage by taking SOLARC observations with the reference from EUV wavelength images to obtain useful information for the study of local coronal magnetic fields. A general tendency can be seen from the comparison between IR and EUV emission sources: the higher the percentage of EUV bright points for a field of view, the more concentrated the distributions of the IR emission sources. In the next step, we plan to compare the inferred 3-D IR source regions with the 3-D structure of the bright EUV features with the help of coronal tomography technology or the 3-D STEREO EUV data.

Acknowledgements

YL is grateful to National Natural Science Foundation of China 10843010 and 10933003.

References

Judge, P. G., Low, B. C., & Casini, R. 2006, *ApJ*, 651, 1229

Kuhn, J. R. 1995, in: J. R. Kuhn & M. J. Penn (eds.), *IR Tools for Solar Astrophysics: What's Next?*, (Singapore: World Scientific), p. 89

Lin, H., Penn, M. J., & Tomczyk, S. 2000, *ApJ*(Letters), 541, L83

Lin, H., Kuhn, J. R., & Coulter, R. 2004, *ApJ*(Letters), 613, L177

Liu, Y. & Zhang, H. 2002, *Publ. Yunnan Obs.*, 92, 1

Liu, Y. & Lin, H. 2008, *ApJ*, 680, 1496

Liu, Y. 2009, *AnGeo*, 27, 2771

Penn, M. J., Lin, H., Tomczyk, S., Elmore, D., & Judge, P. 2004, *Solar Phys.*, 222, 61

Tomczyk, S., Card, G. L., Darnell, T., Elmore, D. F., Lull, R., Nelson, P. G., Streander, K. V., Burkepile, J., Casini, R., & Judge, P. G. 2008, *Solar Phys.*, 247, 411

Solar and Stellar Variability: Impact on Earth and Planets
Proceedings IAU Symposium No. 264, 2009
A. G. Kosovichev, A. H. Andrei & J.-P. Rozelot, eds.

doi:10.1017/S1743921309992481

Observation of interactions between two erupting filaments

Yu Liu[1], **Jiangtao Su**[2], **Yuandeng Shen**[1] **and Liheng Yang**[1]

[1]National Astronomical Observatories,
Yunnan Astronomical Observatory, Kunming 650011, China
email: lyu@ynao.ac.cn

[2]National Astronomical Observatories, Beijing 100012, China

Abstract. Following the first observational study of the interaction between two distinct filaments (Su *et al.* 2007; hereafter, event 1), we present another interesting case observed by SMART telescope on 2005 June 25 with higher spatial resolution (hereafter, event 2). The two events are compared with each other. In event 1 the two filaments erupted subsequently and obvious mass flow was observed to be transferred from one erupting filament to one stable filament which triggered its eruption. On the contrary, in event 2, the two filaments erupted simultaneously and there was no transfer of material noticed between them during the initial stage. The two filaments merged together along the ejection path, indicating the bodily coalesce between the two interacting flux ropes. Moreover, event 1 was associated with a coronal mass ejection (CME), while event 2 was a failed filament eruption, thus without CME association.

Keywords. Sun: corona; Sun: magnetic fields; Sun: filament

1. Introduction

Generally, filament instabilities can be triggered by interior and exterior agents. The interior agents may be magnetic field flux evolution and cancelation on the photosphere. The exterior agents include fast running Moreton waves, remote extending flare ribbons and nearby repetitive surge injections (Liu *et al.* 2005, Jiang *et al.* 2009). Many previous studies have shown that filament disturbances are closely related to pore birth and movements which are results of magnetic flux emergence and cancelation. Moreover, the impact from an erupting filament can also be an exterior trigger for the eruption of a neighbor filament. Su *et al.* (2007) presented such a case well observed in three wavelengths of Hα line (i.e. event 1). In the case, both filaments were successfully erupted and accompanied by a CME. Fig.1 shows the carton describing the process of two filament merge due to magnetic field reconnection between them. The reconnection can be directly caused by one filament impact onto another filament body. This model is consistent with many aspects of the Hα and EUV observations.

2. New observations: event 2

Two filament eruptions were observed on 2005 June 25 in the solar active region NOAA 10780 (S08,W30). They were associated with a C1.3–class flare (03:30–04:30 UT), but there was no radio burst or coronal mass ejection reported for the event. The ground-based observations were taken by the SMART telescope at Hida Observatory. The SMART telescope was operated to take data at the Hα wavelength center and several off-bands, with a full disk mode and a local-monitoring mode in two observing channels. The time cadence was 7–30 s for most Hα data whose resolution was better

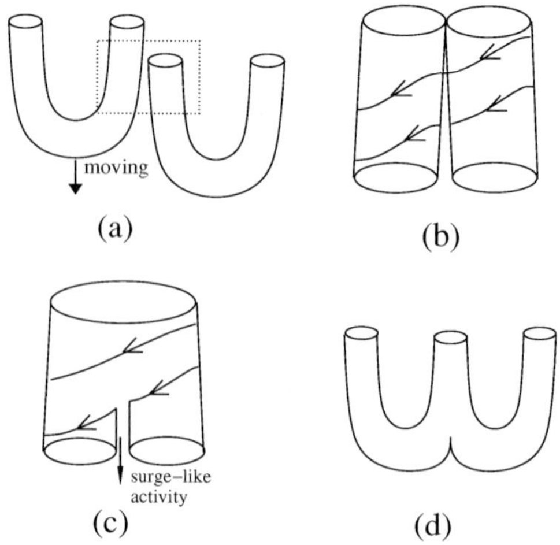

Figure 1. A schematic model showing the interaction and merge of two flux ropes located closely (Su *et al.* 2007). The helical magnetic field components in the filaments are thought to play an important role during the interaction and merge process for two approaching flux ropes.

than 1 arcsec for good seeing conditions. We briefly show the Hα observations in Fig. 2 which presents the main development for the two filament eruptions as well as their morphology evolution. The time and the off-band information are supplied at the bottom for each frame. After the initial stage (Fig. 2*a*–2*b*), the two filaments merged totally (Fig. 2*c*–2*f*).

3. Conclusion

Su *et al.* (2007) reported event 1 as the first filaments interaction and eruption event. In this paper we present event 2 as new observations. In event 1, obvious mass flow was found to be injected from one erupting filament to another stable one, triggering its eruption subsequently. In event 2, however, filament merging process was observed soon after their interaction. The two filaments merged as one moving along the overlying coronal fields (Liu *et al.* 2009). In event 2, the north filament eruption was closely relevant to the underlying flux cancelation due to the collision of two magnetic fluxes with opposite polarity on the photosphere. The event was companied by a C1.3-class flare with two spreading ribbons, the EUV dimmings and the overlying post-flare arcades. Moreover, the event had no associated CME because it was a failed filament eruption and the filaments material returned back to solar surface. Single loop-like bright ejections were observed in Nobeyama radio flux data (17 GHz), TRACE EUV and SXI SXR images. The event clear shows that the disturbing agent of a filament disturbance can be the impact from another filament, which is different from the other exterior agents, *e.g.*, flare ribbons, surges and shock waves. The interaction and mergence between two filaments should involve complicated dynamic and thermal process. The detailed mechanism is beyond the scope of our current observations. We speculate that the interaction and mergence should experience a violent mass and magnetic field flux transfer and exchange. Numerical simulations are needed to reveal more details for understanding the interaction process.

Furthermore, both the filaments in event 2 were sinistral and thus their mergence was consistent with the fact found by Martin *et al.* (1994) that filaments or filament channels of same magnetic handedness may merge with each other while those of opposite handedness may not join.

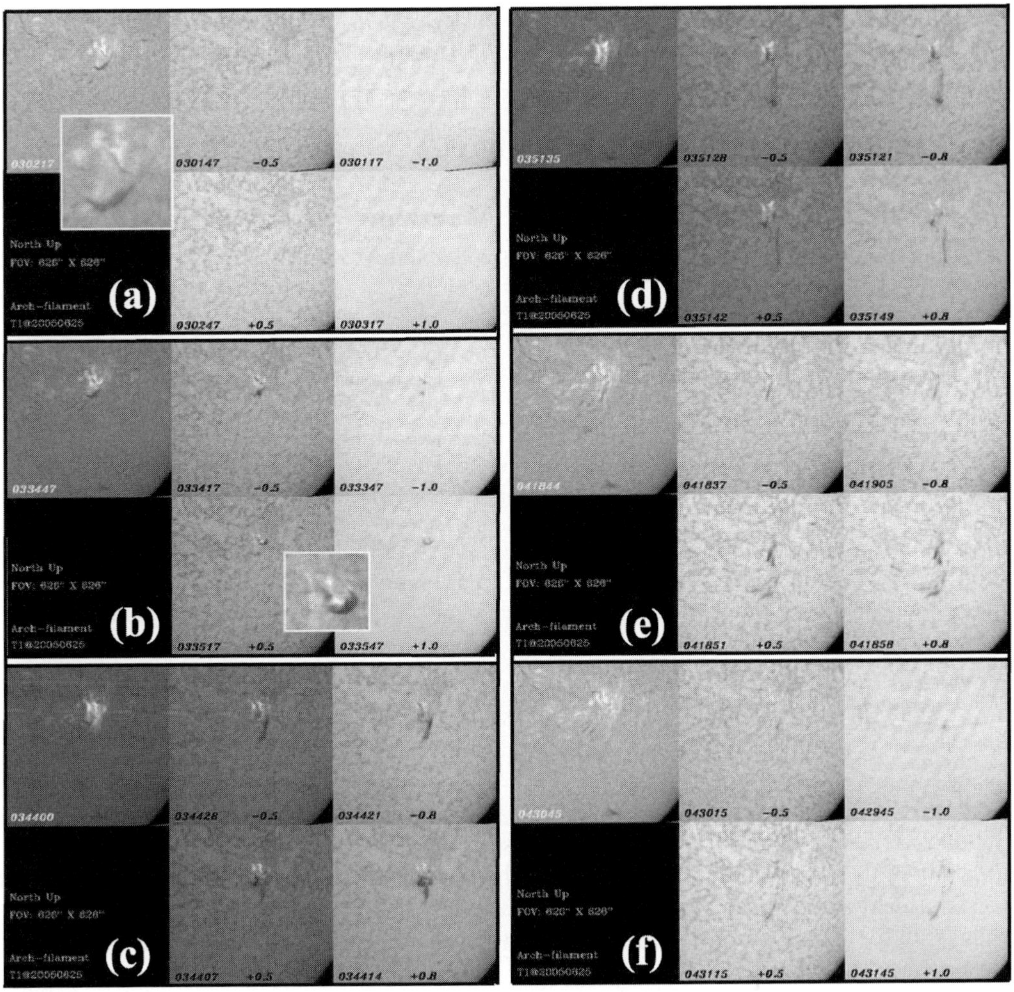

Figure 2. Multi-wavelength observations at Hα line for event 2 of two filaments interaction on 2005 June 25. (a) before eruption. (b) start of eruption and the north untwisting filament. (c–d) filaments interaction during the eruption. (e–f) filament material returned back to the surface and their eruptions were failed.

Acknowledgements

We are grateful to National Natural Science Foundation of China 10843010 and 10933003.

References

Jiang, Y., Bi, Y., Yang, J., Zheng, R., & Wang, J. 2009, *Research in Astro. & Astrophys.*, 9, 603

Martin, S., Billimoria, R., & Tracadas, P. 1994, in: R. Rutten & C. Schrijvers (eds.), *Solar Surface Magnetism*, (Kluwer Acad. Publ.), p. 303

Liu, Y., Kurokawa, H., & Shibata, K. 2005, *ApJ* (Letters), 631, L93

Liu, Y., Su, J., Xu, Z., Lin, H., Shibata, K., & Kurokawa, H. 2009, *ApJ* (Letters), 696, L70

Su, J., Liu, Y., Kurokawa, H., Mao, X., Yang, S., Zhang, H., & Wang, H. 2007, *Solar Phys.*, 242, 53

Solar and Stellar Variability: Impact on Earth and Planets
Proceedings IAU Symposium No. 264, 2009
A. G. Kosovichev, A. H. Andrei & J.-P. Rozelot, eds.

doi:10.1017/S1743921309992493

Study of helicity properties of peculiar active regions

M. C. López Fuentes[1], C. H. Mandrini[1] and P. Démoulin[2]

[1]Instituto de Astronomía y Física del Espacio (CONICET-UBA), CC 67, Suc 28, 1428 Buenos Aires, Argentina
e-mail: lopezf@iafe.uba.ar

[2] Observatoire de Paris, LESIA, UMR 8109 (CNRS), F-92195, Meudon Principal Cedex, France

Abstract. Peculiar solar active regions (ARs), such as δ-islands and other high tilt bipoles, are commonly associated with the emergence of severely deformed magnetic flux tubes. Therefore, the study of these ARs provides valuable information on the origin and evolution of magnetic structures in the solar interior. Here, we infer the magnetic helicity properties of the flux tubes associated to a set of peculiar ARs by studying the evolution of photospheric magnetograms (SOHO/MDI) and coronal observations (SOHO/EIT and TRACE) in combination with force-free models of the magnetic field. We discuss how our results relate to different models of the evolution of emerging magnetic flux tubes.

Keywords. Sun: magnetic fields, Sun: activity, Sun: photosphere, Sun: corona

1. Introduction

Solar active regions (ARs) are the manifestation of the buoyant emergence of magnetic flux tubes from the solar interior. Since these magnetic structures are related to the toroidal field at the base of the convection zone, they are approximately parallel to the solar equator. For this reason, most ARs are bipolar configurations oriented in the east-west direction, obeying what is known as Hale's law (e.g., Schrijver & Swaan 2000). AR bipoles are generally slightly inclined, so the preceding polarities tend to be closer to the solar equator (Joy's law). This inclination ("tilt" angle) is thought to be due to the Coriolis force acting on the flux tubes during emergence (see Fisher *et al.* 2000). A number of ARs have tilts that deviates from Hale's and Joy's laws. A typical case are the "δ-islands"; sunspots formed by two or more umbrae of different polarities within the same penumbra. These regions are associated to flux tubes that suffered a deformation with respect to the normal "Ω" loop shape. Among the possible causes of the deformation are the interaction of the flux tubes with the convection zone plasma and the development of the kink instability (López Fuentes *et al.* 2003, Tian *et al.* 2005). The deformation of the flux tube axis (the tube as a whole) is associated to the magnetic helicity component named writhe. The other component, called twist, corresponds to the rotation of the individual magnetic field lines around the main axis of the tubes. It is expected that, as the flux tube emerges through the photosphere, the peculiar AR bipole rotates towards the normal east-west direction. Therefore, from the rotation of the polarities during flux emergence we can infer the left (negative writhe) or right (positive writhe) chirality of the tubes. The sign of the flux-tube twist, on the other hand, can be inferred comparing force-free models of the coronal magnetic field with EUV and soft X-ray observations. The relation between writhe and twist provides information on the mechanism that produces the flux tube deformation. The kink instability, for instance, implies that both

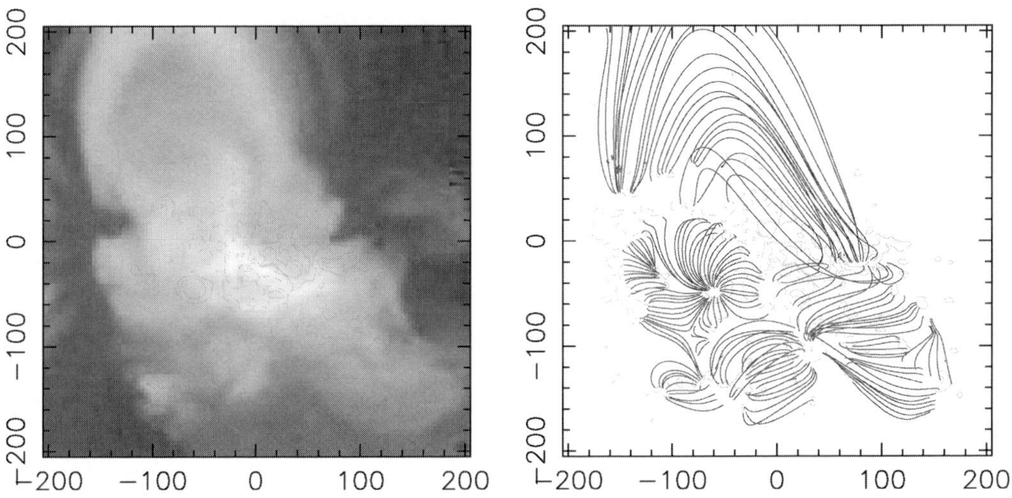

Figure 1. Left panel: Yohkoh/SXT image of AR 9026 observed on 2000 June 7 (14:52UT). Right panel: linear force-free magnetic model of the AR's coronal field extrapolated from SOHO/MDI magnetogram data. Continuous/dashed contours correspond to positive/negative longitudinal component of the photospheric magnetic field of levels ±100 and ±500 Gauss.

components must have the same sign, because it is due to the internal transfer of magnetic helicity from twist to writhe (Fan *et al.* 1999). Here, we analyze a set of peculiar ARs and we obtain the twist and writhe signs of the associated magnetic flux tubes to constrain the possible mechanisms at the origin of their deformation.

2. Analysis and results

We study 4 peculiar ARs from a larger set presented in López Fuentes & Mandrini (2008). Using SOHO/MDI magnetograms we compute the mean position of the positive and negative polarities from which we obtain the tilt angle. From the tilt rotation during the AR evolution across the solar disk we infer the writhe of the associated magnetic flux tubes. It can be demonstrated that counterclockwise rotation (CCW) of the polarities corresponds to positive writhe, while clockwise (CW) rotation indicates a negative writhe. In Table 1 we show the properties of the studied ARs. The first 5 columns correspond to: AR number, AR type, solar hemisphere to which the AR belongs, sense of rotation of the tilt during evolution, and inferred writhe sign.

To infer the sign of the AR twist we compare EUV and soft X-ray observations with models of the coronal magnetic field extrapolated from MDI magnetograms using the linear force-free equation:

$$\nabla \times \mathbf{B} = \alpha \mathbf{B} \tag{2.1}$$

We use the sign of the parameter α as a proxy for the twist sign. Figure 1, left panel, shows a soft X-ray image of AR 9026 obtained on 2000 June 7 with Yohkoh/SXT. In the right panel we show a model with $\alpha = -0.006$ Mm^{-1} extrapolated from a SOHO/MDI magnetogram obtained on 2000 June 7 (11:12UT). The comparison clearly shows that the AR has negative twist. For a more quantitative determination of the twist, we compute the mean distance of model field lines to a series of coronal loops from TRACE and Yohkoh/SXT. The model that best approaches the observed loops is considered the best fitting for the AR twist. In Figure 2 we plot the mean field line distance to TRACE loops versus the alpha parameter for AR 9026 (left panel) and AR 10808 (right panel).

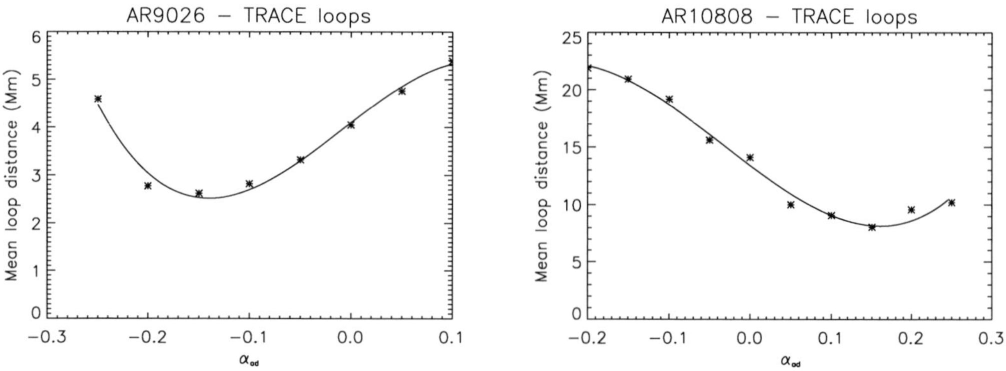

Figure 2. Mean distance between modeled magnetic field lines and observed coronal loops versus the force free parameter α for ARs 9026 and 10808. Curves are polynomial fittings.

Table 1. Properties of the studied ARs.

AR Number	AR Type	Hem.	Rotation	W sign	T sign	Kink?
9026	δ-island	N	CW	Neg	Neg	√
9165	High tilt	N	CW	Neg	Neg	√
9591	Non-Hale	S	CCW	Pos	Neg	–
10808	δ-island	S	CCW	Pos	Pos	√

The fitting procedure shows that AR 9026 has negative twist and AR 10808 has positive twist. Similar analysis for AR 9165 and AR 9591 indicate that both ARs have negative twist (see Table 1). The last column of the table indicates whether the relation between twist and writhe signs is consistent with the kink instability or not.

3. Conclusion

We determine the magnetic helicity properties (twist and writhe) of 4 peculiar active regions that depart appreciably from Hale's law. Our main goal is to constrain the mechanisms proposed to explain magnetic flux-tube deformations. According to their twist and writhe signs, 3 of the studied cases are consistent with the development of a kink instability. Other mechanism, such as the interaction with the convection zone plasma during emergence, should be invoked to explain the remaining case.

References

Fan, Y., Zweibel, E. G., Linton, M. G., & Fisher, G. H. 1999, *ApJ*, 521, 460

Fisher, G. H., Fan, Y., Longcope, D. W., Linton, M. G., & Pevtsov, A. A. 2000, *Solar Phys.*, 192, 119

López Fuentes, M. C. & Mandrini, C. H. 2008, Boletín de la Asociación Argentina de Astronomía La Plata Argentina, 51, 31

López Fuentes, M. C., Démoulin, P., Mandrini, C. H., Pevtsov, A. A., & van Driel-Gesztelyi, L. 2003, *A&A*, 397, 305

Schrijver, C. J. & Zwaan, C. 2000, Solar and stellar magnetic activity/Carolus J. Schrijver, Cornelius Zwaan. New York: Cambridge University Press, 2000. (Cambridge astrophysics series; 34)

Tian, L., Alexander, D., Liu, Y., & Yang, J. 2005, *Solar Phys.*, 229, 63

Solar and Stellar Variability: Impact on Earth and Planets
Proceedings IAU Symposium No. 264, 2009
A. G. Kosovichev, A. H. Andrei & J.-P. Rozelot, eds.

doi:10.1017/S174392130999250X

Automatic detection method, forecast and alert of solar proton events

Ganghua Lin

Key Laboratory of Solar Activity, National Astronomical Observatories,
Chinese Academy of Sciences,
Datun Rd. 20A, Chaoyang District, Beijing, 100012, China
email: lgh@bao.ac.cn

Abstract. The methods of automatic solar active phenomenon or event detection have been researched and explored by people for many years, which have gone into actual services. The paper analyzes the relationship between these methods of automatic detection and the forecast or alert, using the solar short-term proton events predictions as an example. Using automatic method to conduct forecast or alert is under thinking.

Keywords. Sun: activity, particle emission, magnetic fields, solar-terrestrial relations; methods: data analysis, statistical

1. Introduction

The forecasts and alerts of space environment for live of human being, development of society and aerospace profession are vital and necessary. At the same time, the automatic detection methods have characteristics of reliability and rapidity. With the growing demand for forecasts of solar activity, there is great need for these methods. In the past many years, methods of automatic solar active phenomenon or event detection have been researched and explored by people, and gone into actual service. That whether these methods can be used systematically in the forecasts and alerts of space environment or not has a preliminary analysis in the paper.

2. Forecast or alert factors

The forecast is to predict solar event or phenomenon that is going to occur several days later. The alert is to warn an event that is going to occur in dozes of minutes or dozes of hours. They have both similarities and differences. To facilitate the further analysis, table 1 presents the forecast or alert factors of proton event that are frequently used, the obtained method and the automatic detection method. In Table 1, the forecast factors are main clues of propagation process of proton and potential factors of occurrence of the proton event; the obtained method refers to the observing equipments by which the factors can be obtained; the automatic detection method refers to the explored method that is used for detecting the corresponding phenomenon or event.

In the proton event forecast system, the main factors can be roughly separated into four category wang 1999.

Among them, sunspots related factors are : Zürch classification and area of sunspot groups, magnetic classification and magnetic density of sunspot groups, McIntosh classification, special sunspot groups and transverse field structure, sunspot volution structure and movement, alternation of the area of sunspot groups 3 days before the explosion of flare event.

Table 1. Forecast or alert factors of solar proton event.

Forecast or alert factors	Obtained method	Automatic detection method
Soft x-ray flare or H*alpha* flare	X-ray or full disk Ha telescope	explored. Ming *et al.* (2003)
sunspot:		explored. Curto *et al.* (2008), Benkhalil *et al.* (2006), Ming *et al.* (2003)
1 The McIntosh classification 2 The magnetic classification 3 The recent flaring activity 4 The level of flaring activity 5 Special magnetic classification of sunspot group	1 The first McIntosh parameter z, the second McIntosh parameter p, the third McIntosh parameter c. 2 SOHO MDI 6767Å magnetograms, CaII 8542 Å magnetogram(Kitt Peak), the FeI line magnetogram (Mt Wilson). 3 The recent X-ray flares observed on GOES, event lists and the current movies of SOHO EIT FeXII 195Å 4 The X-ray background level on GOES.	
10.7 cm radio flux	10.7 cm radio flux at frequency 2800MHz 2695MHz is an alternative. The latest SOHO MDI continuum or Catania observation	downloading
Proton event of impulse type Proton event of progression type(be rich in proton)	It is corresponding to type III and V of radio eruption. STEREO It is corresponding to type II and IV of radio eruption. STEREO	
Frequency of flare activity		explored. Ming *et al.* (2003), improvable output
CME-driven interplanetary shock waves	LASCO, STEREO, EIT FeXII 195Å, the list of radio bursts prepared by SEC or/and the latest Culgoora PRESTO message.	
Coronal magnetic field	LASCO, EIT 284Å, TRACE 195Å fulldisk GOES/SXI, SOHO/MDI(NSO/KPVT), STEREO	
Interplanetary magnetic field	ACE,the CELIAS/MTOF/PM sensor on the SOHO spacecraft	
Flux density of proton. The distance between the flaring region and root segment region of open magnetic field line in the interplanetary.	GOES 11 STEREO	downloading
Solar disk location of flare eruption		Explored. Ming Qu, *et al.* (2003)
Proton peak flux levels	GOES11	downloading

Active area related factors are: the shape of magnetic neutral line, magnetic flux emerge and longitudinal field gradient, flare activities frequency, similarity between current active area and old proton active area.

The location related factors are: the location of flaring on the solar disk, reproducibility between active region longitude and proton flare, the active area and the foot points of line of magnetic connection of solar-terrestrial are at the same magnetic polarity area, the active area and disk center locate at the same magnetic polarity area, active area and foot points on disk of line of magnetic connection of solar-terrestrial are in the same sector area of interplanetary magnetic field.

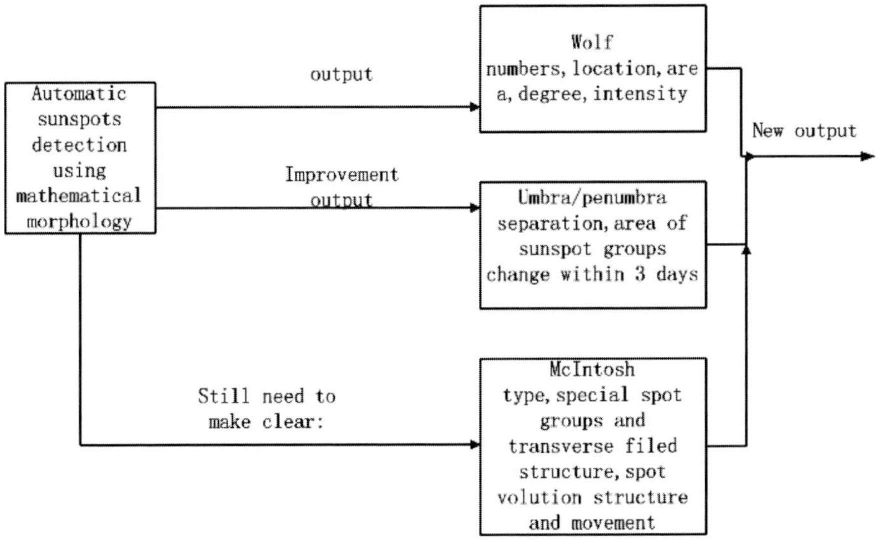

Figure 1. Sunspots automatic detecting and related forecast factors.

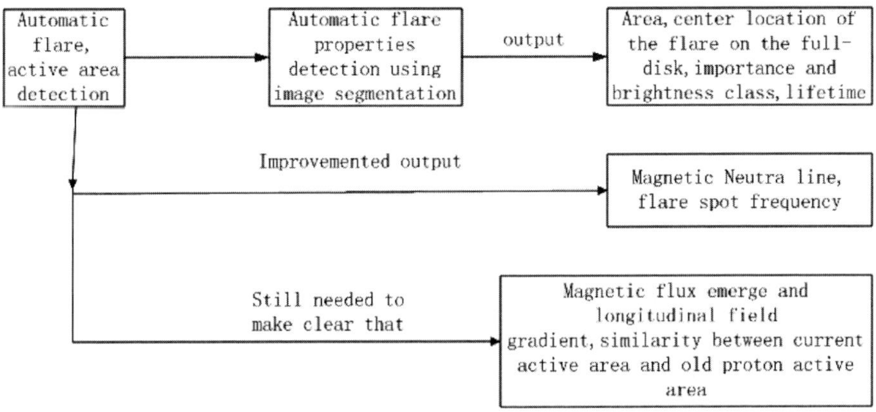

Figure 2. Active area automatic detecting and related forecast factors.

The soft X-ray related factors are: soft x-ray flux, integral peak flux of 1-8Å, flare duration time of x-ray.

3. Method of automatic detecting and forecasts or alerts of flare proton event

The flare proton event has been introduced. We build relationship between the methods of automatic detecting on solar events or phenomenons and their forecasts or alerts. Figure 1 presents possible output and possible output after improvement by means of automatic detection. Some factors may need to be made clear by further human interference. Figures 2 and 3 have similar structures and meanings.

4. Discussion

From Figures 1-4 we can see that half of all the factors can be automatically detected. Sixty percents of them can be automatically gained after the improvement. One third of

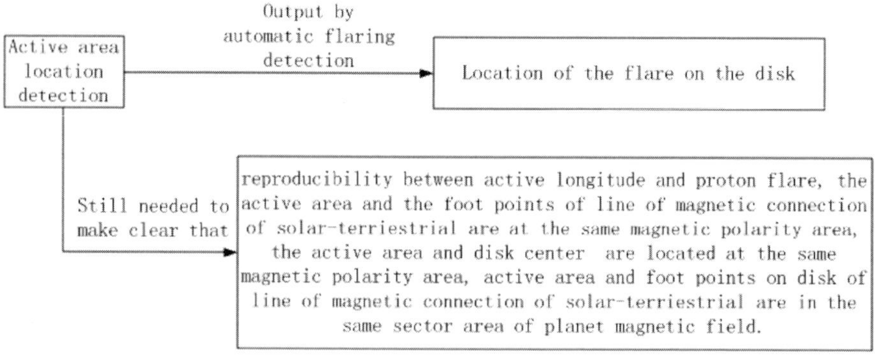

Figure 3. Active area location detecting and related forecast factors.

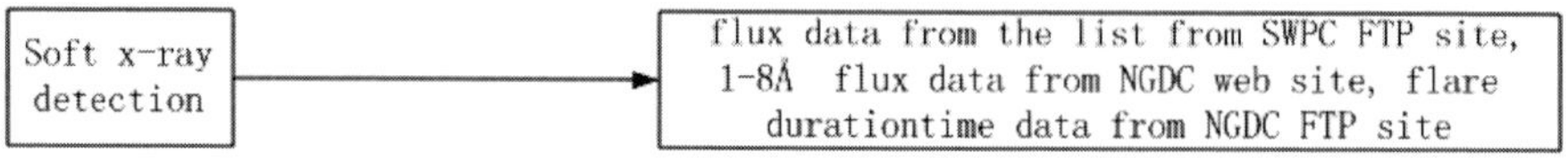

Figure 4. X-ray detecting and related forecast factors.

the remains still need to be make clearer. We can find that some factors of the one third can be automatically obtained by downloading data, computed, others still need to be further analyzed by human.

I thank the editor for his careful and earnest. The work is supported by National Natural Science Foundation of China under grants 60673158.

References

Benkhalil, A., Zharkova, V. V., Zharkov, S., & Ipson, S. 2006, *Solar Phys.*, 235:87-106
Curto, J. J., Blanca, M., & Martinez, E. 2008, *Solar Phys.*, 250:411-429
Ming Qu, Frank Y. Shih, Ju Jing, & Haimin Wang 2003, *Solar Phys*, 217:157-172
Wang Jialong 1999, *Chinese Journal of Astronomy and Astrophysics*, 19, 4, 412:418

Session III

Solar and stellar cycles and variability on century timescale

Chair: J. Stenflo

Solar and Stellar Variability: Impact on Earth and Planets
Proceedings IAU Symposium No. 264, 2009
A. G. Kosovichev, A. H. Andrei & J.-P. Rozelot, eds.

doi:10.1017/S1743921309992511

Sunspot cycles and Grand Minima

Dmitry Sokoloff[1], Rainer Arlt[2], David Moss[3], Steven H. Saar[4] and Ilya Usoskin[5]

[1]Department of Physics, Moscow State University, 119991, Moscow, Russia
email: sokoloff@dds.srcc.msu.su

[2]Astrophysical Institute Potsdam, An der Sternwarte 16, D-14482 Potsdam, Germany.Potsdam
email: rarlt@aip.de

[3]School of Mathematics, University of Manchester, Manchester M13 9PL, UK
email: David.Moss@manchester.ac.uk

[4]Harvard-Smithsonian Center for Astrophysics, Cambridge, MA, USA
email: saar@cfa.harvard.edu

[5]Sodankylä Geophysical Observatory (Oulu Unit), 90014 University of Oulu, Finland
email: ilya.usoskin@oulu.fi

Abstract. Observational data concerning the long-term history of cyclic solar activity as recorded in sunspot and isotopic data are discussed in the context of solar dynamo theory. In particular, a simple dynamo model based on differential rotation and the mirror asymmetry of convection with random fluctuations of dynamo governing parameters is shown to reproduce some basic features of the solar magnetic activity evolution.

Keywords. Sun: magnetic fields, (Sun:) sunspots, stars: magnetic fields.

1. The general form of the record of solar activity

The solar activity cycle is usually considered as a quasiperiodic process with cycles relatively similar to each other. The standard physical explanation for the solar activity cycle is that the phenomenon underlying the cycle is the propagation of a wave of quasi-stationary magnetic field which in turn is excited by the solar dynamo. The classical explanation of the solar dynamo (Parker, 1955) is that solar differential rotation produces toroidal magnetic field from poloidal, while mirror-asymmetric convective flows produce poloidal field from toroidal (the "α-effect"), thus closing the chain of magnetic field self-excitation. This scheme results in latitudinal propagation of a wave of toroidal magnetic field (the dynamo wave). Appropriate tuning of the dynamo governing parameters allows equatorialward propagation of the dynamo wave that fits various observational feature of the activity wave. However, the link between the concepts of solar dynamo and helioseismological data remains an area of intensive discussion in scientific community (e.g. Kosovichev, 2008). Another disputable point here is to what extent a particular parametrization in terms of the α-effect, originating from Parker's paper and developed as a physical concept by Steenbeck, Krause and Rädler (Krause & Rädler, 1980), covers the range of possible physical dynamo mechanisms. For example, should the idea of a transport dynamo based on meridional circulation (e.g. Dikpati & Charbonneau, 1999) be considered as an independent option? In any case, the physical nature of solar activity on the time-scales of several activity cycles can be considered as a relatively understood phenomenon.

Of course, successive solar cycles differ slightly from one another, in amplitude, length and other characteristics. A physical explanation of the corresponding dynamics as well as the ability to predict the forthcoming cycle based on knowledge of recent cycles is being addressed by a number of approaches (see e.g. Kitiashvili & Kosovichev, 2008). This remains however a challenge for solar physics, but it looks plausible that the topic can be developed based on available ideas from solar dynamo theory and helioseismology.

Solar activity is a much more complicated phenomenon than just a quasi-regular cycle. The famous Maunder Minimum interrupted the normal sequence of activity cycles from the mid-XVIIth to the beginning of the XVIIIth century. Fortunately, it was the age of the birth of modern astronomy, and the Maunder Minimum was relatively well observed, and the records survived in the astronomical archives. A personal important scientific contribution from the King of France, Louis XIV, should be mentioned here as a rare example of this kind. For a review of the archive findings in the context of the solar dynamo see e.g. Sokoloff (2005). It is important that cyclic activity is visible, to some extent at least, at the end of the Maunder Minimum and that the activity wave then becomes substantially asymmetric, being visible almost solely in the Southern solar hemisphere, in contrast to the symmetric modern strong cycles. Another occurrence of an asymmetric activity wave is known from Gassendi's observations for the epoch just before the Maunder Minimum. Solar activity looks asymmetric also during the lost cycle however sunspots occurred that time preferably in the Northern hemisphere.

In an analysis of the cosmogenic isotope data, which provides a standard proxy for solar activity in the past, Usoskin *et al.*, 2007 showed that phenomena comparable with the Maunder Minimum occurred from time to time during the history of solar activity. A general name for the phenomenon is now Grand Minima of solar activity. Cosmogenic isotope data give some hints that, apart from Grand Minima, epochs of Grand Maxima of solar activity can be also observed, a typical example being the high solar activity in the second half of the 20th century. Of course, the further back in time a Grand Minimum is, the less precise our knowledge concerning its properties. As far as can be seen from the data available (Miyahara *et al.*, 2006) the Maunder Minimum represents a typical Grand Minimum in its general properties.

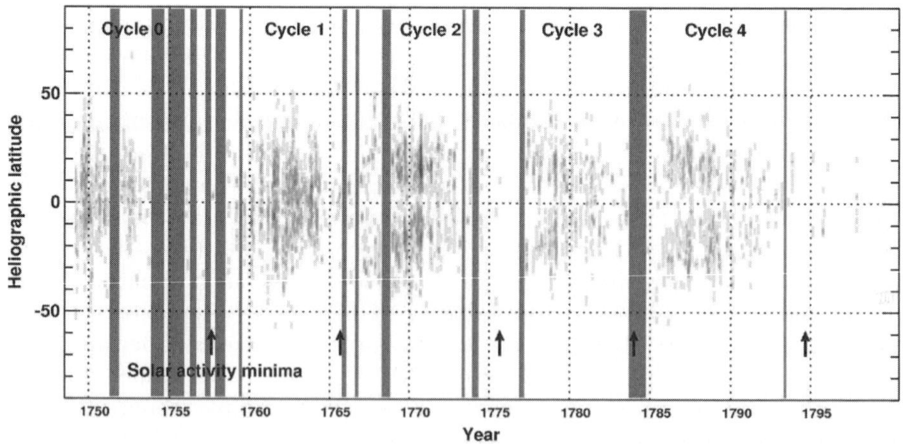

Figure 1. The historical butterfly diagram for XVIII century obtained from the Staudacher data. Sunspot density is given in scales of gray. Black vertical strips indicates epochs for which the data are insufficient to get the butterfly diagram.

The physical nature of Grand Minima in general, and the Maunder Minimum in particular, is much less clear than the nature of the solar cycle itself. Indeed, solar dynamo models (Brandenburg *et al.*, 1989, Jennings & Weiss, 1991, Covas *et al.*, 1998) as well as first laboratory dynamo experiments (Ravelet *et al.*, 2008) show that activity minima occur from time to time in the temporal evolution of particular dynamos, although a physical scenario for such events that is more or less generally accepted in the solar physics community is still lacking. An important point here is that as far as it follows from the available archive data the beginning of the Maunder Minimum was quite abrupt. It remains unclear to what extent a prediction, based on the previous record of solar activity, of a future Grand Minimum can be made. We stress that the prediction of the parameters a new solar cycle and those of a new Grand Minimum may be physically different undertakings. The role of a *local* minima of solar activity such as the Dalton Minimum, as possible precursors of Grand Minima, appears an attractive topic in this field.

One more point is that the temporal sequence of Grand Minima appears quite random; at least it is difficult to isolate a periodic behaviour. On the other hand, the intervals between successive minima vary too strongly to be referred as a realization of a Poisson random process (Usoskin *et al.*, 2007).

In the above context, the XVIIIth century was usually considered as a normal epoch with a typical record of solar activity. Some doubts in this respect came from the concept of a lost activity cycle which seems to have occured at the very end of the century (Usoskin *et al.*, 2003, Usoskin *et al.*, 2009b). Recent archive findings (Arlt, 2009) have confirmed that this impression of normality needs to be revised substantially. First of all, the butterfly diagram (Fig. 1) reconstructed from the Staudacher data for the XVIIIth century shows some features which looks peculiar to this century and are absent in the contemporary record of solar activity. In particular, butterflies for some cycles are concentrated near to the solar equator while others are separated from the equator by a strip of low activity. We discuss below how these features can be embedded in the framework of solar dynamo theory.

2. Dipolar versus quadrupolar modes of solar dynamo

A magnetic field of dipole-like symmetry is not the only magnetic configuration which can be excited by solar and stellar dynamos. Brandenburg *et al.* (1989) were the first to demonstrate that alpha-quenched mean field dynamos in spheres can move through regions with stable dipole-like and quadrupole-like, or even mixed, parity solutions as the dynamo parameters are changed. A rich dynamical behaviour of solar dynamos was discussed in context of zero or one dimensional models by e.g. Weiss *et al.* (1984) and Tobias *et al.* (2005), while experiments with more realistic solar dynamos show that the preferred symmetry of solutions near marginal excitation can be sensitive to quite small changes in the model (e.g. Moss, 1999, Dikpati & Gilman, 2001, Bonnano *et al.*, 2002, Jouve & Brun, 2007), or at least the excitation conditions for even and odd modes are similar (e.g. Moss & Brooke, 2000).

Recently Moss *et al.* (2008) summarized evidence that neither dynamo theory nor the observational data give strong support to the idea that stellar magnetic fields must have dipolar rather than quadrupolar symmetry with respect to the stellar equator. They demonstrated the spontaneous transition of the dynamo-excited magnetic field from one symmetry type to another and explored observational tests to distinguish between the two types of magnetic field symmetry, and thus detect the presence of quadrupolar magnetic symmetry in stars. A complete absence of quadrupolar symmetry would present a distinct challenge for contemporary stellar dynamo theory.

In our opinion, the archival butterfly diagram reconstructed by Arlt (2009) for the XVIIIth century from the Staudacher data (Fig. 1) gives a reasonably meaningful hint that the solar magnetic configuration demonstrates from time to time substantial excursions towards quadrupole-like parity. Of course, the historical data do not contain information concerning the polarity of sunspot groups, and so a direct verification of the Hale polarity law is impossible. The point is, however, that any toroidal magnetic field with dipole-like symmetry has to vanish at the solar equator, while a quadrupole-like field usually has a maximum of toroidal field at the equator, and is very unlikely to have zero amplitude there (Fig. 2). Such a maximum can be compared with the maximum at the solar equator which is visible in Fig. 1 for cycle I (and possibly for cycle 0), in contrast to a clear minimum at the equator visible for cycles 3 and 4.

Moss *et al.* (2008) argue that information on stellar cycle properties may offer another window on magnetic symmetry. Dynamos operating in stars in significantly supercritical regimes can give rise to complex phenomena, such as multiply periodic solutions and beating between modes with different symmetry properties, for example. Beating between dipolar and quadrupolar components might explain the apparent multiple periods observed in some stars. A cyclic quadrupolar component could show up as an amplitude and/or period modulation superimposed on a main dipole cycle period, or as the primary period itself if the quadrupolar symmetry dominates. Using cycle information, there is some limited evidence that fields with quadrupolar symmetry *may* have already been observed in some stars. A subclass of moderate-activity cool dwarfs exhibit multiple cycle periods in long-term Ca II HK measurements (Baliunas *et al.*, 1995). Similar secondary periods $P_{\rm cyc}(2)$ are seen in long-term photometric data of active single dwarfs as well (e.g. Messina & Guinan, 2002, Oláh & Strassmeier, 2002. These $P_{\rm cyc}(2)$ *may* represent the cycle period of the quadrupolar dynamo component $P_{\rm cyc}({\rm Q})$ (if the dipole is dominant), a beat frequency between $P_{\rm cyc}({\rm Q})$ and $P_{\rm cyc}({\rm D})$, the dipole period (if the quadrupole is dominant), or merely a long-term modulation of the amplitude of $P_{\rm cyc}({\rm D})$ due to non-linear effects.

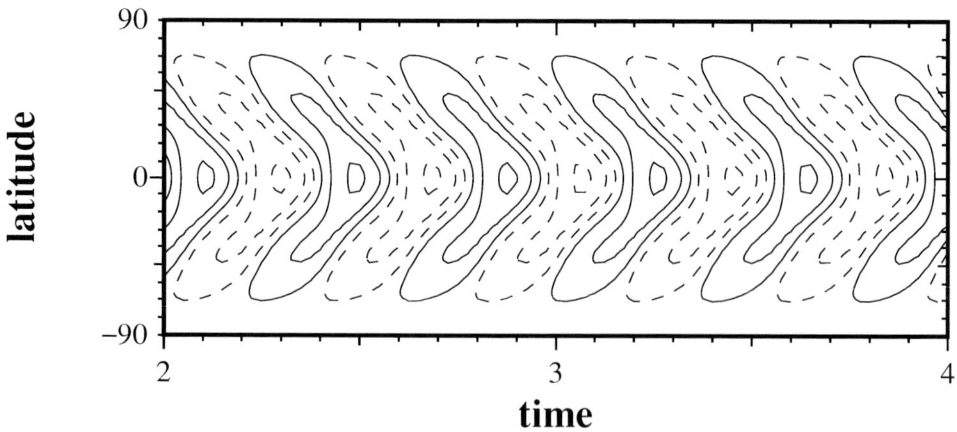

Figure 2. A simulated butterfly diagram for the toroidal magnetic field in a simple Parker migratory dynamo with algebraic α-quenching and quadrupole-like symmetry. Contours show toroidal field strength, with solid contours indicating positive values and broken contours negative values. The contours clearly show a maximum at the solar equator. Time is given in arbitrary dimensionless units not years.

3. Fluctuations in the solar governing parameters and Grand Minima

We stressed above that dynamo theory provides several options that might explain the features observed in the long-term evolution of solar activity including Grand Minima, asymmetric and quadrupole magnetic configurations, beating, lost cycles etc. One could expect that each particular phenomenon of that kind would require a specific explanation in terms of this theory.

The most straightforward idea here is to recognize that the α-effect, being the result of the electromotive force averaged over turbulent vortices, can contain a fluctuating contribution (Hoyng, 1993, Hoyng *et al.*, 1994, Ossendrijver & Hoyng, 1996). The idea can lead to events similar to the Maunder Minimum on the timescale of centuries (see e.g. Brandenburg & Spiegel, 2008).

Moss *et al.* (2008), Usoskin *et al.* (2009a) investigated the long-term dynamics of solar activity by confronting the predictions of a Parker migratory dynamo model containing a random contribution to the α-coefficient with the available data concerning the sequence of Grand Minima and Maxima, to recognize that this simple model provides in a proper parametric range all bulk of the phenomena under discussion similar at least qualitatively to that one observed on the Sun. The intention was to test whether a simple physical model can reproduce the basic phenomena of the long-term solar dynamics. Of course, a detailed explanation of the phenomena needs a much more realistic model, including at least a 2D description of the solar magnetic field, realistic solar rotation curve, etc. Moreover, it is not excluded *a priori* that the phenomena could have some alternative explanation. The analysis was based on a simple illustrative model, rather than on something more realistic, in order to isolate physical phenomena and to take into account the quite limited status of the actual observational information.

Random fluctuations of the dynamo governing parameters can be instructive in explaining the stochastic features of short-term dynamics of solar activity, on the timescale of a few solar cycles (e.g. Moss *et al.*, 1992, Hoyng *et al.*, 1994). In contrast, Moss *et al.* (2008), Usoskin *et al.* (2009a) considered global fluctuations of α on the temporal scales of order the cycle length and spatial (latitudinal) scale of the whole solar hemisphere. The analysis did not include variations of short time and latitudinal extent on the scales

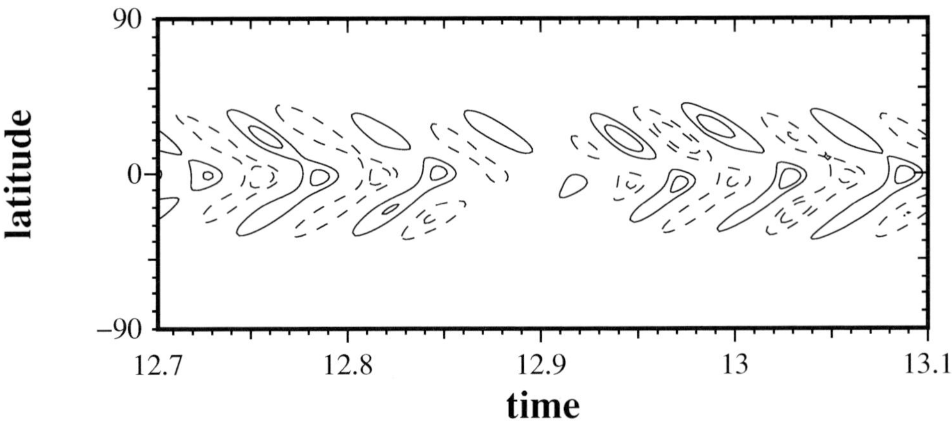

Figure 3. A simulated butterfly diagram for the toroidal magnetic field in a simple Parker migratory dynamo with algebraic α-quenching and fluctuating $\alpha(\theta, t)$ and basically quadrupole symmetry. Contours show toroidal field strength, with solid contours indicating positive values and broken contours negative values. The diagram suggests the presence of various types of complicated dynamics, such as transition to mixed parity and quadrupole-like configurations, lost cycles etc. Time is given in arbitrary dimensionless units not years.

of turbulent vortices, which are obviously important for the short-term dynamics of solar activity. The presence of long-term variations in the alpha-coefficient has been reported from analysis of direct numerical simulations by Brandenburg & Sokoloff (2002) and Otmianowska-Mazur *et al.* (2006).

Here we present some new results for solar dynamo models with fluctuating dynamo parameters, that are intended to be directly related to the current observational situation. Fig. 3 presents the butterfly diagram for a model for a particular choice of parameters which demonstrates various transitions in parity, quadrupole-like configurations and lost cycles. This example gives the evolution of a model which demonstrates strict quadrupole-like symmetry in the absence of fluctuations. Fig. 4 gives a similar example for a model with basically dipole-like symmetry. Fig. 5 shows the corresponding evolution of parity and magnetic energy.

We stress that the above figures significantly exploit the concept that sunspot formation is a threshold phenomenon; the butterfly diagrams from toroidal field contours are only plotted for fields above a certain threshold strength. Cyclic behaviour absent in a particular hemisphere during an activity cycle in the above figures persists still in form of a very weak magnetic field oscillations which are not strong enough to appear in the butterfly diagrams. Possibly, this is an option to explain that some tracers of cyclic sunspot behaviour (Frick *et al.*, 1997), and even of apparent solar diameter (Nesme-Ribes *et al.*, 1995), are recognizable from the archive data for the Maunder Minimum.

4. The record of solar activity on the time-scale of millennia

Moss *et al.* (2008) discuss the possibility that solar activity cannot be considered as a stationary random process even on the timescale of [tens of thousands of] years, as suggested by the following test. Let us consider a time series $f_n = f(t_n)$ where the instants of observations $t_i = n\tau$. If the length of the time series is sufficient to consider it as a stationary random process then the random quantity $g_n = \Sigma_0^n f_i$ is expected to behave as $g_n = \langle f \rangle n + h(n)$ where $\langle f \rangle$ is the mean of the random process f_n and $h(n)$ is a function that grows more slowly then n. By plotting g_n against n and comparing the results with a linear trend we can check whether the time series is long enough to be consider as a realization of a stationary random process (for applications of this

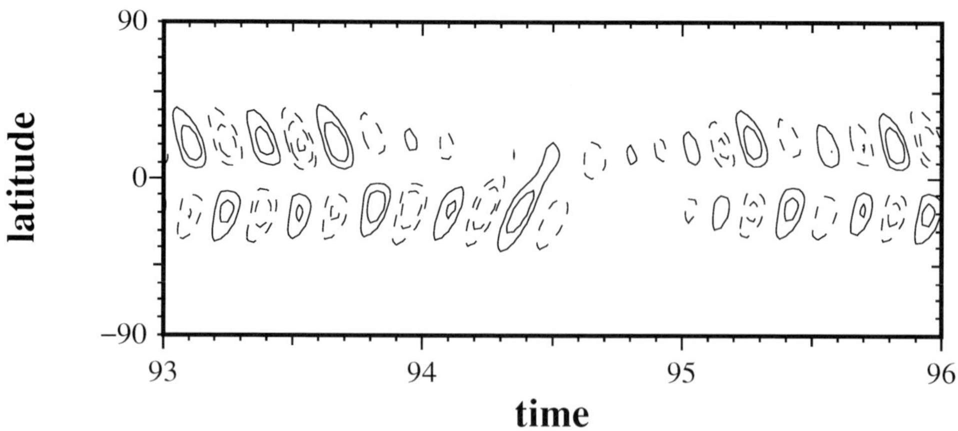

Figure 4. Simulated butterfly diagram for toroidal magnetic field in simple Parker migratory dynamo with algebraic α-quenching and fluctuating $\alpha(\theta, t)$ and basic dipole-like symmetry. Contours show toroidal field strength, with solid contours indicating positive values and broken contours negative values. Time is given in arbitrary dimensionless units not years.

test in solar physics, see Kutvitskii *et al.*, 2009). The test when applied to the record of solar activity that was exploited by Usoskin *et al.* (2009a) to provide a sequence of Grand Minima, demonstrates substantial deviations from [a linear behaviour]. The corresponding plot for dynamo models with fluctuating dynamo governing parameters give similar deviations from a straight line (Moss *et al.*, 2008). This result appears quite unexpected because the underlying physics do not contain any timescale longer than the cycle length, and thus we are faced with a nontrivial behaviour of a noisy system. Unfortunately, the solar activity record based on isotopic data is too short to claim that these dynamics inevitably follow from the observations.

Conclusion of this paper can be summarized as follows. Observational data concerning cyclic solar activity as recorded in sunspot and isotopic data demonstrate a complicated long-term history which includes Grand Minima, asymmetric cycles or even cycles of quadrupole symmetry and lost cycles. A simple dynamo model based on differential rotation and the mirror asymmetry of convection with random fluctuations of dynamo

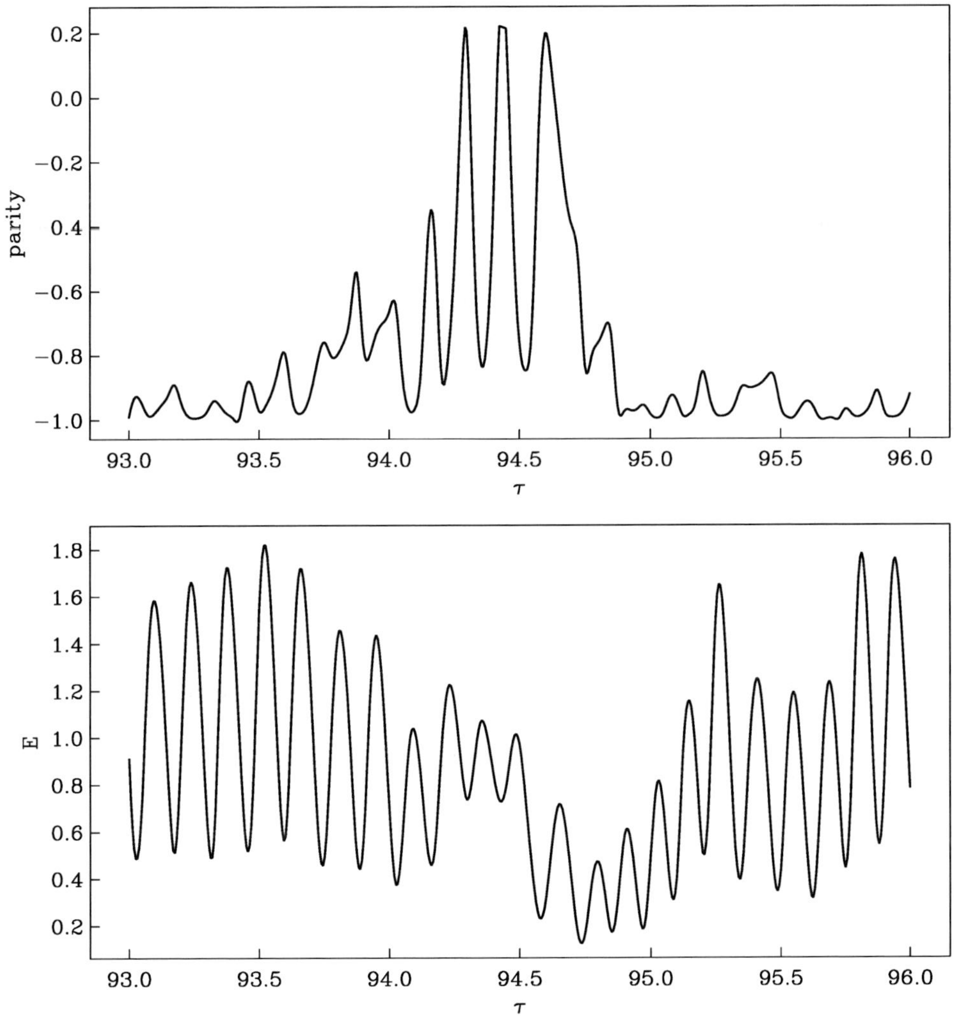

Figure 5. Temporal evolution of magnetic field parity (upper) and energy (lower panel) for the model illustrated in Fig. 4. Parity $P = -1$ means a dipole-like configuration and $P = +1$ corresponds to the quadrupole-like case. Energy is measured in units of equipartition energy.

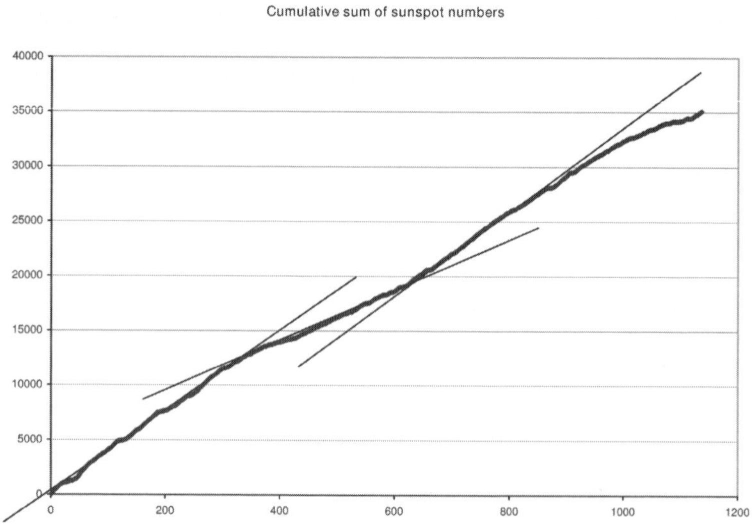

Figure 6. Cumulative sunspot number against time. The straight lines approximate the data locally.

governing parameters is able to reproduce some basic features of the solar magnetic activity evolution.

Acknowledgements

DS is grateful to RFBR under grants 09-02-08154 and 07-02-00127, Universidade do Estado do Rio de Janeiro and the Organizing Committee for financial support for his participation in the meeting. D.M. and D.S. thank the Finnish Academy of Science and Letters (Väisälä foundation) for supporting their visits to the University of Oulu.

References

Arlt, R. 2009, *Solar Phys.*, 255, 143

Baliunas, S. L., Donahue, R. A., Soon, W. H., Horne, J. H., Frazer, J., Woodard-Eklund, L., Bradford, M., Rao, L. M., Wilson, O. C., Zhang, Q., Bennett, W., Briggs, J., Carroll, S. M., Duncan, D. K., Figueroa, D., Lanning, H. H., Misch, T., Mueller, J., Noyes, R. W., Poppe, D., Porter, A. C., Robinson, C. R., Russell, J., Shelton, J. C., Soyumer, T., Vaughan, A. H., & Whitney, J. H. 1995, *ApJ*, 438, 269

Bonanno, A., Elstner, D., Rüdiger, G., & Belvedere, G. 2002, *A&A*, 390, 673

Brandenburg, A., Krause, F., Meinel, R., Tuominen, I., & Moss, D. 1989, *A&A*, 213, 411

Brandenburg, A & Sokoloff, D. 2002, *Geophys. Astrophs. Fluid Dyn.*, 96, 319

Brandenburg, A. & Spiegel, E. A. 2008, *AN*, 329, 351

Covas, E., Tavakol, R., Tworkowski, A., & Brandenburg, A. 1998, *A&A*, 329, 350

Dikpati, M. & Charbonneau, P. 1999, *ApJ*, 518, 508

Dikpati, M. & Gilman, P. A. 2001, *ApJ*, 559, 428

Frick, P., Galyagin, D., Hoyt, D., Nesme-Ribes, E., Shatten, E., Sokoloff, D., & Zakharov, V. 1997, *A&A*, 328, 670

Hoyng, P. 1993, *A&A*, 272, 321

Hoyng, P., Schmitt, D., & Teuben, L. J. W. 1994, *A&A*, 289, 265

Jennings, R. L. & Weiss, N. O. 1991, *Mon. Not. Roy. Astron. Soc.*, 252, 249

Jouve, L. & Brun, A. S., 2007, *A&A*, 474, 239

Kitiashvili, I. & Kosovichev, A. G. *ApJ*, 688, L49

Kosovichev, A. G. 2008, *Adv. Sp. Res.*, 41, 830

Krause, F. & Rädler, K.-H. 1980, *Mean-Field Magnetohydrodynamics and Dynamo Theory* (Oxford: Pergamon Press)

Kutvitskii, V. A., Semikoz, V. B., & Sokoloff, D. D. 2009, *Astron. Rep.*, 53, 166

Messina, S. & Guinan, E. F. 2002, *A&A*, 393, 225

Miyahara, H., Sokoloff, D., & Usoskin, I. G. 2006, in: W.-H. Ip & M. Duldig (eds.), *Advances in Geosciences 5* (Singapore: World Scientific), p. 1

Moss, D. 2008, *MNRAS*, 306, 300

Moss, D., Brandenburg, A., Tavakol, R. K., & Tuominen, I. 1992, *A&A*, 265, 843

Moss, D. & Brooke, J. 2000, *MNRAS*, 315, 521

Moss, D., Saar, S. Y., & Sokoloff, D. 2008, *MNRAS*, 388, 420

Moss, D., Sokoloff, D., Usoskin, I., & Tutubalin, V. 2008, *Solar Phys.*, 250, 221

Nesme-Ribes, E., Frick, P., Zakharov, V., Ribes, J.-C., Sokoloff, D., Vigouroux, A., & Laclare, F. 1995, *C. R. Paris*, 321, 525

Oláh, K. & Strassmeier, K. G. 2002 *AN* 323, 361

Ossendrijver, A. J. H. & Hoyng, P. 1996, *A&A*, 313, 959

Otmianowska-Mazur, K., Kowal, G., & Hanasz, M. 2006, *A&A*, 445, 915

Parker, E. N. 1955, *ApJ*, 122, 293

Ravelet, F., Berhanu, M., Monchaux, R., Aumaitre, S., Chiffaudel, A., Daviaud, F., Dubrulle, B., Bourgoin, M., Odier, Ph., Plihon, N., Pinton, J.-F., Volk, R., Fauve, S., Mordant, N., & Petrelis, F. 2008, *PRL*, 101, 074502

Sokoloff, D. D. 2005, *Solar Physics*, 224, 145

Tobias, S. M., Weiss, N. O., & Kirk, V. 2005, *MNRAS*, 273, 1150

Usoskin, I. G., Sokoloff, D., & Moss, D. 2009a, *Solar Phys.*, 254, 345

Usoskin, I. G., Mursula, K., Arlt, R., & Kovaltsov, G. A. 2009b, *ApJ*, 700, L154

Usoskin, I. G., Mursula, K., & Kovaltsov, G. A. 2003, *A&A* 403, 743

Usoskin, I. G., Solanki, S. K., & Kovaltsov, G. A. 2007, *A&A*, 471, 301

Weiss, N. O., Cattaneo, F., & Jones, C. A. 1984, *Geophys. Astrophys. Fluid Dyn.*, 30, 305

Solar and Stellar Variability: Impact on Earth and Planets
Proceedings IAU Symposium No. 264, 2009
A. G. Kosovichev, A. H. Andrei & J.-P. Rozelot, eds.

doi:10.1017/S1743921309992523

Stellar magnetic cycles

A. F. Lanza

INAF-Osservatorio Astrofisico di Catania, Via S. Sofia, 78 – 95123 Catania, Italy
email: nuccio.lanza@oact.inaf.it

Abstract. The solar activity cycle is a manifestation of the hydromagnetic dynamo working inside our star. The detection of activity cycles in solar-like stars and the study of their properties allow us to put the solar dynamo in perspective, investigating how dynamo action depends on stellar parameters and stellar structure. Nevertheless, the lack of spatial resolution and the limited time extension of stellar data pose limitations to our understanding of stellar cycles and the possibility to constrain dynamo models. I briefly review some results obtained from disc-integrated proxies of stellar magnetic fields and discuss the new opportunities opened by space-borne photometry made available by MOST, CoRoT, Kepler, and GAIA, and by new ground-based spectroscopic or spectropolarimetric observations. Stellar cycles have a significant impact on the energetic output and circumstellar magnetic fields of late-type active stars which affects the interaction between stars and their planets. On the other hand, a close-in massive planet could affect the activity of its host star. Recent observations provide circumstantial evidence of such an interaction with possible consequences for stellar activity cycles.

Keywords. Sun: activity, stars: late-type, stars: activity, stars: magnetic fields, planetary systems.

1. Introduction

The eleven year cycle of solar activity manifests its effects in the circumsolar and interplanetary environments and has a remarkable impact on the Sun-Earth relationship. Understanding stellar magnetic cycles will allow us to look at this important phenomenon from a wider perspective. In this review, I shall focus on a few aspects of solar and stellar cycles that I deem broadly interesting in this context, although without discussing their impact on the circumstellar environments and planets. Several reviews have recently covered the topics that I shall not treat here; among others, Strassmeier (2005), Berdyugina (2005), Hall (2008), Rempel (2008), and Strassmeier (2009). For stellar cycles from X-ray proxies, see, e.g., Güdel (2004) and Favata *et al.* (2008). For the impact of stellar activity on planets see the review by Ribas (2010) in the proceedings of this Symposium.

2. The solar cycle and the solar dynamo

Magnetic fields produce localized perturbations of the flux of energy and momentum in the solar atmosphere that are called *active regions* (hereinafter ARs). They are characterized by different phenomena, according to the mean intensity of the magnetic field. Fields of the order of 0.1 T produce cool features in the photosphere, called sunspots, while fields of the order of 0.01 T are associated with a heating of the atmosphere leading to bright photospheric facular patches and chromospheric plages. Transient releases of magnetic energy on timescales from a few minutes to hours manifest themselves as solar flares. While the formation and evolution of individual active regions are largely unpredictable, their global behaviour reveals a remarkable regularity. The total area of the sunspots varies in a cyclic manner with a period of ~ 11 yr. At the beginning of each cycle, sunspots appear at latitudes of $\approx 35^{\circ}$ in two localized belts, one in each hemisphere, that migrate toward the equator, reaching a latitude of $15^\circ - 20^\circ$ at the maximum of the total sunspot area and of $5^\circ - 10^\circ$ at the next minimum, when a new couple of active belts appears at high latitudes marking the beginning of a new cycle. In addition to the strong magnetic fields localized inside ARs, the Sun displays a global dipolar magnetic field with an intensity of $\sim 10^{-4}$ T, symmetric around the solar rotation axis, which cyclically reverses its polarities close to the maximum of sunspot activity.

The rotation of the solar convection zone reacts in a cyclic way to the variation of the intensity and spatial distribution of the magnetic field, leading to the phenomenon of the *torsional oscillations* (Howe 2009). The frequencies of the p-mode oscillations are also affected by the magnetic field and display relative variations of the order of 10^{-4} along the 11-yr cycle whose origin is not completely understood yet (cf., e.g., Salabert *et al.* 2009).

The time series of sunspot data, which dates back to the XVII century, has revealed that the amplitude and length of the solar cycle are modulated on different timescales, among which the Gleissberg cycle is notable. It consists of a modulation of the amplitude of the 11-yr cycles on a century timescale (e.g., Kollath & Olah 2009). Investigating time series of proxies extending back to 10^4 yr ago, such as the cosmogenic isotopes ^{10}Be or ^{14}C, it is possible to investigate the level of solar activity in the distant past. In combination with sunspot data, these studies have revealed intervals of very low solar activity, called *grand minima* (see Usoskin *et al.* 2007; Sokoloff *et al.* 2010). At the other end of the time scale, there are short-term cyclic variations of the total sunspot area with periods of $150 - 160$ days that sometimes appear close to the maximum of the 11-yr cycle (e.g., Oliver *et al.* 1998; Krivova & Solanki 2002). They were first detected in the occurrence of major flares, which are closely associated to the emergence of new magnetic flux and the formation of new sunspots, by Rieger *et al.* (1984) and are therefore called *Rieger cycles.*

A theoretical framework to interpret the solar cycle is offered by dynamo models based on mean-field magnetohydrodynamics (see Charbonneau 2005; Weiss & Thompson 2009; and references therein). The global solar dipole field is stretched by the differential rotation giving rise to an intense field in the azimuthal direction which emerges under the action of magnetic buoyancy and leads to the formation of sunspots in the photosphere and magnetic loops in the outer atmosphere. However, the poloidal and azimuthal fields are subject to a turbulent diffusion, hence the poloidal field will decay, leading eventually to the decay of the azimuthal field, unless some mechanism can regenerate the poloidal field itself. Such a mechanism is much more subtle than differential rotation and is thought to be associated to the action of the Coriolis force on turbulent convection which imparts a non-vanishing mean helicity to the turbulent velocity field of the plasma. Since the magnetic field is frozen into the plasma, such a mean helicity is capable to regenerate a poloidal field from an azimuthal field, thus allowing a dynamo to operate (see, e.g., Weiss & Thompson 2009, and references therein, for details). This is the so-called α effect.

The role of the different effects mentioned above is summarized in the definition of the non-dimensional *dynamo number*. Indicating with $\Omega' = \partial\Omega/\partial r$ the typical amplitude of the differential rotation as specified by the radial gradient of the angular velocity (see below), and by α_0 and $\eta_{\rm t}$ the typical amplitudes of the α effect and the turbulent diffusivity, respectively, we define the dynamo number as: $D \equiv \alpha_0 \Omega' R^4/\eta_{\rm t}^2$, where R is the radius of the star. In order for the induction effects represented by the parameters α_0 and Ω' to overcome the turbulent diffusivity $\eta_{\rm t}$, the dynamo number must be greater than some critical value $D_{\rm c}$, which depends on the boundary conditions, the geometry of the dynamo shell and the spatial distribution of the induction effects. In other words, only when $D \geqslant D_c$ the dynamo can work and we observe a large-scale magnetic field, otherwise any large-scale field is simply diffused away on a time scale $\tau_d \approx R^2/\eta_t$. When the magnetic field is amplified enough, it starts to react back on the small scale turbulent motions and the large scale velocity fields. Such a feedback tends to reduce both the α effect (α quenching) and the differential rotation, thus opposing any further amplification of the field. When a feedback is included, a dynamo model is said to be magnetohydrodynamic or non-linear, in constrast to the simpler linear (or kinematic) models that neglect it and assume a fixed α effect and differential rotation.

Kinematic dynamo models were proposed long ago for the Sun and account for the oscillation of the sunspot area and the cyclic reversal of the global poloidal field. They predict that the migration of the azimuthal field as the cycle progresses is in the equatorward direction if the product $\alpha\partial\Omega/\partial r$ is negative in the Northern hemisphere, while it is in the poleward direction if that quantity is positive. At the base of the solar convection zone, helioseismic measurements tell us that $\partial\Omega/\partial r > 0$ at latitudes lower than about 40°, while $\partial\Omega/\partial r < 0$ at higher latitudes (Thompson *et al.* 2003). This accounts for the observed direction of migration of the activity belts, if a negative α is assumed close to the base of the solar convection zone in the Northern hemisphere (see, e.g. Schlinchenmaier & Stix 1995). The scenario is consistent with a localization of the dynamo in the overshoot layer immediately below the solar convection zone, where a strong shear, mainly in the radial direction, has been detected by helioseismology (e.g., Thompson *et al.* 2003). Another property of such a layer is its subadiabatic stratification that keeps the magnetic field stable until it reaches a critical intensity of $\approx 1-10$ T, beyond which it emerges toward the photosphere (e.g., Caligari *et al.* 1995). Nevertheless, the possibility of a dynamo action distributed throughout the convection zone or even localized in the subphotospheric shear layer where $\partial\Omega/\partial r < 0$, is not ruled out. It is possible that the activity observed at the surface of the Sun results from the contributions of different kinds of dynamos which differ for their location, the storage mechanism of their azimuthal field and the relative amplitude of the field fluctuations with respect to the mean field (cf. Durney *et al.* 1993; Brandenburg 2005; Kosovichev 2009). Finally, it is important to mention the role of the meridional circulation that may act as a sort of conveyor belt transporting the magnetic field between regions where different induction processes take place, e.g., from the overshoot layer, where the shear is mainly localized, into the convection zone, where the α effect is possibly dominant. Therefore, the meridional circulation affects the basic properties of dynamo models including the period of the cycle and the direction and rate of migration of the mean azimuthal field.

A fundamental question concerns the possibility of understanding a complex and highly non-linear system such as the solar (or a stellar) dynamo by means of simple parametric models like mean-field models. They drastically reduce the number of degrees of freedom of the system which are of the order of 10^{27} (Canuto 2000) and apply highly simplified parametric dependences to account for non-linear effects. Yet, they seem to be quite successful in explaining the main characteristics of the solar cycle. The reason for their success and the degree up to which we may hope to extend the comparison between models and observations has been recently discussed by Spiegel (2009).

3. Stellar activity cycles

3.1. *Stellar cycles from chromospheric proxies*

Stellar discs are not resolved by current observations, therefore, we need suitable proxies for stellar magnetic fields that can be derived from disc-integrated photometric or spectroscopic measurements. In the Sun, the transits of ARs across the disc during its rotation produce characteristic flux variations. Cool spots produce flux decreases while faculae induce an increase of the flux with a remarkable centre-to-limb dependence because their contrast is greater close to the limb. The Total Solar Irradiance, i.e., a measure of the solar bolometric flux at the distance of 1 AU, is a good proxy of the optical flux modulation induced by ARs. Its variations have now been monitored for almost three activity cycles. Owing to the predominant effect of the faculae, the Sun at the maximum of activity is brighter by ~ 0.1 percent than at the minimum (Fröhlich & Lean 2004). In the case of solar-like stars, it has been difficult to measure variations of the order of 0.001 mag from the ground over timescales of decades. Therefore, the use of long-term series of optical photometry to study activity cycles in stars with an activity level comparable to that of the Sun is still quite limited. The situation is different in the case of chromospheric proxies, notably the excess flux in the core of the Ca II H&K lines, which varies by ≈ 10 percent along the solar cycle (Hall & Lockwood 2004). Such a variation is measured by means of the so-called S index (Vaughan *et al.* 1978), whose systematic monitoring in a sample of ~ 100 late-type main-sequence single stars started ~ 40 years ago. This is the famous H&K project dedicated to the detection and study of stellar activity cycles. Its main results have been summarized by Baliunas *et al.* (1995, 1998). They find that about 60 percent of main-sequence stars with spectral types from F to early M show cycles. Cycles shorter than ~ 7 yr are never observer among stars with a rotation period longer than $\approx 15 - 20$ days. Several stars show two independent periodicities. About 25 percent of the sample show a variable S index, but without any obvious periodicity. This behaviour is more frequent among younger and more rapidly rotating stars, while stars with an age comparable to the Sun tend to show a cyclic behaviour or an almost constant flux. The latter have been dubbed *flat activity stars* and amount to ~ 15 percent of the sample. It was suggested that they could be in a phase of extended activity minimum, akin the Sun during its grand minima. However, recent studies have questioned such an interpretation, also because they span a range of chromospheric flux from below the solar minimum to near the higher levels of the present solar cycle (Hall & Lockwood 2004). Moreover, the S index is not a good proxy of magnetic activity at low levels of magnetic flux because the core fluxes of the Ca II H&K lines are dominated by the basal chromospheric heating due to acoustic waves produced by photospheric convection, irrespective of the magnetic flux. It is only when the chromospheric flux is significantly higher than the basal level for a given spectral type (and luminosity class) that the Ca II H&K lines are useful to study magnetic activity and cycles. Such considerations are important also for subgiant and giant stars which show a smaller frequency of cyclic variations (≈ 40 percent) and a more variable level of activity (Baliunas *et al.* 1998), possibly because their chromospheric flux is dominated by acoustic heating.

Chromospheric proxies can be used to measure the rotation period of the ARs in a star, when they have a sufficiently long lifetime to produce a unambiguous rotational modulation of the flux. Therefore, it is possible to correlate the level of chromospheric activity with the rotation period. Observations of the Sun as a star showed a variation of the rotation period in phase with the solar cycle, i.e., a decrease of the rotation period as the cycle progressed from minimum to maximum and then a further reduction followed by an abrupt increase when the activity went through the next minimum and a new cycle began (Donahue & Keil 1995). This was consistent with a migration of the activity belts from the higher latitudes toward the equator, followed by the appearance of new high latitude belts at the beginning of a new cycle. Nevertheless, the solar-type correlation is not the only one observed among the H&K project stars. It has been quite surprising to find that several stars show an *anti-solar* behaviour, i.e., the rotation period of their

ARs increases as the cycle progresses from minimum to maximum activity (see, e.g., the case of HD 114710; Donahue & Baliunas 1992). It is unlikely that such stars possess an anti-solar pattern of surface differential rotation, i.e., with the poles rotating faster than the equator (e.g., Rüdiger *et al.* 1998). Therefore, a possible explanation is that their ARs migrate poleward instead than equatorward. This may happen if stellar activity is not confined to latitudes close to the equator, but is well extended toward the poles. If the dynamo operates in a stellar overshoot region at high latitudes, the direction of migration of the dynamo wave is expected to be poleward in the case of an internal rotation profile similar to that of the Sun because $\partial\Omega/\partial r < 0$ at latitudes higher than $\sim 40°$ (Thompson *et al.* 2003). Of course, there can be another dynamo wave propagating at low latitudes toward the equator, but the poleward wave may dominate the observed stellar activity if the inclination of the stellar rotation axis is low, i.e., the star is viewed approximately pole-on.

As a development of such studies, Baliunas *et al.* (2006) proposed an interpretation for the different shapes of the modulation of the S index vs. time in stars showing cycles. Stars with a low level of activity often show an almost sinusoidal modulation, while more active objects display anharmonic variations. Baliunas *et al.* (2006) attribute such a difference to the different dynamo regimes occurring in different stars. Objects with a dynamo number close to the critical value $D \sim D_c$ are characterized by an almost sinusoidal variation of the S index and by a migration of the activity belts in latitude. On the other hand, stars with $D \gg D_c$ show cycles with a markedly non-sinusoidal modulation and their magnetic field does not migrate appreciably in latitude as their cycle progresses, i.e., there is a weak correlation between the rotation period of their ARs and the phase of the cycle. This interpretation is based on a dynamo model including an α-quenching mechanism. It assumes that in stars with a high level of activity the α effect is negligible for most of the cycle because the magnetic field intensity is so high as to strongly quench the mean helicity of the turbulence. In this regime, the dynamo wave does not migrate appreciably because its rate and direction of migration are controlled by the quantity $\alpha\partial\Omega/\partial r$ which is close to zero. Only when the turbulent diffusion of the poloidal field eventually affects the strength of the azimuthal field, the α effect is resumed to allow a regeneration of the former. However, this occurs only for a limited interval of time during each cycle and has no consequences for the migration of the dynamo wave which stays almost stationary. The prolonged suppression of the α effect leads to a remarkably anharmonic variation of the mean magnetic field which implies an anharmonic variation of the S index.

Finally, interesting results have been derived by a comparison of the chromospheric long-term variations with the seasonal optical magnitudes of several H&K project stars for which time series of $10 - 20$ yr are now available (Lockwood *et al.* 2007; Hall *et al.* 2009). They show two different kinds of behaviour: a) there are stars similar to the Sun which become brighter in the optical (and probably in the bolometric passband) when they are more chromospherically active; and, b) there are stars which become fainter when they are more chromospherically active. The former have a chromospheric activity level comparable to that of the Sun, so their long-term optical variations are dominated by faculae. The latter are more active than the Sun and their ARs are dominated by cool spots, so they get fainter when their activity increases.

3.2. *Stellar cycles from photospheric proxies*

The variability of the optical flux can be easily monitored from the ground to search for activity cycles if its amplitude is greater than a few 0.01 mag. In this way, Messina & Guinan (2002) studied cycles in a sample of young solar analogues with spectral types

G0V-G5V and ages ranging from ≈ 130 to ≈ 700 Myr. Although the sample was limited, it was very interesting to study the early evolution of solar activity. They found that the fastest rotating stars tended to display longer cycles, with the range of the observed cycle periods apparently converging from a large dispersion at ages of $\approx 100-200$ Myr toward the solar cycle period at the age of the Sun. Messina & Guinan (2003) investigated the correlation between the period of the rotational modulation and the phase of the cycle in the above sample of stars finding a clearly significant correlation in two of them, and a less significant correlation in the other four. Both solar-like and anti-solar behaviours were observed, as in the case of the chromospheric cycles of the H&K project sample. The amplitude of the optical variability on timescales from the rotation period to the cycle period, was well correlated to the inverse of the *Rossby number* of each star. The Rossby number is defined as: $Ro \equiv P_{\rm rot}/\tau_{\rm c}$, where $P_{\rm rot}$ is the rotation period of the star and $\tau_{\rm c}$ the correlation time of the turbulent motions at the base of the stellar convection zone, usually set equal to the convective turnover time one mixing length above the base of the convection zone, as computed by means of the standard mixing-length theory. Assuming that $\alpha_0 = \Omega l$, $\Omega' \propto \Omega$, and $\eta_{\rm t} = l^2/\tau_c$, where l is the mixing length at the base of the stellar convective envelope, we find $D \propto Ro^{-2}$. Therefore, the Rossby number is a measure of the dynamo number which can be derived by combining observations of the stellar rotational modulation with a model of its internal structure (see, e.g., Messina & Guinan 2003).

Late-type stars in close binary systems are forced into fast rotation by tidal interactions and show a very high level of magnetic activity. Among those systems, the most active are the RS Canum Venaticorum binaries. They are detached systems consisting of a G or K subgiant, and a dwarf or subgiant of spectral type F or G. Their phenomenology has been described in several reviews (e.g., Rodonò 1992; Guinan & Gimenez 1993). Long-term optical monitoring has revealed cycles of $10-20$ yr in several systems (e.g., Rodonò *et al.* 1995; Henry *et al.* 1995; Oláh *et al.* 1997; Lanza *et al.* 1998; Rodonò *et al.* 2000; Lanza *et al.* 2001; Lanza *et al.* 2002; Lanza *et al.* 2006; Berdyugina & Henry 2007). Recently, Buccino & Mauas (2009) have investigated chromospheric cycles in some RS CVn systems using long-term UV data from the IUE satellite; and Messina (2008) has discussed the long-term variability of the optical colours of several systems finding in some cases evidence of a predominance of the faculae at the maximum of activity. The long-term optical variability of RS CVn systems is generally anharmonic, in agreement with the dynamo models by Baliunas *et al.* (2006) for very active stars. Moreover, changing of the cycle period as well as multiperiodic variations are often observed (cf. Oláh *et al.* 2009), which may be related to the complex feedback mechanisms operating in a highly non-linear stellar dynamo (cf. Spiegel 2009).

The binary nature of RS CVn stars allows us to look for other proxies of magnetic field variations. A very interesting candidate is the long-term variability of their orbital period which shows cycles with relative amplitudes of the order of 10^{-5} on time scales from a few decades up to a few centuries. The same phenomenon is observed in semi-detached binaries, such as Algols and cataclysmic variables. The latter are particularly interesting because they show the phenomenon also when their secondary components are fully convective (Ak *et al.* 2001; Borges *et al.* 2008), thus providing a circumstantial evidence of magnetic cycles in fully convective dwarf stars with a sufficiently fast rotation. For more information on the connection between orbital period modulation and dynamo action see, e.g., Lanza *et al.* (1998), Lanza & Rodonò (1999), Lanza & Rodonò (2004), Lanza (2005), and Lanza (2006a).

3.3. *Dependence of the cycle period on stellar parameters*

Since stellar cycles are manifestations of stellar dynamos whose regimes mainly depend on the dynamo number D, most of the studies have focussed on the relationship between the period of the cycle and the rotation period, or the Rossby number, assumed to be a better proxy for D. A plot of the cycle period $P_{\rm cyc}$ vs. the rotation period $P_{\rm rot}$ shows a large scatter, particularly for $P_{\rm rot} < 10 - 20$ days and a generally weak correlation between $P_{\rm cyc}$ and $P_{\rm rot}$ (cf., e.g., Saar & Brandeburg 1999; Oláh *et al.* 2009). When the cycle period is measured in units of the rotation period of the star and plotted vs. the Rossby number, the correlation improves if the stars are subdivided into three branches (cf. Fig. 5 of Saar & Brandenburg 1999). Those with a low or intermediate level of activity, i.e., pratically all the targets of the H&K project, occupy two roughly parallel branches in the plot and show a $P_{\rm cyc}/P_{\rm rot}$ decreasing with the increase of the inverse Rossby number. On the other hand, very active stars, such as those in RS CVn binaries, show an increase of $P_{\rm cyc}/P_{\rm rot}$ with the increase of the inverse Rossby number. Their branch is connected to that of the intermediate active stars by several objects that may be regarded as being in a transitional evolutionary stage from very high activity to intermediate activity. The six young solar analogues investigated by Messina & Guinan (2002) fall on the branches corresponding to intermediate or highly active stars, with one object in the transitional stage.

This complex scenario is not satisfactorily understood yet. Although some conjectures have been proposed to account for the different branches, such as changes in the α effect with the field strength (cf. Saar & Brandenburg 1999), or different locations and aspect ratios of the dynamo shell (e.g., Bohm-Vitense 2007), it is difficult to constrain the dynamo models.

4. New perspectives opened by the latest and forthcoming observations

The recent launch of the satellite CoRoT has opened a new perspective for the study of stellar activity. The MOST satellite had already shown the potentiality of space-borne photometers in terms of precision, stability and high-duty cycle (cf., e.g., Walker *et al.* 2007; 2008), but CoRoT with its extended time series allowed us to find more interesting results, notably in the case of active stars accompanied by hot Jupiters. Spot modelling techniques based on high-precision photometry have been validated through an extensive application to solar irradiance time series (cf. Lanza *et al.* 2007).

The young active G7V star hosting CoRoT-2b has $P_{\rm rot} \sim 4.52$ days and a spot modelling of its light curve has revealed a short-term spot cycle of ~ 29 days, akin the solar Rieger cycles (Lanza *et al.* 2009a). The shorter period may be a consequence of the faster stellar rotation in the framework of the model by Lou (2000) that interprets Rieger cycles as a manifestation of Rossby-type waves trapped in the outer layers of the convection zone. An alternative explanation calls into play a possible influence of the close-in hot Jupiter because the cycle period is very close to ten synodic periods of the planet with respect to the mean stellar rotation period. Indeed, MOST and CoRoT are providing hints of a possible influence of close-in giant planets on stellar magnetic activity, revealing persistent ARs that rotate synchronously with the planet in systems such as τ Bootis (Walker *et al.* 2008) and CoRoT-4 (Lanza *et al.* 2009b). More on these intriguing phenomena can be found in Lanza (2008), Shkolnik *et al.* (2009), and Lanza (2009).

CoRoT time series will allow us several more detections of Rieger-like cycles and possible interactions between stars and their planets because more than 100,000 late-type stars will be monitored by the end of the mission for time intervals ranging from 20 to 150 days. To date, apart from the Sun and CoRoT-2a, only the active close binary UX Arietis has been suggested to show Rieger cycles (Massi *et al.* 2005). On the other hand, the recently launched Kepler satellite will allow us to extend continuous photometric monitoring to time intervals of 3-4 yr, improving the photometric precision by at least a factor of 3. It should be capable of monitoring the activity cycle variations of p-mode eigenfrequencies in stars brighter than $V \sim 9$, hopefully providing us with information for a better understanding of the phenomenon (e.g., Karoff *et al.* 2009). Another interesting perspective shall be opened by GAIA which will observe the field of Kepler for $2-4$ years after the end of the Kepler mission, although with a very sparse time sampling (on the average 90 points in five years), allowing us to get some information on decade-long magnetic cycles in stars brighter than $V \approx 15$.

Rich opportunities are opened by the new ground-based spectroscopic and spectropolarimetric observations. A database of Doppler Imaging maps of very active stars is now available, in some cases with an extension of $10-15$ years, allowing us an interesting comparison with long-term photometry (see, e.g., Strassmeier 2009 for a discussion of Doppler Imaging results). Among the many studies based on Doppler Imaging, I would like to mention the detection of long-term oscillations of the amplitude of surface differential rotation in the very young (≈ 30 Myr) K0 dwarf AB Dor (Collier Cameron 2007). Such studies allow us to investigate the back reaction of the Lorentz force on stellar rotation in highly non-linear dynamo regimes and can be analyzed on the basis of general models for the angular momentum transport inside a convection zone to derive quantitative estimates of the internal torques and magnetic fields (see Lanza 2006b, Lanza 2007).

Spectropolarimetric observations of τ Bootis have revealed two consecutive reversals of its photospheric magnetic field, suggesting a solar-like magnetic cycle with a period of only ~ 2 yr, i.e., ten times shorter than the solar magnetic cycle (Fares *et al.* 2009). Such a very short cycle could be related to the influence of a close-in giant planet orbiting τ Boo, as conjectured by Lanza (2008; 2009). So far τ Boo is the only star to have shown a complete cycle of polarity reversals, apart from the Sun. A long-term monitoring of the highly active K1IV component of the RS CVn system HR 1099 did not show such a phenomenon, although the observations cover almost 15 years (Donati *et al.* 2003; Petit *et al.* 2004).

5. Conclusions

A major limitation to our understanding of stellar magnetic cycles is the lack of spatial resolution of stellar discs and asteroseismic measurements of internal rotation. Therefore, our picture is not as detailed as in the case of the Sun. In spite of that, the detection and study of stellar cycles by proxies is now a well established field of research and long-term time series have revealed the existence of a complex scenario. Stellar cycles can be associated with quasi-sinusoidal or highly anharmonic variations of the proxies, they show single or multiple periodicities, and their periods may change versus time (cf. Oláh *et al.* 2009). The relationship of the cycle period to the stellar parameters is by no means a simple one. Even considering the correlation with the Rossby number, which provides us with a measure of the dynamo number, no simple scenario emerges. Present dynamo models have difficulties in accounting for such a complex correlation, so they are not significantly constrained by it. A significant progress may come from the new long-term spectroscopic and spectropolarimetric observations that can measure surface differential rotation and photospheric magnetic fields in highly active stars, i.e., young main-sequence stars or late-type components of close active binaries. Such observations provide a direct view of the large scale topology of the magnetic field in the photosphere and can be used to estimate the intensity of the internal field by measuring its back-reaction on stellar rotation. For slowly rotating, sun-like stars a new perspective is opened by CoRoT and Kepler high-precision photometric time series. They have already allowed us to detect Rieger-like cycles in the planet-hosting star CoRoT-2a and have provided hints of a possible influence of close-in giant planets on the activity of their host stars.

References

Ak, T., Ozkan, M. T., & Mattei, J. A. 2001, *A&A*, 369, 882
Baliunas, S. L., Donahue, R. A., Soon, W. *et al.* 1995, *ApJ*, 438, 269
Baliunas, S. L., Donahue, R. A., Soon, W., & Henry, G. W. 1998, in Cool Stars, Stellar Systems and the Sun, ASP Conf. Ser. vol. 154, R. A. Donahue and J. A. Bookbinder (Eds.); p. 153
Baliunas, S., Frick, P., Moss, D., Popova, E., Sokoloff, D., & Soon, W. 2006, *MNRAS*, 365, 181
Berdyugina, S. V. 2005, *Living Rev. Sol. Phys.*, vol. 2, no. 8
Berdyugina, S. V. & Henry, G. W. 2007, *ApJ*, 659, L157
Böhm-Vitense, E. 2007, *ApJ*, 657, 486
Borges, B. W., Baptista, R., Papadimitriou, C., & Giannakis, O. 2008, *A&A*, 480, 481
Brandenburg, A. 2005, *ApJ*, 625, 539
Buccino, A. P. & Mauas, P. J. D. 2009, *A&A*, 495, 287
Caligari, P., Moreno-Insertis, F., & Schussler, M. 1995, *ApJ*, 441, 886
Canuto, V. 2000, *A&A*, 357, 177
Collier Cameron, A. 2007, *AN*, 328, 1030
Donahue, R. A. & Baliunas, S. L. 1992, *ApJ*, 393, L63
Donahue, R. A. & Keil, S. L. 1995, *Solar Phys.*, 159, 53
Donati, J.-F., Collier Cameron, A., Semel, M., *et al.* 2003, *MNRAS*, 345, 1145
Durney, B. R., De Young, D. S., & Roxburgh, I. W. 1993, *Solar Phys.*, 145, 207
Fares, R., Donati, J.-F., Moutou, C., *et al.* 2009, *MNRAS*, 398, 1383
Favata, F., Micela, G., Orlando, S., *et al.* 2008, *A&A*, 490, 1121
Fröhlich, C. & Lean, J. 2004, *A&AR*, 12, 273
Guinan E. F. & Gimenez, A. 1993, in Sahade J., McCluskey G. E., Jr., Kondo Y. (Eds.), The Realm of Interacting Binary Stars, Kluwer Ac. Publ., Dordrecht; p. 51
Güdel, M. 2004, *A&AR*, 12, 71
Hall, J. C. 2008, *Living Rev. Sol. Phys.*, vol. 5, no. 2
Hall, J. C. & Lockwood, G. W. 2004, *ApJ*, 614, 942
Hall, J. C., Lockwood, G. W., & Skiff, B. A. 2007, *AJ*, 133, 862
Hall, J. C., Henry, G. W., Lockwood, G. W., Skiff, B. A., & Saar, S. H. 2009, *AJ*, 138, 312
Henry, G. W., Eaton, J. A., Hamer, J., & Hall, D. S. 1995, *ApJS*, 97, 513
Howe, R. 2009, *Living Rev. Sol. Phys.*, vol. 6, no. 1
Karoff, C., Metcalfe, T. S., Chaplin, W. J., *et al.* 2009, *MNRAS*, 399, 914
Kolláth, Z. & Oláh, K. 2009, *A&A*, 501, 695
Kosovichev, A. G. 2009, *Space Sci. Revs.*, 144, 175
Krivova, N. A. & Solanki, S. K. 2002, *A&A*, 394, 701
Lanza, A. F. 2005, *MNRAS*, 364, 238
Lanza, A. F. 2006a, *MNRAS*, 369, 1173
Lanza, A. F. 2006b, *MNRAS*, 373, 819
Lanza, A. F. 2007, *A&A*, 471, 1011
Lanza, A. F. 2008, *A&A*, 487, 1163
Lanza, A. F. 2009, *A&A*, 505, 339
Lanza, A. F., Aigrain, S., Messina, S., *et al.* 2009b, *A&A*, 506, 255
Lanza, A. F., Bonomo, A. S., & Rodonò, M. 2007, *A&A*, 464, 741
Lanza, A. F., Catalano, S., Cutispoto, G., Pagano, I., & Rodonò, M. 1998a, *A&A*, 332, 541
Lanza, A. F., Catalano, S., Rodonò, M., *et al.* 2002, *A&A*, 386, 583

Lanza, A. F., Pagano, I., Leto, G., *et al.* 2009a, *A&A*, 493, 193
Lanza, A. F., Piluso, N., Rodonò, M., Messina, S., & Cutispoto, G. 2006, *A&A*, 455, 595
Lanza, A. F. & Rodonò, M. 1999, *A&A*, 349, 887
Lanza, A. F. & Rodonò, M. 2004, *AN*, 323, 424
Lanza, A. F., Rodonò, M., Mazzola, L., & Messina, S. 2001, *A&A*, 376, 1011
Lanza, A. F., Rodonò, M., & Rosner, R. 1998b, *MNRAS*, 296, 893
Lockwood, G. W., Skiff, B. A., Henry, G. W., *et al.* 2007, *ApJS*, 171, 260
Lou, Y.-Q. 2000, *ApJ*, 548, 460
Massi, M., Neidhöfer, J., Carpentier, Y., & Ros, E. 2005, *A&A*, 435, L1
Messina, S. 2008, *A&A*, 480, 495
Messina, S. & Guinan, E. F. 2002, *A&A*, 393, 225
Messina, S. & Guinan, E. F. 2003, *A&A*, 409, 1017
Oláh, K., Kolláth, Z., Granzer, T., *et al.* 2009, *A&A*, 501, 703
Oláh, K., Kövari, Zs., Bartus, J., *et al.* 1997, *A&A*, 321, 811
Oliver, R., Ballester, J. L., & Baudin, F. 1998, *Nature*, 394, 552
Petit, P., Donati, J.-F., Wade, G. A., *et al.* 2004, *MNRAS*, 348, 1175
Rempel, M. 2008, *Journal of Physics: Conf. Ser.*, 118, 012032
Ribas, I. 2009, these proceedings, p. 3
Rieger, E., Kanbach, G., Reppin, C., *et al.* 1984, *Nature*, 312, 623
Rodonò, M. 1992, in IAU Symp. 151, Y. Kondo, R. F. Sistero, & R. S. Polidan (Eds.), Kluwer Ac. Publ., Dordrecht; p. 71
Rodonò, M., Lanza, A. F., & Catalano, S. 1995, *A&A*, 301, 75
Rodonò, M., Messina, S., Lanza, A. F., Cutispoto, G., & Teriaca, L. 2000, *A&A*, 358, 624
Rüdiger, G., von Rekowski, B., Donahue, R. A., & Baliunas, S. L. 1998, *ApJ*, 494, 691
Saar, S. H. & Brandenburg, A. 1999, *ApJ*, 524, 295
Salabert, D., García, R. A., Pallé, P. L., & Jiménez-Reyes, S. J. 2009, *A&A*, 504, L1
Shkolnik, E., Aigrain, S., Cranmer, S., *et al.* 2009, in Proceedings of the 15th Cambridge Workshop on Cool Stars, Stellar Systems and the Sun, AIP Conference Proceedings, Vol. 1094, p. 275
Schlichenmaier, R. & Stix, M. 1995, *A&A*, 302, 264
Sokoloff, D., Arlt, R., Moss, D., Saar, S. H., & Usoskin, I. 2010, these proceedings, p. 111
Spiegel, E. A. 2009, *Space Sci. Revs.*, 144, 25
Strassmeier, K. G. 2005, *AN*, 326, 269
Strassmeier, K. G. 2009, *A&AR*, 17, 251
Thompson, M. J., Christensen-Dalsgaard, J., Miesch, M. S., & Toomre, J. 2003, *ARAA*, 41, 599
Usoskin, I. G., Solanki, S. K., & Kovaltsov, G. A. 2007, *A&A*, 471, 301
Vaughan, A. H., Preston, G. W., & Wilson, O. C. 1978, *PASP*, 90, 267
Walker, G. A. H., Croll, B., Kuschnig, R., *et al.* 2007, *ApJ*, 659, 1611
Walker, G. A. H., Croll, B., Matthews, J. M., *et al.* 2007, *A&A*, 482, 691
Weiss, N. O. & Thompson, M. J. 2009, *Space Sci. Revs.*, 144, 53

Solar and Stellar Variability: Impact on Earth and Planets
Proceedings IAU Symposium No. 264, 2009
A. G. Kosovichev, A. H. Andrei & J.-P. Rozelot, eds.

doi:10.1017/S1743921309992535

Do young Suns undergo magnetic reversals?

Stephen C. Marsden[1], Sandra V. Jeffers[2], Jean-Francois Donati[3], Matthew W. Mengel[4], Ian A. Waite[4] and Brad D. Carter[4]

[1]Anglo-Australian Observatory, P.O. Box 296, Epping, NSW 1710, Australia
email: scm@aao.gov.au
[2]Sterrekundig Instituut, Universiteit Utrecht, P.O. Box 80000, NL-3508 TA Utrecht, The Netherlands
email: s.v.jeffers@uu.nl
[3]Laboratoire d'Astrophysique de Toulouse-Tarbes, Observatoire Midi-Pyrénées, 14 avenue Edouard Belin, F-31400 Toulouse, France
email: donati@ast.obs-mip.fr
[4]Faculty of Sciences, University of Southern Queensland, Toowoomba 4350, Australia
email: mengel@usq.edu.au, waite@usq.edu.au, carterb@usq.edu.au

Abstract. A key part of the modern-day regenerative solar magnetic dynamo is the reversal of the Sun's global magnetic field every eleven years. However, recent theoretical models indicate that young-rapidly rotating Sun-like stars may not always undergo full magnetic reversals, but instead may sometimes undergo "attempted" reversals where the magnetic field declines in strength only to return with the same polarity. Using the technique of Zeeman Doppler imaging we have mapped the magnetic field topology of a small sample of young Sun-like stars at multiple epochs, and present tentative evidence of an "attempted" magnetic field reversal on one of our stars.

Keywords. stars: late-type, stars: imaging, stars: magnetic fields

1. Introduction

The generation of magnetic fields is arguably one of the most important process operating inside a star, influencing everything from the angular momentum evolution of the star through to the habitibility of planets. On the Sun, the reversal of the magnetic field every ~11 years is a crucial element of the cyclic solar dynamo, however, what of young solar-type stars? Such stars are often rapidly-rotating and show activity phenomena, such as spots and flares, an order of magnitude or more larger than that of today's Sun, indicating powerful magnetic dynamos. Do such stars have a similar dynamo operation to that of the modern-day Sun or are they fundamentally different? Determining if young Sun-like stars have magnetic cycles is an important step in our understanding of how they produce magnetic fields.

Photometry of young solar-type stars as well as emission in the Calcium H and K lines from such stars, show variations and sometimes cycles in their emission level (i.e. Lanza 2009 and Baliunas *et al.* 1995). While the photometry of young solar-type stars often shows cyclic variations, the variations in the Calcium H and K flux from such young stars appears to be for the most part chaotic or irregular (Baliunas *et al.* 1995). Do these cycles/variations involve reversals of the magnetic field as in the solar case? To determine this we need to be able to observe and monitor the evolution of magnetic fields on young solar-type stars.

2. Detecting and mapping magnetic fields

We can use the fact that starlight will be polarised in the presence of stellar magnetic fields in order to detect magnetic fields on the surfaces of stars. High-resolution spectropolarimeters, such as ESPaDOnS (on the Canada-France-Hawaii Telescope), NARVAL (on the Télescope Bernard Lyot) and SEMPOL (on the Anglo-Australian Telescope) split the starlight into right- and left-handed circularly polarised light. A Stokes V profile can then be reconstructed from the polarised light to reveal the presence of magnetic fields on the stellar surface.

Zeeman signatures in cool stars are usually extremely small, often with amplitudes <0.1% of the continuum level. Thus spectra with signal-to-noise > 1000 are required to even detect these signatures. Unfortunately, we usually cannot simply extend the exposure time in order to increase the signal-to-noise as the young stars we wish to observe are often rapidly-rotating (some with rotational periods of a fraction of a day). If the exposure time is increased too much the magnetic signatures will be smeared out as the star rotates during the exposure. However, when observing with an echelle, each spectrum contains several thousand photospheric lines which are all affected by the presence of surface magnetic fields in a similar way. Thus we can use the technique of Least-Squares Deconvolution (LSD, Donati *et al.* 1997) to combine these lines into one high signal-to-noise LSD profile. Using LSD can provide a multiple gain of 30+ in signal-to-noise over that of a single line.

This allows us to detect the presence of magnetic fields on the surfaces of young solar-type stars, but in order to determine if such stars are undergoing magnetic reversals we need to be able to map the surface magnetic topology of such stars. Magnetic regions on the stellar surface cause distortions in the star's Stokes V LSD profile which travel through the profile as the star rotates. Radial magnetic field (field in/out of the stellar surface) will produce a Stokes V profile that retains its sign as it passes across the stellar disk. In contrast, a region of azimuthal field (field wrapped around the star's rotation axis) will produce a Stokes V profile that changes sign as it passes across the stellar disc. Thus the orientation of the field can be determined. The longitude of a magnetic region is easily determined by the timing of the region passing through the centre of the profile. The latitude of a magnetic region on the stellar surface can also be determined as those regions near the equator will pass through the entire LSD profile while those regions nearer the pole will be restricted to the central part of the profile. Thus the position and orientation of magnetic regions on the stellar surface can be determined.

A time series of Stokes V LSD profiles taken over the stellar rotation period (or a significant part of it) is inverted using Zeeman Doppler imaging (ZDI, Brown *et al.* 1991, Donati & Brown 1997) to reproduce the magnetic topology of the star. Unfortunately, this inversion is an ill-posed problem with an infinite number of solutions that can be found to fit the data to the noise level. In order to choose a unique solution some additional constraint, called regularisation, is usually applied. Our code uses the Skilling & Bryan (1984) maximum entropy optimisation scheme which produces an image with the minimum amount of information required to fit the data.

3. The evolution of magnetic topologies

The technique of ZDI has been around for a number of years now and applied to a number of stars. So has a magnetic reversal actually been seen? Donati *et al.* (2003a) studied the magnetic topologies of several K-stars over a 7-11 year period and did not find any evidence for a magnetic reversal. However, all these stars have low levels of differential rotation (Donati *et al.* 2003b) and this may mean they have long magnetic cycles (see Section 3.2).

The only stars (outside of the Sun) so far to have evidenced a reversal of their magnetic field are the late-F star Tau Boo (Donati *et al.* 2008, Farès *et al.* 2009) and the Sun-like star HD 190771 (Petit *et al.* 2009). Tau Boo has shown two reversals in the polarity of its magnetic field in the space of two years leading to the belief that this star has a cycle length of ~2 years. This is significantly shorter than that of the Sun (~22 years). What could be causing this? There are several possibilities. Firstly, Tau Boo has an extremely high level of surface differential rotation at almost 10 times the solar value and Donati *et al.* (2008) suggest that such a high level of differential rotation could well be responsible for the rapid magnetic cycle of the star. However, Tau Boo is also the host of a "Hot Jupiter" with a period of a few days (Butler *et al.* 1997) and Farès *et al.* (2009) suggest that this could be responsible.

3.1. *Young solar-type stars*

Both Tau Boo and HD 190771 are older than ~1 Gyr and may well already have started to develop magnetic cycles similar to the Sun (although possibly much more rapid), but what of younger solar-type stars? Do such active, rapidly-rotating stars show evidence of magnetic reversals?

HD 171488 is a young (30 - 50 Myr old) early-G star which has had its magnetic topology mapped at four epochs from September 2004 through to November 2007 (see Figure 1). The most striking feature on the star is the ring of azimuthal magnetic field around the pole which is seen in all epochs. There appears to be little evidence of a reversal of the star's magnetic field over the 3 year period.

HR 1817 is a young (10 - 30 Myr old) late-F star which has had its magnetic topology determined at three epochs, December 2001, December 2007, and December 2008 (see Figure 2). A polar ring of azimuthal field is also in evidence at all three epochs, however, in 2007 the ring is significantly weaker and increased regions of negative azimuthal field are seen on the lower-latitudes of the star at this epoch. Again there is little evidence of a reversal of the magnetic field on the star.

3.2. *The role of differential rotation*

If we observe a star over several rotations the magnetic features will evolve under the influence of differential rotation. A simplified solar-like differential rotation law can be included in the imaging process:

$$\Omega(l) = \Omega_{\rm eq} - d\Omega sin^2 l,$$

where $\Omega_{\rm eq}$ is the equatorial rotation rate and $d\Omega$ is the shear from equator to pole (the differential rotation). By treating $\Omega_{\rm eq}$ and $d\Omega$ as free parameters we can determine the differential rotation on a star using the χ^2 minimisation method of Petit *et al.* (2002).

This has been done for both HD 171488 and HR 1817 (Marsden *et al.* 2006, Jeffers & Donati 2008, Marsden *et al.* in prep.) and both have been shown to have extremely high levels of differential rotation, similar to that evidenced by Tau Boo at up to 10 times the solar value. In contrast to Tau Boo, neither star shows any evidence of a magnetic reversal over a 3 year period for HD 171488 and a 7 year period for HR 1817, although the long gap in observations between 2001 and 2007 could well mean that a magnetic reversal (or two) may well have been missed on HR 1817. Does this mean that differential rotation does not play a strong role in the length of a star's magnetic cycle? Does it mean that young solar-type stars do not in fact undergo magnetic reversals or are other factors at work?

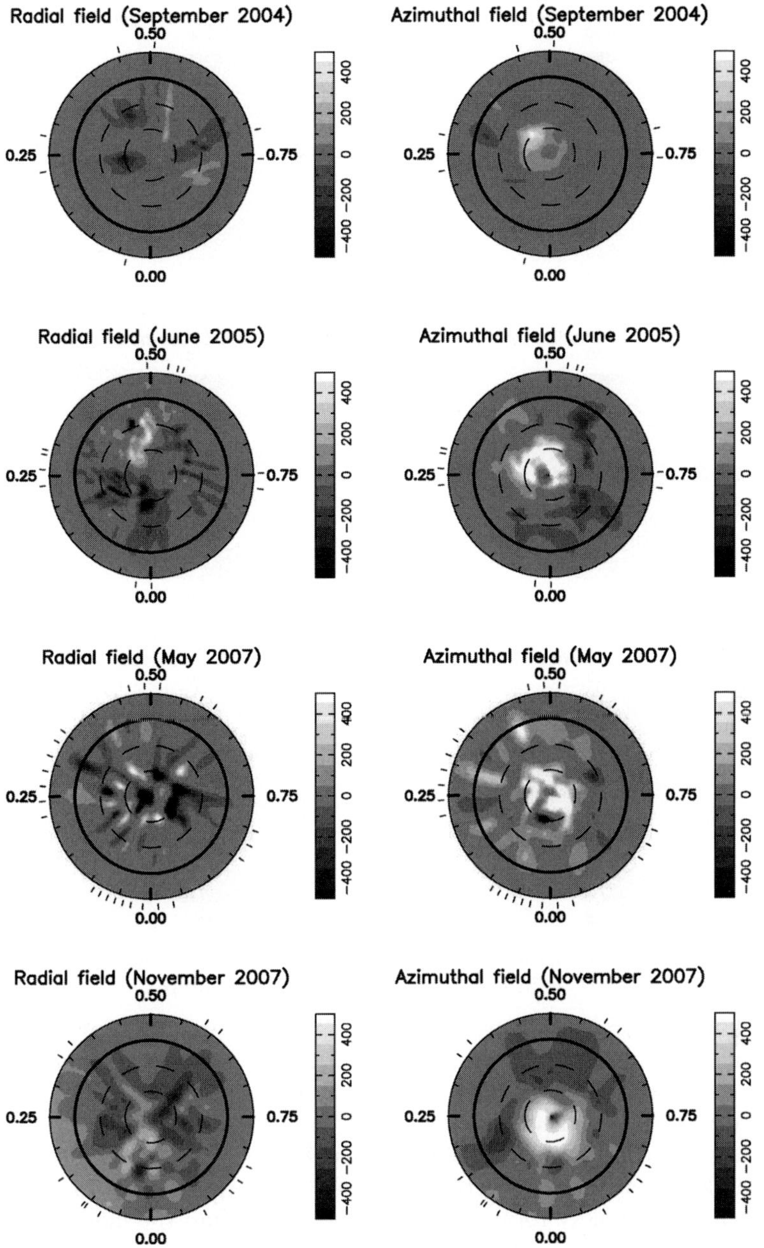

Figure 1. Radial (left) and azimuthal (right) magnetic topologies for HD 171488. The images are flattened polar projections extending down to -30° latitude. The bold line denotes the equator and the dashed lines are +30° and +60° latitude parallels. The radial ticks outside the plots indicate the rotational phases at which the star was observed. The magnetic scale is in gauss (Marsden *et al.* 2006, Jeffers & Donati 2008, Jeffers *et al.* in prep.).

4. Comparison with theory

Recent theoretical models of rapidly-rotating (up to 5 times solar) young Suns by Brown *et al.* (2009) show that the dynamo of such stars appears to be distributed throughout the convection zone (rather than restricted to the interface layer as in the

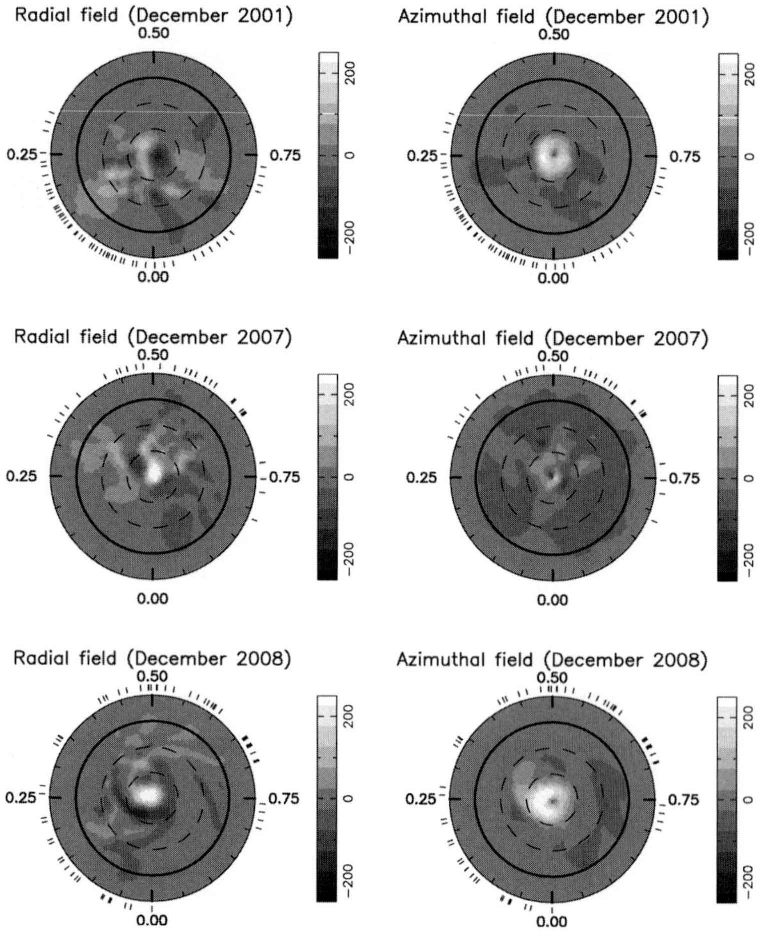

Figure 2. Radial (left) and azimuthal (right) magnetic topologies for HR 1817. The images are as in Figure 1 (Marsden *et al.* in prep.).

solar case). This appears to support the finding of large regions of near-surface azimuthal field on young solar-type stars which should not appear if a solar-type interface-layer dynamo is in operation. Futhermore the models of Brown *et al.* (2009) show that young Suns produce "wreaths" of azimuthal magnetic field near the stellar surface. These "wreaths" appear to be somewhat similar to those shown in our magnetic topologies (Figures 1 and 2) but at lower latitudes than those we see. Could this be due to the fact that our stars are much more rapidly rotating (~20 times solar) than the models of Brown *et al.* (2009), with the Coriolis force deflecting the wreaths to higher latitudes?

In addition, the models of Brown *et al.* (2009) show another interesting aspect. They show that not only do rapidly rotating Suns undergo full magnetic reversals but also undergo a number of "attempted" reversals, where the magnetic field strength in both the Northern and Southern hemispheres decreases but then strengthens with the same polarity rather than undergoing a reversal. Could this be what happened on HR 1817 in December 2007 with the weakening of its polar azimuthal ring?

5. Conclusions

We have used Zeeman Doppler imaging to reconstruct the magnetic field topology of two young solar-type stars at multiple epochs. Both stars show an intense ring of azimuthal field around the poles at almost all epochs with little evidence of a magnetic reversal. Such rings of azimuthal field are seen on a number of stars and appear similar to the "wreaths" of magnetic field produced by the models of Brown *et al.* (2009) except they are at higher latitudes, perhaps due to higher rotation rates. The young late-F star HR 1817 shows a significant weakening of its polar azimuthal ring during one epoch. We postulate that HR 1817 may have undergone an "attempted" reversal, as predicted by the models of Brown *et al.* (2009), during this epoch.

Only one solar-type star (outside of the Sun) has shown a possible magnetic cycle (Tau Boo) with a cycle length an order of magnitude shorter than that of the Sun. Tau Boo has both a high level of differential rotation and hosts a "Hot Jupiter", either of which may be influencing the magnetic cycle of the star. However, both of the young Sun-like stars we have studied (HD 171488 and HR 1817) have similarly high levels of differential rotation, but do not show any evidence of magnetic reversals, even though the observations cover a period similar to the cycle length of Tau Boo. Does this imply that the "Hot Jupiter" of Tau Boo, rather than high differential rotation, is somehow influencing its magnetic cycle? If so, how? Does the fact that both the stars that have shown magnetic reversals are older than ∼1 Gyr mean that solar-type stars have developed regular magnetic cycles by this age?

So, do young Suns undergo magnetic reversals? The jury is still out. Theory says they should, along with undergoing "attempted" reversals, but it will take a dedicated observing program to monitor such stars in order to determine if they do and if so, how often and regularly they do.

References

Baliunas, S. L., Donahue, R. A., Soon, W. H., Horne, J. H., Frazer, J., Woodard-Eklund, L., Bradford, M., Rao, L. M., Wilson, O. C., Zhang, Q., Bennett, W., Briggs, J., Carroll, S. M., Duncan, D. K., Figueroa, D., Lanning, H. H., Misch, A., Mueller, J., Noyes, R. W., Poppe, D., Porter, A. C., Robinson, C. R., Russell, J., Shelton, J. C., Soyumer, T., Vaughan, A. H., & Whitney, J. H. 1995, *ApJ*, 438, 269

Brown, S. F., Donati, J.-F., Rees, D. E., & Semel, M. 1991, *A&A*, 250, 463

Brown, B. P., Browning, M. K., Brun, A. S., Miesch, M. S., & Toorme, J. 2009, *ApJ*, submitted (arXiv:0906.2407)

Butler, R. P., Marcy, G. W., Williams, E., Hauser, H., & Shirts, P. 1997, *ApJ*, 474, L115

Donati, J.-F. & Brown, S. F. 1997, *A&A*, 326, 1135

Donati, J.-F., Semel, M., Carter, B. D., Rees, D. E., & Cameron, A. C. 1997, *MNRAS*, 291, 658

Donati, J.-F., Collier Cameron, A., Semel, M., Hussain, G. A. J., Petit, P., Carter, B. D., Marsden, S. C., Mengel, M., López Ariste, A., Jeffers, S. V., & Rees, D. E. 2003a, *MNRAS*, 345, 1145

Donati, J.-F., Collier Cameron, A., & Petit, P. 2003b, *MNRAS*, 345, 1187

Donati, J.-F., Moutou, C., Farès, Bohlender, D., Catala, C., Deleuil, M., Shkolnik, E., Cameron, A. C., Jardine, M. M., & Walker, G. A. H. 2008, *MNRAS*, 385, 1179

Farès, R., Donati, J.-F., Moutou, C., Bohlender, D., Catala, C., Deleuil, M., Shkolnik, E., Cameron, A. C., Jardine, M. M., & Walker, G. A. H. 2009, *MNRAS*, 398, 1383

Jeffers, S. V. & Donati, J.-F. 2008, *MNRAS*, 390, 635

Lanza, A. F. 2009, these proceedings, p. 120

Marsden, S. C., Donati, J.-F., Semel, M., Petit, P., & Carter, B. D. 2006, *MNRAS*, 370, 468

Petit, P., Donati, J.-F., & Collier Cameron, A. 2002, *MNRAS*, 334, 374

Petit, P., Dintrans, B., Morgenthaler, A., Van Grootel, V., Morin, J., Lanoux, J., Aurière, M., & Konstantinova-Antova, R. 2009, *A&A*, 508, p. L9

Skilling, J. & Bryan, R. K. 1984, *MNRAS*, 211, 111

Solar and Stellar Variability: Impact on Earth and Planets
Proceedings IAU Symposium No. 264, 2009
A. G. Kosovichev, A. H. Andrei & J.-P. Rozelot, eds.

doi:10.1017/S1743921309992547

Long-term stellar variability

Isabella Pagano[1]

[1]INAF - Osservatorio Astrofisico di Catania, v. Santa Sofia 78, p.le M. Rodonò, 95123 Catania
email: isabella.pagano@oact.inaf.it

Abstract. Stars with significant subsurface convection zones develop magnetic loop structures that, arising from the surface upward to the external atmospheres, cause flux variability detectable throughout the whole electromagnetic spectrum. In fact, diagnostics of magnetic activity are in radio wavelengths, where gyrosincrotron radiation arises from the quiescent and flaring corona; in the optical region, where important signatures are the Balmer lines, the Ca II IRT and H&K lines; in the UV and X ray domains, the latter mainly due to coronal thermal plasma. The zoo of different magnetic features observed for the Sun – spots, faculae, flares, CMEs – are characterized by different temporal evolution and energetics, both in quantity and quality. As a consequence, the time scale of variability, the amount of involved energy and the quality of the involved photons are used as fingerprints in interpreting the observed stellar variability in the framework of the solar-stellar analogy. Here I review main results from long-term multiwavelength observations of cool star atmospheres, with emphasis to similarities and differences with the solar case.

Keywords. stars: activity, stars: spots, stars: chromospheres, stars: coronae, stars: mass loss

1. Introduction

Active regions in the atmosphere of late-type stars are the obvious manifestations of surface magnetic activity. Strong magnetic flux ropes are generated by the coupling between differential rotation and turbulent convection in the subphotospheric layers thanks to a "dynamo" mechanisms (see Parker 1955). The flux tubes buoyantly rise through the convection zone and penetrate the stellar surface. By interacting with the surface velocity fields, they become braided, twisted, and ultimately, reconnect releasing magnetic energy that power the atmospheric activity.

The whole phenomenology is higly variable in time. Active regions grow and evolve in days-weeks, but are also site of transient phenomena like flares, microflares, and coronal mass ejections (CMEs). The global level of magnetic activity changes in time following cycles as it happens for the Sun.

The covering factor of active regions, as well as the enery output of stellar flares, can be order of magnitude more intense in stars than on the Sun. This is particularly true for very fast rotators such as young stars and stars locked in close binary systems. The time scale of variability, the amount of involved energy and the quality of the involved photons are used as fingerprints in interpreting the observed stellar variability in the framework of the solar-stellar analogy. Multiwavelength observations are mandatory tools for these investigations. Here, I review some of the results obtained by long-term multiwavelength observations of cool star activity.

2. Sources of long-term data on magnetic activity

Stellar magnetic activity has been detected in all late type stars cooler than $T_{eff} \sim$ 8 300 K, i.e. later than A4 V (Simon *et al.* 2002). For typical characteristics these can be grouped as following:

- *Sun-like stars*: this is a very broad class including from late F to late G type dwarfs and subgiants. Following Cayrel de Strobel & Bentolila (1989), solar-like stars are usually classified as:
  - *Solar analogs*: these are pop I dwarfs having main properties similar to those of the Sun;
  - *Solar twins*: which have all physical parameters almost identical to those of the Sun.
- *M-type stars*;
- *Rapidly rotating stars*: i.e., stars in close-binaries (e.g., BY Dra, RS CVn, FK Com) and young stars.

How magnetic activity changes during the evolution of solar-like stars and K-M dwarfs is of paramount interest to understand properties and evolution of exo-planetary systems. Stellar magnetic activity studies require long term monitoring projects both from the ground and space. Among the observing programs providing bulks of fundamental data for stellar magnetic activity studies I recall the Mount Wilson Observatory Ca II H&K (e.g., Baliunas *et al.* 1998), and the Lowell & Fairborn Observatories (e.g., Radick *et al.* 1998, Messina & Guinan 2002 & 2003, Lockwood *et al.* 2007) projects, both aimed to the detection of chromospheric variability in solar-like stars, the latter dedicated also to observations of the photospheric signatures of magnetic activity. Photometric and spectroscopic monitoring have been carried out in several European observatories and institutes to get maps of spotted photospheres (e.g., Rodonò 1992, Rodonò & Cutispoto 1994, Lanza *et al.* 1998, Rodonò *et al.* 2000, Messina 2008, Vida *et al.* 2007; Strassmeier 2009), and of magnetic field patterns (e.g., Donati *et al.* 2003a, Donati *et al.* 2003b, Donati *et al.* 2007). Recently, space observations from MOST, CoRoT, and Kepler are providing very accurate optical variability data with high cadence and months duration (e.g., Walker *et al.* 2007 & 2008, Lanza *et al.* 2009a&b). Photometric and spectroscopic patrol of stellar flares from the ground have been carried out since the sixties in many places (cf., Gershberg 2005 and references therein).

Finally, the active regions and flares in the upper chromosphere, transition regions and coronae as well as solar-like winds have been studied thanks to the UV, and EUV data from, e.g., IUE, HST, GALEX, EUVE and FUSE (e.g., Jordan *et al.* 1987; Ayres 1995, Pagano *et al.* 2004, Ribas *et al.* 2005, Wood *et al.* 2005), X-ray data from, e.g., Einstein, Rosat, ASCA, Beppo-SAX, Chandra, and XMM (e.g., Favata & Micela 2003 and reference therein, Telleschi *et al.* 2005, Stelzer 2007), and radio observations from VLA, VLBI, VLBA, and single-dishes (e.g., Güdel & Benz 1993, Benz & Güdel 1994, Umana *et al.* 1995, Trigilio *et al.* 1998).

3. Results from long-term observations

3.1. *Photospheres and Chromospheres*

3.1.1. *Solar-like stars*

Ca II H&K emission data from Mount Wilson Observatory (MWO) have been analyzed together with Strömgren b, y photometry from Lowell and Fairborn Observatories (L&FO) to investigate the photosphere vs. chromosphere variability of a wide sample of solar-like stars over 1-2 decades time extension (e.g., Lockwood *et al.* 2007, Hall *et al.*

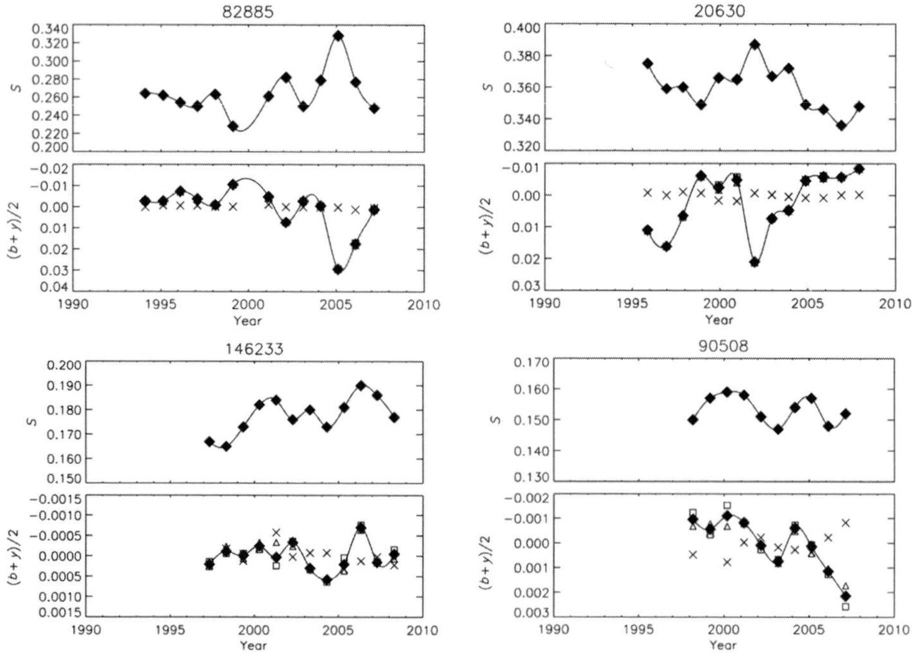

Figure 1. Ca II H&K (*top panels*) and photometric (*bottom panels*) series for HD 82885 (G8 IVV), HD 20630 (G5 V), HD 146233 (G2 V), HD 90508 (G0 V) from Hall *et al.* (2009). Diamonds are the seasonal means. The differential series of the comparison stars are shown by × symbols.

2009). An important result from these investigations is that, on a year-to-year timescale, young active stars become fainter when their Ca II emission increases; i.e., their chromospheric and photospheric light curves are unticorrelated as in the case of HD 82885 and HD 20630 (c.f., Figure 1). Instead, older less active stars, such as the Sun, become brighter when their Ca II emission increases, i.e., their chromospheric and photospheric light curves are correlated as in the case of HD 146233, and HD 90508 (cf. Figure 1). In other words, young active solar-like stars are spot-dominated while older less active stars, like the Sun, are faculae dominated. Hall *et al.* (2009) highlight that the Sun is not peculiar among other solar-like stars; its median photospheric variability is above the one of the solar analogs having chromospheric activity given by $\log R_{HK} < -4.8$. For instance, 18 Sco (G2 V), one of the closest solar twins, with a photometric activity cycle of 7 yr, exhibits a chromospheric variability very similar to the solar one, with $\sigma(\log R_{HK}) = 0.043$ from 1994 to 2008, that is comparable to 0.040 for the Sun over the same period. Data from MWO and L&FO have been used to search for stars in Mounder-like magnetic minimum state. Saar (2006) found that the minimum R_{HK} observed in dwarfs is a decreasing function of stellar metallicity. This effect is well shown in Figure 2-*left* where the chromospheric activity indicator R_{HK} vs. [M/H] is plotted for the sample of solar like stars analyzed by Hall *et al.* (2009).

3.1.2. *Rapidly rotating stars*

Rapidly rotating stars show optical light curves changing on time scales of the order of their rotational period due to rotational modulation of starspots and on longer times due to magnetic cycles. From rotational modulated light curves and from the analysis of changing line profiles during the stellar rotation, maps of the spotted photospheres

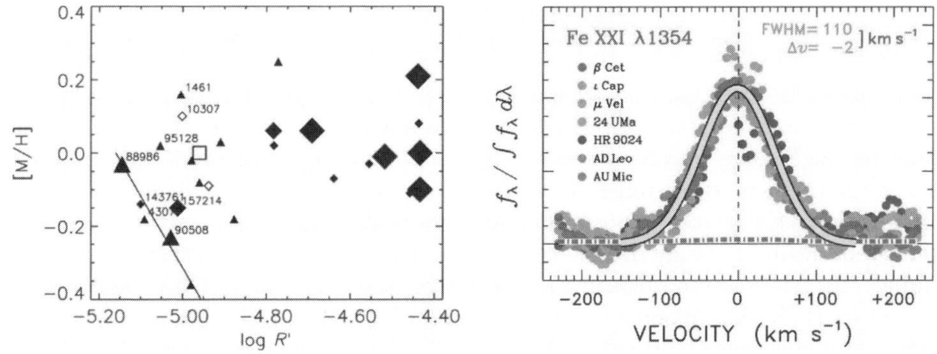

Figure 2. *Left:* R_{HK} vs. metallicity for a sample of solar-like dwarfs from Hall *et al.* (2009). The most inactive stars fall on a locus of increasing chromospheric activity with decreasing metallicity. *Right:* Superposition of scaled Fe XXI λ 1354 profiles of seven stars, mostly G1-K0 giants but including two M dwarfs. The individual line shapes collectively define a "universal" profile that can be matched by a Gaussian with FWHM 110 km s^{-1} and an essentially negligible 2 km s^{-1} blueshift with respect to the predicted average photospheric velocities of the stars and the adopted wavelength of the Fe XXI λ 1354 (from Ayres *et al.* 2003).

and of the magnetic field distribution have been derived by several authors. Among the most noticeable results there are the observations of high-latitude spots (see Güdel 2008, and references therein), and the high-latitude rings of toroidal field detected by Zeeman Doppler imaging (e.g., Donati *et al.* 2003b,Catala *et al.* 2007). Also very solar un-like behavior, such as migration of active regions toward the poles, preferred active longitudes, and "flip-flop" effect (e.g., Rodonò *et al.* 2000, Berdyugina & Henry 2007) are typically observed.

Messina (2008) has analyzed brightness and color variations in the U, B and V bands of a sample of magnetically active close binary stars on a long time scale, finding that some of them become redder when they are fainter, i.e., as is expected from the presence of active regions mainly consisting of cool spots. Others become bluer when they become fainter, i.e., when spots reach the maximum coverage. In such a case hot photospheric faculae dominate the color variation. Again, this could be interpreted in the framework of the scenario proposed by Radick *et al.* (1998), Lockwood *et al.* (2007), and Hall *et al.* (2009) in which young and rapidly rotating stars arrange their surface magnetic flux predominantly into dark spots, whereas, when stars age and their rotation slows down, bright facula-like structures are favored.

Faculae, plage and supergranulation on the Sun are present even at the minimum of the cycle when few or no spots are on the disk. These features carry a large fraction of the global magnetic flux, but can be elusive at optical wavelength while are best detected in UV where, especially plages, show very large intensity contrasts. Chromospheric and TR maps have been indirectly obtained thanks to time series high-resolution spectroscopy in the UV (e.g. Walter *et al.* 1987, Neff *et al.* 1989, Busà *et al.* 1999, Pagano *et al.* 2001). One of the main results from these studies is the presence of T=10^4 K plasma at distances of 1–2 stellar radii from the stellar photospheres. On the other hand, X-ray observations of the same systems reveal that their coronae are filled in with warm ($\sim 10^6$ K) and hot ($\sim 10^7$ K) plasma components. Hence, the outer atmosphere of rapidly rotating cool stars are site of multi-temperature plasma structures possibly thermally confined by magnetic fields.

3.2. *Transition Regions and Coronae*

3.2.1. *Plasma dynamics in the transition regions*

Observations of emission lines originating from the TR (10^5 K) to the corona (10^7 K) of magnetically active stars have been used to shed light on plasma dynamics, a by-product of the heating mechanisms at work in the outer atmospheric layers. TR lines observed for the Sun show redshifts increasing with temperature of line formation till about 1-1.2$\times 10^5$ K (active regions and quiet Sun, respectively). The redshift decreases at higher temperature, crossing over to blueshift at T$\sim 10^6$ K (see e.g., Peter & Judge 1999). A similar effect is observed also for many active stars: α Cen A (G2 V) (Pagano *et al.* 2004), α Cen B (K1 V) (Wood *et al.* 1997), and EI Eri (G5 IV) (Jordan *et al.* 2001) which show the same behavior as the Sun, Procyon (F5 IV-V) that shows a maximum redshift at somewhat higher temperatures (Wood *et al.* 1997), and ξ Boo A (G8 V) that shows a maximum redshift at temperature lower then typically observed on the Sun (Pagano *et al.* 2006). The reason of the persistent line shifts has been a challenge for theoreticians; it was generally argued that the dependence of line shifts from temperature of line formation is due to the dynamic response of the atmosphere to the energy deposition after magnetic reconnection events, but several different models have been proposed (e.g., Peter 2006, Spadaro *et al.* 2006). On the other hand, TR lines of very active dMe stars, like AU Mic (dM1e), certainly show no conspicuous trend of line shift versus formation temperature (Pagano *et al.* 2000, Redfield *et al.* 2002). A model capable to represent the case of the Sun and solar analogs, should also explain whether the absence in AU Mic is a consequence of the strong magnetic field, high activity, high gravity, or late spectral type of this target.

Transition region lines are not only redshifted with respect to the photospheric lines, but they also show very broad wings in MS stars, young rapid rotators, and stars in close binaries. For rapidly rotating stars – e.g., AB Dor (Vilhu *et al.* 1998), or AR Lac (Pagano *et al.* 2001) – broad wings can arise from rings of hot gas at 2-3 stellar radii, like the "slingshot prominences" seen in H-alpha or the extended loops discussed in Section 3.1.2. For other cases, broad wings can be interpreted as a signature of "microflare" heating (Wood *et al.* 1997). Redfield *et al.* (2002) showed that there is a trend of decreasing relative importance of the broad components with increasing line formation temperature, implying that the spectroscopic impact of the mechanism that gives origin to the broad wings might be more important in the lower transition region than at 10^5 K temperature. The extension of these investigations to coronal temperatures is challenging, as discussed in section 3.2.2.

3.2.2. *Hi-res spectroscopy of coronal plasma*

Coronal spectroscopy at high resolution has been possible up to date only thanks to the observations of coronal forbidden lines in the UV and FUV spectrum, e.g., Fe XVIII λ 974, Fe XIX λ 1118, Fe XII λ 1242&1349, and Fe XXI λ 1354 (c.f., Pagano *et al.* 2000, Ayres *et al.* 2003, Redfield *et al.* 2003). As shown in Figure 2-*right* for the Fe XXI λ 1354, centroids of coronal lines of a number of late type giants and M dwarfs fall at rest with respect to the photosphere, hence the high-temperature plasma is not participating in any large systematic mass motions, but it is likely confined in stable magnetic structures. Moreover, the measured widths of the coronal forbidden lines are generally close to their thermal values (FWHM 40–90 km s^{-1} at T$\sim 10^{6.2}$–$10^{7.0}$ K), except for some of the fastest rotating stars in the observed sample (AB Dor, 31 Com, and the G1 star of Capella) all showing broadening exceeding the convolution between the thermal and the rotational widths, consistently with excess turbulence or with emitting regions at $\sim$0.4-1.3 stellar radii from the photosphere.

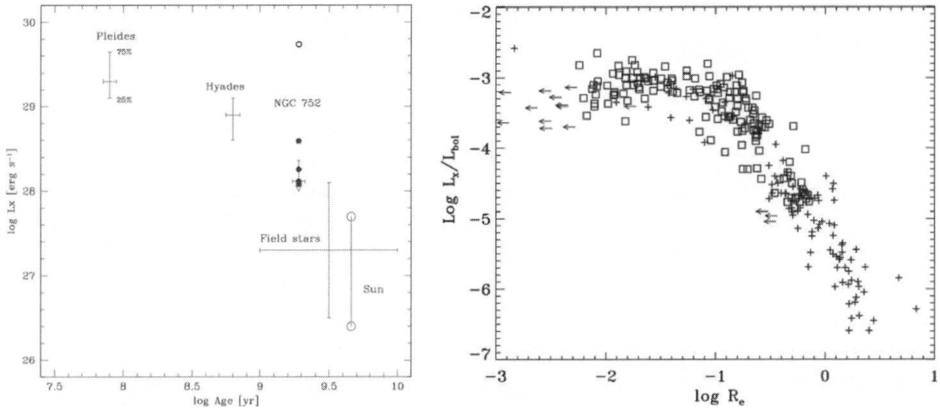

Figure 3. *Left:* X-ray luminosity of stars with mass in the range 0.8–1.2 $M_\odot$ vs. age. The line at age 4.5 Gyr connects the minimum and maximum of the Sun (from Giardino *et al.* 2008). *Right:* X-ray to bolometric luminosity ratio vs. empirical Rossby number including field dwarfs (crosses) and cluster stars (squares) (from Pizzolato *et al.* 2003).

3.2.3. *X-ray emission from coronae*

The soft X-ray flux from stellar coronae is correlated with the Ca II H&K flux over a factor 100,000 in the X-ray flux (e.g., Schrijver 1987, Ayres 1997), with the mean Sun which obeys the relation and moves along it during the solar cycle. During the stellar life, rotation slows down and consequently the X-ray luminosity decreases by 3-4 orders of magnitude (e.g., Giardino *et al.* 2008, Güdel 2008). This is not the case of stars in close binaries that, being tidally locked, maintain their fast rotation even at old ages, and therefore have high X-ray luminosity. Rotation period is therefore a good predictor of the total X-ray luminosity, till a critical value below which stars of different masses (or spectral types) enter a saturated regime, corresponding to $L_X/L_{bol} \sim 10^{-3}$ at a rotation velocity $\gtrsim$ 15 km s^{-1} (cf. Figure 3). Pizzolato *et al.* (2003) have demonstrated that the critical rotation period, below which saturation occours, increases with decreasing stellar mass, suggesting that the X-ray emission in late-type main sequence stars can be considered as an effect depending on the properties of the convective region as well as an effect depending on the total energy budget available in the star.

X-ray luminosity of very fast rotators (v sin i > 100 kms^{-1}) can be a factor of 3-5 below the saturated level (Randich 1998). The mechanisms responsible for this regime, defined as supersaturation, are not thoroughly understood.

3.2.4. *Rotational modulation of coronal emission*

Stellar rotation periods are generally longer than typical duration of X-ray observations, for this reason rotational modulation of X-ray emission has been detected only sporadically. Unambiguous detections of rotational modulation have been reported by Güdel (1995) for the young solar analogue EK Dra, by Marino *et al.* (2003) for VXR45, a G9 V star in IC 2391 (cf. Figure 4), by Hussain *et al.* (2005) for the K0 V AB Dor, and by Stassun *et al.* (2006) for a number of young stars in the Orion Nebula Cluster. AB Dor as well as many of the stars in ONC showing rotational modulation are fast rotators with an X-ray emission in the saturation regime. VXR45 is even in supersaturation regime. The presence of X-ray rotational modulation in these stars indicates a substantial inhomogeneity in the spatial distribution of active regions. Hence, saturation (and

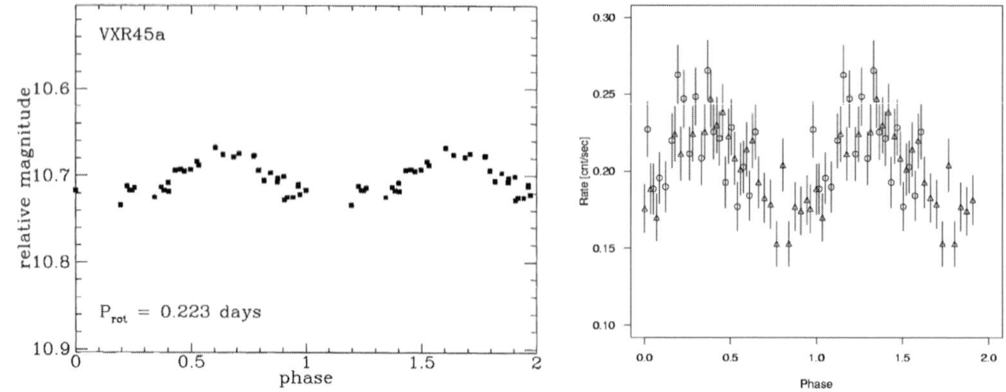

Figure 4. *Left*: Optical light curve of VXR45a obtained by Patten & Simon (1996) folded with a period of 0.223 days. *Right:* X-ray light curve of VXR45a acquired with XMM-Newton in the 0.3-7.8 keV band (see Marino *et al.* 2003), folded with the same rotation period used for the optical light curve. Circles and triangles have used for data acquired during the first and second covered rotations, respectively.

supersaturation) are not related, at least for these objects, to the filling of the stellar surface with active regions.

3.2.5. *Long-term observations of coronae*

The solar X-ray luminosity follows a cycle with the same periodicity as the chromospheric cycle, but with a much larger amplitude, up to a factor of 100 in the Yohkoh 0.73 – 2.5 keV band (Acton 1996). However, cycles are elusive in other stars. In fact, these have been detected only in α Cen B (Ayres 2009), 61 Cyg (Hempelmann *et al.* 2006), and HD 81809 (Favata *et al.* 2008), all solar-like stars with moderate activity. HD 81809 is a close visual binary of slow rotating solar-like stars (G2 V+G9 V) with a well defined chromospheric cycle (Baliunas *et al.* 1995). XMM-Newton observations have shown a well-defined coronal cycle that has been modeled by Favata *et al.* (2008) with a simple extension of the solar case in terms of varying coverage of solar-like active regions, but

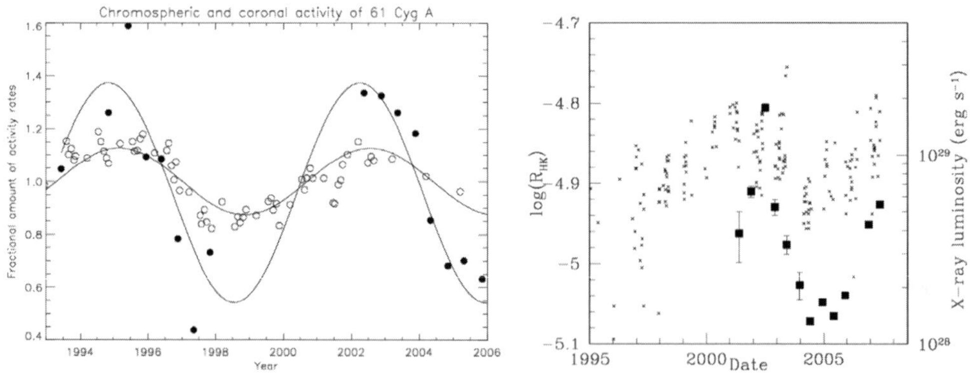

Figure 5. *Left*: The coronal and chromospheric cycles of 61 Cyg A observed by XMM-Newton (0.2-2.0 keV band) and Mt. Wilson & Lowell Observatory projects, respectively (see Hempelmann *et al.* 2006 for details) *Right:* The coronal and chromospheric cycles of HD 81809 measured thanks to XMM-Newton monitoring in the 0.2-2.5 keV band (*squares, right hand scale*) and the monitoring of the chromospheric activity index R_{HK} from Mt. Wilson project (*crosses, left hand scale*). See Favata *et al.* (2008) for details.

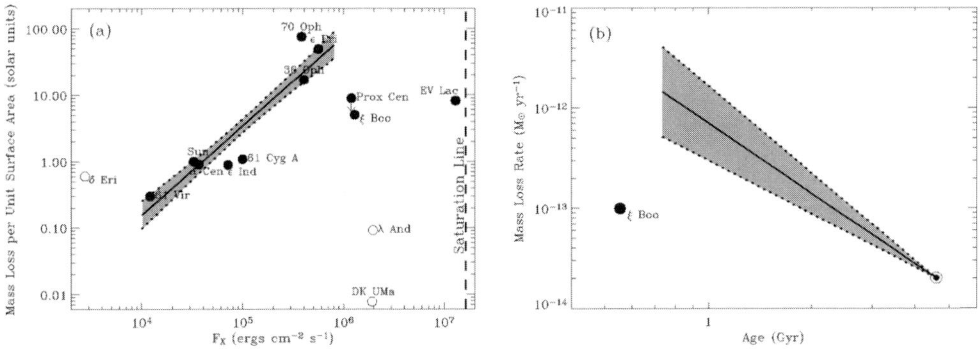

Figure 6. *Left*: Measured mass loss rates (per unit surface area) plotted versus X-ray surface flux. *Right:* The mass loss history of the Sun suggested by the power law relation shown in the left panel. Both figures are from (Wood *et al.* 2005).

with a larger coverage than for the Sun (cf. Figure 6-*right*). The binary stars in 61 Cyg - K5 V (A)+ K7 V (B) - observed by XMM-Newton in the X-rays (0.2–2.0 keV) show both short time scales and long-term variations (Hempelmann *et al.* 2006). In particular, 61 Cyg A has a clear coronal cycle phased with the chromospheric activity cycle (cf. Figure 6-*left*). The changes of coronal properties during the cycle resemble the solar behaviour. The coronal activity of 61 Cyg B is more irregular, but also follows the chromospheric activity (Robrade *et al.* 2007). The close-by α Cen triple system, made by a G2 V (A), a K1 V (B) and the M5 V Proxima Centauri, has been monitored by all available X-ray observatories since the '80. Ayres (2009) confirmed the presence for α Cen B of a coronal cycle with period of nine years (as the chromospheric cycle) and a depth of a factor $\sim$6, comparable to the depth of the solar coronal cycle ($\sim$ 5), while α Cen A did not significantly vary in the 0.2–2.0 keV "ROSAT" band, despite its dramatic fainting in the high energy tail of the coronal spectrum observed by XMM-Newton since 2003 (Robrade *et al.* 2005 & 2006). The average coronal luminosity of α Cen B is a factor of 2 higher than that of the Sun, while α Cen A is a similar factor lower, making this star a good candidate as a star in a Mounder-like magnetic minimum. More details on the coronal cycle of α Cen are given by Ayres in this proceedings.

Coronal cycles are not detected in more active stars, possibly because stochastic variability is prevalent in these objects.

3.3. *Winds from late-type dwarfs*

Spectroscopic analyses of stellar H I Ly-α lines have proven to be the best way so far to clearly detect and measure winds of MS stars like the Sun (Wood *et al.* 2004). In fact, the weak solar-like winds are too hot and thin to provide detectable UV or X-ray absorption (or emission) signatures. Also the wind radio emission is below the sensitivity of radio telescopes in use up to date. In a magnetic active star, Ly-α is in emission and its observed profile is the result of the intrinsic stellar line profile with superimposed the absorptions occurring in the stellar astrosphere (the environment around the star similar to the heliosphere that surrounds the Sun), in the LISM, and finally in the heliosphere. By modeling these absorption features, particularly from the properties of the astrosphere model required to fit the line profile, Wood *et al.* (2002) performed the first quantitative measurements of mass loss rates for G and K dwarf stars, and found that the mass loss rates increase with activity (from X-ray surface flux) and thus with decreasing stellar age. From the observed trend of mass loss rate vs. stellar age it results that the solar wind might have been 1000 times stronger when the Sun was very young, and thereby likely played a major role in the evolution of planetary atmospheres, particularly the stripping of volatiles from primitive Mars.

4. Conclusions

Stellar variability data tell us that the Sun behaves as a typical solar-likes star in the same phase of evolution. In fact, the Sun activity level and its activity modulation fit very well the behavior of late-type MS stars, in every kind of diagnostic (optical continuum, optical and UV emission lines, X-ray emission). On the other hand, rapidly rotating late-type stars such as young stars and stars in close binary systems show magnetic activity features never observed on the Sun; e.g., polar spots, spot migration towards the poles, and bolometric emission dominated by the cool spots. Preliminary measurements of stellar solar-like winds show that the solar wind could have had a role in the evolution of planetary atmospheres.

New exciting results are expected in the near future from the refurbished HST and EVLA. For instance, the access to UV hi-res (R∼20 000) spectra of stars in open clusters will be eventually possible with HST-COS spectrograph, and solar like stellar winds could become directly detectable with the EVLA.

Finally, the understanding of stellar activity phenomena is becoming a must to characterize extrasolar planetary systems, from their discovery to the analysis of their evolution.

References

Acton, L. 1996, *ASP Conf. Ser.* 109, p.45

Ayres, T. R., Fleming, T. A., Simon, T., *et al.* 1995, *APJS*, 96, 223

Ayres, T. R. 2009, *ApJ*, 696, 1931

Ayres, T. R. 1997, *J. Geophys. Res.*, 102, 1641

Ayres, T. R., Brown, A., Harper, G.M., et al. 2003, *ApJ*, 583, 963

Baliunas, S. L., Donahue, R. A., Soon, W. H., *et al.* 1995, *ApJ*, 438, 269

Baliunas, S. L., Donahue, R. A., Soon, W., & Henry, G. W. 1998, *ASP Conf. Ser.*, 154, p. 153

Benz A. & Güdel, M. 1994, *A&A*, 285, 621

Berdyugina, S. V. & Henry, G. W. 2007, *ApJ*, 659, L157

Busà, I., Pagano, I., Rodonò, M., Neff, J. E., & Lanzafame, A. C. 1999, *A&A*, 350, 571

Catala, C., Donati, J.-F., Shkolnik, E., Bohlender, D., & Alecian, E. 2007, *MNRAS*, 374, L42

Cayrel de Strobel, G. & Bentolila, C. 1989, *A&A*, 211, 324

Collier Cameron, A., Bedford, D. K., Rucinski, S. M., Vilhu, O., & White, N. E. 1988, *MNRAS*, 231, 131

Donati, J.-F., Collier Cameron, A., & Petit, P. 2003a, *MNRAS*, 345, 1187

Donati, J.-F., Collier Cameron, A., Semel, M. *et al.* 2003b, *MNRAS*, 345, 1145

Donati, J.-F., Jardine, M. M., & Gregory, S. G. 2007, *MNRAS*, 380, 1297

Favata, F. & Micela, G. 2003, *Space Science Reviews*, 108, 577

Favata, F., Micela, G., Orlando, S. *et al.* 2008, *A&A*, 490, 1121

Gershberg, R. E. 2005, *Solar-type Activity in Main-Sequence Stars*, Springer

Giardino, G., Pillitteri, I., Favata, F., & Micela, G. 2008, *A&A*, 490, 113

Güdel, M. & Benz, A. 1993, *ApJ 405*, L63

Guedel, M., Schmitt, J. H. M. M., Benz, A. O., & Elias, N. M. 1995, *A&A*, 301, 201

Güdel, M. 2008, *Living Reviews in Solar Physics*, vol. 4, no. 3

Hall, J. C., Henry, G. W., Lockwood, G. W., Skiff, B. A., & Saar, S. H. 2009, *AJ* 138, 312

Hempelmann, A., Robrade, J., Schmitt, J. H. M. M., *et al.* 2006, *A&A*, 460, 261

Hussain, G. A. J., Brickhouse, N. S., & Dupree, A. K. 2005, *ApJ*, 621, 999

Jordan, C., Ayres, T. R., Brown, A., Linsky, J. L., & Simon, T. 1987, *MNRAS*, 225, 903

Jordan, C., Sim, S. A., McMurry, A. D., & Aruvel, M. 2001, *MNRAS*, 326, 303

Kürster, M. & Dennerl, K. 1993, in *Physics of Solar and Stellar Coronae*, J. F. Linsky & S. Serio (eds.), Kluwer Acad. pub., p. 443
Lanza, A. F., Catalano, S., Cutispoto, G., Pagano, I., & Rodonò, M. 1998, *A&A*, 332, 541
Lanza, A. F., Pagano, I., Leto, G., *et al.* 2009a, *A&A*, 493, 193
Lanza, A. F., Aigrain, S., Messina, S., *et al.* 2009b, *A&A*, in press (arXiv:0901.4618)
Lockwood, G. W., Skiff, B. A., Henry, G. W., *et al.* 2007, *ApJS*, 171, 260
Marino, A., Micela, G., Peres, G., & Sciortino, S. 2003, *A&A*, 407, L63
Messina, S. & Guinan, E. 2002, *A&A*, 393, 225
Messina, S. & Guinan, E. 2003,*A&A*, 409, 1017
Messina, S. 2008, *A&A*, 480, 495
Neff, J. E., Walter, F. M., Rodonò M., & Linsky, J. L. 1989, *A&A*, 215, 79
Pagano, I., Linsky, J. L., & Carkner, L. 2000, *ApJ*, 532, 497
Pagano, I., Rodonò, M., Linsky, J. L., *et al.* 2001, *A&A*, 365, 128
Pagano, I., Linsky, J. L., Valenti, J., & Duncan, D. K. 2004, *A&A*, 415, 331
Pagano, I., De Martino, C., Lanza, A. F., Spadaro, D., & Linsky, J. L. 2006, *ESA-SP* 617, 91
Parker, E. N. 1955 *ApJ*, 122, 293
Patten, B. M. & Simon, T. 1996, *ApJSS*, 106, 489
Peter, H. 2006, *AdSpR*, 39, 1814
Peter, H. & Judge, P. G. 1999, *ApJ*, 522, 1148
Pizzolato, N., Maggio, A., Micela, G., Sciortino, S., & Ventura, P. 2003, *A&A*, 397, 147
Radick, R. R., Lockwood, G. W., Skiff, B. A., & Baliunas, S. L. 1998, *APJSS*, 118, 239
Randich, S. 1998, *ASP Conf. Ser.*, 154, p. 501
Redfield, S., Linsky, J. L., Ake, T. B., *et al.* 2002, *ApJ*, 581, 626
Redfield, S., Ayres, T. R., Linsky, J. L., *et al.* 2002, *ApJ*, 585, 993
Ribas, I., Guinan, E. F., Güdel, M., & Audard, M. 2005, *ApJ*, 622, 680
Robrade, J., Schmitt, J. H. M. M., & Favata, F. 2005, *A&A*, 442, 315
Robrade, J., Schmitt, J. H. M. M., & Hempelmann, A. 2007, *MSAIt*, 78, 311
Rodonò, M. 1992 *LNP*, 397, 201
Rodonò, M. & Cutispoto, G. 1994, *ASP Conf. Series*, 64, 459
Rodonò, M., Messina, S., Lanza, A. F., Cutispoto, G., & Teriaca L. 2000, *A&A*, 358, 624
Saar, S. 2006, *BAAS* 38, 240
Schrijver, C. J. 1987, *ApJ* 172, 111
Simon, T., Ayres, T. R., Redfield, S., & Linsky, J. L. 2002, *ApJ*, 579, 800
Spadaro, D., Lanza, A. F., Lanzafame, A. C. *et al.* 2006, *ApJ*, 582, 486
Stassun, K. G., van den Berg, M., Feigelson, E., & Flaccomio, E. 2006, *ApJ*, 649, 914
Stelzer, B., Flaccomio, E., Briggs, K. *et al.* 2007, *A&A*, 468, 463
Strassmeier, K. G. 2009, *IAU Symp*, 259, 363
Telleschi, A., Güdel, M., Briggs, K., *et al.* 2005, *ApJ*, 622, 653
Trigilio, C., Leto, P., & Umana, G. 1998, *A&A*, 330, 1060
Umana, G., Trigilio, C.,Tumino, M., Catalano, S., Rodonò, M. 1995, *A&A*, 298, 143
Vida, K., Kovári, Zs., Ŝvanda, M. *et al.* 2007, *AN*, 328, 1078
Vilhu, O., Muhli, P., Huovelin, J., *et al.* 1998, *AJ* ,115, 1610
Walker, G. A. H., Croll, B., Kuschnig, R., *et al.* 2007, *ApJ* 659, 1611
Walker, G. A. H., Croll, B., Matthews, J. M., *et al.* 2008, *A&A* 482, 691
Walter, F. M., Neff, J. E., Gibson, D. M., *et al.* 1987, *A&A*, 186, 241
Wood, B. E., Linsky, J. L., & Ayres, T. R. 1997, *ApJ*, 478, 745
Wood, B. E., Müller, H.–R., Zank, G. P., & Linsky, J. L. 2002, *ApJ*, 574, 412,
Wood, B. E. 2004, *Living Reviews in Solar Physics*, Vol. 1, no. 2
Wood, B. E., Müller, H.-R., Zank, G. P., Linsky, J. L., & Redfield, S. 2005, *ApJ*, 628, L143

Solar and Stellar Variability: Impact on Earth and Planets
Proceedings IAU Symposium No. 264, 2009
A. G. Kosovichev, A. H. Andrei & J.-P. Rozelot, eds.

doi:10.1017/S1743921309992559

The Cycles of Alpha Centauri

Thomas R. Ayres[1]

[1]Center for Astrophysics & Space Astronomy, University of Colorado,
389-UCB (CASA), Boulder, Colorado, USA
email: Thomas.Ayres@Colorado.edu

Abstract. The main AB pair of the nearby Alpha Centauri triple system has one of the most extensive X-ray records of any cosmic object, stretching over three decades. The primary, α Cen A (G2 V), is a near twin of the Sun, with a similarly soft (1–2 MK) corona. The secondary, α Cen B (K1 V), is more active than the Sun, with a generally harder coronal spectrum. Here, spatially resolved measurements of the pair by *Chandra's* High Resolution Camera are compared, on a common basis, with previous pointings from *ROSAT* and *XMM-Newton*.

Keywords. X-rays: stars, stars: individual (HD 128620, HD 128621), stars: activity

1. Introduction

One of the remarkable properties of the Sun is the 11-year sunspot cycle, which in fact is a 22-year magnetic oscillation ("Hale Cycle") when polarity reversals of the mean field are taken into account. This signature property of solar magnetic activity is attributed to a global "dynamo" (e.g., Parker 1970). Because differential rotation acts as a catalyst, and because overturning turbulent gas flows are the ultimate power source of the dynamo, magnetic activity is crucially dependent on the stellar spin rate and whether the star is cool enough to support vigorous surface convection (see review by Güdel 2004).

The outward simplicity of the solar sunspot cycle is deceptive, however, and it has remained surprisingly resistant to theoretical modeling. The stubbornness of the solar example has encouraged the examination of analogous phenomena on other cool stars, to gain a broader perspective. The best known effort in this regard is the long term Ca II HK monitoring program begun by O.C. Wilson in the 1960's (Wilson 1978), to track cycles through subtle changes in chromospheric emission as the population of surface active regions waxes and wanes over long time scales. Most of the stellar cycles recovered to date are strikingly solar-like in HK amplitude and decadal duration (e.g., Radick *et al.* 1998). Unlike the activity itself, however, the stellar cycle periods do not seem to depend strongly on surface temperature or spin rate, except that the more active, fast rotators avoid cycles altogether, in favor of short-term stochastic chromospheric variability.

Surprisingly though, a debate still simmers over important observational properties of the *solar* cycle, especially its amplitude in soft X-rays associated with high energy *coronal* ($T \sim$ 1–10 MK) phenomena. The corona, in fact, is much more closely tied to the magnetic field than the Ca II chromosphere: the coronal gas is heated by a magnetic agency and bottled up in fine scale loop-like magnetic structures, while the chromosphere is heated partly by acoustic shocks. Consequently, the solar HK index varies only about 25% over the solar cycle, while coronal X-rays can change by factors of several to a hundred, depending on the energy band, although the details remain in some dispute. For example, Judge, Solomon, & Ayres (2003) deduced a modest factor of ~6 change in the solar 0.1–2.4 keV X-ray luminosity over the rise of current Cycle 23, based on ultra-soft X-ray photometry from the SNOE satellite; compared with the more extreme $\sim 30\times$ variations inferred from higher energy experiments like *Yohkoh* SXT (Acton 1996).

The X-ray cycle amplitude is a pivotal aspect of solar activity, because coronal ionizing radiations strongly influence the heliosphere in general, and Earth's upper atmosphere ("ionosphere") in particular. Indeed, the whole question of the evolution of the solar ionizing flux (and coronal wind) is central to the development of planetary atmospheres in the early solar system (e.g., Ayres 1997, and references to previous work therein), as well as young exoplanets around other stars; main topics of this Symposium.

Fundamentally, our few decades view of the Sun's X-cycle represents just a glimpse of what that variability might be over the longer term, in the past and into the future. Again, one is tempted to turn to the stars for insight. Discouragingly little is known about stellar X-cycles, however. The best examples, in terms of sample sizes, typically are two or three well separated surveys of young open clusters like the Hyades, where the inter-epoch variations appear to be dominated by flares (e.g., Stern 2001).

There are, however, a few cases of dedicated coronal monitoring programs on solar-age field stars including the G subgiant HD 81809 (Favata *et al.* 2004), 61 Cyg AB (K5 V + K7 V; Hempelmann *et al.* 2006), and most notably the close-by α Centauri triple system (G2 V [A], K1 V [B], M5 V [C, also known as Proxima]). At a distance of just over a parsec, the main AB pair of α Cen has received considerable attention from four generations of X-ray observatories, beginning with its discovery (albeit unresolved) by *HEAO-1* (Nugent & Garmire 1978); separation into distinct coronal sources by *Einstein* (Golub *et al.* 1982); several campaigns with *ROSAT* (Schmitt & Liefke 2004); and continuing into the contemporary eras of *XMM–Newton* (Robrade, Schmitt, & Favata 2005) and *Chandra* (Ayres *et al.* 2008; Ayres 2009). The Robrade *et al.* study reported a sharp decline in the X-ray count rate of the primary in early 2005—a "fainting" of its corona, as the authors put it—which motivated the subsequent *Chandra* work. The α Cen AB orbit has been closing rapidly over the past decade, and in 2005 was only barely resolvable by *XMM–Newton,* although even at minimum separation ($4''$, circa 2016) still will be easy for the sharper $1''$ gaze of *Chandra.* The latter campaign included direct imaging of the pair at roughly six month intervals from 2005 October to the present (most recent observation: 2009 May; one more scheduled for 2009, and two additional pointings in 2010). The *Chandra* campaign also included a pivotal deep transmission grating spectrum in mid-2007. It showed clearly that the steep decline of α Cen A seen by *XMM–Newton* mainly was the result of a loss of the highest temperature components of the primary's corona rather than a truly complete collapse of its hot outer atmosphere.

The present report reviews the progress made to date since the Ayres (2009) study.

2. Observations

Figure 1 illustrates time-lapse *Chandra* imaging of α Centauri over the period 2000–2010. α Cen A (G2 V) is to the upper left in each image pair, while its optically fainter companion (B: K1 V) is to the lower right. The peak intensity of α Cen B is about 2 counts s^{-1}, a relatively bright *Chandra* source. Thin squiggly curves are combined orbital, parallactic, and proper motion tracks. Conspicuous 'wobbles' in the observed X-ray spots represent the $0.8''$ *annual parallax,* easily captured by the high astrometric precision of *Chandra.* Note the rapid closing of the orbital separation in recent years. In the gap between the first *Chandra* pointing in 1999.98, and the beginning of the new HRC-I series in 2005.81, the *XMM–Newton* monitoring program, mentioned earlier, reported the surprising, steep decline in A's coronal X-rays.

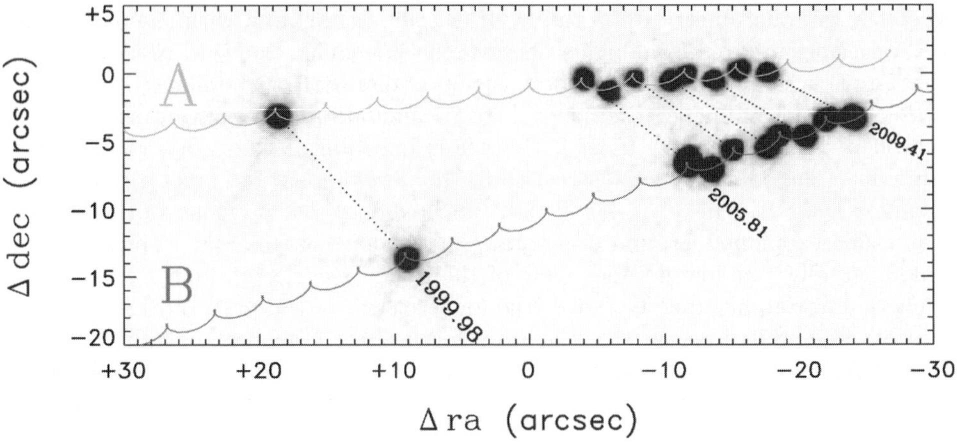

Figure 1. *Chandra* HRC imaging of α Centauri.

Figure 2 is the full X-ray time series of α Cen from *ROSAT* (1995–98 [upper panel: 1996]), and more recently *XMM–Newton* and *Chandra* (lower panel: the two dense *ROSAT* campaigns now are represented by time averages) for the reference energy band 0.2–2 keV. Blue points (lower) are for α Cen A, red points (upper) for B. Note the occasional "flares" in the B record during the two dense *ROSAT* sequences in 1996. The jagged hatched tracks from 2003–2007 are for *XMM–Newton* (semiannual measurements). Energy conversion factors were based on specialized DEM models derived for a range of possible activity states of the two stars. Furthermore, the soft response of the *XMM–Newton* EPIC cameras was progressively suppressed until a reasonable agreement was achieved with the overlapping *Chandra* points. Since EPIC must use the thick optical blocking filter for stars as bright as the α Cen pair (to control CCD "red leak"), it is plausible that the true soft response might be overestimated.

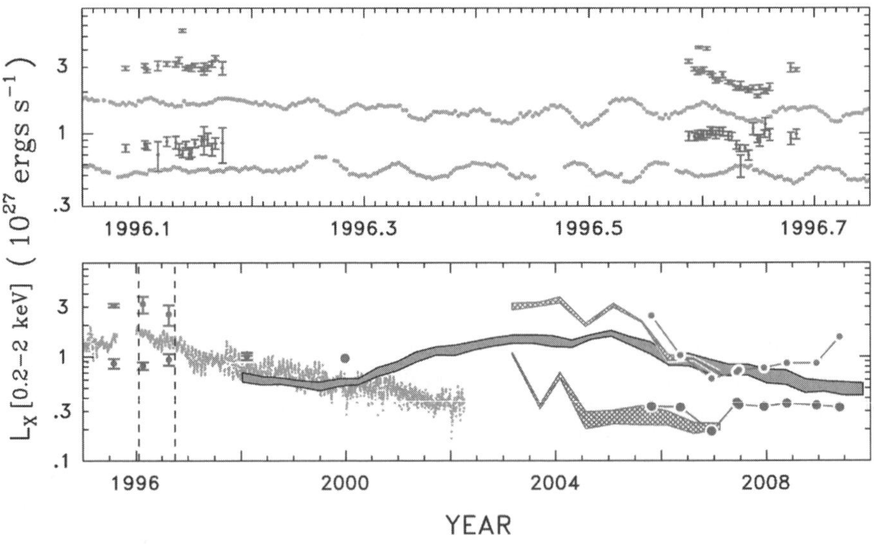

Figure 2. Long term history of α Cen X-rays.

Gray points and the smooth shaded trace depict the estimated behavior of solar X-rays (during previous Cycle 23) in this energy band, shifted in time to roughly match various features of the α Cen series. In the top panel, the upper solar trace is for cycle maximum, the lower for minimum: rotational modulations are clearly visible at both extremes. Although *XMM–Newton* witnessed what appeared to be a dramatic 'fainting' of α Cen A in late 2004, the contemporary *Chandra* measurements, interpreted in light of the LETGS spectra of the AB pair, show that the drop of the primary from its historical high in 2000 has been relatively modest, smaller in fact than typical solar cycle changes.

3. Conclusions

The combined time series suggests that α Cen B was near X-ray maximum in the mid-1990's, minimum in the late-1990's, peaked again in 2004–2005, then was declining in 2006–2007, bottomed out in 2008, and by mid-2009 appears to be rising again. B's roughly 8-year X-ray modulation is shorter than the solar 11-year sunspot period, although the cycle depth is similar (factor of 5 in 0.2–2 keV band). Meanwhile, α Cen A showed minimal variability during 1995–2000, a decline in 2003–2005, and presently is mired in an activity lull, somewhat down from the mid-1990's level. α Cen A is behaving like the Maunder minimum Sun, a "flat activity" star if you will (Judge & Saar 2007).

Aside from the details of the twin X-cycles, the "smoking gun" *Chandra* transmission grating spectrum in mid-2007 emphasized that much of the coronal radiative output of cool-corona objects like the Sun falls at longer wavelengths than are recorded efficiently by contemporary stellar instruments, rendering L_X and cycle depth strongly dependent on the exact energy span of the measurements. This has broad-reaching impact on any discussions of coronal heating requirements, and the changes over starspot cycles.

Acknowledgements

This work was supported by the *Chandra* Guest Investigator program.

References

Acton, L. 1996, in: R. Pallavicini & A. K. Dupree (eds.), *Ninth Cambridge Workshop on Cool Stars, Stellar Systems, and the Sun*, ASP Conf. Proc. 109, (San Francisco: ASP), p. 45
Ayres, T. R. 1997, *JGR*, 102, 1641
Ayres, T. R., Judge, P. G., Saar, S. H., & Schmitt, J. H. M. M. 2008, *ApJ*, 678, L121
Ayres, T. R. 2009, *ApJ*, 696, 1931
Favata, F., *et al.* 2004, *A&A*, 418, L13
Golub, L., Harnden, F. R., Jr., Pallavicini, R., Rosner, R., & Vaiana, G. S. 1982, *ApJ*, 253, 242
Güdel, M. 2004, *A&AR*, 12, 71
Hempelmann, *et al.* 2006, *A&A*, 460, 261
Judge, P. G. & Saar, S. H. 2007, *ApJ*, 663, 643
Judge, P. G., Solomon, S. C., & Ayres, T. R. 2003, *ApJ*, 593, 534
Nugent, J. & Garmire, G. 1978, *ApJ*, 226, L83
Parker, E. N. 1970, *ApJ*, 162, 665
Radick, R. R., Lockwood, G. W., Skiff, B. A., & Baliunas, S. L. 1998, *ApJS*, 118, 239
Robrade, J., Schmitt, J. H. M. M., & Favata, F. 2005, *A&A*, 442, 315
Schmitt, J. H. M. M. & Liefke, C. 2004, *A&A*, 417, 651
Stern, R. A. 2001, *Advances in Space Research*, 26, 1723
Wilson, O. C. 1978, *ApJ*, 226, 379

Solar and Stellar Variability: Impact on Earth and Planets
Proceedings IAU Symposium No. 264, 2009
A. G. Kosovichev, A. H. Andrei & J.-P. Rozelot, eds.

doi:10.1017/S1743921309992560

Harmonic analysis approach to solar cycle prediction and the Waldmeier effect

K. Petrovay

Eötvös University, Department of Astronomy,
Pf. 32, H-1518, Budapest, Hungary
email: `K.Petrovay@astro.elte.hu`

Abstract. It is demonstrated that, in addition to the precursor method, harmonic analysis approach to solar cycle prediction is also strongly conditioned by the Waldmeier effect.

Keywords. Sun: activity

---

1. Introduction

The current unusually deep and extended solar minimum has rekindled interest in solar cycle prediction. The apparently most successful approach to prediction has been the precursor method, based on the correlation of certain solar parameters around or before solar minimum with the amplitude of the following maximum. However, in an important paper Cameron & Schüssler (2007) recently argued that most of these methods may be ultimately a simple consequence of Waldmeier's well known effect consisting in a correlation between the rise rate of a cycle and its maximal amplitude. This casts doubt on the feasibility of true cycle-to-cycle prediction, carried out without relying on indicators of the incipient new cycle.

These problems with the precursor methods suggest that it might be worth taking a second look at an alternative, time-honoured approach to the problem of solar cycle prediction: harmonic analysis. The first attempt to decompose the sunspot number series into a superposition of periodic functions, and then use it to extrapolate the series into the future was due to Kimura (1913). His list of periods already include a large number of values in the range 7 to 15 years; the physical reality of these periods, however, was doubted already by contemporary researchers. The approach most commonly used in harmonic analysis today, the power spectrum, was first taken for the sunspot number series by Cole (1973). In the following we take a fresh look at the power spectrum of the sunspot number series including data for the last few cycles and consider the question to what extent the resulting periods can be considered real physical periods and whether they are stable enough to serve as a basis for solar cycle prediction. Our findings indicate that harmonic analysis results are not free either from the strong influence of the Waldmeier effect.

2. Power spectra of the sunspot number record

Figure 1 presents the power spectrum of the smoothed annual sunspot number series for the whole period covered by the "official" record, i.e. 1749–2008.

Conspicuous peaks can be noticed at the first harmonic (5.5 years) and subharmonic (22 years) of the dominant 11-year peak. This is of course expected as in any series of asymmetrical cycles of variable amplitude such harmonics will naturally appear.

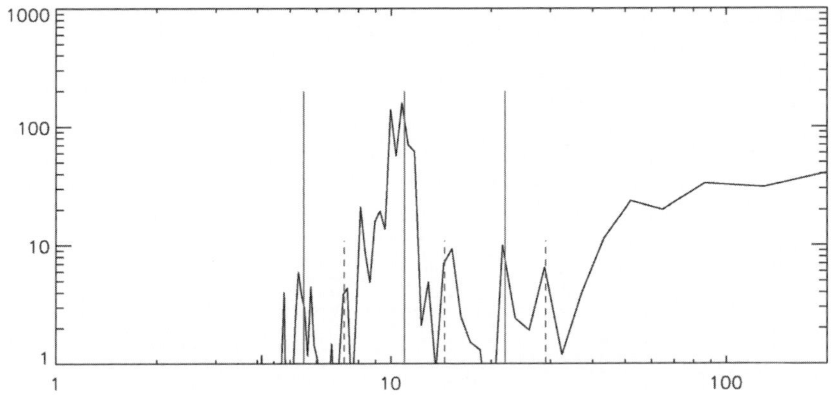

Figure 1. Power spectrum of the smoothed monthly sunspot number series for the period 1749–2008. Solid vertical bars mark the 11-year period, its first harmonic and subharmonic; dashed vertical bars are drawn at a fiducial period of 14.5 years, its harmonic and subharmonic.

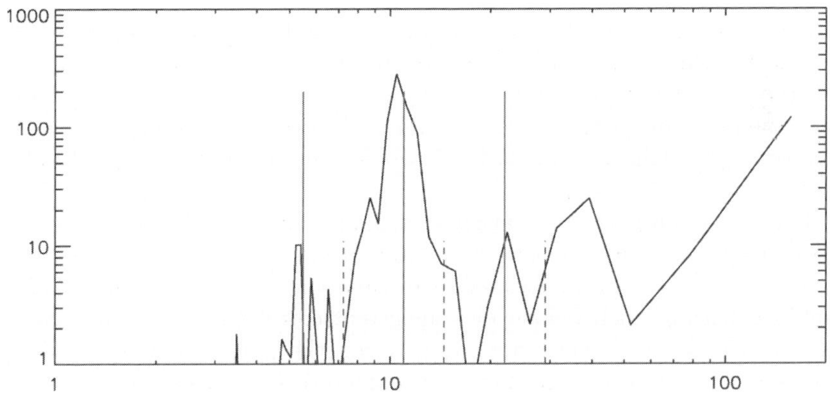

Figure 2. Same as Fig. 1 for the period 1850–2008.

What is more surprising is that, in addition to the finite natural width of the 11-year peak, with sidelobes, a separate secondary peak appears near a period of 14.5 years. The reality of this secondary period seems to be confirmed by the presence of further distinct peaks at the corresponding harmonic and subharmonic periods.

At first sight this finding might be taken as evidence for the physical reality of some decadal periods additional to the 11-year base period. However, constraining the time interval considered to data more recent than 1850, from which time the sunspot number series is generally considered to be more reliable, the 14.5-year secondary peak and its harmonics completely disappear (Fig. 2).

Does this mean that the apparent secondary period is due to unreliable early observations? Probably not: Figure 3 indicates that the appearance of the 14.5-year secondary peak in the complete series is almost entirely due to the strong predominance of this period (and its harmonic) in the interval 1783–1835, covering the Dalton minimum, consisting of three consecutive unusually weak cycles, when the "normal" 11-year mode of operation was completely suppressed.

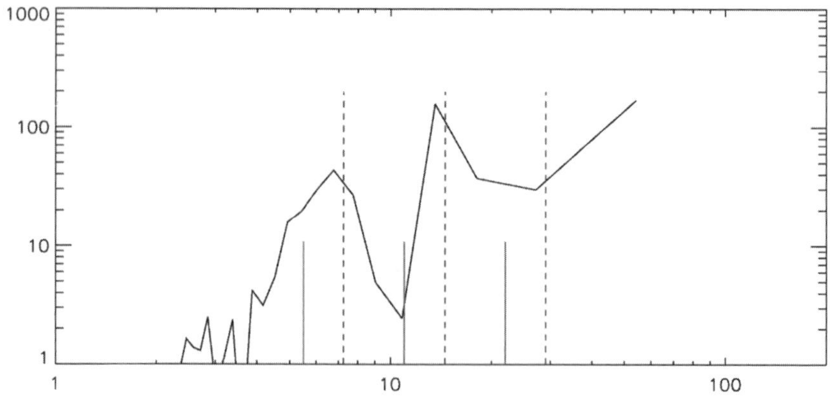

Figure 3. Same as Fig. 1 for the period 1783–1835.

3. Interpretation

Two possible interpretations arise regarding the findings of the previous section. On the one hand, the Sun may have been operating in a different mode during the Dalton minimum, the cycle length being 14.5 years instead of the usual 11 years. Indications that during the much deeper Maunder minimum the cycle period may have been as long as 22 years or so (Usoskin & Mursula 2003) would seem to lend some plausibility to this assumption.

Figure 4, however, supports an alternative explanation. This plot shows the well known inverse correlation between cycle length and amplitude, which in turn is the consequence of the strong inverse correlation between rise time and cycle amplitude (Waldmeier effect), combined with a much weaker or nonexistent correlation between decay time and amplitude. It is clear that cycles 4 to 7, covering the period in Fig. 3, all fall in the right hand side of the plot and are therefore characterized by a longer than average cycle length. The effect is even stronger in the case (not plotted here) when, instead of the usual minimum to minimum definition, cycle length is defined as the interval between successive maxima: indeed, the interval between the maxima of cycles 4 and 5 exceeded 17 years (Usoskin & Mursula 2003, see Table 1 in Usoskin & Mursula 2003).

The cycles around the Dalton minimum, then, seem to lie at the low amplitude (or long period) end of a continuum representing the well known cycle length–amplitude relation, ultimately explained by Waldmeier effect. There seems to be no evidence for a separate mode of operation of the dynamo during the Dalton minimum.

4. Conclusion

The predictive skill of harmonic analysis is well known to be limited but this is usually attributed to secular changes in the periods of modes and their amplitudes in the harmonic spectrum (e.g. Kane 2007). Recently there has been significant advance in the understanding of such secular changes for centennial cycles. For the Gleissberg cycle the results of Kolláth & Oláh (2009) indicate a slow linear secular increase in the period. For the 210-year Suess cycle McCracken & Beer (2008) present further evidence for the temporally intermittent nature of this marked peak in the spectrum of solar proxies. The Suess cycle seems to have a role in regulating the recurrence rate of grand

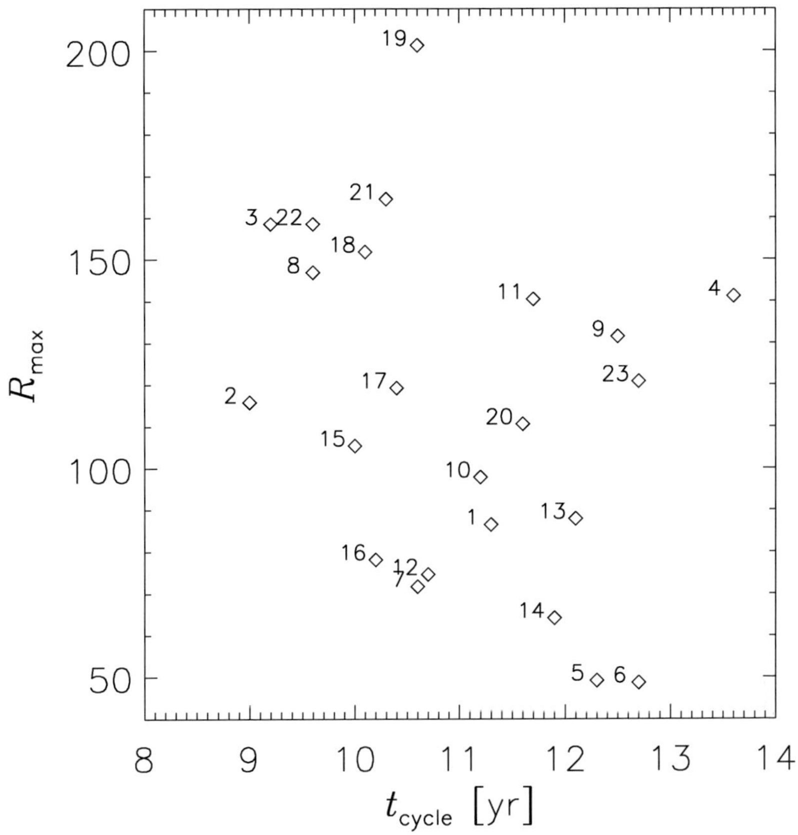

Figure 4. Plot of cycle length vs. cycle amplitude for solar cycles 1 to 23.

minima. Grand minima, in turn, only seem to occur during < 1 kiloyear intervals ("Spörer events") around the minimum of the ∼ 2400-year Hallstatt cycle.

These results may raise the hope that with a better understanding of centennial and millennial variations of solar activity long term activity changes may ultimately be successfully predicted on the basis of the harmonic analysis approach. However, our findings in this paper regarding the role of the Waldmeier effect imply that even if centennial/millennial scale variations may be statistically predicted, there may still be fundamental obstacles in the way of a successful prediction of decadal scale cycle-to-cycle variations that are mankind's prime concern.

Acknowledgement

This research was supported by the Hungarian Science Research Fund (OTKA) under grant no. K67746 and by the European Commission through the RTN programme SOLAIRE (contract MRTN-CT-2006-035484).

References

Cameron, R. & Schüssler, M. 2007, ApJ, 659, 801
Cole, T. W. 1973, Solar Phys., 30, 103

Kane, R. P. 2007, Solar Phys., 246, 487
Kimura, H. 1913, MNRAS, 73, 543
Kolláth, Z. & Oláh, K. 2009, A&A, 501, 695
McCracken, K. G. & Beer, J., 2008, in Proc. 30th International Cosmic Ray Conference, Vol. 1, UNAM, Mexico, p. 549
Usoskin, I. G. & Mursula, K. 2003, Solar Phys., 218, 319

Solar and Stellar Variability: Impact on Earth and Planets
Proceedings IAU Symposium No. 264, 2009
A. G. Kosovichev, A. H. Andrei & J.-P. Rozelot, eds.

doi:10.1017/S1743921309992572

Solar cycles: the past evolution influence

Alexis Klutsch[1,2] and Rubens Freire Ferrero[2]

[1]Universidad Complutense de Madrid, Departamento de Astrofísica, Facultad C.C. Físicas, 28040 Madrid, Spain, email: `klutsch@astrax.fis.ucm.es`

[2]Observatoire Astronomique, Université de Strasbourg & CNRS, UMR 7550, 11 rue de l'Université, 67000 Strasbourg, France, email: `rubens.freire@astro.unistra.fr`

Abstract. The so-called solar cycle is generally characterized by the quasi-periodic oscillatory evolution of the photospheric spots number. This quasi-periodic pattern has always been an intriguing question. Several physical models were proposed to explain this evolution and many mathematical data analysis were employed to determine the principal frequencies noticeable in the measured data. Both approaches try to predict the future evolution of the solar activity and to understand the physical phenomena producing these cycles. Here we present the analysis of the sunspots number evolution using the time-delay approach. Our results show than the solar cycle can also be characterized by this behavior implying the influence of the past evolution over the present one, suggesting an histeresis mechanism, linked probably with magnetic activity.

Keywords. Sun: sunspots, Sun: activity, Sun: evolution

1. Introduction

Solar activity can be seen through the evolution of sunspots number in quasi-periodic oscillatory series with periods going from 8 to 15 years and with a mean period of 11 years. Due to the change of magnetic field polarity in solar hemispheres alternatively each cycle, the period is rather 22 years. The quasi-periodic evolution of many activity phenomena is still an unsolved key problem in solar physics (along with, e.g., heating of the solar chromosphere and corona, and solar flares). An important issue of this understanding is due to the influence of solar activity over the terrestrial climate (Archibald 2006).

Many physical models were proposed to understand the basic mechanisms (e.g., Benevolenskaya 1998) producing solar/stellar activity and its quasi-cyclic evolution. From sunspots time series, mathematical approaches have highlighted several other hidden periods other that the well-apparent 11-years period. These works contributed to provide some observed parameters to constraint theoretical models, and to predict the future evolution of the sunspots number (e.g., Clilverd *et al.* 2003; Sello 2003).

In this work we follow another mathematical approach, investigating the influence of the past evolution of the sunspots number over the present one. A natural and simple way to carry out this analysis is to assume that the susnpots number have a temporal delay behaviour. Thus, the relationship between present and past values becomes non-linear and could prove that the past amplitude influence the present one. As for the abovementioned methods, this one also allows some predictions about future solar cycles by using intermediate parameters to characterize the general past evolution.

2. The solar cycle as a temporal delay phenomena

We consider a temporal delay behavior (Eq. 2.1) to rely the present evolution of a phenomena with some particular events in its past evolution. The variation of N in time is not only related with its current value $N(t)$ but also with its past values $N(t-T)$, where T is our temporal delay. The simplest way is to assume only one past temporal

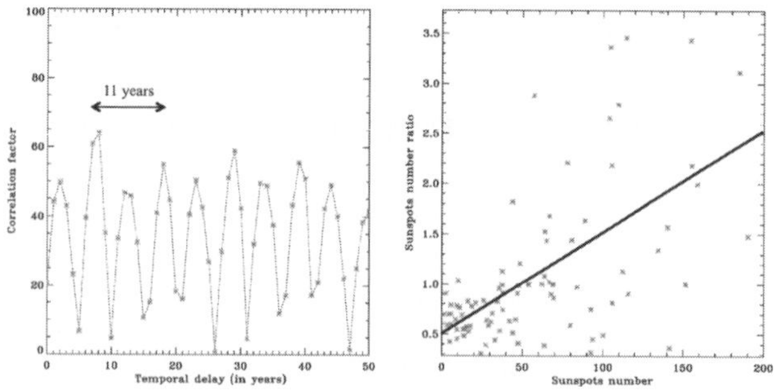

Figure 1. *Left panel:* Correlation factor obtained by chi-square minimization versus the temporal delay. *Right panel:* Sunspots number ratio versus sunspots number with a 8-years delay. The blue line shows a linear fit.

influence and some proportionality between the variables (e.g., Murray 1993), as follows:

$$\frac{dN(t)}{dt} = a\ N(t)\,\{1 - b\ N(t-T)\} \quad \rightarrow \quad \frac{\Delta N_i}{\Delta t} = a\ N_i\,\{1 - b\ N_{i-J}\} \tag{2.1}$$

Because we use yearly values of the relative sunspots number, given by SDIC (http://www.sidc.be/sunspot-data/), Δt is equal to 1. The present and past values of N are already known and we assumed some values for the parameter T ($T = 0, 1, 2, \ldots$; Fig. 1, left panel). So we can determine constants a and b (Eq. 2.2) from the correlation between the sunspots number ratio and the past sunspots number (Fig. 1 right panel).

$$N_i + \Delta N_i = a\ N_i\,\{1 - b\ N_{i-J}\} + N_i \quad \rightarrow \quad \frac{N_{i+1}}{N_i} = \{a+1\} - ab\ N_{i-J} \tag{2.2}$$

3. Results and conclusions

The T values giving the best correlation between the sunspots number ratio and the past spots number, are 7 or 8 years (Fig. 1). To test the accuracy of our method, we applied it on the solar cycle 23 (Table 1 and Fig. 2, left panel). Our predictions are consistent with the observations, except for the unusually long period of low activity at the end of this cycle. We also characterized the next solar cycle 24 (Table 2 and Fig. 2) using our modelling of the solar cycle 23 as well as the spots number observed until 2006.5 and 2007.5. The next maximum sunspots number would take place between 2011 and 2012 and should be close to 60 (Fig. 2, right panel) as for the solar cycle 14. Using the solar minimum occurred in 2008.5 ($R_{min} = 2.9$ from SDIC), the predicted maximum is consistent with the value from the linear least-square fit on the both axis (red line on Fig. 3) and is a few smaller than the range given by Brajsa *et al.* (2009). Moreover the next solar minimum should not occur before 2019 or even 2020.

Table 1. Observed and predicted values for the cycle 23. Results showed in Fig. 2 (left panel).

Observed parameters of solar minimum		Epoch of solar maximum		Maximum sunspot number		Epoch of the end of cycle	
Epoch	Number	Observed	Predicted	Observed	Predicted	Observed	Predicted
1996.4	8.0	2000.3	2001.5	120.8	106.4	2009.?	2006.5

These preliminary results are encouraging because we find a similar delay as that observed between the geomagnetic activity and solar cycle peaks (Hathaway & Wilson 2006) linked probably with magnetic activity by some kind of histeresis mechanism. At present, we can successfully reproduce some previous solar cycles (e.g., epoch and solar maximum). We plan to improve our predictive method including the influence of delays for which there is a good correlation. This more detailed description could allow

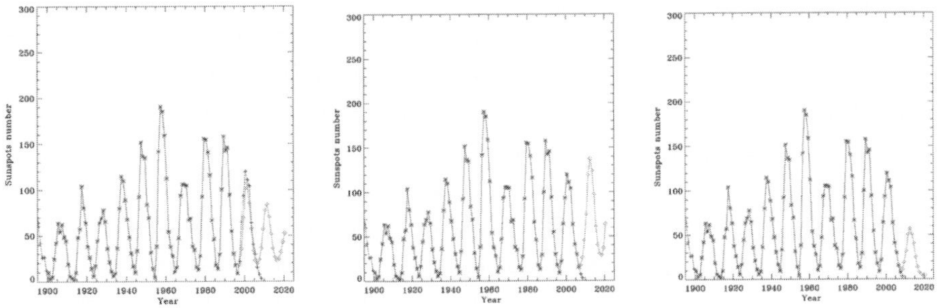

Figure 2. The observed (*plus symbols and dotted line*) and selected (*asterisks and blue line*) yearly values of the relative sunspots number. Our predictions using the temporal delay method are also marked (*diamonds and red line*). *Left panel:* Prediction for the solar cycle 23 and 24. *Middle and Right panels:* Prediction of the solar cycle 24 using the sunspots number observed until 2006.5 and 2007.5 as input.

Table 2. Predicted values for the solar cycle 24 using various data set in input.

Panel of Fig. 2	Input parameters of solar minimum		Predicted parameters of solar maximum		Predicted parameters of next solar minimum	
	Epoch	Number	Epoch	Number	Epoch	Number
Left	2006.5	19.6	2011.5	84.1	2016.5	23.0
Middle	2006.5	15.2	2012.5	137.3	2018.5	28.8
Right	2007.7	7.5	2012.5	57.1	2020.5	4.2

Figure 3. Maximum (R_{max}) versus minimum (R_{min}) sunspots number of a given solar cycle (*dots*). The cycle number is also marked. Blue, green and red diamonds show the locus of our predicted maxima using our modelling of the solar cycle 23 as well as the spots number observed until 2006.5 and 2007.5, respectively (Table 2). The predicted range of Brajsa *et al.* (2009) is plotted as well (*vertical line*). The blue and red lines show linear fits using data of both axis and the linear least-square fit like one given by Brajsa *et al.* (2009, Fig. 6), respectively.

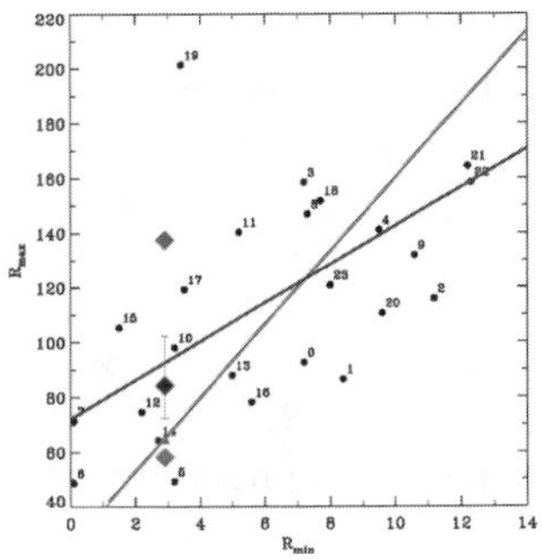

to solve the overestimation of the next solar minimum and to reproduce the asymmetries observed during the previous cycles. Our final aim is to obtain a reliable prediction of the whole solar activity by identifying the fundamental lower and high activity precursors of its present and future evolution. All these improvements are still needed to be able to predict future low solar activity periods (e.g., Maunder/Dalton Minimum).

References

Archibald, D. 2006, Energy & Environment, 17, 29

Benevolenskaya, E. E. 1998, *ApJ*, 509, L49

Brajša, R., Wöhl, H., Hanslmeier, A., *et al.* 2009, *A&A*, 496, 855

Clilverd, M. A., Clarke, E., Rishbeth, H., *et al.* 2003, *Astronomy and Geophysics*, 44, 5.20

Hathaway, D. H. & Wilson, R. M. 2006, Geophys. Res. Lett., 33, L18101

Murray, J. D. 1993, *Mathematical Biology, 2nd corr. ed.* (Springer)

Sello, S. 2003, *A&A*, 410, 691

Session IV

Magnetic activity and dynamo mechanisms

Chair: V. Martinez-Pilet

Solar and Stellar Variability: Impact on Earth and Planets
Proceedings IAU Symposium No. 264, 2009
A. G. Kosovichev, A. H. Andrei & J.-P. Rozelot, eds.

doi:10.1017/S1743921309992584

Towards understanding the global magnetism of the Sun and solar-like stars

Allan Sacha Brun
DSM/IRFU/SAp & UMR AIM, CEA-CNRS-Université of Paris 7, France

Abstract. The Sun and solar-like stars possess intense and cyclic magnetic activity. In order to understand how this comes about we have developed series of 2-D and 3-D models in order to simulate their global dynamics and magnetism. We here report on our latest findings.

Keywords. Sun: activity, interior, magnetic fields; stars: activity; convection, MHD, turbulence

1. Simulating the magnetism of the Sun and solar-like stars

The Sun possesses striking magnetic and dynamical properties, such as its turbulent convective envelope, large-scale surface differential rotation, 22-yr cycle of magnetic activity, butterfly diagram of sunspot emergence, hot corona, etc. (Stix 2002). Understanding how the physical processes operating in the solar turbulent plasma nonlinearly interact to yield this wide range of dynamical phenomena is very challenging. One successful and powerful approach is to rely on multi-dimensional magnetohydrodynamics (MHD) numerical simulations. Today, despite tremendous advances in building powerful supercomputers, it is still not possible to compute a fully integrated 3-D MHD model of the Sun starting from its core up to its corona. One is thus forced to study individually complementary pieces of the full solar MHD puzzle and to progressively incorporate them in a more nonlinearly coupled model. One important characteristic of the Sun that needs to be understood is the origin of its magnetic activity because it has direct societal impact by impairing satellites, damaging electric power grids, interfering with high frequency radio communications and radars. It is currently believed that the solar magnetism is linked to an internal "fluid" dynamo (Parker 1955). More precisely the Sun is the seat of both a small scale and irregular dynamo and a large scale and cyclic dynamo that generate and maintain its magnetic field and lead to the various magnetic phenomena observed at its surface (Parker 1993; Cattaneo & Hughes 2001; Ossendrijver 2003). Developing numerical models of the solar dynamo has thus been a very active field of research. This has mainly involved two types of numerical experiments:

• *kinematic solar dynamo models* that solve only the induction equation in its mean field approximation and assume the velocity field as given (Stix 1976; Moffat 1978; Krause & Radler 1980; see Charbonneau 2005 and Solanki *et al.* 2006 for recent reviews). These models rely on the parametrization of two important effects that are thought to be at the origin of the solar global dynamo, the α and Ω effects. They provide a useful and fast tool to model the solar 22-yr magnetic cycle and its associated butterfly diagram since no feedback from the Lorentz force on the motion is accounted for.

• or *dynamical solar dynamo models* that solve explicitly the full set of MHD equations (see for instance Gilman 1983; Glatzmaier 1985; Cattaneo 1999; Brun *et al.* 2004; Vögler & Schüssler 2007). These models self-consistently compute all the physical processes in three dimensions allowing significant progress to be made on the intricate interactions operating in a turbulent magnetized plasma. The cost of 3-D models and the large number of degrees of freedom needed to model the whole Sun make it difficult, as of today, to provide quantitative predictions such as the cycle period.

Clearly, both methods (2-D vs 3-D) are complementary and are needed to better understand the magnetic solar activity. They have also often been extended to other stellar spectral type (see for instance Charbonneau & Saar 2001; Brun 2009). We propose in this paper to report on our latest findings using this complementary approach.

2. Solar-stellar connection

Another important information to better constraint the magnetic Sun is to compare its activity to other stars. Observations of the magnetism of 'solar type' stars, i.e. stars possessing a deep convective envelope and a radiative interior (late F, G, K and early M spectral type) are becoming more and more available (Giampapa 2005) and may provide additional constraints on our understanding of global scale stellar dynamos. One difficulty of such observational programs is that they require long term observations since stellar cycle periods are likely to be commensurate to the solar 11-yr sunspots cycle period (or 22-yr for a full cycle including two polarity reversals of the global poloidal field). Thanks to the data collected at Mount Wilson Observatory since the late 60's, such data is available (Wilson 1978; Baliunas *et al.* 1995). Among the sample of 111 stars (including the Sun as a star) originally observed between the F2 to M2 spectral type, it is found that about 50% of the stars possess a cyclic activity, with cycle (starspot) periods varying roughly between 5 to 25 yrs, i.e. between half to twice the sunspot cycle period. They further indicate that among the inactive stars of the sample some are likely to be in a quiet phase (as was the Sun during the Maunder minimum). Overall activity cycles seem to be more frequent for less massive stars K than for F stars. More recent observational programs have been pursued that now even provide information on the field topology as a function of the rotation rate, such as the one using the Espadons and Narval instruments and the Zeeman Doppler Imaging technique (Donati *et al.* 2006). Applying this observational technique over a sample of four solar analogues with rotation rate varying from one to three times solar, Petit *et al.* (2008) have shown that the field amplitude increases as a function of the star's rotation rate and, more importantly, becomes more and more dominated by its toroidal component (modulo possible bias in the observational technique used). If such a trend is confirmed, i.e. that the field topology is becoming more toroidal with increasing rotation rate, it is a very important and instructive result that puts strong constraints on the dynamo models.

The systematic analysis of stellar magnetism data revealed that for solar type stars there is a good correlation between the cycle and rotation periods of the stars. This correlation is even stronger when using the Rossby number ($Ro = P_{rot}/\tau$) that takes into account the convection turnover time τ at the base of the stellar convective envelope (Noyes *et al.* 1984; Baliunas *et al.* 1996). As the star rotates faster, its cycle period is found to be shorter. Typically Noyes *et al.* (1984) found that $P_{cyc} \propto P_{rot}^n$, with $n = 1.25 \pm 0.5$. Based on an extended stellar sample Saar & Brandenburg (1999) and Saar (2002) have argued that there is actually two branches when plotting the cycle period vs the rotation period of the stars. They make the distinction between the primary (starspot) cycle and Gleissberg or grand minima type modulation of the stellar activity. For the active branch they found an exponent $n \sim 0.8$ and for the inactive stars $n \sim 1.15$ (Charbonneau & Saar 2001). It is also found that this correlation breaks at high rotation rate with the possible appearance of a super active branch. Further at very high rotation rate saturation of the X-ray luminosity seems to limit the validity of the scaling found at more moderate rotation rates (Pizzolato *et al.* 2003). For G type stars this saturation is found for rotation rate above 35 kms^{-1}, for K type stars at about 10 kms^{-1} and for M dwarfs around 3–4 kms^{-1} (Browning 2008; Reiners, Basri & Browning 2009). How stellar magnetic flux scales with rotation rate is thus also important to understand since it is telling us how the magnetic field generated by dynamo action inside the stars emerges and imprints the stellar surface (Rempel 2008) and if it actually saturates. Both the Corot and Kepler satellites should further improve our understanding of stellar magnetism and convection (we advise the reader to read the contributions related to both satellites in these proceedings).

3. 2-D mean field models

Theoretical considerations to interpret stellar magnetism based on classical 2-D mean field α-ω dynamo models (Durney & Latour 1976, Knobloch *et al.* 1981; Baliunas *et al.* 1996) naturally find correlations between rotation rate and stellar activity. In particular it is found that both magnetic field generation and the dynamo number D (i.e. a Reynolds number characterizing the mean field α and ω dynamo effects used in the models) vary with the rotation period of the star. This is due to the fact that in these models both effects are sensitive to the rotation rate of the star. The ω-effect is a direct measure of the differential rotation $\Delta\Omega$ established in the star. It is well known both theoretically and observationally that the differential rotation in the convective envelope of solar-type stars is directly connected to the star's rotation rate Ω_0 (Donahue *et al.* 1996; Barnes *et al.* 2005; Ballot *et al.* 2007; Brown *et al.* 2008; Küker & Rüdiger 2008). However, the exact scaling n_r (i.e. $\Delta\Omega \propto \Omega_0^{n_r}$) is still a matter of debate among the observers and theoreticians, being sensitive to the observational techniques used and to the modelling approach.

The α-effect used in mean field theory is related to helical turbulence and represents a measure or parameterization of the mean electromotive force (emf) (Moffatt 1978; Pouquet *et al.* 1976). It is thus also linked to the rotation rate of the star and the amount of kinetic helicity present in its convective envelope. However in Babcock-Leighton flux transport dynamo models the standard α-effect is replaced by a surface term linked to the tilt of the active regions with respect to the east-west direction (i.e. Joy's law, Kosovichev & Stenflo 2008). This tilt is thought to be due to the action of the Coriolis force during the rise of the toroidal structures that emerge as active regions (D'Silva & Choudhuri 1993). Recent 3-D simulations in spherical shells (Fan 2008; Jouve & Brun 2007b, 2009) seem to indicate that this is not the only effect responsible for the observed tilt and that the twist and arching of the toroidal structures as well as the continuous action of the surface convection during the emergence have some influence on the resulting tilt.

Another important ingredient in flux transport models is the large scale meridional circulation (MC) used to connect the surface source term generating the poloidal field to the region of strong shear at the base of the convection zone (i.e. the tachocline) where it will be subsequently sheared by the ω-effect in order to close the global dynamo loop (i.e. $B_{pol} \rightarrow B_{tor} \rightarrow B_{pol}$) . The meridional flow (or "conveyor belt") thus plays an important role in setting the cycle period of the global dynamo in this class of flux transport model. As a direct consequence it is natural to ask how the meridional circulation amplitude and profile change with the rotation rate and how these may influence the magnetic cycle period.

Jouve & Brun (2007a) have done a systematic study of the influence on the cycle period of the various parameters of Babcock-Leighton flux transport models. For the reference model possessing only one meridional circulation cell per hemisphere they confirm the results of Dikpati & Charbonneau (1999) that the cycle period is strongly dependent on the meridional circulation amplitude. A least square fit through the ensemble of models they computed leads to the following dependency:

$$P_{cyc} = v_0^{-0.91} s_0^{-0.013} \eta_0^{-0.075} \Omega_0^{-0.014}, \tag{3.1}$$

with v_0, s_0 and η_0 being respectively, the amplitude of the meridional circulation at the surface, that of the Babcock-Leighton source term and of the magnetic diffusivity (see Jouve & Brun (2007a), Jouve, Brown & Brun (2009) for more details). We can thus expect that as we increase the rotation rate the cycle period will shorten if all the other parameters are kept constant but unfortunately not as fast as what is observed (Noyes *et al.* 1984; Saar 2002). However we may also expect the meridional circulation to depend on the rotation rate, and the sign of the variation of v_0 with Ω_0 will most certainly dominate over the weak dependence of P_{cyc} on Ω_0. It is thus crucial to assess how the meridional circulation amplitude and profile vary with the rotation rate Ω_0.

Early work by Dikpati *et al.* (2001) have assumed that the amplitude of the meridional flow is proportional to the rotation rate of the star. With such scaling they seem to be able to reproduce some of the rotation vs cycle period correlation observed in solar type stars. Charbonneau & Saar (2001) have also studied the sensitivity of various mean field dynamo models in reproducing stellar activity observations assuming for the Babcock-Leighton type that the meridional flow amplitude increases either as Ω_0 or as $\log(\Omega_0)$. Similarly Nandy & Martens (2007) have explored the solar-stellar connection with B-L mean field dynamo models and have reached the same conclusion: only a positive scaling of the amplitude of meridional flows with the rotation rate can reconcile the models with observations of magnetism of solar-like stars. However recent theoretical work by Ballot *et al.* (2007) and Brown *et al.* (2008) indicate that this is unlikely to be the right scaling. Instead 3-D simulations reveal that the amplitude of the meridional flow weakens as the rotation rate is increased.

In Jouve *et al.* (2009) we have explored how mean field Babcock-Leighton dynamo models may be used to explain the observational relation between the rotation and magnetic activity period of solar-like stars by incorporating scaling law for the amplitude of the meridional circulation coming from the 3-D simulations of Brown *et al.* (2008). In this 3-D study they found that the energy and amplitude of the meridional circulation decreases with the rotation rate.

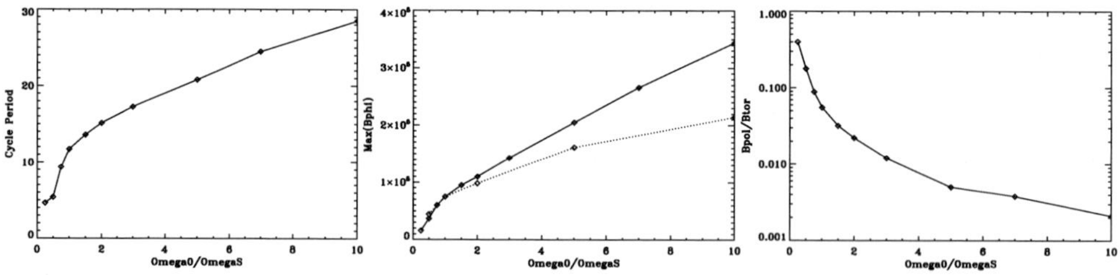

Figure 1. Left panel: Variation of the magnetic cycle period (in years) wih respect to the solar rotation rate for the cases published in Jouve *et al.* (2009). On the 2nd panel we display the variation of the maximum of toroidal field at the base of the CZ for two series of models. The solid line corresponds to the models using the scaled down meridional flow as a function of the rotation rate deduced from Brown *et al.* (2008) whereas the dotted line corresponds to another set of models for which v_0 was kept constant. On the third panel, we show the ratio Bpol/Btor wih respect to the solar rotation rate for the scaled cases.

On Figure 1, we display the results of our series of mean field stellar dynamo models at various rotation rate. First we find that indeed the magnetic field topology tends to

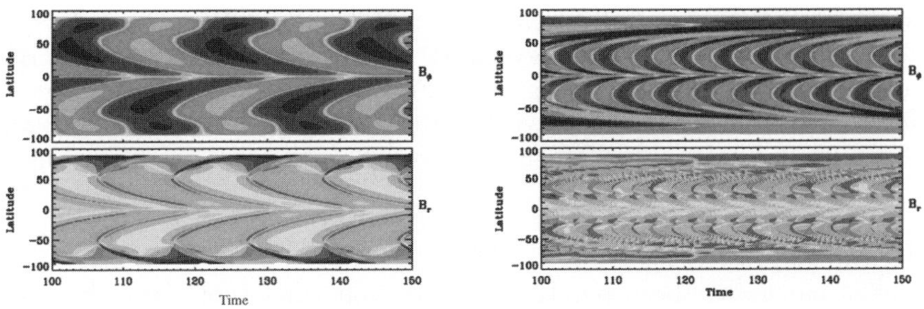

Figure 2. The top panel shows the time-latitude cut of the toroidal field at the base of the convection zone and the bottom one represents the evolution of the radial field at the surface for two cases rotating at 5 times the solar rate. Both models possess a scaled down MC flow but in the right panels the meridional flow is multicellular, with the poleward cell extending from the equator up to 20 degree. Time is in years. The scale of the color table is the same for all: between -5×10^3 G and 5×10^3 G for the radial field and between -5×10^5 G and 5×10^5 G for the toroidal field.

be more and more toroidal, with B_ϕ increasing in strength due to a larger ω-effect and a slower flow at the base of the convection zone, that allows the fields to be stretched further more. Unfortunately we find that Babcock-Leighton flux transport dynamo model based on single cell meridional circulation does not reproduce the P_{cyc} vs P_{rot} relation when using scaling for the MC deduced from 3-D global simulations.

However the strong relation between the meridional circulation speed and the cycle activity is an indication that shortening the advection path may result in shorter cycle period. Dikpati *et al.* (2004), Bonanno *et al.* (2005) and Jouve & Brun (2007a) have studied in details the influence of various meridional circulation profiles on the butterfly diagram and other large scale magnetic properties of the Sun. In their study Jouve *et al.* (2009) demonstrate that adding cells in latitude may resolve the apparent disagreement between B-L dynamo models and observations of stellar activity cycles.

On the butterfly diagrams shown on Figure 2, we clearly see that we have significantly reduced the length of the activity cycle: the period is now 5.2 yr instead of 22 yr. So shortening the advection path seems to be an interesting solution but it also modifies the butterfly diagram. The cyclic activity tends now to be more concentrated at low latitudes. The radial field appears to be busy, exhibiting a periodic but yet complex pattern at low latitudes. We moreover note that above 55 degree, the cyclic activity is lost and we are left with some regions where the magnetic field evolves much more slowly than the time scale of the main cycle.

Another way to reconcile B-L models and observations may be to replace the meridional circulation as the main process to transport the field from the surface down to the tachocline. We refer to the work of Guerrero & Gouveia Dal Pino (2008) and Yeates *et al.* (2008) that have considered respectively turbulent pumping or magnetic diffusion as the dominant transport process. The weaker dependancy of the cycle period to those alternative transport processes may result to a better agreement between B-L dynamo models and stellar cycle observations. We intend to explore further these alternatives. We now turn to discussing our recent nonlinear 3-D MHD studies of solar and stellar convection, rotation and magnetism.

4. 3-D global MHD models

The Sun and solar-like stars are complex magnetohydrodynamic objects that require state-of-the-art observations and numerical simulations in order to pin down the physical processes at the origin of such diverse activity and dynamics. We propose here to make a brief summary of the recent advances made with the Anelastic Spherical Harmonic (ASH) code (Clune *et al.*(1999); Brun *et al.* 2004) in modelling in 3-D the global non-linear solar magnetohydrodynamics. We refer to Ballot *et al.* (2007) and Brown *et al.* (2008, 2009) for the study of the dynamics of solar-like stars rotating at various rotation rate.

- Solar Global Convection

A series of papers has been published on this important topic (Miesch *et al.*(2000); Elliott *et al.*(2000); Brun & Toomre 2002), the most recent being by Miesch *et al.*(2008). In that later paper for the first time a global model of the solar convection with a density contrast of 130 from top to bottom and a resolution equivalent to 1500^3 has been obtained. This has lead to significant results regarding the turbulent convection spectra from large scale (giant cells like) down to supergranular like convection patterns and their correlation with the temperature fluctuations leading to a large (150% $L_\odot$) convective luminosity.

- Solar Differential Rotation and Meridional Circulation

A very recent review by Brun & Rempel (2008) discusses the respective role of Reynolds stresses, latitudinal heat transport and baroclinic effect in setting the peculiar conical differential rotation profile observed in the solar convection zone. Indeed basic rotating fluid dynamic considerations imply that the differential rotation should be invariant along the rotation axis thus yielding a cylindrical rotation profile. Since this is not observed it is necessary to find the source of the breaking of the so called Taylor-Proudman constraint. In particular in a recent paper by Miesch *et al.*(2006) we have been able to show that baroclinic effects are associated to temperature latitudinal variation and that convection by transporting heat poleward contributes for a significant part of that variation but not all. A temperature contrast of about 10 K is compatible with helioseismic inference of the inner solar angular velocity profile. Meridional flows in most cases are found to be multi cellular and fluctuate significantly over a solar rotation. This flow contributes little to the heat transport and to the kinetic energy budget (accounting for only 0.5% of the total kinetic energy). However it plays a pivotal role in the angular momentum redistribution by opposing and balancing the equatorward transport by Reynolds stresses (Brun & Toomre 2002, Brun & Rempel 2008).

- Solar Global Dynamo

In the continuation of the work by Gilman (1983) and Glatzmaier (1985), we have studied at much higher resolution dynamo action in turbulent convective shells (Brun 2004; Brun *et al.* 2004). We have found that dynamo action is reached above a critical magnetic Reynolds number and that the magnetic field is mostly intermittent and small scale (see Figure 3 (right panel)), the large scale axisymmetric field only contributing for about 3% of the total magnetic energy. Reversals of the field occur on a time scale of about 1.5 yr as opposed to the observed 11 yr cycle of solar activity. This is in part due to the absence of a tachocline at the base of the convective envelope. In an attempt to resolve that issue, we have in Browning *et al.* (2006), computed the first 3-D MHD model of a convection zone with an imposed stable tachocline. In that layer the field that has been transported or pumped down from the above turbulent convection zone, is organized in strong axisymmetric toroidal ribbons with a dominant antisymmetry with respect to the equator (see Figure 3 (left panel)). The poloidal field in the convection zone is stabilized by the presence of that layer with much less frequent, if any,

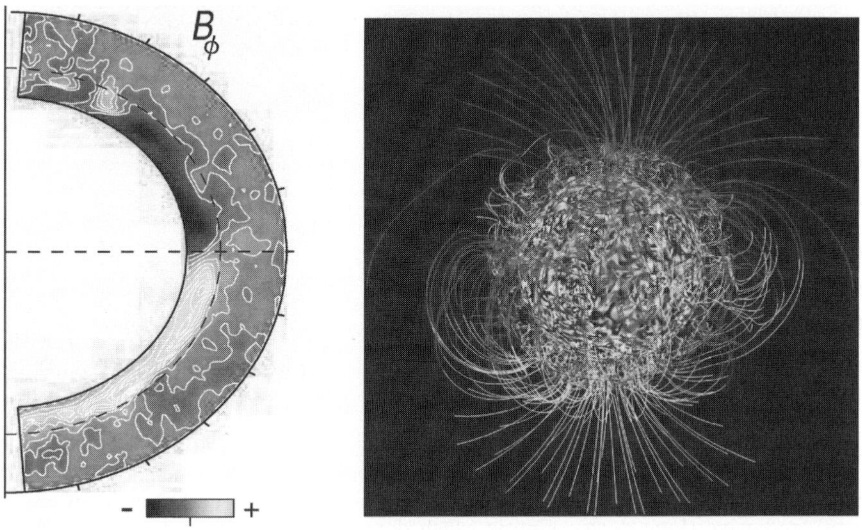

Figure 3. Left panel: Azimuthal and temporal average of the toroidal magnetic field in a 3-D MHD simulation of the solar convective envelope coupled to a stable sheared tachocline (Browning *et al.* 2006). We note the antisymmetric magnetic layers at the base of the convection zone represented by the dash line. Right panel: 3-D magnetic field lines rendering (yellow corresponds to positive Br, i.e the line is directed toward the observer) inside the convection zone (below the transparent sphere) and in a current-free corona (potential field line reconstruction) for the purely convective dynamo case M3 of Brun *et al.* (2004). We note the intricacy of the field lines in the convection zone, often located in strong downflows and the connectivity of the field over large distances at the surface.

reversal. The magnetic energy reaches in both cases about 10% of the total kinetic energy. We also find that the differential rotation is reduced in amplitude due to the nonlinear feedback of the field on the flow via the Lorentz force. In a recent study by Jouve & Brun (2007b), Jouve & Brun (2009), we have also studied flux emergence in isentropic and turbulent rotating convection zone. We confirmed that a minimum amount of field concentration and twist is required for the structure to emerge at the surface and that convection may pin down and recycle weak field. At the surface in the convective case, horizontal converging motions modify the orientation of the emerging structure. We find that Maxwell stresses associated with the emerging structure modify locally the horizontal flows, changing their amplitude and direction.

• Solar Radiation Zone and its coupling to CZ

The thinness of the solar tachocline ($h < 0.05R_\odot$) implies the presence of processes that transport angular momentum horizontally in order to prevent its viscous and thermal spread along the solar evolution. Horizontal turbulence in stratified layers, the presence of a fossil magnetic field or internal waves may provide such a transport (Spiegel & Zahn 1992, Gough & McIntyre 1998). Brun & Zahn (2006) have studied for the first time in 3-D the nonlinear evolution of an horizontal shear (similar to the solar angular profile) in a stable radiative zone and the role of a fossil field. They found that such field by connecting to the surface shear will ease rather than oppose the spread of the tachocline. It is thus difficult to invoke only the magnetic field to explain the solid body rotation of the radiative zone and the thin tachocline. Another important finding in Brun & Zahn (2006) is that simple magnetic configurations undergo non axisymmetric MHD instabilities first discovered by Talyer in the 70's. In particular a purely poloidal (dipolar) fossil

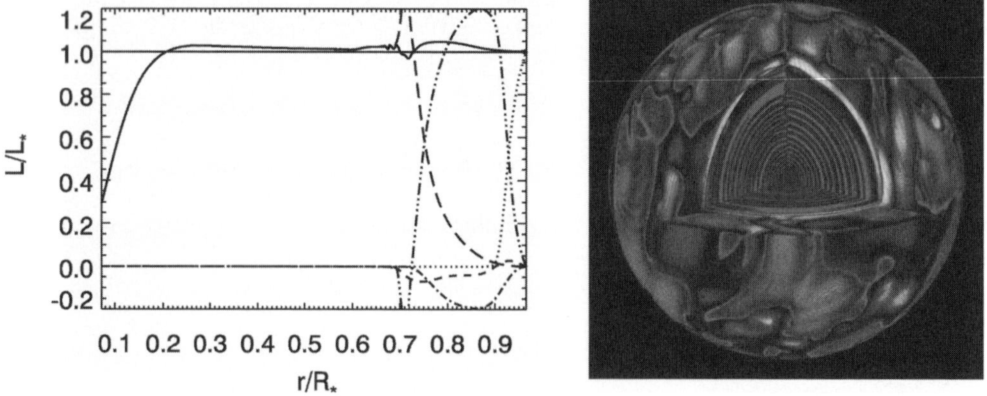

Figure 4. 3-D integrated solar model coupling nonlinearly the convective envelop to the radiative interior (Brun *et al.* 2009, in preparation). The base of the convection zone is located at 0.72 R. Left panel: Radial energy flux balance. Shown are the enthalpy (convection) (dash three dots), radiative (long dash), kinetic (dash dot), viscous (dash), unresolved (dotted) and total (solid) fluxes converted to luminosity and normalized to the solar luminosity. We note the large enthalpy flux in the convection zone and the sharp penetration at the case of the CZ. Right panel: Shown is the a 3-D rendering of the density perturbations with red corresponding to positive fluctuations. We have omitted an octant in order to be able to see the equatorial and meridional planes within the domain. We note the clear presence of internal waves in the radiative zone.

field would undergo high azimuthal wavenumber instabilities. In order to obtain a stable configuration for a magnetic field burried in the solar radiative interior it is necessary to have a coupled poloidal/toroidal topology. Such non axisymmetric instabilities may lead to dynamo action in radiative interiors has also been advocated by Spruit (2002) and discussed in details by Zahn *et al.* (2007).

- Towards a 3-D integrated model of the Sun and solar-like stars

Coupling nonlinearly the convection zone with the radiative interior is key to understand the solar global dynamo and inner dynamics. Brun *et al.* (2009, in preparation) have developped the first 3-D solar integrated model from $r = 0.07R_{\odot}$ up to $0.97R_{\odot}$. We show on Figure 4 (left panel) the energy flux balance realized in such coupled models. We note the dominant role played by the convective flux in transporting most of the energy in the convective envelope, while the radiative flux dominates in the deep interior. A thin convective penetration region exist at the base of the convection zone (as indicated by the negative convective flux around 0.7 R), overlapping with the tachocline of shear self consistently generated in the simulation. On Figure 4 (right panel), we show a 3-D rendering of the density fluctuations over the whole computational domain. The presence of internal waves is obvious in the radiative interior. The penetrative convection is at the origin of these gravito-inertial waves. We are currently studying in details the source function at every depth in the model and the resulting power spectrum at different locations in the radiative interior and find that a large spectrum near the base of the convection zone is excited. The tachocline is kept thin in this model by using a step function at the base of the convection zone for the various diffusion parameters making the thermal and viscous spread of the latitudinal shear imposed by the convective envelope slow with respect to the convective overturning time. We intend in the near future to reproduce the study of Brun & Zahn (2006) by introducing in the integrated model a fossil field taking advantage of the more realistic boundary conditions realized in this new class of models (Strugarek *et al.* 2010, in prepration). We will also adapt our latest solar models to solar-like and more massive stars in order to span the H-R diagram (see for instance Brun 2009).

Acknowledgement

I am thankful to my friends and colleagues J. Toomre, J.-P. Zahn, M. Miesch, M. Derosa, M. Browning, B. Brown & L. Jouve without whom the results reported in this paper would not have been obtained. I acknowledge funding by the European Community via ERC-StG grant 207430 STARS2 (www.stars2.eu). Finally, I am grateful to the organizers of the IAU Symposium 264 for a wonderful time in Rio.

References

Baliunas, S. L. *et al.* 1995, *ApJ*, 438, 269
Baliunas, S. L. *et al.* 1996, *ApJ*, 460, 848
Ballot, J., Brun, A. S., & Turck-Chièze, S. 2007, *ApJ*, 669, 1190
Barnes, J. R., Cameron, A. C., Donati, J.-F., James, D. J., Marsden, S. C., & Petit, P. 2005, *MNRAS*, 357, L1
Bonanno, A., Elstner, D., Belvedere, G., & Rüdiger, G. 2005, *Astron. Nachr.*, 326, 170
Brown, B. P., Browning, M. K., Brun, A. S., Miesch, M. S., & Toomre, J. 2008, *ApJ*, 689, 1354
Brown, B. P., Browning, M. K., Brun, A. S., Miesch, M. S., & Toomre, J. 2009, *ApJ*, accepted
Browning, M., Miesch, M. S., Brun, A. S., & Toomre, J. 2006, *ApJ*, 648, L157
Browning, M. K. 2008, *ApJ*, 676, 1262
Brun, A. S. 2004, *Solar Phys.*, 220, 333
Brun, A. S. 2009, *EAS Publication Series*, 39, 153
Brun, A. S., Miesch, M. S., & Toomre, J. 2004, *ApJ*, 614, 1073
Brun, A. S. & Rempel, M., *Space Sci. Rev.*, 144, 151
Brun, A. S. & Toomre, J. 2002, *ApJ*, 570, 865
Brun, A. S. & Zahn, J.-P. 2006, *A&A*, 457, 665
Cattaneo, F. & Hughes, D. W. 2001, *A&G*, 42, 18
Cattaneo, F. 1999, *ApJ*, 515, L39
Charbonneau, P. 2005, *Living Rev. Solar Phys.*, 2
Charbonneau, P. & Saar, S. H. 2001, in Magnetic Fields across the H-R diagram, eds G. Mathys, S. K. Solanki and D. T. Wickamasinghe, *ASP Conf. Series*, 248, 189
Clune, T. L., *et al.* 1999, *Parallel Comput.*, 25, 361
Dikpati, M. & Charbonneau, P., 1999, *ApJ*, 518, 508
Dikpati, M., Cally, P. S., & Gilman, P. A. 2004, *ApJ*, 610, 597
Dikpati, M., Saar, S. H., Brummell, N., & Charbonneau, P. 2001, in Magnetic fields across the H-R diagram, eds, G. Mathys, S. K. Solanki & D. T. Wickramasinghe, *ASP Conf. Series*, vol. 248, 235
Donahue, R. A., Saar, S. H., & Baliunas, S. L. 1996, *ApJ*, 466, 384
Donati, J.-F., Howarth, I. D., Jardine, M. M., Petit, P. *et al.* 2006, *MNRAS*, 370, 629
D'Silva, S. & Choudhuri, A. R. 1993, *A&A*, 272, 621
Durney, B. R. & Latour, J. 1976, *Geophys. Astrophys. Fluid dynamics*, 9, 241
Elliott, J. R., Miesch, M. S., & Toomre, J. 2000, *ApJ*, 533, 546
Fan, Y. 2008, *ApJ*, 676, 680
Giampapa, M. 2005, in The Sun, solar analogs and the climate, Saas-Fee Advanced Course 34, eds I. Rüedi, M. Güdel and W. Schmutz. Berlin: Springer, 307
Gilman, P. A. 1983, *ApJS*, 53, 243
Glatzmaier, G. A. 1985, *ApJ*, 291, 300
Gough, D. O. & McIntyre, M. 1998, *Nature*, 394, 755
Guerrero, G. & de Gouveia Dal Pino, E. 2008, *A&A*, 485, 267
Jouve, L. & Brun, A. S. 2007a, *A&A*, 474, 239

Jouve, L. & Brun, A. S. 2007b, *Astron. Nachr.*, 328, 1104
Jouve, L. & Brun, A. S. 2009, *ApJ*, 701, 1300
Knobloch, E., Rosner, R., & Weiss, N. O. 1981, *MNRAS*, 197, 45
Kosovichev, A. G. & Stenflo, J. O. 2008, *ApJ*, 688, L115
Krause, F. & Raedler, K. H. *Mean Field Magnetohydrodynamics and Dynamo Theory*, Pergamon Press (Oxford)
Küker, M. & Rüdiger, G. 2008, *Journal of Phys.: Conf. Ser.* 118, 012029
Miesch, M. S. *et al.* 2000, *ApJ*, 532, 593
Miesch, M. S., Brun, A. S., & Toomre, J. 2006, *ApJ*, 641, 618
Miesch, M. S., Brun, A. S., Derosa, M., & Toomre, J. 2008, *ApJ*, 673, 557
Moffat, H. K. 1978, *Magnetic field Generation in Electrically Conducting Fluids*, Cambridge University Press
Nandy, D. & Martens, P. C. H. 2007, *AdSpR*, 40, 7, 891
Noyes, R. W., Weiss, N. O., & Vaughan, A. H. 1984, *ApJ*, 287, 769
Ossendrijver, M. 2003, *ANS*, 324, 64
Parker, E. N. 1955, *ApJ*, 121, 491
Parker, E. N. 1993, *ApJ*, 408, 707
Petit, P., Dintrans, B., Solanki, S. K., Donati, J.-F. *et al.* 2008, *MNRAS*, 388, 80
Pizzolato, N., Maggio, A., Micela, G., Sciortino, S., & Ventura, P. 2003, *A&A*, 397, 14
Pouquet, A., Frisch, U., & Léorat, J. 1976, *Journal of Fluid Mech.*, 77, 321
Reiners, A., Basri, G., & Browning, M. 2009, *ApJ*, 692, 538
Rempel, M. 2008, *Journal of Phys.: Conf. Ser.* 118, 012032
Saar, S. H. & Brandenburg, A. 1999, *ApJ*, 524, 295
Saar, S. H. 2002, in the 11th Cool Stars: Stellar System and the Sun, eds R. J. Garcia Lopez, R. Rebolo and M. R. Zapatero, *ASP Conf. Series*, 223, 292
Solanki, S. K., Inhester, B., & Schuessler, M. 2006, *Rep. Prog. Phys.*, 69, 563
Spiegel, E. A. & Zahn, J.-P. 1992, *A&A*, 265, 106
Spruit, H. C. 2002, *A&A*, 381, 923
Steenbeck, M. & Krause, F. 1969, *Astron. Nachr.*, 291, 49
Stix, M. 1976, *A&A*, 47, 243
Stix, M. 2002, *The Sun: an introduction*, Springer
Vögler, A. & Schüssler, M. 2007, *A&A*, 465, L43
Wilson, O. C. 1978, *ApJ*, 226, 379
Yeates, A. R., Nandy, D., & Mackay, D. H. 2008, *ApJ*, 673, 544
Zahn, J.-P., Brun, A. S., & Mathis, S. 2007, *A&A*, 474, 145

Solar and Stellar Variability: Impact on Earth and Planets
Proceedings IAU Symposium No. 264, 2009
A. G. Kosovichev, A. H. Andrei & J.-P. Rozelot, eds.

doi:10.1017/S1743921309992596

The stellar magnetic dynamo during the evolution across the main sequence

Swetlana Hubrig
Astrophysikalisches Institut Potsdam
An der Sternwarte 16
14482 Potsdam, Germany
email: shubrig@aip.de

Abstract. In comparison with stellar magnetic field studies in the eighties and nineties, where most effort was focused on A, B, and active solar-type stars, magnetic fields are currently directly measured in massive early B and O stars with radiative envelopes, as well as in the lower mass tail consisting of fully convective late-M dwarfs. Knowledge of the magnetic field topology of stars of different mass and at different evolutionary stages is important to understand the underlying magnetic field generation mechanisms. I review the present status of magnetic field studies along with the results of theoretical modeling.

Keywords. MHD, instabilities, techniques: polarimetric, binaries: spectroscopic, stars: chemically peculiar, stars: evolution, stars: formation, stars: magnetic fields, stars: pre–main-sequence, stars: spots

---

1. Introduction

The presence of a convective envelope is a necessary condition for significant magnetic activity. Magnetic activity is found all the way from the late A-type stars (e.g. in Altair: Robrade & Schmitt 2009) with very shallow convective envelopes down to the coolest fully convective M-type stars. Among the early-type main sequence A and B stars, fields with large-scale organised structure consistent with the oblique rotator model have been detected in the subgroup of the Ap and Bp stars. Although these stars have the best studied fields among the early-type stars, the origin of their magnetic fields is still poorly understood. The presence of magnetic fields in massive early B and O stars is known in a very small sample only, but advances in instrumentation over the past decade have led to magnetic field detections in a gradually growing subset of massive stars, which frequently present cyclic wind variability, Hα emission variations, non-thermal radio/X-ray emission and transient features in the absorption line profiles.

2. The presence of a convection zone in massive stars

Recent theoretical developments have highlighted the potentially significant influence of magnetic fields on the structure, evolution and environment of massive stars. The presence of a convection zone in the outer envelope of hot massive stars was recently studied by Cantiello *et al.* (2009). The authors used a stellar evolution code to compute a grid of massive star models at different metallicities. A small convection zone in the envelopes of sufficiently luminous massive main-sequence stars due to the so-called iron peak in stellar opacities was previously predicted by Slothers & Chin (1993).

Cantiello *et al.* mapped the strength of the iron convection zone (FeCZ) in the H-R diagram showing its prominence as a function of stellar parameters. Since in their models the FeCZ has a spatial extent similar to the solar convection zone, the authors suggest

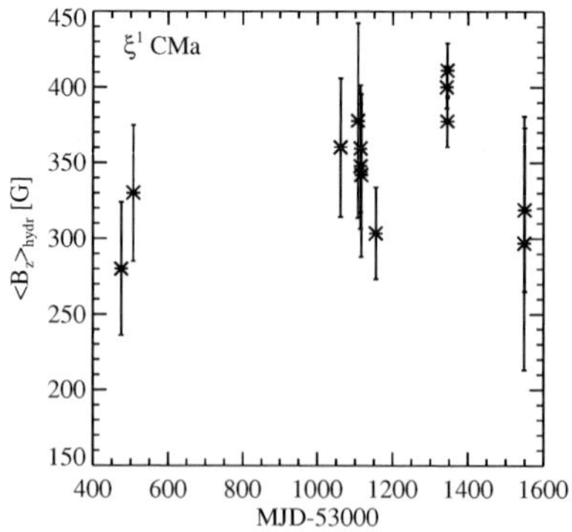

Figure 1. Magnetic field measurements for one of the hottest β Cep star, ξ^1 CMa, during the last 4.4 years.

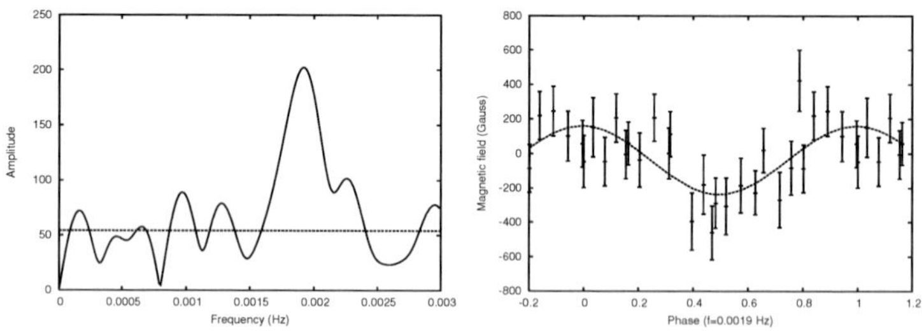

Figure 2. A study of the magnetic field of the B0.2V star θ Car. Left: Fourier spectrum for the magnetic field data. The horizontal line gives the noise level. Right: Phase diagram for the same dataset for the best candidate frequency of 0.0019 Hz, i.e. about 8.8 minutes.

that a dynamo may also work in rapidly rotating OB stars. The rise time of buoyant magnetic flux tubes from the FeCZ to the stellar surface will then be then much shorter than in the model by MacGregor & Cassinelli (2003), who considered the rise of flux tubes from the convective core. The interaction of the stellar wind with the localized surface magnetic field could increase the loss of stellar angular momentum and could probably explain co-rotating density patterns in the outflowing wind. No definitive magnetic field was ever detected in Wolf-Rayet stars and currently only half a dozen of O stars have published magnetic fields (e.g., Donati *et al.* 2002; Donati *et al.* 2006a; Hubrig *et al.* 2008a). Among the hottest early B-type stars, a weak ($\sim$ 40 G) magnetic field has been discovered in the B0.2V star τ Sco (Donati *et al.* 2006b) and in one of the hottest β Cephei stars, the B0.7IV star ξ^1 CMa, with a rather large longitudinal magnetic field of up to 400 G (Hubrig *et al.* 2006a). The amplitude of the observed magnetic field variations in the B0.7IV star ξ^1 CMa is rather low: The obtained values of the longitudinal magnetic field are in the range from 280 to 410 G (see Fig. 1). Our search for a variation period of the magnetic field in this star using a Fourier analysis did not reveal any significant frequency (Hubrig *et al.* 2009a).

The results of the search for a magnetic field in another hot star, θ Car, with spectral type B0.2V, using FORS 1 at the VLT were rather inconclusive (Hubrig *et al.* 2008b). To monitor the behaviour of the magnetic field over at least a part of the stellar surface, we carried out a time series of exposures with short integration time over the time span of $\sim$1.2 h. The observed magnetic field changed several times from positive to negative values over the observing time span. To find the period of variations of the magnetic

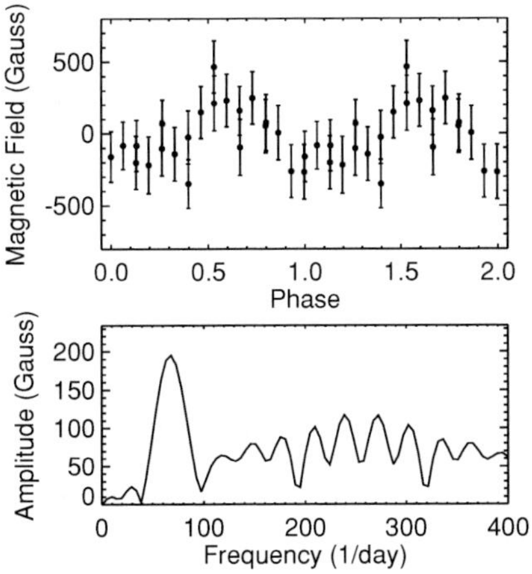

Figure 3. Phase diagram and amplitude spectrum for the magnetic field measurements of the Be star λ Eri in August 2006.

field, we performed a Fourier analysis of the magnetic field measurements. The Fourier spectrum with an amplitude of the magnetic field variation of 202$\pm$43 Gauss and the phase diagram for $f = 0.0019$ Hz are presented in Fig. 2, indicating a possible cyclic variability of the magnetic field.

A number of physical processes in classical Be stars (e.g., angular momentum transfer to a circumstellar disk, channeling stellar wind matter, accumulation of material in an equatorial disk, etc.) are more easily explainable if magnetic fields are invoked. For ten Be stars we performed time-resolved magnetic field measurements over one hour (corresponding to a time series of 20 to 30 measurements per star; Hubrig *et al.* 2009b). In this way, similar to the study of θ Car, we obtained information on the behaviour of the localized transient magnetic field over a part of the stellar surface. In the obtained amplitude spectra a 2.4σ peak corresponding to a period of 21.12 min was detected in the data set of measurements carried out using hydrogen lines in the Be star λ Eri (see Fig. 3). Our observations indicate that a strong field could possibly exist locally, but with a topology such that its net effect can appear only sporadically in disk-integrated variations. Our time-resolved magnetic field measurements of the remaining Be stars indicate that four other Be stars may display a magnetic cyclic variability on a time scale of minutes or tens of minutes. However, since the mechanism of the generation and maintenance of magnetic fields in massive stars is not well understood yet, we are cautious in drawing any conclusion on the behaviour of a magnetic field in θ Car or the studied Be stars.

3. Stars of spectral type B6-A0 with HgMn peculiarity and Ap/Bp stars with large-scale organized magnetic fields

Our recent studies of spectroscopic binaries with B6-A0-type suggest that the majority of slowly rotating late B-type stars formed in binary systems with certain orbital parameters become HgMn stars. The aspect of inhomogeneous distribution of some chemical elements over the surface of HgMn stars has been, for the first time, discussed by Hubrig & Mathys (1995) and has been confirmed in recent studies (e.g. Hubrig *et al.* 2006b). In Fig. 4 we present a typical Doppler Imaging reconstruction of Y on the surface of a HgMn star, in this case the eclipsing SB2 HgMn binary AR Aur (e.g. Savanov *et al.* 2009).

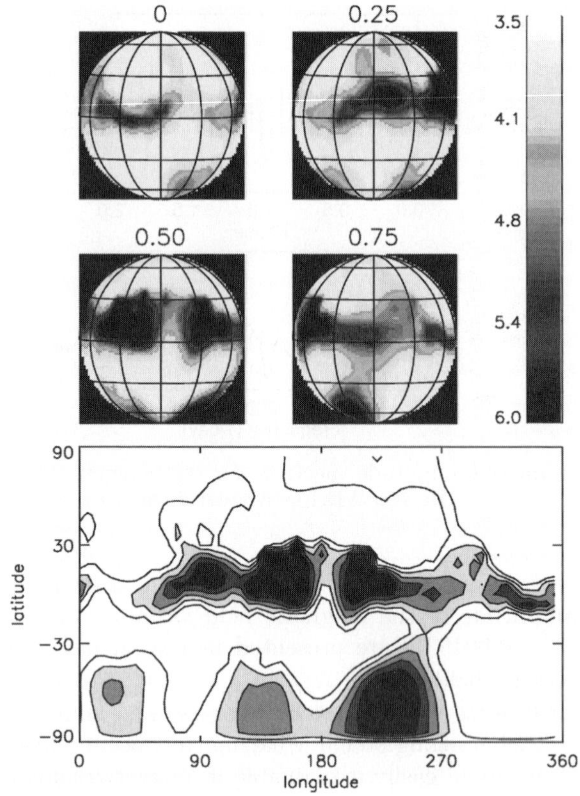

Figure 4. Results of the Doppler Imaging reconstruction of Y on the surface of the eclipsing binary AR Aur. Spherical Doppler Imaging maps are presented at four equidistant rotational phases (upper panel). The rectangular plot (lower panel) shows a pseudo-Mercator projection of the surface maps.

The distribution of some elements was found to present fractured rings around the stellar rotational equator with some features close to the rotational pole.

For one SB1 system, HD 11753, it was possible to gather a large number of spectra with the CORALIE echelle spectrograph attached to the 1.2 m Leonard Euler telescope on La Silla in Chile. In total, we obtained 113 spectra at a spectral resolution of 50,000 (Briquet *et al.*, in preparation). Also for this system many spectral lines exhibit clear variations. In Fig. 5 we present stacked images combining the observed spectra in phase bins with steps of 0.04.

It is a fundamental question whether magnetic fields play a significant role in the development of anomalies in binary B-type stars. Answering this question is also important for the understanding of the processes taking place during the formation and evolution of B stars in general. A scenario how a magnetic field can be built up in binary systems has been presented some time ago by Hubrig (1998) who suggested that a tidal torque varying with depth and latitude in a star induces differential rotation. Differential rotation in a radiative star can be prone to magneto-rotational instability (MRI). Magnetohydrodynamical simulations by Arlt *et al.* (2003) revealed a distinct structure for the magnetic field topology similar to the fractured elemental rings observed on the surface of HgMn stars. The initial model differential rotation was hydrodynamically stable (Taylor-Proudman flow), but the introduction of a magnetic field excites the MRI on a very short time-scale compared to the time-scale of microscopic magnetic diffusion.

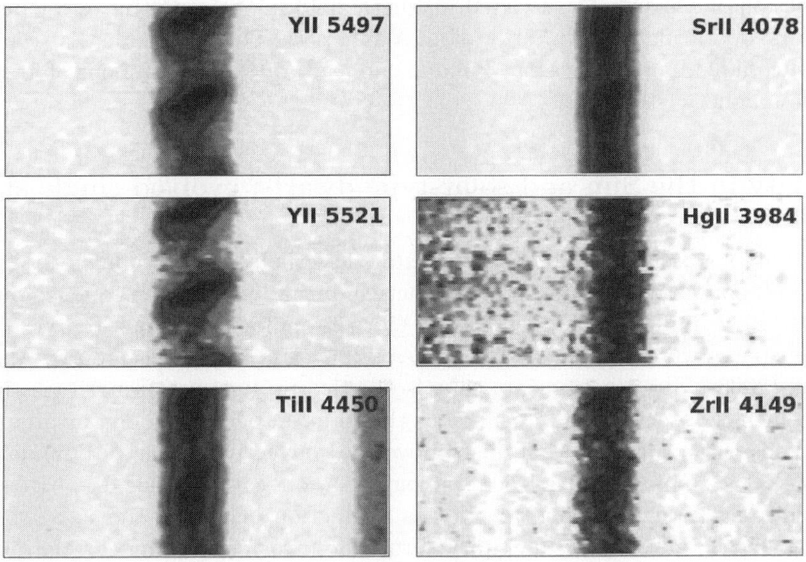

Figure 5. Variations of line profiles of different elements in the CORALIE spectra of the SB1 system HD 11753 as a function of rotation phase.

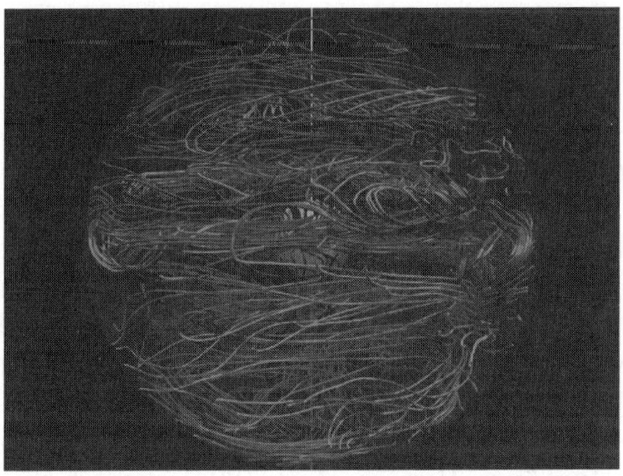

Figure 6. 3-D simulation of magnetic field lines after the onset of MRI. Blue shows radial magnetic field $B_r < 0$, while red stands for $B_r > 0$. Courtesy: R. Arlt.

Although the fields are not very strong, complex surface patterns can be obtained from the nonlinear, nonaxisymmetric evolution of the MRI. In Fig. 6 we present a 3-D simulation of magnetic field lines after the onset of MRI.

The magnetic fields of Ap and Bp stars have a simple non-axisymmetric topology and exhibit virtually no intrinsic variability on short and long time scales. The origin of the magnetic fields of Ap and Bp stars and details of their peculiar properties are still mysterious in spite of large efforts in their study. It is not clear yet whether their magnetism is related to the fossil magnetic fields from the interstellar medium, or results from some dynamo action capable of generating fields in radiative zones. Rüdiger & Kitchatinov (2009) argued that non-axisymmetric poloidal fields of Ap and Bp stars can result from a kink-type instability of their internal toroidal fields. Using this mechanism, they explain the relative slow rotation and observational low boundary of magnetic field strengths of these stars. Featherstone *et al.* (2007) used detailed 3-D simulations to study the interaction of a primordial twisted magnetic field in the radiative envelope with the core-dynamo operating in the interior of an A-type star. They conclude that the interplay between such a field and the stellar dynamo can lead to the development of an organized large-scale magnetic structure.

4. Activity in the Sun and solar-type dwarfs, evolved single stars, and fully convective stars

It has been long assumed that solar-type stars lose their angular momentum via a magnetized wind, spin down, and become less active during their main sequence phase (e.g., Skumanich 1972; Soderblom *et al.* 1991). Gyrochronology relations (curves in colour-period-age space) were introduced by Barnes (2007) and improved upon by Mamajek & Hillebrand (2008). The rotation rate can also be tied to dynamo strength via the Rossby number, which is observationally defined as the rotation period divided by an estimate of the local convective turnover time just above the convective-radiative boundary (Noyes *et al.* 1984). By combining a gyrochronology relation with the $\log R'_{HK}$–Rossby number correlation, one can predict activity as a function of colour and age (Mamajek & Hillebrand 2008). Another useful empirical relation involves a strong correlation between coronal X-ray activity and rotation via the Rossby number (e.g. Hempelmann *et al.* 1995). One of the goals of research on stellar magnetic activity is to understand, from a theoretical point of view, how these empirical relations constrain stellar dynamo theory. Despite considerable efforts in theoretical stellar dynamo studies, there are still fundamental open questions concerning the deep meaning of these relations (e.g. Montesinos *et al.* 2001).

In contrast to solar-type stars, the Ca II H+K emission from evolved single stars with low mass is weak. The drop in chromospheric emission from giants with decreasing $T_{\rm eff}$ corresponds to an even sharper drop in emissions from the transition region and corona. This has led to the discovery of so-called dividing lines in the H-R diagram. HST imaging of the M2 supergiant Betelgeuse has shown a bright spot in ultraviolet (Mg II H&K) wavelengths (Gilliland & Dupree 1996), i.e. located in the star's dynamic chromosphere above the photosphere. It was the first direct detection of stellar surface inhomogeneity on a star other than the Sun. Also some other observations, such as maser polarization in circumstellar envelopes of AGB stars (e.g. Sivagnanam 2004) or X-ray emission observed from some cool giant stars (e.g. Hünsch *et al.* 1998) suggest the existence of magnetic activity in late-type giants. The first systematic search for the presence of magnetic fields in bright giants of different spectral types and with different X-ray luminosities was carried out ~15 years ago by Hubrig *et al.* (1994) who reported 3σ detections for four giants. In the last years, using the LSD technique, Konstantinova-Antova *et al.* (2008) detected Zeeman signatures of the magnetic field in several other giants. Numerical MHD simulations of non-linear dynamo action in Betelgeuse were presented by Dorch (2004). Three different modes of dynamo action were recognized in his computations, but it was not possible to conclude on the presence of a magnetic field since such a field is not observed.

Recent observations of active mid-M dwarfs confirmed that dynamo processes in fully-convective stars with masses of about $0.3\,M_\odot$ are very successful at generating strong poloidal fields with simple axisymmetrical configurations (e.g. Morin *et al.* 2008). For lower mass stars strong and simple large-scale fields were detected, while some stars display weaker and more complex magnetic field topologies. In Fig. 7 we present $\alpha\Omega$ dynamo simulations applied to the rapidly rotating fully convective M4 dwarf V374 Peg, for which an axisymmetric field was reported (M. Küker, private communication).

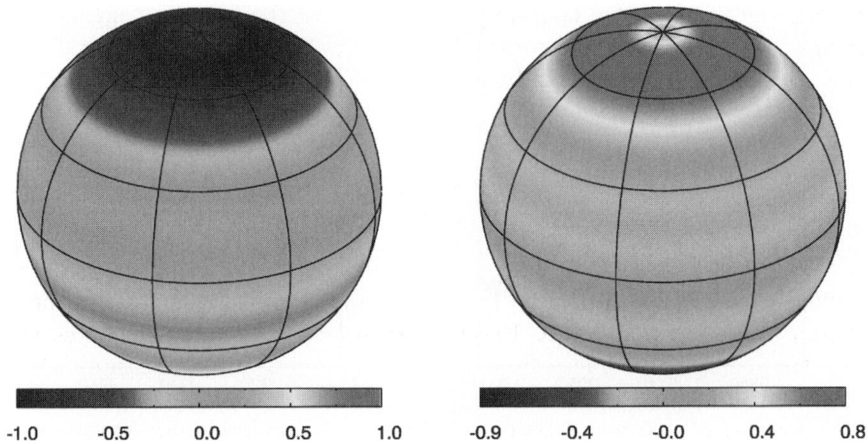

Figure 7. $\alpha\Omega$ dynamo model simulations: The star has solar-type differential rotation, i.e. it rotates faster at the equator than at the poles. The resulting magnetic field is axisymmetric and (essentially) dipolar. This model can be applied to the rapidly rotating fully convective M4 dwarf V374 Peg, for which an axisymmetric field was reported. Courtesy: M. Küker.

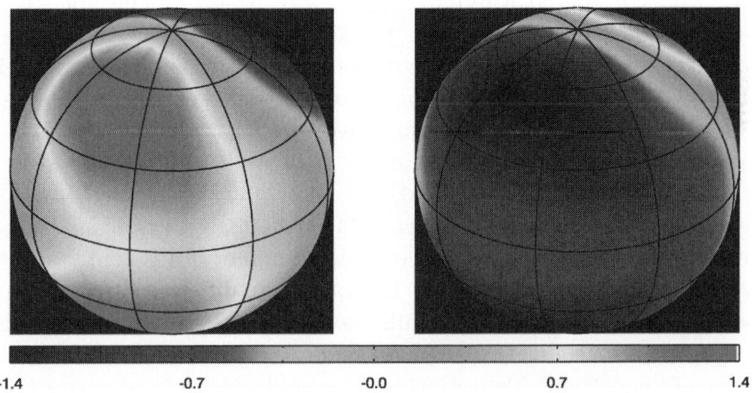

Figure 8. α^2-dynamo model simulations: This dynamo works without differential rotation. All field components are generated by the helicity of the convective gas motions (α–effect). This type of dynamo does not prefer axisymmetric fields and the presented field is purely non-axisymmetric. The surface field shows large regions of strong vertical fields. Courtesy: M. Küker.

Large-scale magnetic topologies and surface differential rotation seem to change at spectral type ~M3.5 at the boundary to full convection (e.g. Reiners & Basri 2009). 3-D nonlinear MHD simulations of the interiors of fully convective M-dwarfs of about 0.3 $M_\odot$ were recently presented by Browning (2008). Despite of the lack of strong differential rotation, the magnetic fields realized in his simulations possess significant axisymmetric components attributed to the strong influence of rotation on the slowly overturning flows.

5. Magnetic fields in star formation: T Tau stars and Herbig Ae stars

Magnetic fields are important ingredients of the star formation process (e.g. McKee & Ostriker 2007). Accumulated direct and indirect evidence points to a much higher level activity in the young Sun (e.g. meteoritic traces and studies of young solar analogs). T Tauri stars stand out by their very large emissions in the chromospheric and transition-region lines. A model for the generation of magnetic fields in fully convective pre-main sequence stars was presented by Küker & Rüdiger (1999), Chabrier & Küker (2006), and

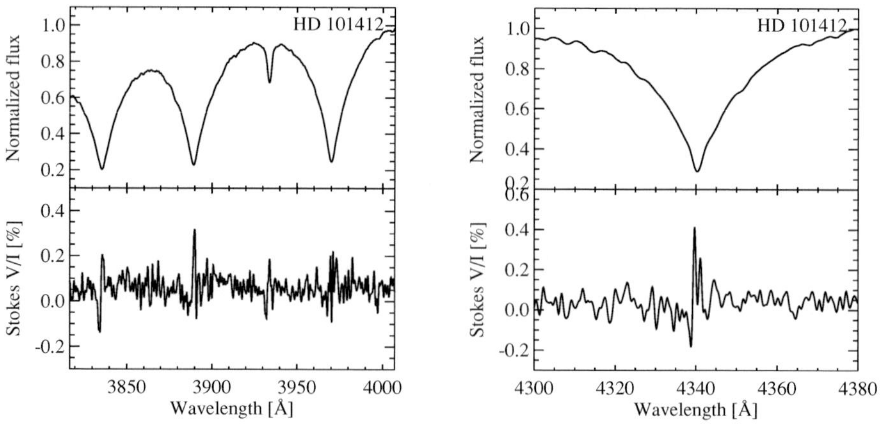

Figure 9. Stokes I and V spectra of the Herbig Ae/Be star HD 101412, with the largest detected magnetic field. Left panel: Zeeman features in H_9, H_8, Ca II H&K, and Hϵ profiles. Right panel: Stokes I and V spectra in the vicinity of the Hγ line.

Küker (2009), who considered a dynamo process of α^2-type. The non-linear induction equation was solved in three dimensions to allow for non-axisymmetric solutions. The resulting magnetic field was found purely non-axisymmetric and steady (see Fig. 8). Two years later Kitchatinov (2001) carried out simulations of a convection zone dynamo and an internal field captured by a growing radiative core for a pre-main sequence solar mass star. His computations suggest that an infant Sun with its core just starting to grow generates a non-axisymmetric global field.

The quite recent work of Sokoloff *et al.* (2008) was based on a representation of the stellar magnetic field as a superposition of a finite number of poloidal and toroidal free damping modes. Assuming a "realistic" law of stellar differential rotation the authors found that a dynamo starts to work in fully convective stars when the dynamo number $D > 5\times10^3$.

The presence of magnetic fields in Herbig Ae stars has long been suspected, in particular on account of Hα spectropolarimetric observations pointing out the possibility of the existence of a physical transition region in the H-R diagram from magnetospheric accretion, similar to that of classical T Tau stars (e.g. Vink *et al.* 2002). In the course of our spectropolarimetric studies of Herbig Ae stars (Hubrig *et al.* 2004; Hubrig *et al.* 2006c; Hubrig *et al.* 2007; Hubrig *et al.* 2009c) we have for the first time examined the relation of the measured field strengths to various parameters that characterize the star-disk system.

In our sample, the Herbig Ae star HD 101412 exhibited the largest magnetic field strength: $\langle B_z \rangle = -454\pm42$ G. In Fig. 9 we present distinct Zeeman features detected at the positions of the hydrogen Balmer lines and the Ca II H&K lines.

In Fig. 10, on the left side, we present the strength of the magnetic field plotted versus $\log L_X$. We find a hint for an increase of the magnetic field strength with the level of the X-ray emission, also supported by the correlation analysis which yields a >90 % probability that a correlation is present between $\langle B_z \rangle$ and $\log L_X$. This could suggest a dynamo mechanism responsible for the coronal activity in Herbig Ae stars.

There are also clear indications for a trend towards stronger magnetic fields in younger Herbig Ae stars (Fig. 10, right side). The magnetic fields become very weak at the end of their PMS life. Similarly, strong X-ray sources are only found at the youngest ages, in qualitative agreement with the predictions of a shear dynamo that decays within a few

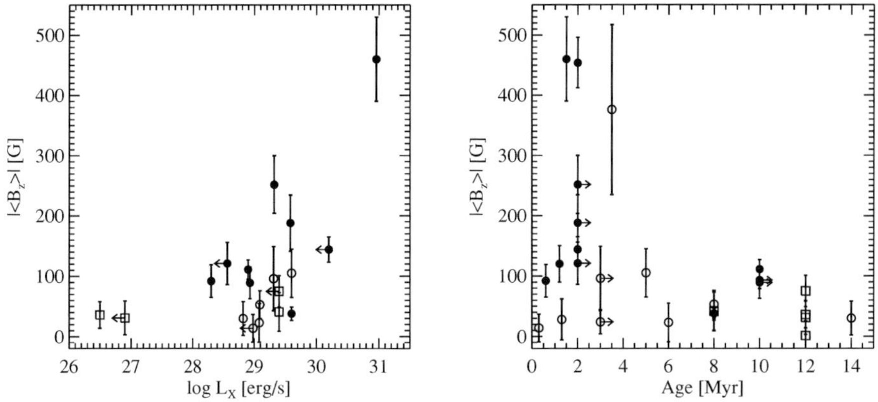

Figure 10. Left panel: The strength of the longitudinal magnetic field plotted over the X-ray luminosity. Right panel: The strength of the longitudinal magnetic field as a function of age. Filled circles denote Herbig Ae stars with a 3σ magnetic field detection, while open circles denote Herbig Ae stars with non-detections. Squares denote stars with debris disks, of which none has a 3σ magnetic field detection.

Myrs as the rotational energy of the star decreases (Tout & Pringle 1995). Furthermore, the stars seem to obey the universal power-law relation between magnetic flux and X-ray luminosity established for the Sun and main-sequence active dwarf stars (Pevtsov *et al.* 2003).

6. Summary

Dynamo models have undergone considerable progress in recent years, evolving from the solar dynamo to the larger class of stellar dynamos. To better understand the magnetic field generation in various types of stars it is necessary to obtain in future observations more information on activity, magnetic fields and magnetic field topology in stars of different mass and at different age. Testing the present dynamo models against observational constraints is crucial to identify necessary steps for the further development of dynamo theories.

References

Arlt, R., Hollerbach, R., & Rüdiger, G. 2003, *A&A*, 401, 1087
Barnes, S. A. 2007, *ApJ*, 669, 1167
Browning, M. K. 2008, *ApJ*, 676, 1262
Cantiello, M., Langer, N., Brott, I., *et al.* 2009, *A&A*, 499, 279
Chabrier, G. & Küker, M. 2006, *A&A*, 446, 1027
Donati, J.-F., Babel, J., Harries, T. J., *et al.* 2002, *MNRAS*, 333, 55
Donati, J.-F., Howarth, I. D., Bouret, J., *et al.* 2006a, *MNRAS*, 365, 6
Donati, J.-F., Howarth, I. D., Jardine, M. M., *et al.* 2006b, *MNRAS*, 370, 629
Dorch, S. B. F. 2004, *A&A*, 423, 1101
Featherstone, N. A., Browning, M. K., Brun, A. S., *et al.* 2007, *AN*, 328, 1126
Gilliland, R. L. & Dupree, A. K. 1996, *ApJ* (Letters), 463, L29
Hempelmann, A., Schmitt, J. H. M. M., Schultz, M., *et al.* 1995, *A&A*, 294, 515
Hubrig, S., Plachinda, S. I., Hünsch, M., *et al.* 1994, *A&A*, 291, 690
Hubrig, S. & Mathys, G. 1995, *ComAp*, 18, 167
Hubrig, S. 1998, *CoSka*, 27, 249
Hubrig, S., Schöller, M., & Yudin, R. V. 2004, *A&A* (Letters), 428, L1

Hubrig, S., Briquet, M., Schöller, M., *et al.* 2006a, *MNRAS*, 369, 61
Hubrig, S., González, J. F., Savanov, I., *et al.* 2006b, *MNRAS*, 371, 1953
Hubrig, S., Schöller, M, Yudin, R. V., *et al.* 2006c, *A&A*, 446, 1089
Hubrig, S., Pogodin, M., Yudin, R. V., *et al.* 2007, *A&A*, 463, 1069
Hubrig, S., Schöller, M., Schnerr, R. S., *et al.* 2008a, *A&A*, 490, 793
Hubrig, S., Briquet, M., Morel, T., *et al.* 2008b, *A&A*, 488, 287
Hubrig, S., Briquet, M., Morel, T., *et al.* 2009a, *AN*, 330, 317
Hubrig, S., Schöller, M., Savanov, I., *et al.* 2009b, *AN*, 330, 708
Hubrig, S., Stelzer, B., Schöller, M., *et al.* 2009c, *A&A*, 502, 283
Hünsch, M., Schmitt, J. H. M. M., & Voges, W. 1998, *A&AS*, 127, 251
Kitchatinov, L. L. 2001, *ARep*, 45, 816
Konstantinova-Antova, R., Auriére, M., Iliev, I. Kh., *et al.* 2008, *A&A*, 480, 475
Küker, M. & Rüdiger, G. 1999, *A&A*, 346, 922
Küker, M. 2009, *AIPC*, 1094, 77
MacGregor, K. B. & Cassinelli, J. P. 2003, *ApJ*, 586, 480
Mamajek, E. E. & Hillebrand, L. A. 2008, *ApJ*, 687, 1264
McKee, C. F. & Ostriker, E. C. 2007, *ARAA*, 45, 565
Montesinos, B., Thomas, J. H., Ventura, P., *et al.* 2001, *MNRAS*, 326, 877
Morin, J., Donati, J.-F., Petit, P., *et al.* 2008, *MNRAS*, 390, 567
Noyes, R. W., Weiss, N. O., & Vaughan, A. H. 1984, *ApJ*, 287, 769
Pevtsov, A. A., Fisher, G. H., Acton, L. W., *et al.* 2003, *ApJ*, 598, 1387
Reiners, A. & Basri, G. 2009, *A&A*, 496, 787
Robrade, J. & Schmitt, J. H. M. M. 2009, *A&A*, 497, 511
Rüdiger, G. & Kitchatinov, L. L. 2009, *astro-ph/0908184l*
Savanov, I. S., Hubrig, S., González, J. F., *et al.* 2009, *IAUS*, 259, 401
Sivagnanam, P. 2004, *MNRAS*, 347, 1084
Skumanich, A. 1972, *ApJ*, 171, 565
Stothers, R. B. & Chin, C. 1993, *ApJ* (Letters), 408, L85
Soderblom, D. R., Duncan, D. K., & Johnson, D. R. H. 1991, *ApJ*, 375, 722
Sokoloff, D. D., Nefedov, S. N., Ermash, A. A., *et al.* 2008, *AstL*, 34, 761
Tout, C. A. & Pringle, J. E. 1995, *MNRAS*, 272, 528
Vink, J. S., Drew, J. E., Harries T. J., *et al.* 2002, *MNRAS*, 337, 356

Solar and Stellar Variability: Impact on Earth and Planets
Proceedings IAU Symposium No. 264, 2009
A. G. Kosovichev, A. H. Andrei & J.-P. Rozelot, eds.

doi:10.1017/S1743921309992602

Helicity of solar magnetic field from observations

Hongqi Zhang

National Astronomical Observatories, Chinese Academy of Sciences, Beijing 100012
email: hzhang@bao.ac.cn

Abstract. The helicity is an important quantity to present the basic topological configuration of magnetic field transferred form the solar subatmosphere into the interplanetary space. In this paper, we present the observational solar magnetic field and the relationship with the magnetic helicity.

Keywords. Sun: magnetic fields, helicity, interior; turbulence, MHD

1. Importance and definition of magnetic helicity

Helicities are topologically a measure of the structural complexity of the corresponding fields. Woltjer (1958) presented the basic properties of magnetic helicity firstly. As indicated by Taylor (1986) that the topological invariants of idcal plasma so that only total magnetic helicity survives. Helicity is described in terms of the internal structure of a flux tube and the external relations between flux tubes. The helicity of open field configurations not bounded by magnetic surfaces is discussed by Berger and Field (1984), and a measure of the propagation of topological structures across open boundaries is given also.

Pouquet, Frisch, and Leorat (1976) indicated that the turbulence and magnetic fields are a common feature of many celestial bodies. The study confirms the importance of helicity both in its kinetic and in its magnetic form for generation of large-scale magnetic fields by turbulence. Keinigs (1983) and Keinigs and Gerwin (1986) analyzed that in a magnetized plasma the alpha effect represents a turbulently generated emf directed along the mean magnetic field. The alpha effect is reevaluated in terms of ensemble-averaged properties of the magnetic fluctuation spectrum. The results indicate that the turbulent current helicity must be opposite in sign to the mean-field current helicity in order for the alpha effect to play a role in overcoming the resistive diffusion of large-scale magnetic fields. Kleeorin and Rogachevskii (1999) analyzed the evolution of the magnetic helicity tensor for a nonzero mean magnetic field and for large magnetic Reynolds numbers in an anisotropic turbulence.

The magnetic helicity density $h_m = \mathbf{A} \cdot \mathbf{B}$, with $\mathbf{A}$ the vector potential for magnetic field $\mathbf{B}$, measures the chirality of magnetic lines of force. The magnetic helicity is defined as

$$H_m = \int_V h_m d^3x = \int_V \mathbf{A} \cdot \mathbf{B} d^3x, \tag{1.1}$$

where the vector potential $\mathbf{A}$ can not be observed immediately. It is noticed that the magnetic helicity is a relative quantity and depends on the selection of gauge of magnetic field $\mathbf{B}$ (i.e. $\mathbf{B} = \bigtriangledown \times (\mathbf{A} + \bigtriangledown\phi)$. It is conserved in a close volume when small resistivity is present.

According to the definition (Berger and Field, 1984), the magnetic helicity can be separated into two kinds. One is the self helicity, which relates to the magnetic flux tubes twisted themselves. This helicity may be used to analyze the twisted magnetic flux loops. Another is the mutual helicity, which relates to the different magnetic flux tubes linked each other. As the helicity contains both, the total helicity can be written in the form

$$H_m = T\Phi^2 + 2L\Phi_1\Phi_2, \tag{1.2}$$

where the T is the twisted number of magnetic flux Φ and the L is the linkage number of different magnetic flux Φ_1 and Φ_2.

The relative change of magnetic helicity in the solar atmosphere can be inferred by the magnetic field across the boundary surface (Berger and Field, 1984)

$$\frac{dH_m}{dt} = -2\oint_S [(\mathbf{V}_t \cdot \mathbf{A}_p)B_n - (\mathbf{A}_p \cdot \mathbf{B}_t)V_n]ds, \tag{1.3}$$

where the magnetic field $\mathbf{B}$ and velocity field $\mathbf{V}$ are observable in the solar atmosphere. The subscripts have their normal meanings. The first term in eq. (1.3) provides the contribution from the twisted motion of footpoints of magnetic field in the solar surface, while the second term does that from the emergence of twisted magnetic flux from the subatmosphere.

An important improvement in this project was demonstrated by Demoulin and Berger (2003) with a relative simple form, who presented that the horizontal motions, deduced by tracking the photospheric cut of magnetic flux tubes, include the effect of both the emergence and the shearing motions whatever the magnetic configuration complexity is. According to the analysis of Demoulin and Berger (2003), one can obtain that

$$\frac{dH_m}{dt} = -2\oint_S (\mathbf{U} \cdot \mathbf{A}_p)B_n ds, \tag{1.4}$$

where

$$\mathbf{U} = \mathbf{V}_t - \frac{V_n}{B_n}\mathbf{B}_t. \tag{1.5}$$

As the linkaged or twisted magnetic flux bundles emerge from the subatmosphere, their footpoints show the sheared or twisted motion and it can be analyzed by the equivalent form in eqs. (1.4) and (1.5). This implies that one can not exclude the contribution of emerging flux in the horizontal motion of magnetic footpoints in the solar surface.

The current helicity density h_c ($h_c = \mathbf{B} \cdot \nabla \times \mathbf{B}$) is another important physical quantity for the measure of the magnetic field in the solar atmosphere. It is noticed that only as $\nabla \times \mathbf{A}$ is parallel to $\mathbf{A}$ the relationship of both helicity densities becomes simple, and both helicity density show the same sign constantly (Zhang, 2001). In the statistical analysis of magnetic fields, the mean magnetic and current helicities would probably show the same sign $< h_c > \sim < h_m >$.

The current helicity is defined in the form

$$H_c = \int_V h_c d^3x = \int_V \mathbf{B} \cdot \nabla \times \mathbf{B} d^3x. \tag{1.6}$$

If neglect the coefficient $\frac{c}{4\pi}$, we can obtain

$$H_c = 2I_1 I_2. \tag{1.7}$$

The similar relationship on the linkage and twist of current helicity relative to eq. (1.2) can be inferred also

$$H_c = TI^2 + 2LI_1 I_2, \tag{1.8}$$

where T is the twisted number of current system I and L is the linkage number of different current system I_1 and I_2.

Moreover, as comparing with the magnetic helicity density, it is found that only a part (vertical component) of current helicity density in the photosphere (Abramenko *et al.*, 1996; Bao and Zhang, 1998),

$$h_{cz} = \mathbf{B} \cdot (\bigtriangledown \times \mathbf{B})_z \tag{1.9}$$

can be inferred from the photospheric vector magnetograms, due to the observational limitation. A similar limitation can be found also on the analysis by the force free factor (Pevtsov *et al.*, 1994)

$$\alpha = \frac{\mu J_z}{B_z}, \tag{1.10}$$

which also does not contain any information on the horizontal part of current helicity density. The mean photospheric current helicity density $\overline{h}_c$ (or mean force free factor $\overline{\alpha}$) is normally used to infer the handedness of magnetic field quantitatively in active regions.

It is suggested a dynamo mechanism in the solar interior based on the combined action of differential rotation and cyclonic convective vortices (Paker 1955) as a viable way to generate magnetic fields capable of driving the activity cycle. We are able to quantify the differential rotation from the motion of large-scale magnetic fields at the solar surface and from helioseismology in the solar convection zone. However, because of the opacity of the solar atmosphere, knowledge of the action of convective vortices can only be obtained from available observations of helical magnetic fields. We can define the magnetic field as a combination of mean and fluctuate field $\mathbf{B} =< \mathscr{B} > +\mathbf{b}$. According to mean field dynamo theory, the electromotive force $\mathbf{E}$ averaged over convective eddies has a component parallel to the magnetic field, $\mathbf{E} = \alpha < \mathscr{B} > + \cdots$, where the pseudoscalar α is related to kinetic and electric current helicities. The determination of the kinetic helicity in the solar atmosphere is difficult, while the twist of magnetic fields, can be estimated from photospheric vector magnetograms of solar active regions (Abramenko *et al.* 1996; Bao and Zhang 1998). The mean current helicity density can be written in the form

$$h_c =< \mathbf{b} \cdot \bigtriangledown \times \mathbf{b} > . \tag{1.11}$$

According to Kleeorin and Rogachevskii (1999) and Brandenburg and Subramanian (2005), the change of magnetic (current) helicity can been inferred in the form

$$\frac{\partial h_c}{\partial t} \approx -\frac{2}{l^2}[< \mathscr{B} > \cdot \bigtriangledown \times < \mathscr{B} > -(\alpha_k + h_c) < \mathscr{B} >^2], \tag{1.12}$$

where the symbols have their normal meanings. It means that the observational solar vector magnetograms can be statistically used to get the possible message on the generation of magnetic field inside of the Sun due to the solar dynamo. In the normal analysis of vector magnetic field of solar active regions, the symbol difference between $\mathbf{B}$ and $\mathbf{b}$ has been neglected.

2. The transfer of magnetic helicity and solar eruptive phenomena

2.1. *Helicity transfer in solar active regions*

Helicity is a notable quantity in the study of solar active regions (Wang, 1996). Chae (2001) shown how to observationally determine the rate of magnetic helicity transport via photospheric footpoint shuffling from a time series of line-of-sight magnetograms. Démoulin and Berger (2003) pointed out horizontal motions include the effect of both the emergence and the shearing motions whatever the magnetic configuration complexity is.

From a series of photospheric-vector magnetograms and corresponding soft X-ray images, Zhang (2001) and Zhang (2006a) found that the newly emerging magnetic flux associates the current helicity from the subatmosphere in the active regions with the redistribution of the current helicity density in the upper atmosphere, i.e. it provides observational evidence that flux and helicity emerge together. Because the injection rate of magnetic helicity and photospheric current helicity density have different means in the solar atmosphere, a combined analysis of the observational magnetic helicity parameters actually provides a relative complete picture of magnetic helicity and its transfer in the solar atmosphere. Liu and Zhang (2002) demonstrated that helicity reversal in magnetic features of a delta -configuration is likely to destabilize the compact structure, as well as to re-organize the magnetic field configuration, and, hence, is important for the rapid disintegration of a delta-spot during major flares. its chirality. Liu and Zhang (2006) found that the rotation of photospheric footpoints forms in the earlier stage of magnetic flux emergence and the relative shear motion of different magnetic flux systems appears later in an active region. The strong shear motion between the new emerging flux system and the old one brings more magnetic helicity into the corona than the twisting motions. Zhang, Liu and Zhang (2008) found the rapidly rotating positive polarity of an extensive δ sunspot in Active Region (AR) NOAA 10486, it produced several powerful flare-CMEs. They found the fastest of them is about 220° for six days. The helicity injection inferred from such rotational motion is about $-3.0 \times 10^{43} Mx^2$, which is comparable that calculated by the local correlation tracking (LCT) method $-5.2 \times 10^{43} Mx^2$ in the whole AR.

Su *et al.* (2009) studied the distributions of local twist α_z and current helicity hc on the active region of NOAA 10930. They found the patches of positive and negative helicities were intermixed showing a mesh pattern in the umbra and a thread pattern in the penumbra. The fine distributions of α_z and hc on a penumbral filament indicated that it may be possible for the two opposite helicities to coexist in a filament and their magnitudes were nearly equivalent. Moreover, Tian and Alexander (2009) separately calculated the distribution of the helicity flux injected in the leading and following polarities of 15 emerging bipolar ARs, using the Michelson Doppler Image 96 minute line-of-sight magnetograms and a local correlation tracking technique. They found from this statistical study that the leading (compact) polarity injects several times more helicity flux than the following (fragmented) one (typically 3–10 times).

2.2. *Helicity and solar flare-CMEs*

Rust and Kumar (1996) found that images of the X-ray corona near the solar disk's center were examined for large, transient brightenings of the type known to be associated with Hα filament eruptions and coronal mass ejections. Many of the brightenings were sigmoid (S-shaped). López Fuentes *et al.* (2000) described the long-term evolution of a bipolar non-Hale active region that was observed from 1995 October to 1996 January. They compared their results with those of other related studies, and they discussed, in particular, whether the kink instability is relevant to explain the peculiar evolution of some active regions. Nindos and Zhang (2002) and Nindos, Zhang and Zhang (2003) investigated whether the bulk of magnetic helicity carried away from the Sun by CMEs comes from helicity injected to the corona by such motions or by emerging magnetic flux. Green *et al.* (2002) studied the magnetic helicity evolution in an active region (NOAA 8100) in which the main photospheric polarities rotate around each other during five Carrington rotations. It is proposed that the ejected helicity is provided by the twist in the sub-photospheric part of the magnetic flux tube forming the active region. Green *et al.* (2007) found the helicity sign of the erupting field and the direction of filament rotation to be consistent with the conversion of twist into writhe under the ideal MHD constraint of helicity conservation. For positive (negative) helicity the filament apex rotates clockwise (counterclockwise), consistent with the flux rope taking on a reverse (forward) S shape, which is opposite to that observed for the sigmoid. This result is incompatible with two models for sigmoid formation: one identifying sigmoids with upward arching kink-unstable flux ropes and one identifying sigmoids with a current layer between two oppositely sheared arcades.

Zhang, Flyer and Low (2006) have pointed out that the accumulation of magnetic helicity in the corona plays a significant role in storing magnetic energy. They propose a conjecture that there is an upper bound on the total magnetic helicity that a force-free field can contain. LaBonte, Georgoulis and Rust (2007) surveyed magnetic helicity injection into 48 X-flare-producing active regions recorded by the MDI between 1996 July and 2005 July. They found that an empirical fit to the data shows that the injected helicity over the range $10^{39} - 10^{43} Mx^2 s^{-1}$ is proportional to magnetic flux squared. Similarly, over a range of 0.3–3000 days, the time required to generate the helicity in a CME is inversely proportional to the magnetic flux squared. Most of the X-flare regions generated the helicity needed for a CME in a few days to a few hours. Kazachenko *et al.* (2009) used the Michelson Doppler Imager and TRACE observations of photospheric magnetic and velocity fields in NOAA 10759 to build a three-dimensional coronal magnetic field model. Their combined analysis yields the first quantitative picture of the helicity and energy content processed through a flare in an active region with an obviously rotating sunspot and shows that rotation dominates the energy and helicity budget of this event

3. Relationship between magnetic helicity and solar cycles

3.1. *Hemispheric rule of magnetic (current) helicity*

Hale *et al.* (1919) firstly found that Hα penumbral features show the direction of whirl in the Northern hemisphere is left-handed or anti-clockwise, while in the Southern hemisphere it is right-handed or clockwise. Ding, Hong, and Wang (1987) found the distribution of spiral patterns in the southern and northern hemispheres shows that the differential rotation may be a fundamental solar dynamo for the formation of the spiral spots. The statistical directions of the emerging twisted magnetic vectors in the active regions in the southern and northern hemispheres are synchronously inverse with a period of about two years.

Seehafer (1990) demonstrated that the electric current helicity is predominantly negative in the northern hemisphere and positive in the southern hemisphere. He pointed out that this finding contradicts the standard dynamo theory according to which the helicity of the large-scale currents is generated by the alpha effect in the convection zone, since it shows helicity of opposite sign. Pevtsov, Canfield and Metcalf (1995) found in their data set, 76% of the active regions in the northern hemisphere have negative helicity, and 69% in the southern hemisphere, positive. Although the data show considerable variation from one active region to the next, the data set as a whole suggest that the magnitude of the average helicity increases with solar latitude, starting at zero near the equator, reaches a maximum near 15 deg–25 deg in both hemispheres, and drops back toward smaller values avove 35 deg–40 deg.

Rust and Kumar (1996) found that reverse-S brightenings outnumbered forward-S brightenings by six to one in the northern hemisphere. Forward-S brightenings were similarly predominant in the south. This hemispherical segregation suggests that the magnetic fields in the transient features are systematically twisted.

Hagino, M. & Sakurai, T. 2005, *PASJ*, 57, 481
Hale, G. E., Ellerman, F., Nicholsin, S. B., & Joy, A. H. 1919, *Astrophys. J.*, 49, 153
Jiang, J., Choudhuri, A. R., & Wang, J. 2007, *Solar Phys.*, 245, 19
Kazachenko, M. D., Canfield, R. C., Longcope, D. W., & Qiu, J. 2009, *Astrophys. J*, 704, 1146
Keinigs, R. K. 1983, *PhFl*, 26, 2558
Keinigs, R. & Gerwin, Richard A. 1986, *IEEE Transactions*, 14, Issue 6, 858
Kleeorin, N. & Rogachevskii, I. 1999, *PhRvE*, 59, 6724
Kleeorin, N., Kuzanyan, K., Moss, D., Rogachevskii, I., Sokoloff, D., & Zhang, H. 2003, *A&A*, 409, 1097
Kuzanyan, K. M., Lamburt, V. G., Zhang, H., & Bao, S. 2003, *ChJAA*, 3, 257
Kuzanyan, K. M., Pipin, V. V., & Zhang, H. 2007, *AdSpR*, 39, 1694
LaBonte, B. J., Georgoulis, M. K., & Rust, D. M. 2007, *Astrophys. J*, 671, 955
Liu, Y. & Zhang, H. Q. 2002, *A&A*, 386, 646
Liu, J. & Zhang, H. 2006, *Solar Phys.*, 234, 21
Longcope, D. W., Fisher, G. H., & Pevtsov, A. A. 1998, *Astrophys. J*, 507, 417
López Fuentes, M. C., Démoulin, P., Mandrini, C. H., & van Driel-Gesztelyi, L. 2000, *Astrophys. J*, 544, 540
Moffatt, H. K. Magnetic field generation in electrically conducting fluids, Cambridge, England, Cambridge University Press, 1978, 353
Nindos, A. & Zhang, H. 2002, *Astrophys. J*, 573, L133
Nindos, A., Zhang, J., & Zhang, H. 2003, *Astrophys. J*, 594, 1033
Parker, E. 1955, *Astrophys. J*, 122, 293.
Pevtsov, A. A. 2000, *Astrophys. J*, 531, 553
Pevtsov, A. A., Canfield, R. C., & Metcalf, T. R. 1995, *Astrophys. J*, 440, L109
Pevtsov, A. A. & Latushko, S. M. 2000, *Astrophys. J*, 528, 999
Pevtsov, A. A., Dun, J. P., & Zhang, H. 2006, *Solar Phys.*, 234, 203
Pevtsov, A. A., Canfield, R. C., Sakurai, T., & Hagino, M. 2008, *Astrophys. J*, 677, 719
Pouquet, A., Frisch, U., & Leorat, J. 1976, *JFM*, 77, 321
Rust, D. M. & Kumar, A. 1996, *Astrophys. J*, 464, L199
Rüdiger, G., Pipin, V. V., & Belvedére, G. 2001, *Solar Phys.*, 198, 241
Seehafer, N. 1990, *Solar Phys.*, 125, 219
Seehafer, N. 1994, *A&A*, 284, 593
Su, J. T., Sakurai, T., Suematsu, Y., Hagino, M., & Liu, Y. 2009, *Astrophys. J*, 697, L103
Taylor, J. B. 1986, *RvMP*, 58, 741
Tian, L., Bao, S., Zhang, H., & Wang, H. 2001, *A&A*, 374, 294
Tian, L. & Alexander, D. 2009, *Astrophys. J*, 695, 1012
Wang, J. 1996, *Solar Phys.*, 163, 319
Woltjer, L. 1958, *PNAS*, 44, 489
Xu, H., Gao, Y., Zhang, H., Sakurai, T., Pevtsov, A. A., & Sokoloff, D. 2007, *AdSpR*, 39, 1715
Xu, H., Gao, Y., Popova, E. P., Nefedov, S. N., Zhang, H., & Sokoloff, D. D. 2009, *ARep*, 53, 160
Yang, S., Zhang, H., & Büchner, J. 2009, *A&A* 502, 333
Yeates, A. R., Mackay, D. H., & van Ballegooijen, A. A. 2008, *Astrophys. J*, 680, L165
Yokoyama, M. & Masuda, S. 2009, *Solar Phys.*, 254, 285
Zhang, H. 2001, *MNRAS*, 326, 57
Zhang, H. 2006a, *ChJAA*, 6, 96
Zhang, H. & Bao, S. 1998, *A&A*, 339, 880
Zhang, H., Sokoloff, D., Rogachevskii, I., Moss, D., Lamburt, V., Kuzanyan, K., & Kleeorin, N. 2006, *MNRAS*, 365, 276
Zhang, M. 2006b, *Astrophys. J*, 646, L85
Zhang, M., Flyer, N., & Low, B. C. 2006, *Astrophys. J*, 644, 575
Zhang, Y., Liu, J., & Zhang, H. 2008, *Solar Phys.*, 247, 39

Solar and Stellar Variability: Impact on Earth and Planets
Proceedings IAU Symposium No. 264, 2009
A. G. Kosovichev, A. H. Andrei & J.-P. Rozelot, eds.

doi:10.1017/S1743921309992614

Probability distribution functions for solar and stellar magnetic fields

Jan O. Stenflo

Institute of Astronomy, ETH Zurich, CH-8093 Zurich, and
Istituto Ricerche Solari Locarno (IRSOL), Via Patocchi, CH-6605 Locarno-Monti
email: stenflo@astro.phys.ethz.ch

Abstract. Magneto-convection structures the magnetic field in solar and stellar atmospheres over scales that for the Sun span about 8 orders of magnitudes, down to the magnetic diffusion scale of order 10 m. The statistical properties of this structuring are governed by probability density functions (PDFs), for the vertical and transverse field components as well as for the field inclination. Due to the fractal nature of the field pattern these PDFs appear to have a high degree of scale invariance. There are serious pitfalls in the derivations of empirical PDFs, pitfalls that are particularly severe in the case of the field inclination. This explains the fragmentary and rather unreliable PDF information available in the published literature. Here we discuss the nature of these pitfalls and indicate how they may be avoided, using Hinode quiet-sun Stokes vector data to derive PDFs for the field strength and field inclination.

Keywords. Sun: magnetic fields, Sun: photosphere, techniques: polarimetric, line: formation

1. Resolved and unresolved scales

As we zoom in on ever smaller scales in solar magnetograms (cf. Fig. 1), we find a fractal-like pattern with a high degree of self-similarity (Stenflo & Holzreuter 2002, 2003). It is in fact hard to guess the spatial scale of a magnetogram without tick marks, due to this scale invariance.

As ever more powerful solar telescopes come on line, we will be able to resolve smaller and smaller magnetic features, but due to the self-similarity, the appearance of the magnetic pattern would be expected to remain pretty much the same, except that the new tick marks will represent smaller scales.

As we go down in scale, the magnetic Reynolds number decreases until it reaches a value of unity around the 10 m scale (de Wijn *et al.* 2009). Below a value of unity the field ceases to be frozen in, because the time scale of magnetic diffusion becomes smaller than the dynamic time scale. As the viscous diffusion limit for the turbulent spectrum lies at even much smaller scales, we expect that magneto-convection will structure the field all the way down to the 10 m scale. This is 4 orders of magnitude smaller than the angular resolution limit of the best current telescopes.

On the quiet Sun we will therefore remain very far from resolving any of the magnetic structures in the foreseeable future. Analysis of Hinode data in different ways (cf. Stenflo 2009) reveals subpixel magnetic structuring far beyond the Hinode 0.3 arcsec angular resolution (see more about this below). For observed magnetic fields, which always refer to a certain finite angular resolution, we should therefore always avoid using the term "field strength" and instead use the term "flux density". Although both these quantities have the same dimension (G), confusion can be avoided if the term "field strength" is only used for resolved fields. In the asymptotic limit of infinite resolution the two concepts "field strength" and "flux density" become identical.

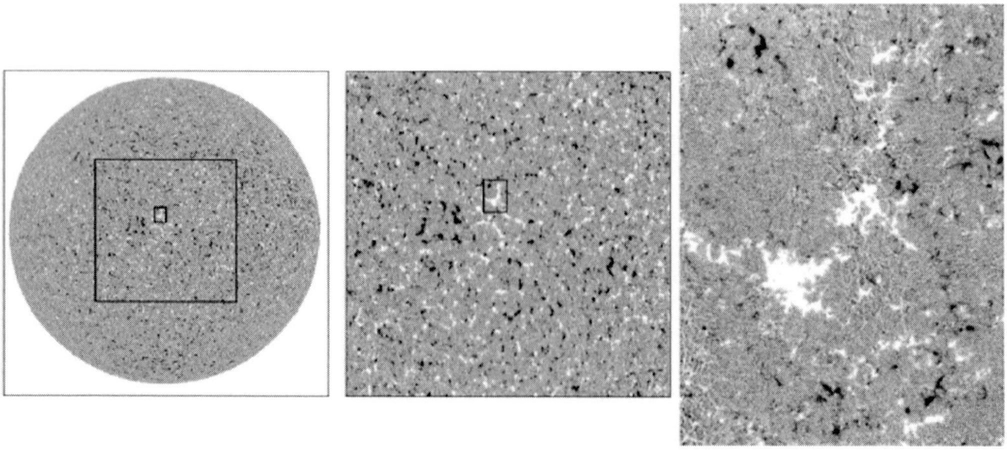

Figure 1. Fractal-like magnetic fields of the quiet Sun, observed on 9 February 1996. The two maps to the left are from a Kitt Peak full-disk magnetogram, while the right, high-resolution magnetogram was recorded the same day at the Swedish La Palma Observatory (courtesy Göran Scharmer). It covers an area that is only 0.35 % of the map next to it.

2. The concept of "intrinsic field strength"

As it was realized already in the 1960s that the true magnetic structures were far smaller than the available resolution, indirect methods were developed to diagnose the nature of the magnetic field beyond the spatial resolution limit. The Stokes V line ratio technique makes use of the differential non-linearity in the relation between field strength and circular polarization measured simultaneously in different spectral lines (Stenflo 1973). With a very special choice of line pair it is possible to isolate the magnetic effects from the thermodynamic effects. This led to the discovery that more than 90 % of the magnetic flux seen with moderate angular resolution is due to subresolution kG fields. Since however the measured flux densities on the quiet Sun are only a few G, the high field strengths imply that the corresponding magnetic elements occupy on average only about 1 % of the quiet-Sun photospheric volume. The theoretical counterpart of these seemingly intermittent strong-field structures became the magnetic flux tubes, which usually were assumed to be embedded in a non-magnetic environment. Ever more sophisticated flux tube models were then developed.

To determine the kG field strength of the spatially unresolved field structures an interpretational model based on a 2-component concept was found to be effective and to give consistent results. The concept is to assume that we have one magnetic component with a certain "intrinsic" field strength and filling factor, and one "non-magnetic" component. This 2-component model became the standard interpretational tool that was applied universally not only in solar but also in stellar physics.

With the success of this approach it tended to be forgotten that it represents an extreme idealization that filters out a certain property of the underlying reality, and that the true nature of this reality may be very different. If we would describe the 2-component model in terms of probability density functions (PDFs) that are frequently used to characterize the statistical properties of the fields found in numerical simulations of magneto-convection, the 2-component PDF would appear very strange indeed: it would consist of two δ function peaks, one near a field strength of 1 kG, representing the magnetic component, while the second peak, located at zero field strength, would be about 100 times taller (since the magnetic filling factor is only of order 1 %). The huge and the tiny δ function peaks would be separated by a vast, empty "desert" in between. In contrast, all numerical simulations of magneto-convection produce continuous PDFs, smoothly connecting the weakest and the strongest fields.

We therefore realize that the 2-component model is a crude interpretational filter that projects out some global property of the strong-field tail of the actual PDF in terms of a single number that we call the "intrinsic field strength". What this number physically really means depends on the nature of the underlying PDF and on the functional relation between field properties and polarization. One should be careful in assigning too much physical meaning to this "intrinsic field strength" as representing the field strength of a discrete flux tube. The underlying fractal reality is likely to be different.

After the kG values for the magnetic component of the quiet Sun had been discovered, the attention turned towards exploring the magnetic nature of the so-called "non-magnetic" component. It was from the start clear that it could not be truly non-magnetic, since a major field-free region in the highly conducting turbulent solar plasma is a physical impossibility. If the field is weak, the frozen-in field lines must be tangled up and amplified by the turbulent motions. The question was how strong this tangled field is.

This question could not be answered with the Zeeman effect (due to cancellation of the mixed-polarity signatures), but the introduction of the Hanle effect opened a window to this aspect of solar magnetism (Stenflo 1982), since the Hanle effect with its different symmetry properties is not subject to the mentioned cancellations. Interpreting the observed Hanle signatures in terms of a single-valued field strength, values of 30–60 G were found (Faurobert-Scholl 1993; Trujillo Bueno, Shchukina & Asensio Ramos 2004).

Even when the 2-component model is extended by including the Hanle effect, it would, if interpreted literally in terms of a PDF, mean that the huge δ function peak at zero field strength is shifted to a location somewhere in the interval 30–60 G (while the tiny 1 kG peak for the "magnetic component" remains). As we however have all reasons to expect that the actual PDF is a smooth function connecting in a continuous fashion the weakest with the strongest (kG) fields, the single numbers extracted from the observed Hanle signatures must represent some filtered global property of the weak-field portion of the underlying PDF. What this single number really means depends on the nature of the PDF, the angular distribution of field vectors, and the non-linear functional relation between the Hanle polarization signature and the field properties (strength and orientation).

It is therefore clear that there is now a big need to transcend the 2-component approach and move in the direction of interpretational models based on PDFs.

3. Angular distribution of field vectors

We have seen that the "intrinsic properties" of PDFs cannot be determined (in any direct way at least) in the spatially unresolved domain, since the unavoidably very idealized interpretational model used gives us a filtered version of the underlying reality characterized by a couple of discrete parameters. While such parameters can provide important constraints on the nature of the underlying PDF, a full reconstruction of the PDF would heavily rely on guessed-at model assumptions. As an alternative we suggest to approach this problem by instead exploring the PDFs in the spatially resolved domain and the scaling laws that they obey.

It might seem that it would in the spatially resolved domain be a straightforward matter to extract the PDFs for the flux densities as averaged over the spatial resolution element. This averaging is however highly non-linear and therefore far from trivial, as we will illustrate for the case of the angular distribution of field vectors. Lites *et al.* (2008)

analysed Hinode SOT/SP Stokes vector spectra of the Fe I 6301 and 6302 Å lines and came to the conclusion that there is 5 times more horizontal than vertical flux in the quiet photosphere. Subsequently Schüssler & Vögler (2008) reported that their numerical simulations of magneto-convection also showed a similar preponderance of horizontal flux. In contrast, I have analysed the identical Hinode data set and come to the opposite conclusion, namely that it is the vertical flux that dominates.

The main reason why it is possible to come to such opposite conclusions from the identical data set has to do with the way one deals with the influence of the magnetic structuring on subresolution scales. Due to non-linearities in the relation between polarization and the field properties, the averaging process affects the polarization and the field differently (cf. Stenflo 1994). The non-linearities in the linear polarization are much larger than in the circular polarization. The way in which they affect the spatial averaging over the Hinode 0.3 arcsec resolution element depends critically on the nature of the field structuring on subpixel scales. If this subpixel structuring is dealt with incorrectly (or ignored, as is often the case), large differential scaling errors for the transverse and longitudinal fields will occur, resulting in completely wrong field inclinations. Thus, if the subpixel structuring is neglected, the fields would look much more horizontal than they actually are.

It is beyond the scope of the present paper to go into the quite involved procedure of how to account for the subpixel structuring, but it is based on information extracted from the observed differential Stokes I and V line profile effects in the two Fe I 6301 and 6302 Å lines, modeled with synthetic Stokes spectra representing various realistic, hot and cold atmospheres. Some results of this rather complex analysis are illustrated in Fig. 2 for the angular distribution of field vectors. Since the angular distribution is a steep function of flux density, we have binned the data in terms of apparent flux density in the 6301 Å line. The upper panel of Fig. 2 shows the distribution for the 100–120 G bin. An isotropic distribution of field vectors would be flat in this representation (histogram in terms of $\mu = \cos\gamma$ bins, γ being the inclination angle). If horizontal fields would dominate, the distribution would have a negative slope. The steep positive gradient shows that the fields for this flux density bin are strongly peaked around the vertical direction.

To characterize each histogram with a single parameter, we have fitted it with the function μ^α, where $\alpha = 0$ represents the isotropic case, and the distribution gets increasingly peaked around the vertical direction as α increases. The lower panel in Fig. 2 shows the fitted α values as a function of flux density. For flux densities below 15 G α becomes negative, which would suggest that the weakest flux densities are predominantly horizontal. This is however entirely an artefact of noise, as has been verified by Monte-Carlo simulations using the actual noise values extracted from the Hinode data.

We find from this analysis that the angular distribution is nearly isotropic for the smallest flux densities, but becomes rapidly very peaked around the vertical direction as the flux density increases. It is natural to expect this to happen as a result of the increasing influence of the buoyancy forces.

4. Probability distributions for the flux densities

While the subpixel structuring has a profound effect on the extracted transverse flux densities, its influence on the line-of-sight flux densities is much more modest, although it needs to be accounted for. In Fig. 3 we show the histogram or probability density function (PDF) for the vertical flux densities, extracted from the mentioned Hinode data set for the quiet disk center (dashed curve). It can be fitted well by the sum of a symmetric Voigt function and an anti-symmetric line dispersion function, characterized by a "damping

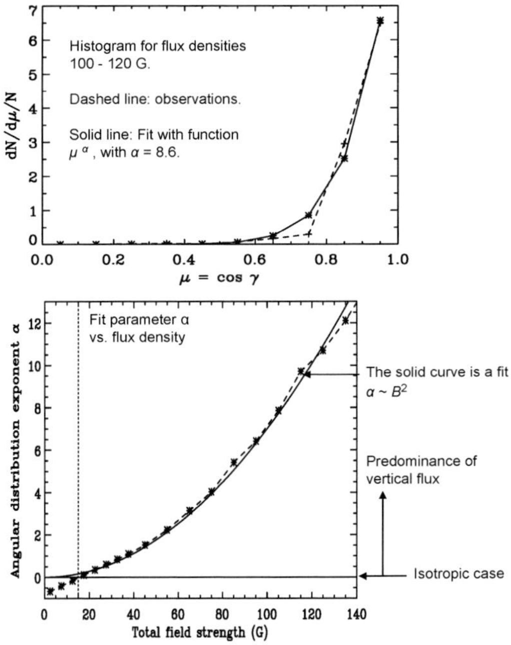

Figure 2. Angular distributions extracted from the Hinode data for the quiet-Sun disk center. Upper panel: Histogram representing the flux density interval 100–120 G. Lower panel: Variation with flux density of fit parameter α, showing how the fields get strongly peaked around the vertical direction as the flux density increases. The negative values of α below 15 G can be shown to be an artefact of noise. The Hinode data are therefore consistent with an isotropic distribution for the weakest fluxes.

constant" (to use the terminology from spectral line analysis) and a "Doppler width" characterizing the shape of the inner core. Since the value of the Doppler width closely agrees with its value for a pure Gaussian noise distribution, using the measured value of the noise in the Hinode data, the noise-free PDF is well represented by the mentioned sum of the Voigt and line dispersion functions, if we simply set the Doppler width to zero. The noise-deconvolved PDF (solid curve in Fig. 3) is proportional to the function

$$\frac{(\gamma/2)^2 + 0.038B}{B^2 + (\gamma/2)^2}\,, \tag{4.1}$$

where B is the flux density in G, and $\gamma = 8\,\mathrm{G}$.

The symmetric (Lorentzian) part of this function is represented by the dotted curve in Fig. 3. The anti-symmetric part accounts for the imbalance between the positive and negative polarities in the solar region that is being sampled.

We note that the PDF has its peak at zero flux density. A non-zero location of the histogram peak that has previously been reported (cf. Lites *et al.* 2008) is an artefact of the way in which the measurement noise was dealt with. The influence of the noise in the data becomes much more dramatic in the case of the horizontal fields derived from the linear polarization, with the result that the apparent PDFs for the horizontal and total flux densities drop towards zero for the weakest fields. Careful analysis of how the influence of the noise propagates, combined with Monte-Carlo simulations, show however that the apparent, shifted location of the PDF maxima is an artefact of noise. Also in the case of the horizontal and total flux densities, the noise-corrected PDFs all have their maxima at zero flux density.

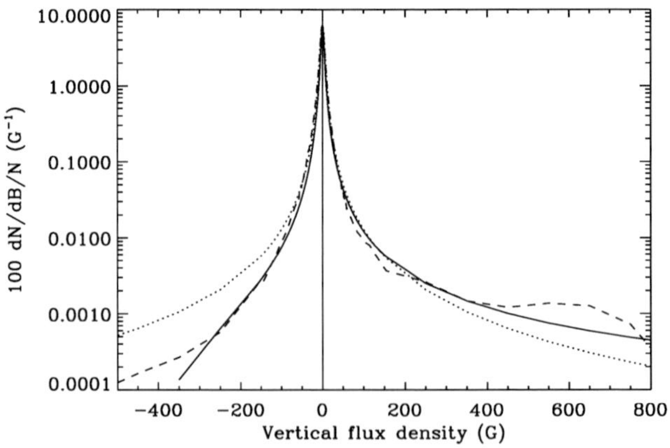

Figure 3. PDF for the vertical flux density, as extracted from the quiet-Sun Hinode data at disk center (dashed curve). The solid curve is a noise-deconvolved analytical representation in terms of a sum of a Lorentz and a line dispersion function, as given by Eq. (4.1). The symmetric part of this function is given by the dotted curve.

5. Concluding remarks

It is clear that we need to transcend the 2-component approach in the direction of continuous PDFs. This however requires some modelling assumptions for the behavior of the PDFs in the spatially unresolved domain, where they are not directly observable. To obtain empirical guidance we need to explore the scaling laws of the PDFs in the resolved domain, where we notice a high degree of fractal-like scale invariance, which is likely to continue into the unresolved domain. Even in the resolved domain, however, the polarimetric quantities averaged over the resolution element are strongly influenced by the nature of the subpixel magnetic structuring. Here we have tried to cast some light on these problems and indicate how they may be overcome.

Acknowledgements

Hinode is a Japanese mission developed and launched by ISAS/JAXA, with NAOJ as domestic partner and NASA and STFC (UK) as international partners. It is operated by these agencies in co-operation with ESA and NSC (Norway).

References

de Wijn, A. G., Stenflo, J. O., Solanki, S. K., & Tsuneta, S. 2009, *Space Science Reviews*, 144, 275
Faurobert-Scholl, M. 1993, *A&A*, 268, 765
Lites, B. W., Kubo, M., Socas-Navarro, H., *et al.* 2008, *ApJ*, 672, 1237
Schüssler, M. & Vögler, A. 2008, *A&A*, 481, L5
Stenflo, J. O. 1973, *Solar Phys.* 32, 41
Stenflo, J. O. 1982, *Solar Phys.* 80, 209
Stenflo, J. O. 1994, *Solar Magnetic Fields — Polarized Radiation Diagnostics.* (Kluwer)
Stenflo, J. O. 2009, in: K. N. Nagendra, H.-G. Ludwig, & P. Bonifacio (eds.), *3D views on cool stellar atmospheres: theory meets observation*, Proc. JD10 at IAU General Assembly, Rio, Aug. 10–11, 2009, *MemSAI*, 80, 690
Stenflo, J. O. & Holzreuter, R. 2002, in: H. Sawaya-Lacoste (ed.), *Magnetic Coupling of the Solar Atmosphere*, ESA Publ. SP-505, 101
Stenflo, J. O. & Holzreuter, R. 2003, in: A. A. Pevtsov & H. Uitenbroek (eds.), *Current Theoretical Models and High Resolution Solar Observations*, Proc. 21st International NSO/SP Workshop, ASP Conf. Ser. (San Francisco: ASP), vol. 286, 169
Trujillo Bueno, J., Shchukina, N., & Asensio Ramos, A. 2004, *Nature*, 430, 326

Solar and Stellar Variability: Impact on Earth and Planets
Proceedings IAU Symposium No. 264, 2009
A. G. Kosovichev, A. H. Andrei & J.-P. Rozelot, eds.

doi:10.1017/S1743921309992626

Oscillatory migratory large-scale fields in mean-field and direct simulations

Dhrubaditya Mitra[1], Reza Tavakol[1], Axel Brandenburg[2,3] and Petri J. Käpylä[2,4]

[1]Astronomy Unit, School of Mathematical Sciences, Queen Mary University of London, Mile End Road, London E1 4NS, UK
emails: dhruba.mitra@gmail.com; r.tavakol@qmul.ac.uk

[2]NORDITA, Roslagstullsbacken 23, SE-10691 Stockholm, Sweden
email: brandenb@nordita.org,

[3]Department of Astronomy, Stockholm University, SE 10691 Stockholm, Sweden,
[4]Observatory, Tähtitorninmäki (PO Box 14), FI-00014, University of Helsinki, Finland
email: petri.kapyla@helsinki.fi

Abstract. We summarise recent results form direct numerical simulations of both non-rotating helically forced and rotating convection driven MHD equations in spherical wedge-shape domains. In the former, using perfect-conductor boundary conditions along the latitudinal boundaries we observe oscillations, polarity reversals and equatorward migration of the large-scale magnetic fields. In the latter we obtain angular velocity with cylindrical contours and large-scale magnetic field which shows oscillations, polarity reversals but poleward migration. The occurrence of these behviours in direct numerical simulations is clearly of interest. However the present models as they stand are not directly applicable to the solar dynamo problem. Nevertheless, they provide general insights into the operation of turbulent dynamos.

Keywords. MHD, convection, Sun: magnetic fields

1. Introduction

One of the major challenges in astrophysics is to understand the generation of large-scale magnetic fields and their range of variations in a variety of settings, including the Sun, stars and galaxies. An important shared feature of such variations is that they often have characteristic time scales that are much shorter than the corresponding microscopic diffusion times. This would require the presence of relatively rapid mechanisms both for generation and dissipation of large-scale magnetic fields. An important candidate for the generation of such fields is a turbulent magnetohydrodynamic dynamo.

Here we shall concentrate on three such variations, namely oscillations, migration–especially equatorward migration as observed for example in the Sun–and the approximately periodic reversal of field polarity.

A great deal of effort has gone into the study of these types of behaviour in dynamo models over the last few decades. These studies have employed both the mean-field (MF) approach (see e.g. Krause & Rädler 1980; Moffatt 1978; Ossendrijver 2003) as well as direct numerical simulations (DNS) (see e.g. Brandenburg & Subramanian 2005, and references therein). In the simplest MF approach the effects of turbulence are parametrised in terms of the turbulent magnetic diffusion and an α effect.

MF models can be classified according to the relative magnitudes of α effect and differential rotation $|\nabla\Omega|$, where Ω is the angular velocity in the convection zone. In general the α effect contributes to the generation of both poloidal and toroidal field components (the $\alpha^2\Omega$ dynamo). Often the $\alpha\Omega$ approximation (appropriate when the differential rotation is relatively large) and α^2 models (which are appropriate when the differential rotation is negligible) are used. It is well known that $\alpha\Omega$ models possess oscillating spatiotemporal (wave–like) solutions (Parker 1955; Krause & Rädler 1980). These oscillations propagate equatorward when the product $\alpha\frac{\partial\Omega}{\partial r}$ is negative in the northern hemisphere (Parker 1955; Yoshimura 1975) (see also Choudhuri 1998, for a recent account).

Mean field models have been phenomenologically successful in accounting for a number of features seen in the Sun and stars (see e.g. Ossendrijver 2003; Charbonneau 2005). For example, nonlinear $\alpha^2\Omega$ MF models which include the back reaction of Lorentz force on the azimuthal flow have been shown to be capable of producing many important features of the spatiotemporal variations observed in the Sun, including the torsional oscillations that penetrate the convection zone, as indicated by recent helioseismological inversions (see e.g. Covas *et al.* 2000a,b, 2004). Similar MF models have also been successfully used in order to account for a number of observed features in late-type rapidly rotating stars (Covas *et al.* 2005). So far, however, these successes have relied on rather *ad hoc* assumptions. For example to obtain the correct equatorward migration in the case of the Sun, the sign of α needs to be taken to be negative in the northern hemisphere. This follows from Parker's result (Parker 1955) linking the direction of migration with the sign of $\alpha\partial\Omega/\partial r$, and conflicts with contemporary results for the sign of α in the bulk of the convection zone in each hemisphere (Yoshimura 1975).

Other ways of producing the correct sense of migration in the Sun includes the incorporation of a meridional circulation (see e.g. Choudhuri *et al.* 1995) by assuming that the Ω effect operates at the bottom of the convection zone in order to generate a toroidal field, which in turn gives rise to a poloidal field by an α effect operating near the surface, and these are coupled by a meridional circulation which is assumed to have a poleward direction near the solar surface and an equatorward direction at the base of the convection zone. This circulation, which has not been directly observed in helioseismological inversions except near the surface, would need to be strong enough to make the equatorward migration possible. It is fair to say that despite their successes, all MF models require a number of *ad hoc* assumptions and choice of values of solar/stellar parameters that are only known very approximately at present (see e.g. Jouve *et al.* 2008; Chatterjee *et al.* 2004) for a list of such models.

The above shortcomings (as well as other fundamental questions such as the debate regarding the 'catastrophic quenching' of nonlinear dynamos) and the need to test various assumptions made in MF models, have in parallel motivated the study of turbulent dynamos using direct numerical simulations. Despite much effort and a number of important successes, however, direct numerical simulations of the full MHD equations in spherical shells have so far had limited success in generating large-scale magnetic fields on dynamical time scales (Gilman 1983; Glatzmaier 1985; Brown *et al.* 2009).

In the following we summarise some of our recent results, concerning oscillations, migrations of large-scale magnetic fields and polarity reversals, obtained using DNS in (i) helically forced MHD (Mitra *et al.* 2009) and (ii) convection-driven dynamos in a rotating frame (Käpylä *et al.* 2009).

2. Oscillation and migration of large-scale fields in DNS

In contrast to the MF studies, direct numerical simulations of turbulent MHD equations have seldom shown large-scale oscillations and migration. Two noteworthy exceptions are Brandenburg *et al.* (2001) and Gilman (1983). The former is a DNS with forced turbulence in Cartesian domains with imposed shear which gave rise to dynamo waves on time scales determined by the turbulent magnetic diffusivity (as opposed to the molecular diffusivity, which would imply that the oscillations were on dissipative time scales). The latter is a convection simulation in a spherical wedge-shaped domain, which showed poleward migration of magnetic activity–as opposed to equatorward migration seen in the Sun–and cylindrical contours of mean angular velocity–as opposed to cone-shaped contours in most of the solar convection zone. In the following we briefly review two recent migratory oscillating solutions obtained by DNS of the MHD equations (see Mitra *et al.* 2009; Käpylä *et al.* 2009, for details). In these studies we perform simulations in a spherical wedge domain as a compromise between requirements of high resolution and globality. We choose our domain to be distributed symmetrically about the equator.

2.1. *Forced DNS of the MHD equations*

We first describe the results of direct numerical simulations of the MHD equations with external forcing in a spherical wedge-shaped domain. The external force is random and white-in-time and injects negative helicity in the northern hemisphere and positive in the southern. We do not explicitly include convection, stratification and rotation, which are known to be present in solar and stellar convection zones, but the helical forcing used here does implicitly model these features (Krause & Rädler 1980). We use perfect-conductor boundary conditions (i.e. set the normal component of the magnetic field to be zero on the boundaries) on all boundaries except the boundary with constant azimuth where we impose periodic boundary conditions. The large-scale magnetic field is defined by averaging over the azimuthal and radial directions. Figure 1a depicts a 'butterfly diagram', showing oscillation, polarity reversals and equatorward migration of large-scale magnetic field.

Further investigations have shown that these results are well reproduced by a MF model of an α^2 dynamo where the sign of α is positive in the northern hemisphere and negative in the southern (see Mitra *et al.* 2009, for details). The characteristic time scales of oscillations are well approximated by $T_{\rm cyc} = 1/\eta_{\rm T} k_\theta^2$ with $\eta_{\rm T} = (1/3)u_{\rm rms}/k_{\rm f}$, $k_{\rm f} = W_{\rm rms}/U_{\rm rms}$, $W = \boldsymbol{\nabla} \times \mathbf{U}$, $k_\theta = 2\pi/L_\theta$, $L_\theta = R_\odot \Delta\theta$ where $\Delta\theta$ is the range of colatitude (θ) in our wedge-shaped domain and $R_\odot = 1$ in our simulations. Furthermore our MF studies have shown that the ratio between the maximum amplitude of the large-scale magnetic field and the equipartition magnetic field ($B_{\rm eq}$) decreases as a function of the magnetic Reynolds number, $\mathrm{Re_M} = u_{\rm rms}/\eta k_{\rm f}$ as $\mathrm{Re_M^{-0.4}}$. This is related to the well known phenomenon of catastrophic quenching. This catastrophic quenching is expected to be alleviated by an exchange of magnetic helicity across the equator. Such a helicity flux, which is thought to be mediated by diffusion, has not yet been verified in DNS. Brandenburg *et al.* (2009) recently incorporated such a flux in a MF model and have shown that at high enough magnetic Reynolds numbers the diffusive flux is able to alleviate the quenching. We refer the reader to Mitra *et al.* (2009) for details of the model, methods of simulations and elaboration of the results reported above.

Finally we note that oscillations in one or two dimensional models of α^2 dynamos, where the sign of α changes within the domain are well known literature (Baryshnikova & Shukurov 1987; Rädler & Bräeuer 1987; Stefani & Gerbeth 2005; Rüdiger & Hollerbach 2004), but migrational properties have not been investigated earlier.

2.2. *Rotating, convective DNS in spherical wedge-shaped domain*

Next we summarise results obtained using simulations of convective dynamos in spherical wedge-shaped domains, that include rotation explicitly. We consider a computational setup where the initial stratification in the region below $r = 0.7R_\odot$ is convectively stable, whereas the region above is unstable. The convection is driven self-consistently by a heat flux imposed at the lower boundary. The radii of the bottom of the domain, bottom of the convectively unstable layer, and the top of the domain are given by $(r_1, r_2, r_3) = (0.6, 0.7, 1)R_\odot$. We impose stress-free boundary conditions for the velocity and 'open' vertical field boundary condition (i.e. set the tangential component of magnetic field to zero at the boundary) at $r = R_\odot$. Starting from an uniformly rotating initial state, rotational shear is generated, and we again obtain large-scale magnetic fields which show oscillations, polarity reversals and migration, but the migration in this case is poleward, see Fig. 1b. This behaviour can be reproduced by an $\alpha\Omega$ dynamo. An accurate measurement of α is difficult in the present case because the test-field method has yet to be applied to this problem. However we use the approximation $\alpha \propto -\overline{\boldsymbol{\omega}\cdot\mathbf{u}}$, where $\boldsymbol{\omega} = \boldsymbol{W} - -\overline{\mathbf{W}}$, $\mathbf{u} = \boldsymbol{U} - -\overline{\mathbf{U}}$, $W = \boldsymbol{\nabla}\times\mathbf{U}$ and an overbar denotes averaging over the azimuthal direction. A contour plot of the product $\alpha \mathrm{d}\Omega/\mathrm{d}r$ is given in Fig. 1c. This shows that in clear agreement with the result of Parker (1955), that the sign of $\alpha \mathrm{d}\Omega/\mathrm{d}r$ determines the direction of migration. The details of the model, methods of simulation and elaboration of the results given in this Section can be found in (Käpylä *et al.* 2009).

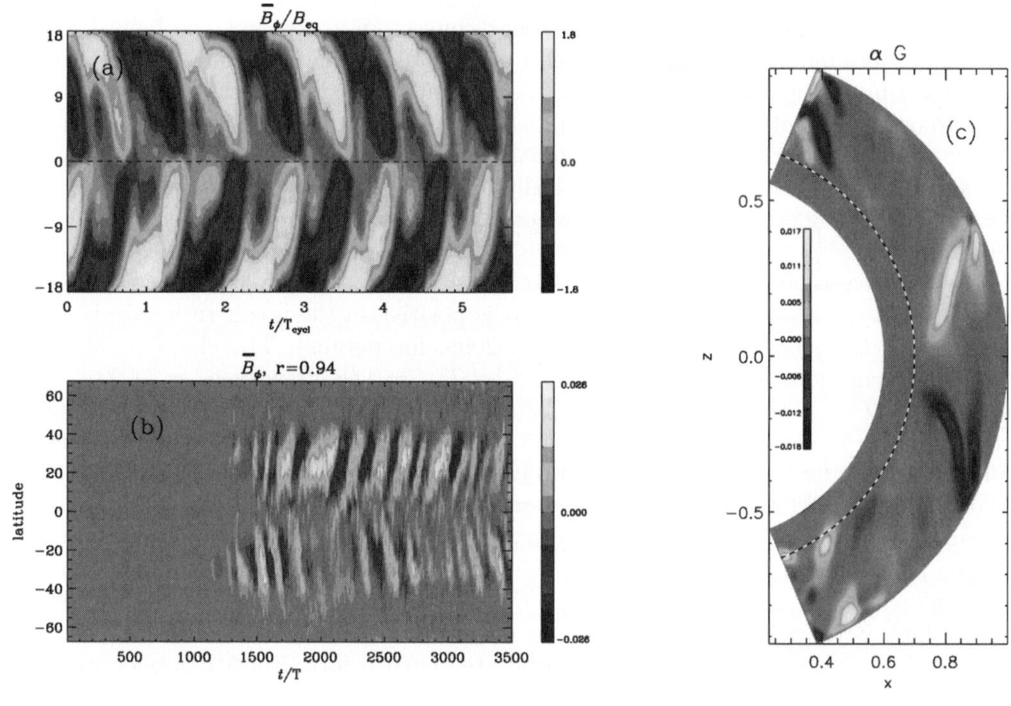

Figure 1. (a) Butterfly diagram for the toroidal field from the simulations of helically forced MHD equations in spherical wedge-shaped domain. (b) Butterfly diagram of the toroidal field from the simulations of rotating, convective MHD equations in spherical wedge-shaped domain. (c) A contour plot of a the product αG in the meridional plane from our convective simulations.

3. Discussion

Motivated by the observations of solar magnetic fields, we have considered oscillations, polarity reversals and migration of large-scale magnetic fields using direct numerical simulations We have summarised our recent results obtained using DNS of both non-rotating helically forced and rotating convection driven MHD equations in spherical wedge-shaped domains.

In both cases we find large-scale magnetic fields are generated. In the non-rotating helically forced models, using perfect-conductor boundary conditions along the latitudinal boundaries, we find the field shows oscillations, polarity reversals and equatorward migration. In the case where convection and rotation are present we also find oscillations and polarity reversals, but the migration is in this case poleward.

Using analogous mean field models we have found that these results are well described by α^2 and $\alpha\Omega$ mean field models respectively.

Clearly the occurrence of these features in direct numerical simulations is of interest. Nevertheless, the present models as they stand are not directly applicable to the solar dynamo problem, even though they provide insight into the operations of turbulent dynamos.

Acknowledgement

We would like to thank David Moss for very helpful discussions.

References

Baryshnikova, I. & Shukurov, A. 1987, *AN*, 308, 89

Brandenburg, A. 2005, *ApJ*, 625, 539

Brandenburg, A., Bigazzi, A., & Subramanian, K. 2001, *MNRAS*, 325, 685

Brandenburg, A., Candelaresi, S., & Chatterjee, P. 2009, *MNRAS*, 398, 1414

Brandenburg, A., Rädler, K.-H., Rheinhardt, M., & Subramanian, K. 2008, *ApJ Lett.*, 687, L49

Brandenburg, A. & Subramanian, K. 2005, *Phys. Rep.*, 417, 1

Brown, B. P., Browning, M. K., Miesch, M. S., Brun, A. S., & Toomre, J. 2009, *ApJ* (submitted), arXiv:0906.2407; see also, Brown, B. P., Browning, M. K., Brun, A. S., Miesch, M. S. *et al.* 2007, *AIPC*, 948, 271; Browning, M. K., Miesch, M. S., Brun, A. S., & Toomre, J. 2006, *ApJ*, 648, L157; Brun, A. S., Miesch, M. S., & Toomre, J. 2004, *ApJ*, 614, 1073

Charbonneau, P. 2005, *Living Reviews in Solar Physics*, 2

Chatterjee, P., Nandy D., & Choudhuri, A. 2004, *A&A*, 427, 1019

Covas, E., Tavakol, R., Moss, D., & Tworkowski, A. 2000, *A&A*, 360, L21

Covas, E., Tavakol, R., & Moss, D. 2000, *A&A*, 363, L13

Covas, E., Moss, D., & Tavakol, R. 2004, *A&A*, 416, 775

Covas, E., Moss, D., & Tavakol, R. 2005 *A&A*, 429, 657

Choudhuri, A. 1998, The physics of Fluids and Plasmas: An introduction for Astrophysicists. Cambridge University Press, Cambridge

Choudhuri, A. R., Schussler, M., & Dikpati, M. 1995, *A&A*, 303, L29+

Gilman, P. A. 1983, *ApJS*, 53, 243

Glatzmaier, G. A. 1985, *ApJ*, 291, 300

Jouve, L., Brun, A. S., Arlt, R., Brandenburg, A., Dikpati, M., Bonanno, A., Käpylä, P. J., Moss, D., Rempel, M., Gilman, P., Korpi, M. J., & Kosovichev, A. G. 2008, *A&A*, 483, 949

Käpylä, P. J., Korpi, M. J., Brandenburg, A., Mitra, D., & Tavakol, R. 2009, arXiv:0909.1330

Krause F. & Rädler, K.-H. 1980, Mean-field magnetohydrodynamics and dynamo theory. Pergamon Press, Oxford

Mitra, D., Tavakol, R., Käpylä, P. J., & Brandenburg, A. 2009, arXiv:0901.2364

Moffatt, K. 1978, Magnetic Field Generation in Electrically Conducting Fluids. Cambridge University Press, Cambridge.

Ossendrijver, M. 2003, *Astron. Astrophys. Rev*, 11, 287–367 (2003).

Parker, E. 1955, *ApJ*, 1122, 293

Parker, E. 1987, *Sol. Phys.*, 110, 11

Rädler, K.-H. & Bräeuer, H.-J. 1987, *AN*, 308, 101

Rüdiger, G. & Hollerbach, R. 2004, in Rüdiger, G., Hollerbach, R., eds, The Magnetic Universe: Geophysical and Astrophysical Dynamo Theory. Wiley-VCH, p. 343

Stefani, F. & Gerbeth, G. 2005, *Phys. Rev. Lett*, 94, 184506

Yoshimura, H. 1975, *ApJ*. 201, 740.

Solar and Stellar Variability: Impact on Earth and Planets
Proceedings IAU Symposium No. 264, 2009
A. G. Kosovichev, A. H. Andrei & J.-P. Rozelot, eds.

doi:10.1017/S1743921309992638

Prediction of solar activity cycles by assimilating sunspot data into a dynamo model

Irina N. Kitiashvili[1,3] and Alexander G. Kosovichev[2,3]

[1]Center for Turbulence Research, Stanford University, Stanford, CA 94305, USA
email: irinasun@stanford.edu

[2]Hansen Experimental Physics Laboratory, Stanford University, Stanford, CA 94305, USA
email: sasha@sun.stanford.edu

[3]NORDITA, Dept. of Astronomy, AlbaNova University Center, SE 10691 Stockholm, Sweden

Abstract. Solar activity is a determining factor for space climate of the Solar system. Thus, predicting the magnetic activity of the Sun is very important. However, our incomplete knowledge about the dynamo processes of generation and transport of magnetic fields inside Sun does not allow us to make an accurate forecast. For predicting the solar cycle properties use the Ensemble Kalman Filter (EnKF) to assimilate the sunspot data into a simple dynamo model. This method takes into account uncertainties of both the dynamo model and the observed sunspot number series. The method has been tested by calculating predictions of the past cycles using the observed annual sunspot numbers only until the start of these cycles, and showed a reasonable agreement between the predicted and actual data. After this, we have calculated a prediction for the upcoming solar cycle 24, and found that it will be approximately 30% weaker than the previous one, confirming some previous expectations. In addition, we have investigated the properties of the dynamo model during the solar minima, and their relationship to the strength of the following solar cycles. The results show that prior the weak cycles, 20 and 23, and the upcoming cycle, 24, the vector-potential of the poloidal component of magnetic field and the magnetic helicity substantial decrease. The decrease of the poloidal field corresponds to the well-known correlation between the polar magnetic field strength at the minimum and the sunspot number at the maximum. However, the correlation between the magnetic helicity and the future cycle strength is new, and should be further investigated.

Keywords. Sun: activity, sunspots; methods: data analysis

1. Introduction

The interest to the problem of the solar cycle prediction has recently increased because of the unusually long minimum of solar activity. This minimum illustrates deficiencies in our understanding of the global processes in the Sun. The 400-year sunspot number record shows a chaotic behavior but also regular properties of the solar cycles, like the Waldmeier effect, which tells us that the stronger cycles have shorter razing times than the weaker cycles. Also, statistical analyses show the existence of long quasi-periodic variations of the solar activity amplitude in addition to the 11-years period (e.g. Miyahara *et al.* 2010). A number of methods for predicting solar cycles has been developed (e.g. Pesnell 2008), but the forecasts for the next solar cycle are widely different.

It is well-accepted that the variations of solar activity are a result of a complicate dynamo process in the convection zone. However, because of the imperfect dynamo models and deficiency of the necessary observational data, despite the known general properties of the solar cycles, a reliable forecast of the 11-year sunspot number is still a problem.

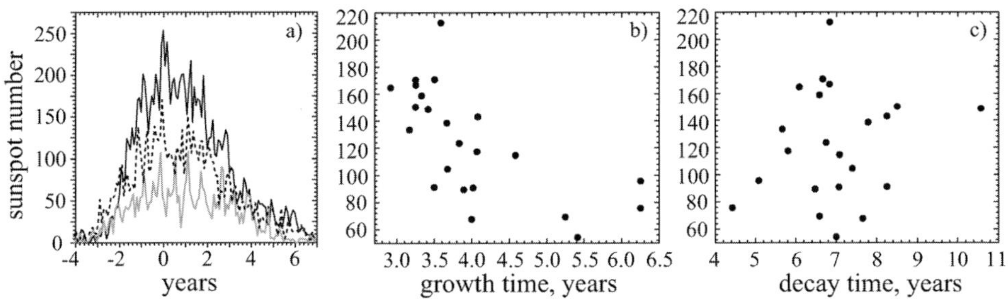

Figure 1. Basic properties of the solar cycles. Asymmetry of mean profile of sunspot number variations (a) shown for three solar cycles 14 (grey curve), 19 (black curve) and 23 (dotted curve), which are aligned according to their maxima ($t = 0$). Relationships between the amplitude of the sunspot number and the growth (b) and decay time (c) for the real solar cycles in 1755–2007.

The great variety of the predictions of the upcoming cycle 24 is caused by uncertainties in models and model parameters, and errors in both models and observations (Pesnell 2008). For this reason, we propose to use the data assimilation approach, which allows us to successively correct the model state according to observational data, and also take into account uncertainties of the model and data in the forecast solution.

Our idea is to combine observational data (in a first approximation, the sunspot data) and a dynamo model by a data assimilation method, and analyze the model and observational uncertainties. For application of the data assimilation approach we use the Ensemble Kalman Filter (EnKF) method, which works well for dynamical systems (Evensen 2007). In the first approximation, we consider the phenomenon of solar activity in the context of sunspot number variations, which have observational data during the past 23 solar cycles. In order to relate these data to dynamo models we propose to use the dependence of the sunspot number, W, on the toroidal component of magnetic field, B, in the form: $W \sim\mid B \mid^{3/2}$ (Bracewell 1953, 1988), for representation of the solar cycles.

For implementation of the EnKF method it is very important to reproduce in the dynamo model the basic properties of the solar cycle: in particular, the mean profile of the sunspot number variations (Fig. 1a), which is characterized by fast growth and slow decay, and the relationship between the cycle amplitude and the growth and decay times, (Fig. 1b and c). In the next section we consider a dynamo model, which includes the Parker's migratory dynamo model (Parker 1955) and an equation for magnetic helicity (Kleeorin & Ruzmaikin 1982; Kleeorin *et al.* 1995), which provides dynamical quenching of the dynamo process.

2. Parker's model with magnetic helicity and Waldmeier effect

Currently, there is no generally accepted model of the solar dynamo. However, most of the models are based on the Parker's oscillatory $\alpha\Omega$-dynamo mechanism (Parker 1955), which includes a turbulent helicity and magnetic field stretching by the differential rotation. Recent observational and theoretical investigations (e.g. Brandenburg & Subramanian 2005; Sokoloff 2007) revealed an important role of magnetic helicity (Pouquet *et al.* 1976). Thus, for this investigation we added to the original Parker's model an equation describing the evolution of the magnetic helicity, α_m. This equation was derived by Kleeorin and Ruzmaikin (1982) from a balance of the total magnetic helicity. Then, the dynamo model in local Cartesian coordinates can be written as

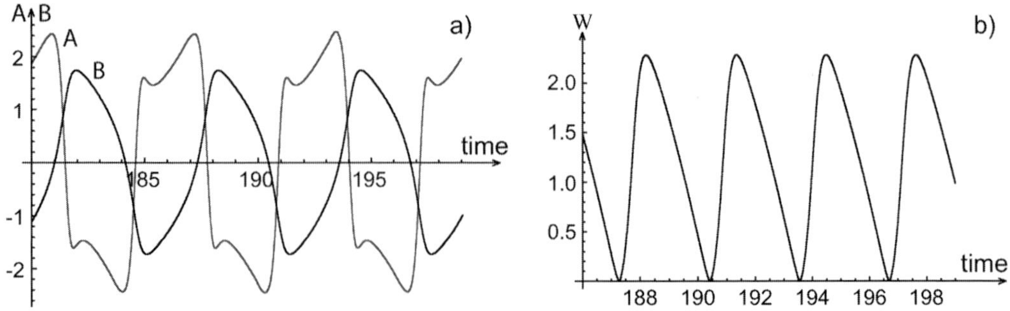

Figure 2. Periodical solutions of the dynamo model with included magnetic helicity for the middle convective zone parameters (panel a). Black and grey curves show variations for the toroidal component of magnetic field, B, and the vector-potential, A, of the poloidal magnetic field. Simulated variations of the sunspot number (panel b) obtained by applying the Bracewell's law: $W \sim\mid B \mid^{3/2}$ to the dynamo solution shown in panel a.

(Kitiashvili & Kosovichev 2009)

$$\begin{aligned} \frac{\partial A}{\partial t} &= \alpha B + \eta \nabla^2 A, \quad \frac{\partial B}{\partial t} = G\frac{\partial A}{\partial x} + \eta \nabla^2 B, \\ \frac{\partial \alpha_m}{\partial t} &= \frac{Q}{2\pi\rho}\left[\langle\vec{B}\rangle(\nabla \times \langle\vec{B}\rangle) - \frac{\alpha}{\eta}\langle\vec{B}\rangle^2\right] - \frac{\alpha_m}{T}, \end{aligned} \tag{2.1}$$

where B is the toroidal component of magnetic field, A is the vector potential of the poloidal component of the mean magnetic field, $\langle\vec{B}\rangle = \vec{B}_P + \vec{B}_T$ ($\vec{B}_P = \mathrm{curl}(0,0,A)$, $\vec{B}_T = (0,0,B)$ in the spherical coordinates); η describes the total magnetic diffusivity, which is the sum of the turbulent and molecular magnetic diffusivity; $\eta = \eta_t + \eta_m$ (usually $\eta_m << \eta_t$); $G = \partial\langle v_x\rangle/\partial y$ is the rotational shear; coordinates x and y are in the azimuthal and latitudinal directions respectively, parameter α is the total helicity represented in the form $\alpha = \alpha_h/(1+\xi B^2) + \alpha_m$, α_h and α_m are the kinetic and magnetic parts; ξ is a quenching parameter, ρ is density, T is a characteristic time of dissipation magnetic helicity (which includes dissipation though helicity transport) and, $Q \sim 0.1$.

Following the approach of Weiss *et al.* (1984) we average the system of equations (2) in a vertical layer to eliminate z-dependence of A and B and consider a single Fourier mode propagating in the x-direction assuming $A = A(t)e^{ikx}$, $B = B(t)e^{ikx}$. This dynamo model has been investigated in detail by Kitiashvili & Kosovichev (2009).

Figure 2 shows a typical nonlinear periodic solution (left panel) for the toroidal component of magnetic field, B (black curve), and vector-potential A (grey curve). We found that the dynamo model can reproduce the typical observed solar cycle profiles characterized by fast growth and slow decay (right panel). Figure 3 shows the relationships between of the amplitude of the sunspot number parameter, the growth and decay times and the cycle duration, for different values of α_h and D_0. These characteristic times were determined from the points of minima and maxima of the model sunspot number, W. The crosses represent the amplitude of the periodic field variations for $D_0 = -0.82$ and different values of kinetic helicity, α_h. The circles correspond to the case of constant $\alpha_h = 2.44$ and different values of dynamo number D_0. In the first case, the relationship between the cycle amplitude and the growth time is well-defined and monotonic. However, in the case of a fixed α_h and varying D_0 (circles), the amplitude initially, at small D_0, follows the same sequence as in the variable α_h case, but then at higher values of $|D_0|$ (shown by bigger circles) the amplitude decreases. The decay time is longer for the higher amplitude cycles (Fig. 3b).

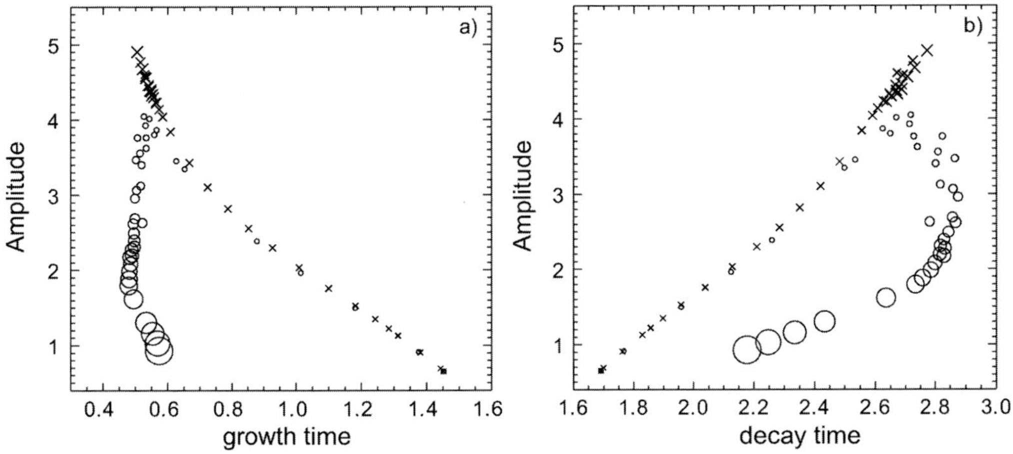

Figure 3. Relationships between the model sunspot number amplitude, W, and a) the cycle growth time and b) the decay time. The circles show a sequence for a fixed value of the kinetic helicity, $\alpha_h = 2.44$ and the dynamo number varying from -7 to -0.82. The crosses show the case of fixed $D_0 = -0.82$ and varying α_h, from 2.44 to 3. The size of the crosses and circles is proportional to the corresponding values of D_0 and α_h. Model parameters correspond to the middle convective zone: $\nu = 1.28, \lambda = 1.23 \times 10^{-6}$.

Thus, we have constructed the dynamo model, which reproduces qualitative properties of the sunspot cycles; and then using the data assimilation approach we can estimate the system state according to the observed sunspot number variations. In the next section, we briefly discuss the basic principles of the data assimilation approach.

3. Data assimilation methods

A primary target of any research is prediction of a physical system behavior. Often this is an extremely difficult problem, because the models are constructed with some approximations and assumptions, contain uncertainties, and deviate from the reality. Therefore, a theoretical model cannot describe the true condition of a system. On the other hand, observational data also include errors, which are often difficult to estimate.

The idea of data assimilation is in making continuous estimates of the system state and corrections to the initial conditions at subsequent moments of time according to the available observational data, and in creating a prediction of a future model state. Thus, the advantage of data assimilation methods is in their ability to combine the observational data and the models for possible efficient and accurate estimations of the physical properties, which cannot be observed directly. The data assimilation methods such as the Kalman Filter (Kalman 1960) allow us, with the help of an already constructed model and observational data, to determine the state of the model that is in agreement with a set of observations, and using this state as the initial conditions to obtain a forecast of future observations and error estimates (Evensen 2007; Kitiashvili & Kosovichev 2008). In our case, we know from observations the sunspot number (with some errors) and want to estimate the parameters of solar magnetic fields and helicity, described by the dynamo model given by equations (2.1).

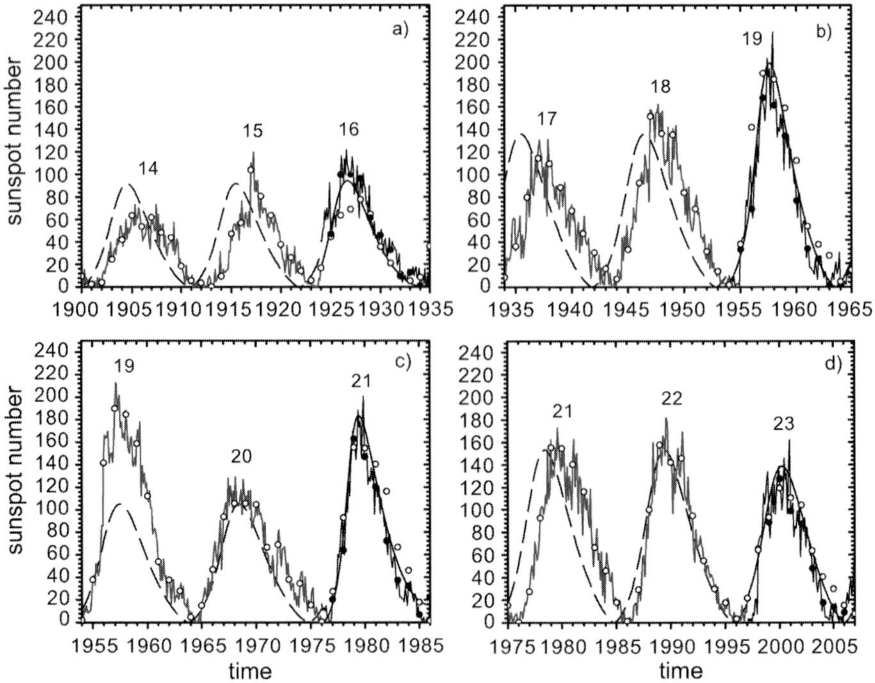

Figure 4. Predictions for solar cycles 16, 19, 20 and 23. Black dashed curves show the model reference solution. Gray curves show the best estimate of the sunspot number using the observational data (empty circles) and the model, for the previous cycles. Filled circles are simulated observational data. Black thick and thin curves show the reference solutions according new initial conditions and prediction results correspondingly.

4. Ensemble Kalman Filter in the solar cycle prediction problem

For predicting the solar cycle properties we use the Ensemble Kalman Filter (EnKF), a data assimilation method, which is effective for non-linear models formulated as dynamical systems and provides a statistical analysis of an ensemble of possible fluctuations (Evensen 2007). This method is tested by calculating predictions of the past cycles using only the observational data (annual sunspot numbers) until the start of these cycles.

Figure 4 shows examples of the EnKF implementation of the forecasting for solar cycles 16, 19, 20 and 23. In this approach, the exact model solution is corrected according to the previous observational data. This allows us to redefine the initial conditions of the model for the magnetic field components and helicity, and construct a model solution for the next time interval.

The experiments with the previous sunspot data show that this approach can provide reasonable forecasts of the strength of the following solar cycles. However, there are significant discrepancies between the forecasts and the actual data. For instance, the strength of cycle 16 is over-estimated, and the strength of cycle 19 is under-estimated. The main uncertainties are caused by inaccuracies in determining the time of the end of the previous cycle from the sunspot number data, and, of course, by the incompleteness of the model and insufficiency of the sunspot number data. In particular, we found the forecast can be inaccurate when the sunspot number changes significantly from the value of the previous cycle (Kitiashvili & Kosovichev 2008).

In our forecast experiments, we found a strong dependence on the phase relationship between the reference model solution and the observations. The phase difference appears

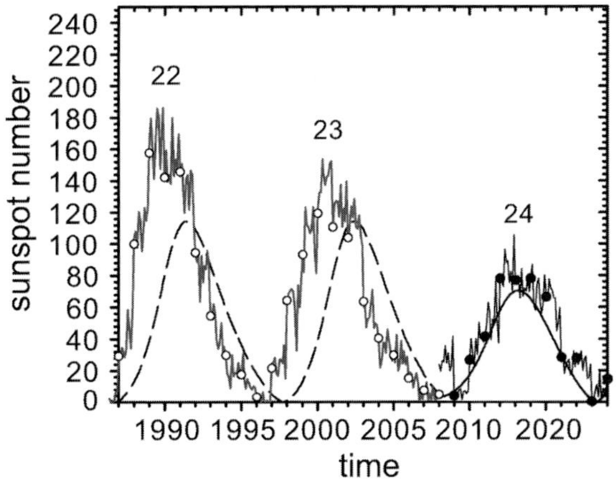

Figure 5. Prediction of the solar cycle 24. Notations are the same as in Fig. 4

to be due to the constant period of the model solution. Curiously, when the model phase is ahead of the solar cycle phase, adding a data point at the start of the cycle substantially improves the forecast. However, when the model phase lags, this improvement does not happen. This effect is taken into account by correcting the phase of a reference solution that it is slightly ahead of the solar cycle phase.

We used the same analysis scheme for predicting of the upcoming solar cycle 24 (Fig. 5). According to our analysis, the solar cycle 24 which starts in 2009 will be weaker than the current cycle by approximately 30% (Kitiashvili & Kosovichev 2008). The estimated formal error of our prediction is $\sim 10\%$.

5. Estimation of the state of the solar dynamo

The basic data assimilation procedure in our forecast scheme is in estimating the properties (or state) of the solar dynamo model at the solar minima by using the sunspot data available up to this minimum. In our simple model these properties include the toroidal component of magnetic field, B, the vector-potential of the poloidal component, A, and the magnetic helicity, α_m. These properties are estimated by using the Ensemble Kalman Filter method. Then, they are used as the initial conditions for calculating the properties of the next solar cycle by solving Eqs (2.1). Thus, it is interesting to see how these initial conditions correlate with the predicted maxima of the sunspot number.

In Figure 6 (top panel) we shows the "predicted" maximal annual sunspot number (squares) for solar cycles from 16 to 24, with an estimated the forecast uncertainty and the actual observed values for cycles 16–23 (circles). In the bottom panel, we show the corresponding initial conditions for the preceding solar minima. The variations of the toroidal field, B, do not show a particular pattern, and are close to zero as expected for during the solar minima. The vector-potential of poloidal field, A, shows changes the sign changes corresponding to the polar field reversal. The amplitude at the start of cycles 20 and 24 is substantially lower than during the other minima. This may correspond to the well-known correlation between the strength of the polar magnetic field and the following sunspot number (Schatten 2005; Svalgaard *et al.*, 2005). However, there is no such correlation for weak cycles 16 and 17. The variations of the magnetic helicity, α_m, (solid curve with triangles) shows significantly better correlation with the future sunspot numbers, indicating that the magnetic helicity substantially decreases prior the weak sunspot cycles. Perhaps, the combination of two factors, the strength of the poloidal

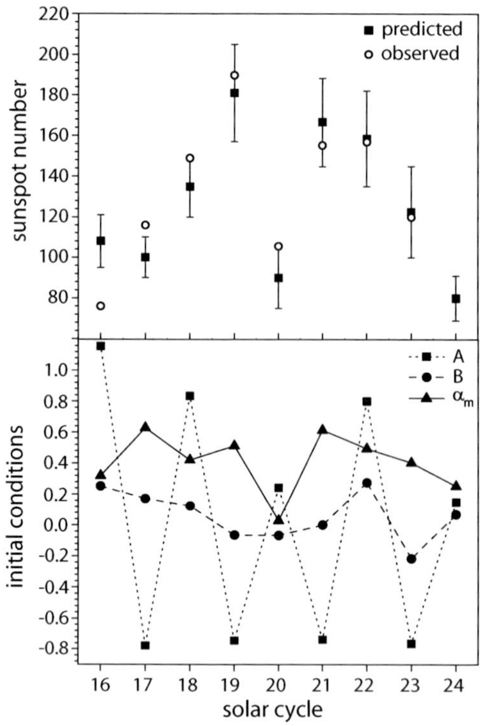

Figure 6. Correlation between the predictions of the sunspot maxima (squares with errorbars), the actual observed maximum sunspot numbers (open circles), shown in the top panel, and the initial conditions of the dynamo model (vector-potential of the poloidal magnetic field component, A, toroidal magnetic field component, B, and the magnetic helicity, α_m) at the preceding minima, obtained by the EnKF method for the solar cycle forecasts (bottom panel).

magnetic field and the level of the magnetic helicity determines the strength of the future cycles. Of course, this requires further investigation.

6. Conclusions

We have presented a numerical analysis of a simple non-linear dynamical model, which includes the classical Parker's dynamo equations (Parker 1955) with the standard α-quenching and an equation for the magnetic helicity evolution (Kleeorin & Ruzmaikin 1982). The formulation for the magnetic helicity is based on the balance between the large-scale and turbulent magnetic helicities. Using a low-order dynamical system approach we have examined the influence of the kinetic and magnetic helicities on the non-linear fluctuations of the dynamo-generated magnetic field in the conditions of the solar plasma, and compared the model solutions with the sunspot number variations observed during the solar 11-year cycles. The analysis of the model showed the existence of non-linear periodic and chaotic solutions (Kitiashvili & Kosovichev 2009). For this model we obtained profiles of the sunspot number variations, which qualitatively reproduce the typical profile of the solar cycles.

The results of assimilation of the annual sunspot number data into the solar dynamo model and the prediction of the previous solar cycles (Fig. 4 and 5) demonstrate a new method of forecasting the solar activity cycles. This method predicts a weak solar cycle 24 with a maximum of the smoothed annual sunspot number of approximately 80 (Fig. 5).

It is interesting to note that the simulations show a delay of the upcoming cycle that is observed at the present time. According to the prediction, the maximum of the next cycle will be reached approximately in 2013.

Using the data assimilation results we have estimated the parameters of the dynamo model during the solar minima and found that the vector-potential of the toroidal component substantially decreased prior the weak sunspot cycles 20 and 23, and also at the present time, prior cycle 24. This may corresponds to the observed weak polar magnetic fields. The level of the magnetic helicity seems to show a reasonably good correlation with the future sunspot maxima. This may have important implications for solar dynamo models, but, of course, requires further investigation.

Future investigations of the data assimilation approach will include more complete dynamo models and assimilation of the solar synoptic magnetic field data, which are available for the past 3 cycles.

Acknowledgements

IK would like to acknowledge the American Astronomical Society for the travel grant and the local organizing committee of GA IAU for the financial support. The paper was completed during the Nordita program "Solar and stellar dynamos and cycles". The authors thanks the participants for useful discussions of this work, and Nordita for financial support.

References

Bracewell, R. N. 1953, *Nature*, 171, 649.
Bracewell, R. N. 1988, *Mon. Not. R. Astr. Soc.*, 230, 535.
Brandenburg, A. & Subramanian, K. 2005, *Physics Reports*, 417, 1.
Evensen, G. 2007, *Data assimilation. The Ensemble Kalman Filter* (Springer, Germany 2007).
Kalman, R. E. 1960, *J. Basic Engineering*, 82 (series D), 35.
Kitiashvili, I. N. & Kosovichev, A. G. 2008, *ApJ* (Letters), 688, L49.
Kitiashvili, I. N. & Kosovichev, A. G. 2009, *Geophys. Astrophys. Fluid Dyn.*, 103, 53.
Kleeorin, N., Kuzanyan, K., Moss, D., Rogachevskii, I., Sokoloff, D., & Zhang, H. *Astron. Astrophys.*, 409, 1097.
Kleeorin, N., Rogachevskii, I., & Ruzmaikin, A. 1995, *A&A*, 297, 159.
Kleeorin, N. & Ruzmaikin, A. 1982, *Magnetohydrodynamics*, 18, 116 (in russian).
Miyahara, H., Yokoyama, Yu, & Yamaguchi, Ya. T. 2010, *this Proceedings*.
Parker, E. N. 1955, *ApJ*, 122, 293.
Pesnell, W. D. 2008, *Solar Physics*, 252, 209.
Pouquet, A., Frisch, U., & Léorat, J. 1976, *J. Fluid Mech.*, 77, 321.
Schatten, K. 2005, *Geophys. Res. Lett.*, 32, 21106.
Sokoloff, D. 2007, *Plasma Phys. Control. Fusion*, 49, B447.
Svalgaard, L., Cliver, E. W., & Kamide, Y. 2005, in: *Large-scale Structures and their Role in Solar Activity*, ASP Conf. Ser., Vol. 346, K. Sankarasubramanian, M. Penn, and A. Pevtsov. eds, p. 401.

Solar and Stellar Variability: Impact on Earth and Planets
Proceedings IAU Symposium No. 264, 2009
A. G. Kosovichev, A. H. Andrei & J.-P. Rozelot, eds.

doi:10.1017/S174392130999264X

Transport of open magnetic flux between solar polar regions

Alexei A. Pevtsov[1] **and Valentyna I. Abramenko**[2]

[1]National Solar Observatory, Sunspot, NM 88349, U.S.A.
email: apevtsov@nso.edu

[2]Big Bear Solar Observatory, New Jersey Institute of Technology, Big Bear City, CA 92314-9672, U.S.A.
email: avi@bbso.njit.edu

Abstract. We present the observations of coronal hole that has originated at the polar hole in one hemisphere, extended to equatorial region, got disconnected and transported across the equator to polar region of opposite hemisphere.

Keywords. Sun: magnetic fields, Sun: activity, Sun: corona, Sun: photosphere

1. Introduction and data

Earlier studies of coronal holes (CHs) have found near rigid rotation, which initially was interpreted as the consequence of CHs magnetic field rooted deep in the convection zone. Later, Wang & Sheeley (2004) have introduced the footpoint switching (or interchange reconnection), when the rigid rotation is maintained by continuous reconnection between closed (bipoles) and open (CH) field lines. Karachik, Pevtsov, & Sattarov (2006) used coronal bright points (CBPs) to trace solar rotation inside of three coronal holes. They found no difference in solar rotation for CBPs inside and outside of CHs. At the same time, the average rotation of studied CHs did not show presence of the differential rotation in agreement with the interchange reconnection model. On the other hand, Jones (2005) had pointed out that the internal rotation derived by helioseismology methods indicates that CHs may be rooted as deep as the base of the convection zone. Harvey & Recely (2002) had classified CHs on three category: polar CHs (latitude $> 60°$), isolated non-polar CHs ($\pm$ 60°), and transient CHs. Their study had revealed that polar CHs evolve from high-latitude isolated CHs. During their evolution, polar CHs may connect to low latitudes through polar lobes – a narrow equatorward extension of polar CH. However, the previous studies indicate that although polar holes in opposing hemispheres may extend to low latitudes, there is no direct interaction between two polar regions. Here we provide the first observational evidence of a complete transport of one CH across the equator, from one solar polar region to the other.

The data used in this study include full disk observations from SOHO/EIT (286Å) and SOHO/MDI. Our observations cover period of nine solar rotations, starting from Carrington Rotation (CR) 1982. Figure 1-left shows evolution of the total magnetic flux at high latitudes ($>$70°) computed from MDI Carrington maps. Using the intersection of linear polynomial with zero flux line, one can estimate that the Northern (Southern) polar field had reversed its sign around CR1965 (CR1976). Reversal dates for N-Pole/S-Pole were also reported as CR1972/CR1976 (Pevtsov *et al.* 2003) and CR1975/CR1981 (Durrant & Wilson 2003). Despite some disagreement in dates, we conclude that the transport of open magnetic flux (CH) across solar equator described below had occurred after both solar poles had reversed their polarities.

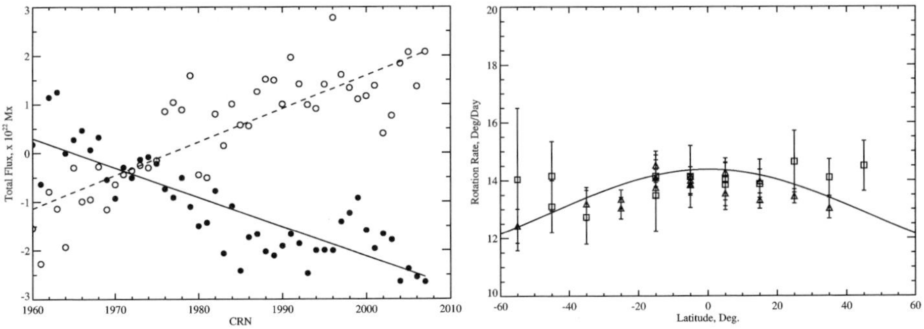

Figure 1. Left: Magnetic flux of polar regions at the time of polarity reversal. Open/filled circles corresponds to Southern/Northern polar regions. Dashed and solid lines are least squares best-fits to the data by first degree polynomial. Right: Mean rotation rate of CH boundaries. Triangles (squares) denote measurements with low (high) errors in rotation rates. Solid line is a typical sunspot rotation rate.

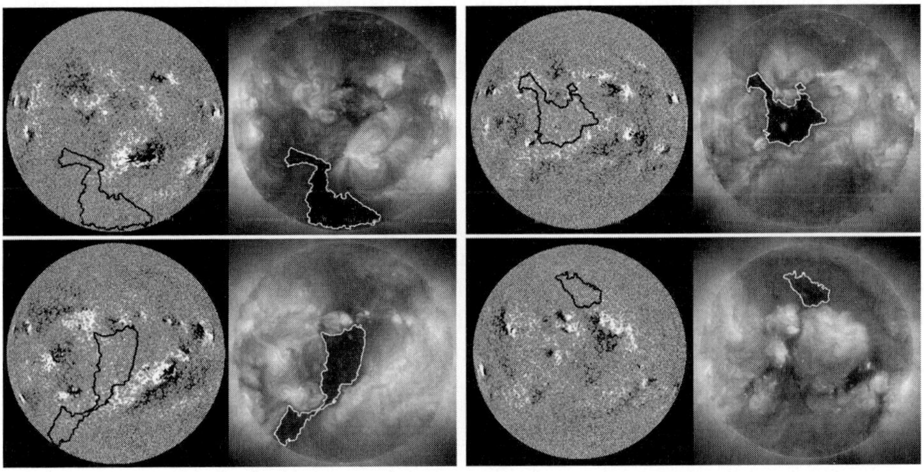

Figure 2. Coronal hole and its magnetic field. Left: CR1982, CR1985. Right: CR1987, CR1990.

Table 1. Flux, area, and Mean Field of CH.

CRN	M-Flux, Mx	Area, Pix	B_{AVG}, G
1982	3.8×10^{21}	7.9×10^{5}	2.3
1983	9.9×10^{21}	11.9×10^{5}	4.0
1984	11.2×10^{21}	9.0×10^{5}	6.1
1985	11.0×10^{21}	6.4×10^{5}	8.3
1986	19.0×10^{21}	7.6×10^{5}	12.2
1987	9.4×10^{21}	4.9×10^{5}	9.3
1988	4.0×10^{21}	2.1×10^{5}	9.1
1989	9.6×10^{21}	4.4×10^{5}	10.5
1990	0.7×10^{21}	1.0×10^{5}	3.6

2. Evolution of the Coronal Hole over Nine Solar Rotations

Examination of SOHO/EIT images starting with CR1982 indicates that although both poles had reversed their polarity, no polar CHs had developed. On CR1982, southern polar CH is situated very asymmetrically in respect to S-pole, and later, (CR1991) it

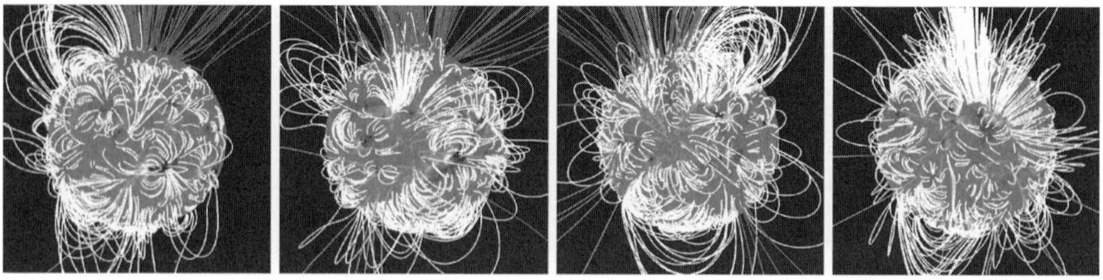

Figure 3. PFSS modelling of magnetic field shown on Fig. 2. White indicates closed filed lines; purple and green are opened field lines of opposite polarity.

disappears completely. CH on the North pole is also intermittent: it exists in CR1982, but is absent on later rotations. The studied coronal hole had originated as an extension of the South polar CH (Fig. 2). Once the extension reached the equator, the CH got disconnected from the southern polar hole. Next, it had travelled across the equator, and crossed to high latitudes in the Northern hemisphere. At the time of the event, the North polar region has had prevailing negative polarity field, while the newly transported CH was associated with the positive magnetic polarity. Table 1 and Fig. 2 demonstrate lack of continuity in total magnetic flux and area of the CH as it travels across the equator. Total magnetic flux associated with this CH varies between $\approx 4\%$ and 50% of total magnetic flux of polar field at the time of the event ($2\text{–}4\times10^{22}$ Mx). Significant variations in total flux and area indicate that these key properties of CH are largely determined by its photospheric roots. In that respect, the CH is analogous to a grass fire, which supports itself by continuously propagating from one patch of dry grass to the other. Such behavior is consistent with the interchange reconnection model. The lack of continuity and significant variations in flux and area of CH suggests that although morphologically it appears as the same coronal hole, its photospheric/magnetic footprint is constantly changing, thus suggesting that CHs are not a deep-rooted phenomena. Rotation profile of CH (Fig. 1, right) was computed by tracking the position of the mid-point between East and West boundaries of CH at a given latitude. Disregarding extreme latitudes, where the errors of this method are high, the average rotation profile is in a good agreement with the photospheric rotation. A gradual distortion of shape of CH (Fig. 2) also indicates that its photospheric footprint follows closely to the solar differential rotation profile. Travel "path" of the CH does not appear to be random. Despite the presence of weak magnetic fields of right polarity in vicinity of CH, it does not always move to these polarity patches. Analysis of EIT images suggests that CH avoids areas of closed magnetic fields, and its path follows the large-scale boundaries between separate flux systems. Potential Field Source Surface (PFSS, Fig. 3) modelling indicates that overall motions of the coronal hole are governed by location of magnetic separatrices in the corona.

References

Durrant, C. J. & Wilson, P. R. 2003, *Solar Phys.*, 214, 23

Harvey, K. L. & Recely, F. 2002, *Solar Phys.*, 211, 31

Jones, H. P. 2005, in: Sankarasubramanian, K., Penn, M., & Pevtsov, A. (eds.) Large-scale Structures and their Role in Solar Activity, Astron. Soc. Pac. CS-346, 229

Karachik, N., Pevtsov, A. A., & Sattarov, I. 2006, *ApJ*, 642, 562

Pevtsov, A. A., Balasubramaniam, K. S., & Rogers, J. W. 2003, *ApJ*, 595, 500

Wang, Y.-M. & Sheeley, N. R. 2004, *ApJ*, 612, 1196

Solar and Stellar Variability: Impact on Earth and Planets
Proceedings IAU Symposium No. 264, 2009
A. G. Kosovichev, A. H. Andrei & J.-P. Rozelot, eds.

doi:10.1017/S1743921309992651

Surface temperature maps for II Peg during 1999–2002

M. Lindborg[1], M. J. Korpi[1,4], I. Tuominen[1], T. Hackman[1], I. Ilyin[2], and N. Piskunov[3]

[1]Observatory, PO BOX 14, FI-00014 University of Helsinki, Finland

[2]Astrophysikalisches Institut Potsdam, An der Sternwarte 16, 14482 Potsdam, Germany

[3]Department of Astronomy and Space Physics, Uppsala University, SE-751 20, Uppsala, Sweden

[4]NORDITA, Roslagstullsbacken 23, SE-10691 Stockholm, Sweden

Abstract. The active RS CVn star II Peg has been spectroscopically monitored for almost 18 years with the SOFIN spectrograph at NOT, La Palma, Spain. In this paper we present five new surface temperature maps of the object for the years 1999 (two maps), 2001 (one map) and 2002 (two maps).

Keywords. techniques: spectroscopic, stars: late-type, stars: spots

1. Introduction

II Peg is one of the very active RS CVn stars and it is the brightest X-ray star with 50pc of the Sun. RS CVn stars are closely detached binaries where the more massive component is a G-K giant or subgiant and the secondary a subgiant or dwarf of spectral class G to M. Because of the low luminosity of the secondary many RS CVn systems appear as single-line binaries which are suitable for spectral analysis (Berdyugina *et al.* 1998a). In close binaries the rapid rotation is maintained by tidal forces due to the close companion. Large amplitude brightness variation of RS CVn stars imply the presence of enormous star spots on their surfaces covering up to 50% of the visible disk. Also coronal X-ray and microwave emissions, strong flares in the optical, UV, radio and X-ray are seen. Cool spots on the stellar surface will locally alter the photospheric absorption lines and continuum intensities.

Previous investigations on the temperature distribution over the surface of II Peg include the study of Berdyugina *et al.* (1998b), who presented surface temperature maps for 1992-1996, and Bergyugina *et al.* (1999c) with surface maps for 1996–1999, both studies were based on observations with the SOFIN-spectrograph at NOT. Gu *et al.* (2003) presented surface images of II Peg for 1999-2001 based on observations with the Coude echelle spectrograph at the Xinglong station of the National Astronomical Observatories, China, but the spectral lines used for inversions were different to that of SOFIN observations. Photometric light curve variations of the object were analysed by Berdyugina & Tuominen (1998c), and by Rodonò *et al.* (2000). The results of Berdyugina *et al.* (1998a,b, 1999a,b,c) and Berdyugina & Tuominen (1998c) consistently show that there are two active longitudes separated approximately by 180°, migrating in the orbital reference frame, and that a switch of activity level occurs periodically with a period of 4.65 years. In the surface images of Gu *et al.* (2003) the general spot pattern was quite similar, but the drift with respect of the orbital frame was less obvious, and the switch of the activity level appeared to occur earlier than predicted by Berdyugina

Table 1. Summary of the observations.

HJD 2450000+	Phase	S/N	Label	HJD 245000+	Phase	S/N	Label
July-August 1999				September 1999			
1383.7017	0.7998	248	9	1443.4928	0.6915	179	4
1384.7178	0.9509	238	11	1443.6184	0.7102	164	5
1385.7232	0.1004	217	1	1444.4788	0.8382	230	6
1386.7124	0.2475	229	3	1445.5656	0.9998	194	7
1387.7094	0.3958	200	5	1447.6084	0.3036	174	1
1388.7278	0.5472	181	7	1448.6212	0.4542	184	2
1389.7031	0.6923	178	8	1449.5979	0.5994	178	3
1390.7198	0.8435	151	10				
1391.7045	0.9899	243	12				
1392.7108	0.1396	258	2				
1393.7275	0.2908	189	4				
1394.7386	0.4411	206	6				
August 2001				August 2002			
2120.5503	0.3796	192	3	2507.6078	0.9405	139	9
2121.5642	0.5303	225	4	2508.6582	0.0967	208	1
2122.7102	0.7008	185	6	2509.5877	0.2349	200	3
2123.6083	0.8344	196	8	2510.5909	0.3842	216	4
2124.5644	0.9765	193	10	2511.5413	0.5255	195	5
2125.6758	0.1418	177	1	2512.6316	0.6876	212	7
2126.7146	0.2963	189	2	2513.5494	0.8242	179	8
2128.6067	0.5777	166	5	2514.5854	0.9782	206	10
2129.6470	0.7324	190	7	2515.5768	0.1257	207	2
2130.6415	0.8804	208	9	2518.5749	0.5715	129	6
November 2002							
2588.3964	0.9549	207	7				
2589.3764	0.1007	160	1				
2591.3567	0.3951	241	3				
2592.3705	0.5459	219	4				
2597.4333	0.2988	252	2				
2599.4798	0.6031	275	5				
2600.4896	0.7532	259	6				

et al. (1999c). Rodonò *et al.* (2000) found a much more complicated spot pattern from their analysis of photometry: they report on the existence of a longitudinally uniformly distributed component together with three active longitudes, with complicated cyclic behavior. Carroll *et al.* (2009) have also applied a Zeeman Doppler imaging technique to derive the magnetic field configuration on the surface of II Peg during 2007 using spectropolarimetric (Stokes I and V) observations with SOFIN. Their maps show a very similar spot pattern as found by Berdyugina *et al.* (1998b, 1999c); moreover, the radial field direction is opposite on different active longitudes.

2. Observations

When a star rotates rapidly star spots modify the observed spectral line profiles. As the star rotates, these bumps will move across the absorption line profiles (Hackman *et al.* 2001; Kurster, M. 1993). The surface imaging techique is basically to trace these distortions and create a surface map of the star. We use the surface imaging technique developed by Piskunov (the code INVERS7, Piskunov *et al.* 1990; Piskunov 1991).

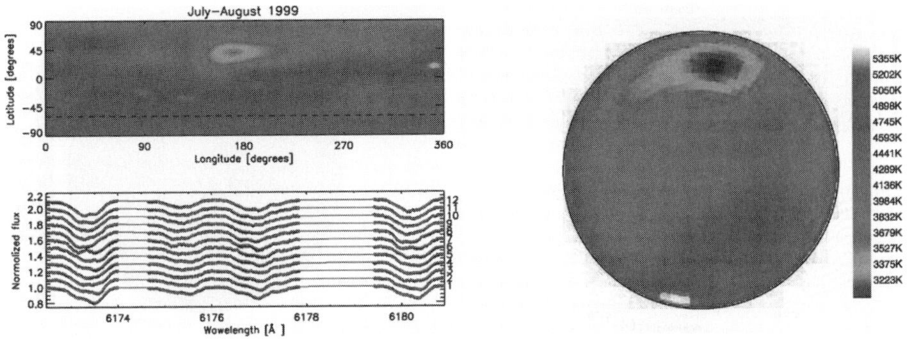

Figure 1. Observing run during July - August 1999. Upper left panel: Mercator projection of the obtained surface temperature distribution. Lower left panel: observed (crosses) and calculated line profiles (solid lines). Panel on the right: Pole-on projection of the surface temperature distribution.

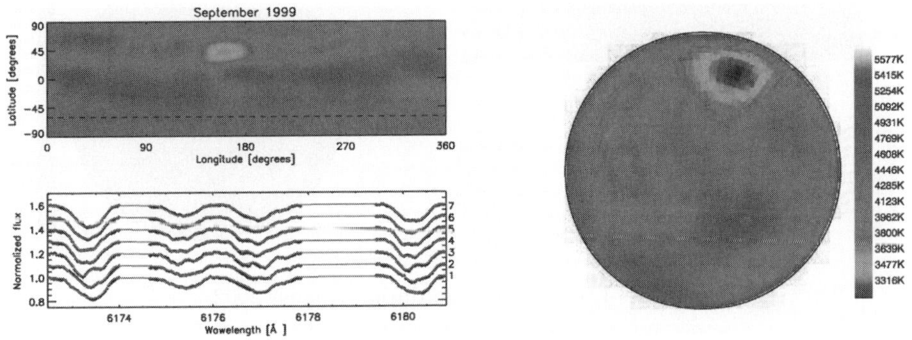

Figure 2. Observing run during September 1999.

We use the stellar model atmospheres of Kurucz (1993) for local line profile calculations for a set of temperatures and limb angles. This table is then used for the disk integration of a given surface temperature distribution. The surface imaging problem can be solved by searching for a such surface temperature distribution that minimizes the discrepancy function between the observations and the calculated line profiles (Hackman *et al.* 2001).

High resolution spectra of II Peg were measured in July-August 1999, September 1999, August 2001, August 2002 and November 2002. All the observations were made using the SOFIN high resolution echelle spectrograph at the 2.6 m Nordic Optical Telescope (NOT), La Palma, Spain. The data were acquired with the 2nd camera equipped with a CCD detector of 1152×298 pixels, and the spectral region 6160 – 6210Å was chosen for surface imaging. The observations are summarized in Table 1.

The spectral observations were reduced with the 4A software system (Ilyin 2000). Bias, cosmic ray, flat field and scattered light corrections, wavelength calibration and normalization were included in the reduction process.

The stellar parameters used in the inversions were chosen according to Berdyugina *et al.* (1998a), and read T_{eff}=4600 K, $\xi_t = 2.4$ kms^{-1}, ζ_t=3.5 kms^{-1}, $P_{\rm orb}$=6.724333 d, $v \sin i$=22.6 kms^{-1}, $i = 60°$.

3. Results and conclusions

The following spectral lines were used in the surface temperature inversions: Fe I 6173 Å, Ni I 6175 Å, Ni I 6177 Å, and Fe I 6180 Å.

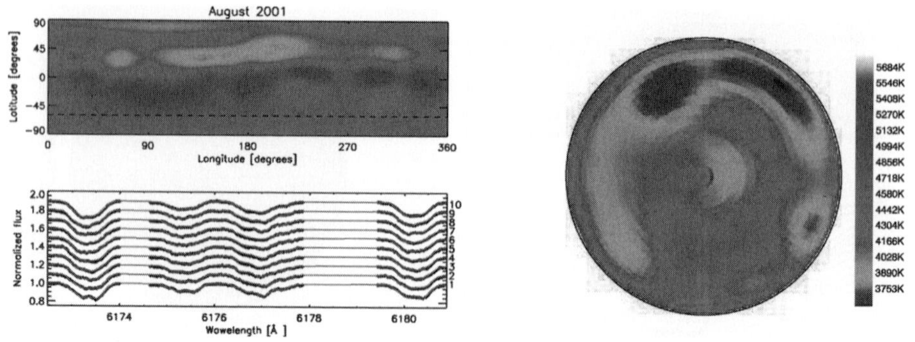

Figure 3. Observing run during August 2001.

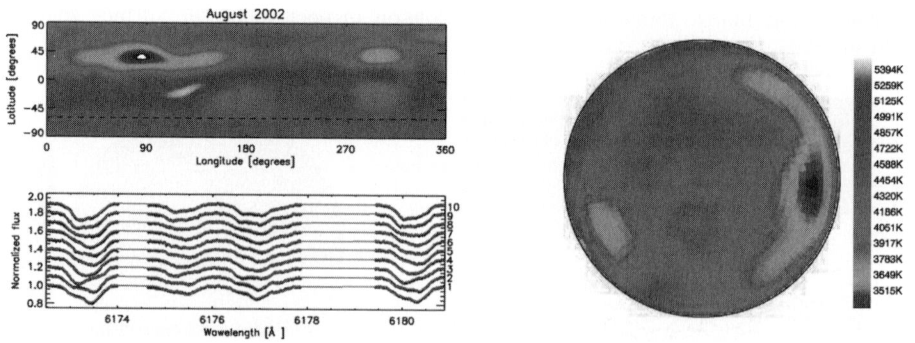

Figure 4. Observing run during August 2002.

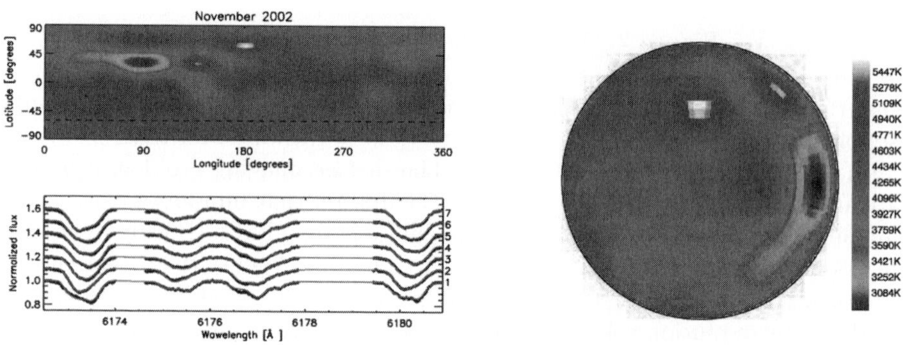

Figure 5. Observing run during November 2002.

Spectral line parameters were obtained from the Vienna Atomic Line Database (Kupka *et al.* 1999); the $log(gf)$ values of the two included Ni I lines were modified from the standard value of -0.53 to -0.2 as the standard values were found to produce much weaker absorption lines than the actually observed ones. This is equivalent to an increase in the Ni abundancy, which is probably the real reason for the observed strong Ni lines.

Figures 1–5 show the results of the inversions. During July-August 1999 (Fig. 1) only one spot is seen approximately at latitude 40°. The minimum temperature inside the spot was measured to occur at the approximate longitude of 170° or phase 0.47. Our image is quite different from the one obtained by Gu *et al.* (2003) for almost a simultaneous observing period, but a different spectral region. Their inversions gave much larger, longitudinally extended, spot structures around 170–290° at roughly 60° of latitude.

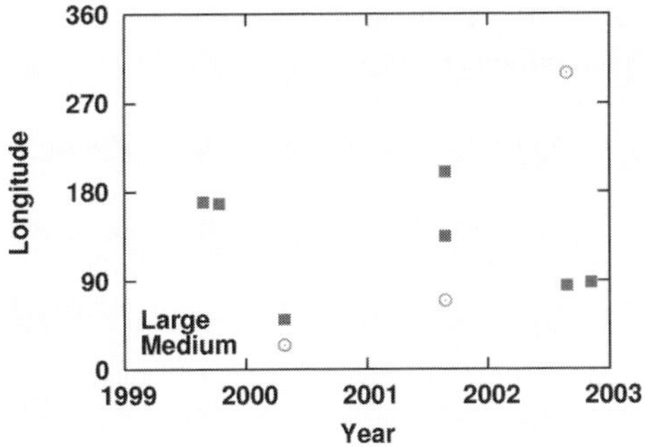

Figure 6. The spot longitudes of II Peg versus time. Square symbols represent larger active regions and circles mark weak structures.

In September 1999 our inversions also show only one strong spot that has barely moved in the orbital reference frame (latitude 44° and longitude 160° or phase 0.44). There is a weaker, longitudinally extended feature visible between 220–270° or phase 0.61–0.75.

Almost two years later, in August 2001 (Fig. 3), the star exhibits much more surface structures: three spots are visible in our image (longitudes 70°, 140°, and 200° or phases 0.19, 0.39 and 0.56) at an approximate latitude of 40°. The inversions of Gu *et al.* (2003) for an observing run 5 months later show much less structure, and the maximal activity seems to have moved roughly to the other side of the star than what was seen in their images during 1999 and 2000.

One year later, in August 2002 (Fig. 4) our maps show one strong spot at the latitude 40° and longitude 80° or phase 0.22, and a weaker one at 300° or phase 0.83. In November 2002 (Fig. 5) only the stronger spot is seen approximately at the same location. Comparing the surface temperature distribution during the observing seasons 1999 and 2002, the maximal spot activity seems to have moved by roughly 100° in the orbital reference frame, while very little drift of the spots can be seen during the consecutive observing runs during 1999 and 2002. In between these two 'states' of only one strong spot at different location on the surface, a much more complex distribution could be seen during August 2001. Our images give some support to the conclusion of Gu *et al.* (2003) of a significant change of the longitudinal spot distribution having occurred sometime during 2001 (Fig. 6).

We plan to continue to study the spot evolution on II Peg by analysing photometric and spectropolarimetric observations of the object, both to invert the surface temperature distribution, but also the magnetic field configuration of the object.

Acknowledgements

The authors acknowledge the hospitality of NORDITA during the programme 'Solar and stellar dynamos and cycles', and useful discussions with Dr. Oleg Kochukhov.

References

Berdyugina, S. V., Berdyugin, A. V., Ilyin, I., & Tuominen, I. 1998a, *A&A*, 340, 437
Berdyugina, S. V., Jankov, S., Ilyin, I., Tuominen, I., & Fekel, F. C. 1998b, *A&A*, 334
Berdyugina, S. V. & Tuominen, I. 1998c, *A&A*, 336, L25
Berdyugina, S. V., Ilyin, I., & Tuominen, I. 1999a, *A&A*, 347, 932
Berdyugina, S. V., Ilyin, I., & Tuominen, I. 1999b, *A&A*, 349, 863
Berdyugina, S. V., Berdyugin, A. V., Ilyin, I., & Tuominen, I. 1999c, *ApJ*, 350, 626
Carroll, T., Kopf, M., Strassmeier, K., Ilyin, I., & Tuominen, I. 2009, *IAU Symposium*, 259

Gu, S.-H., Tan, H.-S., Wang, X.-B., & Shan, H.-G. 2003, *A&A*, 405, 763
Hackman, T., Jetsu, L., & Tuominen, I. 2001, *A&A*, 374, 171
Ilyin, I. 2000, *PhD thesis, University of Oulu, Finland*
Kupka, F. & Ryabchikova, T. A. 1999, *VALD-The Vienna Atomic Line Database: A Survey*
Kurster, M. 1993, *A&A*, 274, 851
Kurucz, R. L. 1993, *VizieR On-line Data catalog: VI/39*
Piskunov, N. E., Tuominen, I., & Vilhu, O. 1990, *A&A*, 230, 363
Piskunov, N. E. 1991, *Proceedings of Colloquium No. 130 of the International Astronomical Union*
Rodonò, M., Messina, S., Lanza, A. F., Cutispoto, G., & Teriaca, L. 2000, *A&A*, 358, 624

Solar and Stellar Variability: Impact on Earth and Planets
Proceedings IAU Symposium No. 264, 2009
A. G. Kosovichev, A. H. Andrei & J.-P. Rozelot, eds.

doi:10.1017/S1743921309992663

'Negative' surface differential rotation in stars having low Coriolis numbers (slow rotation or high turbulence)

Kwing L. Chan[1]

[1]Department of Mathematics,
Hong Kong University of Science and Technology, Hong Kong, China
email: maklchan@ust.hk

Abstract. A general picture of differential rotation in cool stars is that they are 'solar-like', with the equator spinning faster than the poles. Such surface differential rotation profiles have also been demonstrated by some three-dimensional simulations. In our numerical investigation of rotating convection (both regional and global), we found that this picture is not universally applicable. The equator may spin substantially slower than the poles ($(\Omega_{equator} - \Omega_{pole})/\Omega$ can reach -50%). The key parameter that determines the transition in behavior is the Coriolis number (inverse Rossby number). 'Negative' differential rotation of the equator (relative to the mean rotation) occurs if the Coriolis number is below a critical value.

Keywords. stars: interiors, rotation; convection

1. Introduction

The interaction of convection and rotation can produce different zonal flow patterns. A switch in the sign of the vertical shear in the zonal flow at the equator has been discussed in an earlier conference paper (Chan 2003). The pattern transition was linked to the magnitude of the Coriolis parameter (inverse of the Rossby number). The process was demonstrated by both regional f-box calculations and global-scale spectral calculations. In recent years, 'anti-solar' differential rotations have been reported in a number of observations (Kovain *et al.* 2007, Erdem *et al.* 2009) and models (Kitchantinov & Rudiger 2004, Steffen & Freytag 2007, Palacios & Brun 2007). The phenomenon, we believe, reflects the situation of low Coriolis number (high Rossby number) as discussed in the earlier paper. In this paper, we present a more systematic (and higher resolution) investigation of the flow field transition in a global-scale shell model.

2. Model

The model solves the stratified form of the compressible hydrodynamics equations (Chan 1994) for an ideal gas in a rotating convection zone with a thickness of 30% outer radius. The stratification contains 6.3 pressure scale heights. Four calculations with different values of angular velocity Ω have been made. The rest of the model setup is not changed. Values of the Coriolis number ($\mathrm{Co} = \Omega d/v$) for the four cases are 4.6, 9.6, 22, and 39. The length scale d is taken to be the depth of the layer and the velocity scale v is the root-mean-square total velocity in the domain. The three-dimensional numerical code uses the transform spectral method, and the spatial resolution is T84 $\times$ 116 where T84 stands for triangular truncation with maximum zonal degree 84 and 116 stands for the number of vertical grid levels. The models took long time to relax.

3. Results

The latitude-radius distributions of the longitudinally and temporally averaged relative angular velocity (in comparison with the mean rotation of the shell) are shown on the panels of Figure 1 in an order of increasing Co. The light/dark shadings represent positive/negative values. Figure 2 show the latitudinal profiles of the surface differential angular velocity for the cases arranged in the same order. A critical value of Co exists between 9.6 and 22. For lower Co, the radial shear of the mean zonal velocity at the equator is negative; for higher Co, the shear is positive. When the shear is positive, the surface differential rotation is 'solar-like'; when the shear has the opposite sign, the differential rotation is 'anti-solar'. Besides this qualitative change, the profiles show further variations for different Co values.

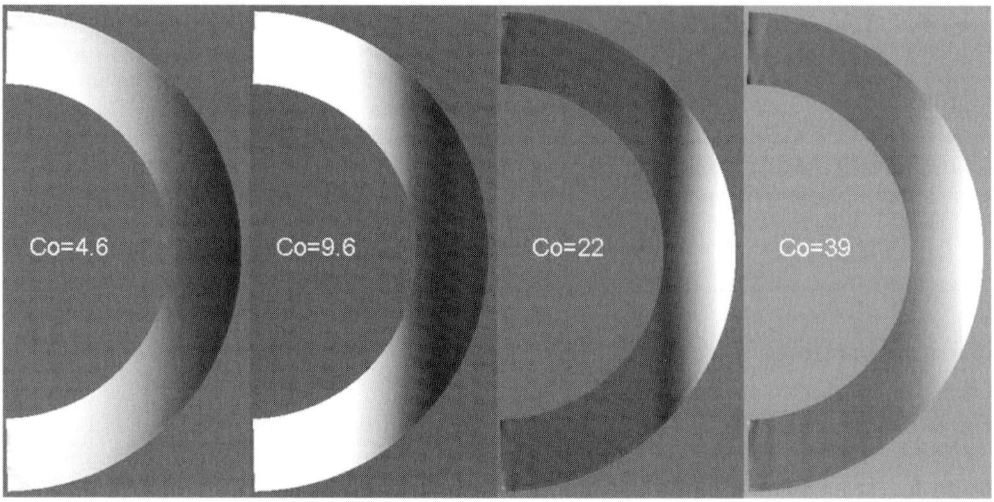

Figure 1. Differential angular velocity distributions

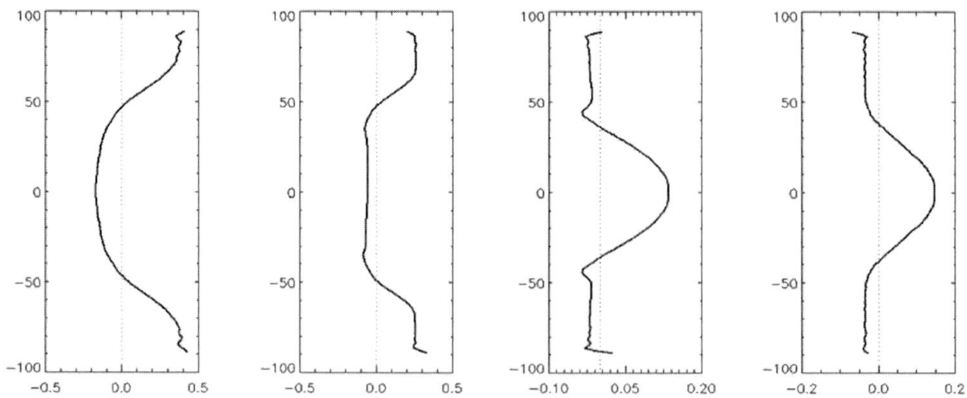

Figure 2. Latitudinal profiles of the surface differential angular velocity

4. Conclusion

The Coriolis number determines whether differential rotation at the surface of a cool star is 'solar-like' or 'anti-solar'. Observation of this surface characteristic may help to impose constraints on the properties of the convection zone (e.g. depth, turbulence strength).

Acknowledgement

KLC thanks the Hong Kong Research Grants Council for support (600309).

References

Chan, K. L. 2003, in S. Turcotte, S. C. Keller & M. Cavallo (eds.), *3D Stellar Evolution* (Ann Arbor: Sheridan), p. 168 (can be found at http://www.math.ust.hk/~maklchan)
Chan, K. L., Mayr, H. G., Mengel, J. G., & Harris, I. 1994, *J. Comput. Phys.*, 113, 165
Erdem, A. *et al.* 2009, *New Astron.*, 14, 545
Kitchatinov, L. L. & Rudiger, G. 2004, *AN*, 325, 496
Palacios, A. & Brun, A. S. 2007, *AN*, 328, 1114
Steffen, M. & Freytag, B. 2007, *AN*, 328, 1054

magnitude also occurs around the maximum of the solar cycle when the slowly varying polar field is near zero and the variable equatorial component has a minimum. Like the sunspot number, the axial dipole is much weaker during the current solar minimum and the duration of the minimum is much longer. The current value is about −0.26 G compared with values of 0.6, −0.7, and 0.55 in the minimum years 1976, 1986, and 1996.

The reduced value of the polar cap field changes the configuration of the large-scale field in the overlying corona and heliosphere, especially at minimum. It may also presage a smaller maximum for Solar Cycle 24.

3. The Effect of the Polar Field on the Corona and Heliosphere

The complexity of the photospheric field pattern simplifies in the overlying corona and the heliosphere. Coronagraph observations at various heights show this, as do *in situ* measurements of solar wind structure throughout the heliosphere. One convenient way to model the large-scale structure of the solar atmosphere during the solar cycle is to use a potential field approximation applied to the photospheric field in conjunction with an imposed source surface located at some height at which the field becomes radial and no longer changes with height, [e.g. Schatten, 1968 and Hoeksema *et al.*, 1982]. Of course much more sophisticated force-free field and MHD models can be applied to calculate more quantities, predict events, refine the details, and investigate time variations, but the basic large-scale configuration of the heliospheric field remains the same.

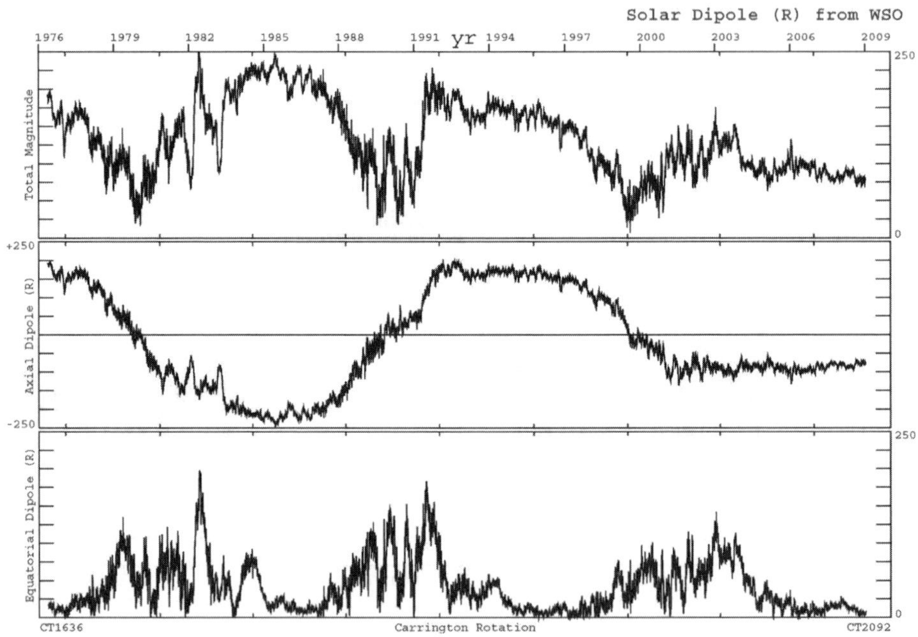

Figure 2. The top panel shows the total magnitude of the solar dipole field from May 1976 to November, 2009, as in Figure 1. The center and bottom panel show its two constituents the axial dipole and the magnitude of the equatorial dipole. The variation is relatively smooth near cycle minima when the axial polar dipole dominates. The dominant feature of the axial dipole is the reversal of the polar field every 11 years about the time of solar maximum. The equatorial dipole is strongly driven by patterns due to the alignment of active regions and their remnants at mid and low latitude. Such alignments have a lifetime of 12–18 months.

Figure 3 compares the evolution of the source surface field during CYCLES 21 and 23. The left column shows the 10 years of Cycle 21, from 1976 to 1986. The right column shows 13 years of Cycle 23, from 1996 to 2009. Each panel is two rotations wide to show more clearly the field configuration. The large-scale field usually evolves on a relatively slow time scale, so showing one panel every 12 rotations provides a reasonable view of

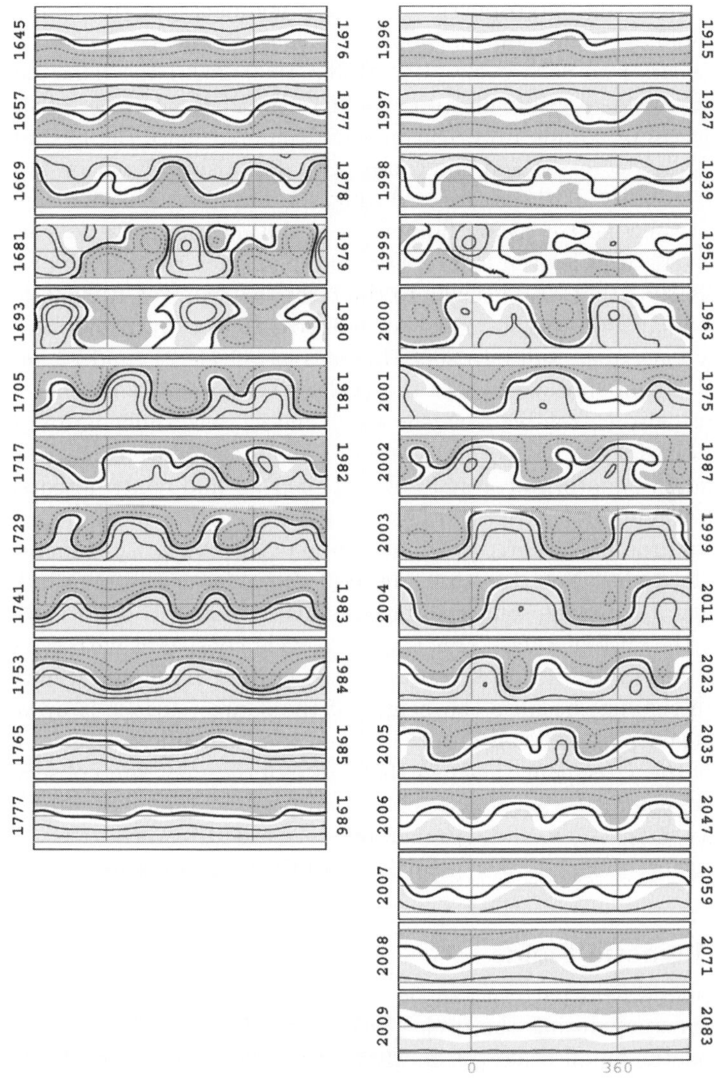

Figure 3. The panels on the left show the Solar Cycle 21 variations of the magnetic field configuration at the base of the solar wind computed with a potential field source surface model. Each panel shows the field at 2.5 solar radii in a two-rotation wide window. The Carrington Rotation is indicated on the left and the year on the right. The vertical axis is sine latitude from pole to pole and data are shown between 70N and 70S. The dark line is the neutral line where the polarity reverses. Light shaded areas with solid contours show positive (out of the Sun) regions. The panels are spaced by 12 rotations, about 11 months, to show the entire 11-year cycle in 12 panels from 1976 to 1986. The panels on the right show the same for Solar Cycle 23. Here the Rotation is indicated on the right and the year on the left, near the center. Cycle 23 lasted longer so more panels are shown, but the polarity was the same as Cycle 21. The neutral line at the base of the heliospheric current sheet is nearly flat at minimum.

how the field varies. The dominant polar field at solar minimum confines the neutral line to the equator at the beginning and end of each cycle. The Earth, never more than 7 degrees from the equator, encounters the interplanetary extension of the neutral line, the heliospheric current sheet (HCS), every rotation. At higher latitude during long intervals near minimum the polarity of the heliospheric field does not reverse [McComas *et al.*, 2000]. The HCS becomes increasingly distorted during the rise of the cycle as active regions make the photospheric field more complex. As the polar fields weaken in response to old active region flux moving poleward, the HCS crosses all latitudes. Three or four years into each cycle (1979–1980; 1999–2000) the polar fields reverse in the corona. This reversal gradually propagates outward to the edge of the heliosphere, affecting the access of cosmic rays. During the declining phase the latitudinal extent of the HCS gradually decreases and the complexity diminishes. Each cycle is unique, but the multi-year stability of the features during the declining phase is quite impressive. Because so little new active region flux emerged during the last few years of Cycle 23, the lifetime of structures is even longer. For example, track the lobe of positive flux near 0 degrees in CR 1987 that drifts slowly to the right through to CR 2059, five years later.

Because the polar field was so weak during Cycle 23, the warps of the HCS remained large for a very long time, well into 2008. In contrast, Cycle 21 (and 22, not shown here) flattened just eight or nine years into the cycle. The coronal holes that are the source of the open flux, also calculable with the PFSS model, also differ late in Cycle 23. The polar coronal holes for the last several years are smaller, and even though the flux at low latitude was small, equatorial coronal holes opened up and contributed significantly to the solar wind. The unipolar caps did not dominate the heliospheric magnetic structure. For several years around the three previous minima virtually no open flux emanated from low or middle latitudes. Note too that Figure 3 shows fewer contours in Cycle 23. This means that with a constant source surface radius, less flux is open to the heliosphere during most of the current cycle. This is qualitatively consistent with the observation of values of open flux in the heliosphere during the current minimum [Smith and Balogh, 2008]. While intriguing, because the height of the source surface is an adjustable and somewhat arbitrary parameter, it is not clear how much significance should be attributed to the details of the agreement. The fact that much of the solar wind observed near Earth at this minimum emerged from smaller near-equatorial holes may explain why solar wind velocity, density, magnitude, and dynamic pressure are all lower.

What is the cause of the weaker polar fields during this minimum?

4. Zonal Magnetic Fields

The photospheric field pattern can be explained by surface flux transport acting on the flux that emerges during the cycle [DeVore *et al.*, 1985]. Even though more total flux emerges on the small scale [Zirin, 1987], that flux is not organized and does not contribute significantly to the evolution of the pattern. The magnetic flux distribution in active regions is locally imbalanced systematically because of the differences between leading and following polarity concentrations and because of the inclination specified by Joy's law. Differential rotation, diffusion, and meridional flow act on flux elements to generate the patterns. Much of the emergent flux in active regions cancels locally, but what remains eventually makes its way poleward to form the new polar cap or travels equatorward where it cancels with flux from the other hemisphere.

Considering the entire solar surface, one can investigate the characteristics of flux in each latitude zone during each rotation. Figure 4 shows the zonal magnetic field versus time for the last three solar cycles. The lower panel is the familiar butterfly diagram and shows that the total flux is concentrated in the active region belts during most of the cycle. Flux emerges at middle latitude shortly after minimum and gradually moves equatorward. The total polar flux vanishes at solar maximum and is strongest at solar minimum. Cycle 23 is longer and weaker than either Cycle 21 or 22 and there is a long period in 2007 and 2008 where there is very little flux, except in the weak polar regions.

The upper panel shows the net flux. Flux emerges at middle latitudes shortly after minimum, with leading polarity appearing at lower latitude than following polarity. The regions of following polarity expand gradually poleward, cancelling the flux of the old cycle poles and ultimately forming the new polar cap. The surges sometimes move rapidly (in one or two years) from low latitude to high. The surges can be of either polarity, see for example the alternating polarity stripes in the southern hemisphere in 1981 and 1982. When surges reach the poles, the strength of the cap is affected. So, when the leading-polarity positive surge reaches the south pole in late 1990, the polar field reversal temporarily stalls (see Figure 1). Emerging active regions and remnants of decaying spots contribute to these surges. The surges often occur simultaneously in the two hemispheres, but can be symmetric, antisymmetric, or asymmetric about the equator.

The net zonal polarity looks weaker and significantly less organized in Cycle 23. The onset of the cycle looks fairly typical, but the pattern is soon disrupted, with the polarity even at low latitude reversing more than once. This happens in the south in 2002 and in the north in 2003. There seem to be more leading polarity surges, particularly in the south in 1999, 2001, 2002, 2003, and 2005. The north is less clear, but weaker surges of old polarity can be detected in 2000, 2002, and 2004. This apparently has the effect of

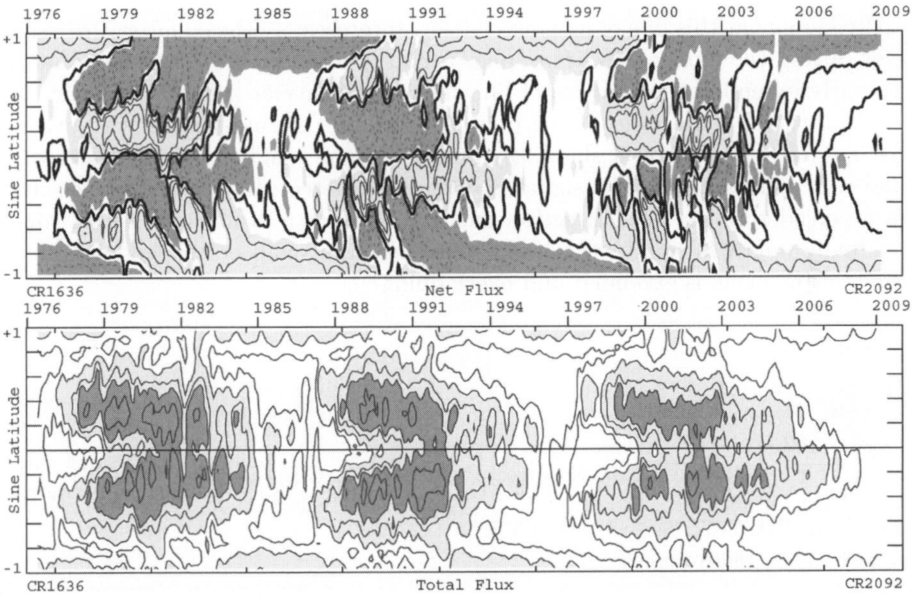

Figure 4. The time evolution of the total flux in each latitude zone is shown in the lower panel for Solar Cycles 21-23. For each Carrington Rotation in each bin in sine latitude the absolute value at each longitude is summed to find the total flux. A contour plot of the result versus time reveals the familiar butterfly diagram. Flux associated with active regions dominates during most of the cycle at lower latitudes. The polar field is strong at higher latitude at other times. The flux emergence moves equatorward through each cycle. The upper panel shows the net flux in each zone, i.e. the sum is performed on the signed values. Structure with the new polar field polarity emerges at mid latitudes shortly after minimum and gradually moves poleward. Each cycle is quite different.

lowering the polar field strength. This explanation of the weaker polar field obviously begs the question: what about the emergence of active regions led to a different pattern during this cycle?

5. Discussion

Long-term observations of the large-scale field are essential to understanding the solar cycle. While the transition from Cycle 23 to 24 is unusual compared to recent cycles, Cycle 14, about 100 years ago, was not that different. Lanza [2009] and others at this meeting show that stars like the Sun have similar dynamo-driven cycles with similar variations in characteristics, but observational limitations make it difficult to be certain of the details. Even solar dynamo models have given very divergent predictions for the magnitude of the upcoming cycle [Pesnell, 2008]. Improved measurements of meridional and other flows in the photospheric and below should help resolve the solar uncertainties. The Kepler and CoRot missions will greatly increase our knowledge of cyclic variations on other stars.

Historically sunspot cycles have overlapped and the current transition is no different, just longer and weaker. Spots from both cycles have been observed from 2007 2009. The curious extended 18-year overlapping solar cycle has not been discussed here, but we can point to three important indications. Altrock [2009] sees the emergence of precursors of the cycle at high latitude in low-coronal Fe XIV emission several years before the minimum in three solar cycles. Torsional oscillations in the upper convection zone also indicate that a zonal flow of enhanced shear appears at high latitude well in advance of the cycle minimum and makes its way equatorward [Howe *et al.*, 2009]. Sunspots eventually emerge at this line of increased shear. Interestingly the equatorward migration is slower during the current declining phase than it was one cycle ago. Lo *et al.* [2009] have recently extended the analysis of Shrauner and Scherrer [1994] to measure the net large-scale toroidal component of the photospheric field during the last two cycles. The toroidal field is revealed by a slight asymmetry in the magnetic field observed in the east and west hemispheres. The signal is difficult to detect, but also indicates that flux with a reversed toroidal field direction emerges at high latitude well before minimum and gradually migrates equatorward. Future observations and better modeling may help explain why the cycle is extended and overlapping.

References

Altrock, R. C. 2009, *B. A. A. S.* 41, 861

Devore, C. R., Sheeley, N. R., Jr., Boris, J. P., Young, T. R., Jr., & Harvey, K. L. 1985, *Solar Physics* 102, 41

Hoeksema, J. T., Wilcox, J. M., & Scherrer, P. H. 1972, *J. Geophys. Res.* 87, 10331

Howe, R., Christensen-Dalsgaard, J., Hill, F., Komm, R., Schou, J., & Thompson, M. J. 2009, *ApJ. Lett.* 701, L87, 10.1088/0004-637X/701/2/L87

Lanza, A. F. 2009, this volume

Lo, L., Hoeksema, J. T., & Scherrer, P. H. 2009, SOHO-23 Workshop, sun.stanford.edu/~todd/posters/SOHO23.Lo.3CyclesToroidal.pdf

McComas, D. J. *et al.* 2000, *J. Geophys. Res.* 105, 10419.

Pesnell, W. D. 2008, *Solar Physics*, 252, 209, 10.1007/s11207-008-9252-2.

Schatten, K. 1968, *Nature*, 220, 1211

Shrauner, J. A. & Scherrer, P. H. 1994, *Solar Physics*, 153, 131

Smith, E. J. & Balogh, A. 2008, *Geophys. Res. L.* 35, 10.1029/2008GL035345

Svalgaard, L., Duvall, T. L., Jr., & Scherrer, P. H. 1978, *Solar Physics*, 58, 255

Zirin, H. 1987, *Solar Physics*, 110, 101, 10.1007/BF00148205

Session V

Physical mechanisms of solar and stellar variability

Chair: H. Shibahashi

Solar and Stellar Variability: Impact on Earth and Planets
Proceedings IAU Symposium No. 264, 2009
A. G. Kosovichev, A. H. Andrei & J.-P. Rozelot, eds.

doi:10.1017/S1743921309992687

Mechanisms for total and spectral solar irradiance variations

Margit Haberreiter

Laboratory for Atmospheric and Space Physics, University of Colorado, 1234 Innovation Drive, Boulder, CO, 80303, USA
email: haberreiter@lasp.colorado.edu

Abstract. The total and spectral irradiance varies over short time scales, i.e. from days to months, and longer time scales from years to decades, centuries, and beyond. In this talk we review the current understanding of irradiance changes from days to decades. We present the current status of observations and discuss proposed reconstruction approaches to understand these variations. The main question that ultimately needs to be answered is what are the physical processes that could explain the enhanced heating of the photosphere, chromosphere, transition region, and corona, leading to a change in the solar radiative output at various wavelengths. As semi-empirical models allow us to reproduce the solar spectrum over a broad wavelength range, they offer a powerful tool to determine the energy necessary to heat certain layers and at the same time balance the radiative losses.

Keywords. Sun: activity, Sun: atmosphere, Sun: magnetic fields, radiative transfer

1. Introduction

The incoming solar radiation is the main energy driving the Earth's climate system. This radiation varies as a function of wavelengths on timescales of days to months, and years to decades and centuries. While the variations of the total solar irradiance (TSI) cannot be the reason for the increase of the mean global temperature over the past 20 years (Lockwood & Fröhlich 2007), the variations of the solar spectral irradiance (SSI) are considered to have an effect on the Earth's climate system. For example, Egorova *et al.* (2004), and Austin *et al.* (2008) demonstrate that the Earth's atmosphere shows an effect on the solar cycle variability in the UV spectral range. The effect of a long-term trend of the solar spectral variability on the Earth's climate system are however not yet fully understood. This is partly due to the lack of a sufficiently long and precise time series of the SSI that is needed as input for Global Circulation Models (GCMs). Therefore, reconstructions of the TSI and SSI for time scales of decades and longer are essential for fully understanding of the Sun's role in the Earth's climate system.

Another important aspect of the SSI is the varying EUV radiation that influences the thermosphere and ionosphere (Fuller-Rowell *et al.* 2004). Due to the variations of the incoming EUV radiation, the neutral density in the Earth's upper atmosphere changes. As the motion of satellites depends on the neutral density, being able to nowcast and forcast the density allows a more precise localization of the satellites. This clearly improves the reliablity of satellite operation.

Finally, understanding the effects of the varying SSI on the Earth's atmosphere and climate system has received increasing interest not only for the Earth, but also with respect to the conditions of the atmospheres of other planets.

In the next section the latest observations are briefly mentioned. In Sec. 3 the current status of the reconstruction models for the TSI and SSI is presented. Then, in Sec. 3.3

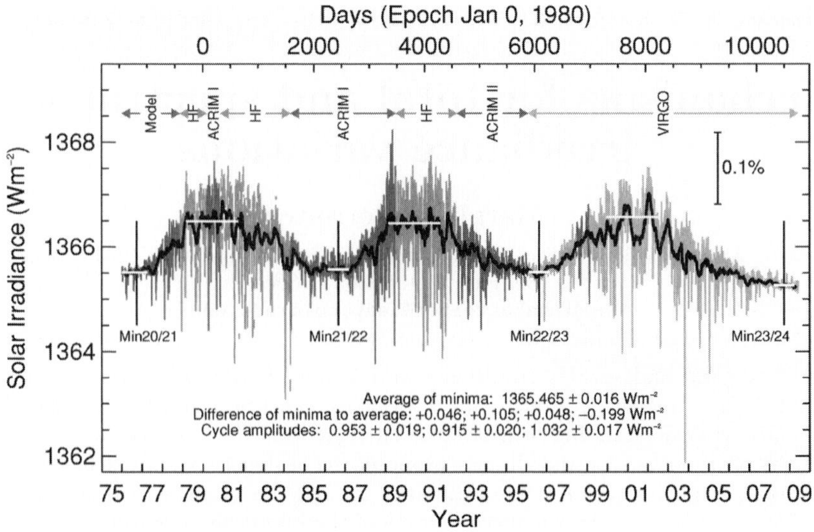

Figure 1. Latest available update of the TSI composite since the start of its measurements from space. The colored lines indicate daily measurements taken by different instruments the composite is based on; for details see Fröhlich (2009).

we present latest calculations for the EUV spectral range. Finally, in Sec. 4 we discuss the latest reconstruction approaches in a broader perspective.

2. Observations

2.1. *Total Solar Irradiance Variations*

The TSI is the wavelength integrated radiation emitted by the Sun measured at a distance of 1 AU. For more than three decades it has been measured from space. The measurements were taken by a number of different instruments; for details see e.g. Domingo *et al.* (2009), Sec. 2.1. Fig. 1 shows the composite compiled by Fröhlich (2009) showing a clear 11-year solar cycle variability. This variability is considered to be a result of the changing contribution of surface features, such as the quiet and active network, sunspots, plage, and faculae, to the overall emergent intensity.

2.2. *Spectral Solar Irradiance Variations*

The variability of the SSI is a strong function of wavelength. Therefore, in order to study the response of the Earth's atmosphere over the solar cycle the SSI has to be continuously observed over decades with sufficient spectral resolution and over a broad wavelength range. For a detailed list of past and future missions dedicated to observing the SSI see Domingo *et al.* (2009), Sec. 2.3. A few of the most recent instruments are listed below. In the UV, a continuous time series from space was only started in 1991 with the launch of the Upper Atmosphere Research Satellite (Willson 1994, UARS), having the Solar Ultraviolet Spectra Irradiance Monitor (Brueckner *et al.* 1993, SUSIM) and the Solar Stellar Irradiance Comparison Experiment (Rottman *et al.* 1993, SOLSTICE) onboard. With the launch of the SOlar Radiation and Climate Experiment (Rottman 2005, SORCE) the spectral range of the SSI from 1 nm to 2,700 nm was for the first time covered for continuous observations. The Solar Auto-Calibrating EUV/UV Spectrometer (SolACES) and SOLar SPECtrum (SOLSPEC) experiments on SOLAR/ISS (Thuillier *et al.* 2003; Schmidtke *et al.* 2006) that was launched in February 2009 measure the solar spectrum from 16 to 226 nm (in 4 separate channels) and 180 to 3000 nm.

Recently, Harder *et al.* (2009) discovered from the SORCE/SIM (Harder *et al.* 2005b) observations that different wavelength intervals of the solar spectrum show opposite trends in the solar cycle variability. These new findings show that the TSI variability is the sum of a SSI variability, which partly cancels due to its opposite sign. These findings pose a challenge for recontruction models that need to be able to reproduce this effect.

The continuous time series of SSI provided by SORCE is extremely valuable for the study of the solar radiation effects on the Earth's climate. For climate studies the ultimate goal is to reconstruct solar irradiance as far back in time as possible to study its long-term effects on the Earth's climate. Thus, precise reconstructions of the SSI over decades and beyond are necessary. In the next section the available reconstruction models are discussed.

3. Models for irradiance variations

3.1. *Total Solar Irradiance Variations*

With the availability of precise TSI and SSI measurements from space since the 1970s, modeling these variations became a challenging task. The early models were based on indices such as e.g. the sunspot number (SSN), the Mg II core-to-wing ratio, and the solar F10.7 cm radio flux. These indices are associated with solar activity, but not directly with the emitted radiation. Fröhlich & Lean (2004) could explain the TSI variations by a combination of the sunspot darkening derived from the SSN and the facular brightening based on a long-term and short-term analysis of the Mg II index. The SATIRE model (Solanki *et al.* 2005, and references therein) uses intensity spectra calculated with the ATLAS9 code by Kurucz *et al.* (1991) to describe the intensity emitted by different solar surface features, i.e. quiet Sun, sunspot umbra and penumbra, network, and faculae. SOHO/MDI and Kitt Peak magnetograms are employed to derive the surface area covered by the features as a function of position on the disk. For the calculation of the intensity spectra, local thermodynamic equilibrium (LTE) is assumed. These models are very successful and can explain the TSI variability almost entirely due to the changes of the solar surface magnetic field.

Evidence that the Sun varies on longer time scales is seen in radionuclide data, e.g. ^{10}Be from ice cores (Beer 2000). Radionuclides are produced when cosmic ray particles enter the Earth's atmosphere and interact with the atoms in the atmosphere, thereby also producing neutrons. The flux of the cosmic ray particles is controlled by the heliospheric magnetic field (HMF) and the geomagnetic field. In fact, the production rate of the radionuclides is anti-correlated with the extent of the HMF. It has been shown by Beer (2000) that the rates of neutrons and ^{10}Be behave very similar. This fact allowed Schöll *et al.* (2007) to reconstruct the variability of the TSI based on neutron monitor data and the SSN. Based on ^{10}Be data, recently Steinhilber *et al.* (2009) presented a TSI reconstruction for the past 9300 years.

3.2. *Spectral Solar Irradiance Variations*

3.2.1. *Reconstruction from magnetograms*

The calculation of solar intensity spectra based on the assumption of LTE, as described in Sec. 3.1, is not suitable for the UV and shorter wavelengths. Moreover, the formation of negative hydrogen, the main continuum opacity source in the visible and IR, depends on the electron density, which in turn depends on the ionizing UV radiation. As the latter

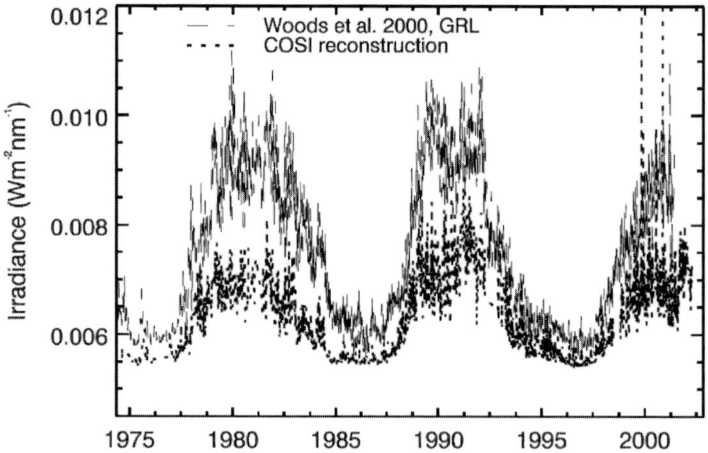

Figure 2. Reconstruction of Lyman α based on NLTE spectral synthesis with the COSI code and the feature identification from magnetograms as employed in the SATIRE model. Adopted from Haberreiter *et al.* (2005).

shows non-LTE (NLTE) effects, so does the formation of negative hydrogen; for details see e.g. Shapiro *et al.* (2009). Thus, accounting for the NLTE effects is essential for the synthesis of the correct solar spectrum, from the UV to the IR. These effects clearly also influence the solar cycle variation of the SSI.

Nevertheless, within the SATIRE model a method has been developed that allows extrapolating the SSI for wavelengths down to 115 nm. Based on this method, Krivova *et al.* (2009) present the reconstruction of the TSI and SSI for the years from 1974 to 2007. Again, the SATIRE model is very successful in describing the variability of the TSI and SSI. However, one has to keep in mind that the physical processes involved in the irradiance changes are very complex, and, in order to understand them they need to be accounted for in the models in full detail.

In order to improve the spectral synthesis Haberreiter *et al.* (2005) extended the work by Krivova *et al.* (2003) and calculated solar spectra with the NLTE radiative transfer COde for Solar Irradiance (Haberreiter *et al.* 2008a,b, COSI) using the atmosphere structures by Fontenla *et al.* (1999). Based on the magnetogram analysis used in SATIRE they reconstructed the SSI for the wavelength range from 100 to 400 nm. Fig. 2 shows their reconstruction for Lyman α and compares it to the composite by Woods *et al.* (2000). Following the SATIRE approach, the reconstruction is based on the changing intensity contributions for the quiet Sun, sunspots, and faculae. From Fig. 2 it is clear that the reconstruction underestimates the solar cycle variability by up to a factor of 2. This reconstruction however did not include the quiet and active network contribution. Whether the additional quiet and active network contribution is sufficient to explain the observed solar cycle variability still remains open. Another important point is that all reconstruction approaches discussed above assume a constant quiet Sun intensity over the solar cycle. However, work by Schühle *et al.* (2000) indicates that the quiet Sun radiance changes over the solar cycle. If this is confirmed the reconstruction approaches certainly need to be revised.

3.2.2. *Reconstruction from intensity images*

The analysis of magnetograms is only an indirect way to determine the brightness components responsible for irradiance variations. In fact the direct measurement of the intensity of all of the components allows a better understanding of their nature.

The Presicion Solar Photometric Telescope (Coulter *et al.* 1996; Ermolli *et al.* 2003; White *et al.* 2000, PSPT) at Rome and M. Loa Observatories take images in a narrow-band Ca II filter at 393 nm and a continuum wavelengths at 607 nm. These measurements allow the identification of solar surface features based on the intensity measured in each pixel. For example, the Solar Radiation Physical Modeling (SRPM) system provides tools to identify the surface features, such as intergranular cell, quiet and active network, faculae, plage, sunspot umbra and penumbra, which are identified through their contrast as a function of position on the disk (Fontenla *et al.* 2005, 2009b; Harder *et al.* 2005a).

The latest set of 1D atmosphere structures (Fontenla *et al.* 2009b) facilitates calculating the observed intensity and center-to-limb behavior for each of the surface features as identified from the PSPT observations. Please note that the latest set of models does not represent the quiet Sun with one model. In fact the intensity spectrum for the quiet Sun is considered to be a combination of intergranular cell (Model B), quiet network (Model D) and active network (Model F). Fig. 3 shows the reconstruction based on this scheme for different wavelengths with an earlier set of atmosphere structures. The agreement is very robust, with slightly larger deviations for 430 nm. The approches discussed so far are used to reconstruct the TSI and SSI for past times, but there is also an increased intereset in nowcasting and forecasting the solar irradiance, in particular for shorter wavelengths.

Recently, a lot of progress has been made in forecasting active regions on the far side of the Sun (Lindsey & Brown 2000). Along with the observations of backscattered Lyman α (Quémerais & Bertaux 2002), recently Fontenla *et al.* (2009a) developed a tool that is capable of forecasting the Lyman α radiation for a solar rotation. The great advantage of this method is that it accounts for growing and decaying active regions that have not yet appeared on the near side. This opens the possibility to forecasting the full solar spectrum in the near future. One important application for forecasting the EUV is to be able to determine the neutral density in the ionosphere, which in turn permits a more precise satellite positioning.

3.3. *SSI variations in the EUV*

Similar to the reconstruction approach presented in Sec. 3.2.2, the irradiance in the EUV is considered to be the sum of all intensities emitted by the different features at and beyond the solar disk. Fig. 4 shows the synthetic spectrum calculated with the SRPM system in spherical symmetry compared to the observed spectrum measured with the EVE rocket instrument during a calibration flight on April 14, 2008 (Chamberlin *et al.* 2009). For details on the spectral synthesis see Haberreiter & Fontenla (2009). We are currently in the process of calculating intensity spectra for various types of coronal features. Based on the feature identification from EIT images, this will finally allow us to calculate the irradiance in the EUV.

Depending on the wavelength, the EUV is emitted by different layers, i.e. the chromosphere, transition region, or corona. SOHO/EIT (Delaboudinière *et al.* 1995) observes at four wavelengths, i.e. 17.1 nm, 19.5 nm, 28.4 nm, and 30.4 nm. The wavelength range at 30.4 nm represents the cooler chromosphere, while 28.4 nm represents the hotter corona. It is clear that the feature identification in these regimes has to be different than in the case of the PSPT images. Nevertheless, the intensity distribution in each image is a powerful measure to define thresholds that are characteristic for the chromospheric and coronal features. Fig. 5 shows the possibility to identify coronal features from a SOHO/EIT image observed on July 30, 2007 at 17.1 nm. The gray-scale is the logarithm of the intensity normalized to its median. The colored contour lines are thresholds that define coronal features such as coronal and equatorial holes (blue), quiet corona (green), coronal network (yellow), and hot and super-hot corona (orange and red). As the area covered by

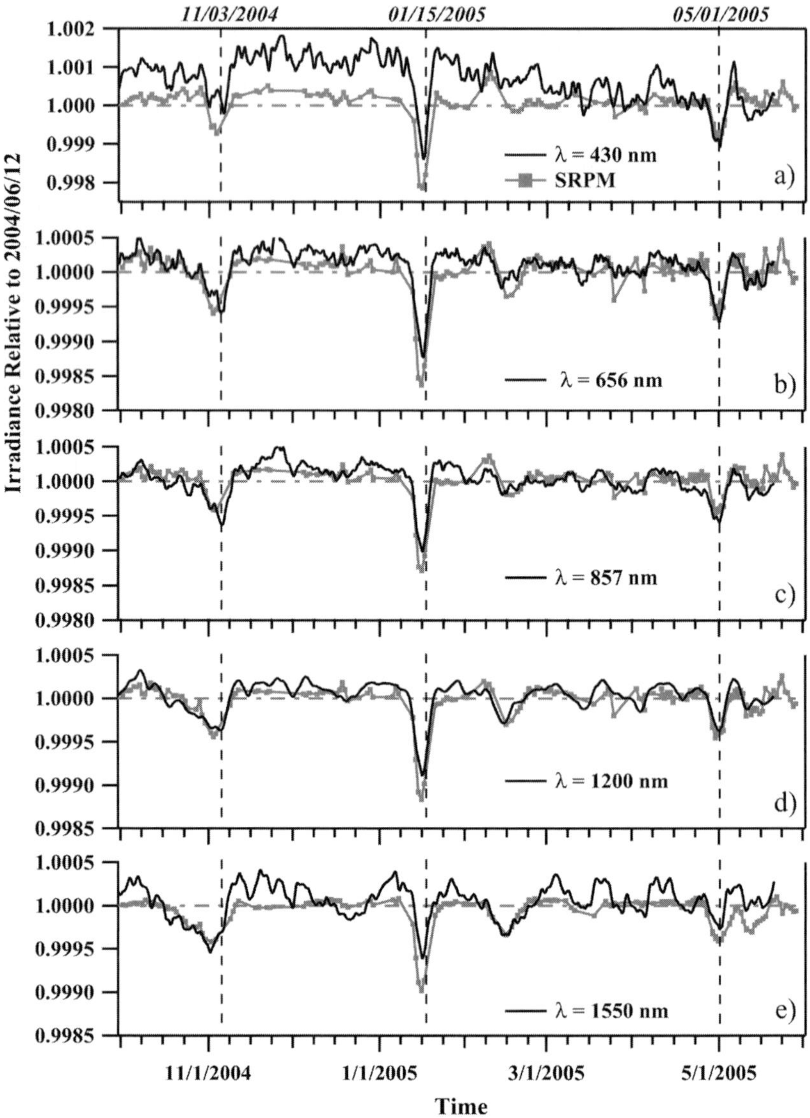

Figure 3. Comparison of the SRPM reconstruction based on PSPT images and SIM/SORCE spectral irradiance measurements for different wavelengths (Harder *et al.* 2005a). The relative changes of the SSI are in solid agreement with the observation, indicating that there is a firm understanding of the SSI variability at these wavelengths.

these features changes over the solar cycle, so does the disk-integrated radiation emitted by the Sun. In the EUV a difficulty arises from the fact that with increasing distance the extended corona gets increasingly optically thin, which makes a feature identification more problematic. Furthermore, it is clear that a spherical line-of-sight integration that takes into account the extended corona is very important.

4. Discussion

Comprehending the irradiance variations ultimately depends on understanding the heating mechanisms that are responsible for an enhanced brightness of the different

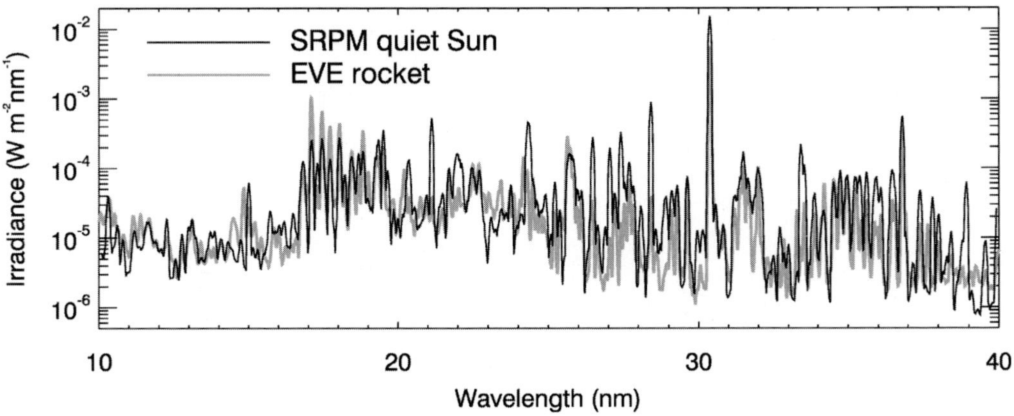

Figure 4. Comparison of the synthetic spectrum (thin black line) with the spectrum observed with the EVE instrument during a rocket calibration flight on April 14, 2008 (thick gray line). Adopted from Haberreiter & Fontenla (2009).

features on the solar disk. It is understood that the reason why sunspots are dark is due to the suppression of convective heat flow in the presence of strong magnetic field. On the other hand quiet and active network, plage, and faculae show an increased contrast. All these features are associated with enhanced magnetic field.

However, the contrast of magnetic elements is a function of wavelength. Sunspots show a negative contrast in the visible wavelength ranges but are typically bright in the UV, in particular in Lyman α. The fact that the contrast of solar surface features changes as a function of wavelength indicates that different heating processes are involved in different layers of the solar atmosphere.

The 1D atmosphere structures by Fontenla *et al.* (2009b) are suitable to reproduce solar spectra at a moderate spacial resolution of 2" from the FUV to the IR. These structures are time-independent representations of an average solar atmosphere, and as such they cannot account for any of the possibly dynamic processes involved in the chromospheric and coronal heating. The additional strength of these models is, however, that they might serve as a diagnostic tool to determine how much heating needs to be provided by reasonable mechanisms in order to sustain certain temperatures in different layers of the solar atmosphere and at the same time balance the radiative losses.

Some of the proposed heating mechanisms are low-frequency acoustic waves (Jefferies *et al.* 2006), the Farley-Buneman instability (Fontenla *et al.* 2008), coronal heating from Alvén waves Cranmer *et al.* (2007), nanoflares Patsourakos & Klimchuk (2009), and heating processes of the corona that are rooted in the chromosphere (De Pontieu *et al.* 2009). Ultimately, to fully understand the physics of the solar atmosphere, the physical processes of various heating mechanisms need to be included in the forward modeling, which eventually will allow us to reproduce solar spectra over a broad wavelength range. Additionally, the heating mechanisms also have to explain the solar cycle variability. Here, the physical properties of the 1D atmosphere structures offer good diagnostics about the temperature and pressure in certain regimes of the solar atmosphere. On the other hand, forward modeling using 3D MHD codes allows accounting for dynamic heating processes. Combining the insight that has been gained from both approaches has certainly great potential to advance our understanding of the physics of the solar atmosphere, and ultimately the variability of the TSI and SSI.

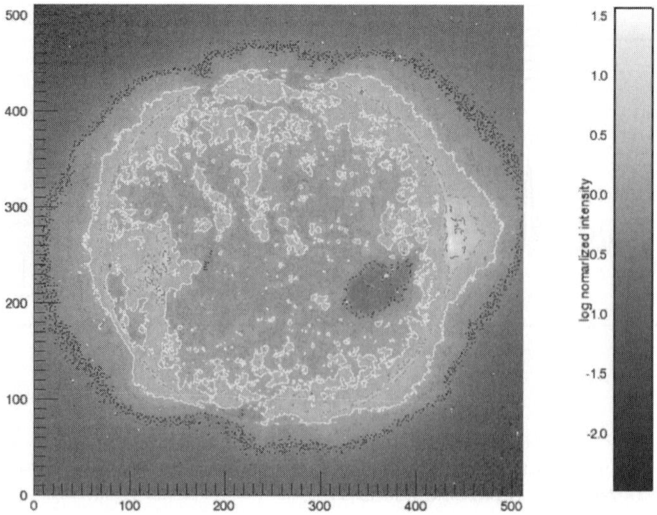

Figure 5. Coronal features identified from a SOHO/EIT image take at 17.1 nm. The contour levels correspond to -0.3 (blue dotted line, coronal hole), 0 (green dashed line, quiet corona), 0.2 (yellow solid line, coronal network), 0.5 (orange dashed-dotted line, hot corona) and 0.75 (red long dashes, super-hot corona) of the logarithm of the intensity normalized to the median. These contour levels provide a good first measure to identify different coronal features.

5. Conclusions

Our understanding of the TSI and SSI variations over shorter and longer timescales has advanced substantially over the past decades. Starting with solar activity proxies to reproduce these variations, we are now able to use detailed radiative transfer calculations to account for the irradiance variations. One of the remaining open questions is whether the quiet Sun radiance shows a long-term trend. If this is confirmed, then the current reconstruction models need to be refined to account for this effect.

The increased precision of the TSI and SSI measurements builds the base for the understanding of the processes that are behind its variability. The upcoming irradiance measurements by Picard/PREMOS (Schmutz *et al.* 2009), PROBA2/LYRA (Hochedez *et al.* 2006), and SDO/EVE (Woods *et al.* 2006) will certainly improve our understanding of the TSI and SSI variability. SDO/AIA will provide images with an unprecedented combination of spatial and temporal resolution, and thus be very valuable for advancing our understanding of the physics behind the irradiance variations.

Acknowledgements

MH gratefully acknowledges travel support by the International Astronomical Union. MH also thanks the Solar Influences Group at LASP for helpful discussions.

References

Austin, J., Tourpali, K., Rozanov, E., Akiyoshi, H., Bekki, S., Bodeker, G., Brühl, C., Butchart, N., Chipperfield, M., Deushi, M., Fomichev, V. I., Giorgetta, M. A., Gray, L., Kodera, K., Lott, F., Manzini, E., Marsh, D., Matthes, K., Nagashima, T., Shibata, K., Stolarski, R. S., Struthers, H., & Tian, W. 2008, *Journal of Geophysical Research (Atmospheres)*, 113, 11306

Beer, J. 2000, *Space Science Reviews*, 94, 53

Brueckner, G. E., Edlow, K. L., Floyd, L. E., Lean, J. L., & Vanhoosier, M. E. 1993, *Journal of Geophysical Research*, 98, 10695

Chamberlin, P. C., Woods, T. N., Crotser, D. A., Eparvier, F. G., Hock, R. A., & Woodraska, D. L. 2009, *Geophysical Research Letters*, 36, 5102.

Coulter, R. L., Kuhn, J. R., & Lin, H. 1996, *Bulletin of the American Astronomical Society*, 28, 912

Cranmer, S. R., van Ballegooijen, A. A., & Edgar, R. J. 2007, *Astrophysical Journal Supplement Series*, 171, 520

Delaboudinière, J. P., Artzner, G. E., Brunaud, J., Gabriel, A. H., Hochedez, J. F., Millier, F., Song, X. Y., Au, B., Dere, K. P., Howard, R. A., Kreplin, R., Michels, D. J., Moses, J. D., Defise, J. M., Jamar, C., Rochus, P., Chauvineau, J. P., Marioge, J. P., Catura, R. C., Lemen, J. R., Shing, L., Stern, R. A., Gurman, J. B., Neupert, W. M., Maucherat, A., Clette, F., Cugnon, P., & van Dessel, E. L. 1995, *Solar Physics*, 162, 291

De Pontieu, B., McIntosh, S. W., Hansteen, V. H., & Schrijver, C. J. 2009, *Astrophysical Journal*, 701, L1.

Domingo, V., Ermolli, I., Fox, P., Fröhlich, C., Haberreiter, M., Krivova, N., Kopp, G., Schmutz, W., Solanki, S. K., Spruit, H. C., Unruh, Y., & Vögler, A. 2009, *Space Science Reviews*, 145, 337

Egorova, T., Rozanov, E., Manzini, E., Haberreiter, M., Schmutz, W., Zubov, V., & Peter, T. 2004, *Geophysical Research Letters*, 31, 6119

Ermolli, I., Berrilli, F., & Florio, A. 2003, *Astronomy and Astrophysics*, 412, 857

Fontenla, J., White, O. R., Fox, P. A., Avrett, E. H., & Kurucz, R. L. 1999 *Astrophysical Journal*, 518, 480

Fontenla, J. & Harder, G. 2005, *Memorie della Societa Astronomica Italiana*, 76, 826

Fontenla, J. M., Peterson, W. K., & Harder, J. 2008, *Astronomy and Astrophysics*, 480, 839

Fontenla, J. M., Quémerais, E., González Hernández, I., Lindsey, C., & Haberreiter, M. 2009a, *Advances in Space Research*, 44, 457

Fontenla, J. M., Curdt, W., Haberreiter, M., Harder, J., & Tian, H. 2009b, *Astrophysical Journal*, accepted for publication

Fröhlich, C. & Lean, J. 2004, *The Astronomy and Astrophysics Review*, 12, 273

Fröhlich, C. 2009, *Astronomy and Astrophysics*, 501, L27

Fuller-Rowell, T., Solomon, S., Roble, R., & Viereck, R. 2004, in: J. M. Pap, P. Fox, C. Fröhlich, H. S. Hudson, J. Kuhn, J. McCormack, G. North, W. Sprigg, & S. T. Wu (eds.), *Solar Variability and its Effects on Climate. Geophysical Monograph 141*, p. 141

Haberreiter, M., Krivova, N. A., Schmutz, W., & Wenzler, T. 2005, *Advances in Space Research*, 35, 365

Haberreiter, M., Schmutz, W., & Hubeny, I. 2008a, *Astronomy and Astrophysics*, 492, 833

Haberreiter, M., Schmutz, W., & Kosovichev, A. G. 2008b, *Astrophysical Journal*, 675, L53

Haberreiter, M. & Fontenla, J. 2009, in: I. Hubeny, J. M. Stone, K. MacGregor, & K. Werner (eds.), *Recent Directions in Astrophysical Quantitative Spectroscopy and Radiation Hydrodynamics*, p. 355

Harder, J., Fontenla, J., White, O., Rottman, G., & Woods, T. 2005a, *Memorie della Societa Astronomica Italiana*, 76, 735

Harder, J., Lawrence, G., Fontenla, J., Rottman, G., & Woods, T. 2005b, *Solar Physics*, 230, 141

Harder, J. W., Fontenla, J. M., Pilewskie, P., Richard, E. C., & Woods, T. N. 2009, *Geophysical Research Letters*, 36, 7801

Hochedez, J.-F., Schmutz, W., Stockman, Y., Schühle, U., Benmoussa, A., Koller, S., Haenen, K., Berghmans, D., Defise, J.-M., Halain, J.-P., Theissen, A., Delouille, V., Slemzin, V., Gillotay, D., Fussen, D., Dominique, M., Vanhellemont, F., McMullin, D., Kretzschmar, M., Mitrofanov, A., Nicula, B., Wauters, L., Roth, H., Rozanov, E., Rüedi, I., Wehrli, C., Soltani, A., Amano, H., van der Linden, R., Zhukov, A., Clette, F., Koizumi, S., Mortet, V., Remes, Z., Petersen, R., Nesládek, M., D'Olieslaeger, M., Roggen, J., & Rochus, P. 2006, *Advances in Space Research*, 37, 303

Jefferies, S. M., McIntosh, S. W., Armstrong, J. D., Bogdan, T. J., Cacciani, A., & Fleck, B. 2006, *Astrophysical Journal*, 648, L151

Krivova, N. A., Solanki, S. K., Fligge, M., & Unruh, Y. C. 2003, *Astronomy and Astrophysics*, 399, L1

Krivova, N. A., Solanki, S. K., Wenzler, T., & Podlipnik, B. 2009, *Journal of Geophysical Research (Atmospheres)*, 114, 1

Kurucz, R. L. 1991, in: *NATO ASIC Proc. 341: Stellar Atmospheres - Beyond Classical Models*, p. 441

Lindsey, C. & Braun, D. C. 2000, *Science*, 287, 1799

Lockwood, M. & Fröhlich, C. 2007, *Proceedings of the Royal Society A*, 463, 2447

Patsourakos, S. & Klimchuk, J. A. 2009, *Astrophysical Journal*, 696, 760

Quémerais, E. & Bertaux, J. L. 2002, *Geophysical Research Letters*, 29, 2

Rottman, G. J., Woods, T. N., & Sparn, T. P. 1993, *Journal of Geophysical Research*, 98, 10667

Rottman, G. 2005, *Solar Physics*, 230, 7

Rozanov, E., Egorova, T., Schmutz, W., & Peter, T. 2006, *Journal of Atmospheric and Solar-Terrestrial Physics*, 68, 2203

Schmidtke, G., Fröhlich, C., & Thuillier, G. 2006, *Advances in Space Research*, 37, 255

Schmutz, W., Fehlmann, A., Hülsen, G., Meindl, P., Winkler, R., Thuillier, G., Blattner, P., Buisson, F., Egorova, T., Finsterle, W., Fox, N., Gröbner, J., Hochedez, J.-F., Koller, S., Meftah, M., Meisonnier, M., Nyeki, S., Pfiffner, D. Roth, H., Rozanov, E., Spescha, M., Wehrli, C., Werner, L., & Wyss, J. U. 2009, *Metrologia*, 46, 202

Schöll, M., Steinhilber, F., Beer, J., Haberreiter, M., & Schmutz, W. 2007, *Advances in Space Research*, 40, 996

Schühle, U., Wilhelm, K., Hollandt, J., Lemaire, P., & Pauluhn, A. 2000, *Astronomy and Astrophysics*, 354, L71

Shapiro, A. I., Schmutz, W., Schöll, M., Haberreiter, M., & Rozanov, E. 2009, *Astronomy and Astrophysics*, submitted

Solanki, S. K., Krivova, N. A., & Wenzler, T. 2005, *Advances in Space Research*, 35, 376

Steinhilber, F., Beer, J., & Fröhlich, C. 2009, *Geophysical Research Letters*, 36, 19704

Thuillier, G., Hersé, M., Labs, D., Foujols, T., Peetermans, W., Gillotay, D., Simon, P. C., & Mandel, H. 2003, *Solar Physics*, 214, 1

White, O. R., Fox, P. A., Meisner, R., Rast, M. P., Yasukawa, E., Koon, D., Rice, C., Lin, H., Kuhn, J., & Coulter, R. 2000, *Space Science Reviews*, 94, 75

Willson, B. C. 1994, in: J. M. Pap, C. Frolich, H. S. Hudson, & S. Solanki (eds.), *Proceedings from IAU Colloquium 143: The Sun as a Variable Star: Solar and Stellar Irradiance Variations*, p. 54

Woods, T. N., Tobiska, W. K., Rottman, G. J., & Worden, J. R. 2000, *Journal of Geophysical Research*, 105, 27195

Woods, T. N., Lean, J. L., & Eparvier, F. G. 2006, in: N. Gopalswamy, & A. Bhattacharyya (eds.), *Proceedings of the ILWS Workshop*, p. 145

Solar and Stellar Variability: Impact on Earth and Planets
Proceedings IAU Symposium No. 264, 2009
A. G. Kosovichev, A. H. Andrei & J.-P. Rozelot, eds.

doi:10.1017/S1743921309992699

Large-scale patterns and 'active longitudes'

Vladimir N. Obridko

Institute of Terrestrial Magnetism, Ionosphere,
and Radio Wave Propagation RAS, Troitsk, Moscow Region, 142190 Russia
email: obridko@izmiran.ru

Abstract. The following aspects of the physics of large-scale solar magnetic fields are discussed: structure of large-scale fields (LSF) and connection with local fields; dynamo and origin of LSF; LSF cycle variation; meridional circulation and LSF; rotation of LSF; fine structure of the field in quiet regions and the concept of the pebble-shaped field; active longitudes, their manifestation in various solar indices, and dependence on the power of solar activity.

Keywords. Sun: magnetic fields, rotation, sunspots

1. Introduction

The majority of up-to-date high-resolution magnetograms (SOHO MDI, SOLIS) display the magnetic field as an aggregate of very fine features. This philamentary structure was foreseen long ago, and its direct observation nowadays is, undoubtedly, a major achievement in the experimental solar physics. The cluster model of sunspots is corroborated by helioseismic data, which has far reaching implications for the theory of magnetic field generation in the Sun.

On the other hand, we have, unexpectedly, lost clear understanding of the large-scale background field. To begin with, its very existence is disputable. There are reasons to believe that the field between the so-called 'kilogauss' tubes is absent. Then, however, the extended regions, obviously, dominated by the fields of one polarity can not be accounted for. High-resolution images only reveal these extended unipolar regions, but if we superimpose a low-resolution map on a high-resolution magnetogram, the polarities will coincide. So, the fine magnetic elements are not randomly distributed over the solar surface, but, rather, form vast regions where one or another polarity is dominating. This is most likely due to the influence of a large-scale (or global) weak magnetic field. Thus, we arrive at the conclusion that a relatively weak global field must exist, which somehow organizes the position of the fine-structure elements.

Both energy and hydrodynamic aspects of this process still remain obscure. We do not know whether background fields are the primary features or are formed as a result of decay of local fields. It is not clear how deep they go down; i.e., are they large-scale all over or only at the surface? Mathematically, the field extrapolation to the upper corona does not make distinction between these cases.

Many properties of the background fields are still vague and even puzzling. Besides, one must discriminate between the notions of background and global fields, which often, is not that simple.

2. Magnetic dipole

The effective magnetic dipole is the base of the large-scale structure. Its magnetic moment and direction are readily calculated from WSO observation data downloaded from (http://quake.stanford.edu/wso/wso.html). The analysis performed by Livshits and Obridko (2006) has shown that the total magnetic moment occasionally decreases significantly, never turning to zero. During one or two years in the declining phase of the activity cycle, the magnetic moments of the vertical (aligned with the solar rotation axis) and horizontal (aligned with the solar equator) dipoles are comparable. This situation is known in astrophysics as the oblique rotator.

During the epoch of the solar minimum, the pole of the dipole executes relatively regular precession motions completing 1–2 circles around the solar rotation axis. This quasi-precession lasts for 1–3 years. Then, the dipole makes a jump to the equatorial zone for 0.7–1.2 years and continues a smooth longitudinal motion for another 1.5–3 years. After that, a new jump occurs, and the 'precession' goes on around the opposite pole of the Sun.

Large-scale fields rotate in anti-correlation with the solar activity indices. The rotation of the global magnetic field decelerates as the activity increases both in the 11-year cycle and in longer term cycles. The maximum rotation periods were recorded in the middle of the XX century, in the extremely powerful cycles 18 and 19 (Obridko and Shelting 2001).

The discrepancy between the rotation of local and global fields may be important in studying the mechanism of dynamo. The Carrington rotation manifests the rotation of the existing sunspots, while the global field may reflect the rotation rate of the sunspot generation region and may affect the distribution of sunspots on the disk.

3. Fine structure of the background field

An interesting feature was revealed lately by the analysis of high-resolution SOHO/MDI observations. It might seem that the background field, which, by definition, lies beyond the local fields, must be quasi-uniform. In fact, it consists of fine elements and looks like a neat pebble beach.

Another peculiarity is that the field in these elements is virtually horizontal (see Fig. 1). The center-to-limb variation for the background field as a whole is positive; i.e., the observed values decrease to the limb as they must do in a quasi-radial field. For the weakest fields considered to be small-scale elements of the background field, the variation is negative suggesting that these fields are mainly tangential (Ioshpa *et al.*, 2009). A similar result was obtained in the very first measurements of the full vector outside the active regions, but it was attributed to low sensitivity in measuring transversal fields. At

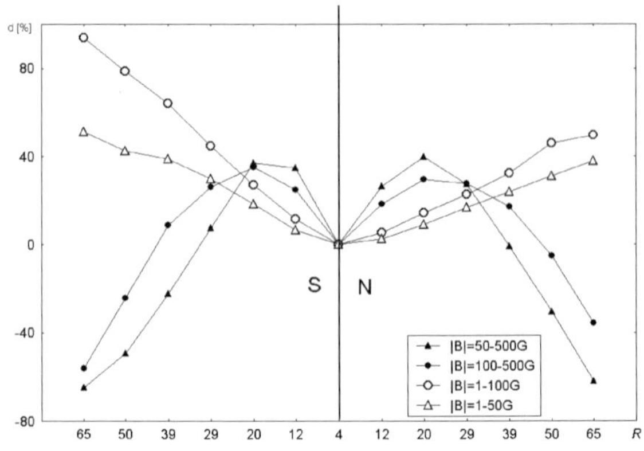

Figure 1. Center-to-limb variation of SOHO/MDI magnetic fields of different scales.

present, it is established for sure that the field in the small-scale elements is tangential. The elements of the background field resemble the windfall of trees in the forest.

The existence of these elements may be due to the fact that some physical processes take place inside a thin double layer located at about 0.995 R, which is called **leptocline** (Lefebvre *et al.* 2007).

4. Meridional motions

Obridko and Shelting (2003) revealed mutually opposite motions of large-scale and local fields, whose rates were not constant. They interpreted these results as follows. Magnetic fields are generated at the base of the convection zone. The wave of generation moves from the mid latitudes to the equator. The rapidly emerging concentrated local fields follow the wave of generation and form a butterfly diagram, being probably, additionally enhanced in the leptocline region. On the other hand, the slowly emerging diffuse large-scale fields loose connection with the wave of generation and are driven to the poles by the meridional drift.

In addition to the mean meridional flow from the equator to the poles, Švanda *et al.* (2007) found extra meridional circulation cells converging towards the activity belts and migrating to the equator as the solar cycle progresses.

The speed of the magnetic flux transport towards the solar poles may differ significantly from the longitudinally averaged speed of the meridional flow. Therefore, the longitudinal structure of these flows should be taken into account.

In the past decades, the study of cyclic variation of solar activity, i.e., regularities that govern the 11-year and 22-year (magnetic) cycles as a whole, is giving place to a detailed study of their internal fine structure. These investigations have revealed a great number of phenomena whose nature is not entirely clear. Besides the cycle variation, all parameters of solar activity experience quasi-periodic oscillations with shorter periods ranging from seconds to years. Of particular interest are oscillations with a period of the order of 1–2 years.

5. Variation of solar indices with a period of 1.3 years

Several studies are devoted to oscillations with a period of 1.3 years revealed at the bottom of the solar convection zone. Komm *et al.* (2002, 2003, 2004) and Howe *et al.* (2000) found out that the 1.3-year periodicity at the base of the convection zone is most pronounced in the solar rotation rate at the equator of the tachocline. The 1.3-year variation of the solar rotation rate in the vicinity of the tachocline was also revealed by Schou (2003). Komm *et al.* (2006) and Kosovichev (2003) did not find evidence of the 11-year cycle near the tachocline, where the 1.3-year oscillations of the solar rotation rate were most noticeable. As shown by Christensen-Dalsgaard (2006), the solar rotation is not stationary: regions of faster and slower rotation alternate in the narrow tachocline region with the mean amplitude increasing towards the solar equator. It is also noted that the amplitudes of variation of the rotation rate and solar activity correlate (see Toomre *et al.*, 2003). A series of work (McDonald *et al.*, 2005; Cadavid *et al.*, 2005) was devoted to the study of quasi-periodicity in fluctuations of the axisymmetric solar magnetic field with the use of the main component and independent component analyses. The oscillation periods of the order of 1.0–1.5 years were considered to be typical of the polar and high-latitude fields and the periods of 1.3 and 1.7 years, to characterize the mid- and low-latitude solar events. The wavelet analysis carried out by Krivova and Solanki (2002) revealed 1.3-year periodicities in the tracers of emerging flux at the solar surface, in particular, sunspots. The power of these variations was observed to vary strongly with time.

Using Mt.Wilson magnetographic data for 1986–1994, in particular, rotation velocity measurements, Javaraiah (2003) found that the 1.3-year periodicity was dominant in the solar surface rotation rate. Özgüç *et al.* (2003) used the Fourier and wavelet transforms to show that variations with a period of 1.3 years were present in the daily flare index. It turned out that the amplitude of short-term oscillations varied strongly with time. Rybák and Karlovský (2003) arrived at a similar conclusion. They isolated a few unstable periodicities in the Wolf number variation for 1850–2002, including the period of 1.3 years. Ikhsanov and Ivanov (2004) revealed the same periodicity in the large-scale solar magnetic field from Stanford (1976–2004) and Kitt Peak (1970–1984) magnetic data. The spherical harmonic coefficients of radial magnetic field for the axisymmetric and non-axisymmetric modes were calculated in (Knaack *et al.*, 2005; Knaack and Stenflo, 2005; Knaack, 2005) from Mt.Wilson and Kitt Peak daily magnetograms, and a few modes were isolated in the variation of the photospheric magnetic field. It was shown that these modes differ in different activity cycles; e.g., the 1.3-year periodicity is better pronounced in the even cycles (20, 22) than in the odd ones (21, 23).

Mendoza *et al.* (2006) studied the periodicities in various types of the solar magnetic flux (total, closed, open, low- and high-latitude open fluxes). All fluxes fluctuate with a period of 1.7 years. The medium-term, one-year fluctuation is significantly present in the total and closed fluxes, but it is less important in the open fluxes. Due to the uncertainties involved in estimating the exact period of the quasi-annual peak, the authors could not differentiate it from the 1.3-year periodicity. In their opinion, this high-frequency component is in phase with the 11-year solar cycle. Wang and Sheeley suggested that 1.3-year oscillations could occur as a result of stochastic interaction of local fields and meridional flows (Wang, 2004; Wang and Sheeley, 2003).

It should be noted that the studies using the data on large-scale magnetic field are necessarily based on relatively short observation series covering 2–3 solar cycles. Direct helioseismic observations are even shorter. An exception is the work by Makarov *et al.* (2002) where the authors calculated the photospheric magnetic field by an original method using a long series of observations of H-alpha filaments and prominences from 1915 to 2000. The revealed 1.3-year periodicity in latitudinal oscillations of the zonal boundaries of the large-scale magnetic field (neutral lines) was associated with oscillations of the solar rotation rate in the tachocline region. It was shown that this latitudinal oscillation was weak in the period 1950–1970 in the northern hemisphere, but was clearly pronounced in 1960 in the southern hemisphere. The authors did not find significant correlation of this periodicity with the 11-year cycle.

Gulyaev (2006) has demonstrated that these oscillations coincide with oscillations of the heliospheric current sheet. This result was corroborated by Obridko and Shelting (2007).

Livshits and Obridko (2006) analyzed the periodic components of the rotation of the effective solar dipole. They found out that the solar dipole can be resolved into the 'vertical' (co-axial) and horizontal components. Only two frequencies can be identified in the variation spectrum of the magnetic moments of these dipoles. The quasi-11-year cycle is absolutely similar for the magnetic moments of the horizontal and vertical dipoles. The situation is quite different if we consider oscillations with a period of 1.3–2.5 years, so called QBO. These periods are only present in the magnetic moment of the horizontal dipole and are absolutely absent in the case of the vertical dipole.

In the recent years, the oscillation periods of 1.3 years are not as clearly identified in helioseismic measurements as they were earlier (Howe, 2009). This raises the question of how reliable they are and how often they occur in the variation spectrum of different solar parameters. Obridko and Shelting (2007) have shown that oscillations with a period of 1.3 years usually vanish in the declining phase of the 11-year cycle. Below, in Section 9, these oscillations are treated as a short-term acceleration of the global field.

6. Tentative conclusions. What we do not know?

We do not fully understand the nature and evolution of large-scale field but it is sure enough that this is not a mere result of low-resolution observations.

It is not clear how the weak large-scale fields can control the organization of stronger local fields.

We do not know whether the fields of different scales are generated at the same or at different depths.

Some physical processes take place inside a thin double layer, located near the photosphere

7. Active longitudes

The active longitudes were first discovered by Wolfer in 1897, but they still remain the object of heated debates. It should be noted that the term 'active longitudes' without the particular index specified is quite indefinite. Obviously, the active longitudes for sunspots are not the same as the active longitudes for coronal holes. In this work, we shall deal mainly with the active longitudes for sunspots.

Below, we shall mention only some publications on active longitudes to illustrate the main contradictions. The tendency of the solar cycle to manifest itself at some preferred longitudes was found by Benevolenskaya *et al.* (1999) and Bumba *et al.* (2000). The data on the occurrence of sunspots confirm the existence of two preferred longitudes separated by 180°, which migrate in a fixed rotation frame but are persistent throughout 120 years. Their differential rotation differs significantly from that of individual spots. This implies that the depth at which sunspots are formed (and affected by the non-axisymmetric component of the field) is not the same as the anchoring depth of developed sunspots (I. G. Usoskin, S. V. Berdyugina and J. Poutanen (2005).

The lifetime of the sunspot formation zones exceeds significantly the lifetimes of individual spots and may reach 15–20 rotations. The stability of the active longitudes is larger for the larger spots (Ivanov, 2007). This is superimposed by a more intricate system of longitudes that rotate at different velocities.

The system of active longitudes with the rotation period $P \sim 27$ days prevails in the most intensive 11-year activity cycles. The system of active longitudes with $P \sim 28$ days is best pronounced at the minimum of the secular cycle. The most stable magnetic structures at a quasi-source surface in the solar corona are separated by $\sim 180°$ in heliographic longitude and are similar to dipolar structures. The nature and behavior of these large-scale magnetic patterns are interpreted as a superposition of cyclic dynamo modes and the nonaxially symmetric relic field of the Sun (Mordvinov and Kitchatinov, 2004; Kitchatinov and Olemskoi, 2005).

This result is in contradiction with the aforementioned deceleration of the global magnetic field in cycles 18–19 (Obridko and Shelting, 2001). However, the rotation periods of global and local fields may depend on activity in different way.

The occurrence of the active longitudes is considered in a number of theoretical papers. Bigazzi and Ruzmaikin (2003) explain the Suns preferred longitudes as a coupling of the magnetic dynamo modes. Elstner and Korhonen (2005) believe that, 'to explain

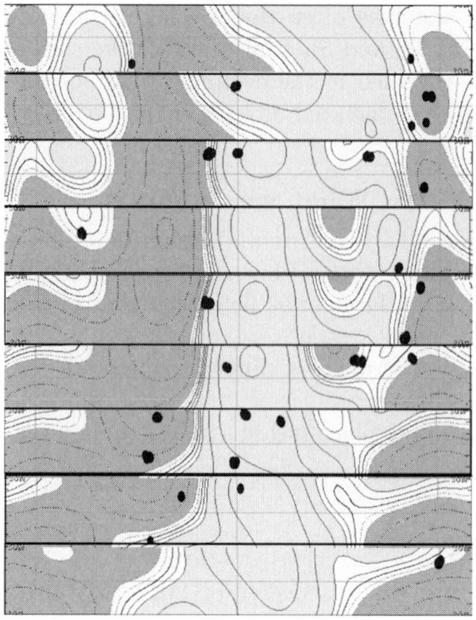

Figure 2. Synoptic maps of large-scale magnetic field (Carrington rotations 1676–1684) with the positions of major active regions marked.

this phenomenon, a non-axisymmetric dynamo mode, giving rise to two permanent active longitudes at opposite stellar hemispheres, is needed together with an oscillating axisymmetric magnetic field'. Axel Brandenburg and Petri J. Käpylä (2005) solve a two-dimensional mean field dynamo model where magnetic helicity conservation is fully included. The model develops longitudinal variability with activity patches traveling in longitude. These patches may be associated with active longitudes.

On the other hand, the existence of two active longitudes was questioned by various authors or was denied altogether, at least, over long time intervals. Losh (1938) found that there is one maximum and one minimum per rotation, and that the phase of the maximum is stable over a solar cycle. Balthasar and Schüssler (1984) investigated sunspot numbers and showed that the phase of the maximum remains more or less the same for two solar cycles and then, it changes by about 180°. Balthasar (2007) revealed only one maximum in the Fourier spectrum, i.e., one active longitude. Knaack & Stenflo (2005) and Knaack *et al.* (2005) used Mt.Wilson and Kitt Peak magnetograms and found that the dominant rotation periods were different for cycles 21, 22 and 23. Bouwer (1992) investigated various indicators of solar activity and found that precise periods between 27 and 28 days persist only for a short time, sometimes only for a few solar rotations. Pelt *et al.* (2006) claim that strong and well substantiated evidence for an essential and century-scale persistent nonaxisymmetry in the sunspot distribution does not exist. Henney and Durney (2007) found a surprisingly non-negligible likelihood, approximately 1 in 3, that observed periodicities from integrated full-disk solar parameters are a chance occurrence for time series of the order of 20 years in duration.

8. Active longitudes and structure of the large-scale field

The success in detecting the active longitudes depends on many factors, in particular, on the intensity of events and on the shift of the large-scale field boundaries. As the latter, we can assume the heliospheric equator. Such a study was performed by Bumba and Obridko (1969). They analyzed the positions of major proton complexes relative to the sector boundaries (Bartels active longitudes) and arrived at the following conclusions:

1. The flare activity and especially the proton flare activity is concentrated in the zones of 'Bartels' active longitudes.

2. The flare activity and especially the proton flare activity is concentrated in the neighborhood closest to the sector boundaries.

3. It seems that the concentration of flare activity around the 'active longitudes' as well as around the sector boundaries increases with the importance of the event. The highest degree of concentration may be seen for the proton-flare regions.

Later, these conclusions were verified more than once and were corroborated on the whole, particularly, when major groups were involved.

Figure 2 shows the position of the heliospheric equator and some major active regions. About 70% of spots with the area > 500 m.p.h are located at a distance less than 20° in longitude from the neutral line of the large-scale field.

However, the longitude of the axis of the effective solar dipole (and the longitude of the associated neutral line) changes with the course of time.

Now, let us consider how the Carrington longitude of the north pole of the effective magnetic dipole changes over a long time interval. Figure 3 illustrates the time variation of this coordinate for 29 years from 1977 to 2005.

One can see that the Carrington longitude of the north pole of the effective magnetic dipole increases gradually by about 1.5° for one Carrington rotation and by 1.5° for a year. This means that the rotation period of the global magnetic field is 27.1621 days, i.e., is by 0.1132 day smaller than the Carrington rotation rate. After approximately one Hale cycle, the system of longitudes returns to its initial position.

This is seen from the longitude-latitude diagram in Fig. 4.

The dotted curve represents the motion of the north pole of the effective solar dipole in the Carrington reference frame. The cycle minima are marked. The black solid curve is approximation by the 5-order polynomial. Note that transition from one hemisphere to another occurs at approximately the same longitudes.

Thus, the magnetic dipole restores both its magnitude and position every 22 years after having passed through all Carrington longitudes and latitudes. The associated reference frame executes a similar periodic motion.

The failure in identifying the active longitudes is often due to the fact that the latter are rather associated with the rotation of the large-scale field than with Carrington

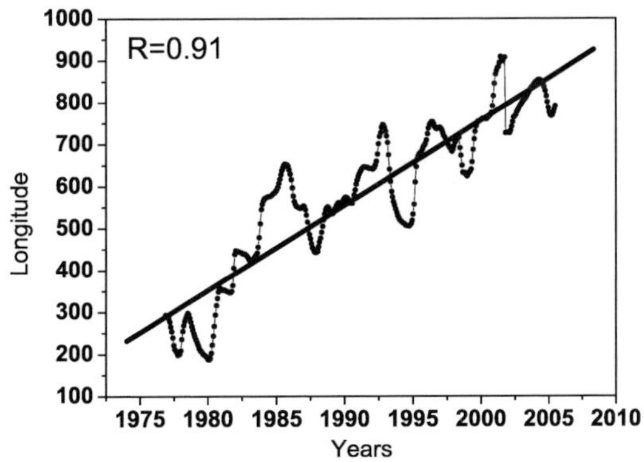

Figure 3. Time variation of the longitude of the north pole of the effective solar dipole

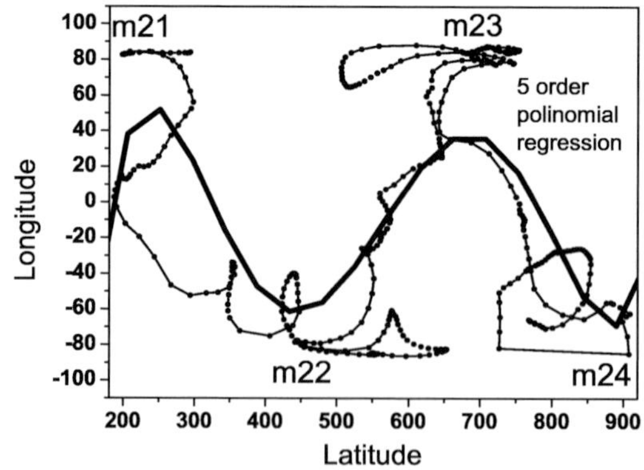

Figure 4. Longitude-latitude diagram of the rotation of the solar dipole

coordinates. Over a few years, the active longitudes in Carrington coordinates become unstable and are restored again in the following cycle.

9. Rotation of the solar effective dipole

Knowing the longitude shift of the pole, we can calculate the time dependence of the dipole rotation rate (Fig. 5). It is seen from the figure that, though the mean rotation rate of the dipole is only by 0.4% larger than the Carrington rotation rate, the dipole is occasionally accelerated to reach the velocity of 14.7°/day, which exceeds the Carrington rotation rate by 13.6% and corresponds to the syndic rotation period of 24 days. The solid curve shows approximation by the 4-order polynomial. As is readily seen, it is precisely these short-term accelerations that make the mean velocity in the relatively low cycles 21 and 23 exceed the mean velocity in the higher cycle 22 in agreement with (Obridko and Shelting, 2001).

It is interesting to note that the accelerations have a periodic character. Figure 6 represents the Fourier power spectrum of the rotation rate of the effective magnetic dipole. One can see that the acceleration occurs with a typical period of 1.4 years. In the declining phase of cycle 23, however, the acceleration is absent in agreement with the result by Howe (2009).

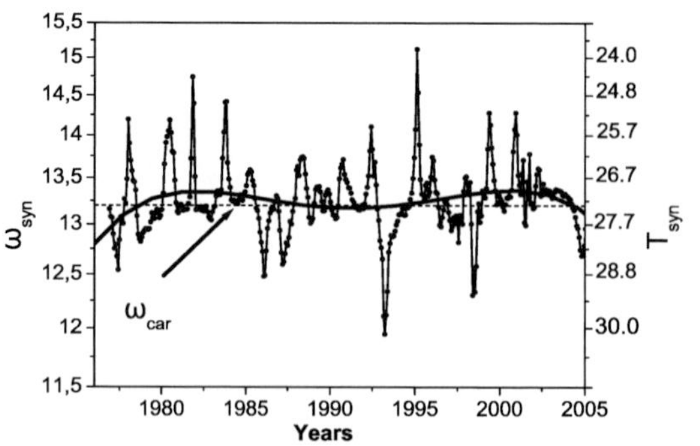

Figure 5. Time dependence of the synodic angular rotation rate of the solar effective dipole (ω_{syn}) and synodic period (T_{syn}). The dotted line shows the angular rate of Carrington rotation (ω_{car}).

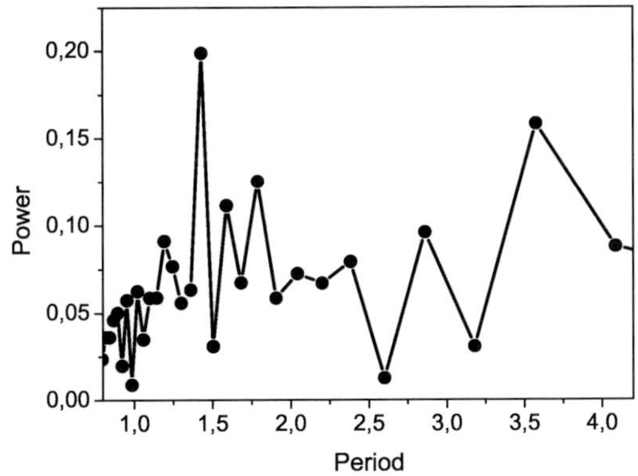

Figure 6. Fourier spectrum of the rotation rate of the solar dipole

10. General conclusions

All active events in the Sun are the result of interaction of the global (possibly poloidal) and axisymmetric (quadrupole-like) fields.

The term 'active longitudes' is not quite correct, since they are not constant either in time or space and are closely related to the structure of the large-scale field.

The active longitudes are connected with the large-scale field and are best pronounced in the most powerful events and major sunspot groups.

The work was supported by the Russian foundation for basic research, project no. 08-02-00070.

References

Balthasar, H. 2007, *A&A*, 471 (1), 281
Balthasar, H. & Schüssler, M. 1984, *Solar Phys.*, 93 (1), 177
Benevolenskaya, E. E., Kosovichev, A. G., & Scherrer, P. H. 1999, *Solar Phys.*, 190 (1/2), 145
Bigazzi, A. & Ruzmaikin, A. 2004, *ApJ*, 604 (2), 944
Bouwer, S. D. 1992, *Solar Phys.*, 142 (2), 365
Brandenburg, A. & Käpylä, P. J. 2005, *eprint arXiv:astro-ph/0512639*
Bumba, V., Garcia, A., & Klvana, M. 2000 *Solar Phys.*, 196 (2), 403
Bumba, V. & Obridko, V. N. 1969, *Solar Phys.*, 6 (1), 104
Cadavid, A. C., Lawrence, J. K., McDonald, D. P., & Ruzmaikin, A. 2005, *Solar Phys.*, 226 (2), 359
Christensen-Dalsgaard, J. 2006, in: *Convection in Astrophysics, International Astronomical Union. Symposium no. 239, held 21–25 August, 2006 in Prague, Czech Republic*, p. 30
Elstner, D. & Korhonen, H. 2005, *Astron. Nachr.* 2005, 326 (3), 278
Gulyaev, R. A. 2006, *Solar System Res.*, 40 (4), 368
Henney, C. J. & Durney, B. R. 2005, *ASP Conference Ser.*, 346, 381
Howe, R. 2009, Solar Phys., 6 (1), ???)
Howe, R., Christensen-Dalsgaard, J., Hill, F., Komm, R. W., Larsen, R. M., Schou, J., Thompson, M. J., & Toomre, J. 2000, *Science*, 287 (5462), 2456
Ikhsanov, R. N. & Ivanov, V. G. 2004, in: Stepanov, A. V., Benevolenskaya, E. E., & Kosovichev, A. G. (eds.) *Multi-Wavelength Investigations of Solar Activity* (Cambridge University Press, Cambridge, UK, IAU Symp., vol. 223), p. 629
Ioshpa, B. A., Obridko,V. N., & Chertoprud, V. E. 2009, *Astron. Lett.*, 35 (6), 424
Ivanov, E. V. 2007 *Adv. Sp. Res.*, 40, 959
Javaraiah, J. 2003, *Bull. Astron. Soc. India*, 31, 317
Kitchatinov, L. L. & Olemskoi, S. V. 2005, *Astron. Lett.*, 31 (4), 280

Knaack, R. 2005, *Global evolution of magnetic fields in the photosphere of the Sun during cycles 20–23* (Ph.D dissertation. 179 pages; Dr.sc.nat. Switzerland: Eidgenoessische Technische Hochschule Zuerich (Switzerland), Publication Number: AAT C820734. DAI-C 66/03), p. 653

Knaack, R. & Stenflo, J. O. 2005, *A&A*, 438 (1), 349

Knaack, R., Stenflo, J. O., & Berdyugina, S. V. 2005, *A&A*, 438 (3), 1067

Komm, R., Howe, R., Durney, B., & Hill, F. 2002, *Bull. American Astron. Soc.*, 34, 644

Komm, R., Howe, R., Durney, B., & Hill, F. 2003, *ApJ*, 586 (1), 650

Komm, R., Howe, R., & Hill, F. 2004, in: *COSPAR Scientific Assembly. Held 18–25 July 2004* Paris, France, p. 1397

Komm, R., Howe, R., & Hill, F. 2006, *Adv. Sp. Res.*, 38 (5), 845

Kosovichev, A. G. 2003, in: Wilson, A. (ed.), *Solar variability as an input to the Earth's environment. International Solar Cycle Studies (ISCS) Symposium, 23–28 June 2003, Tatranska Lomnica, Slovak Republic* (ESA SP-535, Noordwijk: ESA Publications Division, ISBN 92-9092-845-X), p. 795

Krivova, N. A. & Solanki, S. K. 2002, *A&A*, 394, 701

Lefebvre, S., Rozelot, J. P., & Kosovichev, A. G. 2007, *Adv. Sp. Res*, 40 (7), 1000

Livshits, M. A. & Obridko, V. N. 2006, *AZh*, 83 (11), 1031

Losh, H. M. 1939, *Publ. Obs. Univ. Michigan* , 7 (5), 127

Makarov, V. I., Tavastsherna, K. S., Tlatov, A. G., & Callebaut, D. K. 2002, in: Wilson, A. (ed.), *Solar variability: from core to outer frontiers. The 10th European Solar Physics Meeting, 9–14 September 2002, Prague, Czech Republic* (ESA SP-506, Vol. 1. Noordwijk: ESA Publications Division, ISBN 92-9092-816-6), p. 173

McDonald, D. P., Cadavid, A. C., Lawrence, J. K., & Ruzmaikin, A. 2005, in: *American Geophysical Union, Spring Meeting 2005* (abstract SP43B-05)

Mendoza, B., Velasco, V. M., & Valdés-Galicia, J. F. 2006, *Solar Phys.*, 233 (2), 319

Mordvinov, A. V. & Kitchatinov, L. L. 2004, *Astron. Rep.*, 48 (3), 254

Obridko, V. N. & Shelting B. D. 2001, *Solar Phys.*, 201, 1

Obridko, V. N. & Shelting, B. D. 2003, *Astron. Rep.*, 47 (4), 333

Obridko, V. N. & Shelting, B. D. 2007, *Adv. Sp. Res.*, 40 (7), 1003

Özgüç, A., Ataç, T., & Rybák, J. 2003, *Solar Phys.*, 214 (2), 375

Pelt, J., Brooke, J. M., Korpi, M. J., & Tuominen, I. 2006, *A&A*, 460 (3), 875

Rybák, J. & Karlovský, V. 2003, in: Wilson, A. (ed.), *Solar variability as an input to the Earth's environment. International Solar Cycle Studies (ISCS) Symposium, 23–28 June 2003, Tatranska Lomnica, Slovak Republic* (ESA SP-535, Noordwijk: ESA Publications Division, ISBN 92-9092-845-X), p. 145

Schou, J. 2003, *Bull. American Astron. Soc.*, 35, 854

Švanda, M., Zhao, J., & Kosovichev, A. G. 2007 *Solar Phys.*, 241 (1), 27

Toomre, J., Christensen-Dalsgaard, J., Hill, F., *et al.* 2003, in: Sawaya-Lacoste, H. (ed.), *Local and global helioseismology: the present and future. Proc. SOHO 12 / GONG+ 2002, 27 October–1 November 2002, Big Bear Lake, CA, USA* (ESA SP-517, Noordwijk, Netherlands: ESA Publications Division, ISBN 92-9092-827-1), p. 409

Usoskin, I. G., Berdyugina, S. V., & Poutanen, J. 2005, *A&A*, 441 (1), 347

Wang, Y.-M. 2004, *Solar Phys.*, 224 (1–2), 21

Wang, Y.-M. & Sheeley, N. R., Jr. 2003, *ApJ*, 590 (2), 1111.

Solar and Stellar Variability: Impact on Earth and Planets
Proceedings IAU Symposium No. 264, 2009
A. G. Kosovichev, A. H. Andrei & J.-P. Rozelot, eds.

doi:10.1017/S1743921309992705

Is there more global solar activity on the Sun?

J. X. Wang[1], Y. Z. Zhang[1], G. P. Zhou[1], Y. Y. Wen[1] and J. Jiang[2]

[1]Key Laboratory of Solar Activity, National Astronomical Observatories of China, CAS, A20 Datun Road, Chaoyang District, Beijing, 100012
email: wangjx@ourstar.bao.ac.cn

[2]Max Planck Institute for Solar System Research
email: jiangjie@ourstar.bao.ac.cn

Abstract. There appear indications of more global activity on the Sun which is larger, much beyond the scale of solar active regions (ARs). These indications include formation, flaring and eruption of the trans-equatorial loops seen in EUV and X-rays, formation and eruption of trans-equatorial filaments, global magnetic connectivity in EUV dimming associated with halo-coronal mass ejections, wide spread of radio burst sources in meter wavelength in the solar corona, and quasi-simultaneous magnetic flux emergence in both hemispheres seen during some major solar events. With examples of a few major events in the last solar cycle we discuss the possibility that there is large or global-scale activity on the Sun. Its spatial scale is many times larger than that of AR and temporal scale is over 10 hours. The exemplified trans-equatorial loops are anchored in ARs and their activity is temporally associated with flares in ARs too. In some sense the flares in ARs appear either as a part of or a precursor of the more global activity. It is likely that the combination of the flares in ARs and the associated global activity is responsible to the major solar-terrestrial events. More efforts in understanding the global activity are undertaken.

Keywords. Sun: activity, Sun: filaments, Sun: flares, Sun: magnetic fields.

1. Introduction

In the discussions of a few Sun-Earth connection events, there appear two extreme viewpoints to be discerned (see Culhane & Siscoe 2007). On one hand the flare/CME events may be related to quasi-simultaneous magnetic interactions that occur on a global scale given favorable magnetic topology (Wang *et al.*, 2006, 2007; Zhang *et al.* 2007; Zhou *et al.*, 2007). Alternatively, the eruptions may begin on the scale of an individual AR and then become global in nature owing to mechanisms of the kind described by Attrill *et al.* (2006), Harra *et al.* (2007), and Mandrini *et al.* (2007).

Although some large-scale CMEs are showing global signature on the solar surface, the question raised by van Driel-Gesztelyi *et al.* (2008) is penetrating: are these events having intrinsically large or global scale in nature or developing from small or AR scale by nurture. As summarized by these authors, some recent studies suggest that those CMEs are inherently in large-scale, so that their energy supply and final angular width originate from a large-scale region (see Zhang *et al.*, 2007; Wen *et al.* 2007; Wang *et al.*, 2007; Zhou *et al.* 2007; Zhukov and Veselovsky, 2007); while, other approaches consider that CMEs start from AR scale, and then evolve to become large-scale even in the low corona due to interaction between the expanding magnetic structure of the CMEs and other low-coronal magnetic structures (see Maia *et al.*, 1999; Pohjolainen *et al.*, 2001; Bemporad *et al.*, 2005; Attrill *et al.*, 2007a; Mandrini *et al.*, 2007; Moore and Sterling, 2007).

Put aside how small the initiation site of a CME is, our questions are: (1) whether or not there is a large- or global-scale magnetic structure which is involved in the CME process?

(2) Is there more global activity which goes together with flares/filament eruption in AR scale in the CME process?

To be accurate and clear, whenever we say large-scale of the CME source region, we simply mean that the material and magnetic flux of a CME come from global scale coronal structure. A CME may be triggered in an AR or in a filament channel, and in such a "small-scale" complex magnetic structure, it may manifest first as a flare. Moreover, multi-magnetic flux system and complex topology connections may be included in a CME source region, and magnetic interaction among them would, of course, be unavoidable.

In this paper, we exemplify a few major events in the last solar cycle, and try to answer if there are a large-scale magnetic structures involved in flare/CMEs, and if there is more global activity on the Sun which comprises flare and CME as its inseparable manifestation. In section 2 a few flare/CME events are exemplified, and then discussions are given on what we can learn from these observations.

2. Indications To Global Scale Solar Activity

2.1. *On the Halloween events in 2003*

Ceaseless magnetic activity happens on the Sun everywhere. However, is there a hierarchy of the Sun's magnetic activity (Zirin, 1985)? Are the different rank magnetic structures in the hierarchy similar? Is the small-scale magnetic activity the indispensable parts of the AR-scale activity? Similarly, is AR-scale activity a part of the larger scale magnetic activity? Do different levels of magnetic activity simultaneously happen or be inter-coupled or interacted each other? Whether we pay too more attentions to AR phenomena, or we simply isolate an AR from their large-scale field background (see Liu *et al.* 2007)? Here, a few major events are used to examine the possibility of large- or global-scale magnetic structures involved in a CME in the Sun's low corona.

The extraordinary episodes in October – November 2003 marked some of the most intense solar activity in the 23rd solar circle, manifesting extremely powerful X-class flares, fast coronal mass ejections (CMEs), solar energetic particle (SEP) events and so on. The commonly accepted idea is that each of the powerful solar eruptive events in the Halloween epoch of 2003 was related to a single AR, e.g., AR 10484, 10486, or 10488; the contribution of the larger magnetic topologies where an AR located was negligible.

Zhang *et al.* (2007) studied the global magnetic connectivity in CME associated dimmings in one of the major event on October 28. They find that though the CME-related flare happened in one AR, e.g., 10486, the CME-related coronal dimming spread across multiple magnetic topologies as shown in Figure 1a-1c. Brightening and dimming appear in both hemisphere, in ARs as well as under large-scale arcades (see Fig. 1(c)). The background image of Figure 1d is a part of MDI image at 12:48 UT on 28 Oct. 2003, in which eight quasi-simultaneously emerged EFRs are outlined by the pink lines. These EFRs do locate in the places with key magnetic connectivity associated with brightening and dimming. It has been proposed that a large-scale inter-coupled magnetic flux system built a magnetic connection among the different ARs, as well as, some large-scale magnetic structures (Zhou *et al.* 2007). The instability in the interactive multi-flux system is believed to be responsible for the generation of this and other flares/CMEs during this period. It supports the idea of some sub-photospheric link and global activity on the Sun for this Halloween epoch. These results show that the instability of the large or global magnetic structure, in which there was vigorous interaction of multi-scale flux system, would be the true origin of the prolific CME outbursts.

2.2. *On the Bastille Day event in 2000*

Bastille day flare/CME has been well studied. Wang *et al.* (2006) have demonstrated that this major event is not related to only one single AR, but the activation and eruption of a huge transequatorial filament are seen to precede the simultaneous filament eruption and flare in AR 9077, and the full halo-CME in the high corona. Wen *et al.* (2006) illustrate the wide spread of non-thermal radio bursts, i.e., the non-thermal electron sources, based on the NRH (Nancay radio-heliograph) data. The non-thermal electron sources cover almost the full range of longitude and wide range of latitude of the visible hemisphere. Radio Bursts 'B' and 'E' (see Fig. 2) are spatially correlated with the magnetic structure of the transequatorial filament. Moreover, the non-thermal electron sources spread more broadly than the EUV waves shown by SOHO/EIT observations.

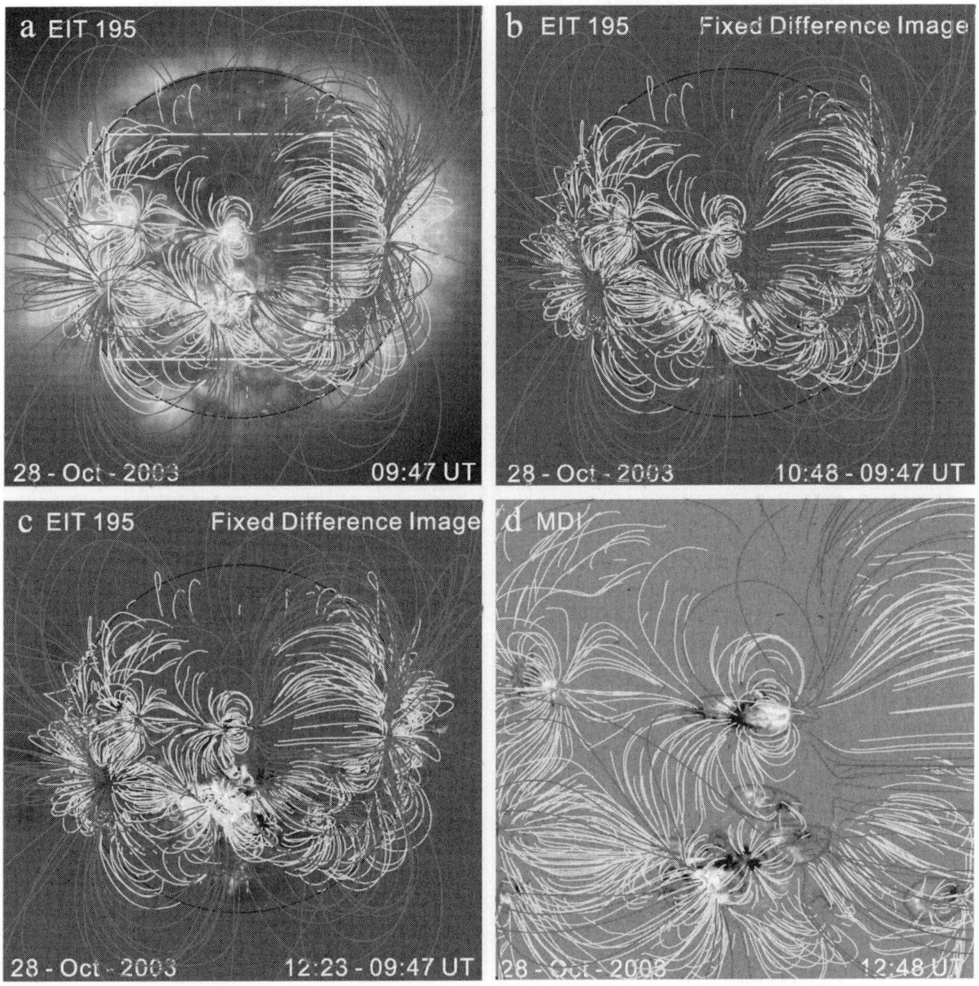

Figure 1. Global magnetic connectivity is found in the CME-associated dimmings and quasi-simultaneous magnetic flux emergence on Oct. 28 2003. Potential extrapolation of observed magnetic fields are overlaid on different background images, i.e., an EIT image at 09:47 UT (a), EIT fixed difference images from 10:48-09:47 UT (b) and from 12:23-09:47 UT (c). Panel (d) is MDI image at 12:48 UT, whose size is indicated by the white square in Panel (a). The contours in pink color denote eight quasi-simultaneous emerging flux regions (EFRs) related to the super activity.

2.3. *On the events in early November of 2004*

The events in the early November of 2004 are characterized by vigorous activity in trans-equatorial loops (TELs) and trans-equatorial filaments (TEFs) (Wang *et al.* 2007). A set of TELs are connecting opposite polarity flux of AR10696 & 695 plus a coronal hole on both hemispheres and laying above the large-scale trans-equatorial magnetic neutral lines, above which there appear an active trans-equatorial filament since November 7, in the interval of November 3-9. During the eruption of the trans-equatorial filament, a beautiful two-ribbon flare was shown by SOHO/EIT 19.5 nm images (also shown by Hα

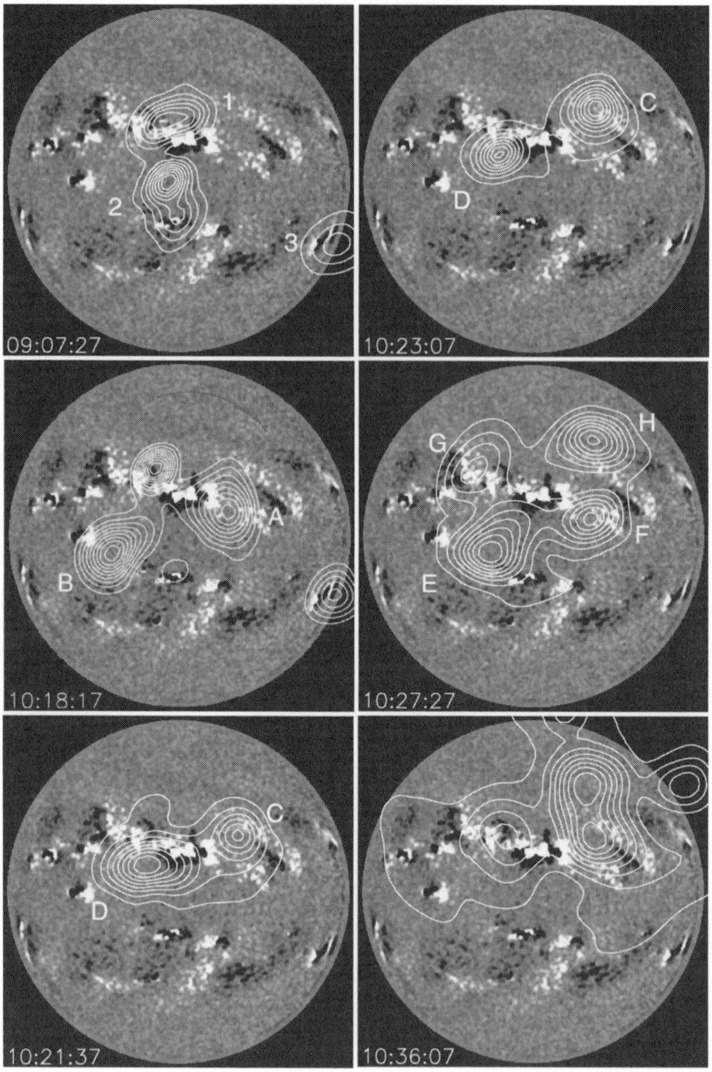

Figure 2. Radio emission sources at 164MHz obtained from NRH superposed on the MDI magnetograms at the closest time. Pre-event Type I storm sources are numbered as '1', '2', '3', and burst sources are denoted by capital letters (see Wen el al. 2006). The wide spreading of non-thermal electron source is shown during CMEs, which show that large-scale magnetic systems were involved in the CME development. The EIT wave front at 10:18 UT is denoted by two red lines in the middle left panel. The angular resolution of the NRH is 5.5'×3.2' at 164MHz.

images, but much more diffuse), which overran the size of AR10696 (See Fig. 8 of Wang *et al.* 2007). Often, a TEL is growing and flaring, coupled with major flare or flares in AR 10696, during which the associated CME was accelerating. The energy release rate of a TEL is much lower than that of a major flare in the AR, but their lifetime was longer than 15 hours (see Figs. 3 and 11 of Wang *et al.* 2007). Jiang *et al.* (2007) have revealed from a theoretical approach that TELs are visible signatures of poloidal field lines across the equator in Babcock-Lighton dynamo process, and an intrinsic component of solar magnetism in solar cycle. However, as an AR not always produces flares, only episodic flaring happens in some TELs.

3. Discussions

The observations, as those described above, support an idea that some CMEs with wide-span may have their source regions embedded in large- or global-scale magnetic structures, though their initiation site is in an AR. A single active region with its flare activity may not be capable of producing a CME with large angular width. Zhukov and Veselovsky (2007) suggest a concept of global CMEs to describe observations of wide-span CMEs whose source regions are truly global in nature. Activity with large spatial scale have been clearly shown in trans-equatorial magnetic structures (Harra *et al.* 2003, 2007; Wang *et al.* 2006, 2007).

Driel-Gesztelyi and her co-authors (2008) suggest on the other hand that the wide-span CMEs, or large- or global-scale activity events, initiate from small scale then become large in the low corona by successive interaction of expanding structures of CME itself with surrounding magnetic structures. For the same events in the 2003 Halloween epoch, they drawn a nice cartoon to illustrate their idea (see their Fig. 12). One possible problem with their cartoon is in the fact that whenever there are multi-polar magnetic configuration, nulls develop and a topology skeleton, e.g., spines, fans, and null-null lines, develop and couple the multi-flux systems (see, Zhao *et al.* 2008). A least-complex configuration which show the basic property of multi-polar magnetic complexity is a quadrupole flux system. MHD numerical simulation shows that for such a multi-flux system, double current sheets develop and catastrophes happen immediately (Zhang *et al.* 2006; Zhang & Wang 2007) to respond the magnetic evolution in the photosphere, so that the magnetic breakout takes place on the current sheet above the global flux system. The magnetic break-out model (Antiochos 1998, Antiochos *et al.* 1999) appears to be a model that relies on or comprise a large or global-scale magnetic source for CMEs.

On the contrary of the idea of large scale CME source, Moore et al. (2007) argued that the final angular width of a CME can be directly related to the flaring AR. It might be true for their examples. However, in pretty rich literatures, large-scale CME sources are shown (see Wang *et al.* 2002). Zhou *et al.* (2006) find that about 40% of earth-directed CMEs are associated with trans-equatorial loops.

However, taking the study of TELs as an example, we still have not fully understand the physics in many aspects, e.g., TEL's footprints in magnetic clouds. For quite long time, our solar astronomers seem to concentrate ourselves on AR and flare activity; on the other hand, the full disk observations with adequate cadence and temporal coverage are still not enough. Future diagnosis of more global magnetism and activity is fundamental in understanding solar and stellar activity.

References

Antiochos, S. K. 1998, *ApJ*, 502, L181–L184
Antiochos, S. K., DeVore, C. R., & Klimchuk, J. A. 1999, *ApJ*, 510, 485–493

Attrill, G., Nakwacki, M. S., Harra, L. K., van Driel-Gesztelyi, L., Mandrini, C. H., Dasso, S., & Wang, J. 2006, *Sol. Phys.*, 238, 117–139

Attrill, G., Harra, L. K., van Driel-Gesztelyi, L., & Demoulin, P. 2007a, *ApJL*, 656, L101–L104

Bemporad, A., Sterling, A. C., Moore, R. L., & Poletto, G. 2005, *ApJL*, 635, L189–L192

Culhane, J. L. & Siscoe, G. L. 2007, *Sol. Phys.*, 244, 3–12

Harra, L. K., Matthews, S. A., & van Driel-Gesztelyi, L. 2003, *ApJ*, 598, L59–L62.

Harra, L. K., Crooker, N. U., Mandrini, C. H., van Driel- Gesztelyi, L., Dasso, S., Wang, J., Elliott, H., Attrill, G., Jackson, B. V., & Bisi, M. M. 2007a, *Sol. Phys.*, 244, 95–114

Harra, L. K., Hara, H., Imada, S., Young, P. R., Williams, D., Sterling, A., Korendyke, C., & Attrill, G. 2006, *Publ. Astron. Soc. Japan*, 59, S801–S806

Liu, Y. 2007, *ApJ*, 654, L171–L174

Maia, D., Vourlidas, A., Pick, M., Howard, R., Schwenn, R., & Magalhães, A. 1999, *J. Geophys. Res.*, 104, 12507–12514

Mandrini, C. H., Nakwacki, M. S., Attrill, G., van Driel-Gesztelyi, L., Demoulin, P. Dasso, S., & Elliot, H. 2007, *Sol. Phys.*, 244, 25–43

Moore, R., Sterling, A. C., & Suess, S. T. 2007, *ApJ*, 668, 1221–1231

Pohjolainen, S., Maia, D., Pick, M., Vilmer, N., Khan, J. I., Otruba, W., Warmuth, A., Benz, A., Alissandrakis, C., & Thompson, B. J. 2001, *Astrophys. J.*, 556, 421–431

van Driel-Gesztelyi, L., Attrill1, G. D. R., Démoulin, P., Mandrini, C. H., & Harra, L. K. 2008, *Ann. Geophys.*, 26, 3077–3088

Wang, J., Zhou, G., Wen, Y., Zhang, Y., Wang, H., Deng, Y., Zhang, J., & Harra, L. K. 2006, *ChJAA*, 6, 247–259

Wang, J.-X., Zhang, Y.-Z., Zhou, G.-P., Harra, L.,Williams, D., & Jiang, Y.-C. 2007, *Sol. Phys.*, 244, 75–94

Wang, T., Yan, Y., Wang, J., Kurokawa, H., & Shibata, K. 2002, *ApJ*, 572, 580–597

Wen, Y., Wang, J., Maia, D. J. F., Zhang, Y., Zhao, H., & Zhou, G. 2006, *Sol. Phys.*, 239, 257–276

Zhang, Y., Wang, J., Attrill, G. D. R., Harra, L. K., Yang, Z., & He, X. 2007, *Sol. Phys.*, 241, 329–349

Zhang, Y., Wang, J., & Hu, Y. Q. 2006, *ApJ*, 641, 572–576

Zhang, Y. & Wang, J. 2007, *ApJ*, 663, 592–597

Zhao, H., Wang, J., Zhang, J., Xiao, C.-J., & Wang, H. 2008, *ChJAA*, 8, 133–145

Zhou, G.-P., Wang, J.-X., Wang, Y.-M., & Zhang, Y.-Z. 2007, *Sol. Phys.*, 244, 13–24

Zhukov, A. N. & Veselovsky, I. S. 2007, *ApJL*, 664, 131–134

Zirin, H. 1985, *Aust J. Phys.*, 38, 961–969

Solar and Stellar Variability: Impact on Earth and Planets
Proceedings IAU Symposium No. 264, 2009
A. G. Kosovichev, A. H. Andrei & J.-P. Rozelot, eds.

doi:10.1017/S1743921309992717

Magnetic energy release: flares and coronal mass ejections

Cristina H. Mandrini

Instituto de Astronomía y Física del Espacio (CONICET-UBA),
CC 67, Suc 28, 1428 Buenos Aires, Argentina
e-mail: mandrini@iafe.uba.ar

Abstract. Free energy stored in the magnetic field is the source that powers solar and stellar activity at all temporal and spatial scales. The energy released during transient atmospheric events is contained in current-carrying magnetic fields that have emerged twisted and may be further stressed via motions in the lower atmospheric layers (i.e. loop-footpoint motions). Magnetic reconnection is thought to be the mechanism through which the stored magnetic energy is transformed into kinetic energy of accelerated particles and mass flows, and radiative energy along the whole electromagnetic spectrum. This mechanism works efficiently at scale lengths much below the spatial resolution of even the highest resolution solar instruments; however, it may imply a large-scale restructuring of the magnetic field inferred indirectly from the combined analysis of observations and models of the magnetic field topology. The aftermath of magnetic energy release includes events ranging from nanoflares, which are below our detection limit, to powerful flares, which may be accompanied by the ejection of large amounts of plasma and magnetic field (so called coronal mass ejections, CMEs), depending on the amount of total available free magnetic energy, the magnetic flux density distribution, the magnetic field configuration, etc. We describe key observational signatures of flares and CMEs on the Sun, their magnetic field topology, and discuss how the combined analysis of solar and interplanetary observations can be used to constrain the flare/CME ejection mechanism.

Keywords. Sun: magnetic fields, Sun: activity, Sun: flares, Sun: coronal mass ejections (CMEs)

1. Introduction

Solar flares and CMEs are the most violent and energetic events in the solar system. They represent the most dynamic aspect of the Sun's corona and involve a fast and large-scale destabilization of its magnetic field. As a result, large amounts of magnetic free energy contained in the coronal field is released and transformed into thermal energy, radiation, accelerated particles, and large scale flows of plasma (see Forbes 2000 for quantitative energy estimations), all of which can directly affect the Earth's environment.

Though, in general, the words "flare" or "CME" are used to refer to the radiative manifestation of energy release and the coronagraph-recorded ejection of plasma, respectively; flares and CMEs can occur separately or together in any order (Aschwanden *et al.* 2001), and they may, in fact, be manifestations of the same underlying physical process. In this sense, we will frequently use the term flare/CME to stress this fact. Figure 1 (top and bottom left panels) illustrates three stages of a large-scale eruptive event, observed by the Transition Region and Coronal Explorer (TRACE), during which a flare/CME occurs in close association with a filament eruption. The bottom right panel shows the chromospheric impact of energy release, but for a different event.

In this paper we will address key aspects common to flares and CMEs, such as: the origin of the energy released, the observational signatures of the energy release mechanism and the magnetic field topology where it occurs. Finally, we will comment on how we can

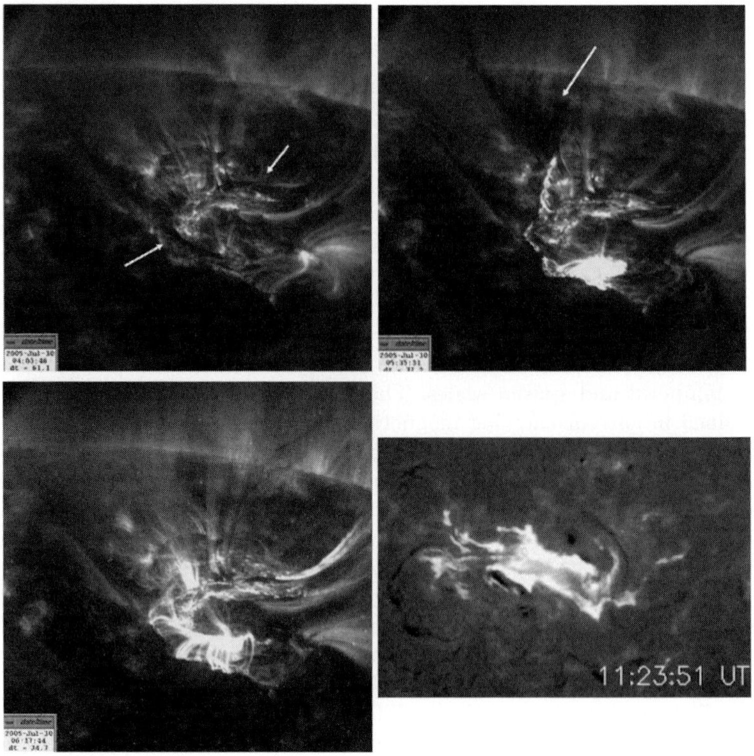

Figure 1. Top and bottom left panels show the temporal evolution of a flare/CME (TRACE, 171 Å) on 30 July 2005. The top left image (04:30 UT) shows a cool-dark curved filament (indicated by white arrows) extending along the line of sight magnetic inversion line (IL) before eruption. The top right panel (05:35 UT) depicts the filament undergoing eruption (white arrow) while flare brightenings develop at lower coronal heights. The bottom left image (06:17 UT) shows a cooling arcade of flare loops with footpoints at both sides of the IL. The bottom right panel is an Hα image taken by the Hα Solar Telescope for Argentina (HASTA) on 28 October 2003; this image shows the chromospheric response to coronal energy release during an event similar to the one in previous panels. All together, these panels illustrate the evolution of a typical two-ribbon flare.

profit from the combined study of observations from different physical regimes (*i.e.*, the solar corona and the interplanetary (IP) space) to constrain our ideas on the flare/CME ejection mechanism. More extended reviews on flare observations and models are those by Benz (2008) and Schrijver (2009); while for CMEs, those by Schwenn (2006), Forbes (2000), Klimchuk (2001), and Forbes *et al.* (2006).

2. Solar flares and coronal mass ejections: observations and models

2.1 General characteristics

Energy released during large flares/CMEs can reach values in excess of 10^{33} erg. Many minor events (microflares, nanoflares) occur at energies in the range of 10^{24}–10^{25} erg. Historically, flares have been identified and, therefore, defined as localized brightenings observed in Hα images (*e.g.* Svestka 1976). These brightenings or ribbons typically appear in pairs, at both sides of the IL, and the event is called a two-ribbon flare. The standard empirical model that was built to explain the sequence of observed flare/CME manifestations is the so-called CSHKP model (Carmichael 1964, Sturrock 1966, Hirayama

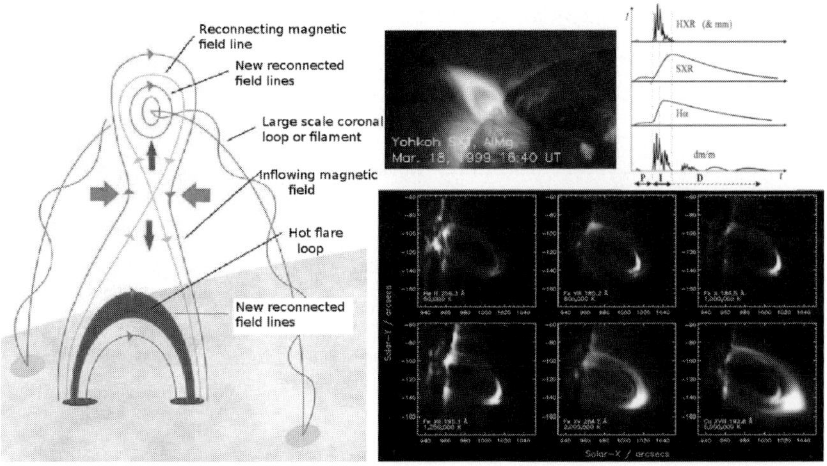

Figure 2. On the left, a cartoon representing a two-ribbon flare (adapted from Shibata 1998). On the top right, typical light-curves corresponding to emissions from the chromosphere to the corona. The image to the left is a cusp-shaped loop observed at the limb by SXT after a flare/CME on 18 March 1999 (Yokoyama *et al.* 2001). The bottom images show a system of loops observed at the west solar limb on 17 December 2006 by EIS. The images were taken simultaneously and show an increase with height of the loops' temperature as expected after a two-ribbon flare (figure by P. Young, see also Hara *et al.* 2008).

1974, Kopp & Pneuman 1976). A cartoon of the model, typical flare light-curves, and other plasma responses to energy release are shown in Figure 2. These events occur in a dominantly bipolar magnetic configuration in which a filament, contained in a twisted flux tube, is embedded. As the configuration evolves, *e.g.* the shear increases close to the IL, the filament goes through a series of equilibria reaching higher coronal heights. The overlying arcade is stretched, oppositely directed field lines are brought together, and a current sheet is formed. Magnetic reconnection (see 2.3) sets in below the filament. This process implies a change of connectivity during which more flux is added to the rising filament increasing its magnetic pressure, while a lower arcade is formed below the reconnection region. As the filament continues rising, it reaches a point where no equilibrium is possible and reconnection, which is slow at first, becomes fast. The filament accelerates and erupts. Subsequent in-flowing plasma brings the surrounding field into the dissipation region and the stored energy (see 2.2) is transformed into heat and kinetic energy of accelerated particles. Large electric fields, as well as shock waves, are created in the reconnection region. Heat conduction fronts and particle beams are channeled along the newly reconnected field lines. Accelerated electrons emit gyrosynchrotron radiation as they spiral along the field lines and, as particles reach the chromosphere, hard X-rays (HXR, $\succeq$ 20 keV) are emitted by bremsstrahlung. The impulsively heated chromosphere is seen in emission in the UV and optical ranges (i.e. Hα). The heated plasma expands upward increasing the density and temperature of the newly reconnected loops, which are observed in EUV and soft X-rays (SXR). These loops cool down and are seen later in Hα. Reconnection continues between field lines which are anchored farther from the IL; newer reconnected loops are higher and the flare ribbons at their footpoints separate. Several tenths of the total flare/CME energy are released in the first few minutes during the impulsive phase. The emission in HXR, EUV, and several radio wavelengths is very structured during this phase, indicating that the energy release is intermittent. The SXR and Hα emission may rise sharply, but their maximum occurs later. During the decay phase reconnection still continues but at a decreasing rate.

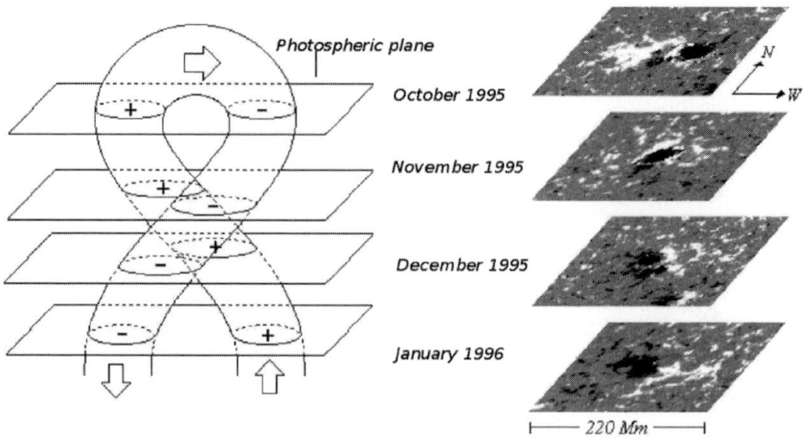

Figure 3. An example of the emergence of a twisted (kinked) flux tube. The right panel shows the photospheric evolution of AR 7912 along four rotations (line of sight magnetograms close to central meridian passage) positive/white (negative/black) values of the field are saturated above (below) 50G (-50G). On the left, we depict a sketch of the flux tube, as deduced from observations. The cuts by four horizontal planes show the approximate locations of the photosphere at the times of the magnetograms (adapted from López-Fuentes *et al.* 2000).

Since the CSHKP model is basically empirical, innumerable observations support the scenario; a few of the highlights are: 1) Masuda *et al.* (1994) observed a HXR source above a SXR loop suggesting that reconnection occurs at coronal heights. 2) Yokoyama *et al.* (2001) observed clear evidence of the existence of reconnection inflows in Soft X-ray Telescope (SXT) data on board Yohkoh. 3) Forbes & Acton (1996) showed that newly formed cusped loops shrink and relax due to magnetic tension. 4) McKenzie & Hudson (1999) observed X-ray voids in SXT and interpreted them as fast downflows due to intermittent reconnection. 5) Asai *et al.* (2004) showed that the start of these downflows is associated with HXR emission and microwave bursts. 5) Czaykowska *et al.* (1999) and Harra *et al.* (2005) found upflowing plasma along the outer edges of flare ribbons, while downflows were seen in the inner ribbon edges.

The events just described are associated with CMEs with cores formed by the ejected filaments. The ejected mass ranges from some 10^{13} g to 10^{16} g, this plasma is embedded in a magnetic flux of 10^{20}–10^{23} Mx (Schwenn 2006). Because the coronal pressure decreases with height, the eruptive configuration expands and, as suggested by Attrill *et al.* (2007) (see Mandrini *et al.* 2007), it may find favourably oriented surrounding fields with which reconnection takes place; then, large coronal areas may provide mass to and become part of the CME (van Driel-Gesztelyi *et al.* 2008).

2.2 The origin of the energy released

There is strong evidence showing that magnetic flux emerges twisted from the solar interior, i.e. carrying free magnetic energy to be released (Leka *et al.* 1996, Lites 2009). It is very difficult for a flux tube to traverse the dense photosphere, since in the concave-up parts under the flux tube axis plasma accumulates leading to its fragmentation (*e.g.* Manchester *et al.* 2004); therefore, flux tubes can only emerge through many small-scale reconnections (Pariat *et al.* 2004). Nevertheless, magnetic patterns, due to the contribution of the field azimuthal component, are observed as the flux tube crosses the photosphere (López-Fuentes *et al.* 2000); this indicates that there is an overall organization in the emerging flux which is compatible with a global twist. Furthermore, a rotation of the main active region (AR) polarities (see Figure 3) can be also a signature of the emergence of highly twisted flux tubes (López-Fuentes *et al.* 2000, 2003, Brown *et al.* 2003). Besides, computations of the magnetic helicity balance in ARs show that helicity has to be brought up mainly by flux emergence to explain the helicity shed via CMEs (Démoulin *et al.* 2002, Green *et al.* 2002). From a theoretical and numerical point of view, early 2D and 2.5D MHD simulations showed that untwisted flux tubes could not survive their transit across the convective zone since they are eroded by the vortices formed in their wake (Schüssler 1979); however, a certain amount of twist could prevent the tube fragmentation (Moreno-Insertis & Emonet 1996). Recent 3D simulations indicate that emerging flux is less twisted than previously thought. Fan (2008) found that to comply with Joy's law, the twist-induced tilt in rising flux ropes should be less than half of that needed for a cohesive rise. These simulations and observations imply that all the large-scale flux that has crossed the convective zone must be twisted.

Before the finding that flux emerges twisted, the generally accepted idea was that free magnetic energy is built up by shearing motions. Computations were done to estimate the time these motions should be imposed to store enough energy to power a flare (see e.g McClymont & Fisher 1989). Large-scale flows are present in the solar photosphere (e.g. differential rotation) and local deviations from the mean differential rotation rate are seen in flow-maps of the solar surface (*e.g.* Meunier 2005). Leka *et al.* (1996) clearly showed, however, that shearing footpoint motions were insufficient to create all the electric currents observed in an emerging AR. Furthermore, apparent surface shearing flows can result from the emergence of a flux tube (Démoulin & Berger 2003). In addition to organized patterns, random footpoint motions are also present.

2.3 The energy release mechanism and the field topology

Magnetic field reconnection, that can be defined as a topological restructuring of the field which leads to a change of its connectivity, has been the energy release mechanism proposed at the origin of flare and some CME models. From a theoretical point of view, magnetic configurations with a complex topology, i.e. having separatrices (places where the field connectivity is discontinuous), are the locations where current sheets can form in 2D. When going to 3D, and if the photospheric magnetic field is described by a series of isolated polarities (surrounded by field-free regions), a complete topological description of the magnetic configuration is given by the skeleton formed by null points, spines, fans, separators, and associated separatrices (Longcope 2005 and examples therein).

However, if the photosphere is fully magnetized, most of the above topological structures disappear; only separatrices associated to coronal magnetic nulls remain. Separatrices of a different origin are linked to concave-up field lines tangential to the photosphere (defining bald-patch locations) (Titov *et al.* 1993). For some observed configurations, those structures are sufficient to explain the location of flare kernels resulting from reconnection. Examples have been found where coronal null points, computed using either subphotospheric sources or extrapolations to represent the coronal field, were associated with coronal activity (Mandrini *et al.* 1991, Gaizauskas *et al.* 1998, Aulanier *et al.* 2000, Fletcher *et al.* 2001a, Mandrini *et al.* 2006, Luoni *et al.* 2007). Concerning bald patches, they have been only found in connection with low energy release events: small flares (Aulanier *et al.* 1998), EUV brightenings (Fletcher *et al.* 2001b), surges (Mandrini *et al.* 2002), and Ellerman bombs (Pariat *et al.* 2004).

The analysis of the topological structure of numerous ARs has shown that flares occur in a variety of configurations larger than just discussed (see e.g. Démoulin *et al.* 1994 and the reviews by Démoulin 2006, 2007). Quasi-separatrix layers (QSLs) (Priest & Démoulin 1995, Démoulin *et al.* 1996, Titov *et al.* 2002), which are regions where there is a

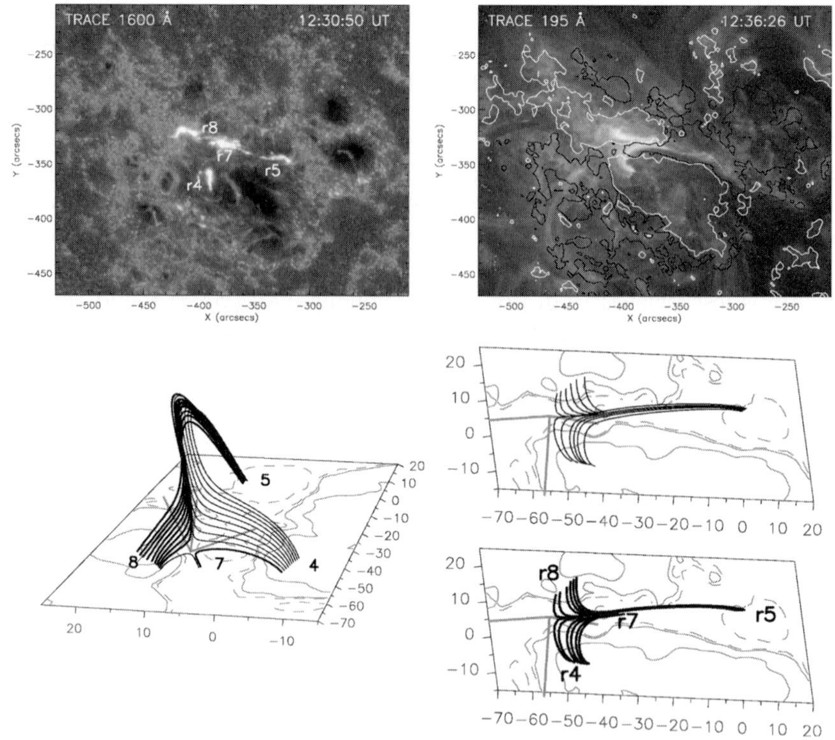

Figure 4. Top panels: TRACE images of AR 10486 on 27 October, 2003, (1600 Å (left) and 195 Å (right)) showing the locations of flare kernels (r4, r5, r7 and r8). The bottom left panel depicts the location of a 3D coronal null point (three lines) with field lines computed starting integration at finite distances from the null. A set of thin continuous lines have footpoints in polarities 4 and 5. These lines could reconnect at the null with lines linking 7 to 8, represented by only one short and thin continuous line. Reconnection forms the set of thick continuous field lines with footpoints at 8 and 5, and those that connect 4 and 7 (the latter are represented by only one thick line). The two bottom-right panels show the coronal field model in the vicinity of the null point. Field lines are the same as in the left panel but as seen from the observer's point of view. The upper of these panels shows two sets of lines representing the pre-reconnected loops. The lower one corresponds to field lines after reconnection (notice the similarity between their shape and loops in 195 Å). Several short field lines have been added as compared to the left panel. Negative (positive) field isocontours are shown in continuous (dashed) thin lines, their values are ± 100 and ± 1000 G. The axes are labeled in Mm (adapted from Luoni *et al.* 2007).

drastic change in field-line linkage, generalize the concept of separatrices to configurations without null points and/or bald patches. Using coronal field models, QSLs have been found to exist in the broadest variety of observed magnetic configurations (Démoulin *et al.* 1997, Fletcher *et al.* 2001b, Bagalá *et al.* 2000). QSLs have been shown to coincide with flare kernels in the chromosphere and coronal loop brightenings of various intensity. These brightenings were connected by field lines as expected from the magnetic reconnection theory. Moreover, in cases where vector magnetograms were available, photospheric current concentrations were also located at the photospheric trace of QSLs. Recently, to relate QSLs with the formation of strong current concentrations and study the characteristics of the reconnection process, 3D MHD simulations of typical observed field distributions and photospheric motions have been developed by Aulanier *et al.* (2005, 2006). The results of these simulations imply that electric currents at QSLs may amplify in time only if the QSLs are broader than the dissipative scale length, as it was suggested in previous observational studies. Related to magnetic reconnection, the simulations show

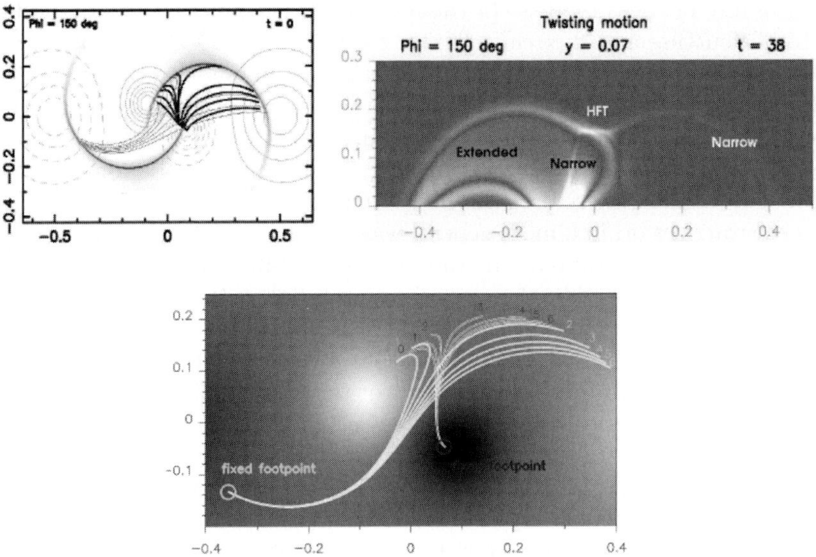

Figure 5. Top left panel: Location of QSLs in a simple magnetic configuration formed by two magnetic bipoles. Two sets of field lines are added to illustrate the drastic change in connectivity, both sets have footpoints starting along a segment crossing a QSL at the lower boundary. Continuous (dashed) contours show positive (negative) values of the vertical field component. Top right panel: Formation of strong electric currents at QSLs after the configuration on the left has evolved in a MHD simulation (twisting motions of the smallest positive polarity). The figure is a vertical cut of the 3D configuration along a horizontal line at y=0.07 in the top left panel. The current scales linearly with the grey-scale intensity (white is the highest value). Extended and narrow currents layers coexist. Bottom panel: Magnetic reconnection after the same twisting motions are applied. The evolution of two slip-running field lines with fixed footpoints in the negative polarities is shown with a fixed time step; as reconnection proceeds, each line changes its connectivity, shifting along the QSL. Figures are adapted from Aulanier *et al.* (2005, 2006).

that it occurs only due to the self-pinching and dissipation of narrow current layers previously formed at QSLs. A property of reconnection is the continuous slippage of field lines along each other as they pass through the current layers. Observational evidence of this slip-running reconnection has been found in flares seen by the X-ray Telescope on board Hinode (Aulanier *et al.* 2007) and by TRACE (Masson *et al.* 2009). The events analyzed in the papers mentioned previously are confined flares, the connectivity change occurs between two closed field configurations, since this is linked to the intrinsic definition of QSLs. However, energy release by reconnection may occur in more general topologies (see 2.1). QSLs have been recently found associated with intense outflows observed by the Extreme-ultraviolet Imaging Spectrometer (EIS) on board Hinode at the border of ARs (between closed and large-scale open-like field lines) by Baker *et al.* (2009), indicating that signatures of energy release along QSLs is even broader. Figures 4 and 5 show the topology of an observed AR and a MHD simulation.

2.4 CMEs initiation models and the link to interplanetary events

CME initiation models can be divided in different ways, i.e. taking into account the presence of non-ideal or ideal instabilities as the drivers (Forbes 2000), or in a tutorial way emphasizing their basic physical differences (Klimchuk 2001). We will follow a different approach and will separate models according to characteristics that may be useful when comparing solar to IP associated events; that is, models in which the ejected flux tube is mainly formed during the eruption from models in which it is already present in the pre-eruptive configuration. Furthermore, it is not our intention to make a detailed description of any of these models. In the first group, we put two sets of models: a) the classical tether cutting and flux cancellation models (Moore & Rumeolitis 1992, Moore *et al.* 2001, Mikic & Linker 1994), and b) the breakout model (Antiochos *et al.* 1999, Lynch *et al.*, 2004, 2008). This group of models requires the presence of sheared arcades and reconnection to start the eruption. The fundamental difference between both sets is that reconnection initially occurs in the core sheared field in the former and above it in the breakout model. Therefore, the latter requires the presence of a null point and associated separatrices or, in a more general way, QSLs in the corona; the basic magnetic field configuration is quadrupolar. In the first set the flux tube forms from the initial arcade reconnection process; in the second set, the flux tube forms when reconnection below the rising arcade sets in. For the second group of models, the eruption is driven by an ideal MHD process: a) a catastrophe (Forbes & Isenberg 1991, Isenberg *et al.* 1993, Lin *et al.* 1998), or b) an instability (kink and/or torus instabilities) (Fan & Gibson 2003, 2004, 2007, Török & Kliem 2003, 2005, Kliem *et al.* 2004, Török *et al.* 2004, Kliem & Torok 2006). Reconnection is not required initially but may help to trigger the eruption.

Following a flare/CME, plasma and magnetic field expand into the IP space. The resulting structures are called interplanetary CMEs (ICMEs). The term magnetic cloud (MC), where more restricted characteristics are present, is used for a subset of ICMEs. We can relate MC physical parameters to those of the solar eruptive region; a relevant one is the magnetic flux, since the magnetic flux involved in the eruption should equate the flux in the MC. On the Sun, surface phenomena, such as coronal dimmings, can be identified and associated with the footpoints of the erupting structure. Since the usual interpretation of dimmings is that they are the sites of material outflow (*e.g.* Harra & Sterling 2001) and two dimmings are often located on both sides of the eruption site over opposite magnetic polarities, it has frequently been assumed (Webb *et al.* 2000) that the dimmings form at the ejected flux tube footpoints. Magnetic flux in the dimmings may be measured from magnetograms and compared with the flux in the related MC. Axial and azimuthal fluxes are computed from *in situ* magnetic data by fitting a model (Dasso *et al.* 2005). Comparisons often show rough agreements between flux in dimmings and cloud axial fluxes (Lepping *et al.* 1997, Webb *et al.* 2000). However, MCs are highly twisted structures and may have substantial azimuthal flux. Mandrini *et al.* (2005) and Attrill *et al.* (2006) have found an excellent agreement between flux in dimmings and the azimuthal MC flux. Twist and flux (in an overlying arcade) may be added via reconnection to an expanding tube, such that the flux from dimmings can contribute to the cloud azimuthal flux component; therefore, these results (see also Qiu *et al.* 2007) suggest that the ejected flux tube is mainly formed during the eruption.

3. Discussion and conclusions

From this concise overview of flare/CME characteristics, we can conclude that the magnetic free energy, which is released during these events, is mostly brought up from the solar interior by twisted flux emergence. Though all observational evidence is indirect, there is a consensus that the energy release mechanism at work is magnetic reconnection. This is at the origin of flares and CMEs, and it even contributes to the acceleration of the eruption in some cases. The observational examples showing various levels of activity originating from different topologies, teach us that reconnection can occur in configurations with much wider topological characteristics than was traditionally considered. QSLs, a generalization of separatrices to configurations in which the magnetic connectivity is not discontinuous (but changes drastically), are the most common locations where reconnection takes place. Three-dimensional MHD simulations have shown that magnetic reconnection proceeds in such way that field lines slip-run along

QSLs as they reconnect. Combined quantitative studies of solar events and their IP counterparts can greatly contribute to our understanding of the flare/CME eruption mechanism. The few examples analyzed so far indicate that the flux tube observed in the IP space is mainly formed during the eruption.

To further understand the details of solar activity events (flares/CMEs) from the build-up of magnetic energy to its effective release and evolution in the IP space, we need both high-resolution measurements of all three components of the magnetic field vector at different solar atmospheric heights, improved 3D views (measurements) of the events from the Sun through the IP space (as the STEREO spacecraft are providing), as well as the development of more realistic 3D models of the coronal and IP magnetic field. Obvious though these assertions may sound, we all know that they are not easy to achieve.

References

Asai, A., Yokoyama, T., Shimojo, M., & Shibata, K. 2004, *ApJ*, 605, L77

Attrill, G. D. R., Harra, L. K., van Driel-Gesztelyi, L., & Démoulin, P. 2007, *ApJ*, 656, L101

Attrill, G., Nakwacki, M. S., & Harra, L. K., *et al.* 2006, *Sol. Phys.*, 238, 117

Aulanier, G., *et al.* 2007, *Science*, 318, 1588

Aulanier, G., DeLuca, E. E., Antiochos, S. K., McMullen, R. A., & Golub, L. 2000, *ApJ*, 540, 1126

Aulanier, G., Démoulin, P., Schmieder, B., Fang, C. & Tang, Y. H. 1998, *A&A*, 183, 369

Aulanier, G., Pariat, E., & Démoulin, P. 2005, *A&A*, 444, 961

Aulanier, G., Pariat, E., Démoulin, P., & Devore, C. R. 2006, *Sol. Phys.*, 238, 347

Antiochos, S. K., DeVore, C. R., & Klimchuk, J. A. 1999, *ApJ*, 510, 485

Aschwanden, M. J., Poland, A. I., & Rabin, D. M. 2001, *Ann. Rev. Astron. Astrophys.*, 39, 175

Bagalá, L. G., Mandrini, C. H., Rovira, M. G., & Démoulin, P. 2000, *A&A*, 363, 779

Baker, D., van Driel-Gesztelyi, L., Mandrini, C. H., Démoulin, P., *et al.* 2009, *ApJ*, 705, 926

Benz, A. O. 2008, *Liv. Rev. Sol. Phys.*, 5, 1

Brown, D. S., Nightingale, R. W., Alexander, D., Schrijver, C. J., *et al.* 2003, *Sol. Phys.*, 216, 79

Carmichael, H. 1964, *NASA SP*, 50, 451

Czaykowska, A., de Pontieu, B., Alexander, D., & Rank, G. 1999, *ApJ*, 521, L75

Dasso, S., Mandrini, C. H., Démoulin, P., Luoni, M. L., *et al.* 2005, *Adv. Spa. Res.*, 35, 711

Démoulin, P. 2006, *Adv. Spa. Res.*, 37, 1269

Démoulin, P. 2007, *Adv. Spa. Res.*, 39, 1367

Démoulin, P., Bagalá, L. G., Mandrini, C. H., Hénoux, J. C., & Rovira, M. G. 1997, *A&A*, 325, 305

Démoulin, P. & Berger, M. A. 2003, *Sol. Phys.*, 215, 203

Démoulin, P., Hénoux, J. C., & Mandrini, C. H. 1994, *A&A*, 285, 1023

Démoulin, P., Hénoux, J. C., Priest, E. R., Mandrini, C. H. 1996, *A&A*, 308, 643

Démoulin, P., Mandrini, C. H., & van Driel-Gesztelyi, L., *et al.* 2002, *A&A*, 382, 650

Fan, Y. 2008, *ApJ*, 676, 680

Fan, Y. & Gibson, S. E. 2003, *ApJ*, 589, L105

Fan, Y. & Gibson, S. E. 2004, *ApJ*, 609, 1123

Fan, Y. & Gibson, S. E. 2007, *ApJ*, 668, 1232

Fletcher, L., López Fuentes, M. C., Mandrini, C. H., *et al.* 2001b, *Solar Phys.*, 203, 255

Fletcher, L., Metcalf, T. R., Alexander, D., Brown, D. S., & Ryder, L. A. 2001a, *ApJ*, 554, 451

Forbes, T. G. 2000, *JGR*, 105, 23153

Forbes, T. G. & Acton, L. W. 1996, *ApJ*, 459, 330

Forbes, T. G. & Isenberg, P. A. 1991, *ApJ*, 373, 294

Forbes, T. G., Linker, J. A., Chen, J., Cid, C., Kóta, J., *et al.* 2006, *Spa. Scien. Rev.*, 123, 251

Gaizauskas, V., Mandrini, C. H., Démoulin, P., Luoni, M. L., & Rovira, M. G. 1998, *A&A* 332, 353

Hara, H., Watanabe, T., Matsuzaki, K., & Harra, L. K., *et al.* 2008, *PASJ*, 60, 275

Harra, L. K., Démoulin, P., Mandrini, C. H., Matthews, S. A., *et al.* 2005, *A&A*, 438, 1099
Harra, L. K. & Sterling, A. C. 2001, *ApJ*, 561, L215
Hirayama, T. 1974, *Sol. Phys.*, 34, 323
Isenberg, P. A., Forbes, T. G., & Démoulin, P. 1993, *ApJ*, 417, 368
Kliem, B., Titov, V. S., & Török, T. 2004, *A&A*, 413, L23
Kliem, B. & Torok, T. 2006, *PRL*, 96, 255002
Klimchuk, J. A. 2001, in: P. Song, H. Singer & G. Siscoe (eds.), *Geophys. Mon.*, 125, 143, AGU
Kopp, R. A. & Pneuman, G. W. 1976, *Sol. Phys.*, 50, 85
Leka, K. D., Canfield, R. C., McClymont, A. N., & van Driel-Gesztelyi, L. 1996, *ApJ*, 462, 547
Lepping, R. P., Szabo, A., DeForest, C. E., & Thompson, B. J. 1997, *ESA SP*, 415, 163
Lin, J., Forbes, T. G., Isenberg, P. A., & Démoulin, P. 1998, *ApJ*, 504, 1006
Lites, B. W. 2009, *Spa. Sci. Rev.*, 144, 197
López Fuentes, M. C., Demoulin, P., Mandrini, C. H., & van Driel-Gesztelyi, L. 2000, *ApJ*, 544, 540
López Fuentes, M. C., Démoulin, P., Mandrini, C. H., *et al.* 2003, *A&A*, 397, 305
Longcope, D. W. 2005, *Liv. Rev. Sol. Phys.* 2, 7
Luoni, M. L., Mandrini, C. H., Cristiani, G. D., & Démoulin, P. 2007, *Adv. Spa. Res.*, 39, 1382
Lynch, B. J., Antiochos, S. K., DeVore, C. R., Luhmann, J. G., *et al.* 2008, *ApJ*, 683, 1192
Lynch, B. J., Antiochos, S. K., MacNeice, P. J., Zurbuchen, T. H., & Fisk, L. A. 2004, *ApJ*, 617, 589
Manchester, W., IV, Gombosi, T., DeZeeuw, D., & Fan, Y. 2004, *ApJ*, 610, 588
Mandrini, C. H., Démoulin, P., Hénoux, J.-C., & Machado, M. E. 1991, *A&A* 250, 541
Mandrini, C. H., Demoulin, P., Schmieder, B., Deluca, E. E., *et al.* 2006, *Sol. Phys.*, 238, 293
Mandrini, C. H., Démoulin, P., Schmieder, B., Deng, Y. Y., & Rudawy, P. 2002, *A&A*, 391, 317
Mandrini, C. H., Nakwacki, M. S., Attrill, G., *et al.* 2007, *Sol. Phys.*, 244, 25
Mandrini, C. H., Pohjolainen, S., Dasso, S., Green, L. M., *et al.* 2005, *A&A*, 434, 725
McClymont, A. N. & Fisher, G. H. 1989, in: J. Waite, J. Burch, & R. Moore (eds.), Solar System Plasma Physics, 219, AGU
McKenzie, D. E. & Hudson, H. S. 1999, *ApJ*, 519, L93
Masson, S., Pariat, E., Aulanier, G., & Schrijver, C. J. 2009, *ApJ*, 700, 559
Masuda, S., Kosugi, T., Hara, H., Tsuneta, S., & Ogawara, Y. 1994, *Nature*, 371, 495
Meunier, N. 2005, *A&A*, 443, 309
Mikic, Z. & Linker, J. A. 1994, *ApJ*, 430, 898
Moreno-Insertis, F. & Emonet, T. 1996, *ApJ*, 472, L53
Moore, R. L. & Roumeliotis, G. 1992, IAU Colloq. 133: Eruptive Solar Flares, 399, 69
Moore, R. L., Sterling, A. C., Hudson, H. S., & Lemen, J. R. 2001, *ApJ*, 552, 833
Pariat, E., Aulanier, G., Schmieder, B., Georgoulis, M. K., *et al.* 2004, *A&A* 614, 1099
Priest, E. R. & Démoulin, P. 1995, *JGR*, 100, 23443
Qiu, J., Hu, Q., Howard, T. A., & Yurchyshyn, V. B. 2007, *ApJ*, 659, 758
Schrijver, C. J. 2009, *Adv. Spa. Res.*, 43, 739
Schrijver, C. J., De Rosa, M. L., Metcalf, T., Barnes, G., *et al.* 2008, *Astrophys. J.*, 675, 1637
Schwenn, R. 2006, *Liv. Rev. Sol. Phys.*, 3, 2
Shibata, K. 1998, *Ap&SS*, 264, 129
Sturrock, P. A. 1966, *Nature*, 211, 695
Schüssler, M. 1979, *A&A*, 71, 79
Svestka, Z. 1976, *Solar Flares*, Springer-Verlag Berlin Heidelberg
Titov, V. S., Priest, E. R., & Démoulin, P. 1993, *A&A*, 276, 564
Titov, V. S., Hornig, G., & Démoulin, P. 2002, *JGR*, 107, SSH 3, 1–13
Török, T. & Kliem, B. 2003, *A&A*, 406, 1043
Török, T. & Kliem, B. 2005, *ApJ*, 630, L97
Török, T., Kliem, B., & Titov, V. S. 2004, *A&A*, 413, L27
van Driel-Gesztelyi, L., Attrill, G., Démoulin, P., Mandrini, C., *et al.* 2008, *An. Geo.*, 26, 3077
Webb, D. F., Lepping, R. P., Burlaga, L. F., DeForest, C. E., *et al.* 2000, *JGR*, 105, 27251
Yokoyama, T., Akita, K., Morimoto, T., Inoue, K., & Newmark, J. 2001, *ApJ*, 546, L69

Solar and Stellar Variability: Impact on Earth and Planets
Proceedings IAU Symposium No. 264, 2009
A. G. Kosovichev, A. H. Andrei & J.-P. Rozelot, eds.

doi:10.1017/S1743921309992729

RS CVn binary IM Peg – investigation of stellar activity and surface flows

Heidi Korhonen[1], Michael Weber[2], Markus Wittkowski[1], Thomas Granzer[2] and Klaus G. Strassmeier[2]

[1]European Southern Observatory, Karl-Schwarzschild-Str. 2, D-85748 Garching bei München, Germany

[2]Astrophysical Institute Potsdam, An der Sternwarte 16, D-14482 Potsdam, Germany

Abstract. We have obtained high resolution, high S/N spectra of the RS CVn binary IM Peg using UVES spectrograph at Kueyen 8.2m telescope of ESO VLT. We use Doppler imaging techniques to obtain stellar surface temperature maps from the UVES data. The TempMap code allows us to use surface differential rotation as an input parameter and thus to try to construct the rotation pattern on the stellar surface as part of the inversion process. The UVES observations are combined with spectroscopic observations from another time period obtained at the STELLA observatory. We obtain stellar surface temperature maps also from these spectra. These Doppler images are used to study the magnetic activity and surface differential rotation on IM Peg.

Keywords. stars: activity, chromospheres, spots

1. Target and observations

IM Peg is an active RS CVn binary with an orbital period of 24.45 days. The secondary showing signs for both photospheric and chromospheric activity is of spectral type K2III, and has rotational velocity of $v\sin i = 28$km/s. IM Peg is a quite well studied system of which several surface temperature maps showing spots at very high latitudes have been obtained (e.g., Berdyugina *et al.* 2000).

Very high S/N and high resolution spectroscopic observations were obtained at the European Southern Observatory with UVES (Ultraviolet and Visual Echelle Spectrograph) mounted on the 8-m Kueyen telescope of the VLT. The observations obtained in October-December 2008 have a spectral resolution of 110 000 and S/N of 350.

More high resolution spectra were obtained between June 26 and July 17 2009, at the fully robotic 1.2-m STELLA observatory in Tenerife. The fibre-fed SES spectrograph gives a resolving power of 55 000 and a wavelength coverage of 3900–9000 Å. The typical S/N of the observations was 250 at the peak, but approximately 50% less in the spectra region used for Doppler imaging.

2. Spots and surface flows

The Doppler images were obtained using the TempMap inversions code (Rice *et al.* 1989). The observations are compared to a grid of local line profiles calculated using SPECTRUM spectral synthesis code (Gray & Corbally 1994) and Kurucz model atmospheres (Kurucz 1993). Before inversions the STELLA data were de-noised using 15 of 401 components of a PCA analysis using 747 spectral lines (Carroll *et al.* 2007). In the inversion results shown here only the Fe I 6430 Å line has been used.

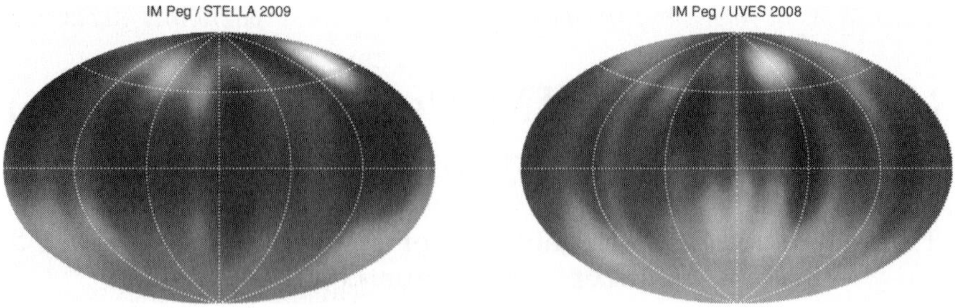

Figure 1. Temperature maps obtained from the UVES and STELLA data using Doppler imaging techniques and Fe I 6430 Å line.

The surface flows on stars can be investigated from Doppler images. In one commonly used method two temperature maps close in time are cross-correlated to study the shift of the surface features between the maps. Another way is to introduce the flow pattern in the Doppler imaging code and look for the parameter set that minimises the deviation between the observations and the model (so-called χ^2-landscape). Fig. 2 shows the χ^2-landscape for the UVES dataset of IM Peg. The result shows quite a shallow minimum indicating anti-solar surface differential rotation, i.e., with the pole rotating faster than the equator. This result is similar to the one presented by Weber (2007).

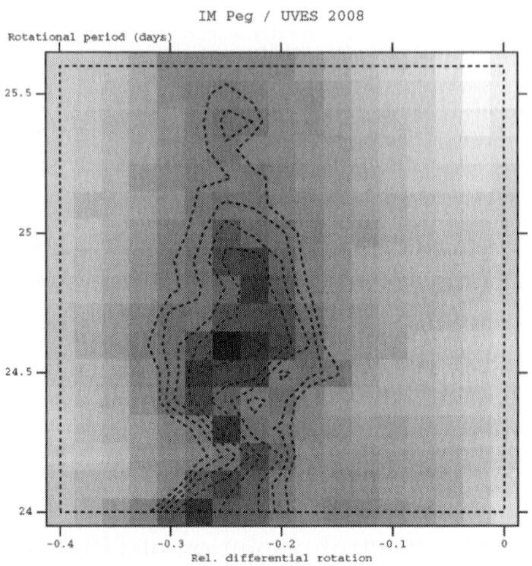

Figure 2. The goodness of fit parameter for the UVES dataset with varying period and differential rotation.

3. Chromospheric activity

The Hα profiles from all the epochs obtained with UVES are plotted in Fig. 3. In IM Peg the Hα line is in absorption, and shows clear variations with time. For studying in detail the activity in the upper atmosphere a dynamic spectrum, also shown in Fig. 3, was obtained from the Hα profiles. For showing the variations better an average spectrum (shown as a thick line in the profile plot) was first removed from all the profiles. The dynamical spectrum shows enhanced emission around phase 1.0 (0.0), most likely from plages. Signs of a flare are seen in the phase 0.98. The extra absorption at the phase 0.43 gives some indication of cool chromospheric clouds.

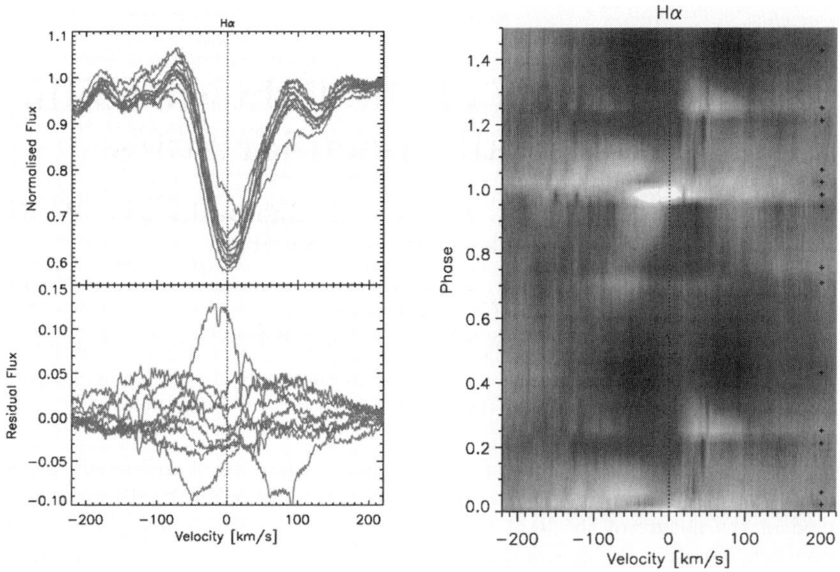

Figure 3. Left: All the Hα profiles from the UVES data. Upper plot shows the profiles normalised to the continuum, and the lower plot the residual profiles after subtracting the mean profile (thick line in the upper plot). Right: Dynamic spectrum of the Hα line based on the residual profiles. The x-axis gives the velocity and the y-axis the phase. The crosses on the right hand side of the plot give the phases of the observations and the dotted line the zero velocity. Brighter the colour the more emission is observed. The phases go from 0.0 to 1.5 to better show the flare occurring near the phase 0.0 (1.0).

References

Berdyugina, S. V., Berdyugin, A. V., Ilyin, I., & Tuominen, I. 2000, *A&A*, 360, 272
Carroll, T. A., Kopf, M., Ilyin, I., & Strassmeier, K. G., 2007 *AN*, 328, 1043
Gray, R. O. & Corbally, C. J. 1994, *AJ*, 107, 742
Kurucz, R. L. 1993, *Kurucz CD No.* 13
Rice, J. B., Wehlau, W. H., & Khokhlova, V. L. 1989, *A&A*, 208, 179
Strassmeier, K. G., Granzer, Th., Weber, M., *et al.* 2004, *AN*, 325, 527
Weber, M., Granzer, Th., Strassmeier, K. G., & Woche, M. 2008, *SPIE*, 7019, 70190L
Weber, M. 2007 *AN*, 328, 1075

Solar and Stellar Variability: Impact on Earth and Planets
Proceedings IAU Symposium No. 264, 2009
A. G. Kosovichev, A. H. Andrei & J.-P. Rozelot, eds.

doi:10.1017/S1743921309992754

The mechanism of the light variability of chemically peculiar stars

J. Krtička[1], Z. Mikulášek[1], J. Zverko[2], J. Žižňovský[2], G. W. Henry[3], J. Skalický[1] and P. Zvěřina[1]

[1]Institute of Theoretical Physics and Astrophysics, Masaryk University, CZ-611 37 Brno, Czech Republic

[2]Astronomical Institute, Slovak Academy of Sciences, SK-059 60 Tatranská Lomnica, Slovakia

[3]Center of Excellence in Information Systems, Tennessee State University, Nashville, Tennessee, USA

Abstract. Until recently, the mechanism of the light variability of chemically peculiar (CP) stars was unclear. To improve this situation, we started a theoretical and observational campaign aimed at the nature of the light variability of these stars. We use the TLUSTY model atmospheres calculated for the appropriate surface chemical composition to obtain the emergent flux and to predict the rotationally modulated light curves. We show on example of several well-studied CP stars that their light variations can be explained as a result of i) the uneven surface distribution of the elements (creating overabundant regions), ii) the flux redistribution from the ultraviolet to the visible part of the spectrum (in the overabundant regions), and iii) rotation of the star. We show that the silicon and helium bound-free transitions and iron bound-bound transitions provide the main contribution to the flux redistribution. This result is also a very precise test of modern stellar model atmospheres. We conclude that the mentioned mechanism is a very promising explanation for the light variations in CP stars of earlier spectral types.

Keywords. stars: atmospheres, stars: chemically peculiar, stars: spots, stars: variables: other

1. Magnetic chemically peculiar stars

Magnetic chemically peculiar stars belong to one of a few groups of variable stars whose light variations are not very well understood. These stars show inhomogeneous surface distribution of different elements, including helium, silicon or iron (see Fig. 1). Krtička *et al.* (2007), and Krtička *et al.* (2009) showed that the redistribution of the emergent flux due to the bound-free (continuum) and bound-bound (line) transitions in the stellar atmosphere (Fig. 3) is a mechanism explaining the observed light variations (Fig. 2). Moreover, the comparison of observed and simulated light variations can provide an important new test of modern model stellar atmospheres.

2. Light curve calculation

The models of the stellar atmospheres are calculated using the code TLUSTY (Hubeny & Lanz 1995, Lanz & Hubeny 2007) assuming a fixed stellar effective temperature and surface gravity. The abundance of chemical elements concerned is set in agreement with the maps of surface elemental distribution derived from spectroscopy (Fig. 1).

The emergent fluxes from individual surface elements are calculated using the code SYNSPEC. From these fluxes the magnitudes of the star in individual filters of Strömgren photometric system are derived (Fig. 3). Finally, these simulated light variations are compared with the observed ones (Fig. 2). We do not use any free parameter to fit the observed light curves.

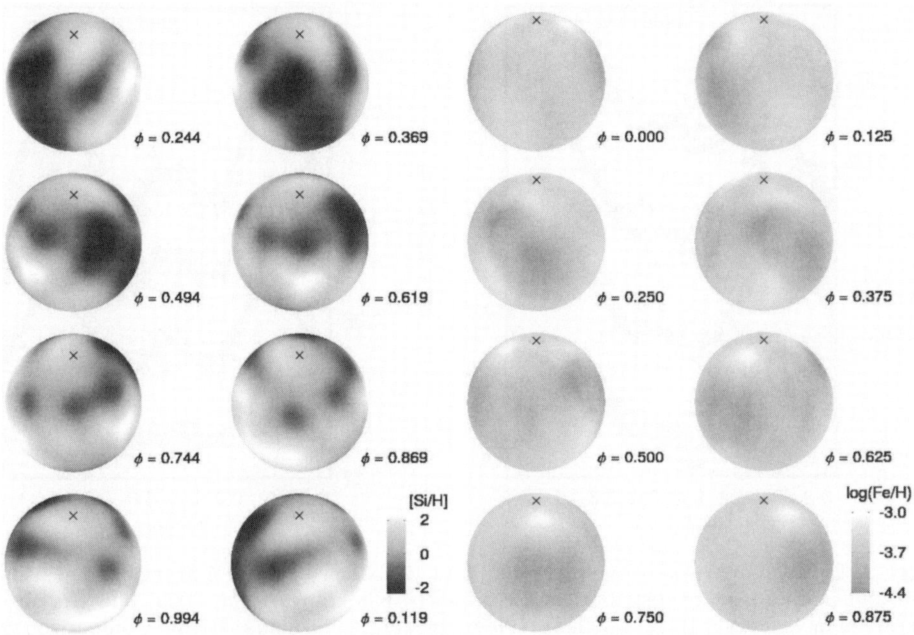

Figure 1. Left: The variation of silicon abundances on the visible disc of HD 37776 for different rotational phases after Khokhlova *et al.* (2000). Silicon is overabundant on the visible disc around the phase $\phi = 0$. Right: The variation of iron abundances on the visible disc of HR 7224 for different rotational phases after Lehmann *et al.* (2007). Iron is overabundant on the visible disc around the phase $\phi = 0$.

Here we use the ephemeris derived by Mikulášek *et al.* (2008) for HD 37776 and by Krtička *et al.* (2009) for HR 7224.

3. Conclusions

We conclude that the observed light variations of HR 7224 and HD 37776 can be explained as a result of inhomogeneous distribution of helium, silicon and iron on the surface of these stars. This causes the redistribution of the flux from the ultraviolet to the visible part of the spectrum, and, due to the stellar rotation, the light variations. Our models are also able to reproduce the Hβ line variability of HR 7224.

A very good agreement of the observed and simulated light variations serves as an independent test of modern model stellar atmospheres. The successful comparison of the predicted and observed light curves would not be possible without using the calculations of radiative ionization cross-section done by the Opacity Project team (Seaton *et al.* 1992).

Acknowledgement

This work was supported by grant GAAV IAA301630901 and VEGA 2/0074/09.

References

Adelman, S. J. 1997a, *A&AS*, 122, 249

Adelman, S. J. 1997b, *A&AS*, 125, 65

Adelman, S. J. & Pyper, D. M. 1985, *A&AS*, 62, 279

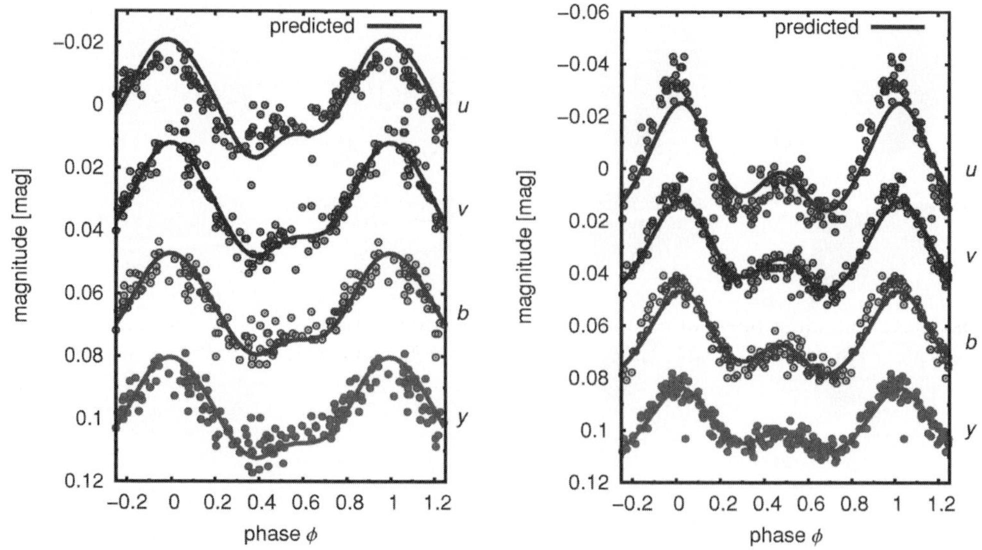

Figure 2. Left: Comparison of the predicted light variations of HD 37776 (calculated using the silicon and helium surface distribution according to Khokhlova *et al.* 2000) and the observed one (Adelman & Pyper 1985, Adelman 1997b) in different colours. Right: Comparison of the predicted light variations of HR 7224 (calculated using the iron and silicon surface distribution according to Lehmann *et al.* 2000) and the observed one (Adelman 1997a) in different colours.

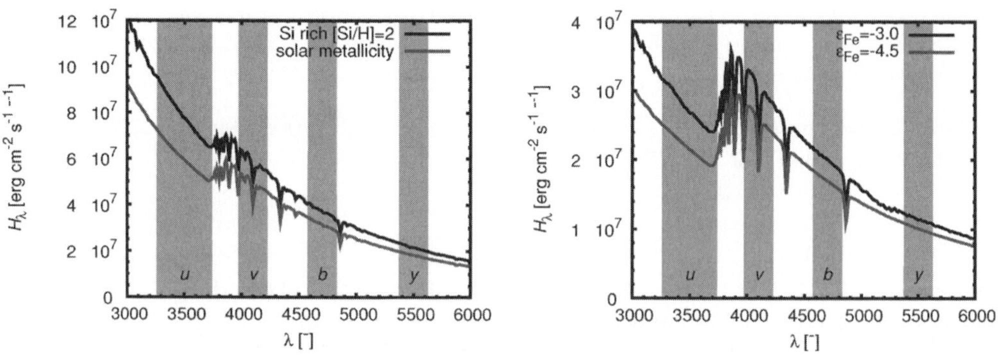

Figure 3. Left: Calculated fluxes from HD 37776 atmosphere with different silicon abundances. Theoretical flux was smoothed to better demonstrate the influence of silicon abundance. Silicon rich spots are brighter in *uvby* colours than the regions with low silicon abundance. Right: Calculated radiative fluxes from HR 7224 atmosphere with different iron abundances. Theoretical flux was smoothed to better demonstrate the influence of iron abundance. Iron rich spots are evidently brighter in *uvby* colors than the regions with low iron abundance.

Hubeny, I. & Lanz, T. 1995, *ApJ*, 439, 875

Khokhlova, V. L., Vasilchenko, D. V., Stepanov, V. V., & Romanyuk, I. I. 2000, *AstL*, 26, 177

Krtička, J., Mikulášek, Z., Zverko, J., & Žižňovský, J. 2007, *A&A*, 470, 1089

Krtička, J., Mikulášek, Henry, G. W., Zverko, J., Žižňovský, J., Skalický, J., & Zvěřina, P. 2009, *A&A*, 499, 567

Lanz, T. & Hubeny, I. 2007, *ApJS*, 169, 83

Lehmann, H., Tkachenko, A., Fraga, L., Tsymbal, V., & Mkrtichian, D. E. 2007, *A&A*, 471, 941

Mikulášek, Z., Krtička, J., Henry, G. W., Zverko, J., *et al.* 2008, *A&A*, 485, 585

Seaton, M. J., Zeippen, C. J., Tully, J. A., *et al.* 1992, *Rev. Mexicana AyA*, 23, 19

Solar and Stellar Variability: Impact on Earth and Planets
Proceedings IAU Symposium No. 264, 2009
A. G. Kosovichev, A. H. Andrei & J.-P. Rozelot, eds.

doi:10.1017/S1743921309992766

Analysis of the event of 2004 November 10

Yuan Ma and LiYing Zhu

Yunnan Astronomical Observatory/National Astronomical Observatories, Chinese Academy of Sciences, Kunming, 650011
email: mayuan@ynao.ac.cn

Abstract. The large burst of complex type which occurred at the metric, decimetric and microwave wavelengths from 0203 to 0304 UT on 2004 November 10 is analyzed in this present article, some prominent features of this event are obtained from the correlative analysis of this event with the Hα flare and coronal mass ejection (CME) and the radiation process of this event is discussed based on these observational features.

Keywords. Sun: radio radiation, flares, coronal mass ejections (CMEs), activity

1. Introduction of the active region

At 0203–0304 UT of 2004 November 10 the radio burst occurred in the active region AR10696, which has been one of the comparatively strong energetic active regions since the 23rd solar cycle and during its stay on the solar disc there occurred 45 Hα flare, located at (N09, W49), among which there were one 3B flare, 2 flares of Importance 2, 4 flares of Importance 1 and 38 subflares, and in the radio wave band (the meter-wave, decimetric wave and microwave wavelengths) there also occurred the large bursts of complex type and CMEs many times.

2. Observation and analysis

2.1. *Observed data*

The enhanced radio radiation of the event of 2004 November 10 was observed in the frequency range from 34000 MHz to 18 MHz in the lower coronal. The observed data of the 14 wave bands (245, 410, 500, 610, 986, 1161, 1336, 1415, 2695, 4995, 8800, 15400, 17000 and 34000 MHz) selected from the frequency range are used to make the analysis with those of the Hα flares and CMEs. This event was the great burst of multispike and complex type during the whole spectrum. The detailed analysis of a great burst of 47 GB type recorded at the National Astronomical Observatories/Yunnan Observatory in the frequency range of 625 to 1500 MHz is principally made in this present article (see Fig. 1). From Fig. 1 it can be seen that the burst begins in the vicinity of 0202 UT, reaches its maximum at 0206 UT and ends near 0233 UT, with the peak flux being 1776 sfu.

The Hα flares and radio materials used in this paper are take from the data of Solar Geophysical Data (SGD).

2.2. *Tim distribution of the burst*

From the data we analyze it can be seen that the meter-wave burst started at 0206–0207 UT, the decimetric wave burst began at 0203-0204 UT and the microwave burst commenced at 0203-0205 UT. The Hα flare began at 0204 UT, reached its maximum at

0210UT and ended at 0315 UT, located at (N09, W49). The CME occurred at 0226 UT, Accompanied by the type II burst. Based on the difference of the starting time of the bursts at different frequencies it can be reckoned that the place of the initial electron acceleration and energy release of the event was in the vicinity of the decimetric waveband (986–1336 MHz).

2.3. *Intensity distribution of the burst*

The radiant intensity of the burst at 610 MHz firstly reached a peak value, 6100 sfu, and two minutes later, the peak value at 15400 MHz reached the maximum, 6800 sfu, which occurred in the impulsive phase of the 3B flare. The CME which occurred at 0226 UT was in the full halo state (the angular breadth was 360 degrees), with the velocity reaching 3387 km/s. The slow-drift type II burst was also reached at Culgoora, Astralia, in the low frequency range.

2.4. *Fine structures of the decimetric waves*

From Figure 1, we can see that there are several groups of normal and reverse type III bursts with the duration of hundreds of ms and drift rate of hundreds of MHz/s. This type III bursts have a starting frequency of about 1128 MHz and ending frequency of about 1210 MHz. There are no type III burst beyond the frequency range from 1000–1215 MHz.

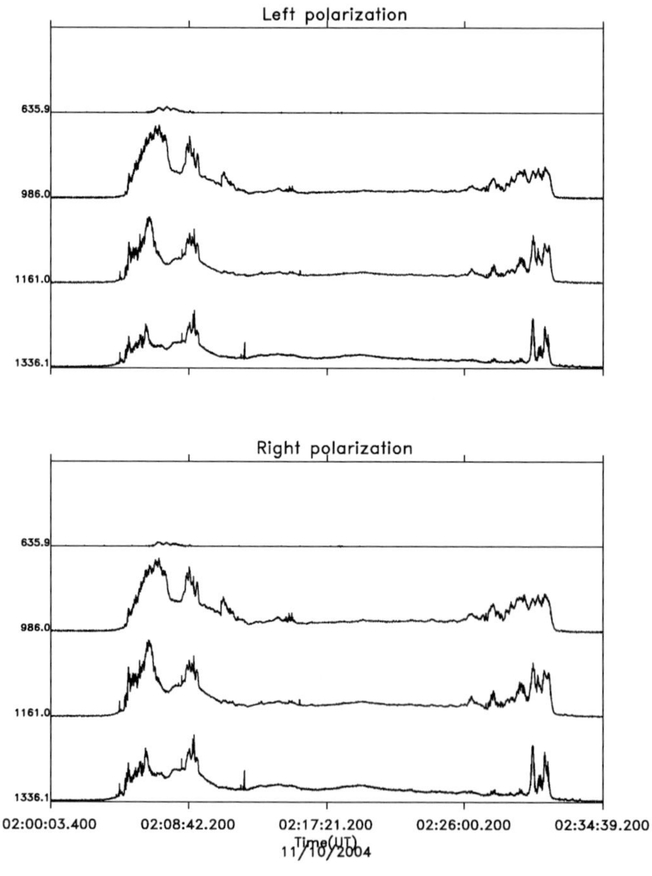

Figure 1. The solar radio burst on 2004 November 10.

3. Discussions

This burst had quite broad radiant frequency range, showing that there should exist a very large radiant region. The broad band or impulsive microwave bursts with long or short duration generally stem from the closed magnetic field in the low corona active regions and are all attributed to the synchro-cyclotron radiation of non-thermal energetic electrons (Vourlidas, 2004, Yan *et al.*, 2007, Huang and Nakajima, 2009), while the bursts at metric and decimetric wavelengths possibly originate from plasma radiation. In addition, in the coronal the propagation of the flare blast wave and the instability of the ensuing interaction of it with the prestored large-scale magnetic structure probably lead to the type II radio bursts observed in the range of metric waves (Wang *et al.*, 2002, Pohjolainen *et al.*, 2001). The meter-wave and decimetric wave type II bursts most probably result from the plasma radiation on the wave front of the MHD shock waves (Pohjolainen *et al.*, 2005).

We hold that the electron acceleration of the flare impulsive phase is caused by the electric field or electrostatic wave in the magnetic reconnection. During the impulsive phase the reconnection may be composed of lots of small current sheets and in the wave bands near the reconnection the peak value of the burst may be the multiple superposition of minimal spikes while at the place far away from the acceleration because of the degradation and diffusion of the accelerated electrons in the propagation these small spikes will be smoothed more and more, showing that the decay phase after the magnetic reconnection is the turbulence acceleration process. Thus, in the burst flux cross-sections of different wavelengths in the impulsive phase, the place where there are much more spikes may be that much closer to the magnetic reconnection electron acceleration in the flare. For example, at 986-1336 MHz, in the burst curves of the three wave bands there are the most burst peak values at 1194 MHz which should be much closer to the electron acceleration region.

In a word, the radiation process of a great burst with a long duration which occurs at metric, decimetric and microwave wavelengths is very complicated, not a homogeneous radiation mechanism but the radiation process where there exists the combination simultaneously.

Acknowledgements

This research is supported by the NSFC grants (Nos. 10473020 and 10333030) and the foundation of the Chinese Academy of Sciences. It is also supported by the Project of Cultivating Western Talents of China.

References

Huang, G. L. & Nakajima, H. 2009, *ApJ*, 702, 19
Kliem, B., Karlicky, M., & Benz, A. O. 2000, *A&A*, 360, 715
Pohjolainen, S., Vilmer, N., & Khan, J. I. 2005, *A&A*, 434, 329
Pohjolainen, S., Maia, D. J. F., Pick, M. *et al.*, 2001, *ApJ*, 556, 421
Vourlidas, A. 2004, *In Solar and Space Weather Radiophysics*, 223
Willson, R. F. 2004, *Solar Physics*, 197, 399
Wang, H. M., Gallagher, P., & Yurchyshyn, V. 2002, *ApJ*, 69, 26
Yan, Y. H., Huang, J., Chen, B., & Sakurai, T. 2007, *PASJ*, 59, 815

Solar and Stellar Variability: Impact on Earth and Planets
Proceedings IAU Symposium No. 264, 2009
A. G. Kosovichev, A. H. Andrei & J.-P. Rozelot, eds.

doi:10.1017/S1743921309992778

Study of the structures of the explosive events in the UV

R. T. Niembro-Hernandez[1], **J. E. Mendoza-Torres**[2] **and K. Wilhelm**[3]

[1]Fac. de Ciencias, UNAM, Mexico
email: tat17v@gmail.com

[2]Instituto Nacional de Astrofisica, Optica y Electronica, INAOE, Mexico
email: mend@inaoep.mx

[3]Max-Planck-Institute for Solar System Research, MPS, Germany
email: wilhelm@linmpi.mpg.de

Abstract. We analyze Si IV 139.37 nm emission line during solar Explosive Events (EE) near the center of the solar disk with the aim to study the structure of the sources of EEs observed at the VUV. The observations were made by SUMER, on board SoHO, with a raster regime of six EW positions that allowed us to identify the times and EW, NS positions of the maximum amplitude of each EE. Based on one dimensional NS distributions at three different wavelengths (blue, central and red) for times around each maximum we have identified three different shapes of the sources. Also, It was found that the maximum at the blue wing is attained after the maxima at red and central wavelengths.

Keywords. Sun: UV radiation, corona, activity, techniques: spectroscopic

1. Introduction

A class of some sudden increase of VUV emission in small spatial scale (2–4 arcsec) that take place at the solar TR are referred to as Explosive Events (EEs). EEs occur in $1-4$ minutes (Kirk *et al.* 1994), sometimes with line wing increases that denote flows of $100-250$ km/s (Innes & Toth 1999).

EEs have been observed with a rate of occurrence of more than 1000 per second all over the solar disk (Brueckner & Bartoe 1983) which could give a non negligible input of energy for Coronal Heating.

The extension of the EEs has been studied considering the region where the amplitude of the area covered by the slit was maximum, but it has not been verified if the observed increase corresponds to the maximum attained over the extension of the EE (Mendoza-Torres, Torres-Papaqui & Wilhelm 2005).

In studying the structure (projected on the disk) of EEs it is important to know the actual extension. In order to do this, it is necessary to identify the events that really where sampled all the way during the increasing and decreasing phases and whose maximum became at the central regions of the observed field. Then, we are able to study the structure over such field.

A better knowledge of the source structure can help us to identify the energy transport between the layers of the Sun around the TR, (Chae *et al.* 1998a).

2. Observations

We selected SUMER data for the Si IV line at 139.37 nm (Wilhelm *et al.* 1995 in a raster regimen that produces spectra for a field of $120''\mathrm{NS}\times 6''\mathrm{EW}$ of June 20th

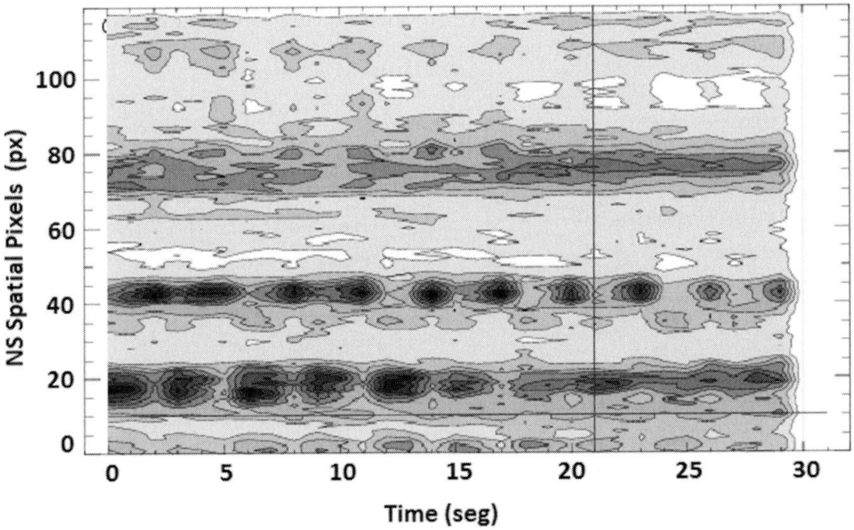

Figure 1. Plot of amplitude as function of NS position and time. It may be seen that EEs take place at some preferential locations.

1996, where Coronal Holes and Active Regions were not observed, with an integration of 5 seconds at each position.

3. Data analysis and results

From the 2.5 h of SUMER data for June 20th, 1996 we selected EEs based on some criteria. In such a way we selected EEs for which both the rising and the decreasing phases were observed and whose maximum was not observed at the extremes of the field. With these restrictions we get a sample of 41 EEs (Table 1). For the selected EEs, one dimensional NS distribution profiles were made for the amplitude at the central wavelength of the line and for wavelengths at both wings (blue and red) for times around the maximum. This method does not require to fit Gaussians in order to identify flows. We have seen that EEs are observed at some given preferential locations (Fig. 1).

The extensions of the sources during the maximum at each of the wavelengths were estimated at 0.5 and at 0.2 of the maximum amplitude.

Table 1. Amount of EEs at the different wavelengths and combinations of them

Component	Amount	Combination	Amount
All	41	Blue-Central-Red	18
		Blue-Central	2
Blue	26	Blue-Red	5
		Central-Red	5
Central	26	Blue	1
		Central	1
Red	33	Red	5

Based on the NS distributions we noticed that one after another maxima can take place around a given location which means that an EE could be related to more than one energy release. Most of the EEs have a component of low amplitude but large extension sorrounding the maximum amplitude as shown in Fig. 2c.

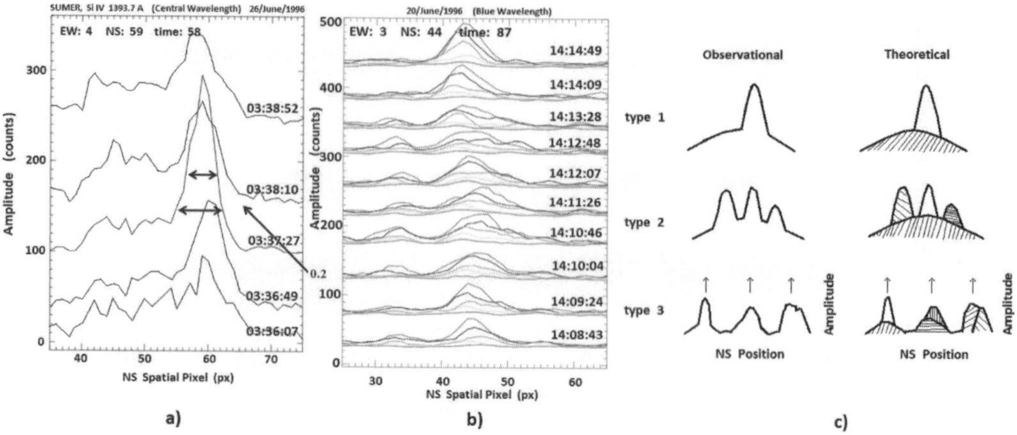

Figure 2. a) NS Distributions for the growing, the maximum and the decreasing phase of an EE. An extended region with increased amplitude is observed during the EE, local variations at this region occur during the EE. b) Amplitude distributions around the time and location of the maximum amplitude of an EE. Increases at near locations are seen. c) Schematic representation of the possible sources that could be taking place for the different types of distributions observed to occur in the studied EEs.

We found out that 18 EE showed increases of the three wavelengths and fulfilled the selection criteria.

Expansion and displacement of the sources are sometimes seen after the occurrence of the maximum amplitude (Fig. 2b).

In most of the cases the blue maxima occurred after the central maxima (11 cases of 18 EEs). For the red component we found that in 7 of 18 ocassions it took place as the central maxima (at the same time) and 4 of 18 red was attained before the central maxima (Table 1).

This behaviour and the spatial one (Fig. 2) around maxima seem to indicate that more than one energy releases take place or that it happens several up and down flows in one EE, possible in different locations and in different phases of the EE.

References

Innes, D. E. & Toth, G. 1999, Solar Physics, vol. 185, 127–141

Wilhelm, K. *et al.* 1995, Solar Physics, vol. 162, 189–231

Chae *et al.* 2003, Journal of the Korean Astronomical Society, vol. 36, 13–20

Brueckner, G. E. & Bartoe J.-D. F. 1983, Astrophysical Journal, vol. 272, 329–348

Chae, J., Wang, H., Lee, C., Goode, P., & Schuhle, U. 1998, Astrophysical Journal, vol. 504, 123

Mendoza-Torres, J. E., Torres-Papaqui, J. P., & Wilhelm, K. 2005, Astronomy and Astrophysics, vol. 431, 339–244

Brekke, P., Hassler, D. M., & Wilhelm, K. 1997, Solar Physics, vol. 175, 349–374

Dere, K. P. 1994, Advances in Space Research, vol. 14, 413–422

Solar and Stellar Variability: Impact on Earth and Planets
Proceedings IAU Symposium No. 264, 2009
A. G. Kosovichev, A. H. Andrei & J.-P. Rozelot, eds.

doi:10.1017/S174392130999278X

Nonlinear analysis of decimetric solar bursts

Reinaldo R. Rosa[1], Mauricio J. A. Bolzan[2], Francisco C. R. Fernandes[2], H. S. Sawant[1] and Marian Karlický[3]

[1]LAC and DAS, Instituto Nacional de Pesquisas Espaciais (INPE), São José dos Campos, Brazil
email: reinaldo@lac.inpe.br

[2]IPD, Universidade do Vale do Paraiba, São José dos Campos, SP, Brazil
email: mauricio.bolzam@pq.cnpq.br

[3]Ondrejov Observatory, Czech Republic

Abstract. The solar radio emissions in the decimetric frequency range (above 1 GHz) are very rich in temporal and spectral fine structures due to nonlinear processes occurring in the magnetic structures on the corresponding active regions. In this paper we characterize the singularity spectrum, $f(\alpha)$, for solar bursts observed at 1.6, 2.0 and 3 GHz. We interpret our findings as evidence of inhomogeneous plasma turbulence driving the underlying plasma emission process and discuss the nonlinear multifractal approach into the context of geoeffective solar active regions.

Keywords. turbulence, Sun: radio radiation, solar-terrestrial relations; methods: data analysis

1. Introduction

Solar energetic events, as solar flares, produce typical radio emission patterns throughout the whole radio emission spectrum range. In particular, several physical aspects of the active region are related to the high resolution and high sensitivity data observed in the lowest microwave range (1–3 GHz), usually reported as the decimetric range (Aschawanden 2005). Usually the solar flare process is related to a magneto-hydrodynamic (MHD) instability taking place in strongly anisotropic turbulent plasma (Kuperus, 1976). The importance of this MHD scenario has been investigated, for example, from nonlinear analysis of decimetric bursts at 3 GHz observed during the June 6, 2000 flare (Rosa *et al.* 2008). It was found that the 3 GHz radio burst power spectrum exhibits a power-law which is an evidence of stochastic intermittency due to a self-affine dynamics as found in the MHD turbulence theory. Intermittent energetic process implies that the fluctuations are correlated without a dominant characteristic time scale, as predicted in the models for multi-loop interactions (Tajima 1987). However, in order to characterize more precisely the nature of such inhomogeneous self-affine turbulent process, a complementary analysis, based on the so-called singularity spectrum technique, is required.

As known from the turbulence theory, the intermittency leads to deviation from usual Kolmogorov energy structure functions and its main signature are the singularity spectra exponents, $f(\alpha)$, which represent a power-law scaling-free dependence (Frisch 1995). For inhomogeneous plasma turbulence, the so-called multi-fractal *p-model* describes how the energy can be distributed among scales following a multiplicative rescaling structure (Halsey *et al.* 1986). In this paper, taking into account the *p-model* singularity spectra, we obtain the $f(\alpha)$ for typical decimetric bursts, improving our search for a robust technique by which inhomogeneous plasma turbulence process might be identified.

2. Data Analysis, Interpretation and Discussion

Fig. 1 shows the intermittent time series observed at 1.6 GHz (a), 2 GHz (b) and 3 GHz (c). The decimetric bursts at 1.6 (Fig. 1a) and 2 GHz (Fig. 1b) were observed by using the Brazilian Solar Spectroscope (BSS) with time and frequency resolutions of 100ms and 10 MHz, respectively (Madsen *et al.* 2004). Figure 1c shows the 3 GHz burst observed employing the Ondrejov radiospectrograph with time resolution of 0.01s (Jiricka *et al.*, 1993). These three decimetric bursts were related to geoeffective energetic solar events as flares and Coronal Mass Ejections.

Here we have considered, in order to obtain the singularity spectrum $f(\alpha)$ from the SFU-component time series, the Wavelet Transform Modulus Maxima (WTMM) (Mallat 1989). The basic idea behind the WTMM method is to describe a nonlinear partition function over only the modulus maxima of the wavelet transform of a signal SFU(t). Recently, the robustness of this methodology has been tested for intermittent geomagnetic fluctuations using, as a reference nonlinear pattern, the so-called *p-model* (Halsey *et al.* 1986). The *p-model* is a canonical mathematical system that describes nonhomogeneous energy cascade processes in turbulent-like flows (e.g., Rodrigues Neto *et al.* 2001). In this process the flux density is transferred to the two smallest eddies with the same length but different flux probabilities p_1 and p_2 ($p_1 + p_2 = 1$). The process is repeated several times with p_1 and p_2 randomly distributed, being the asymmetric breakdown in the fragmentation process driven by the parameter $p = p_1 = 1 - p_2$. The common value of $p_1 = p_2 = 0.5$ corresponds to the homogeneous energy transfer rate with no intermittency effects. The values of $p > 0.5$ correspond to an intermittent nonlinear dynamics found in nonhomogeneous turbulence and chaotic advection. In the present work, we use a maximum likelihood algorithm to fit the *p-model* in each characteristic solar burst scale. Hence we have performed the theoretical multiplicative cascade *p-model* as a canonical reference in our analysis.

Choosing typical parameters values for the *p-model* (Bolzan *et al.* 2009) we use an algorithm to fit the experimental and theoretical data. The results are shown in Fig. 2 from where the SFU complex pattern variability can be interpreted as the result of a possibly multifractal process related to the presence of a nonhomogeneous asymmetric energy cascade during the energy release process in the respective active regions. Characteristic values of the α for each $f(\alpha)_{max}$ show that the relation between the radio frequency and the multifractal spectrum characterized by means of α for each $f(\alpha)_{max}$ is nonlinear. This result makes evident the importance of studying in detail the presence of intermittent nonlinear phenomena driving the solar decimetric fluctuations during flares.

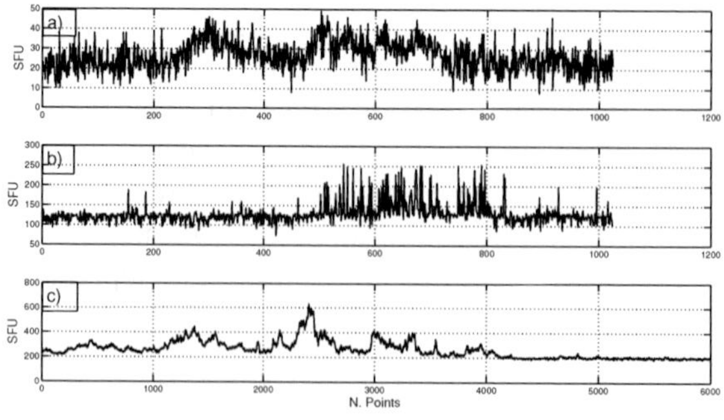

Figure 1. SFU times series for decimetric solar bursts observed at 1.6 GHz (a), 2 GHz (b) and 3 GHz (c).

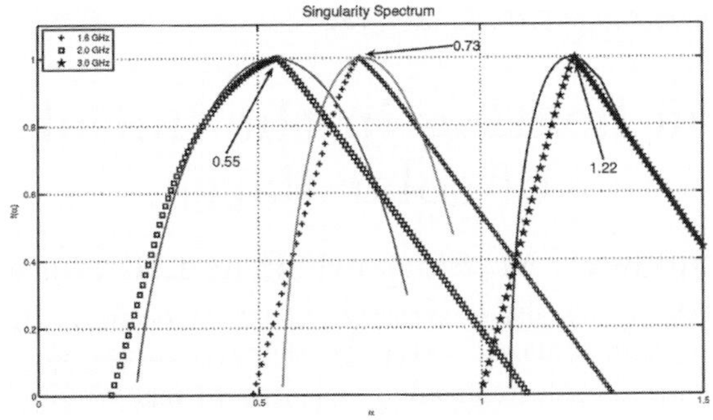

Figure 2. The singularity spectra for each SFU time series showed in Figure 1 and their corresponding *p-model* fitting. The values of the respective characteristic α for each $f(\alpha)_{max}$ are shown.

The inhomogeneous nature of the decimetric solar radio emission was successfully detected by using the WTMM. The decimetric SFU time series can be interpreted as being the response of an out-of-equilibrium process, possibly related to the particle acceleration from a transversal loop-loop interaction, where MHD oscillations can play an important role and a counterpart phenomena as turbulent interaction between electron beams and evaporation shocks can act as a secondary source. In order to check such hybrid mechanism composed by a inhomogeneous decimetric extended source higher spatial resolution data are required. Once the decimetric solar bursts analysed here are related to geoeffective events, the microphysics processes as particle acceleration and magnetic reconnection can be addressed in a more general framework involving the plasma solar-terrestrial environment.

3. Acknowledgements

The Brazilian agencies FAPESP, CNPq and CAPES.

References

Aschwanden, A. 2005, Springer, Berlin, 2005

Bolzan, M. J. A., Rosa, R. R., & Sahai, Y. 2009, *Annales Geophysicae*, 27(2), 569–576

Enescu, B., Ito, K., & Struzik, Z. 2006, *Geoph. Journal International*, 164(1), 63–74

Frisch, U. 1995, in: *Turbulence*, Cambridge University Press, New York

Halsey, T. C., Jensen, M. H., Kadanoff, L. P., Procaccia, I., & Shraiman, B. I. 1986, *Phys. Rev. A*, 33, 1141

Kuperus, M. 1976, *Solar Phys.*, 47(1), 7990

Madsen, F. R. H., Sawant, H. S., Fernandes, F. C. R., & Cecatto, J. R. 2004, *Braz. J. Phys.*, 34(4b)

Mallat, S. 1989, *IEEE Trans. Pat. An. Mach. Intel.* 11, 674–693

Oswiecimka, P., Kwapien, J., & Drozdz, S. 2006, *Phys. Rev. E*, 74, 016103

Rodrigues Neto, C., Zanandrea, A., Ramos, F. M., Rosa, R. R., Bolzan, M. J. A., S, L. D. A. 2001, *Physica A*, 295(1–2), 215–218

Rosa, R. R., Karlicky, M., Veronese, T. B., Vijaykumar, N. L., Sawant, H. S., Borgazzi, A. I., Dantas, M. S., Barbosa, E. B. M., Sych, R. A., & Mendes, O. 2008, *Adv. Sp. Res.*, 41(5), 844–851

Struzik, Z. R. 2000, *Fractals*, 8, 163–179.

Tajima, T., Sakai, J., Nakajima, N., Kosugi, T., Brunel, F., & Kundu, M. R. 1987, *ApJ*, 321, 1031

Solar and Stellar Variability: Impact on Earth and Planets
Proceedings IAU Symposium No. 264, 2009
A. G. Kosovichev, A. H. Andrei & J.-P. Rozelot, eds.

doi:10.1017/S1743921309992791

Plasma heating in the initial phase of solar flares

P. Rudawy[1], **M. Siarkowski**[2] **and R. Falewicz**[1]

[1]Astronomical Institute, University of Wrocław, 51-622 Wrocław, ul. Kopernika 11, Poland
email: rudawy; falewicz@astro.uni.wroc.pl

[2]Space Research Centre, Polish Academy of Sciences, 51-622 Wrocław, ul. Kopernika 11, Poland
email: ms@cbk.pan.wroc.pl

Abstract. In this paper we analyze soft and hard X-ray emission of the 2002 September 20 M1.8 *GOES* class solar flare observed by *RHESSI* and *GOES* satellites, where soft X-ray emission precedes the onset of the main bulk hard X-ray emission by ~5 min. This suggests that an additional heating mechanism may be at work at the early beginning of the flare. However *RHESSI* spectra indicate presence of the non-thermal electrons also before impulsive phase. So, we assumed that a dominant energy transport mechanism during rise phase of solar flares is electron beam-driven evaporation. We used non-thermal electron beams derived from *RHESSI* spectra as the heating source in a hydrodynamic model of the analyzed flare. We showed that energy delivered by non-thermal electron beams is sufficient to heat the flare loop to temperatures in which it emits soft X-ray closely following the *GOES* 1–8 Å light-curve.

Keywords. Sun: flares, Sun: X-rays, gamma rays

1. Introduction

There are several papers that investigate temporal dependencies between the beginnings of SXR and HXR emissions. On average the SXR emission starts 3 min before the hard X-ray emission (see Machado *et al.* 1986, and Veronig *et al.* 2002).

An enhanced thermal emission preceding the onset of the hard X-rays may be indicative of a thermal preheating phase prior to the impulsive electron acceleration. Current-sheet models (e.g., Heyvaerts *et al.* 1977) of solar flares predict a preheating phase. Li *et al.* (1987) discussed the preheating phase of solar flares triggered by new emerging magnetic flux.An alternative mechanism to electron beam-driven evaporation, namely conducted driven evaporation, was developed lately by Battaglia *et al.* (2009). They explain the time evolution of the observed emission for all analyzed events as an effect of saturated heat flux. The tendency of the soft X-ray flux to appear before hard X-rays emission can be attributed also to the sensitivity threshold of the hard X-ray detectors Dennis (1988).

In this paper we show, using unprecedented high sensitivity of the RHESSI detectors Lin *et al.* (2002) and a numerical model of flare, that early SXR emission observed prior to impulsive phase could be fully explained without any ad-hoc assumptions (at least for analyzed event). All necessary energy to explain the soft emission could be derived from observed HXR spectra.

2. Observations and modeling of the heating of the loop

For our work we selected RHESSI observations obtained without the activation of the attenuators in order to prevent discontinuities in fitting parameters. A solar flare with a simple single-loop structure was then chosen for convenience in numerical modeling. The investigated flare occurred in active region NOAA 10126 (S23E69) on 2002 September 20. It was classified as an M1.8 *GOES* class flare. The SXR emission started to increase slowly at 09:18:15 UT (see Fig. 1). The *RHESSI* data were analyzed using RHESSI OSPEX package of the SolarSoftWare system. Data were summed over the front segments of the seven detectors with detectors number 2 and 7 excluded. The spectra were measured with 4 sec temporal resolution in 158 energy bands from 4 to 300 keV. We applied the energy widths $dE = 0.3$ keV within the range 4–15 keV, $dE = 1.0$ keV in the range 15–100 keV, and $dE = 5.0$ keV above 100 keV. The analyzed spectra were corrected for pulse pile-up, decimation and albedo effects.

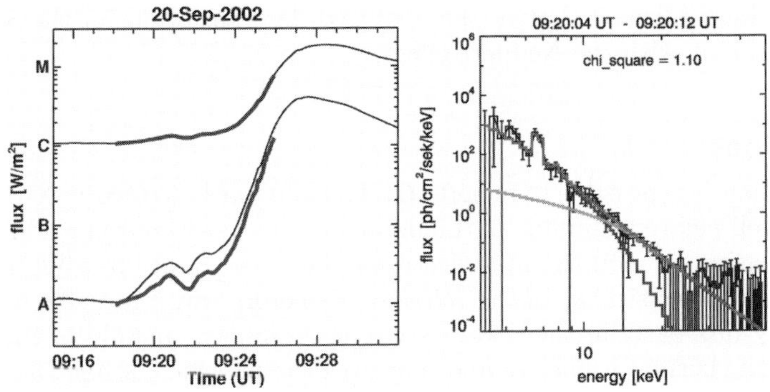

Figure 1. Left panel: *RHESSI* spectral fit results for data accumulated with a 8 sec time interval between 09:20:04-09:20:12 UT. The spectrum was fitted with one temperature thermal model (blue color) and thick target model (green color) with energy cut-off $E_c = 15.8$ keV. Total fitted spectrum is shown in red color. **Right panel:** Comparison of the observed (black thin lines) and calculated (thick blue line) *GOES* fluxes in 1–8 Å (upper curves) and 0.5–4. Å (lower curves) energy bands during initial phase of the analyzed flare.

The spectra were fitted with single temperature thermal plus thick-target models. The thermal model was based on the X-ray continuum and line emission calculated by the CHIANTI atomic code. Figure 1 shows the fit of the spectrum, accumulated between 09:20:04 UT and 09:20:12 UT. In addition to the thermal component, there is an important power law shape for emission between 15 keV and 20 keV which can be recognized and fitted as thick-target emission of the non-thermal electrons.

In this work we used 1D Naval Research Laboratory Solar Flux Tube Model code by Mariska and his co-workers (Mariska *et al.* 1982, Mariska *et al.* 1989), slightly modified by us. We included: new radiative loss and heating functions, the VAL-C model of the initial structure of the lower part of the loop, and double precision calculations. For details see the paper by Falewicz *et al.* (2009).

We modeled the evolution of the analyzed flare as follows: an initial, quasi-stationary pre-flare models of the flaring loop was built using geometrical and thermodynamic parameters estimated from *RHESSI* and *GOES* data. These initial parameters of the flaring loop were as follows: semi-length 9500 km, radius 1900 km, pressure in feet 22 dyn/cm^2. Small volumetric heating was used to keep this model in the quasi stationary state on the pre-flare level of activity before the start of the non-thermal heating. Then we started to model the heating of the loop by non-thermal electrons adding a dose of energy and calculating the resulting *GOES* flux. We used thick target parameters N_{nth}, δ and E_c obtained from fits of consecutive *RHESSI* spectra for each time-step as input into the Fisher's heating function. However, an acceptable fit can be obtained for all E_c values in the range 5–30 keV for all following spectra. Because the low energy cut-off determines an amount of energy delivered to the loop, this non-uniqueness could be limited using an independent energetic condition, like observed 1–8 Å GOES flux. Indeed, for each time step we adjusted E_c values in order to achieve conformity of the observed and modeled fluxes in *GOES* 1–8 Å band. Such use of GOES 1–8 Å flux put important limitation on the allowed low energy cut-off values and reduces importantly the non-uniqueness problem.

The final result of our modeling is presented in Figure 1 - right panel. The synthesized *GOES* 1–8 Å light-curve closely follows the observed one. The correspondence between observed and calculated fluxes in the case of the 0.5–4 Å band is not so ideal but as a result we fully restored the observed slow increase of the SXR flux recorded before impulsive heating using only non-thermal electrons characteristics derived from the observed HXR spectra. Thus we confirmed that in the analyzed flare variations of the SXR and HXR fluxes are consistent with the Neupert effect.

3. Conclusions

We showed that it is possible to fit soft (*GOES*) and hard (*RHESSI*) X-ray emissions of a solar flares well before beginning of the impulsive phase without any additional heating besides the heating by non-thermal electrons. This was made possible because of the unprecedented high sensitivity of the *RHESSI* detectors which are able to measure very low hard X-ray flux early in the flare. Part of the emission, mainly in the energy range $< 25keV$, is non-thermal in nature and indicates the presence of non-thermal electrons. In this case of a M1.8 GOES class flare, the non-thermal electron energy fluxes of the order of 10^{26} $ergs/sec$, derived under the thick target interpretation, fully explains the required heating of the plasma and resulting increase in SXR emission.

Our results extend the standard model of SXR and HXR relationship to the early phases of solar flares and thus expands the number of flares consistent with the Neupert effect. These results also indicates that the process of electrons acceleration appears during the early stage of the flare, well before the impulsive phase.

Our method of adjustment of the low energy cut-off E_c in order to equalize synthesized and observed *GOES* fluxes in 1–8 Å channel can be considered as a new method of E_c determination.

4. Acknowledgments

The authors acknowledge the *RHESSI* consortium. This work was supported by the Polish Ministry of Science and Higher Education, grant number N203 022 31/2991.

References

Battaglia, M., Fletcher, L., & Benz, A. O. 2009, *A&A*, 498, 891
Dennis, B. R. 1988, *Solar Phys.*, 118, 49
Falewicz, R., Rudawy, P., & Siarkowski, M. 2009, *A&A*, 500, 901
Heyvaerts, J., Priest, E. R., & Rust, D. M. 1977, *ApJ*, 216, 123
Li, H.-W., Pallavicini, R., & Cheng, C. 1987, *Solar Phys.*, 107, 271
Lin, R. P., Dennis, B. R., Hurford, G. J., *et al.* 2002, *Solar Phys.*, 210, 3
Machado, M. E., Orwig, L. E., & Antonucci, E. 1986, *Adv. Space Res.*, 6(6), 101
Mariska, J. T., Boris, J. P., Oran, E. S., Young, T. R. Jr., & Doschek, G. A. 1982, *ApJ*, 255, 738
Mariska, J. T., Emslie, A. G., & Li, P. 1989, *ApJ*, 341, 1067
Neupert, W. M. 1986, *ApJ*, 153, L59
Veronig, A., Vrsnak, B., Temmer, M., & Hanslmeier, A. 2002, *Solar Phys.*, 208, 297

Solar and Stellar Variability: Impact on Earth and Planets
Proceedings IAU Symposium No. 264, 2009
A. G. Kosovichev, A. H. Andrei & J.-P. Rozelot, eds.

doi:10.1017/S1743921309992808

The influence of spicules in the solar radius at multiple radio wavelengths

Caius L. Selhorst[1,2], Adriana Silva-Válio[2], Priscila A. Martins[2], Daiane B. Seriacopi[2], Pierre Kaufmann[2] and Hugo Levato[3]

[1]IP&D - Universidade do Vale do Paraíba - UNIVAP, Av. Shishima Hifumi, 2911
Urbanova, São José dos Campos, SP 12244-000, Brasil
email: caius@univap.br

[2]CRAAM, Universidade Presbiteriana Mackenzie, São Paulo, SP, Brasil
[3]CASLEO, San Juan, Argentina

Abstract. In this work, we analyze observations of the solar radius at 22 and 43 GHz obtained with the 13.7 m antenna of the Itapetinga Radio Observatory (Atibaia, Brazil) and at submillimeter-wave frequencies, 212 and 405 GHz, obtained by the Solar Submillimeter-wave Telescope (SST) (El Leoncito, San Juan, Argentina). The radius is defined as the limb position where the intensity is equal to half of the quiet Sun value. These measured radii are then compared with those predicted by a model of the solar atmosphere proposed by Selhorst, Silva, and Costa (2005). The results show that at 22 and 43 GHz, the emission comes from regions high in the chromosphere. Furthermore, the Itapetinga observations yield radii of 985" ± 5" and 981" ± 6", at 22 and 43 GHz respectively, consistent with the theoretical positions in the atmosphere. On other hand, the submillimeter observations resulted in a mean radius of 972" ± 3" and 975" ± 5" at 212 and 405 GHz, respectively, considered equal within the uncertainties. The latter results can be explained by the origin of the emission being very close to the region of minimum temperature, between the photosphere and chromosphere. This is a dynamic region largely affected by many solar features, like spicules and plages.

Keywords. Sun: radio radiation, Sun: radius, Sun: spicules

1. Introduction

The observations of solar radius at multiple radio wavelengths provide an important tool to study the structure of the solar atmosphere, since different frequencies are formed at distinct atmospheric heights. Moreover the height, with respect to the solar surface, where the emission originates increases with wavelength.

Costa *et al.* 1999 and Selhorst *et al.* 2004 showed that the solar radius measured at the chromosphere and layers above it, reflects the solar atmospheric heating during the solar cycle. In the atmospheric model SSC proposed by Selhorst *et al.* 2005, the authors used the observations of solar radius at 17 GHz to adjust the model, and concluded that the presence of spicules increase the the solar radius at this frequency.

In this work, we observe the solar radius at 4 distinct frequencies (22, 43, 212, and 405 GHz) studding the solar emission at different atmosphere heights.

2. Observations

The observations at 22 and 43 GHz were obtained with the 13.7 m antenna of the Itapetinga Radio Observatory (Atibaia, Brazil) between 10 and 14 March of 2006, which resulted in 37 maps.

A total of 92 azimuth maps, obtained by the Solar Submillimeter-wave Telescope (SST) (El Leoncito, Argentina) at 212 and 405 GHz (June - August/2005) were analyzed here.

We defined the solar radius as follow:

- The quiet Sun was take as the most common intensity in the solar disk;
- For each intensity profile the limb was defined as the position where the intensity is half the quiet Sun value taken in two points for each scan;
- These points are then fit by a circumference, which yields a mean radius value.

Table 1. Observational Radius

Frequency (GHz)	Radius (arcsec)
22	985 ± 6
43	981 ± 6
212	972 ± 3
404	975 ± 5

The results of mean solar radius are summarized in Table 1. Except for the 405 GHz radius, the observations show that the radius reduces with the increase of the frequency, which agrees with the theory.

3. Atmospheric Model

The SSC is a numerical atmospheric model (Selhorst, Silva & Costa 2005) proposed to reproduce the radio frequencies observations. The authors suggested that an extended chromosphere (up to 3500 km) is necessary to reproduce the radio observation, which represents the average effect of small size atmospheric features, such as spicules, for example.

Table 3, summarizes the radius and limb brightening results at the frequencies studied here, after the convolution with a Gaussian beam similar to the antenna resolution. A small limb brightening is predict in the simulations, however, it is not clearly defined by the observations. The simulated radius at 22 and 405 GHz are compatible with the observations, while the radius at 43 and 212 GHz are 1 arcsec outside the measurements even when the error bars are considered.

Table 2. Simulated Radius

Frequency (GHz)	HPBW (arcmin)	Quiet Sun (K)	Limb Brightening (% above the Quiet Sun)	Simulated Radius (arcsec)	Observed Radius (arcsec)
22	4	9890	5	983	985 ± 6
43	2	8750	7	974	981 ± 6
212	4	6680	5	976	972 ± 3
404	3	5690	7	976	975 ± 5

Although the SSC model represents a mean atmosphere, including the influence of atmospheric features, the spicules influence close to the limb differs from those at disc center, since they can reach coronal heights.

The spicules characteristics (density, temperature, height, inclination, and position in the surface) are attributed using the Monte Carlo Method (for more details see Selhorst2005). Figure 1 shows the simulation result for an atmosphere with spicules calculated at 22 GHz. The black dots represent the simulation with 1 arcsec resolution, whereas the black curve is the result convolved with a 4' Gaussian beam.

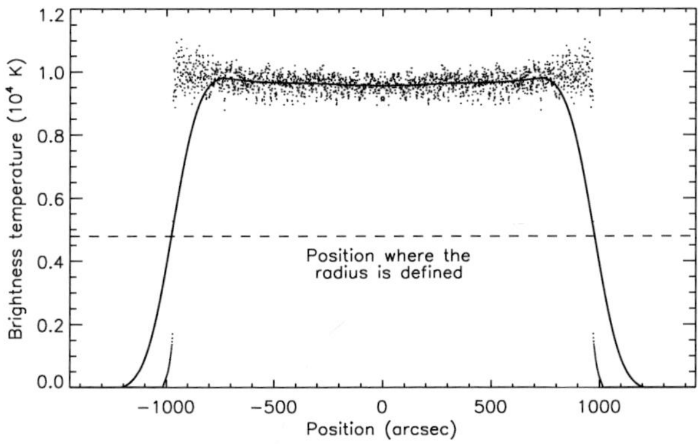

Figure 1. Simulation result for an atmosphere with spicules calculated at 22 GHz. The black dots represent the simulation with 1 arcsec resolution, whereas the solid black curve is the result convolved with a 4' Gaussian beam.

In the simulations, we used the spicules physical parameters used in Selhorst *et al.* 2005 at 17 GHz. For these type of spicules the results are:

- reduction of the limb brightening at all frequencies studied;
- increase of the radius at submillimeter waves (212 and 405 GHz);
- the radius at 22 and 43 GHz were smaller than in the simulations without spicules.

Table 3. Simulated Radius with spicules

Frequency (GHz)	HPBW (arcmin)	Quiet Sun (K)	Limb Brightening (% above the Quiet Sun)	Simulated Radius (arcsec)	Observed Radius (arcsec)
22	4	9600	2	977	985 ± 6
43	2	8690	3	973	981 ± 6
212	4	6740	4	979	972 ± 3
404	3	5790	7	980	975 ± 5

4. Summary

The results of the simulations can be separated in two parts:

- Model without spicules:

- agrees with the observations of radius at 22 and 405 GHz, while at 43 and 212 GHz are 1 arcsec outside of the measured with the error bars;

- presented small limb brightening, between 5 and 7% above the quiet Sun. These values are within the noise level of the data.

- Spicules influence:

- the spicules reduce the limb brightening in the simulations, except for 405 GHz;

- spicules increase the radius at submillimeter frequencies, and reduce the microwave ones. However, the results without spicules are in better agreement with the observations.

References

Costa, J. E. R., Silva, A. V. R., Makhmutov, V. S., Rolli, E., Kaufmann, P., & Magun, A. 1999, *ApJ*, 520, L63

Selhorst, C. L., Silva, A. V. R., & Costa, J. E. R. 2004, *A&A*, 420, 1117

Selhorst, C. L., Silva, A. V. R., & Costa, J. E. R. 2005, *A&A*, 433, 365

Solar and Stellar Variability: Impact on Earth and Planets
Proceedings IAU Symposium No. 264, 2009
A. G. Kosovichev, A. H. Andrei & J.-P. Rozelot, eds.

doi:10.1017/S174392130999281X

Stellar flare diagnostics from multi–wavelength observations

Alexander V. Stepanov[1], **Yuri T. Tsap**[2,1] **and Yulia G. Kopylova**[1]

[1]Pulkovo Observatory
196140 S.–Petersburg, Pulkovskoye chaussee 65/1, Russia
email: stepanov@gao.spb.ru; yulia00@mail.ru

[2]Crimean Astrophysical Observatory,
98409 Nauchny, Crimea, Ukraine
email: yur@crao.crimea.ua

Abstract. Quasi–periodic pulsations in various wavebands are natural manifestations of emission of stellar flares. We suggest a diagnostic tool of stellar flares based on the coronal seismology and the solar–stellar analogy. Two approaches are used: (I) flare loop as a resonator for MHD oscillations and (II) flare loop as an equivalent electric circuit. Using optical, X–ray, and radio data we obtained flare plasma parameters for the red dwarfs EQ Peg, AT Mic, and AD Leo. The characteristic length of stellar flare loops $l \sim R_\star$ and their electric currents turned out to be one–two orders of magnitude lager than the solar ones. Advantages of proposed diagnostics in comparison to the scaling law methods are given.

Keywords. stars: late–type, flare, plasmas, magnetohydrodynamics, oscillations, coronae

1. Introduction

Cool dwarf stars are the dominant stellar component of the Galaxy. According to the Sloan Digital Sky Survey more than 50% of stars between M4–M9 have the high level of the magnetic activity (West *et al.* 2004).

Observations show that in contrast to solar flares the visual luminosity of such stars can increase by several orders of magnitude and their emission is often at a maximum in the optical wavelength range. Nonetheless, there is currently much evidence for a common origin of the flare energy release on the Sun and these stars (Gershberg 2005).

The main structural elements of the solar corona are magnetic loops. According to the universally accepted scenario of solar flares, accelerated particles precipitate at the footpoints of coronal loops (magnetic traps). As a result, the heated dense plasma of the lower atmosphere emits in Balmer lines and in the optical continuum, whereas its hottest part with a temperature $T = 10^7 - 10^8$ K, while evaporating, fills the loop structure and radiates in the UV and soft X–ray ranges. Microwave emission is determined by non–thermal electrons of the coronal magnetic trap.

Scaling law methods based on the consideration of the energy balance in a flare loop are usually used for the flare plasma diagnostics (e.g., Haisch 1983). Meanwhile such approach has serious disadvantages. For example, it is usually suggested that the thermal flux at the footpoints is equal to zero. As a result of this crude approximation as well as a large number of unknown parameters the estimates obtained by many authors differ markedly. Therefore, the methods of coronal loop diagnostics on flare stars should be developed.

2. Flare plasma diagnostics

Two approaches are used for diagnostics of flare loops based on coronal seismology. According to the first approach a flare loop is considered as a resonator for MHD oscillations. The second one results from the analogy between a flare loop and an equivalent electric RLC–circuit. Let us consider these models in more detail.

Sausage modes and phase relations Equations described the linear oscillations in of the magnetic tube are (Roberts, 1995)

$$\frac{\partial \delta P}{\partial t} = \rho v_A^2 \frac{\partial \delta v_z}{\partial z} - \rho_0 (v_A^2 + c_s^2) \nabla \cdot \delta \mathbf{v}, \tag{1}$$

$$\rho \left(\frac{\partial^2}{\partial t^2} - v_A^2 \frac{\partial^2}{\partial z^2} \right) \delta \mathbf{v}_\perp + \nabla_\perp \frac{\partial \delta P}{\partial t} = 0, \tag{2}$$

$$\rho \left(\frac{\partial^2}{\partial t^2} - c_T^2 \frac{\partial^2}{\partial z^2} \right) \delta v_z + \frac{c_T^2}{v_A^2} \frac{\partial}{\partial z} \left(\frac{\partial \delta P}{\partial t} \right) = 0, \tag{3}$$

where the standard notation is used, $\delta P = \delta p + B \delta B / 4\pi$ is the disturbance of the total pressure and c_T is the cusp velocity.

In the case of harmonic oscillations

$$\left(\frac{\partial}{\partial t^2} - v_A^2 \frac{\partial^2}{\partial z^2} \right) \delta \mathbf{v}_\perp = \alpha_1 \delta \mathbf{v}_\perp, \quad \left(\frac{\partial}{\partial t^2} - c_T^2 \frac{\partial^2}{\partial z^2} \right) \delta v_z = \alpha_2 \delta v_z,$$

where α_1 and α_2 are real numbers, using (2) and (3), we have

$$\delta \mathbf{v}_\perp = -\frac{1}{\alpha_1 \rho} \nabla_\perp \frac{\partial \delta P}{\partial t}, \quad \delta v_z = -\frac{1}{\alpha_2 \rho} \frac{c_T^2}{v_A^2} \frac{\partial}{\partial z} \left(\frac{\partial \delta P}{\partial t} \right). \tag{4}$$

For standing waves, which are formed due to reflection of eigen–modes of coronal loops from the photosphere, assuming $\delta P \propto f(r) \exp(-i\omega t + im\varphi) \sin kz, \quad m = 0, 1, 2, \ldots,$ equations (4) give

$$\delta v_r \propto \frac{i\omega}{\alpha_1} \frac{\partial f(r)}{\partial r} \sin kz, \quad \delta v_\varphi \propto -\frac{\omega m}{\alpha_1} f(r) \sin kz, \quad \delta v_z \propto -\frac{\omega k f(r)}{\alpha_2} \cos kz. \tag{5}$$

As follows from equations (5) the phase shift on kz between components of velocities $\delta v_\perp$ and δv_z is $\pi/2$, and when $\delta v_\perp = 0$, then $\delta v_z \neq 0$, and vice versa. It means that sausage modes can not be excited in coronal loops with freezing footpoints, where $\delta v = 0$. However, this problem does not arise, if the ratio of amplitudes $|\delta v_z|/|\delta v_\perp|$ will be negligibly small.

Using equations (1)–(3), the Hain–Lüst equations, and equation for amplitude of sausage modes ($m = 0$), we can obtain (Tsap 2006)

$$\frac{|\delta v_z|}{|\delta v_r|} \approx \frac{ka}{\eta_0} \frac{\beta}{1 + \beta}, \tag{6}$$

where $\eta_0 \approx 2.4$ is the first of the Bessel function J_0, the plasma parameter $\beta = 8\pi n T / B^2$. Equation (6) suggests that excitation of sausage oscillations in the thick ($l \sim a$) magnetic tubes with $\beta \sim 1$ is the problem since oscillations must be damped very quickly because of violation of freezing-in conditions of loop footpoints into the photosphere.

Mathiodakis *et al.* (2006) proposed that ten seconds oscillations from EQ Peg, observed with William Herschel Telescope, are caused by non–leaky sausage modes and got flare parameters using Haisch's scaling laws (Haisch 1983). However, these modes can be excited only in the thick magnetic flux tubes(Stepanov *et al.* 2005), and taking values adopted by Mathiodakis *et al.* (2006), from (6) we obtain $|\delta v_z|/|\delta v_r| \approx 0.2$. These oscillations are impossible due to freezing footpoints of a loop.

From MHD equations for sausage oscillations, assuming that the modulation depth M is determined by the flux of precipitated electrons towards the loop footpoints and Q–factor is determined by the electron thermal conductivity and ion viscosity, we have derived relations for the temperature, plasma density, and magnetic field (Stepanov *et al.* 2005)

$$T \approx 1.2 \times 10^{-8} \frac{\tilde{r}^2 M}{T_p^2 \chi}, \quad n \approx 3.49 \times 10^{-13} \frac{\tilde{r}^3 \tilde{\kappa} M^{5/2} Q \sin^2 \theta}{T_p^4 \chi^{3/2}},$$

$$B \approx 3.81 \times 10^{-18} \frac{Q^{1/2} \tilde{r}^{5/2} \tilde{\kappa}^{1/2} M^{5/4} \sin \theta}{T_p^3 \chi^{5/4}}, \tag{7}$$

where $\theta = \arctan(k_\perp/k_\parallel)$ is the angle between the direction of the magnetic field $\mathbf{B}$ and the wave vector $\mathbf{k}$, $\tilde{r} = 2\pi a/\eta_0$, $\chi = 20/3\beta + 2$, $\tilde{\kappa} = 486\beta \cos^2 \theta + 1$. According to flare optical oscillations observed on EQ Peg by Mathioudakis *et al.* (2006), $T_p \approx 10$ s, $M \approx 0.1$, and $Q \approx 30$. Suggesting the aspect ratio $a/l = 0.1$ and $\theta = \mathrm{arctg}(\eta_0 l/\pi a) = 76°$, radius of the loop cross section $a = 10^9$ cm, according to (7), we have $T \approx 6 \times 10^7$ K, $n \approx 2.7 \times 10^{11}$ cm^{-3} $B \approx 540$ G. These estimates are distinguished from values determined on the basis of Haish's scaling laws and adopted by Mathioudakis *et al.* (2006) ($n = 4 \times 10^{12}$ cm^{-3}, $B = 1100$ G).

Slow magnetoacoustic oscillations and X–ray pulsations Mitra–Kraev *et al.* (2005) presented soft X–ray observations with XMM–Newton, which revealed damped flare oscillations from the red dwarf AT Mic with a period $T_p = 750$ s and an exponential damping time $\tau \approx 2000$ s.

Using X-ray spectra and a multi–temperature model, Raassen *et al.* (2003) estimated for this event a flare temperature of $T \approx 2.4 \times 10^7$ K and an electron density of $n \approx 4 \times 10^{10}$ cm^{-3}. Therefore we can compare these results with results obtained by the coronal seismology method.

It is easy to show that damping of slow magnetoacoustic oscillations is determined by the electron thermal conduction and the damping rate is (Stepanov *et al.* 2006)

$$\frac{1}{\tau} \approx 75.6 \frac{T^{3/2}}{T_p^2 n} \ [\mathrm{s}^{-1}]. \tag{8}$$

At $T \approx 2.4 \times 10^7$ K, $T_p \approx 750$ s, and $\tau = 2000$ s, equation (8) gives the number density $n \approx 3.2 \times 10^{10}$ cm^{-3}. This is in good agreement with estimate of Raassen *et al.* (2003), which were found by an independent technique from X–ray spectra.

Pulsating microwave emission from AD Leo and LRC–circuit Zaitsev *et al.* (1998), taking into account the generalized Ohm's law, have established that a current–carrying magnetic loop can be considered as a LRC–circuit. The global electrodynamics equation, described electric current fluctuations of a loop $|\tilde{I}| \ll |I|$ has the form

$$\frac{1}{c^2} L \frac{\partial^2 \tilde{I}}{\partial t^2} + R(I) \frac{\partial \tilde{I}}{\partial t} + \frac{\tilde{I}}{C(I)} = 0, \tag{9}$$

where the circuit inductance and the effective capacity of a loop are

$$L = 4l \ln\left(\frac{4l}{\pi a} - \frac{7}{4}\right), \quad C(I) = \frac{4\pi a^4 \rho}{l(4I^2 + c^2 a^2 B_z^2)}.$$

When $I \gg caB_z/2$, at $n = 10^{10}$ cm^{-3}, $l = 10^{10}$ cm, $S = \pi a^2 = 10^{18}$ cm^2, from (9) we have

$$T_p = \frac{2\pi}{c}\sqrt{LC(I)} \approx 10\left(\frac{10^{12}A}{I}\right)\ [s]. \qquad (10)$$

Bastian *et al.* (1990) observed on the May 4, 1987 the stellar flare on the AD Leo with the Arecibo radio telescope at 21 cm and revealed quasi–periodic pulsations with $T_p \approx 0.7$ s. Since the Q—factor of this oscillations was very high ($\gtrsim 200$) they could not be caused by magnetoacoustic modes of coronal loop because of strong dissipation. This suggests that the origin of observed pulsations is connected with fluctuations of $\tilde{I}$ in the current–carrying loop and, according to equation (10), the value of electric current $I \approx 10^{13}$ A, which is about two orders of magnitude higher than the typical value on the Sun. It should be noted that the similar high–Q oscillations with $T_p \approx 2$ s were also revealed on the May 19, 1997 at the Effelesberg radio telescope (Zaitsev *et al.* 2004). Using equation (10), we obtain $I \approx 5 \times 10^{12}$ A for this event. Hence, the electric currents in the stellar flares are one–two orders larger compared to the solar ones.

3. Conclusions

• Pulsations in the radio, optical, and X-ray emission are natural manifestations of the stellar flare activity.

• Coronal seismology is a powerful tool for the plasma and magnetic field diagnostics of flaring stars.

• Stars have stronger magnetic fields and more active surface convection in comparison with the Sun.

• Red dwarfs coronae are "dense packed" by hot coronal loops.

• Scale of the flare loops are comparable with the red dwarf radiuses.

• To comprehend physics of stellar activity more multi–wavelength observations are needed.

This study was performed within the framework of Russian–Ukrainian project supported by the the RFBR (9–02–00624–a) and the SFBRU (09–02–90448–a), and by RAS Programm "Origin and evolution of stars and galaxies".

References

Bastian T. S., Bookbinder, J., Dulk, G. A., & Davis, M. 1990, *ApJ*, 353, 265

Gershberg, R. E. 2005, *Solar-type activity in main-sequence stars* (Berlin: Springer)

Haisch, B. M. 1983, in: P. B. Byrne & M. Rodono (eds.), *IAU Colloq. 71: Activity in Red–Dwarf Stars*, (D. Reidel Publishing Co.), 255

Mathioudakis, M., Bloomfield, D. S., Jess, D. B. *et al.* 2006, *A&A*, 456, 323

Mitra-Kraev, U., Harra, L. K., Williams, D. R., & Kraev, E. 2005, *A&A*, 436, 1041

Raassen, A. J. J., Mewe, R., Audard M., & Güdel, M. 2003, *A&A*, 411, 509

Roberts, B. 1995, *in Advances in Solar System Magnetohydrodynamics*, eds. E. R. Priest and A. W. Hood (Cambrige: Cambrige University Press), 105.

Stepanov, A. V., Kopylova Yu. G., Tsap, Yu. T., & Kupriyanova, E. G. 2005, *Astron. Lett.*, 31, 612

Stepanov, A. V., Tsap, Yu. T., & Kopylova, Yu. G. 2006, *Astron. Lett.*, 32, 569

Tsap, Yu. T. 2006, *Kinematika i Fizika Nebesnykh Tel*, 22, 40.

West, A. A., Hawley, S. L., Walkowicz, L. M. *et al.* 2004, *AJ*, 128, 426

Zaitsev, V. V., Stepanov, A. V., Urpo, S., & Pohjolainen, S. 1998, *A&A*, 337, 887

Zaitsev, V. V., Kislyakov, A. G., Stepanov, A. V., Kliem, B., & Furst, E. 2004, *Astron. Lett.*, 30, 319

Solar and Stellar Variability: Impact on Earth and Planets
Proceedings IAU Symposium No. 264, 2009
A. G. Kosovichev, A. H. Andrei & J.-P. Rozelot, eds.

doi:10.1017/S1743921309992821

The non-radial propagation of coronal streamers in minimum activity epoch

Andrey G. Tlatov and Valeria V. Vasil'eva

Kislovodsk Mountain Station, The Central Astronomical Observatory of RAS at Pulkovo, 357700, Box-145, Gagarina st., 100, Kislovodsk, Russia
email: tlatov@mail.ru

Abstract. We have analyzed the shape of the solar corona using the data of daily observations with Mark-3/4 (1980-2008) and SOHO/Lasco-2 (1996–2008) telescopes. The angles of deviation of coronal rays from the radial direction $\Delta\theta$ vary cyclically, reaching the maximum deviation towards the solar equator at the minimum of the solar activity. At the minimum of the 24-th cycle of activity, the $\Delta\theta$ angles were smaller than they were at the minimum of the 22-nd and 23-rd cycles.

We also analyzed of the solar structure corona during eclipses for minimum activity from 1870 till 2008. We examined changes in the index, which characterizes the angle of large coronal streamers to the equatorial plane. It has been shown that the index has been smoothly changing during the last 140 years. The maximal value of an index was during 17–19 activity cycles. The minimal values are reached in the end of 19 centuries and at the present time.

We consider the relations between the angles of deviation of coronal rays at the minimum of activity, the parameters of the global magnetic field of the Sun, and the amplitude of the subsequent cycle of activity, and discuss the hypothesis that the variations of the inclination of coronal rays may affect the parameters of the solar wind and the indices of geomagnetic perturbations at the minima of the solar activity cycles.

Keywords. Sun: corona, activity, fundamental parameters

1. Overview

The solar corona structure corresponds to the configuration of solar magnetic fields. Since the magnetic field of the Sun is subjected to cyclic variations, the corona shape also changes cyclically. The coronal rays are distinctive structures in the solar corona, which propagate at a small angle to the radial direction from the Sun and display the electron density in K-corona enhanced by the factor of 3 to 10. The angle $\Delta\theta$ (Figure 1a) that describes the deviation of the rays from the radial position varies with the phase of the solar cycle Kim *et al.* (2004) and the latitude Eselevich & Eselevich (2002). The regular observations with the SOHO/LASCO and Mark III/IV coronagraphs at the Mauna Loa solar observatory make it possible to analyze the structure of the solar corona from 1980. In order to determine the inclinations of coronal rays, we developed a technique of the identification of coronal streamers in two-dimensional images of the corona obtained with SOHO/LASCO-2 and Mark III/IV in automatic mode. The analysis of the SOHO/Lasco-2 data for 1996–2008 indicates that the inclination angle $\Delta\theta$ varies with the 11-year cycle of the activity, reaching the maximum values at the minimum of the activity. Figure 1b presents the monthly averages for the inclination, averaged along the entire limb. The coronal rays at the minimum of the activity and the phases of the decline are, as a rule, turned towards the solar equator. These conclusions are also confirmed by the analysis of the structure of the solar corona at the heights $1.2-2.5R$ based on the MLSO MarkIII/IV data. Figure 1c presents the angle $\Delta\theta$ for the latitude zone $\pm30^o$ in the time interval

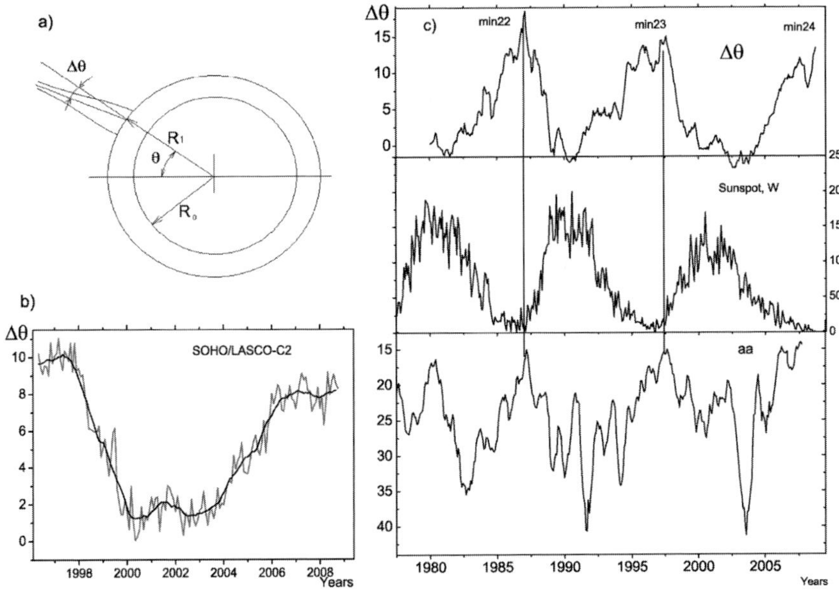

Figure 1. a) The determination of inclination angles for coronal rays. b) The monthly-averaged deviation from the radial direction $\Delta\theta$ derived from the SOHO/Lasco-2 data. The envelope curve averaged from 12 points is also presented. c) The comparison between the deviations of coronal rays $\Delta\theta$ according MLSO- MarkIII/IV data, the Wolf indices for the solar spots W and the geomagnetic index aa. The data are smoothed with a sliding window technique for 6 months.

1980–2008. At the minimum of the 22-nd cycle of the solar activity, the angle $\Delta\theta$ was around 18.2^o, at the minimum of the 23-rd cycle it was 14.7^o, while at the minimum of the 24-th cycle about 9°. This Figure presents the graph of the deviation angle in comparison with sunspot index W and the index of geomagnetic perturbations aa. The non-radial propagation of the solar wind may explain the relationship between geomagnetic indices detected during the minima of activity and the amplitude of a subsequent solar cycle Ohl (1966). Indeed, the relation between the amplitude of the angles $\Delta\theta$ at the minima of the 22-th - 24-th cycles is close to that for the index of the large-scale magnetic field of the Sun in these times Tlatov (2009), and this field is one of prognostic indices of the solar activity.

We can also test changes deviations of coronal streamers from the radial direction for over 100 years (Figure 2). You can use the catalogs of solar eclipses Loucif & Koutchmy (1989), Vsekhsvyatsky *et al.* (1965). Let us introduce index γ that characterizes the angle between high-latitude boundaries of the large coronal streamers at a distance of $2R$. The γ index is the sum of the angles at the eastern and western limbs: $\gamma = \gamma_N + \gamma_S$. In fact, the γ index is a simpler version of the corona flattening indices Nikolskiy (1955) but it is calculated at height R above solar limb.

2. Implications

The awareness of the variations of the angles $\Delta\theta$ and γ with the phase of a cycle is important for theoretical models describing the structure of the corona and the geometry of the magnetic field above the solar limb Wang (1996). The applied aspect of the studies for the deviations of coronal streamers, along with the plasma flows in the solar wind, is also of great importance, since the deviation may affect the geo-efficiency of the solar wind impact. The presence of long-term trends in the solar corona structure can be caused

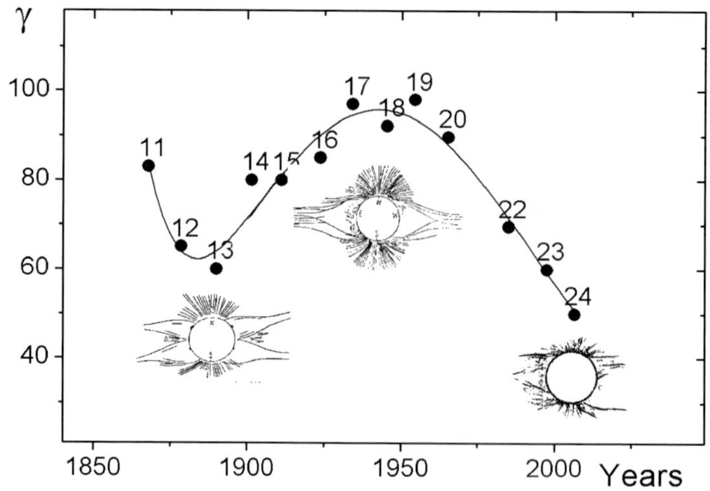

Figure 2. Distribution of parameter γ for the structure of the corona of the minimal type. Numbers of activity cycles are given. The shape of the solar corona to cycles 13,19 and 24 are present.

by changes in the configuration of the global magnetic field of the Sun. The non-radial propagation of the solar wind may explain the relationship between geomagnetic indices detected during the minima of activity and the amplitude of a subsequent solar cycle (Ohl, 1966). Indeed, the relation between the amplitude of the angles at the minima of the 22-th - 24-th cycles is close to that for the index of the large-scale magnetic field of the Sun in these times, and this field is one of prognostic indices of the solar activity. Thereby, we may suggest a link between the large-scale magnetic field and the deviation of coronal rays towards the equator, which in turn affects the level of geomagnetic indices at a solar minimum as well as solar-terrestrial relations.

Thus, analysis of the corona shape has revealed a long-term modulation of the global magnetic field of the Sun (Figure 1b; 1c; 2). Possibly, there exists a secular modulation of the global solar magnetic field which is most pronounced during the solar activity minimum epoch.

References

Eselevich, V. G. & Eselevich, M. V. 2002, *Solar Phys.*, 208, 5

Kim, G.-D., Makarov, V. I., & Tlatov, A. G. 2004, *International Journal of Geomagnetism and Aeronomy*, 5, 2, GI2011

Loucif, M. L. & Koutchmy, S. 1989, *ApJS*, 77, 44

Nikolskiy, G. M. 1955, *Astron. Lett.*, 160, 11

Ohl, A. I. 1966, *Solnycnye Danie*, N9, 84

Tlatov, A. G. 2009,*Solar Phys.*, in pres

Vsekhsvyatsky, S. K., Nikolsky, G. M., Ivanchuk, V. I., Nesmuanovich, A. T., Ponomarev, E. A., Rubo, G. A., & Cherednichenko, V. I. 1965, *The Solar Corona and Corpuscular Emission in Interplanetary Space* (Kiev Univers. Press, Kiev), 293 pp

Wang, Y.-M. 1996, *Apj*, 456, L119

Solar and Stellar Variability: Impact on Earth and Planets
Proceedings IAU Symposium No. 264, 2009
A. G. Kosovichev, A. H. Andrei & J.-P. Rozelot, eds.

doi:10.1017/S1743921309992833

The formation of an equatorial coronal hole

Liheng Yang[1], Yunchun Jiang[1] and Jun Zhang[2]

[1]National Astronomical Observatories/Yunnan Observatory, CAS, Kunming 650011, China
email: yangliheng@ynao.ac.cn

[2]National Astronomical Observatories, CAS, Beijing 100012, China

Abstract. The formation of an equatorial coronal hole (CH) from 2006 January 9 to 12 was simultaneously observed by *GOES-12*/SXI, *SOHO*/EIT and *SOHO*/MDI instruments. The varieties of soft X-ray and EUV brightness, coronal temperature, and total magnetic flux in the CH were examined and compared with that of a quiet-sun (QS) region nearby. The following results are obtained. (1) A preexisting dark lane appeared on the location of the followed CH and was reinforced by three enhanced networks. (2) The CH gradually formed in about 81 hours and was predominated by positive magnetic flux. (3) During the formation, the soft X-ray and EUV brightness, coronal temperature, and total magnetic flux obviously decreased in the CH, but were almost no change in the QS region. The decrease of the total magnetic flux may be the result of magnetic reconnection between the open and closed magnetic lines, probably indicating the physical mechanism for the birth of the CH.

Keywords. Sun: corona, Sun:evolution, Sun: magnetic fields

1. Introduction

Coronal holes (CHs) are large-scale regions of considerably low density and temperature in the solar corona (Krieger *et al.* 1973). In soft X-ray (SXR) and EUV observations, they are identified as dark areas, and their magnetic fields are usually weak, unipolar and open (Joshi 2001). CHs are considered to have a strong association with the fast solar wind, which may induce geomagnetic storms on the Earth (Krieger *et al.* 1973). However, the concrete explanation of their formation and subsequent evolution is still ambiguous.

Until now, only a few works investigated the birth of a single CH. Solodyna *et al.* (1977) showed that a small equatorial CH was formed in less than 9.5 hours and accompanied by a 40% decrease of X-ray emission. However, no obvious change was detected in the photospheric magnetic field. Harvey & Sheeley (1979) then found that a CH rapidly formed in 1 day also without associated change in the photospheric magnetic field.

To unclose the nature of the CH formation, we analyze both the morphological and quantitative varieties of a single equatorial CH in its formation and compare with that of a nearby quiet sun (QS) region.

2. Observations

For the present work, the following data are used:

1. Full-disk longitudinal magnetograms from *SOHO*/MDI (Scherrer *et al.* 1995) with a 96-minute cadence and a pixel size of about 2″.

2. Full-disk EUV images from *SOHO*/EIT (Delaboudinière *et al.* 1995). We primarily examined the data from Fe XII 195 Å and Fe IX/X 171 Å filters, whose ratio can obtain the temperature maps. The EIT obtained 195 Å images roughly every 12 minutes with a scale of 2.6″, but 171 Å images only once in 6 hours.

3. Full-disk CH SXR images from the *GOES-12*/SXI (Hill *et al.* 2005). The images cover a wavelength range of 6-65Å (0.9-2.4 MK), with a 6-hr cadence and a 5″ pixel size.

3. Results

The CH was formed from 2006 January 9 to 12 and centered at E47N02 (E05N02) on January 9 (12). Figure 1a and 1b show its appearance in full-disk MDI magnetogram and EIT 195 Å image when it was fully developed. It is clear that the CH (marked by curves) was a dark, extended region in 195 Å image and lay within a predominantly unipolar magnetic region with positive polarity. Noted that three enhanced networks (marked as R1, R2 and R3, respectively) were around the CH. Figure 1c and 1d display the formation process of the CH in EUV and SXR wavelengths, respectively. Before its formation, a dark lane (marked by the oblique white arrows) was identified on the location of the followed CH. Similar structure, also observed and called as separatrix by Solodyna *et al.* (1977), is thought to be the location of the footprints of open magnetic field lines and pre-existing structure to produce a CH. Then the CH gradually formed around the dark lane and reached its maximum area at about 18:47 UT on January 12.

Figure 2a presents the ratios of EUV brightness in the CH to that in the QS region. Since the formation was slow, the start and end times of the CH formation were difficult to determine only by vision. We thus set the time corresponding to the maximum value of the ratio, 10:00 UT on January 9, as the start time, and the time corresponding to the minimum value, 18:47 UT on January 12, as the end time. Based on the definitions, the formation experienced about 81 hours. Figure 2 also displays the time profiles of the SXR and EUV brightness, coronal temperature, and total magnetic flux in the CH and QS regions. During the whole CH formation, its mean SXR brightness had a relative decrease of 40%, the EUV brightness decreased from 480 to 80 counts pixel $^{-1}$, the temperature decreased about 0.14 MK, and the total magnetic flux decreased from 2.1×10^{21} MX to 1.4×10^{21} MX. By contrast, the SXR and EUV brightness were nearly stable in the QS region, the coronal temperature had a mean value, 1.15MK, and the total magnetic flux was stable at a level of 5.2×10^{21} MX. The decrease of the magnetic flux in the CH probably indicated that magnetic reconnection between open and closed magnetic fields may occur within the CH and represented a potential physical mechanism of the CH formation.

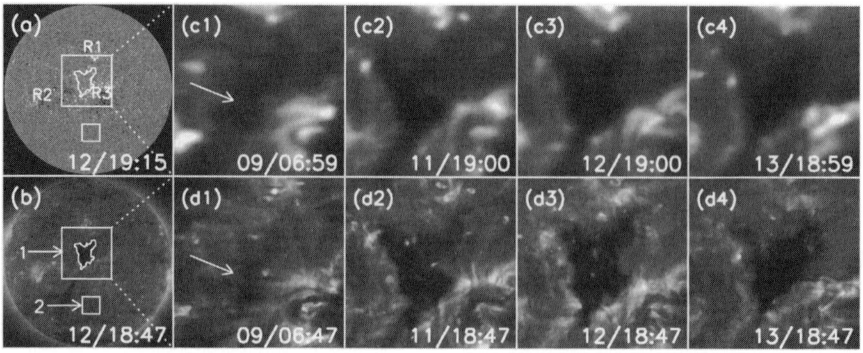

Figure 1. Full-disk magnetogram (a) and EIT 195 Å image (b) showing the CH as it completely formed. EIT 195 Å (c1-c4) and SXI SXR (d1-d4) images showing its formation. The field of view (FOV), outlined by window 1, is 600″×600″. The CH region is encircled by curves. Window 2 denotes the QS region, with a FOV of 218″×203″. Three enhanced networks are marked by R1, R2 and R3.

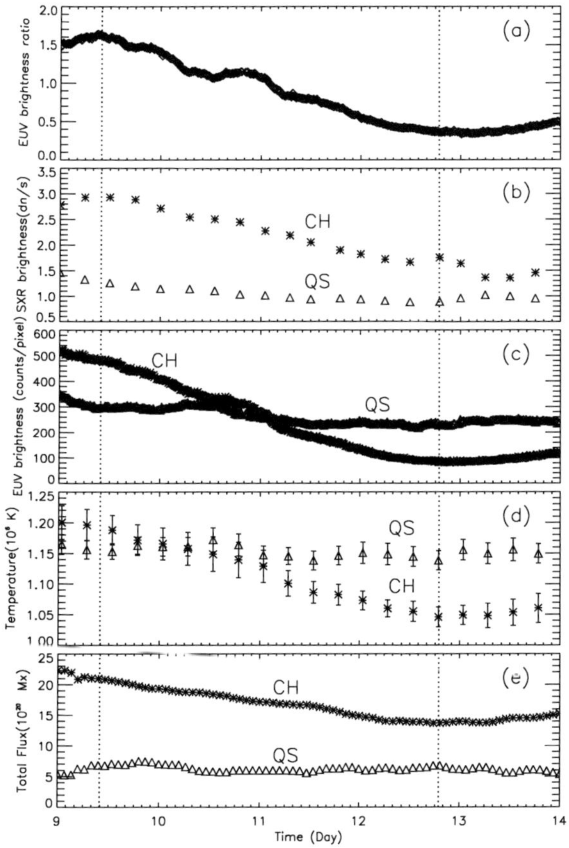

Figure 2. (a) The ratios of EIT 195 Å brightness in the CH (enclosed by curves in Fig. 1) to that in the QS (marked by window 2 in Fig. 1) region as a function of time. Time profiles of SXI SXR brightness (b), EIT 195 Å brightness (c), coronal temperature (d), and total magnetic flux (e) in the CH and QS region. The dashed lines indicate the start and end times of the CH formation, respectively.

Acknowledgements

The authors are indebted to the *SOHO* and *GOES* teams for providing the wonderful data. It is a pleasure to acknowledge many useful advices provided by professors Jingxiu Wang. The work is supported by the 973 program (0797311005).

References

Delaboudiniére, J. P., Artzner, G. E., Brunaud, J. *et al.* 1995, *Solar Phys.*, 162, 291
Harvey, J. W., & Sheeley, Jr, N. R. 1979, *Space Sci. Revs.*, 23, 139
Hill S. M., Pizzo V. J., Balch C. C. *et al.* 2005, *Solar Phys.*, 226, 255
Joshi, A. 2001 *Solar Phys.*, 198, 149
Krieger, A. S., Timothy, A. F., & Roelof, E. C. 1973, *Solar Phys.*, 29, 505
Scherrer, P. H., Bogart, R. S., Bush, R. I. *et al.* 1995, *Solar Phys.*, 162, 129
Solodyna, C. V., Krieger, A. S., & Nolte, J. T. 1977, *Solar Phys.*, 54, 123

Solar and Stellar Variability: Impact on Earth and Planets
Proceedings IAU Symposium No. 264, 2009
A. G. Kosovichev, A. H. Andrei & J.-P. Rozelot, eds.

doi:10.1017/S1743921309992845

New findings increasing solar trend that can change Earth climate

Jean-Pierre Rozelot[1], Cilia Damiani[2] and Sandrine Lefebvre[3]

[1]OCA-Fizeau Dpt, CNRS UMR 6525, & Université de Nice-Sophia-Antipolis,
Av. Copernic, 06130 Grasse, France
email: rozelot@obs-azur.fr

[2]INAF-Osservatorio Astronomico di Roma, via Frascati 33, 00040 Monteporzio Catone (Italia)
email: damiani@oa-roma.inaf.it

[3]Université Versailles St-Quentin; UPMC Univ. Paris 06; CNRS/INSU, LATMOS-IPSL,
Tour 45-46/4ème étage, 4 place Jussieu, 75252 PARIS cedex 05, France
email: sandrine.lefebvre@latmos.ipsl.fr

Abstract. Early attempts to find how solar activity can influence the Earth's climate involved comparison of many physical processes, such as dynamo mechanism, magnetic reconnection and eruptive activity, irradiance, open flux and particles variations, global atmospheric chemistry and dynamics.... However, such direct links seem to be weak even if the solar effects has been found to be stronger during extended maxima or minima of solar activity. Thus, temporal scales ranging from days to thousand of years must be investigated. A description of the most recent results on solar variability and its possible influence on the Earth's climate and atmosphere will be here addressed, with a particular emphasize on modulations of about 120 years (and harmonics). The extrapolation indicates a significant negative decrease of the solar signal, and consequently a decrease of the global Earth's temperature in the forthcoming years. Such a modulation is also testifying by other means, such as spectral observations of temperature sensitive lines indicating a decline of solar activity around 2015 (up to a new prolonged minimum). Prediction of global effects from the Sun's influence over the climate is thus planted in a new way.

Keywords. Sun: general, activity, sunspots, solar-terrestrial relations; Earth

---

1. Introduction

If it is obvious, since the highest Antiquity, that the Sun is the primary source for driving the Earth's climate system, it is not so evident since the middle of the 70's, that the solar changes may have a corresponding response on the climate system. Two major arguments were developed to thwart the assertion that the Sun's output variability may play a key role. The first one concerns the anthropogenic effects and is today widely debated. If a general consensus is reached concerning the increase of global Earth's temperature due to human activities (see reports from the IPCC†), it seems that this increase tends to slow down since a few months: does this break in the positive temperature trend since the years 1970 corresponds today with the low solar activity as seen from the year 2008?

The second argument is linked with the so-called Total Solar Irradiance (TSI). The Earth receives at the top of the atmosphere a solar energy that was recognized as constant, and fixed at about 1366 W/m^2 up to the satellites era. Radiometers on board dedicated satellites record a fluctuation of $\pm$ 1.3 W/m^2 during a solar cycle, a value considered as too low to significantly affect the climate of the Earth (Foukal *et al.*, 2006).

† IPCC: Intergouvernental Panel on Climatic Change, Fifth Assessment Report.

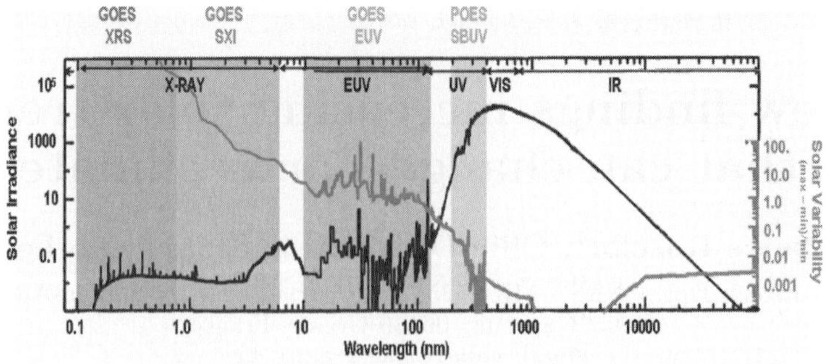

Figure 1. The solar spectrum (black curve) is highly variable (green curve: the spectral variability is the ratio (max-min)/min). The UV and EUV portions of the spectrum have the greatest variability. After NOAA forecast center.

Furthermore, if such an estimate of the "solar constant" may produce a climatic forcing, do its variations are of enough sensitivity to explain the differently equatorial, tropical or polar latitudes perturbations of the Earth's atmosphere? Even by admitting a modulation of the climatic signal on the top of the stratosphere, how changes can be reverberated on the lower layers, seat of our climate? An embryo of answer is suggested for example by Haigh (1994) who emphazised the role of the ozone destructive cycle, or more likely, through an amplification of the solar signal within the UTLS (Upper-Troposphere-Lower-Stratosphere) zone (Kodera and Kuroda, 2002; Kilifarska, 2006)‡.

Recent works has significantly improved our knowledge on the TSI, mainly on the UV part of the spectrum as seen in Fig. 1. The most remarkable issue concerns its variability, reaching a factor 2 peak-to-peak, that was addressed to cause a significant impact on the stratosphere. This finding was followed by a profusion of new studies, which are just beginning to be discussed in many symposia. This may reinforce what Hansen wrote in 2000: *"Even if the solar forcing has been smaller than the anthropogenic forcing, it is incorrect to assume that the Sun necessarily will be an insignificant player in climate change in the 21st century"*. As the interplanetary medium is neither unstructured or quasi-static, nor is a simple magnetic stratified object, thus the interaction of the solar electromagnetic radiation (photons), hot plasma (electrons, protons and other ions), cosmic rays, microscopic dust particles, and magnetic fields (primarily from the Sun) with the upper environment of our Earth leads to a complex physics which is far to be understood; their characterization deserves to be clarified.

The solar output variability which may influence our Earth's environment and more specially our climate is called helioclimatology (Rozelot, 1990, Lefebvre *et al.*, 2007).

2. The key role of time ranging

From the core of the Sun up to the Earth, four main mechanisms generate a variability on the Earth global insolation (Beer *et al.*, 2000): (i) nuclear fusion in the core of the Sun, (ii) transport through the radiative and convective zone of the Sun, (iii) emission of radiation from the photosphere towards the Earth, and (iv) changes in the celestial mechanical parameters (eccentricity, nutation, inclination of orbits). These sources of

‡ Let us quote here Kodera: "it is suggested that the solar influence produced in the upper stratosphere and stratopause region is transmitted to the lower stratosphere through (1) modulation of the internal mode of variation in the polar night jet and (2) a change in the Brewer-Dobson circulation; see also Lefebvre *et al.*, these proceedings.

variability act on different time scales, ranging from billion of years to seasons, sometimes to days or minutes.

The first case (in the range of 0 to 0.3 solar radius R) is studied through different models leading to high luminosity changes through very long periods of time (a doubling of the luminosity over the first 8 billions years, known as "the early faint Sun paradox").

The second case can be divided into two sub-groups: the first one concerns the energy transport through the radiative zone (0.3 to 0.7 R), likely to be stable, and the second one (0.7 to 1R) concerns the convective zone. Although the ultimate source of solar energy is the nuclear reactions taking place in the core, the immediate source of energy is the surface. Nuclear reactions are certainly steady on short time scale, but mechanisms which carry the energy in the convective zone may not be. If the central energy source remains constant while the rate of energy emission from the surface varies, there must be an intermediate reservoir, where the energy can be stored or released depending on the variable rate of energy transport (Pap *et al.*, 1998). The gravitational field (the virial theorem indicates that there is a connection between magnetic energy and gravitational energy), is one such energy reservoir. If energy is stored in this reservoir, it will result a change in the Sun's radius, at least in the uppermost layers of the Sun (the leptocline). Recent works show the compatibility of this theory with irradiance variations (Lefebvre and Kosovichev, 2005; Stothers, 2006; Fazel *et al.*, 2008). These mechanisms act at the level of days to several years (cyclic solar activity). This radius variability will be addressed later on.

The third case is linked with the transport of the radiation from the photosphere up to the Earth. Changes in the global insolation are due to on one hand to anistropic transport, and to the other hand, on changes in the UV part of the spectrum, for which certain spectral bands of the atmosphere are very sensitive (ozone layers for instance). The effects are also of the order of days and years. Finally, the variability due to the movement of the Earth on its orbit and over its axis of rotation, leads to climatic changes that were first studied by Milankovich in 1930. The time scales involved here are of the order of thousands years.

The three last processes show cyclicities and the question, which remains an open question is: do exist resonant phenomena able to amplify changes in the global warming or cooling of our climate? In the same way, we are still ignorant of how feedbacks may act, but we know that such mechanisms may amplify a weak solar signal. In this scope, the UTLS region is a key-layer in the understanding of climatic phenomena. Changing chemistry in this region has lead us to better unravelling the dynamical feedback that can occur here. Undoubtedly, this effect in conjunction with the solar irradiance changes, can influence the interannual variability of temperature in this zone (Kilifarska, 2006, Damiani-Badache *et al,* 2007).

3. Emphazing forecasting solar activity

Pseudo-cyclicity of the solar activity is one of the most intriguing puzzle in solar physics. Why the Sun seems to beat regularly, at around 11-years, but also at around other specific periods, 80-yrs, 211-yrs, 400-yrs, even 2115-yrs (Damon and Jirikowic, 1992). This "around" is the core of the problem. We do not know why the lengths of the solar cycles vary from one cycle to the other, the differences being tiny significant, of the order of 2 to 3-yr for the Schwabbe 11-yr cycle for instance, more for other cycles. This renders the prediction difficult. However, planning for satellite orbits and space missions (especially for man-lived missions) often require knowledge of solar activity levels years in advance. The study of the Lyapunov exponent (Benest *et al.*, 2007), for which the determination is between 3 to 4 years, shows that the solar cycle is not deterministic: it is thus impossible, in the solar case, to accurately determine the date and the level of intensity of the activity signal more than 3 or 4 years ahead. In other words, the forecast is good: the behaviour of a sunspot cycle is fairly reliable once the cycle is well underway, about 3 years after the minimum. For instance, in 2004 it is possible to forecast that the next solar maximum will take place around 2011. However the prediction is poor: inaccuracy on the date, inaccuracy on the estimate of the level of the signal. The "observed" is not enough to deduce what will be the "observable".

There is a great lot of techniques (including neuronal networks), each of them showing advantages and drawbacks. Predictions for solar cycle 24 has been made by several authors, leading to a large range of amplitude of the signal (15 papers are available on the subject for the single year 2008). In spite of the fact that some methods predict a relatively high cycle 24, three different analysis lead to about the same conclusion:

- Kilcik *et al.* (2009) reported that the maximum intensity of cycle 24, will be of 87.4 in December 2012; according to this forecast, the current solar cycle will have a magnitude far lower than any other since 1890–1910.

- Kitiashvili & Kosovichev (2008 and this proceedings) using a nonlinear dynamo model as described by Kleeorin & Ruzmaikin (1982), which takes into account dynamics of the turbulent magnetic helicity, predict that the next sunspot cycle will be significantly weaker (by $\approx 30\%$) than the previous cycle, continuing the trend of low solar activity.

- Using direct polar field measurements, now available for four solar cycles, Svalgaard *et al.* (2005) predict that solar cycle 24 (2011 maximum) will have a peak of 75 ± 8 in the smoothed monthly sunspot number, making it potentially the smallest cycle in the last 100 years.

We need to improve the scientific way of predicting solar activity, both for a better understanding of the heliosphere and dynamo physics, as well as its usefulness to all space agencies interested in solar activity related phenomena, ranging from power grid spikes, to communication blackouts and to satellite orbital dynamics.

4. Understanding dynamical solar internal processes

A priori, it seems that understanding how the core of the Sun rotates is very far from the climatic research. It is not so sure, as one the key is the magnetism: the interplanetary medium is shaped by the solar magnetism. The solar magnetism is no more considered as a purely superficial phenomenon. If it has been shown that the length of the solar cycle depends on the transition region between radiation and convection, the internal solar magnetism is still poorly known. Some main space missions will give us a 3-D vision of the Sun at the horizon of about 2012, which will increase our knowledge on this difficult problem (for instance DynaMICCS, Turck-Chièze *et al.*, 2008).

5. Are we entering a new "blank" area?

Space-based measurements reveal the existence of 11-year cycles in solar radiation upon which are superimposed larger and shorter term changes. Indirect proxies of solar activity† exhibit 11-year solar cycles as well as longer-term changes or cycles that exceed the amplitudes of their 11-yrs cycles. Comparisons of these proxy records with direct

† Such proxies can be listed as: dendochronology, palynology, isotopic variations of marine sediments (foraminifers), ice analyzes (trapping of CO_2, Be^{10}), criticized study of old documents (parochial registers, ship's books, dates of harvests and vintages, wheat prizes, plum blossoming in Japan), glaciers advance and retreats, comets and auroras registers, C^{14} analysis, etc).

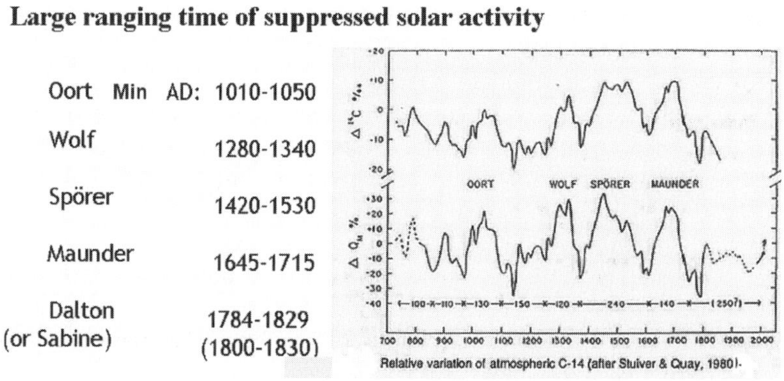

Figure 2. The different past solar minima (inverted from the C^{14} analysis due to the Forbusch effect) show a pseudo long oscillation. An extrapolation shows that the next prolonged minima will occur by around 2020. (After Suiver and Quay, 1980, Sol. Phys., 74, 2, pp. 479–481). See also Hiroko Miyahara, these proceedings.

observations suggest that anomalously low solar activity have existed in the past. These periods are commonly known as the "Dalton", "Maunder", "Spörer", "Wolf" and "Ort" Minima (which occurred from 1784 to 1829, from 1645 to 1715, from 1420 to 1530-1535, from 1280 to 1340 and 1010-1050 respectively: see Fig. 2).

By contrast, since about 1950, solar activity increased steadily, to an amplitude never reached during the first half of the twentieth century. A crude analysis of the frequency at which appeared the above-mentioned minima together with this last remark, lead us to conclude that we are entering a new long solar minima, which will occurred around next cycle 25. If solar activity continues as low as it has been since 2007, 2008 would have racked up a whopping 290 spotless days by the end of December, making it a century-level year in terms of spotlessness (Hataway, 2008). At the end of the year 2009, even if newly emerged Cycle 24 regions were recorded, plages were still of low probability of producing any activity, even if some have been seen (such as on 2009, November, 1st).

Penn and Livingston (2006) observed spectroscopic changes in temperature sensitive molecular lines, mainly in the magnetic splitting of a Fe I line, and in the continuum brightness of over 1000 sunspot umbrae from 1990–2005. There measurements show consistent trends in which the darkest parts of the sunspot umbra have become warmer (45K per year) and their magnetic field strengths have decreased (77 Gauss per year), independently of the normal 11-year sunspot cycle.

Left panel of Figure 3 shows the decrease in field strength now found with respect to time (1992–2009), which shows a linear trend independent of the solar cycle: a linear extrapolation of this trend may suggest that few sunspots will be visible after 2015. Because of the nature of the observing program, the earliest measurements in this plot are probably skewed toward higher magnetic field values (larger sunspots); nonetheless, the linear trend in the magnetic field value is clear, even excluding all pre-1995 data. Is this an omen of long-term sunspot decline, analogous to the Maunder Minimum?

Climatic parameters of many types often exhibit cycles that are also common in solar activity proxies, such as near 11–, 22–, 80– and 210– years. It is well known that times of cooler climate in past millennia usually coincide with reduced levels of solar activity. For instance, during the Little Ice Age, which occurred from about 1450 to 1850, surface temperatures were from 0.6°C to 1°C colder than at present (depending on geographic location, as changes in solar radiation can often result in regional or local responses). Solar activity was lower than at present because of the occurrence of the above-mentioned

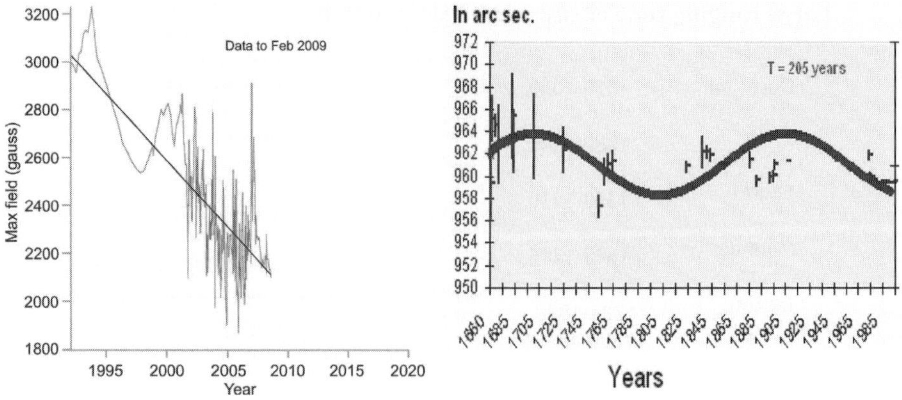

Figure 3. Left: The maximum sunspot field strength is plotted versus time, during the period from 1992 to February 2009; a 12- point running mean is shown, and a linear fit to the data is plotted. Apart from a few measurements, the linear trend has been seen to continue throughout this solar minimum. By courtesy of W. Livingston and M. Penn, 2009. Right: Solar radius variations as observed since the pioneer's work of Picard and La Hire, two french astronomers at the time of the reign of Louis XIV named "Roi-Soleil". In spite of the fact that the error bars were large at that time, a general sinusoidal fit can be adjusted, leading to a next minimum in around 2015–2018.

Minima. Low-frequency signals in long tree-ring chronologies have been used to reconstruct past temperature variability since the year 631 (Esper *et al.*, 2002; see also Fig. 2 in Feynman, 2007). Yearly data has been plotted in Fig. 4 and a sinusoidal fit has been adjusted. In spite of the global warming which occurred these last few years, a general behaviour appears that would lead to a next minimum around the years 2018. The period found, of roughly 425-years, is twice the powerful harmonic (212-yrs) considered as fundamental by Damon and Jirikowic (1992).

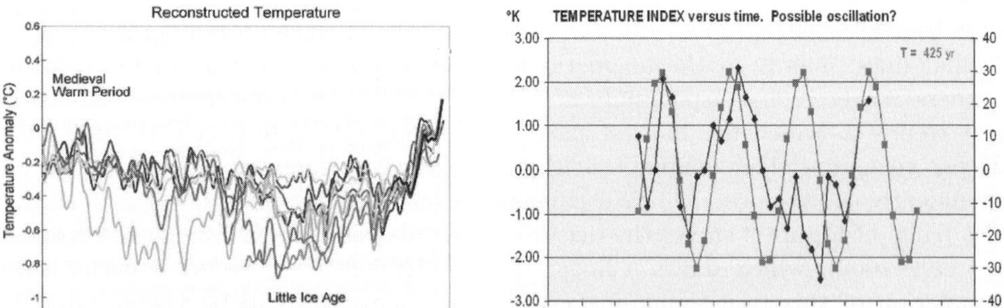

Figure 4. Left: Comparison of 10 different published reconstructions of mean temperature changes during the last 2000 years. More recent reconstructions are plotted towards the front and in redder colors, older reconstructions appear towards the back and in bluer colors. For instance: Red (1-1979): A. Moberg, D.M. Sonechkin, K. Holmgren, N.M. Datsenko and W. Karln (2005). *"Highly variable Northern Hemisphere temperatures reconstructed from low- and high-resolution proxy data"*, Nature, 443: 613–617. Light green (831–1992): J. Esper, E.R. Cook, and F.H. Schweingruber (2002). *"Low-Frequency Signals in Long Tree-Ring Chronologies for Reconstructing Past Temperature Variability"*, Science, 295(5563): 2250–2253. Right: Adjusting a sinusoidal fit which smoothes the data during the ascending phase of the temperature due to men's effects -and bearing in mind the possible pause of Earth's temperature since 2008–, it can be seen that we may enter a new era of low temperature.

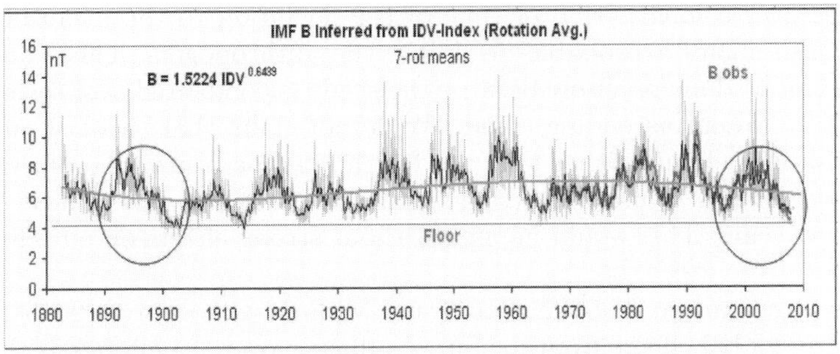

Heliospheric Magnetic Field. Cycles 13 and 23 are circled.
L. Svalgaard, E. Cliver, and P. LeSager, Determination of interplanetary magnetic field strength, solar wind speed, and EUV irradiance 1890-2003, in International Solar Cycle Studies Symposium, June 23-28, 2003, Tatranska Lomnica, Slovak Republic, Proceedings (ESA SP-535), 15, ed. A. Wilson. 2003

Figure 5. The Interplanetary Magnetic Field strength shows a modulation that may reach a minimum in 2015.

An analysis of the solar radius variability, as it appeared for instance in Thuillier *et al.* (2005) or Kilcik *et al.* (2009) indicates that a modulation of amplitude can be seen of about 205-years will lead to a next "bigger" Sun by 2018 (Fig. 3-right). Even if this variability must be taken with cautious, and as the data are certainly not all of the same quality, it can be noticed nevertherless that this periodicity is about 205-yrs, i.e. the fundamental, as above stated.

Finally, the analysis of the interplanetary magnetic field strength as described by Svalgaard *et al.* (2003 -see caption of Fig. 5) shows a similarity of cycles 13 and 23, which will leads again to a minimum after 2013 as seen in Fig. 5.

The authors are perfectly aware of the empirical character of the predictions made here. But it is rather striking, that the four parameters described, temperature index, solar radius variability, umbral-sunspots magnetic fields and interplanetary magnetic field strength, lead to a new great Minima which might occur between 2015 and 2018.

6. Conclusion

Comparison of cycles 4 and 23, then 13 and 23 lead to the conclusion that solar activity will be less intense in the future. This is reinforced by a long-term oscillation in the temperature reconstruction, in the photospheric radius variations, in the decreasing of the solar magnetic field in the quiet solar disc and in the interplanetary magnetic field strength.

It is extremely likely that we are just entering a new Minima. The level of solar activity remained very low during several months in 2008. During that time, high-latitudes region of activity yielded frequently no flares, despite developing mixed-polarity areas indicating a new cycle difficult to start. A lot of predictions based of the past dataset point out that even if the maximum of solar activity has been increasing since the late 19th century, the next maximum will begin to decrease. By looking at all the past Minima deduced from historical records, it can be set forth that solar activity is certainly modulated over long periods of time. This modulation of amplitude is undoubtedly driven through dynamo processes, but the gravitational energy in the most outer layers certainly plays a key role (the "leptocline" shallow layer may be the cradle of turbulent pressure –Lefebvre *et al.*, 2007; Lefebvre *et al.*, 2008–). This gravitational energy expands or shrinks the solar envelope on an antiphasing process with the activity. Lastly, a magnetic field analysis conducted at Kitt Peak on spectral observations of temperature sensitive lines, show a possible long negative trend indicating a decline of solar activity, which could lead also to a suppressed activity in the forthcoming years.

The analysis over a long period of time of the deviation to the mean of the Earth's temperature, also shows a probable declining, the anthropogenic effect not taken into account. This can be done precisely because the study is conducted over several centuries, which smooth the recent ascending trend. We do not claim that this last effect does not exist, and we do not claim neither that the Earth's temperature follows the solar cycle. We merely said that a sinusoidal fitting of the data would yield a next temperature minima in a few years. This would happen without strong effects on our climate, due to man's activity.

Cautiously, we can argue that the next Grand Minima will occur by 2015-2018. Prediction of long-term global effects from the Sun's influence over the climate is thus impinged in a new way.

References

Beer, J., Mende, W., & Stellmacher, R., 2000, "The role of the Sun in climate forcing" Quaterly Science Reviews, 19, pp. 403–415.

Benest, C., Froeschl, C., & Lega, E.: 2007. In "Topics in Gravitational Dynamics", L. N. P., 729, Springer ed., pp.391–405.

Damiani-Badache, C., Rozelot, J. P., Coughlin, K., & Kilifarska, N. 2007, "Influence of the UTLS region on the astrolabes solar signal measurement", Mont. Nont. Roy. Astr., 380, pp. 609–614.

Damon, P. E. & Jirikowic, J. L., 1992, "The Sun as a low-frequency harmonic oscillator", Radiocarbon, 34, pp. 199-205.

Esper, J., Cook, E. R., & Schweingruber, F. H., 2002., Science, 295 (5563): pp. 2250–2253.

Fazel, Z, Rozelot, J. P. Lefebvre, S., Ajabshirizadeh, A., & Pireaux, S., 2008, "Solar gravitational energy and luminosity variations", New Astronomy, 13, pp. 65–72.

Feynman, J., 2007, "Has solar variability caused climate change that affected human culture?", Advances in Space Research, 40, pp. 1173–1180.

Foukal, P., Fröhlich, C., Spruit, H., & Wigley, T. M. L., 2006, "Variations in solar luminosity and their effect on the Earth's climate", Nature, 443/14, pp. 161–166.

Haigh, J. D., 1994, "The Role of Stratospheric Ozone in Modulating the Solar Radiative Forcing of Climate". Nature, 370, 6490/AUG18, p. 544.

Hansen, J. E, 2000, "The Sun's Role in Long-Term Climate Change", Space Science Review, 94, pp. 349–356.

Hataway, D., H., 2008, "Solar Cycle Forecasting", Space Science Reviews, DOI 10.1007/s11214-008-9430-4.

Kodera & Kuroda, 2002, "Dynamical response to the solar cycle" JGR, 107, D24, pp. 4749.

Kilcik, A., Anderson, C. N. K., Rozelot, J. P., Ye, H. Sugihara, G., & Ozguc, A., 2009, "Nonlinear Prediction of Solar Cycle 24", The Astrophysical Journal, 693, pp. 1173–1177.

Kilifarska N. A., 2006, "Solar Variability and Climate-UTLS amplification of Solar Signal", Sun and Geosphere, v.1, No.1, pp. 51–55.

Kitiashvili, I. N. & Kosovichev, A. G. 2008, eprint arXiv:0807.3284

Kleeorin, N. I. & Ruzmaikin, A. A. 1982, Magneto-hydrodynamics, 18, pp. 116–122.

Lefebvre, S. & Kosovichev, A. K., 2005, "Changes in the Subsurface Stratification of the Sun with the 11-Year Activity Cycle", The Astrophysical Journal, 633, Issue 2, pp. L149–L152.

Lefebvre, S., Rozelot, J. P., & Kosovichev, A. G., 2007, "The change of solar shape in time and depth. Some consequences for space climate", Advances in Space Research, 40, pp. 1000–1005.

Lefebvre, S., Nghiem, P. A. P., Turck-Chize, S., 2008, "Impact of a radius and composition variation on the solar subsurface layers stratification, arXiv:0809.1726v1, The Astrophysical Journal, 690, pp. 1272–1279

Lefebvre *et al.*, 2009 "Influence of the solar radiation on Earth's climate using the LMDz-REPROBUS model, these proceedings.

Milankovich, M., 1930, "Theorie der Klimaschankungen", Berlin, pp. 1–176.

Pap, J., Kuhn, J., Frölich, K., Ulrich, R., Jones, A., & Rozelot, J.P., 1998, "Importance of monitoring solar global properties, luminosity, radius and oscillations", ESA-SP-417, pp. 267–271.

Penn, M. J. & Livingston, W., 2006, "Temporal Changes in Sunspot Umbral Magnetic Fields and Temperatures", The Astrophysical Journal, Vol. 649, Issue 1, pp. L45–L48.

Rozelot, J. P.: 1990, "Helioclimatology, a new science?" International Colloquium on Geo-cosmic relation, Amsterdam, 19-22 avril 1989. Published in "The Earth and its environment", ed. Tomassen, Pudoc (NL), 1990, p. 92–100.

Sabatino, S. & Haberreiter, M.: 2005, "Past, present and future measurements of the solar diameter", Advances in Space Research 35, pp. 329 340 and two Corrigenda (i) Advances in Space Research, 37, 8, pp. 1649–1650, and Advances in Space Research, 37, 2, pp. 439–439.

Stothers, R. B. 2006, "A virial theorem investigation of magnetic variations in the Sun". The Astrophysical Journal, 653, pp. L73–L75.

Stuiver, M. & Quay, P. D.: 1980, "Changes in atmospheric Carbon-14 attributed to a variable sun". Science, 207, pp. 11–19.

Svalgaard, L., Cliver E. W., & LeSager, P., 2003, "Determination of interplanetary magnetic field strength, solar wind speed, and EUV irradiance 1890-2003", In Proceedings of International Solar Cycle Studies Symposium held 13-28 June, 2003 in Tatranska Lomnica, Slovak Republik. Edited by A. Wilson.

Svalgaard, L., Cliver, E. W., Kamide, Y., 2005, "Sunspot Cycle 24: Smallest Cycle in 100 Years?"' In Large-scale Structures and their Role in Solar Activity, ASP Conference Series, Vol. 346, Proceedings of the Conference held 18-22 October, 2004 in Sunspot, New Mexico, USA. Edited by K. Sankarasubramanian, M. Penn, and A. Pevtsov, p. 401.

Turck-Chièze, S. & other co-authors, 2008, "Thc DynaMICCS perspective", Experimental Astronomy, Experimental Astronomy, (Special Issue on ESA's Cosmic Vision), Vol. 23, pp. 1017–1055.

Solar and Stellar Variability: Impact on Earth and Planets
Proceedings IAU Symposium No. 264, 2009
A. G. Kosovichev, A. H. Andrei & J.-P. Rozelot, eds.

doi:10.1017/S1743921309992857

Thermospheric temperature and density variations

Hitoshi Fujiwara[1], **Yasunobu Miyoshi**[2], **Hidekatsu Jin**[3], **Hiroyuki Shinagawa**[4], **Yuichi Otsuka**[5], **Akinori Saito**[6] **and Mamoru Ishii**[7]

[1]Department of Geophysics, Tohoku University, Sendai, Japan
email: `fujiwara@pat.gp.tohoku.ac.jp`

[2]Department of Earth and Planetary Sciences, Kyushu University, Fukuoka Japan
email: `miyoshi@geo.kyushu-u.ac.jp`

[3]National Institute of Communication Technology, Tokyo, Japan
email: `jin@nict.go.jp`

[4]National Institute of Communication Technology, Tokyo, Japan
email: `sinagawa@nict.go.jp`

[5]Solar Terrestrial Environment Laboratory, Nagoya University, Nagoya, Japan
email: `otsuka@ste.nagoya-u.ac.jp`

[6]Department of Geophysics, Kyoto University, Kyoto, Japan
email: `saitoua@kugi.kyoto-u.ac.jp`

[7]National Institute of Communication Technology, Tokyo, Japan
email: `mishii@nict.go.jp`

Abstract. The thermosphere is the transition region from the atmosphere to space. Both the solar ultraviolet radiation and the solar wind energy inputs have caused significant thermospheric variations from past to present. In order to understand thermospheric/ionospheric disturbances in association with changes in solar activity, observational and modelling efforts have been made by many researchers. Recent satellite observations, e.g., the satellite CHAMP, have revealed mass density variations in the upper thermosphere. The thermospheric temperature, wind, and composition variations have been also investigated with general/global circulation models (GCMs) which include forcings due to the solar wind energy inputs and the lower atmospheric effects. In particular, we have developed a GCM which covers all the atmospheric regions, troposphere, stratosphere, mesosphere, and thermosphere, to describe variations of the thermospheric temperature and density caused by both effects from the lower atmosphere and the magnetosphere. GCM simulations represent global and localized temperature and density structures, which vary from hour to hour, depending on forcings due to the lower atmosphere, solar and geomagnetic activities. This modelling attempt will enable us to describe the thermospheric weather influenced by solar activity in cooperation with ground-based and satellite observations.

Keywords. solar-terrestrial relations, solar wind, sunspots

1. Introduction

The thermospheric temperature and density have been observed with satellites and ground-based optical and radar techniques. The early satellites provided thermospheric density data obtained from their orbital changes due to the air-drag force (e.g., King-Hele, 1959, 1987). The mass spectrometer measurements by satellites also have revealed composition and density changes in the thermosphere (e.g., Prölss, 1982). The accelerometer measurements by satellites have successfully described spatio-temporal variations of the thermospheric density at present. For example, the double-hump structure of the neutral mass density in the equatorial region (Liu *et al.*, 2005), density enhancements due to traveling atmospheric disturbances (e.g., Forbes *et al.*, 2005; Bruinsma *et al.*, 2006), density enhancements during solar flare events (e.g., Sutton *et al.*, 2006) ere found from accelerometer measurements by he CHAMP satellite.

The Fabry-Perot interferometer (FPI), which can detect the Doppler shifted and broadening emissions from the atmospheric species, is one of the most important instruments to observe the neutral temperature and wind in the thermosphere (e.g., Aruliah *et al.*, 1991; Ishii *et al.*, 1999; Shiokawa *et al.*, 2003; Ford *et al.*, 2007). In addition, the radar techniques enable us to obtain information of the thermospheric wind and temperature. The lower thermospheric wind and temperature and the exospheric temperature have been obtained from the incoherent scatter (IS) radar observations (e.g., Nozawa and Brekke, 1995; Buonsanto *et al.*, 1998).

The global numerical models, which are so-called general/global circulation models (GCMs), of the thermosphere have been developed since the first one by Fuller-Rowell and Rees (1980). Recently, the GCMs have become ones which are coupled with models of the ionosphere, magnetosphere, and the lower atmosphere (e.g., Roble, 2000; Miyoshi and Fujiwara, 2003, 2009; Wang *et al.*, 2004; Toth *et al.*, 2007; Akmaev *et al.*, 2008; Fuller-Rowell *et al.*, 2008). These models are now essential for space weather researches to understand effects of the solar activity on the geospace environment.

The global (large-scale) patterns of the thermospheric wind, temperature, and density during both the solar minimum and maximum periods are almost understood from the observational and modelling studies as mentioned above. Some empirical (statistical) models have been developed based mainly on the ground-based radar and optical observations and satellite and rocket observations. The standard ones are, for example, NRLMSISE-00 (Picone *et al.*, 2002) for the thermospheric temperature and composition (density), HWM (Hedin *et al.*, 1996) for the thermospheric wind, and International Reference Ionosphere (IRI) (Bilitza and Reinisch, 2008) for the ionospheric parameters (ion temperature and composition, electron temperature and density).

Although the empirical models describe well the global patterns of the thermospheric temperature and density depending on the 11-year solar cycle, there are some discrepancies between the localized structures obtained from observations and the empirical models. It is also difficult to describe disturbances of the thermosphere during geomagnetic storms by using the empirical models and GCMs. In addition, previous GCMs cannot predict day-to-day variations of the thermosphere and ionosphere even during geomagnetically quiet periods. In order to understand the above unknown features of the thermosphere, we have developed a GCM which covers all the atmospheric regions, troposphere, stratosphere, mesosphere, and thermosphere. The GCM calculates the thermospheric temperature, density, and wind varying from hour to hour. The thermospheric temperature and density variations due to the solar external forcing and the lower atmospheric effects simulated by the GCM are shown in the following sections.

2. Basic features of the thermospheric temperature and density

The thermospheric temperature and density profiles are determined primarily by the solar energy inputs into the thermosphere. The height of the thermospheric upper boundary also depends on the solar energy inputs; namely, the height depends on the solar and geomagnetic activities, local time, latitude, and season. The temperature profiles obtained from an empirical model of NRLMSISE-00 are shown in Figure 1. The solid and dashed lines indicate the temperatures during the periods of the solar minimum in winter and the solar maximum in summer, respectively. These are the typical ones at mid-latitude at

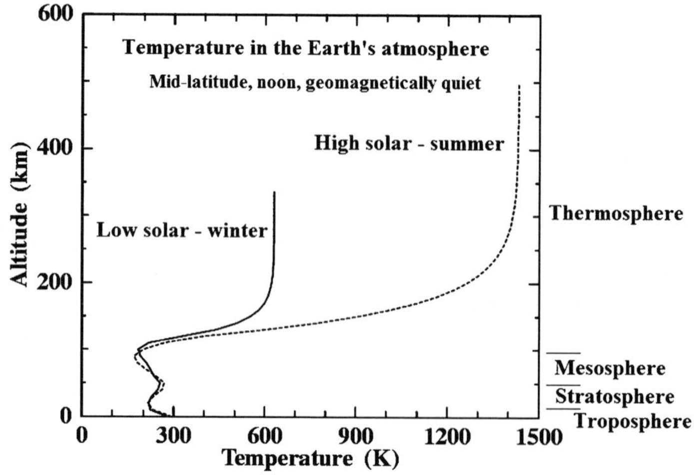

Figure 1. Examples of the temperature profiles obtained from an empirical model of NRLMSISE-00 for solar minimum and maximum conditions at mid-latitude noon.

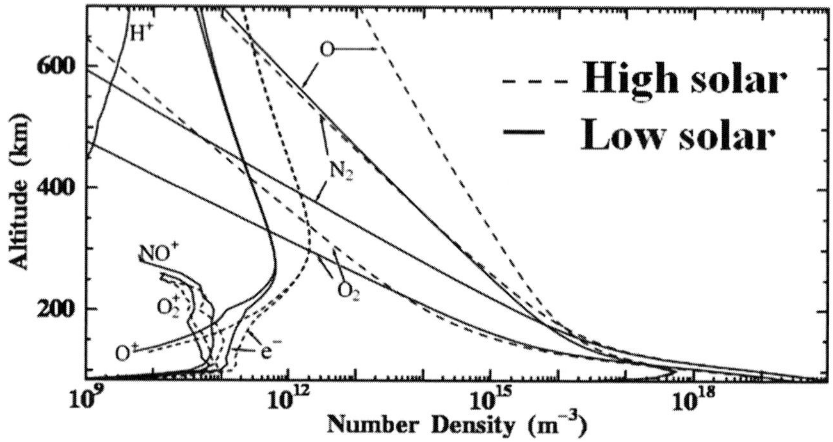

Figure 2. Height profiles of the major constituents in the thermosphere and ionosphere obtained from empirical models of NRLMSISE-00 and IRI for solar minimum and maximum conditions at mid-latitude noon.

noon for geomagnetically quiet condition. In the above case, the temperature at the top of the thermosphere (i.e., the exospheric temperature) changes by about 800 K due to the 11-year solar cycle since the atmospheric species obtain their thermal energy mainly from the solar extreme ultraviolet (EUV) and ultraviolet (UV) radiation which strongly depends on the solar activity.

Figure 2 shows height profiles of the major constituents in the thermosphere and ionosphere obtained from empirical models of NRLMSISE-00 and IRI. The solid and dashed lines indicate profiles during the solar minimum and maximum periods, respectively. These profiles are also typical ones at mid-latitude at noon for geomagnetically quiet condition. The neutral species show significant enhancements at high altitudes during the solar maximum period due to thermal expansion of the air or increase in the scale-height. For example, the number densities of the atomic oxygen and molecular nitrogen at 400 km altitude vary more than one order during the 11-year solar cycle in this case.

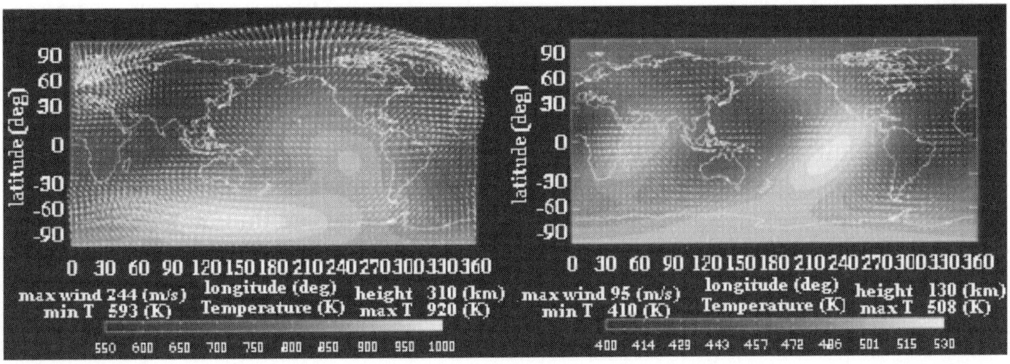

Figure 3. The global distributions of the thermospheric temperature and wind obtained from empirical models of NRLMSISE-00 and HWM in December during the solar minimum and geomagnetically quiet period at 310 (left panel) and 130 km (right panel) altitudes. Note that the strength of wind vector and temperature range are different in each panel.

The global distributions of the thermospheric temperature and wind obtained from empirical models of NRLMSISE-00 and HWM are shown in Figure 3. The left and right panels indicate the global distributions in December during the solar minimum and geomagnetically quiet period at 310 and 130 km altitudes, respectively. The maximum wind vector indicates 244 and 95 m/s in the left and right panels, respectively. At 310 km altitude (left panel), diurnal temperature distribution is remarkable. In addition, high temperature region is clearly seen in the southern (summer) hemisphere. At 130 km altitude (right panel), semi-diurnal temperature variation is clear in the low-latitude region. High temperature regions are also seen at high-latitudes, particularly in the southern (summer) hemisphere.

3. Whole atmosphere GCM

In the present study, we use a whole atmosphere GCM developed by Miyoshi and Fujiwara (2003) as an extension of the middle atmosphere GCM developed at Kyushu University (Miyahara *et al.*, 1993; Miyoshi, 1999). The GCM was originally developed as a tropospheric GCM at the Japan Meteorological Agency (Kanamitsu *et al.*, 1983), and developed as a community climate model at the University of Tokyo. The GCM was extended to include the processes of the middle atmosphere, and then the thermosphere.

The whole atmosphere GCM solves the full nonlinear primitive equations for momentum, thermodynamics, continuity, and hydrostatics. The continuity equation of mass mixing ratio for the major species, N_2, O_2, and O, is also solved taking into account the photo-dissociation of O_2 and oxygen chemistry. The EUVAC spectral model (Richards *et al.*, 1994) is used for the solar EUV flux poured into the thermosphere. This GCM is a global spectral model (triangular truncation of T21 which is equivalent to a spatial resolution of 5.6^o for latitude and longitude) with 75 vertical pressure levels (vertical resolution of 0.4 scale height above the tropopause) and contains all the atmospheric regions from the ground to exobase. The time step for integrating the equations is 100 s. The corrected centered dipole is assumed to describe the Earth's magnetic field. The effects of auroral particle precipitation on heating the neutral gases are evaluated by using an analytical prescription into the auroral oval. The magnetospheric convection electric field and empirical ionosphere are used for calculating Joule heating and ion-drag force. The details of the GCM are described in Miyoshi and Fujiwara (2003, 2006), Fujiwara and Miyoshi (2006), and references therein.

4. Thermospheric temperature and density variations simulated by GCM

Figure 4 shows GCM results of global temperature and horizontal wind distributions at 00:00 UT on March 5, June 5, September 5, and December 5 on a constant-pressure surface of about 302-312 km altitude. Note that the strength of the maximum wind vector is different in each panel to overview global wind patterns. Diurnal (or day-night) temperature variation is clearly seen in all the panels. In the December case, the large-scale (day-night) temperature structure is quite similar to that from an empirical model of NRLMSISE-00 shown in Figure 3; namely, high temperature regions at dayside low-latitudes and southern (summer) high-latitudes. On the other hand, localized temperature structures appear in the simulation results; for example, localized structures are clearly seen from high- to low-latitudes in the southern (summer) hemisphere in the December case. Fujiwara and Miyoshi (2009) showed that the localized structures resulted from the lower atmospheric effects which cause day-to-day variation of the thermospheric temperature. Some localized structures in the simulation results show wavy structures/variations. For example, wavy structures are seen at around the solar terminator and auroral oval. Fujiwara and Miyoshi (2006) made mention of these structures based on their GCM simulations as one of characteristics of the thermospheric temperature structure during geomagnetically quiet periods.

Figure 5 shows GCM results of global density distributions at 00:00 UT on March 5, June 5, September 5, and December 5 at a specific height of 400 km. Although the atmospheric parameters are calculated on constant pressure surfaces because of assumption of the hydrostatics, we can infer the parameters at a specific height from the output data with the spline interpolation. The density variation ranges are 0.93-3.56 $\times$ 10^{-12} kg/m^3, 1.02-4.22 $\times$ 10^{-12} kg/m^3, 0.84-3.47 $\times$ 10^{-12} kg/m^3, and 1.12-4.44 $\times$ 10^{-12} kg/m^3 in March, June, September, and December, respectively. Localized/wavy structures are also seen in the high-latitude region and near solar terminator particularly in the June and December cases. Forbes *et al.* (2008) found solar terminator waves from the neutral mass density variations observed by CHAMP. These waves are closely related to tidal forcings originating from the lower atmosphere (Miyoshi *et al.*, 2009).

Sudden changes in the energy inputs into the thermosphere due to the solar wind variation produce high temperature (pressure) regions at high-latitude. The produced pressure bulges generate traveling atmospheric disturbances (TADs) which propagate globally in the thermosphere. Figure 6 shows examples of TADs simulated with the whole atmosphere GCM. In this case, the cross polar cap potential drop is changed from 30 (initial level) to 60 kV (disturbed level) during 1-hour (00:00-01:00 UT) to produce TADs. The upper panel shows usual simulation results for the thermospheric temperature and wind at 00:00 UT on December 1 on a constant pressure surface of about 310 km altitude. The middle panel shows temperature and wind distributions at 100 minutes after enhancement of the high-latitude energy inputs. High temperature regions, which extend almost in the longitudinal direction, are clearly seen from high- to mid-latitudes. The bottom panel shows differences of temperature and wind between results with TADs (middle panel) and in the usual case. Longitudinally-extent TADs, which are generated and propagate from high- to low-latitudes, are clearly seen in both hemispheres in the bottom panel. Fujiwara and Miyoshi (2006, 2009) also performed TAD simulations with the whole atmosphere GCM. Superposition of TADs and the lower atmospheric effects can produce localized temperature structures globally in the upper thermosphere (Fujiwara and Miyoshi, 2009). Figure 6 also shows localized temperature structures produced by TADs and the lower atmospheric effects. In this case, the amplitudes of TADs are in the

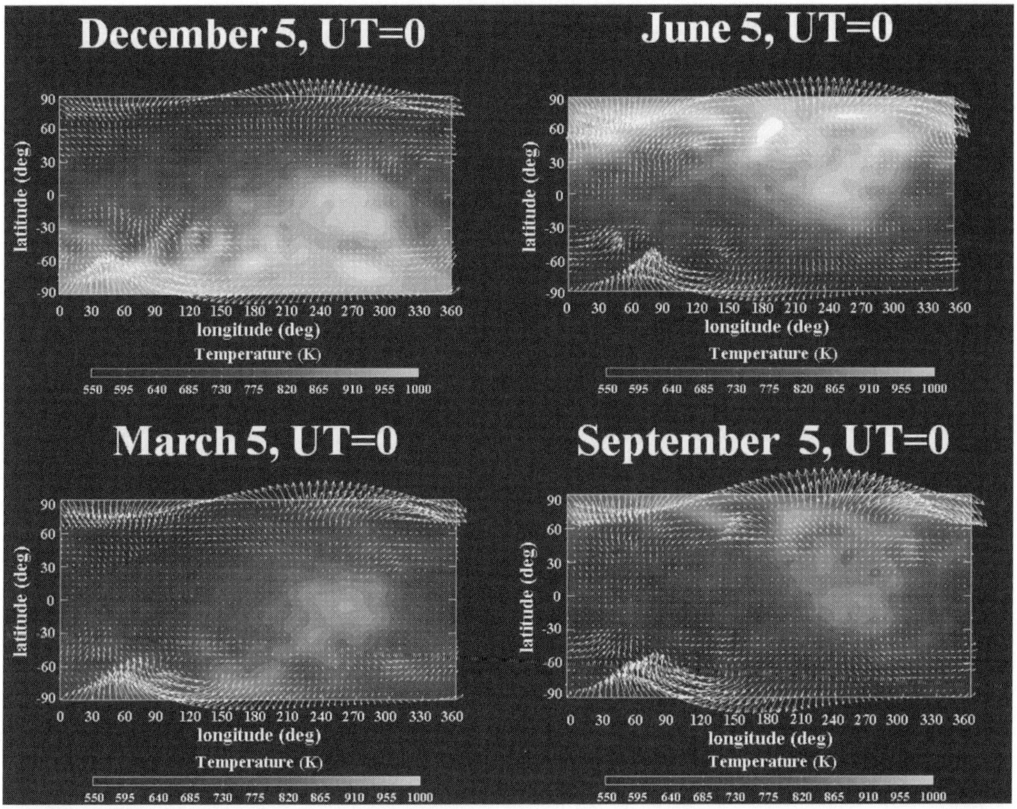

Figure 4. GCM results of temperature and horizontal wind distributions at 00:00 UT on March 5, June 5, September 5, and December 5 on a constant-pressure surface of about 302-312 km altitude. Note that the strength of the maximum wind vector is different in each panel to overview global wind patterns.

temperature range between -50 and 133 K. In addition to TADs propagating from high- to low-latitudes, TADs propagating over the polar cap region are seen in the dayside and nightside regions in the northern and southern hemispheres, respectively (see bottom panel).

5. Future works

In addition to the thermospheric phenomena, the ionospheric ones and coupling processes between the thermospheric neutrals and ionospheric plasmas are quite interesting. Recent observations have revealed ionospheric variations originating from the tropospheric phenomena (e.g., "Coupling Processes in the Equatorial Atmosphere" the special issue of Earth Planets and Space, vol.61(No.4), 2009). In order to describe the coupling results, we have started a modelling study for coupling the whole atmosphere GCM and ionospheric models. As the first step of the work, we investigated low-latitude ionospheric variations (e.g., wavenumber-4 structure) caused by the atmospheric tidal forcings by using an ionospheric dynamo model with use of the GCM output data (Jin *et al.*, 2008).

Some projects for the thermospheric/ionospheric sciences are planned in the Japanese community. For example, our future satellite mission, the Ionosphere, Mesosphere, upper Atmosphere, and Plasmasphere mapping (IMAP), will provide us with important information on coupling between the ionosphere and atmosphere through optical imaging

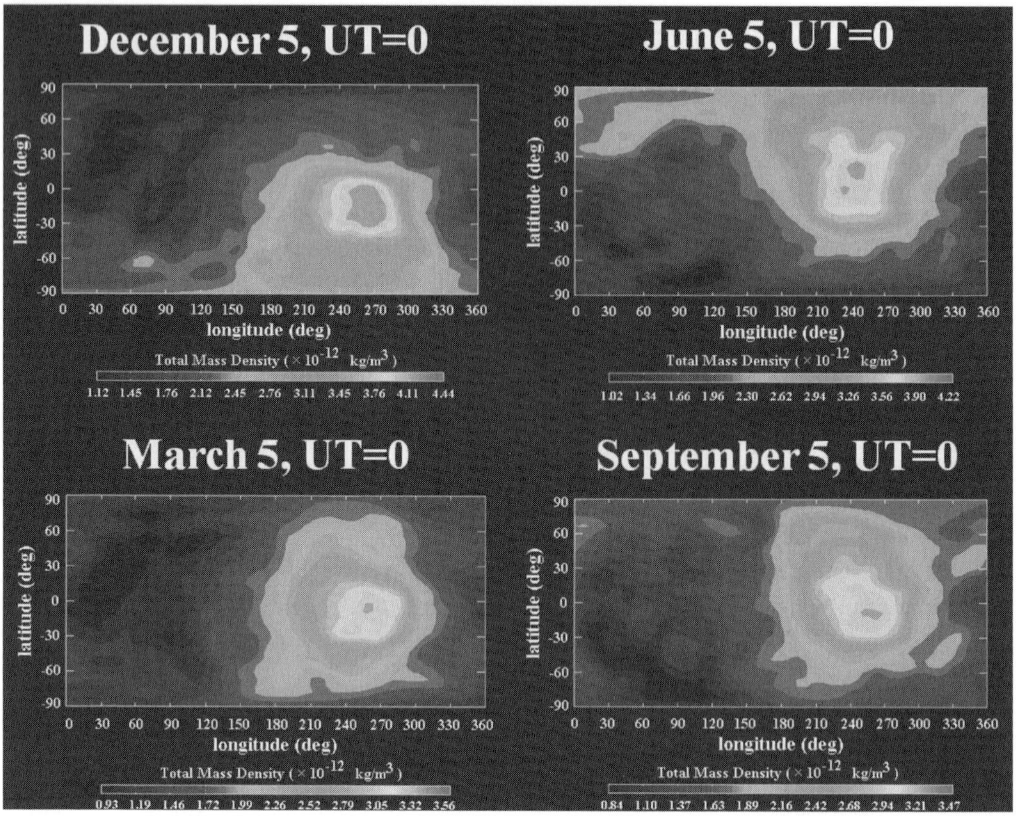

Figure 5. Same as Figure 4 except for neutral mass density at a specific height of 400 km.

observations. The IMAP project is now moving ahead as one of application projects of International Space Station (ISS).

6. Summary

The thermospheric temperature and density variations have been investigated from observational and modelling studies. The general features of thermospheric responses to change in the solar activity due to the 11-year solar cycle have been well understood. However, thermospheric localized structures and their variations from hour to hour are not understood well. In addition, it is also difficult to describe thermospheric responses to sudden changes in the solar radiation and solar wind energy inputs. Modelling efforts of the Earth's and planetary atmospheres are essential to understanding the solar variability impacts on Earth and planets. In particular, general/global circulation model (GCM), which includes all the atmospheric regions, is a very effective tool for studying coupling processes between the lower and upper atmospheres. Our whole atmosphere GCM simulates the thermospheric variations from hour to hour, localized wavy structures, and traveling atmospheric disturbances. The future coupled model of the whole atmosphere and the ionosphere will describe more details of the coupling phenomena between the thermospheric neutrals and the ionospheric plasmas in cooperation with ground-based observations and future missions, e.g., the IMAP project in Japan.

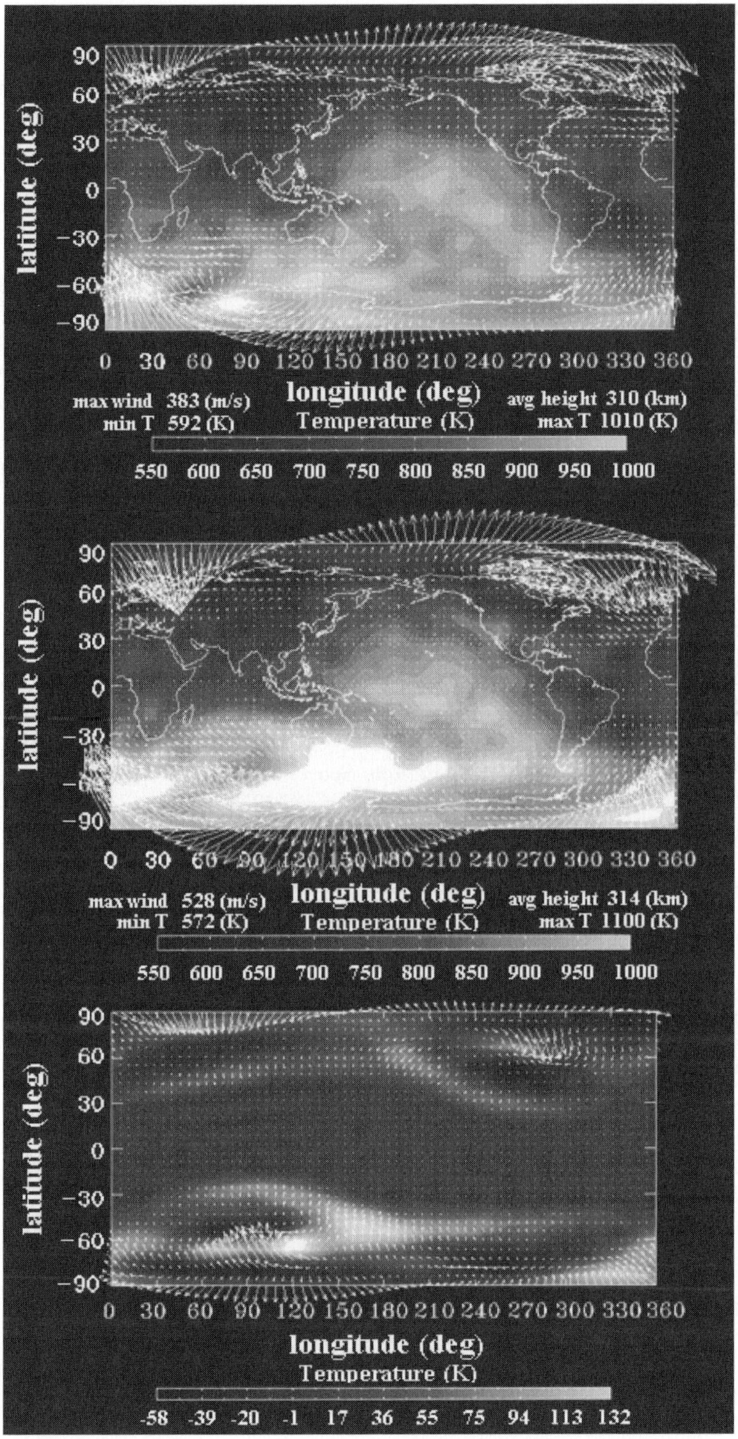

Figure 6. The thermospheric temperature and wind at 00:00 UT on December 1 on a constant pressure surface of about 310 km altitude obtained from GCM simulation in the same condition as that shown in Figure 4 (upper panel). Temperature and wind distributions at 100 minutes after enhancement of the high-latitude energy inputs (middle panel). Differences of temperature and wind between results of usual and TADs simulations (bottom panel). (See text for details)

Acknowledgements

This work was supported in part by Grant-in-Aid for Scientific Research C (20540435), Grant-in-Aid for Scientific Research on Innovative Areas (20200047), and the Global COE program "Global Education and Research Center for Earth and Planetary Dynamics" at Tohoku University by the Ministry of Education, Science, Sports and Culture, Japan. A part of this work was also supported by the joint research program of the Solar-Terrestrial Environment Laboratory, Nagoya University. The empirical models of NRLMSISE-00 and HWM-93 are provided by NSSDC/NASA and IRI is from ModelWeb of CCMC/NASA.

References

Akmaev, R. A., Fuller-Rowell, T. J., Wu, F., Forbes, J. M., Zhang, X., Anghel, A. F., Iredell, M. D., Moorthi, S., & Juang, H.-M. 2008, *Geophys. Res. Lett.*, 35, L03810, doi:10.1029/2007GL032584.

Aruliah, A. L., Fuller-Rowell, T. J., and Rees, D. 1991, *J. Atmos. Terr. Phys.*, 53, 467.

Bilitza, D. & Reinisch, B. 2008, *J. Adv. Space Res.*, 42, 599.

Bruinsma, S., Forbes, J. M., Nerem, R. S., & Zhang, X. 2006, *J. Geophys. Res.*, 111, A06303, doi:10.1029/2005JA011284.

Buonsanto, M. J. & Pohlman, L. M. 1998, *J. Geophys. Res.*, 103 , 23381.

Forbes, J. M., Lu, G., Bruinsma, S., Nerem, R. S., & Zhang, X. 2005, *J. Geophys. Res.*, 110, A12S27, doi:10.1029/2004JA010856.

Forbes, J. M., Bruinsma, S. L., Miyoshi, Y., & Fujiwara, H. 2008, *Geophys. Res. Lett.*, 35, L14802, doi:10.1029/2008GL034075.

Ford, E. A. K., Aruliah, A. L., Griffin, E. M., & McWhirter, I. 2007, *Ann. Geophys.*, 25, 1269.

Fujiwara, H. & Miyoshi, Y. 2006, *Geophys. Res. Lett.*, 33, L20108, doi:10.1029/2006GL027103.

Fujiwara, H. & Miyoshi, Y. 2009, *Earth Planets Space*, 61, 463.

Fuller-Rowell, T. J. & Rees, D. 1980, *J. Atmos. Sci.*, 37, 2545.

Fuller-Rowell, T. J., Akmaev, R. A., Wu, F., Anghel, A., Maruyama, N., Anderson, D. N., Codrescu, M. V., Iredell, M., Moorthi, S., Juang, H.-M., Hou, Y.-T., & Millward, G. 2008, *Geophys. Res. Lett.*, 35, L09808, doi:10.1029/2007GL032911.

Hedin, A. E., Fleming, E. L., Manson, A. H., Schmidlin, F. J., Avery, S. K., Clark, R. R., Franke, S. J., Fraser, G. J., Tsuda, T., Vial, F., & Vincent, R. A. 1996, *J. Atmos. Terr. Phys.* 58, 1421.

Ishii, M., Oyama, S., Nozawa, S., Fujii, R., Sagawa, E., Watari, S., & Shinagawa, H. 1999, *Earth Planets Space.*, 51, 833.

Jin, H., Miyoshi, Y., Fujiwara, H., & Shinagawa, H. 2008, *J. Geophys. Res.*, 113, A09307, doi:10.1029/2008JA013301.

Kanamitsu, M., Tada, K., Kudo, T., Sato, N., & Isa, S. 1983, *J. Meteor. Soc. Japan*, 61, 812.

King-Hele, D. G. 1959, *Nature*, 183, 1224.

King-Hele, D. G. 1987, *Satellite Orbits in an Atmosphere: Theory and application*, Springer, p. 304.

Liu, H., Luhr, H., Henize, V., & Kohler, W. 2005, *J. Geophys. Res.*, 110, A04301, doi:10.1029/2004JA010741.

Miyahara, S., Yoshida, Y., and Miyoshi, Y. 1993, *J. Atmos. Terr. Phys.*, 55, 1039.

Miyoshi, Y. 1999, *Earth Planets Space*, 51, 763.

Miyoshi, Y., & Fujiwara, H. 2003, *Geophys. Res. Lett.*, 30, 1789, doi:10.1029/2003GL017695.

Miyoshi, Y., & Fujiwara, H. 2006, *J. Geophys. Res.*, 111, D14108, doi:10.1029/2005JD006993.

Miyoshi, Y., Fujiwara, H., Forbes, J. M., & Bruinsma, S. L. 2009, *J. Geophys. Res.*, 114, A07303, doi:10.1029/2009JA014110.

Nozawa, S., & Brekke, A. 1995, *J. Geophys. Res.*, 100, 14717.

Picone, J. M., Hedin, A. E., Drob, D. P., & Aikin, A. C. 2002, *J. Geophys. Res.*, 107(A12), 1468, doi:10.1029/2002JA009430.

Prölss, G. W. 1982, *J. Geophys. Res.*, 87, 5260.

Richards, P. G., Fennelly, J. A., & Torr, D. G. 1994, *J. Geophys. Res.*, 99, 8981.

Roble, R. G.. 2000, *AGU monograph*, 123, 53.

Shiokawa, K., Kadota, T., Otsuka, Y., Ogawa, T., Nakamura, T., & Fukao, S. 2003, *Earth Planets Space*, 55, 271.

Sutton, E. K., Forbes, J. M., Nerem, R. S., & Woods, T. N. 2006, *Geophys. Res. Lett.*, 33, L22101, doi:10.1029/2006GL027737.

Toth, G., De Zeeuw, D. L., Gombosi, T. I., Manchester, W. B., Ridley, A. J., Sokolov, I. V., & Roussev, I. I. 2007, *Space Weather*, 5, S06003, doi:10.1029/2006SW000272.

Wang, W., Wiltberger, M., Burns, A. G., Solomon, S. C., Killeen, T. L., Maruyama, N., & Lyon, J. G. 2004, *J. Atmos. Solar-Terr. Phys.*, 66/15–16, 1425.

Solar and Stellar Variability: Impact on Earth and Planets
Proceedings IAU Symposium No. 264, 2009
A. G. Kosovichev, A. H. Andrei & J.-P. Rozelot, eds.

doi:10.1017/S1743921309992869

Are there variations in Earth's global mean temperature related to the solar activity?

Olav Kjeldseth-Moe[1] and Sven Wedemeyer-Böhm[2]

[1]Institute of Theoretical Astrophysics, University of Oslo, Oslo, Norway
email: olavm@astro.uio.no

[2]Institute of Theoretical Astrophysics, University of Oslo, Oslo, Norway
email: sven.wedemeyer-bohm@astro.uio.no

Abstract. We have analyzed the record of Earth's global temperature variations between 1850 and 2007 looking for signals of periodic variations and compared our results with solar activity variations in the same time period. Significant periods are found at 9.4, 10.6 and 20.9 years. These periodic variations may be caused by solar activity. However, and amazingly enough, we also find at least 17 other significant periodic variations in addition to expected variations with periods of 1 year and of half a year. The result is considered in terms of solar related forcing mechanisms. These may be variable solar heating associated with the small changes in solar irradiance over the solar cycle, or direct effects of interactions between variable magnetic fields carried by the solar wind and particles and fields in interplanetary space or in the Earth's ionosphere.

Keywords. Earth: global temperature, Sun: activity

1. Introduction

The role of solar forcing on Earth's global temperature level is a disputed question. In addition to radiative forcing, which is weak, it has been suggested that variations in the influx of cosmic rays into interplanetary space, owing to solar magnetic activity, causes variations in cloud formation on Earth that may account for climate changes on long timescales, from years to millennia (e.g. Svensmark 2007). However, other magnetically related effects are also conceivable.

Looking for periods in the global temperature on the same time scale as solar activity variations may thus be a way of checking on the amount and character of solar forcing, whether thermal or magnetic. Any periodic variations may, so to speak, serve as a calibration for solar effects. We have therefore analyzed the record of Earth's global temperature between 1850 and 2007 looking for periodic signals in the range 1–30 years. To establish a good average value for the solar activity period in the same time interval, we also analyzed the record of sunspot numbers, taking these as a satisfactory indicator of solar magnetic variations, causing variations in the spectrally integrated solar irradiance as well as interacting with magnetic fields on Earth and in space around us.

2. Data sources and data reduction

The data consist of the record of Earth's global temperature variation together with solar activity data for the period 1850 through 2007, i.e. 158 years.

Solar activity. The significance of the solar activity variations is twofold. Firstly, a small variation in the irradiance with solar activity has been found from three decades of space observations (see e.g. Fröhlich 2006, Scafetta 2009 and references therein). An absolute calibration of the variable solar irradiance is problematic, but clearly its peak-to-peak variation is ≈1 W m^{-2}, with an average value near 1366 W m^{-2}, i.e. a total variation of 0.075 percent. Secondly, the transport of magnetic fields from the Sun into interplanetary space increases strongly with increasing solar activity. This may influence cloud formation on Earth, according to Svensmark (2007), or affect tropical electric storm activity with further effects on global temperatures.

To find the length of the solar activity cycle we used daily sunspot numbers listed by the Solar Influences Data Analysis Center, Royal Observatory of Belgium, (see SIDC: http://www.sdic.be/sunspot-data). A recent thorough analysis of these data was made by Gil-Alana(2009). He obtained a value for the period of 130 months, or 10.83 years, for the interval 1749–2008. However, in the first century of this interval the sunspot record is incomplete.

We therefore performed a Fourier analysis of the data in the 158 year interval referred to above and found a value for the period of 10.45 ± 0.6 years. The given uncertainty is simply estimated from the width of a Gaussian fit to the slightly asymmetrically placed power spectrum peak corresponding to the 11 year period.

Global temperatures. Global temperatures are taken from the United Kingdom Met Office Hadley Centre observations datasets. A table of monthly and annual temperature anomalies, i.e. global temperatures in °C relative to a mean over a range of years, is listed in the HadCRUT3 database (see http://hadobs.metoffice.com/hadcrut3/). Formats of the HadCRUT3 tables can be obtained from the same site. Regarding the formats we note that the data tables, in addition to the most reliably estimated anomalies, also contain the upper and lower 95% uncertainty ranges from the combined effects of all uncertainties. The data and uncertainties are described by Brohan *et al.* (2006).

In our investigation we shall use the monthly temperature anomalies. The best monthly anomalies from 1850 through 2007 as listed in the HadCRUT3 database, are shown in the left panel of Figure 1.

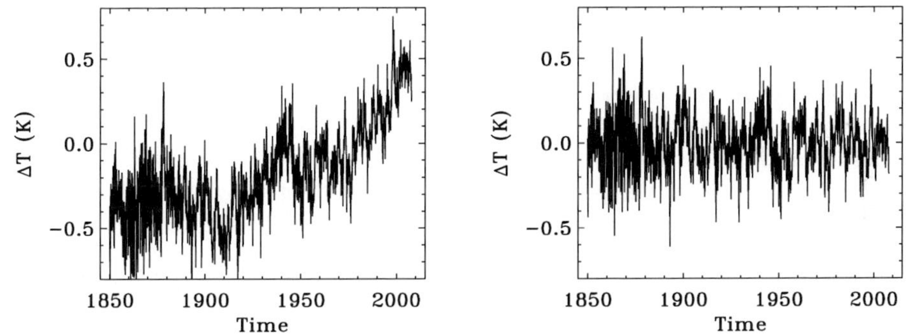

Figure 1. Left panel: Global temperature anomaly 1850–007. Right panel: Temperature anomaly with long term trends subtracted and average equal to zero.

Before we analyzed the data we subtracted long-term trends. For this we fitted a high order polynominal to the observed anomalies and subtracted the time series values obtained from the polynominal. We found that a 10th order polynominal gave the lowest chi-square value for the fit. Afterwards we subtracted the very small average of the resulting anomalies (≈ 10^{-4} °C). This removed the zero frequency power from the power spectrum and facilitated detection of periods of the order 10-20 years. The procedure only adds a small uncertainty to the lowest frequencies, f = 1–5. The result, which we shall call the differential temperature anomaly, is shown in the right panel of Figure 1.

Power spectrum and significance of power signals. The monthly differential temperature anomalies have been Fourier analyzed and the power spectrum calculated. The result is shown in Figure 2. We have limited the plot to the 170 lowest frequencies. A yearly variation is clearly present at f=158. In addition there is a likely significant peak at f=316, i.e. corresponding to a half-yearly variation.

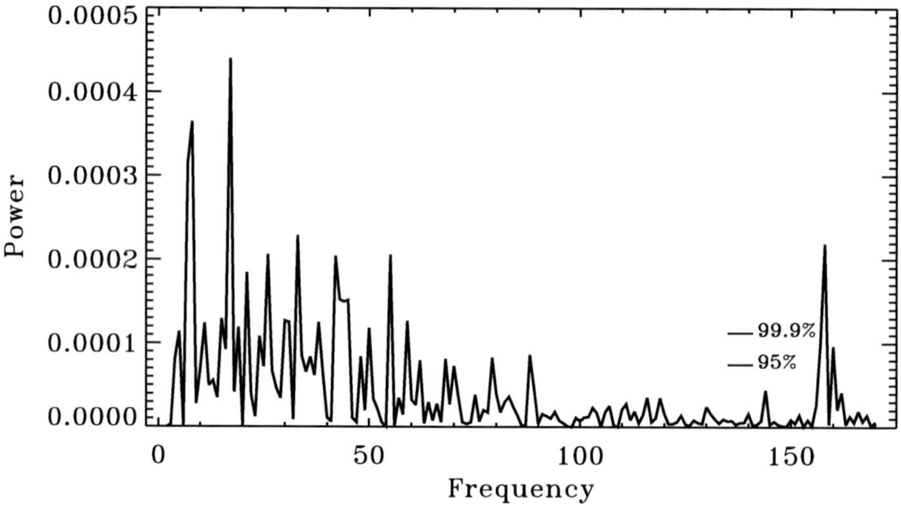

Figure 2. Power spectrum of the differential global temperature anomaly, for f = 0–170.

Two annotated marks in the figure give power levels where we find that given fractions of power values above these levels correspond to real signals and are not caused by noise. Thus, for power values larger than $7.5\ 10^{-5}$ and $1.1\ 10^{-4}$ the fraction of real signal values relative to the total number of noise and signal together is 95 % and 99.9 %, respectively. We shall now describe how we have estimated this significance of any particular power value, something that is needed in order to decide which of the peaks in the power spectrum correspond to real time variations, as distinguished from noise.

The noise level in the power spectrum of a data set may be estimated by shuffling the data randomly and calculate the power spectrum of the shuffled data. If this is done many times, say a 1000 times or more, a significant average is obtained. The procedure gives a reasonable result if the power in the data is dominated by the noise, which is the case here. However, there is always an amount of signal mixed into the noise obtained this way.

The significance of a particular power value may, however, be estimated in another fashion. If we plot the distribution of values in our power spectrum we get a normalized distribution function as shown in the left panel of Figure 3. The normalized distribution, $N(p)/N$, is defined as the number of values per power unit in the power spectrum as a function of power, p. It is obtained by dividing the power spectrum into small sections each covering a range in power from p to p+δp, counting the number of values in each section, and dividing by the interval δp. The crosses in the figure give the counted values while the drawn line is the best estimate of a curve through the points. This curve consists of two exponential functions added together, where the exponents are linear functions of power, p. Thus, they appear as straight lines on the logarithmic scale of the y-axis.

The two separate domains for high and low powers correspond to a noise dominated and a signal dominated part of the distribution function. A synthetic model may easily demonstrate this. Thus, we made a data set of 20 oscillation with different amplitudes and with frequencies close to the probable frequencies in the power spectrum of the real data. Then we added different amounts of noise to the resulting time series values. The result is seen in Figure 3b, where two of the curves have noise levels bracketing what is found in the real data and the third curve (dot-dashed) is from the synthetic data set with no noise at all. We did not include the counted values in the plots in Figure 3b, but they correspond in a general fashion to the counted values in the analysis of real temperature data shown in the left panel, Figure 3a.

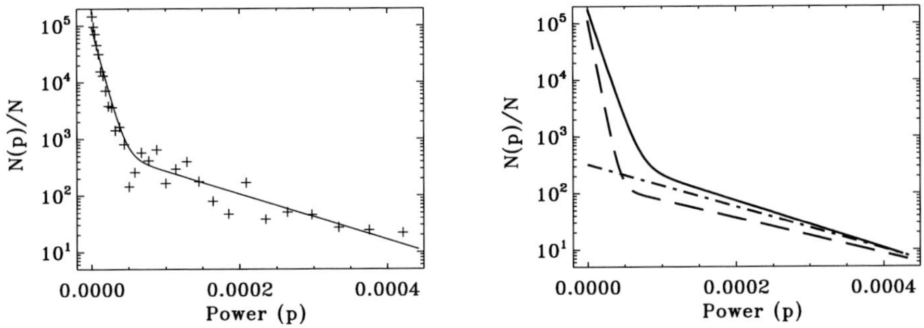

Figure 3. Left panel (a): Normalized distribution of power values in the power spectrum of Earth's differential global temperature anomalies. Right panel (b): Normalized distribution of power values in the power spectrum of a synthetic set of 20 oscillations, with different levels of noise added.

The likelihood of a particular power value being caused by a real signal or by noise may then be derived from the two contributing distributions in Figure 3a. This is how the fractional annotation values in Figure 2 were determined. Uncertainties in the derived likelihood values may furthermore be computed from the uncertainties in the fitting of straight lines to the logarithm of the "noise" and the "signal" distributions, respectively, in the two domains in Figure 3a.

Finally, the values of the power peaks (and indeed any value in the power spectrum) have uncertainties owing to the inherent total uncertainty of the given temperature anomalies. The 95% uncertainty ranges from the combined effects of all the uncertainties are, as mentioned above, included in the data set. The uncertainties may not be randomly distributed, but for our purpose we have assumed this as a sufficiently close approximation. A large number (30000) of time series of possible temperature anomalies were created, where Gaussian randomly distributed increments were added to each temperature value in the original best series. These random temperature increments were all different, had a mean of zero and a standard value corresponding to the given uncertainties in the measured temperature anomalies. Thus, the 95% limits were regarded as 2 × sigma deviations from a mean value. The synthetic temperature anomalies were then treated the same way as the standard set of values and their power spectra were derived. From this we obtained the distribution of the power in the power spectrum. An example, including the two strong power peaks at f=7-8 and f=17 and a selection of four other peaks in the power spectrum, is shown in Figure 4. The values for the 95% and 99% likelihood of the power peak being caused by a real signal is marked in the figures (broken vertical lines).

3. Results

The results of the investigation are presented in Figure 2 and in Table 1. In Table 1 we list parameters for oscillations with power values larger than $0.85{\cdot}10^{-4}$. Power values at this lower limit are significant to 95%, even if the uncertainty in the limit is taken into account. Most of the listed power peaks can, however, be considered as real signals with a significance of well above 99.9%.

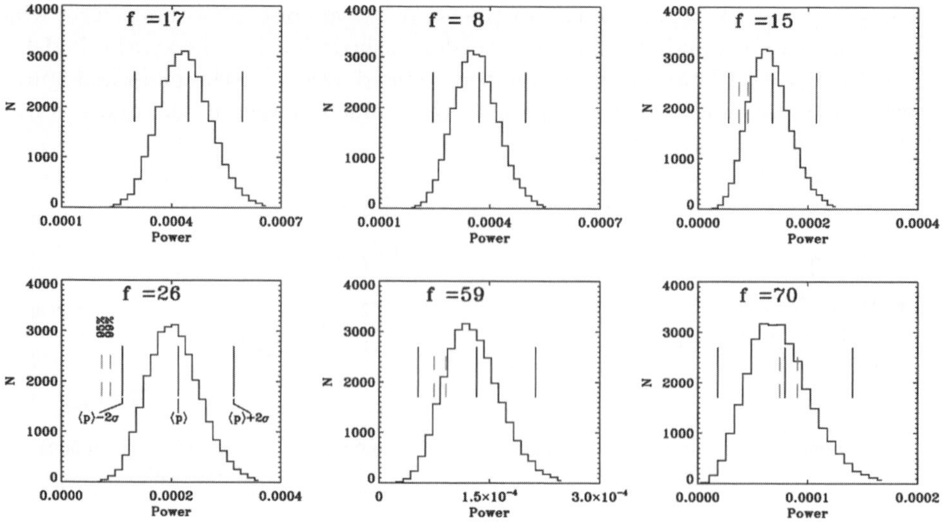

Figure 4. Distribution of power values at 6 selected frequencies derived from the uncertainties in the observed temperature anomalies, 1850–2007. $< p >$ is the average power value.

Table 1. The strongest maxima in the global temperature power spectrum.

Frequency (nominal)	Frequency–f_a (adjusted)	Power value (at peak x 10^4)	Period (years)	Remarks
(4)-5	4.60	1.18 ± 0.30	34.4	Asymmetric feature
(7)-8	7.56	3.70 ± 0.63	20.9	Asymmetric feature
11	10.87	1.30 ± 0.39	14.5	
15	14.90	1.36 ± 0.39	10.6	
17	16.90	4.45 ± 0.72	9.35	
19	18.75	1.25 ± 0.39	8.42	
21	21.21	1.90 ± 0.48	7.45	
24	24.29	1.14 ± 0.37	6.51	
26	25.98	2.13 ± 0.51	6.08	
30-31		1.27	5.2	Blended feature
33	33.24	2.29 ± 0.5	4.75	
38	38.02	1.31 ± 0.40	4.16	
42-45		2.15	3.75-3.5	Broad blended feature
48	48.11	0.91 ± 0.33	3.28	
50	50.07	1.25 ± 0.39	3.15	
55	55.01	2.12 ± 0.51	2.87	
59	59.11	1.32 ± 0.40	2.67	
62	61.80	0.86 ± 0.32	2.56	
68	68.17	0.88 ± 0.32	2.32	
79	79.21	0.89 ± 0.32	1.99	
158	157.68	2.25 ± 0.53	1.00	
160	160.14	1.03 ± 0.35	0.987	
316	315.93	0.90 ± 0.34	0.50	

We start by noting that the two strongest periods in the power spectrum are located at nominal frequencies f=8 and f=17. These peaks are unequivocally caused by a real signal. They are clearly outstanding also if we analyze the annually averaged temperature anomalies. In addition there is a weaker power peak at f=15, blended in with the stronger peak at f=17. Table 1 gives more accurate frequencies, found by fitting Gaussians to the sometimes asymmetrically placed power peaks and deriving the position of these profiles. The periods of oscillations corresponding to f=8, 15 and 17, are 20.9 ± 1.7 years, 10.60 ± 0.42 years, and 9.35 ± 0.35 years, respectively. Periods are determined from the expression $P = 158/f_a$, since the data set is 158 years long. Uncertainties in the periods are estimated from an assumed full width at half maximum (FWHM) of the power peaks of 1 frequency unit. The periods of f=8 and f=15 agree especially well with the solar activity period of 10.45 ± 0.6 years and with twice this period, which correspond to the periodic variation of the changing solar magnetic polarity. The temperature amplitudes of the three oscillations are approximately 0.05 K, 0.035 K and 0.05 K, respectively.

The investigation started as a search for periodicities corresponding to the solar activity period. However, the detection of a number of other persistent and clearly significant periods, came as a serendipitous, amazing and exciting additional result. Periods of half a year, one year, and two years, are clearly related to the Earth, moving in its eccentric orbit around the Sun, and having different distributions of continents and oceans on the two hemispheres. The other oscillations have periods shorter than $\approx$ 9 years, with exception of one period at 34 years. It seems likely that these variations persist over long times, since the global mean temperature data span almost 160 years.

4. Concluding remarks

The closeness in period between the strongest global temperature variations and the period of solar activity points to a clear connection between the two. The 21 year period might possibly be an undertone of the shorter period variation and might thus be caused by solar irradiance variations. However, a mechanism depending on solar magnetic polarity regulating how well solar wind particles penetrate the Earth's magnetosphere, would also have to be considered. Whether solar irradiance variations would be able to drive oscillations with the derived temperature amplitudes is also an open question, but may at a first glance seem likely. But what about all the other persistent oscillations in the global temperatures? They also have to be driven by an external variable energy input. Also here the ultimate driving force may be solar variations, but perhaps the required energies from irradiance variations alone are insufficient. These and other questions will be addressed in ongoing investigations.

References

Brohan, P., Kennedy, J. J.., Harris, I., Tett, S. F. B., & Jones, P. D. 2006, *Geophys. Res.*, 111, D12106

Fröhlich, Claus 2006, *Space Science Reviews*, 125, 531

Gil-Alana, Luis A. 2009, *Solar Phys.*, 257, 371

Scafetta, Nicola 2009, *arXiv:0908.0792S [astro-ph.SR] 6 Aug 2009*

Svensmark, Henrik 2007, *Astronomy and Geophysics*, 48, 1.18

Solar and Stellar Variability: Impact on Earth and Planets
Proceedings IAU Symposium No. 264, 2009
A. G. Kosovichev, A. H. Andrei & J.-P. Rozelot, eds.

doi:10.1017/S1743921309992870

The CME link to geomagnetic storms

Nat Gopalswamy

NASA Goddard Space Flight Center, Greenbelt, MD 20771, USA
email: nat.gopalswamy@nasa.gov

Abstract. The coronal mass ejection (CME) link to geomagnetic storms stems from the southward component of the interplanetary magnetic field contained in the CME flux ropes and in the sheath between the flux rope and the CME-driven shock. A typical storm-causing CME is characterized by (i) high speed, (ii) large angular width (mostly halos and partial halos), and (iii) solar source location close to the central meridian. For CMEs originating at larger central meridian distances, the storms are mainly caused by the sheath field. Both the magnetic and energy contents of the storm-producing CMEs can be traced to the magnetic structure of active regions and the free energy stored in them.

Keywords. Sun: coronal mass ejections (CMEs), solar wind, magnetic fields, interplanetary medium

1. Introduction

A geomagnetic storm occurs when the southward magnetic field component (B_s) of an interplanetary (IP) structure reconnects with Earth's magnetic field thereby allowing the entry of the solar wind energy into the magnetosphere. The intensity of the geomagnetic storm is expressed using a number of indices. Here we use the Dst (disturbance storm time) index, which represents the average change in the horizontal component of Earth's magnetic field (in units of nT) brought about by the geomagnetic storm at four low-latitude stations (see http://swdcwww.kugi.kyoto-u.ac.jp/dstdir/dst2/onDstindex.html). A storm is indicated when $Dst \leqslant -50$ nT (Loewe & Prölss 1997); $Dst \leqslant -100$ nT indicates intense storms. Alfven waves in the solar wind, corotating interaction regions (CIRs) and coronal mass ejections (CMEs) are the sources of B_s in the IP magnetic field (IMF; Lindsay *et al.* 1995; Echer, Gonzalez & Alves 2006; Xu *et al.* 2009). CMEs cause the most intense of the storms (see e.g., Zhang *et al.* 2007; Gopalswamy, Yashiro & Akiyama 2007). The ability of an interplanetary structure in causing a geomagnetic storm is referred to as geoeffectiveness. The geoeffectiveness of IP structures has been well established by the 1990s (e.g., Gosling 1993; Tsurutani *et al.* 1995). The focus in this paper is on the geoeffectiveness of CMEs observed near the Sun.

It is easy to understand the CME link to geomagnetic storms from the following empirical relationship between the Dst index (nT) and the speed (V in km/s) and the magnitude of the out-of-the-ecliptic component of the IMF ($B_s = -B_z$ in nT) of the solar wind structure:

$$Dst = -0.01VB_z - 32. \tag{1.1}$$

This empirical relation was obtained by Gopalswamy *et al.* (2008) for a subset of interplanetary CMEs (ICMEs) known as magnetic clouds (MCs; Burlaga *et al.* 1981). For $V = 1000$ km/s and $B_z = 60$ nT, one gets $Dst = -632$ nT, which is roughly the maximum recorded value of the Dst index since solar cycle 19. The Carrington event (1859 September 1) is supposed to have caused a bigger storm, which is not fully understood (see Siscoe, Crooker & Clauer 2006, and references therein). The dependence of Dst on V and B_z in Equation 1.1 seems to be firmly established (see e.g., Wu & Lepping 2002). We shall discuss the applicability of this relationship to non-cloud ejecta in a later section. The B_z dependence of Dst is fundamental in that only when B_z is negative (southward) that a geomagnetic storm occurs. The speed V is ultimately related to the CME speed at the Sun modified by the solar wind. Statistical studies have shown that the magnetic field strength in ICMEs is weakly correlated with the ICME speed (Gonzalez *et al.* 1998), which seems to hold even when the CME speed is used in place of the ICME speed (Gopalswamy *et al.* 2008). Thus the dependence of Dst on V and B_z can ultimately be traced to the magnetic properties of the source active regions and the kinematics of CMEs. The CME link can be further understood in terms of the CME source location on the Sun, magnetic properties of the source active regions, the spatial and magnetic structures of CMEs, the CME speed evolution, and the phase of the solar cycle. Towards this end, we use geomagnetic storms from solar cycle 23 because only during this cycle we have continuous CME observations that can be correlated with geomagnetic storms. We also consider only intense storms ($Dst \leqslant -100$ nT) because these storms can be directly linked to CMEs (Zhang *et al.* 2007; Gopalswamy, Yashiro & Akiyama 2007).

2. CME source connection

The solar source of an intense geomagnetic storm can be a CME or a CIR. A CME can be traced to a closed magnetic field region on the Sun such as an active region or a quiescent filament region. The CME is often accompanied by a flare in soft X-rays and/or in H_α. The flare location on the disk is usually defined as the solar source of the CME. Imaging observations in EUV indicate the flare location, the surrounding dimming region and wave disturbances associated with the CMEs. Microwave imaging can also indicate the location of the filament eruption associated with the CME and the post-eruption arcade (flare). Thus the CME source can be identified in a number of ways, ultimately giving the heliographic coordinates of the source location, which is the point of maximum flare brightness in H_α, EUV, or X-rays. Details of how one can identify the source regions of ICMEs and geomagnetic storms can be found in Gopalswamy, Yashiro & Akiyama (2007). In the case of CIRs, it is fairly straightforward to identify the underlying coronal hole as the solar source. The central meridian passage of the coronal hole occurs a couple of days before the arrival of the high speed stream at Earth. In the in-situ data, CIRs appear as regions of enhanced density and temperature during the transition from slow to high speed solar wind.

Figure 1 shows the heliographic coordinates of CMEs that produced large geomagnetic storms ($Dst \leqslant -100$ nT) during solar cycle 23. The sizes of the circles denote the intensity of the storms: largest circles correspond to super storms ($Dst \leqslant -300$ nT). The medium circles represent storms with $Dst \leqslant -200$ nT but > -300 nT (second-level storms). The smallest circles correspond to storms with $Dst > -200$ nT (first-level storms). Note that the superstorms are located roughly within a central meridian distance (CMD) of $\sim 15°$ (inner region) with equal number of events in the eastern and western hemispheres. There are only 4 second-level storms in the eastern hemisphere, but 7 in the western hemisphere. Furthermore, three of the western second-level storms have their solar sources beyond the inner region. The first-level storms show western bias both in number and longitudinal spread of source locations. There is only one source beyond a CMD of 75° in each hemisphere. Thus, we see that CMEs originating closer to the disk center have a better opportunity to produce intense magnetic storms (see also Wang *et al.* 2002; Zhang *et al.* 2007; Gopalswamy, Yashiro & Akiyama 2007). The slight western bias

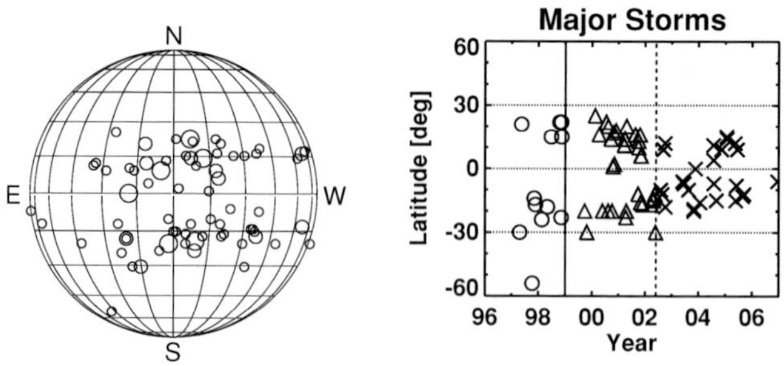

Figure 1. (left) Solar source locations of large geomagnetic storms of solar cycle 23 distinguished according to the storm strength. (right) the source latitudes as a function of time with the three phases of the solar cycle shown separately (circles: rise phase—1996 to 1998; triangles: maximum phase—1999 to 2002.6; crosses: declining phase—2002.6 to 2007).

may be due the fact that CMEs undergo a slight eastward deflection because of the solar rotation as was pointed out by Gosling *et al.* (1987a). Since CMEs are ejected radially, one expects direct impact on the magnetosphere when they originate from close to the disk center.

The latitudes of CME sources plotted as a function of time in Fig. 1 show that the storm-producing CMEs typically originate within $\pm 30°$ latitude (there is only one exception in the beginning of the solar cycle). During the solar minimum, the sources are at slightly higher latitudes compared to the maximum and declining phases. The latitude distribution of the solar sources resembles the sunspot butterfly diagram, suggesting that most of the storm-producing CMEs originate from active regions. Solar rotation brings active regions closer to the disk center; solar activity brings active regions closer to the disk center over a longer time scale. During solar minimum, the global dipolar field also channels CMEs closer to the equator (even though the active regions themselves are located at higher latitudes).

3. The kinematic connection

The kinematic link to the geomagnetic storms (the V part in Equation 1.1) can be understood from Fig. 2, which shows the speed and width distributions of CMEs that produced major geomagnetic storms ($Dst \leqslant -100$ nT) during cycle 23. The speed distribution peaks in the 700 km/s bin and has an average speed of 999 km/s. The width (W) distribution is dominated by the single bin at 360° (halo CMEs). In fact 68% of the CMEs are full halos ($W = 360°$), which increases to 92% when partial halos are included (i.e., all CMEs with $W \geqslant 120°$). Even the small number of non-halos are wide (average width ~88°.7). No CMEs can be found in the bins below the 50° bin. The fraction of halo CMEs in a population is indicative of the fact that these CMEs are very energetic compared to the general population of CMEs. This can be easily seen from the observation that the average speed of halo CMEs is ~1000 km/s (Gopalswamy 2004). Furthermore, faster CMEs are known to be wider (Gopalswamy *et al.* 2001a, 2009a), implying that halo CMEs must be inherently wider and hence more energetic. Faster CMEs also decelerate more within the coronagraphic field of view (Gopalswamy *et al.* 2001a). Accordingly, there is a clear bias towards negative acceleration in Fig. 2. The average acceleration is

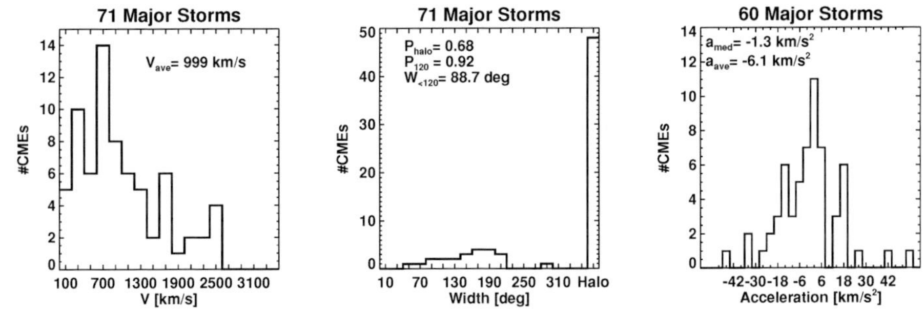

Figure 2. Distributions of speed, width and acceleration of CMEs associated with major geomagnetic storms during cycle 23. The average speed of CMEs is nearly twice that of the general population. Fraction of full halo CMEs in the population (P_{halo}), Fraction of wide CMEs (P_{120}), and non-halo CMEs ($W_{<120}$) are shown on the width plot. Note that the halo CME bin is very tall compared to other bins in the width distribution. The acceleration distribution is clearly skewed toward negative values, consistent with the high speed of the CMEs.

-6.1 ms^{-2} and the median value is -1.3 ms^{-2}. These values are approximate because of projection effects (the height-time measurements used for determining the speed and acceleration are made in the sky plane).

CMEs launched into the interplanetary medium are subject to a drag force that depends on the CME speed and the ambient physical conditions (speed, density and drag coefficient). The drag force tends to slow down fast CMEs and speed up slow CMEs, which can be represented by an interplanetary acceleration, $a = -0.0054(V_{CME} - 406)$, where a is the acceleration in ms^{-2} and V_{CME} is the CME speed (km/s) measured near the Sun (Gopalswamy *et al.* 2001b). The number 406 km/s corresponds to the average speed of the slow solar wind into which the CMEs are launched. For speeds higher than $\sim$1000 km/s, the acceleration is a quadratic function of the initial speed (Gopalswamy 2002). The interplanetary acceleration ranges from ~ -10 ms^{-2} for fast CMEs to $\sim$2 ms^{-2} for slow CMEs. Figure 3 shows the relationship between the speeds measured at the Sun and at 1 AU for a large number (184) of events during cycle 23. The plot includes both types of ICMEs: those with flux rope structure (MCs) and without (non-MCs). The CME speeds at the Sun are in the range $\sim$100 to >3000 km/s, but the corresponding speeds at Earth are in the range $\sim$300 to $\sim$1500 km/s. The narrower range of speeds at Earth is a direct result of the interaction between CMEs and the solar wind. The speed at 1 AU (V) depends on V_{CME} according to $V = 365 + 0.16V_{CME} + 1.4 \times 10^{-5}V_{CME}^2$. When V_{CME} $\sim$400 km/s, we see that $V \sim V_{CME}$ because these CMEs are not affected by the solar wind drag. Lindsay *et al.* (1999) had obtained a linear relationship between V and V_{CME} for a smaller range of speeds.

The ICME speed in Fig. 3 is slightly higher for MCs compared to non-MCs (see Fig.3). This is likely to be due to the fact that the observing spacecraft passes through the nose of the ICMEs when they are MCs and hence record the maximum speed. For non-MCs, the spacecraft may pass through the flanks of the CMEs at varying distances from the nose, so a lower speed is recorded and also varies significantly from event to event. These effects can be directly linked to the CME source location on the Sun: MCs generally result from CMEs originating close to the disk center, while non-MCs result from CMEs originating at larger central meridian distance (Gopalswamy 2006b). V_{CME} is also subject to projection effects, which are more pronounced for MCs than non-MCs, again due to their different locations with respect to the Sun center.

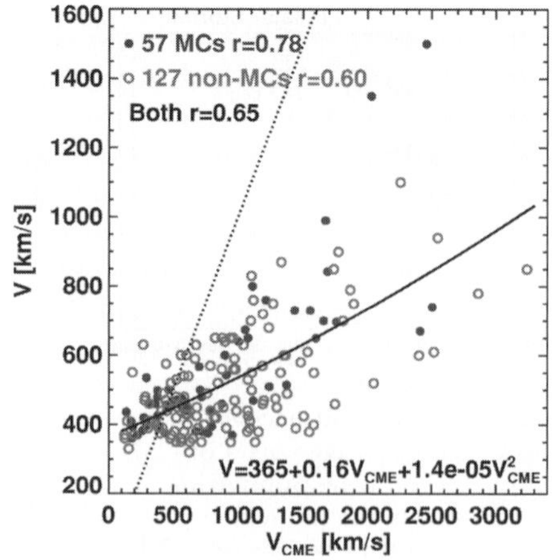

Figure 3. The relationship between the speeds measured at the Sun and at 1 AU for a large number of CMEs during cycle 23. Fifty seven CMEs that resulted in magnetic clouds (MCs, filled circles) and 127 that resulted in non-clouds ICMEs (non-MCs, open circles) are distinguished. The bisector ($V = V_{CME}$) and the quadratic fit to the data points are shown.

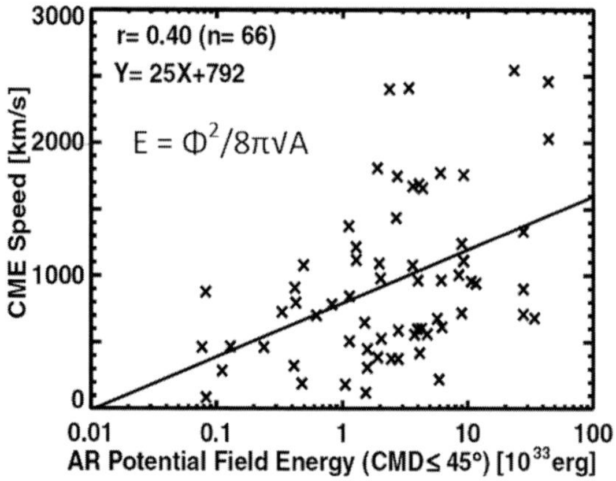

Figure 4. Scatterplot between the active region (AR) potential energy and the CME speed for 66 events observed during cycle 23.

4. The magnetic connection

4.1. *CME speed and active region free energy*

The relation between CME speed and the magnetic potential energy of the source active regions is shown in Fig. 4. The plot includes only those CMEs that resulted in large geomagnetic storms (same as in Fig. 1) and MCs at 1 AU. It is believed that the free energy in an active region is of the order of the magnetic potential energy (MacKay *et al.* 1997). This gives a useful way of estimating the maximum free energy that can be stored in an active region. Fastest CMEs originate from active regions of highest magnetic potential energy, even though the correlation between the two quantities is only moderate (correlation coefficient $r = 0.40$). From Fig. 4 we see that the maximum active region potential energy during cycle 23 did not exceed $\sim 4 \times 10^{34}$ erg, suggesting that the maximum free energy that can be stored cannot be more than this value. If all the free energy were to be released in a single CME, the CME kinetic energy could be as high as $\sim 4 \times 10^{34}$ erg. In reality, the maximum CME kinetic energy observed during cycle 23 was $\sim 1.2 \times 10^{33}$ erg, estimated for the CME on 2003 October 28 (Gopalswamy *et al.*

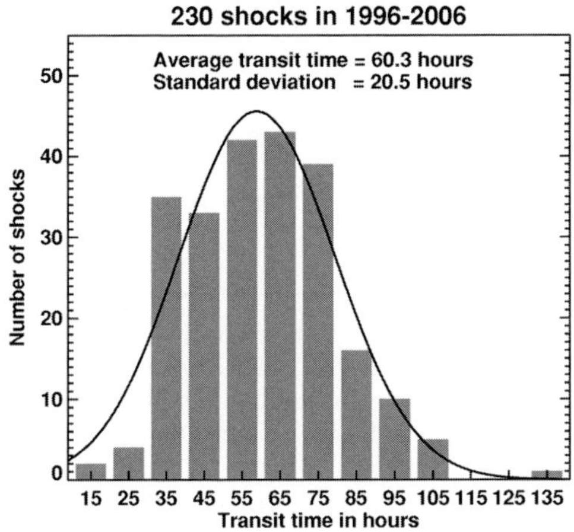

Figure 5. Transit time distribution of CME-driven shocks detected at Sun-Earth L1 during solar cycle 23. A Gaussian fit to the histogram is overlaid on the distribution. The average of the distribution and the standard deviation are noted on the plot.

2005). This means that all the free energy stored in an active region need not be released in a single CME; several CMEs may occur over a period of time to exhaust all the free energy. The CME speed distribution indicates a sharp cutoff beyond ~3500 km/s (see e.g., Gopalswamy 2006a). Assuming that the maximum speed attained by CMEs is ~4000 km/s and using the maximum CME mass of $\sim 10^{17}$ g, one can estimate the maximum kinetic energy to be $\sim 8 \times 10^{33}$ erg. This is again consistent with the observations of magnetic potential energy during cycle 23.

The maximum energy suggested in Fig. 4 has some interesting implications. The active region energy E can be written as $E = \Phi^2/8\pi\sqrt{A}$, where Φ and A are the active region magnetic flux and area, respectively. The maximum energy implies a limit to the maximum field strength (B) and the area of the active regions. The field strengths in sunspots has been estimated to have a maximum value of ~6000 G (Livingston *et al.* 2006). Active region areas as high as 5000 millionths of solar hemisphere (msh) have been reported in the literature (Newton 1955). Combining these two, one can estimate the limit of active region energy as $E \sim 2.7 \times 10^{36}$ erg. This is a factor of ~ 70 higher than the observed value during cycle 23. The maximum active region area during cycle 23 was 2610 msh in AR 10486 (Gopalswamy *et al.* 2005), which is only about half the maximum value reported in Newton (1955). The peak field strength in AR 10486 was ~1700 G (Chumak, Zhang & Guo 2008). Therefore, the maximum active region potential energy is accordingly smaller ($\sim 8. \times 10^{34}$ erg.)

4.2. *CME structure and geomagnetic storms*

CMEs causing geomagnetic storms are generally fast and wide CMEs (see Fig. 2), which means most of them must be shock-driving. The shock surrounds the CME as a bow wave and stands at a distance that depends on the shock strength and the CME size. The average standoff distance corresponds to ~0.5 day between the shock and ICME arrivals at Earth (Gosling *et al.* 1987b; Gopalswamy *et al.* 2008). This means the first thing from a CME that strikes the magnetosphere is the shock. The shock compresses the magnetosphere, which produces a spiky increase in *Dst*, known as the storm sudden commencement (SSC) because a geomagnetic storm usually follows (Curto, Araki & Alberca 2007, and references therein). Shocks take anywhere from ~19 hours to more than four days to reach Earth, as can be seen in Fig. 5, which shows the distribution

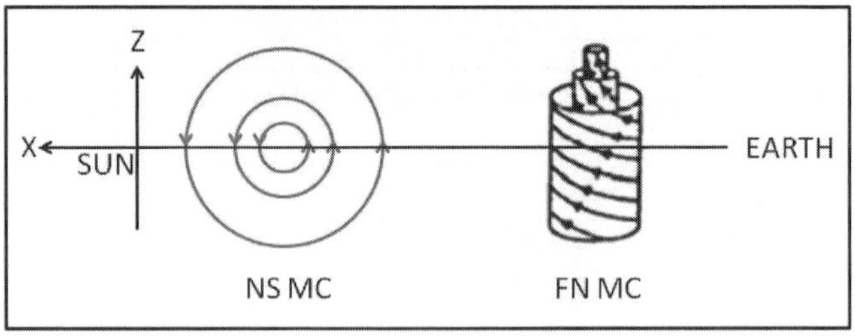

Figure 6. Out-of-the-ecliptic component (B_z) of the interplanetary magnetic field due to NS and FN MCs. For NS MCs, the front has $B_z > 0$ (not geoeffective) and the back has $B_z < 0$ (geoeffective). The FN MCs have $B_z > 0$ throughout the MC interval. For the SN and FS MCs, the arrows point in the opposite direction. The Y-axis points into the plane of the page.

of the transit time of 230 shocks detected during solar cycle 23. The distribution is roughly symmetric with an average transit time of 60.3 h(standard deviation of 20.5 h). Whether a geomagnetic storm would follow the shock depends critically on the existence of B_s somewhere within the interval of interest (IOI), which starts from the shock and ends at the end of the driving ICME. Favorable conditions can occur within the IOI when the sheath and/or the ICME portions contain B_s. While it is straightforward to identify the shocks in the solar wind data, one has to infer from type II radio burst data (Gopalswamy 2006c) or from coronagraphic observations (Gopalswamy *et al.* 2009b; Ontiveros & Vourlidas 2009). Unfortunately, the remote-sensing and in-situ instruments do not detect the same section of the CME-driven shocks.

When the ICME is an MC (flux rope), it is easy to guess the time structure of the *Dst* index. MCs can be classified into four groups: south-north (SN), north-south (NS), fully-south (FS) and fully-north (FN), depending on the direction of B_z in the MC (see Fig. 6). The flux-rope axis of the SN and NS clouds are in the ecliptic plane, so B_z is the azimuthal component; it is perpendicular to the ecliptic for FS and FN clouds, so B_z is the axial component. The SN and NS MCs have B_s in the front and back of the cloud, respectively and hence making them geoeffective. No storm follows FN clouds due to the lack of B_s. On the other hand, FS MCs are always geoeffective because of the axial B_s. In addition, if the sheath contains B_s, then one gets a complex *Dst* profile depending on the distribution of B_s in the IOI. Gopalswamy (2008) reported that the peak of sheath storms precede the MC arrival by ~3.2 h, while the cloud storms peak after ~10.6 h. The delay of course depends on the MC type: the largest is for NS clouds (18.6 h), the shortest for SN MCs (5.5 h) and intermediate (9.3h) for FS MCs.

White-light CMEs associated with the MCs typically erupt from very close to the disk center (CMD $< 30°$). This property is shared by storm-producing CMEs shown in Fig.1. On the other hand, CMEs associated with non-MCs originate at larger CMDs on the average. In the extreme case, one observes shocks at 1 AU without a discernible driver when energetic CMEs originate from close to the solar limb. In these cases, only the shock flank is expected to impact Earth, occasionally producing a geomagnetic storm. One of the implications of this geometrical effect is that all ICMEs may contain flux ropes (MCs) but may or may not be observed as flux ropes: the observing spacecraft passes through the nose of the ICME only for MCs; for non-MCs, the observing spacecraft passes through the edges of the ICMEs; for shock flanks, the spacecraft completely misses the ICME part (Marubashi 1997; Gopalswamy 2006a; Riley *et al.* 2006). The flux rope is thought to be formed due to reconnection in the closed field regions where CMEs originate. The post eruption arcade is supposed to be an indication that a flux rope has formed due to reconnection and erupted (Qiu *et al.* 2007). It is also possible that the flux rope grew

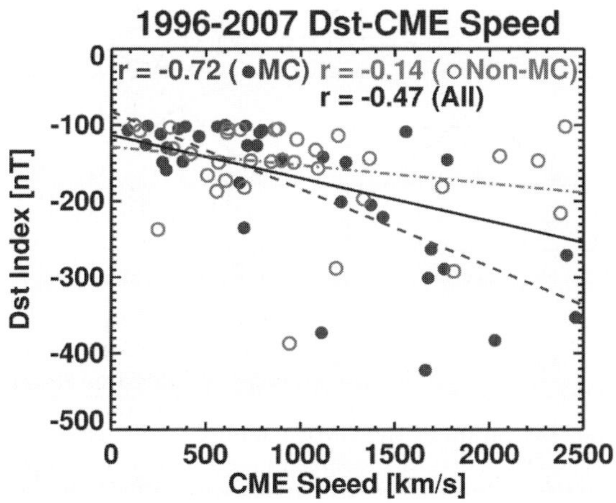

Figure 7. Scatterplot between *Dst* index and the CME speed for large geomagnetic storms during cycle 23. The storms caused by MCs and non-MCs are distinguished. The correlation coefficients for all storms, MC storms and non-MC storms are shown on the plot.

out of a preexisting flux rope by adding flux during the reconnection process. The flux rope derives its kinetic energy from the free energy stored in the active region. The flux rope axis is typically parallel to the neutral line (or filament) in the active region, while the handedness of the flux rope can be inferred from the sign of the active region helicity (Marubashi 1997). Thus one can even infer the B_z part of Equation 1.1, but this is a field of active research.

The glancing impact expected from non-MCs may have consequences for their geoeffectiveness. Figure 7 shows the dependence of *Dst* on the CME speed for major storms of cycle 23. The overall correlation is modest ($r = -0.47$), but the CMEs that resulted in MCs show a much better correlation ($r = 0.72$). On the other hand, CMEs resulting in non-MCs show poor correlation ($r = -0.14$). This can once again be attributed to the fact that non-MCs may strike Earth at different glancing angles. The good correlation between CME speed and *Dst* index for MCs is somewhat reduced ($r = -0.46$) when starting from all MCs and compiling the strength of the resulting storms without any restriction on the *Dst* values (Gopalswamy *et al.* 2008).

5. Summary and Discussion

The primary CME link to the geomagnetic storms arises from the fact that they can introduce an out-of-the ecliptic component to the interplanetary magnetic field that arrives at Earth. Whenever the out-of-the-ecliptic component of the magnetic field in a solar wind structure points to the south, a geomagnetic storm ensues due to a series of physical processes elucidated by Dungey (1961). Figure 8 summarizes the CME connection to geomagnetic storms via Equation 1.1: both the CME speed and the southward magnetic field it contains determine the strength of the storm. The southward component of the magnetic field in a CME arises from their flux-rope structure and in the sheath region due to field line draping around the flux rope and compression. The flux rope originates from the source active region, whereas the sheath field is from the heliosphere. The CME speed depends on the active region free energy, modified by the interaction with the solar wind. Thus both the speed and magnetic field parameters are linked to the free energy and magnetic properties of the source active region. For example, a CME

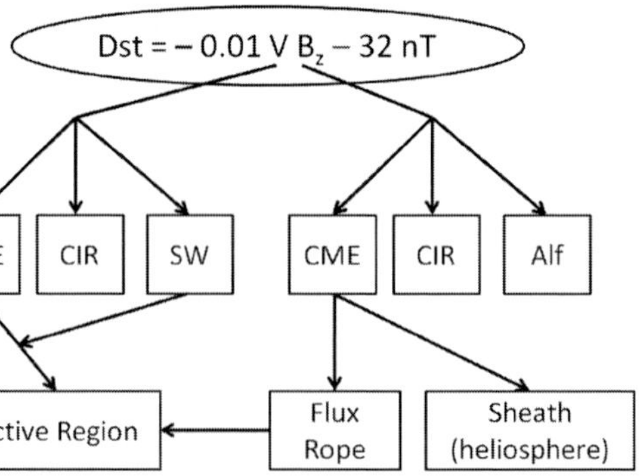

Figure 8. Schematic representation of the CME connection to geomagnetic storms. The empirical relation connecting the Dst index to the speed (V) and southward magnetic field component (B_s) of the interplanetary structures is from Gopalswamy *et al.* (2008). Other sources of V and B_z such as the solar wind (SW), Alfven waves (Alf), CIR, and shock sheath are also shown.

that results in an MC with larger magnetic flux and/or helicity is launched from the Sun with a higher initial speed (Sung *et al.* 2009).

Since CMEs are the main source of major geomagnetic storms, one would like to predict the CME speed and the magnitude of B_s from the observed properties of the active regions (Falconer *et al.* 2009). There have been several attempts to qualitatively link the ICME properties to the active region properties: sign of helicity (Bothmer & Schwenn 1994; Martin & McAllister 1997; Yurchyshyn *et al.* 2001), azimuthal magnetic flux of MCs and the magnetic flux in flare ribbons (Qiu *et al.* 2007), and axial magnetic flux to the flux in coronal dimming regions (Webb *et al.* 2000). A detailed review can be found in Démoulin (2008). Predicting the kinematic and magnetic properties of CMEs near the Sun and in the IP medium based on the active region properties is required to provide advanced warning of potential geomagnetic storms (see e.g., Srivastava 2006).

Finally, we note that the importance of B_s in causing geomagnetic storms is due to the fact that Earth's magnetic field currently points to the north. Paleomagnetic records and numerical experiments indicate that the polarity of the geomagnetic field can be reversed on geological time scales (Roberts & Glatzmaier 2000). When Earth's magnetic field points to the south, one has to worry about the north-pointing interplanetary magnetic structures in causing the geomagnetic storms. Accordingly, the north-south magnetic clouds would cause prompt storms rather than the south-north clouds.

Acknowledgements

The author thanks P. Mäkelä for help with formating and several figures. Work supported by NASA's LWS TR&T program.

References

Bothmer, V. & Schwenn, R. 1994, *Space Sci. Revs* 70, 215
Burlaga, L., Sittler, E., Mariani, F., & Schwenn, R. 1981, *J. Geophys. Res.* 86, 6673
Chumak, O. V., Zhang, H.-Q., & Guo, J. 2008, *Astronomy Reports*, 52(10), 852
Curto, J. J., Araki, T., & Alberca, L. F. 2007, *Earth Planets Space* 59, p. i
Démoulin, P. 2008, *Ann. Geophys.* 26, 3113
Dungey, J. W. 1961, *Phys. Rev. Lett.* 6, 47
Echer, E., Gonzalez, W. D., & Alves, M. V. 2006, *Space Weather* 4, S06001
Falconer, D., Moore, R.L., Gary, G. A., & Adams, M. 2009, *ApJ* 700, L166
Gonzalez, W. D., Gonzalez, A. L. C., Dal Lago, A. *et al.* 1998, *Geophys. Res. Lett.* 25(7), 963

Gopalswamy, N. 2002, *COSPAR Colloquia Series* 14, 157
Gopalswamy, N. 2004, in: G. Poletto & S. T. Suess (eds.), *The Sun and the Heliosphere as an Integrated System*, (Boston: Kluwer), p. 201
Gopalswamy, N. 2006a, *JA&A* 27, 243
Gopalswamy, N. 2006b, *Space Sci. Revs* 124, 145, doi:10.1007/s11214-006-9102-1
Gopalswamy, N. 2006c, in: N. Gopalswamy, R. A. Mewaldt, & J. Torsti (eds.), *Solar Eruptions and Energetic Particles*, Washington, DC: AGU, p. 207
Gopalswamy, N. 2008, *J. Atmos. Sol.-Terr. Phy.*, doi:10.1016/j.jastp.2008.06.010
Gopalswamy, N., Yashiro, S., Kaiser, M. L., Howard, R. A., & Bougeret, J.-L. 2001a, *J. Geophys. Res.* 106, 29219
Gopalswamy, N., Lara, A., Yashiro, S., Kaiser, M. L., & Howard, R. A. 2001b, *J. Geophys. Res.* 106, 29207
Gopalswamy, N., Yashiro, S., Liu, Y., Michalek, G., Vourlidas, A., Kaiser, M. L., & Howard, R. A. 2005, *J. Geophys. Res.* 110, A09S15, doi:10.1029/2004JA010958
Gopalswamy, N., Yashiro, S., & Akiyama, S., 2007, *J. Geophys. Res.* 112, A06112, doi: 10.1029/2006JA012149
Gopalswamy, N., Akiyama, S., Yashiro, S., Michalek, G., & Lepping, R. P. 2008, *J. Atmos. Sol.-Terr. Phy.* 70, 245
Gopalswamy, N., Dal Lago, A., Yashiro, S., & Akiyama, S. 2009a, *Central European Astrophysical Bulletin* 33, 115
Gopalswamy, N., Thompson, W. T., Davila, J. *et al.* 2009b, *Solar Phys.* 259, 227
Gosling, J. T. 1993, *J. Geophys. Res.* 98, 18937
Gosling, J. T., Thomsen, M. F., Bame, S. J., & Zwickl, R. D. 1987a, *J. Geophys. Res.* 92, 12399
Gosling, J. T., Baker, D. N., Bame, S. J., Feldman, W. C., Zwickl, R. D. & Smith, E. J. 1987b, *J. Geophys. Res.* 92, 8519
Lindsay, G. M., Russell, C. T., & Luhmann, J. G. 1995, *J. Geophys. Res.* 100, 16999
Lindsay, G. M., Luhmann, J. G., Russell, C. T., & Gosling, J. T. 1999, *J. Geophys. Res.* 104, 12515
Livingston, W., Harvey, J. W., Malanushenko, O. V., & Webster, L. 2006, *Solar Phys.* 239, 41
Loewe, C. A. & Prölss, G. W. 1997, *J. Geophys. Res.* 102, 14209
Martin, S. F. & McAllister, A. H. 1997, in: N. Crooker, J. A. Joselyn, & J. Feynman (eds.), *Coronal Mass Ejections*, Geophys. Monogr. Ser. 99 (Washington, D.C.: AGU), p. 127
Marubashi, K. 1997, in: N. Crooker, J. A. Joselyn, & J. Feynman (eds.), *Coronal Mass Ejections*, Geophys. Monogr. Ser. 99 (Washington, D.C.: AGU), p. 147
MacKay, D. H., Gaizauskas, V., Rickard, G. J., & Priest, E. R. 1997, *ApJ* 486, 534
Newton, H. W. 1955, *Vistas in Astronomy* 1(1), 666
Ontiveros, V. & Vourlidas, A. 2009, *ApJ* 693, 267
Qiu, J., Hu, Q., Howard, T. A., & Yurchyshyn, V. B. 2007, *ApJ* 659, 758
Riley, P., Schatzman, C., Cane, H. V., Richardson, I. G., & Gopalswamy, N. 2006, *ApJ* 647, 648
Roberts, P. H. & Glatzmaier, G. A. 2000, *Rev. Mod. Phys.* 72, 1081
Siscoe, G., Crooker, N. U., & Clauer, C. R. 2006, *Adv. Sp. Res.* 38(2), 173
Srivastava, N. 2006, *JA&A* 27, 237
Sung, S.-K., Marubashi, K., Cho, K.-S. *et al.* 2009, *ApJ* 699, 298
Tsurutani, B. T., Gonzalez, W. D., Gonzalez, A. L. C. *et al.* 1995, *J. Geophys. Res.* 100, 21717
Wang, Y. M., Ye, P.Z., Wang, S., Zhou, G. P., & Wang, J. X. 2002, *J. Geophys. Res.* 107, 1340
Webb, D. F., Lepping, R. P., Burlaga, L. F., DeForest, C. E., Larson, D. E., Martin, S. F., Plunkett, S. P., & Rust, D. M. 2000, *J. Geophys. Res.* 105, 27251
Wu, C. C. & Lepping, R. P. 2002, *J. Geophys. Res.* 107, 1346
Xu, D., Chen, T., Zhang, X. X., & Liu, Z. 2009, *J. Atmos. Sol.-Terr. Phy.* 57, 1500
Yurchyshyn, V. B., Wang, H., Goode, P. R., & Deng, Y. 2001, *ApJ*, 563, 381
Zhang, J. *et al.* 2007, *J. Geophys. Res.* 112, A10102, doi:10.1029/2007JA012321

Solar and Stellar Variability: Impact on Earth and Planets
Proceedings IAU Symposium No. 264, 2009
A. G. Kosovichev, A. H. Andrei & J.-P. Rozelot, eds.

doi:10.1017/S1743921309992882

Lower and middle atmosphere and ozone layer responses to solar variation

Ana G. Elias[1,2]

[1]CONICET, Consejo Nacional de Investigaciones Cientificas y Tecnicas, Argentina

[2]Lab. Fisica de la Atmosfera, Depto. de Fisica, Facultad de Ciencias Exactas y Tecnologia, Universidad Nacional de Tucuman, Av. Independencia 1800, 4000 Tucuman, Argentina
email: aelias@herrera.unt.edu.ar / anagelias@yahoo.com

Abstract. Global warming in the troposphere and the decrease of stratospheric ozone concentration has become a major concern to the scientific community. The increase in greenhouse gases and aerosols concentration is believed to be the main cause of this global change in the lower atmosphere and in stratospheric ozone, which is corresponded by a cooling in the middle and upper atmosphere. However, there are natural sources, such as the sun and volcanic eruptions, with the same ability to produce global changes in the atmosphere. The present work will focus on solar variation and its signature in lower and middle atmosphere parameters. The Sun can influence the Earth and its climate through electromagnetic radiation variations and also through changes in the solar wind which causes geomagnetic storms. The effects of both mechanisms over the lower and middle atmosphere and ozone layer will be discussed through an overview of selected papers, which by no means cover this subject that is extremely wide and complex. A fundamental understanding of the atmosphere response to solar variations is required for understanding and interpreting the causes of atmospheric variability. This is an essential focus of climate science, which is seeking to determine the extent to which human activities are altering the planetary energy balance through the emission of greenhouse gases and pollutants.

Keywords. Sun: solar-terrestrial relations, Sun: activity, Earth

1. Introduction

A fundamental and challenging component of Earth Science is the proper attribution of observed climate change to variations in solar output. For this purpose, the response of the atmosphere to solar variability must be determined, and the physical mechanism explaining this response should be understood. This is a very complex subject since, among other factors, the effects of the sun occur simultaneously with other natural and anthropogenic influences and present complex spatial patterns.

In general, to analyze a response to any source of variation we can do it in two ways: theory and experiment, and of course a mix of both. In the case of the detection and analysis of atmosphere and ozone layer responses to solar variation these two ways would be: (1) hypothesis based on physical mechanisms and models, like the General Circulation Models (GCMs) in which we can modify the source of variability in order to measure the kind and magnitude of the response, always followed by an experimental proof for validation, and (2) statistical analysis of experimental data using statistical tools such as FFT, wavelet analysis, correlations, filtering techniques, SSA, etc., followed by an "explanation" of the results obtained.

We know that climate is changing in all the atmospheric layers. The external sources responsible for this are anthropogenic sources (i.e. increasing greenhouse gases concentration, deforestation) and natural sources (i.e. solar variation, volcanic activity, astronomical factors). Considering solar variation, which is the concern of this paper, it presents trends and cycles which are not constant in time neither in periodicity. The most commonly used index of solar activity is the sunspot number. Its most prominent variability is the 11-year or Schwabe cycle, which has a variable length of 9-14 years for individual cycles. The background for the 11-year cycle is the 22-year or Hale magnetic polarity cycle, which relates to the reversal of the global magnetic field of the sun. The long-term change in the 11-year cycle amplitude is known as the secular Gleissberg cycle, which is rather a modulation of the 11-year cycle envelope with a varying timescale of 60–120 years. Longer cycles cannot be studied using direct solar observations, but several such cycles have been found in cosmogenic isotope data. A cycle with a period of 200–210 years, called the de Vries or Suess cycle in different sources, is a prominent feature, observed in varying cosmogenic data.

All solar activity parameters (F10.7, group sunspot number, sunspot area, etc.), which include also geomagnetic indices (Kp, Ap, aa, Dst, etc.), present the mentioned cycles. There are also short and mid-term periodicities in the solar output such as the 27-day rotational period and the semiannual variation which is prominent only in geomagnetic data.

In the present work I will emphasize the statistical approach in order to detect the lower and middle atmosphere and ozone layer responses to solar variation through an overview of selected papers that cover only a small part of this extremely wide and complex subject in the field of Sun-Earth interactions.

2. Lower atmosphere

The first papers dealing with climate-sun associations date back to late eighteenth century. For a very good review see Le Treut (2007). Although there were many papers showing statistically significant correlations between troposphere temperature and solar variations, there was not a physical explanation for these statistical results due to the complexity of the climate system. The total solar irradiance, TSI, which corresponds to the wavelength-integrated irradiance reaching the Earth, is the channel through which solar output can impact on troposphere climate. In fact, the incoming solar radiation provides the energy budget and the external forcing for the Earth-atmosphere system. However, the TSI variation in an 11-year solar cycle is 0.1%, which is not enough, according to models, to produce any noticeable effect on climate.

Figure 1 shows global and hemispheric temperature anomaly from Jones *et al.* (2009) for the period 1850-2008, together with Rz and TSI. A 10-year running mean was applied to the time series. The long term variation noticed in Rz, and also in TSI, corresponds to the Gleissberg cycle with a minimum around 1900 and a maximum around 1950. The correlation coefficients between temperature and Rz are 0.77, 0.71 and 0.75 for the Northern Hemisphere, Southern Hemisphere and global respectively. In the case of TSI, these values increase to 0.87, 0.83 and 0.87. That is, Rz can explain between 50% and 60% of temperature variance and TSI between 69% and 76%. But, if the radiative forcing associated to solar variation is not enough according to models to produce the observed temperature changes, how can we obtain such high correlation coefficients?

Another problem is that associations change according to the periods considered, appearing also negative correlations. This can be seen in Figure 2, which shows as an example the running correlation with a 40-year window between global temperature and Rz, and between global temperature and TSI. Also shown, in dashed line, are the correlations considering the whole period. We can think that the response may change in

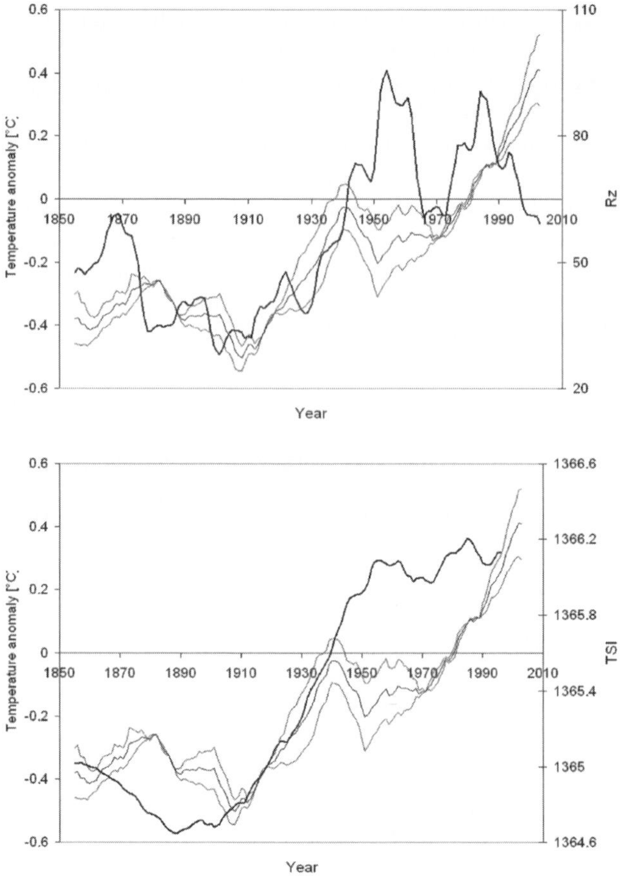

Figure 1. Northern hemisphere (red), Southern hemisphere (green) and global (blue) surface temperature anomalies relative to the 1961-1990 Mean, from Jones *et al.* (2009) and: (a) sunspot number, Rz (black), (b) total solar irradiance, TSI (black). A 10-year running mean was applied to the monthly time series.

time due to interaction with other sources of variation or to environmental changes that would affect sun-Earth association.

A controversial paper by Friis-Christensen and Lassen (1991) intensified the ever existing debate on the role of the sun in climate change. Their achievement was to use the solar cycle length (SCL) instead of Rz, or any other solar proxy. SCL lags ahead Rz around 20 years, and this was enough to match almost exactly the time series of the Northern Hemisphere temperature. Butler (1994) showed later that a single location, Armagh (54.3°N, 6.6W), also presents a high correlation with SCL. The mechanism proposed to explain this association was cloud cover variations caused by cosmic ray flux which is modulated by solar activity (Svensmark and Friis-Christensen 1997; Svensmark 2000; Marsh and Svensmark 2003; Kirby 2007; Carslaw 2009). In this way, a plausible physical mechanism, although still very controversial, was able to explain the statistical results.

Another association between climate and solar activity is linked to climate dynamics and geomagnetic activity, caused by solar variability, mostly by variations of solar-wind properties and the interplanetary magnetic field. In fact, the solar influence on Earth's surface can also consist of a dynamical response to the solar forcing transmitted from the

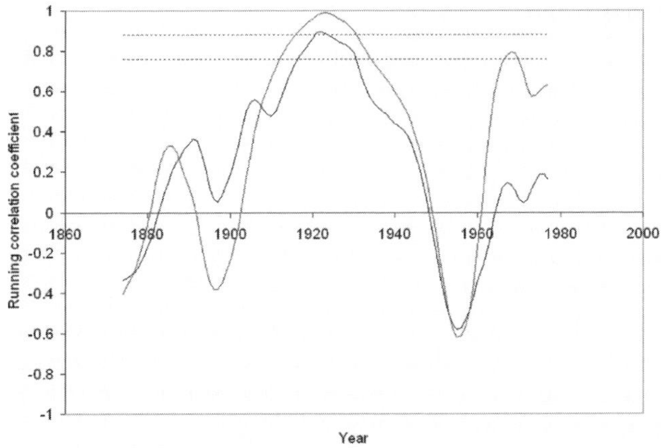

Figure 2. Running correlation with a 40-year window between global temperature and Rz (black solid line) and between global temperature and TSI (red solid line). Dashed lines correspond to the correlation coefficient considering the whole period: 0.75 for temperature vs Rz (black dashed line) and 0.87 for temperature vs TSI (red dashed line). Global temperature from Jones *et al.* (2009).

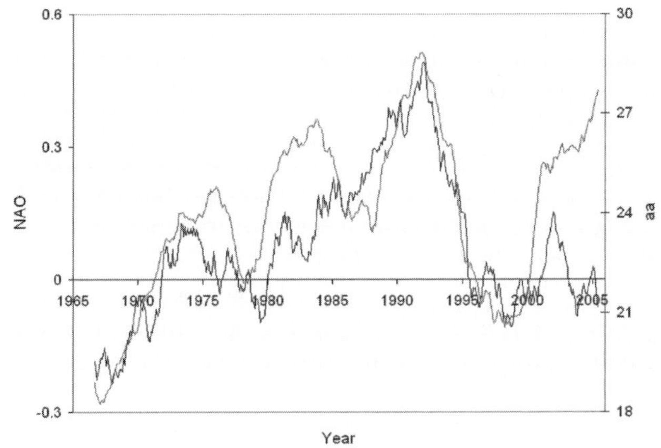

Figure 3. The North Atlantic Oscillation, NAO, index (obtained from ftp://ftp.cpc.ncep.noaa.gov/wd52dg/data/indices) (black) and geomagnetic aa index (red). Both monthly series were smoothed with a 6-year running mean.

stratosphere, rather than a direct radiative effect. As an example Boberg and Lundstedt (2002), Bucha (2002), Thejll *et al.* (2003), Palamara and Bryant (2004), to mention a few, analyzed the North Atlantic Oscillation, NAO, defined as the normalized pressure difference between the Azores and Iceland, which refers to a redistribution of atmospheric mass between the Arctic and subtropical Atlantic producing large changes in the heat and moisture transport between the Atlantic and the neighboring continents. The general conclusion is that enhanced geomagnetic forcing acting on the stratosphere propagates its influence downward and leads to the intensification of the westerly flow affecting the NAO.

Figure 3 shows the NAO index (obtained from ftp://ftp.cpc.ncep.noaa.gov/wd52dg/data/indices) and the geomagnetic aa index, both smoothed with 6-year running mean. The similar behavior of both curves can be clearly noticed. The correlation coefficient between them is 0.68.

3. Middle atmosphere

The middle atmosphere, that is stratosphere and mesosphere, is affected by the UV spectral band of TSI. Temperature and dynamical responses are expected, but as in the lower atmosphere, results are still controversial. In addition, as we go higher in the atmosphere, the period for which measurements exist becomes shorter, making it more difficult to obtain reliable results.

The temperature response to solar UV changes is expected through ozone and oxygen absorption processes in the middle atmosphere. Keckhut *et al.* (2005) obtained a positive correlation between stratosphere temperature and solar activity with an effect of 1-2 K in the upper stratosphere between minimum and maximum solar activity levels. A similar positive response is obtained by Fadnavis and Beig (2006) for the lower and upper stratosphere, but a negative response for the middle stratosphere. For the mesosphere, model studies suggest an in-phase response to the UV flux, peaking in the upper mesosphere, with a 2 K amplitude, and at the stratopause, with a 1-2 K amplitude (Brasseur 1993; Matthes *et al.* 2004). However, experimental results do not completely agree, revealing much smaller solar signatures and, in some cases, no significant solar signal at all. Beig *et al.* (2008) present an excellent overview and up-to-date status of the mesosphere response to solar variation based on recently available observational data, concluding that mesosphere temperature responds to solar variation depending on season, height and location, with an average of in 1-3 K per 100 sfu.

Regarding dynamics, solar UV variation, which is around 6-10% from solar minimum to solar maximum, produce ozone heating variations that affect temprature and zonal winds. Zonal wind anomalies alter planetary waves propagation, which in turn alters tropospheric circulation (Balachandran and Rind 1995; Kodera 2006). Dynamical processes may also mask solar effects, ending in responses that can be detected only after special filtering procedures, as is the case of some stratosphere parameters. The group of Physics of the Middle Atmosphere of the Freie Universitat Berlin in several publications (Labitzke, 1987, 2004, 2005; Labitzke and van Loon, 1988; Labitzke and Kunze, 2009) showed that there exists a strong signal of the 11-year sunspot cycle, but this signal can only be identified, if the data are stratified according to the phase of the quasi-biennial oscillation, QBO.

4. Ozone layer

The solar UV radiation is an essential factor in ozone production, so it directly affects its concentration. Hood (1997), applying multiple regression methods, estimates the solar cycle variation of total ozone using TOMS data. He obtains maximum variations from solar minimum to maximum of approximately 11 Dobson units around 30°N latitude. Labitzke and van Loon (1997), analyzing also TOMS ozone data, obtained the highest correlation coefficients between F10.7 and the total ozone between 5° and 30° latitude in either hemisphere, thus in regions where ozone is transported poleward. Statistically insignificant correlations are obtained for the equatorial regions where ozone is produced, and for the subpolar regions where it is most plentiful. They conclude then, that it is unlikely that the high correlations are a result of direct radiative interaction between the Sun and ozone but may rather be due to a solar influence on the poleward transport of ozone.

Soukharev and Hood (2006) obtained a vertical structure of the tropical ozone response to the 11-year solar cycle characterized by statistically significant positive responses in the upper stratosphere (40-50 km altitude) and lower stratosphere (below 25 km altitude) and by statistically insignificant responses in the middle stratosphere (28-38 km altitude).

Geomagnetic activity also takes its part in the ozone layer response to solar variations. Geomagnetic storms and sub-storms inject energetic particles that penetrate down into the Earth's mesosphere and upper stratosphere ionizing molecules and producing large enhancements of odd nitrogen species at high latitudes that intervene in catalytic ozone destruction. According to Lastovicka and Mitch (1999) and Lastovicka and Krizan (2005) the ozone response can also be a redistribution. They observe significant effects of strong geomagnetic storms on total ozone but only during winter and under specific conditions: high solar activity levels and easterly QBO phase.

5. Conclusions

The response of the atmosphere to solar variations is manifest in several parameters (temperature, precipitation, winds, constituent concentrations, etc.) but in such a complex way, that in most cases cannot be detected in a simple way and with usual statistical tools. Here, I tried to give a general and very compact outline of the subject covering only a small portion of this extremely wide and complex theme.

Why is important to understand and measure the lower and middle atmosphere and ozone layer response to solar variation? We live in the Earth and we want to understand and predict the atmosphere behavior which is essential for human life. And, in the present context, a fundamental understanding of the atmosphere response to solar variations is an essential focus of climate science, which is seeking to determine the extent to which human activities are altering the planetary energy balance through the emission of greenhouse gases and pollutants.

Acknowledgements

I gratefully acknowledge the travel support by the Observatorio Nacional of Rio de Janeiro to attend the IAU Symposium 2009, and also thank to Alexander G. Kosovichev and Jucira L. Penna.

The work was supported by the Projects "Geomagnetic activity and Earth's magnetic field as possible causes of long-term trends in the ionosphere" (CONICET/PIP 114-200801-00470) and "Atmosphere and climate variability and long-term trends" (CIUNT/26-E345)

References

Balachandran, N. K. & Rind, D. 1995, Modeling the effects of UV variability and the QBO on the troposphere-stratosphere system. Part I:The middle atmosphere, *Journal of Climate*, 8, 2058–2079.

Beig, G., Scheer, J., Mlynczak, M. G., & Keckhut, P. 2008, Overview of the temperature response in the mesosphere and lower thermosphere to solar activity, *Reviews of Geophysics*, 46, RG3002, doi:10.1029/2007RG000236.

Boberg, F. & Lundstedt, H., 2002, Solar Wind Variations Related to Fluctuations of the North Atlantic Oscillation, *Geophys. Res. Let.*, 29, doi10.1029/2002GL014903.

Brasseur, G. 1993, The response of the middle atmosphere to long-term and short-term solar variability: A two dimensional model, *J. Geophys. Res.*, 98, 23, 079–23, 090, doi:10.1029/93JD02406.

Bucha, V. 2002, Long-term trends in geomagnetic and climatic variability, *Physics and Chemistry of the Earth*, 27, 427–431.

Butler, C. J. 1994, Maximum and minimum temperatures at Armagh obsservatory, 1844-1992, and the length of the sunspot cycle, *Solar Physics*, 152, 35–42.

Carslaw, K. 2009, Cosmic rays, clouds and climate, *Nature*, 460, 332-333.

Fadnavis, S. & Beig, G. 2006, Decadal solar effects on temperature and ozone in the tropical stratosphere, *Ann. Geophys.*, 24, 2091–2103.

Friis-Christensen, E. & Lassen, K. 1991, Length of the Solar Cycle: An indicator of Solar Activity Closely Associated with Climate, *Science*, 254, 698–700.

Hood, L. L. 1997, The solar cycle variation of total ozone: Dynamical forcing in the lower stratosphere, *J. Geophys. Res.*, 102, 1355–1370.

Jones, P. D., Parker, D. E., Osborn, T. J., & Briffa, K. R. 2009, Global and hemispheric temperature anomalies-land and marine instrumental records. In *Trends: A Compendium of Data on Global Change.* Carbon Dioxide Information Analysis Center, Oak Ridge National Laboratory, U.S. Department of Energy, Oak Ridge, Tenn., U.S.A. doi: 10.3334/CDIAC/cli.002.

Kodera, K. 2006, The role of dynamics in solar forcing, *Space Science Reviews*, 125, 319–330.

Keckhut P., Cagnazzo, C., Chanin, M. L., Claud, C., & Hauchecorne, A., 2005, The 11-year solar-cycle in the temperature in the upper-stratosphere and mesosphere: Part I: Assessment of observations, *J. Atm. Terr. Sol. Phys.*, 67, 940–947.

Kirby, J. 2007, Cosmic Rays and Climate, *Surv. Geophys.*, 28, 333–375.

Labitzke, K. 1987, Sunspots, the QBO, and the stratospheric temperature in the north polar region, *Geophys. Res. Lett.*, 14, 535–537.

Labitzke, K. & van Loon, K. 1997, Total ozone and the 11-yr sunspot cycle, *Journal of Ammpherrc and Solar-Terreslrial Physics*, 59, 9–19.

Labitzke, K. 2004, On the Signal of the 11-Year Sunspot Cycle in the Stratosphere and its Modulation by the Quasi-Biennial Oscillation (QBO), *J. Atmos. Solar-Terr. Phys.*, 66, 1151–1157.

Labitzke, K. & van Loon, H. 1988, Association between the 11-year solar cycle, the QBO and the atmosphere. Part I: The troposphere and stratosphere in the northern hemisphere in winter, *J. Atmos. Terr. Phys.*, 50, 197–206.

Labitzke, K. 2005, On the Solar Cycle-QBO-Relationship: A Summary, *J. Atmos. Solar-Terr. Phys.*, 67, 45–54.

Labitzke, K. & Kunze, M. 2009, Variability in the stratosphere: The sun and the QBO, In: *Climate and Weather of the Sun-Earth System (CAWSES): Selected Papers from the 2007 Kyoto Symposium*, Edited by T. Tsuda, R. Fujii, K. Shibata, and M. A. Geller, pp. 257-278, Terrapub, Tokyo.

Lastovicka, J. & Mlch P. 1999, Is ozone affected by geomagnetic storms?, *Advances in Space Research*, 24, 631–640.

Lastovicka, J. & Krizan, P. 2005, Geomagnetic storms, Forbush decreases of cosmic rays and total ozone at northern higher middle latitudes, *Journal of Atmospheric and Solar-Terrestrial Physics*, 67, 119–124

Le Treut, H., Somerville, R., Cubasch, U., Ding, Y., Mauritzen, C., Mokssit, A., Peterson T., & Prather, M. 2007, Historical Overview of Climate Change. In: *Climate Change 2007: The Physical Science Basis. Contribution of Working Group I to the Fourth Assessment Report of the Intergovernmental Panel on Climate Change*, Cambridge University Press, Cambridge, United Kingdom and New York, NY, USA.

Marsh, N. & Svensmark, H., 2003, Solar influence on Earth's climate, *Space Science Reviews* 107, 317–325.

Matthes, K., Langematz, U., Gray, L. L., Kodera, K., & Labitzke, K. 2004, Improved 11-year solar signal in the Freie Universitat Berlin Climate Middle Atmosphere Model (FUBCMAM), *J. Geophys. Res.*, 109, D06101, doi:10.1029/2003JD004012.

Palamara, D. R. & Bryant, E. A. 2004, Geomagnetic activity forcing of the Northern Annular Mode via the stratosphere, *Annales Geophysicae*, 22, 725–731.

Soukharev, B. E. & Hood, L. L. 2006, Solar cycle variation of stratospheric ozone: Multiple regression analysis of long-term satellite data sets and comparisons with models, *J. Geophys. Res.*, 111, D20314, doi:10.1029/2006JD007107.

Svensmark, H. 2000. Cosmic rays and Earth's climate, *Space Science Reviews* 93, 175–185.

Svensmark, H. & Friis-Christensen, E., 1997, Variation of cosmic ray flux and global cloud coverage-a missing link in solar-climate relationships, *Journal of Atmospheric and Solar-Terrestrial Physics* 59, 1225–1232.

Thejll, P., Christiansen, B., & Gleisner, H. 2003, On correlations between the North Atlantic Oscillation, geopotential heights, and geomagnetic activity, *Geophys. Res. Lett.*, 30, 1347, doi:10.1029/2002GL016598.

Solar and Stellar Variability: Impact on Earth and Planets
Proceedings IAU Symposium No. 264, 2009
A. G. Kosovichev, A. H. Andrei & J.-P. Rozelot, eds.

doi:10.1017/S1743921309992894

Possible traces of solar activity effect on the surface air temperature of mid-latitudes

Ali Kilcik[1], **Atila Özgüç**[2] **and Jean-Pierre Rozelot**[3]

[1]Department of Physics, Faculty of Science, Akdeniz University, 07058 Antalya, Turkey
email: kilcikali@gmail.com

[2]Kandilli Observatory and E.R.I., Bogazici University, Cengelkoy, 34684 Istanbul, Turkey
email: ozguc@boun.edu.tr

[3]OCA-Fizeau Dpt, CNRS UMR 6525, & Université de Nice-Sophia-Antipolis,
Av. Copernic, 06130 Grasse, France
email: rozelot@obs-azur.fr

Abstract. In this study we investigate the effects of solar activity on the surface air temperature of mid-latitudes. This enables us to understand existence of solar activity effects on the temperature. We used surface air temperature and pressure data as climate parameters, and solar flare index data as solar activity indicator, for the 25 - 50 degree longitude and 30 - 70 degree latitude zone, including Turkey and European part of Russia. We considered the parameters temperature, pressure and flare index data for the period ranging from January 1975 to the end of December 2007, which covers almost three solar cycles, namely 21st, 22nd, and 23rd. We found some significant correlations between solar activity and surface air temperature for cycles 22 and 23 for some zones. We applied multitaper method to obtain the cyclic behavior of surface air temperature data sets. The most pronounced power peaks were found by this transform around 1.2 and 2.5 years which were reported earlier for some solar activity indicators. We concluded that signature of solar activity effect exists on surface air temperature of mid-latitudes where we studied.

Keywords. Sun: activity, flares, solar-terrestrial relations, methods: data analysis

1. Introduction

There is a great deal of interest in studying links between solar activity and changes in Earths climate. It has been known for some time that there is a statistical relation between solar activity as expressed, for instance, by the sunspot number or flare index, and the surface air temperature (see Fig. 1). Much scientific effort has been exercised in trying to understand and quantify how much influence the Sun exerts on Earths climate. Here, we highlight some of the recent scientific researches on such solar linkages with surface air temperature. For example, Valev (2006) found statistically significant correlations between global and hemispheric temperature anomalies and solar and geomagnetic indices from 1856 to 2000. Kilcik (2005) compared the solar irradiance model data to surface air temperature variations of two countries: USA and Japan and showed the influence of solar activity variations on the Earth's climate. Recently Kilcik *et al.* (2008) investigated the effects of solar activity on the surface air temperature of Turkey to understand existence of solar activity effects on the temperature. They found significant correlation coefficients between the two parameters. This put in evidence that the relationship between solar activity and surface air temperature is very complicated and varies with time and probably also with the considered geographic area. On a global scale, the correlation between solar activity indices and surface air temperature may be

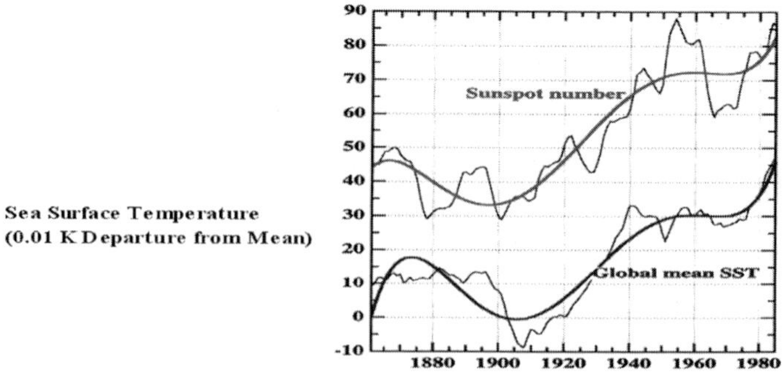

Figure 1. The correlation between the solar and climate records can be seen in this figure comparing polynomial fits to the sunspot record and the global mean sea-surface temperature SST. The Similarity is remarkable (after NOAA Space Environment Center).

positive, negative, or even zero. The correlation is always different for different areas observed (Zhang *et al.* 1989). For example the existence of a positive relation between the surface air temperature of the Northern hemisphere and the solar activity in the period 18811988 has been shown by Georgieva (1998). In this study high positive correlation was found between the geomagnetic activity and the surface air temperature in Middle and Southern Europe However, the same correlation over Canada area is negative (Bucha and Bucha Jr., 1998). In this paper, we attempt to find the effects of a specific solar activity index over a well-defined area. We specify the data used and describe the methods of analysis in Section 2. The results are given in Section 3, while discussion and conclusion remarks are reported in Section 4.

2. Data and Methods

Monthly Solar flare index data set is used as solar activity indicator. The flare index data, which we used, are taken from Kandilli Observatory website† web site. Flare index introduced by Kleczek (1952) is the quantity of Q = i t, where i represents the intensity scale of importance of a flare Ha and t the duration (in minutes) of the flare. The determination of Q has been explained previously (Özgüç *et al.*, 2003, 2004). The anomalies of the temperatures for four zones and these of the solar flare index are shown in Fig. 4.

To investigate the existence of any relation between yearly solar flare index and four zones temperature data, Pearson correlation analyses were applied to the entire yearly data and yearly data set and then we fragmented the temperature and the flare index data according to the solar cycles and again Pearson correlation is applied to these fragmented zonal yearly temperature and flare index data sets. The correlation coefficients for the entire data and for the four zones are presented in Table 2. In order to compare the spectral characteristics of the flare index to the spectral characteristics of surface air temperature we used multitaper method (MTM), which is developed by Thomson (1982) for reducing the variance of spectral estimates by using a set of tapers. In this study we used three sinusoidal tapers. This method has been previously successfully used for the analysis of climate time series of different time scale and resolution (Ghil and Vautard, 1991; Yiou *et al.*, 1996; Anandan *et al.*, 2004; Kilcik *et al.* 2008). Since this method is now standard, and can be found elsewhere, we do not describe it here. To analyze our data the frequency range from 0.00 to 0.078 for solar flare index data, and from 0.005 to

† http://www.koeri.boun.edu.tr/astronomy/astronomy.html

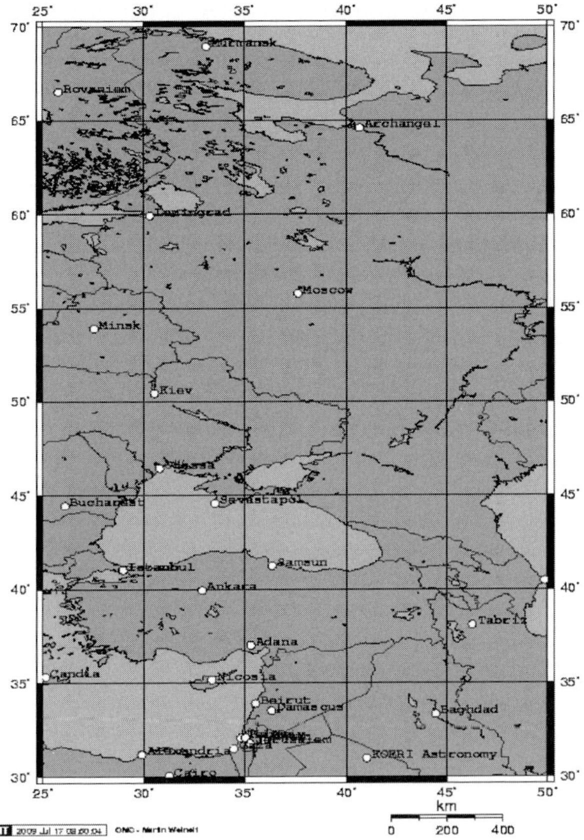

Figure 2. Map of the investigated geographic regions used in this study.

0.078 for all temperature data used in this study were selected. Our significance test is carried out with respect to red noise, since the temperature records, like most climatic and other geophysical time series, have larger power at lower frequencies. All harmonic signals have been obtained by using of 95 % confidence level.

Latitudinal zones	Number of stations
36 – 42 (Turkey)	23
40 – 50	5
50 – 60	15
60 – 70	4

Table 1. Four latitudinal zones and the number of stations for each zone.

3. Analyses and Results

To investigate Sun climate relation for the geographic regions under consideration, we compare the general behavior of solar flare index and four zones temperature anomalies in the first step. These comparisons are given in Figure 4. As shown in this figure, a

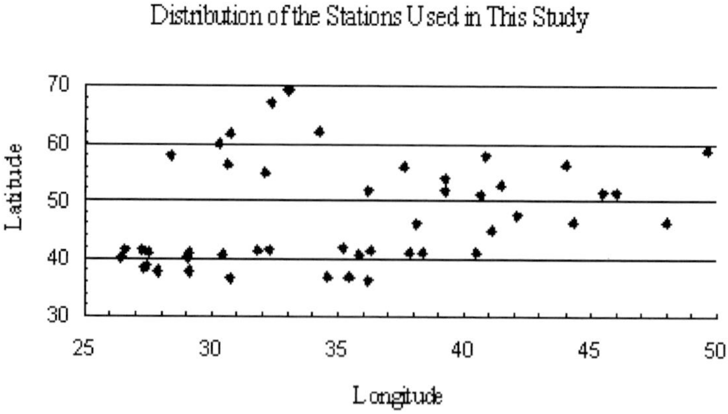

Figure 3. Distribution of the stations used in this study.

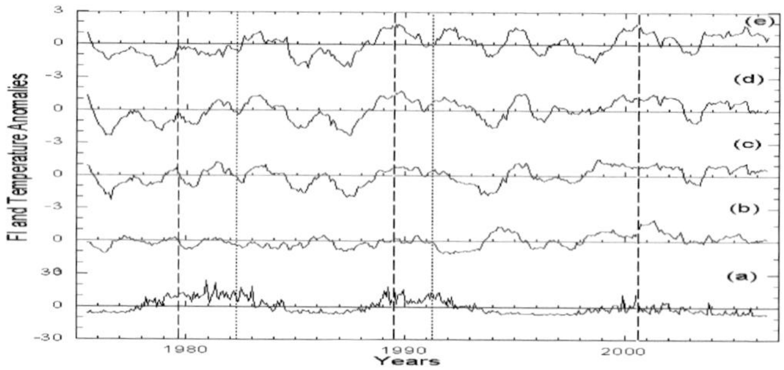

Figure 4. Comparison of the temperature anomalies of four zones with solar flare index anomalies. The temperature anomaly values are given in Celsius. Dashed vertical lines mark the solar activity cycle maxima, according to NOAA mean monthly sunspot numbers. Dotted vertical lines mark the beginning of the two volcanic activities, El Chichon and Pinatuba, respectively. In the figure (a) solar flare index, (b) 30–40 degree (Turkey), (c) 40–50 degree, (d) 50–60 degree, and (e) 60–70 degree zonal anomalies are shown respectively.

good agreement is found between solar flare and four regions temperature anomalies, during the period of cycles maxima especially for cycle 22 and 23. Correlation analyses between yearly solar flare and temperature data were applied to each cycle separately. In the results of our correlation analyses we could not find a statistically significant correlation between yearly mean surface air temperature of all regions and solar flare index data sets from 1976 to 2007. From the correlation analyses results, there is no significant correlation between the two data sets for cycle 21 (1976-1986). We obtained a significant correlations for 50–60 and 60–70 degree of latitude, but we could not obtained a significant correlations for 30 40 and 40–50 degree latitude intervals for cycle 22 (1986-1996). For cycle 23 (1996-2006) we could not obtain any significant correlation for 60–70 degree latitude interval. All data correlation coefficients are given in Table 2.

In order to compare the spectral characteristics of the flare index to the spectral characteristics of surface air temperature we used MTM harmonic test. Our spectral analyses results are given in Figure 5 and 6 for monthly solar flare index and four zonal temperature data respectively. Our periodogram analysis results show that all zonal temperature data sets have all about the same periodicity, showing some small differences.

30° - 40°	40° - 50°	50° - 60°	60° - 70°	
-0.09284	0.12881	0.18595	0.04499	Full data
0.12364	0.22086	0.12770	-0.02867	Cycle 21
-0.17344	0.34993	**0.60899**	**0.64285**	Cycle 22
0.76732	**0.52019**	**0.63700**	0.06950	Cycle 23

Table 2. Correlation coefficients of the entire and cyclic yearly mean data sets. Significant coefficients are given in bold face type.

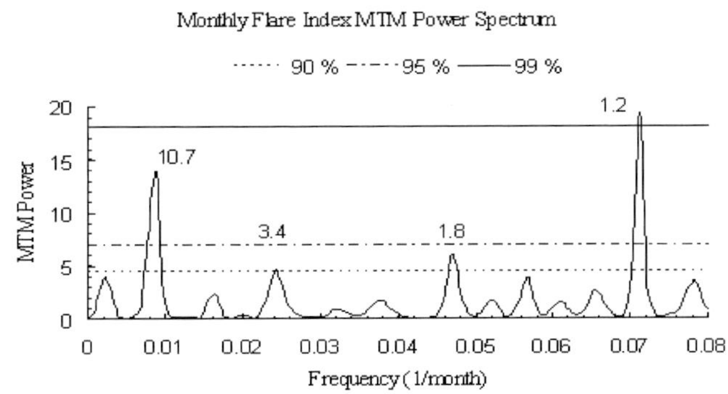

Figure 5. Multitaper method analysis of the 33 years long of the solar flare index from 1975 to 2007. Horizontal lines show the significant levels corresponding to red noise spectrum at 90, 95, and 99 % confidence limits.

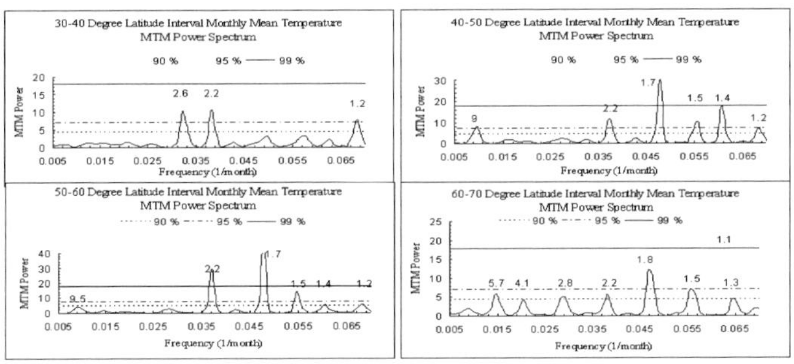

Figure 6. Multi taper method analysis of the 33 years long surface air temperature of the four zones from 1975 to 2007. Horizontal lines show the significant levels corresponding to red noise spectrum at 90, 95, and 99 % confidence limits.

4. Conclusion and discussion

The comparisons we have outlined imply that solar cycle variability may therefore play a significant role in surface air temperature in some mid latitudinal zones. As shown in Table 2 all the considered zones have not any significant correlation for the cycle 21, but for cycles 22 and 23 there are some meaningful correlations; 40–50 and 50–60 degree latitude intervals temperature data show better positive correlations than other two intervals. In results of our correlation analyses we may assume the existence of low degree latitude dependencies of sunclimate relation. As shown in Figure 4, there is an increasing tendency in all zones temperature anomalies for the whole investigated time period point of view. These increases may arise from the enormous increasing of the amount of atmospheric greenhouse gases (GHGs) (CO_2, CH_4, NO_2, etc.) concentrations.

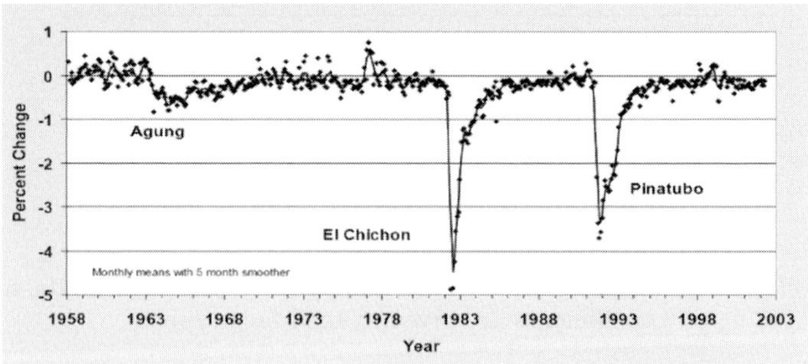

Figure 7. Net solar radiation measured at Mauna Loa Observatory, relative to 1958, showing the effects of major volcanic eruptions. Annual variations are due to the transport of Asian dust and air pollution to Hawaii. Figure is taken from Earth System Research Laboratory web page (http:// www.esrl.noaa.gov/gmd/about/climate.html).

FI and temperature anomalies show an opposite behavior from 1982 to 1985 (see figure 4). It is interesting to note that at the same time period there was strong volcanic activity (March 28, 1982, El Chichon (Mexico); 17.4N, 93.2W) and as seen in Figure 7, the net solar radiation decreased remarkably as measured at Mauna Loa observatory in Hawaii (19.54N, 155.58W). A close look at Figure 4 reveals that the temperature anomalies show really deep valley from 1991 to 1994 and again, this time period is in accordance with another strong volcanic activity (April 2, 1991, Pinatubo (Mexico); 15.13N, 120.35E). Also the net solar radiation decrease is measured at Mauna Loa Observatory and can be seen clearly in Figure 7. As can be seen from Figure 4 it is interesting to note that the volcanic aerosol effects depend on latitude; in the higher latitudes the effect is seen later than the lower ones. These results indicate that GHGs and stratospheric volcanic aerosols have a remarkable effect on temperature behaviors.

There are two meaningful groups of periods for the surface air temperature, which are at 1.3-1.8 and 2.4–2.6 years (see fig.5 and fig.6) The first group found in our study was reported by different authors, such as Kirivova and Solanki (2002) when they analyzed the sunspot areas and the sunspot number data, respectively for the 1880–2000 and the 1750–2000 time periods. Such a period of 1.3-year can be a sub-harmonic of other periodic effects including the well known Schwabe cycle (10.5-year). This issue may be related to the quasi-biennial oscillations occurring in the stratospheric winds. Obridko and Shelting (2007) recently reported that oscillations of 1.3-yr period are closely associated with quasi-biennial oscillations of large-scale solar magnetic field. We may assume this as another possibility. Kilcik et al (2008) also reported same periodicities from surface air temperature of Turkey and flare index for the 1975–2006 time period. In addition our spectral analysis of 60–70 degree zone shows 5.7 years periodicity. This period may be related to geomagnetic activity (5.25-year) and of course indirectly to solar activity. However using the geomagnetic activity index aa, Kane (1997) found a QBO and a 5.4-years periodicity during the period of time 1868–1994. This periodicity was also reported in the biological and biophysical studies (Faraone *et al.*, 2001; Kovac and Mikulecky, 2005; Dimitrova, 2008).

From the correlative and spectral analyses of 33 years data of the flare index and the surface air temperature of some mid latitude zones, conclusions of the present study are:

- Signatures of solar activity effect exist on surface air temperature of some mid-latitude regions according to our statistical analysis and over the considered period of time.

- Change of stratospheric volcanic aerosol concentration has a remarkable effect on surface air temperature.

- Change of atmospheric greenhouse gases concentrations have remarkable effect on temperature.

Acknowledgements

One of the authors (A. Ö.) expresses his thanks to the organizers of the IAU symposium 264 for permitting to attend the IAU General Assembly. This work was supported by the Scientific and Technical Council of Turkey under project number 107T878.

References

Bucha, V., Bucha Jr., & V., 1998, *JASTP* 60, 145
Dimitrova, S., 2008, *JASTP* 70, 420
Faraone, P., Cornelissen, G., Katinas, G. S., Halbergi, G. F., & Siegelova, J., 2001, *Scripta Medica* 74, 107
Georgieva, K., 1998, *Bulg. Geophys.J.* 3, 114
Ghil, M. & Vautard, R., 1991, *Nature* 350, 324
Kane, R. P., 1997, *Ann. Geophys.* 15, 1581
Kilcik, A., 2005, *JASTP* 67, 1573
Kilcik, A., Özgüç, A., Rozelot, J. P., & Yesilyurt, 2008, *JASTP* 70, 1669
Kovac, M., Mikulecky, M. S., 2005, *Bratislavske Lekarske Listy* 106, 423
Krivova, N. A. & Solanki, S. K., 2002, *Astron. Astrophys.* 394, 701
Obridko, V. N. & Shelting, B. D., 2007, *Adv.Space Res.* 40, 1006
Özgüç, A., Atac, T., & Rybak, J. 2003, *Solar Phys.*, 214, 375
Özgüç, A., Atac, T., & Rybak, J. 2004,*Solar Phys.*, 223, 287
Rigozo, N. R., Nordemann, D. J. R., Souza Echer, M. P., Echer, E., da Silva, H. E., Prestes, A., & Guarnieri, F. L., 2007, *JASTP* 69, 1049
Rozelot, J. P., 1990 In "New approaches in Geomagnetism and the Earth's Rotation", ed. S. Flodmark, World Scientific, London, pp. 245–253
Thomson, D. J., 1982, *IEEE Proceedings* 70, 1055
Valev, D., 2006, *Phys. Chem. Earth* 31 109
Yiou, P., Baert, E., & Loutre, M. F., 1996,*Survey in Geophysics* 17, 619
Zhang Y. D. & Li W. B., 1989, Sunspot Numbers, Beijing: Overseas Chinese Press, 139

Solar and Stellar Variability: Impact on Earth and Planets
Proceedings IAU Symposium No. 264, 2009
A. G. Kosovichev, A. H. Andrei & J.-P. Rozelot, eds.

doi:10.1017/S1743921309992900

Influence of the solar radiation on Earth's climate using the LMDz-REPROBUS model

Sandrine Lefebvre[1], **Marion Marchand**[1], **Slimane Bekki**[1], **Philippe Keckhut**[2], **Franck Levèvre**[1], **Chantal Claud**[3], **David Cugnet**[1], **Gérard Thuillier**[2] **and Alain Hauchecorne**[2]

[1]Université Versailles St-Quentin; UPMC Univ. Paris 06; CNRS/INSU, LATMOS-IPSL, Tour 45-46/4ème étage, 4 place Jussieu, 75252 PARIS cedex 05, France
email: sandrine.lefebvre@latmos.ipsl.fr

[2]Université Versailles St-Quentin; UPMC Univ. Paris 06; CNRS/INSU, LATMOS-IPSL, Route des Gatines, BP 3, 91371 Verrières Le Buisson, France

[3]LMD/IPSL, Ecole Polytechnique, 91128 Palaiseau Cedex, France

Abstract. The atmospheric response to the 11-year solar cycle is studied using the fully interactive 3-D coupled chemistry-general circulation model LMDz-REPROBUS with a complete seasonal cycle. We will show results concerning a comparison between two series of 20-year runs, one in maximum of activity and the other in minimum. The stratosphere-troposphere system shows partly significant response to a solar cycle enhancement of UV radiation. We show how the changes in stratospheric ozone, temperature and zonal wind are connected.

Keywords. Methods: numerical, Sun: solar-terrestrial relations

1. Introduction

The impact of solar irradiance variations on the terrestrial atmosphere has long been seen as a significant issue. Understanding and quantifying natural climate variability on decadal to centennial time scales is a prerequisite to project future climate changes. Despite the fact that the solar radiation is one of the main drivers for Earth's climate, the mechanisms by which its short-term variation influences atmospheric parameters are controversial and difficult to identify. From a purely energetic point of view, during the 11-year solar cycle, the Total Solar Irradiance (TSI) varies from less than 0.1%. However, much larger fractional changes take place at short wavelengths: for instance, wavelengths less than 400 nm contribute to about 9% of the change, and 32% of the radiation over a solar cycle occur below 250 nm (Lean 1989). Hence the direct impact of solar irradiance variability is larger in the middle and upper atmosphere than at lower altitudes. Therefore, if there is a solar impact on climate, there must exist processes (maybe depending on the wavelength) that emphasize the effect of the solar cycle.

The interaction of solar radiation with the atmosphere is fundamental in determining its temperature structure and in controlling many of the chemical processes which take place there. The latter depend on the concentrations and spectral properties of the atmospheric constituents. Photodissociation is an essential component of ozone formation, hence most ozone is produced at low latitudes in the upper atmosphere. Due to transport by the mean meridional circulation, observations show however that ozone is also present in considerable quantities in the mid- and high latitude stratosphere. The stratospheric ozone modulated by the solar cycle could be at the origin of a modulation of the Brewer-Dobson circulation (Shindell *et al.* 1999, Haigh 1994 and Haigh 1996). The polar night jet could also be affected as shown by Kodera & Kuroda (2002), Matthes *et al.* (2006),

Matthes *et al.* (2004) and Haigh & Blackburn (2006). Solar radiative heating due to the absorption of solar UV through ozone in the upper stratosphere and stratopause regions creates a large temperature gradient during winter, which consequently induces a strong westerly jet (Kodera & Kuroda 2002).

In this paper, we investigate the atmospheric reponse to the 11-year solar cycle using the fully interactive 3-D coupled chemistry-climate model LMDz-REPROBUS. After briefly describing the model, we will show the first results and a compare with recent analysis of observations.

2. Stratospheric response to solar forcing

2.1. *Description of the model LMDz-Reprobus*

LMDz-Reprobus is a coupled chemistry-climate model which includes full representations of dynamical, radiative, and chemical processes in the atmosphere (full chemistry in the stratosphere only) and their interactions, especially feedbacks of the chemical tendencies on the dynamics : in particular, ozone is strongly affected by dynamics and transport. The horizontal grid resolution is 2.5° x 3.75°, there are 50 vertical levels extending from the surface to 0.07 hPa (~65 km). Details are included in Lefèvre *et al.* (1994), Lefèvre *et al.* (1998), Lott *et al.* (2005) and Jourdain *et al.* (2008). Solar variability is forced explicity in the models through changes in the photolysis rates.

2.2. *Experimental setup*

The goal here is to evaluate the model response to the 11-year solar cycle. We have done simple time-slice experiments with solar fluxes from the SUSIM† data: two series of 20-year runs, one in maximum of activity (chosen as February 2, 1992) and the other in minimum (chosen as October 10, 1996) are computed. Figure 1 shows the irradiance signal for both chosen dates and their ratio: note the non negligeable value of the ratio (up to 20%) in the UV band 140-240 nm.

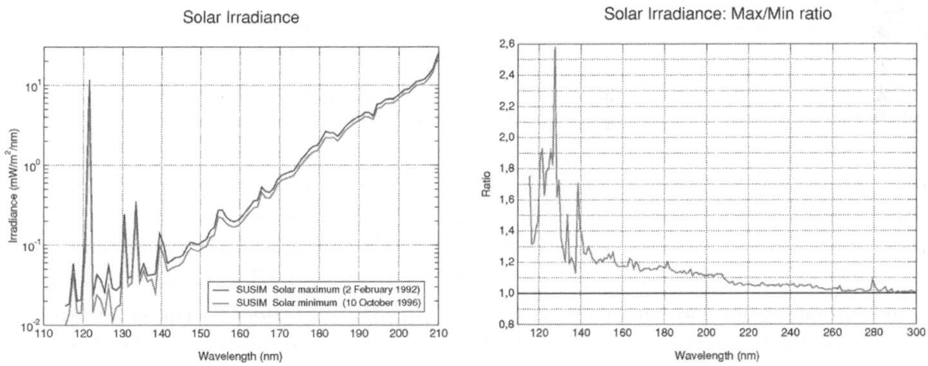

Figure 1. *Left*: Variation of the SUSIM solar irradiance for the chosen dates February 2, 1992 and October 10, 1996. *Right*: Ratio of both signal max/min.

2.3. *Results*

Preliminary results for the temperature, the zonal wind and the ozone concentration are shown in Figures 2 and 3. These graphs present a comparison between both series by

† Solar Ultraviolet Spectral Irradiance Monitor

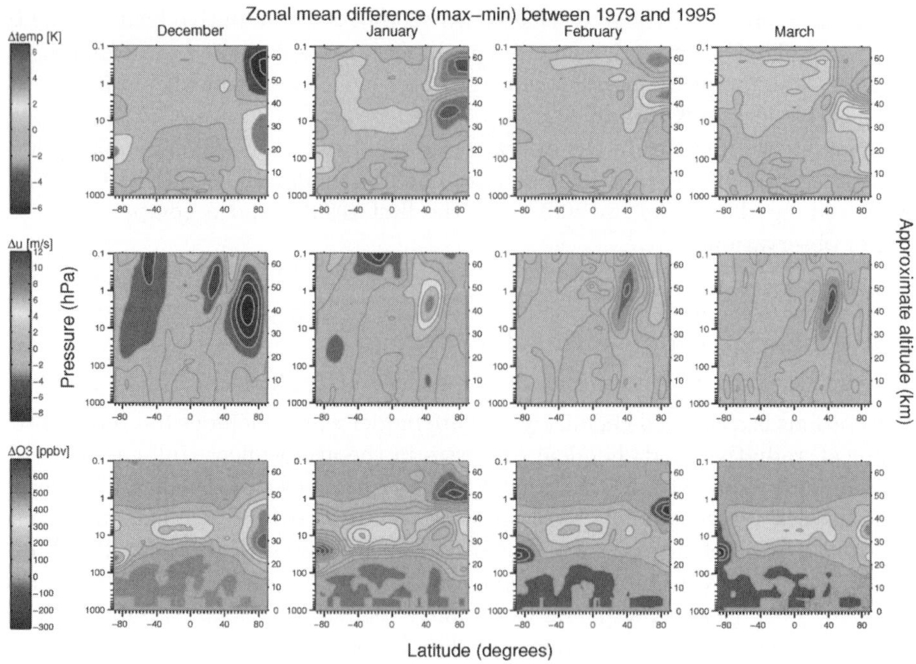

Figure 2. Zonal mean difference (max-min) for the temperature *temp* (top), the zonal wind u (middle) and the ozone concentration O_3 (bottom), in the atmosphere for the northern hemisphere winter months (December to March).

computing the mean difference max-min. We plotted panels respectively for the northern hemisphere winter months where the solar signal is strongly zonally asymmetric, and for the southern hemisphere winter months. The stratosphere is generally warmer of about 0.5 to 1 K at maximum, accompanied by a slight cooling at low latitude in the troposphere, suggesting a dynamical origin. At higher latitudes, the response is well marked with a reversal of the temperature difference near the stratopause level in both hemisphere. The temperature anomalies are associated with zonal wind anomalies for the same months: in austral winter, the zonal wind anomaly peaking at 40° and at 2 hPa is always positive, meaning a reinforcement of the austral polar vortex. This is not the case in boreal winter, where there is a positive anomaly of the zonal wind in January, so a stronger polar vortex during this month, whereas this anomaly becomes strongly negative during the other months, meaning a destabilization of the boreal polar vortex. For clarity of the graphs, we did not plot the significance levels but all these strong anomalies are statistically significant at 95%. A clear ozone increase is visible in the stratosphere at solar maximum, with a peak at 10 hPa in the equatorial band. There are two negative anomalies at high latitude at about 20 hPa and 1hPa in January and February in the winter season, but only one peaking at 2-3 hPa in summer, and a positive one centered around 10 hPa in December and March.

3. Comparison with observations

These results correspond to a reinforcement of the polar winter vortex above the Antarctic but the behavior is different above the Arctic with a stronger polar vortex in January and a destabilization in February, under solar maximum conditions (SMC). Now if we do a comparison with recent observations, first the temperature response is

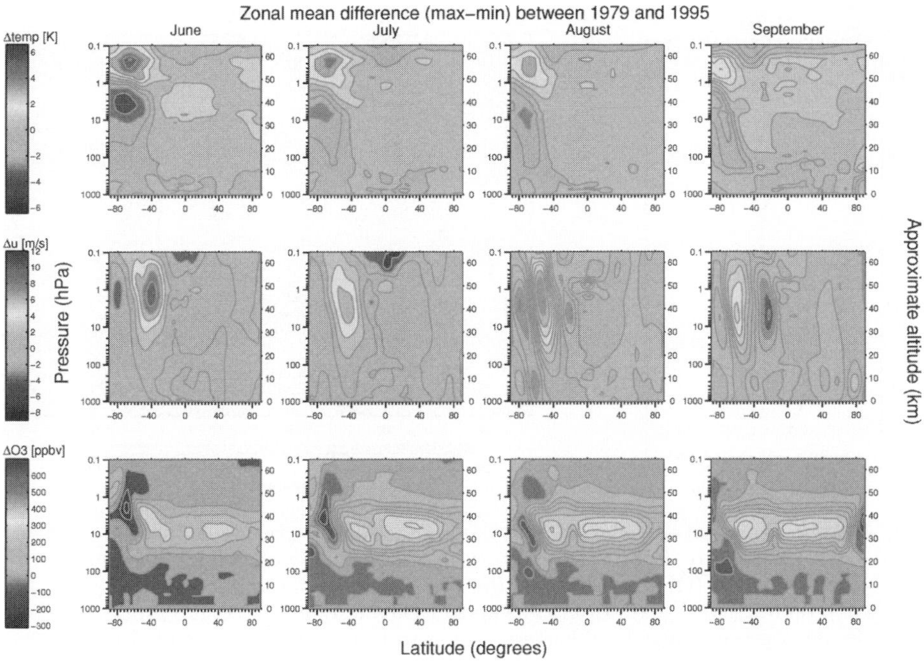

Figure 3. Same as Figure 2 but for the southern hemisphere winter months (June to September).

consistent with the observations in the upper stratosphere-lower mesosphere (direct effect on ozone) as shown in Figure 4 from Keckhut *et al.* (2005). Indeed, curves in left panel confirm a significant sign change near 50 km during winter at high northern latitude. Moreover the right panel clearly shows a strong seasonal behavior at mid-and high-latitude with a maximum response around 50° in both hemispheres while at low latitudes the response is more slowly changing. The high latitude response ranges from a slightly positive reponse to a large negative one. The later extends from October to April in the northern hemisphere with a maximum amplitude of -1.2 K, while it extends from May to september in the Antarctic with an amplitude of −2 K. Our runs exhibit also a strong negative response in the southern hemisphere during austral winter. These results confirm that both hemisphere have a different response in term of amplitude, extension, and timing.

Secondly, by a study using satellite SSU† temperature measurements and ERA-40‡ zonal winds, Claud *et al.* (2008) also demonstrated that temperatures are generally warmer for low- and mid-latitudes under SMC, and that, at northern high latitude, the polar vortex is stronger with the exception of February and to a lesser extent March in the northern hemisphere. Indeed, Figure 5 shows that, for latitudes north of 50°N, there is a significant negative signature in early winter, then in February-March the sign reverses, before returning to negative from April onward. While in December, SMC are associated with colder temperatures practically all along the vertical, in January, a warming in the low stratosphere appears. In February, north of 50°N, temperatures are warmer over the whole vertical range (significantly from 50 to 5 hPa), and temperatures are colder in the equatorial region below 20 hPa. In March, SMC are again associated with

† Stratospheric Sounding Unit

‡ 40 year Re-Analysis Data Archive of the European Centre for Medium-Range Weather Forecasts

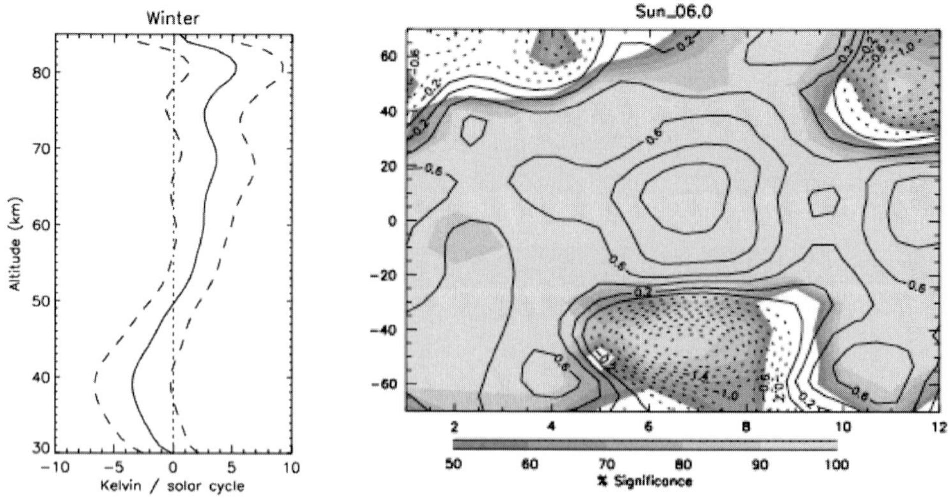

Figure 4. *Left*: Temperature solar response as a function of altitude deduced from lidar data obtained since 1979 up to 2001 at Observatory of Haute Provence (44°N), averaged over the 6 northern winter months (October-March). *Right*: Monthly zonal temperature response to solar activity based on SSU/MSU data from 1979 up to 1998 and represented as a function of latitudes and months for a given height level corresponding to 6 hPa. The gray shaded regions indicate statistically significant signal (from Keckhut *et al.* 2005).

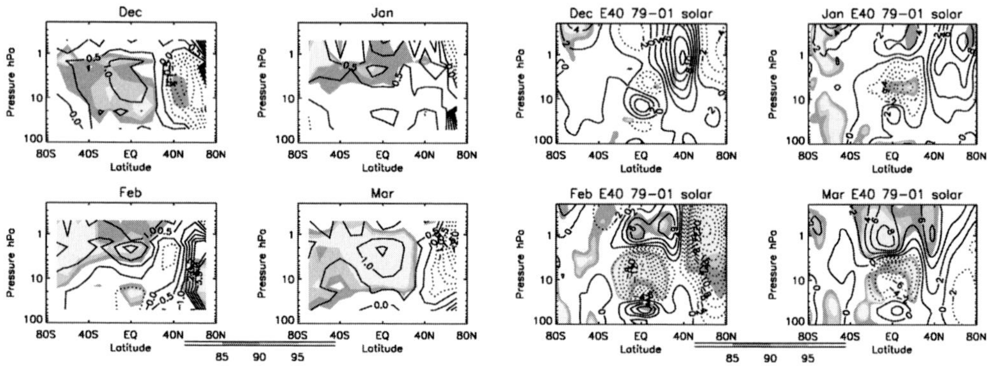

Figure 5. *Left*: SSU temperature measurements from radiometer on board NOAA satellites between 1979 and 2004: zonal temperature response (in K/solar cycle) to solar activity for the winter-early spring of the northern hemisphere. The shades of gray correspond to significance level. Contours are every 0.5 K/solar cycle. *Right*: ERA-40 wind measurements betweeen 1979 and 2001: zonal wind response (in m/s/solar cycle) to solar activity for the winter-early spring of the northern hemisphere. The shades of gray correspond to significance level. Contours are every 2 m/s/solar cycle (from Claud *et al.* 2008).

a cooling of the polar vortex. Figure 5 shows a positive zonal wind anomaly in the northern hemisphere that appears in december above 10 hPa, and developing in January, consistent with a stronger polar vortex in solar maximum. In February, a negative significant anomaly is observed north of 40°N indicating that this month the solar maximum is associated with a weaker vortex. The analysis on the zonal wind then reports a stronger polar vortex during solar maximum, associated with a latitudinal temperature gradient at the stratopause that is evident in December and in January, but not in February and partially in March.

4. Discussion and conclusions

The present study shows how the stratosphere-troposphere system partly responds significantly to a solar cycle enhancement of UV radiation. But the associated mechanism is currently unknown even if ideas has been suggested in the literature.

The main idea is that the mechanism is maybe linked to the interaction of the planetary waves with the main circulation. In particular, Kodera & Kuroda (2002) has argued that changes in the winter polar stratosphere brought about by anomalous solar heating may influence the passage of upward propagating planetary waves and thus their deposition of momentum that influences the strength of the mean overtuning of the stratosphere. Gray *et al.* (2001) also demonstrated that zonal wind anomalies in the sub-tropical upper stratosphere can influence the timing and amplitude of sudden stratospheric warmings, events during the polar winter in which enhanced planetary wave activity disturbs the cold polar vortex.

Perspectives of this work consist in the computation of the Eliassen-Palm fluxes which will permit to investigate the interaction with the planetary waves. Moreover, in a next future is planned the analysis of a run with a real solar signal as input for the irradiance.

This work is led in the framework of the PICARD mission, whose launch is scheduled at the beginning of 2010. This satellite is devoted to a better knowledge of (i) the physics and internal structure of the Sun, and (ii) the solar forcing on Earth's climate (see `http://smsc.cnes.fr/PICARD/index.thm` and Thuillier *et al.* 2006)

Acknowledgements

S. Lefebvre is supported by a CNES post-doctoral grant.

References

Claud, C., Cagnazzo, C., & Keckhut, P. 2008, *JASTP*, 70, 2031
Gray, L. J., Phipps, S. J., Dunkerton, T. J., *et al.* 2001, *Quart. J. R. Meteo. Soc.*, 127, 1985
Haigh, J. D. 1994, *Nature*, 370, 544
Haigh, J. D. 1996, *Science*, 272, 981
Haigh, J. D. & Blackburn, M. 2006, *Space Sci. Revs*, 125, 331
Jourdain, L., Bekki, S., Lott, F., & Lefèvre, F. 2008, *Ann. Geophys.*, 26, 1391
Keckhut, P., Cagnazzo, C., Chanin, M.-L., Claud C., & Hauchecorne A. 2005, *JASTP*, 67, 940
Kodera, K. & Kuroda, Y. 2002, *JGR*, 107(D24), 4749
Lean, J. L. 1989, *Science*, 244, 197
Lefèvre, F., Brasseur, G. P., Folkins, I., Smith, A. K., & Simon, P. 1994, *JGR*, 99, 8183
Lefèvre, F., Figarol, F., Carslaw, K. S., & Peter, T. 1998, *GRL*, 25(13), 2425
Lott, F., Fairhead, L., Hourdin, F., & Levan, P. 2005, *Clim. Dyn.*, 25, 851
Matthes, K., Langematz, U., Gray, L. J, Kodera, K., & Labitzke, K. 2004, *JGR*, 109, D06101
Matthes, K., Kuroda, Y., Kodera, K., & Langematz, U. 2006, *JGR*, 111, D06108
Shindell, D., Rind, D., Balachandran, N., Lean, J., & Lonergan, P. 1999, *Science*, 284, 305
Thuillier, G., Dewitte, S, & Schmutz, W. 2006, *Adv. Sp. Res*, 38, 1792

Solar and Stellar Variability: Impact on Earth and Planets
Proceedings IAU Symposium No. 264, 2009
A. G. Kosovichev, A. H. Andrei & J.-P. Rozelot, eds.

doi:10.1017/S1743921309992912

Solar cycle changes of large-scale solar wind structure

P. K. Manoharan

Radio Astronomy Centre, National Centre for Radio Astrophysics,
Tata Institute of Fundamental Research, Udhagamandalam (Ooty), 643001, India
email: mano@ncra.tifr.res.in

Abstract. In this paper, I present the results on large-scale evolution of density turbulence of solar wind in the inner heliosphere during 1985–2009. At a given distance from the Sun, the density turbulence is maximum around the maximum phase of the solar cycle and it reduces to ~70%, near the minimum phase. However, in the current minimum of solar activity, the level of turbulence has gradually decreased, starting from the year 2005, to the present level of ~30%. These results suggest that the source of solar wind changes globally, with the important implication that the supply of mass and energy from the Sun to the interplanetary space has significantly reduced in the present low level of activity.

Keywords. turbulence, scattering, Sun: corona, Sun: evolution, Sun: solar wind

1. Introduction

The solar wind plasma and its magnetic field undergo significant evolution with the phase of the solar cycle. In this paper, I present new results on the changing of large-scale structure of density turbulence in the solar wind over the solar cycles 22 and 23. The solar wind estimates have been obtained from the interplanetary scintillation (IPS) observations made with the Ooty Radio Telescope, operating at 327 MHz (e.g., Swarup *et al.* 1971). The IPS technique exploits the scattering of radio waves from a compact radio source by the density fluctuations in the solar wind (e.g., Hewish *et al.* 1964) and allows to probe the density turbulence in the solar wind, at a range of distances from the Sun as well as over an extended period of time.

2. Scintillation Index and Solar Wind Density Turbulence

The degree of interplanetary scintillation is given by the scintillation index, m = *rms of intensity fluctuation/mean intensity of the source*, which is a measure of density turbulence ($C_N^2(R) \sim m^2(R)$) in the solar wind. IPS measurements at Ooty can probe the solar wind in the heliocentric distance range of $R = 10 - 250$ solar radii ($R_\odot$) (e.g., Manoharan *et al.* 2000). Figure 1 shows the scintillation index variations with distance from the Sun, for the compact radio quasar 1148-001 between 1985 and 2008. In these plots, the enhanced scintillation caused by intense transient events (e.g., coronal mass ejections, refer to Manoharan *et al.* 2000) have been excluded and plots represent the average ambient turbulence condition of the heliosphere at different phases of the solar cycle.

The level of scintillation, as shown in the figure, increases to a peak value at a distance of $R \approx 40\ R_\odot$, and then decreases for further closer solar offsets, where the scattering becomes strong and saturated (e.g., Manoharan 1993). In the weak-scattering region at $R > 40\ R_\odot$, the gradual decrease in the level of scintillation is linearly related to the fall of density fluctuations ($C_N^2(R) \sim \delta N_e^2(R)$) with distance from the Sun and the radial

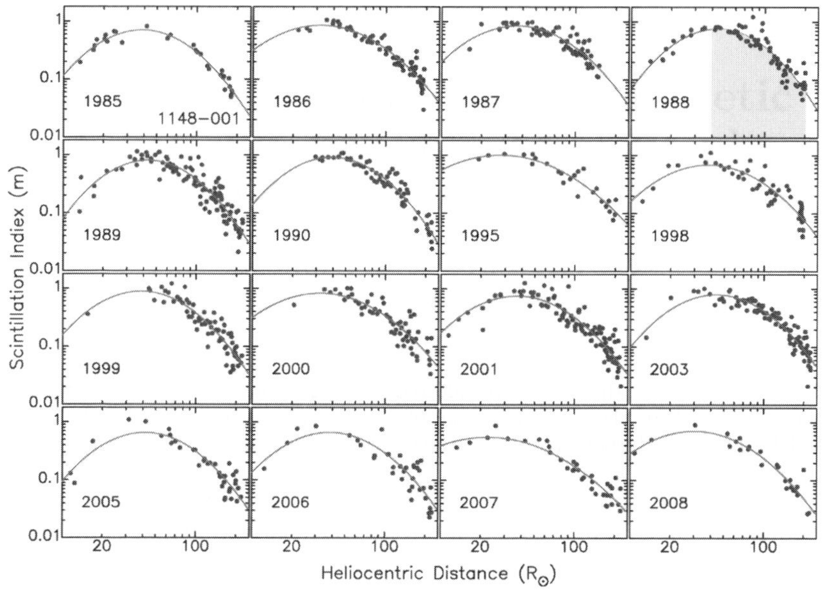

Figure 1. Scintillation index as a function of distance from the Sun for the radio quasar 1148-001 at 327 MHz, over the years 1985–2008. The peak of $m - R$ curve, close to unity, shows that the radio quasar 1148-001 is compact, which in fact has an equivalent diameter of ∼15 milli arcsec. In each plot, the best-fit to the data points is shown by a continuous curve.

dependence of scintillation index, $m = m_0 R^{-\alpha}$, can be related to the density turbulence, where m_0 corresponds to scintillation at unit distance from the Sun. However, an IPS measurement represents the integration along the line-of-sight to the radio source and therefore, the '*distance-density turbulence*' relationship, $C_N^2(R) \sim R^{-\beta}$, is related to the scintillation index by $2\alpha = (\beta - 1)$. In the above plots, the radial dependence of density turbulence can be obtained from the linear slope of the curve at distances >40 $R_\odot$ and slopes vary in the range between $\alpha \approx 2.0$ at the minimum and $\alpha \approx 1.5$ at the maximum of the cycle. These results indicate that the density turbulence falls off rapidly with distance, when the solar activity is low. Whereas it follows a typical $1/R^2$ dependence around the peak of the solar cycle. These results are in agreement with the earlier findings for solar cycles 21 and 22 (Manoharan 1993).

2.1. *Solar Cycle Changes of Large-Scale Turbulence*

During the current minimum of activity, after about the year 2005, the peak of the scintillation curve however seems to move close to the Sun. It suggests that a given level of density turbulence moves closer to the Sun (i.e., a systematic reduction in turbulence at a particular distance from the Sun, from one year to the next). In order to determine the global characteristics of density turbulence, the area under the $m^2(R)$ profile has been calculated as given by,

$$A = \int_{R=40R_\odot}^{R=200R_\odot} m^2(R)\, dR. \tag{2.1}$$

The area has been estimated in the weak-scattering regime ($R \geqslant 40\ R_\odot$), which can provide a qualitative estimate of the large-scale turbulence in the inner heliosphere. Figure 2 shows the area of $m^2(R)$ curve plotted against the year for 10 compact sources. For each source, the area has been normalized by its fraction of scintillating flux density

In this paper, we study the scale dependence of kurtosis and phase coherence in intermittent magnetic field turbulence measured at three different locations of the solar-terrestrial environment: (1) in the solar photosphere, (2) in the solar wind, and (3) on the ground. We investigate two scenarios: a non-ICME event in February 2002 and an ICME event in January 2005.

2. Non-Gaussianity and phase synchronization in the solar-terrestrial environment

Solar images obtained by the SOHO MDI instrument provide the measurement of the magnetic field in the solar photosphere. Near the centre of the solar image the projection effects are negligible, hence the MDI solar image represents the vertical or line-of-sight component of the photospheric magnetic field $B_{\|}$ (Abramenko *et al.*, 2002). Figure 1(a) shows a solar magnetogram taken by SOHO MDI on 1 February 2002. The white color corresponds to positive magnetic polarity, and the black color corresponds to negative magnetic polarity. The two white squares mark two selected areas, one containing the active region AR 09802 (upper) and the other containing a quiet region (lower). Figure 1(b)

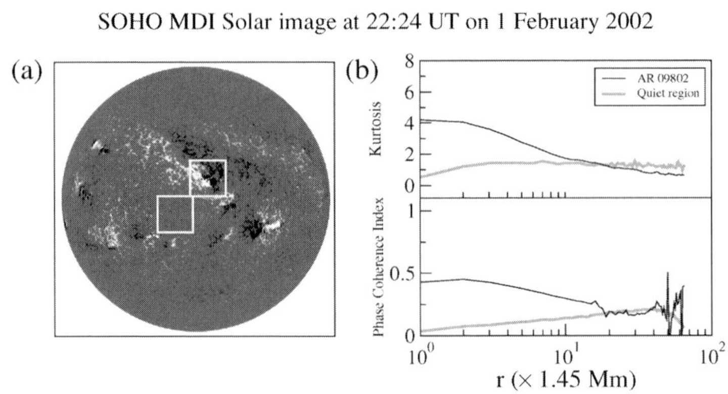

Figure 1. (a) SOHO MDI solar image taken at 22:24 UT on 1 February 2002. The two white squares enclose two areas containing the active region AR 09802 (upper), and a quiet region (lower). (b) Kurtosis (upper panel) and the phase coherence index (lower panel) as a function of spatial scale r computed from AR 09802 and the quiet region.

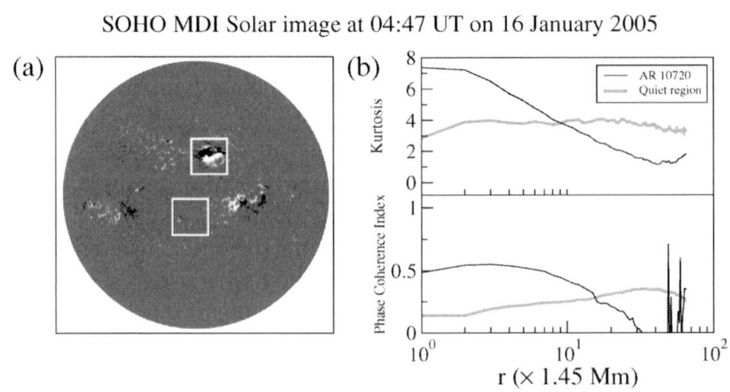

Figure 2. (a) SOHO MDI solar image taken at 22:24 UT on 16 January 2005. The two white squares enclose two areas containing the active region AR 10720 (upper), and a quiet region (lower). (b) Kurtosis (upper panel) and the phase coherence index (lower panel) as a function of spatial scale r computed from AR 10720 and the quiet region.

shows kurtosis and the phase coherence index as a function of spatial scale r computed from the two selected regions. From Figure 1(b) we observe that the variation of kurtosis with r in the quiet region is close to a Gaussian process ($K = 0$), being scale-invariant for all scales $r \gtrsim 2$ pixels ~ 2.9 Mm, consistent with the features of a monofractal process (Abramenko *et al.*, 2002). The active region, on the other hand, displays an increase of kurtosis as the spatial scale r decreases, which is a characteristic of a non-Gaussian process related to nonlinear energy cascade within the inertial subrange, and these values are higher than those obtained from the quiet region for scales $\lesssim 20$ pixels ~ 29 Mm. The degree of phase synchronization measured by the phase coherence index (Hada *et al.*, 2003; Koga *et al.*, 2007; Chian & Miranda, 2009) in the active region increases with decreasing spatial scale r, while the quiet region presents low-degree of synchronization at all scales. Note that at large scales the lack of datapoints introduces big errors in the computation of the phase coherence index.

Figure 2(a) shows a SOHO MDI solar magnetogram obtained on 16 January 2005. The two white squares enclose two selected areas containing the active region AR 10720 (upper) and a quiet region (lower), respectively. On 19-20 January 2005 several flares associated with CMEs were observed in AR 10720, however in this period the active region is too close to the solar limb, and the projection effects cannot be neglected. Hence, we restrict our analysis to this earlier solar image when AR 10720 is near the disk centre. Figure 2(b) shows kurtosis and the phase coherence index as a function of spatial scale r. It shows that the kurtosis of AR 10720 increases as the scale r decreases, while kurtosis of the quiet region displays scale-invariance, similar to Fig. 1(b). Likewise, the phase coherence index in Fig. 2(b) presents similar behavior as Fig. 1(b).

Now we direct our attention to the interplanetary magnetic field data collected *in situ* in the solar wind. Figure 3(a) shows the time series of the modulus of magnetic field $|B|$ obtained by ACE and Cluster from 19:40:40 UT on 1 February 2002 to 03:56:38 UT on 3 February 2002. During this interval Cluster is in the solar wind upstream of the Earth's bow shock (Chian & Miranda, 2009). Although Fig. 1(a) indicates the presence of several solar active regions on 1 February 2002, no M- or X-class solar flares occurred during the selected interval, and strong interplanetary disturbances such as ICMEs were not seen. The upper panel of Figure 3(b) shows the variation of kurtosis as a function of time scale τ for magnetic field fluctuations of ACE and Cluster. For 10 s $\lesssim \tau \lesssim 10^3$ s, kurtosis increases as the time scale decreases which characterizes non-Gaussianity. The lower panel of Fig. 3(b) shows the variation of the phase coherence index with τ which presents similar behavior as kurtosis. From Fig. 3(b) we observe that the behavior of kurtosis and the phase coherence index detected by ACE and Cluster are very similar except for scales around 10 s, where ACE observes a higher level of intermittency and phase synchronization than Cluster.

Figure 4(a) shows the time series of the modulus of magnetic field $|B|$ measured by ACE for the ICME event of 21–22 January 2005 (Foullon *et al.*, 2007). We selected two intervals from this event. The first interval is located upstream of the ICME shock which begins at 06:00:00 UT on 21 January and ends at 16:00:00 UT on 21 January. The second interval is located downstream of the ICME shock which begins at 16:47:19 UT on 21 January, and ends at 21:20:00 UT on 22 January. In order to ensure the stationarity of data we avoid the "foot" transition region associated with the ICME shock front. Figure 4(b) shows kurtosis and the phase coherence index as a function of time scale τ computed from the modulus of magnetic field $|B|$ in upstream and downstream regions of the ICME shock. It is evident, from Figure 4(b), that both upstream and downstream regions indicate the features of intermittency and phase synchronization across scales; moreover, the level of intermittency and phase synchronization are higher in the downstream region than the upstream region for all scales τ.

We analyze next the Earth's geomagnetic field data obtained from two ground magnetometers. The top panel of Fig. 5(a) shows the time series of the modulus of magnetic field $|B|$ observed by ACE at L1 during the non-ICME solar wind event of 1-3 February 2002. The second panel shows the modulus of the Earth's geomagnetic field measured by a ground magnetometer at Ji-Paraná (JPA), Brazil (geomagnetic latitude $\sim 0^{\mathrm{O}}$). The three bottom panels show the time series of Pc3, Pc4 and Pc5 geomagnetic micropulsations, respectively. Each time series of geomagnetic micropulsations is obtained by applying a Fourier band-pass filter. Fig. 5(b) shows kurtosis and the phase coherence index of ACE magnetic field data and JPA geomagnetic field data as a function of time scale τ. Fig. 5(b) indicates that the geomagnetic field fluctuations measured on the ground are intermittent. For almost all scales the level of intermittency and phase coherence of ACE are higher than the JPA.

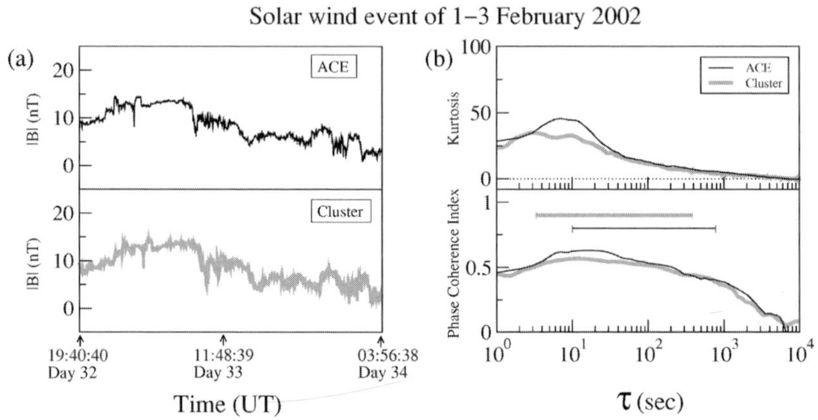

Figure 3. (a) Time series of the modulus of magnetic field $|B|$ measured by ACE (upper panel) and Cluster (lower panel) for the solar wind event of 1-3 February 2002. (b) Kurtosis (upper panel) and the phase coherence index (lower panel) of $|B|$ as a function of time scale τ. Horizontal bars indicate the inertial subranges of ACE and Cluster, respectively.

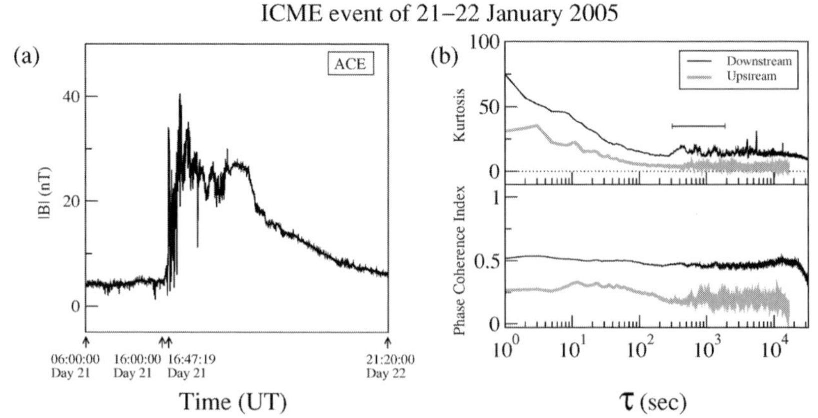

Figure 4. (a) Time series of the modulus of magnetic field $|B|$ measured by ACE for the ICME event of 21–22 January 2005. (b) Kurtosis (upper panel) and the phase coherence index (lower panel) of $|B|$ as a function of time scale τ for the upstream and downstream regions of the ICME shock.

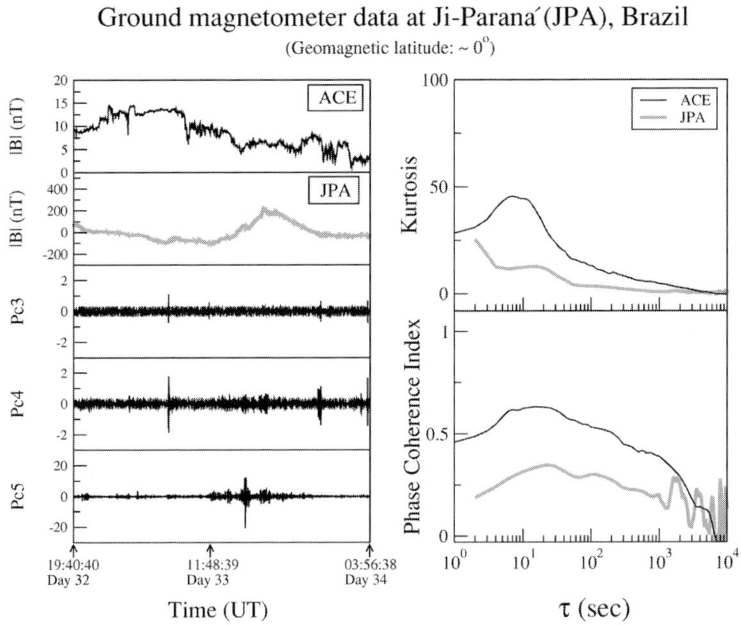

Figure 5. (a) From top to bottom: time series of $|B|$ measured by ACE for the solar wind event of 1-3 February 2002; modulus of the Earth's geomagnetic field $|B|$ measured by a ground magnetometer at Ji-Paraná, Brazil, during the same time interval; time series of Pc3 (10-45 s), Pc4 (45-150 s) and Pc5 (150-1000 s) micropulsations. (b) Kurtosis (upper panel) and the phase coherence index (lower panel) of $|B|$ measured by ACE and the ground magnetometer at JPA as a function of time scale τ.

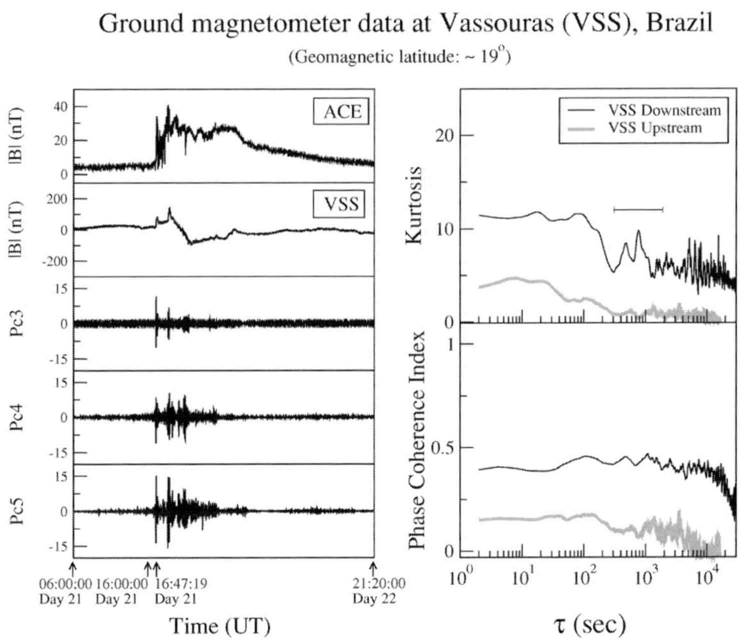

Figure 6. (a) From top to bottom: time series of $|B|$ (nT) measured by ACE for the ICME event of 21–22 January 2005; modulus of the Earth's geomagnetic field $|B|$ (nT) measured by a ground magnetometer at Vassouras, Brazil; time series of Pc3 (10-45 s), Pc4 (45-150 s) and Pc5 (150-1000 s) micropulsations. (b) Kurtosis (upper panel) and the phase coherence index (lower panel) of $|B|$ measured by ACE and the ground magnetometer at VSS as a function of time scale τ.

Finally, for the ICME event of 21–22 January 2005, we plot in the top panel of Figure 6(a) the time series of the modulus of magnetic field $|B|$ observed by ACE and in the second panel the modulus of the Earth's geomagnetic field measured by a ground magnetometer at Vassouras (VSS), Brazil (geomagnetic latitude $\sim 19^{\rm o}$). The three bottom panels show the time series of Pc3, Pc4 and Pc5 geomagnetic micropulsations. All the time series of VSS were shifted by -1462 s to synchronize with the ICME shock arrival at ACE. After shifting, we divide the VSS geomagnetic field time series into "upstream" and "downstream" intervals in analogy with the ACE magnetic field data of the ICME shock. Figure 6(b) shows kurtosis and the phase coherence index of the VSS upstream and downstream intervals as a function of time scale τ. It shows that for all scales the level of intermittency and phase coherence are higher in the "downstream" geomagnetic field-fluctuations after the arrival of ICME than the "upstream" geomagnetic field fluctuations before the arrival of ICME.

A comparison of the variation of kurtosis and the phase coherence index as a function of τ for the downstream intervals observed by ACE in the solar wind (Fig. 4(b)) and the VSS ground magnetometer (Fig. 6(b)) reveals a common feature consisting of three peaks between scales $\tau \sim 304$ s and $\tau \sim 1860$ s, marked with a bar in both figures. This can be interpreted as evidence of a close correlation of the Earth's geomagnetic field with the ICME driver at these scales.

3. Conclusions

In this paper, we studied non-Gaussianity and phase synchronization in solar-terrestrial magnetic field turbulence in two scenarios: a non-ICME event and an ICME event. For each scenario, we calculated first kurtosis and the phase coherence index for an active region and a quiet region, respectively, in the solar photosphere. We then applied the same techniques to solar wind turbulence using *in situ* data. The response of the Earth's geomagnetic field using data from ground magnetometers in Brazil was presented. In particular, we showed that the interplanetary magnetic field turbulence downstream of the ICME shock is closely correlated with the geomagnetic turbulence detected on the ground.

Acknowledgments

R. A. Miranda and A. C.-L. Chian wish to thank Professor Michio Yamada, Dr. Yoshitaka Saiki and RIMS at Kyoto University for their kind hospitality. This work is supported by CAPES and CNPq, Brazil.

References

Abramenko, V. I., Yurchiyshiyn, V. B., Wang, H., Spirock, T. J., & Goode, P. R 2002, *ApJ* 577, 487

Chian, A. C.-L. & Miranda, R. A. 2009, *Ann. Geophys.* 27, 1789

Foullon, C., Owen, C. J., Dasso, S., Green, L. M., Dandouras, I., Elliott, H. A., Fazakerley, A. N., Bogdanova, Y. V., & Crooker, N. U. 2007, *Solar Phys.* 244, 139

Hada, T., Koga, D., & Yamamoto, E. 2003, *Space Sci. Revs* 107, 463

Kamide, Y. & Chian, A. C.-L. (Eds.) 2007, *Handbook of the Solar-Terrestrial Environment* (Berlin, Springer)

Koga, D., Chian, A. C.-L., Miranda, R. A., & Rempel, E. L. 2007, *Phys. Rev. E* 75, 046401

Sahraoui, F. 2008, *Phys. Rev. E* 78, 026402

Telloni, D., Antonucci, E., Bruno, R., & D'Amicis, R 2009, *ApJ* 693, 1022

Solar and Stellar Variability: Impact on Earth and Planets
Proceedings IAU Symposium No. 264, 2009
A. G. Kosovichev, A. H. Andrei & J.-P. Rozelot, eds.

doi:10.1017/S1743921309992948

Observation of magnetic reconnection and current sheets in the solar wind

Pablo R. Muñoz[1], **Abraham C.-L. Chian**[1,2], **Rodrigo A. Miranda**[1,2], **and Michio Yamada**[2]

[1]National Institute for Space Research (INPE), São José dos Campos-SP 12227-010, Brazil
email: pmunoz@dge.inpe.br

[2]Research Institute for Mathematical Sciences (RIMS), Kyoto University, Kyoto 606-8502, Japan

Abstract. We apply single- and multi-spacecraft techniques to search for currents sheets in the solar wind during the ICME event of 21 January 2005, using the Cluster magnetic field data. Two large-scale currents sheets are detected at the leading boundary of the ICME ejecta using the single-spacecraft technique, which exhibit physical characteristics typical of magnetic reconnection exhausts in the solar wind.

Keywords. Sun: solar wind, coronal mass ejections (CMEs), magnetic fields

1. Introduction

Magnetic reconnection in a current sheet is a key dynamical process that converts magnetic energy into particle kinetic energy in space and astrophysical plasmas (Nishida, 2007; Priest, 2007). Retinò *et al.* (2007) developed a systematic technique to search for thin current sheets based on the simultaneous measurement of magnetic field by four Cluster spacecraft. Using this method *in situ* evidence of magnetic reconnection and associated energy release was found in the Earth's magnetosheath (Sundkvist *et al.*, 2007). In addition, Li (2008) proposed a systematic procedure to look for the existence of current sheets in the solar wind based on two-point correlation measurement of magnetic field by a single spacecraft. This single-spacecraft technique succeeded to identify large-scale current-sheet-like structures in the solar wind using one Cluster spacecraft (Li *et al.*, 2008).

In this paper, we study current sheets and magnetic reconnection in the solar wind upstream of the Earth's bow shock during the Interplanetary Coronal Mass Ejection (ICME) event of 21 January 2005, based on Cluster data. Four regions of this ICME are identified by ACE and WIND (Foullon *et al.*, 2007): (1) upstream region, (2) sheath region, (3) non-compressive density enhancement (NCDE) region, and (4) magnetic cloud. First, a systematic search of current sheets is performed in the first three regions detected by Cluster using both single- (Li, 2008) and multi-spacecraft (Retinò *et al.*, 2007; Sundkvist *et al.*, 2007) techniques. Our systematic search is capable of finding both large-scale and small-scale current sheets in all three regions of the ICME. Second, the technique proposed recently by Gosling *et al.* (2005) for observing magnetic reconnection exhaust in a current sheet in the solar wind was applied to two current sheets located at the leading boundary of the ICME ejecta. We analyze these two current sheets in detail and demonstrate their association with magnetic reconnection exhausts in the solar wind, with evidence of bifurcated current sheets and a jet of plasma flow linked to each current sheet.

2. Detection of current sheets

A current sheet is a localized region of an electric current confined to a nearly two-dimensional surface which is ubiquitous in a magnetized plasma. To confine the electric current into a surface, the magnetic field surrounding the current sheet must change its direction significantly across the sheet surface (i.e., with strong shear in the magnetic field vector **B**). Two methods of single-spacecraft (Li, 2008) and multi-spacecraft (Retinò *et al.*, 2007; Sundkvist *et al.*, 2007) are used to systematically search for current sheets in the ICME event of 21 January 2005, based on the Cluster magnetic field data. The interval studied includes the upstream region of the ICME shock, the ICME sheath region (marked by the vertical dashed lines) between the shock arrival (SA) and the leading boundary of the ejecta (SB), and a portion of the non-compressive density enhancement (NCDE) region when Cluster is in the upstream region of the Earth's bow shock (Foullon *et al.*, 2007). In the single-spacecraft method of Li (2008), in order to determine the presence of current sheets in a given time series B(t) it is necessary to prove first that the integrated distribution function F(θ, t) for the interval, representing the probability of finding the angle between $\mathbf{B}(t)$ and $\mathbf{B}(t+\tau)$ larger than θ, scales linearly with the time separation τ when θ is larger than some critical angle θ_0, i.e., $F(\theta, N\tau) \sim NF(\theta,\tau)$ where N is an arbitrary integer greater than zero. The current sheets seen in the upper- and middle-panels of Fig. 1 are detected by following the procedure of Li (2008), assuring that a linear behavior is clearly seen in all scales for different critical angles. Our results show that this single-spacecraft method is capable of detecting current sheets of any scale $\tau \gg \delta$, δ is the resolution of data. The top and middle panels show current sheets of large-scale ($\tau \sim 120$ sec) and small-scale ($\tau \sim 2$ sec), respectively, using the Cluster-1 data. The upper-panel of Fig. 1 shows two large-scale current sheets, SB1 and SB2, associated with the leading boundary of the ejecta SB. For this event, the multi-spacecraft method was able to detect current sheets of small-scale $\tau \sim 1.1$ sec using four Cluster spacecraft

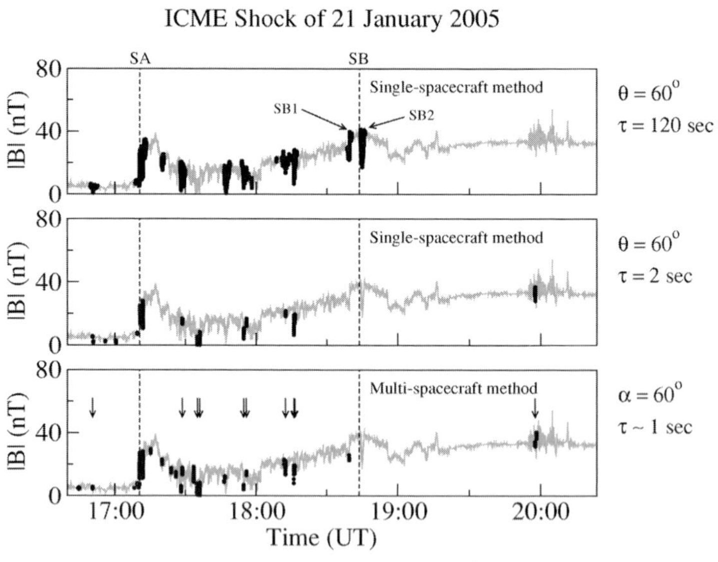

Figure 1. Detection of current sheets by single- and multi-spacecraft techniques using the magnetic field data of Cluster for the ICME shock of 21 January 2005. The gray line denotes the modulus of magnetic field $|B|$, and the black dots denote the current sheets detected. The arrows mark the current sheets detected by both single-spacecraft at $\tau = 2$ sec and multi-spacecraft methods. SA denotes shock arrival, SB denotes the leading boundary of the ejecta where two large-scale current sheets SB1 and SB2 are found.

(lower panel). The scale of the current sheets detectable by the multi-spacecraft method is given by $\tau = \Delta/V_{SW}$, where Δ is the minimum distance between N spacecraft and V_{SW} is the average speed of the solar wind with respect to the spacecraft. For this event, $\Delta \sim 933$ km and $V_{SW} \sim 878$ km/s. The arrows in the lower-panel of Fig. 1 indicate the currents sheets detected by both methods of single-spacecraft at $\tau = 2$ sec (middle-panel) and multi-spacecraft (lower-panel).

3. Observation of magnetic reconnection

We discuss next the observation of magnetic reconnection exhausts in the ICME shock of 21 January 2005. Figure 2 displays the dynamics of magnetic reconnection processes associated with two current sheets, SB1 (left panels) and SB2 (right panels), located at the leading boundary of the ICME ejecta of Fig. 1. From top to bottom: the modulus of magnetic field $|B|$ (nT); three components of magnetic field (B_L, B_M, B_N) in the current sheet (L, M, N) coordinates derived from the Minimum Variance Analysis (MVA) (Sonnerup and Cahill, 1967); the modulus of observed plasma velocity and the plasma velocity predicted by the reconnection theory of Sonnerup *et al.* (1981); three components of observed (V_L, V_M, V_N) plasma velocity in LMN coordinates; the modulus of current density $|J|$ (nA/m^2) computed by the curlometer method based on multi-spacecraft Cluster data (Dunlop *et al.*, 2002); three components of current density (J_L, J_M, J_N). It is seen that V_L is anti-correlated (correlated) with B_L in the leading (trailing) edge of SB1 and V_L is correlated (anti-correlated) with B_L in the leading (trailing) edge of SB2. Such pairs of oppositely coupled changes in V and B are the signatures of magnetic

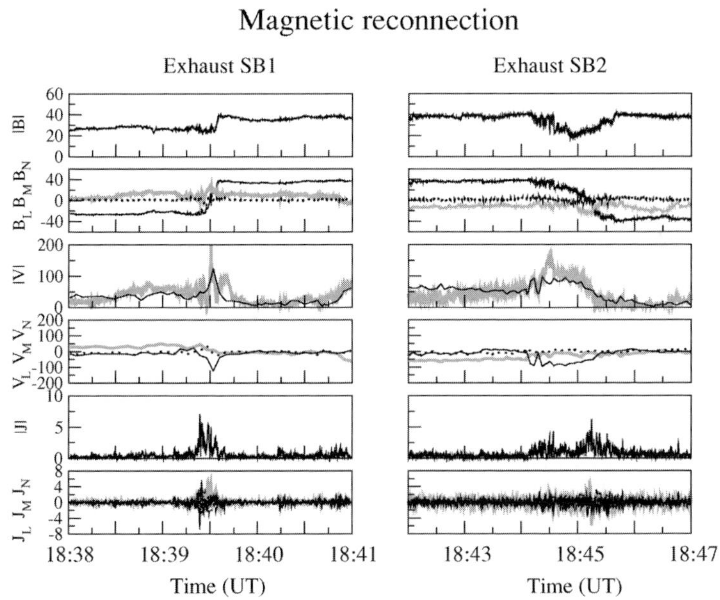

Figure 2. Observation of magnetic reconnection exhausts in two current sheets, SB1 and SB2, located at the leading boundary of the ICME ejecta of 21 January 2005, where $|B|$ (nT) is the modulus of magnetic field, B_L (black), B_M (gray) and B_N (dotted) are the three components of the magnetic field in the (L, M, N) coordinates, V (km/s) is the modulus of observed plasma velocity (black) and the plasma velocity (gray) predicted by the reconnection theory of Sonnerup *et al.* (1981), V_L (black), V_M (gray) and V_N (dotted) are the three components of plasma velocity, $|J|$ (nA/m^2) is the modulus of current density, and J_L (black), J_M (gray) and J_N (dotted) are the three components of the current density.

reconnection exhaust in bifurcated current sheets in the solar wind, as seen in the time series of $|J|$ (Gosling *et al.*, 2005). Our three-dimensional analysis of magnetic field, current density and plasma velocity of two current sheets $SB1$ and $SB2$ are in agreement with the physical features of magnetic reconnection exhausts. The magnetic field vector **B** has a strong shear angle ($\theta_{SB1} \sim 146^{\mathrm{O}}$ and $\theta_{SB2} \sim 150^{\mathrm{O}}$) between the leading and trailing edges of each current sheet. The current density vector **J** computed from the multi-spacecraft curlometer technique (Dunlop *et al.*, 2002) shows a current sheet associated with the region of strong magnetic shear for both $SB1$ and $SB2$. In particular, each current sheet is associated with a strong jet of plasma flow resulting from a magnetic reconnection.

4. Conclusion

The solar wind provides a natural laboratory for observing coherent structures such as current sheets embedded in the interplanetary intermittent turbulence, upstream and downstream of an ICME. Phase synchronization due to nonlinear multiscale interactions is responsible for the formation of these coherent structures (Chian & Miranda, 2009). Magnetic reconnection in current sheets plays an important role in energy dissipation in the solar wind turbulence. The methodology adopted in this paper to systematically detect current sheets and investigate the signatures of magnetic reconnection may deepen our understanding of the chaotic nature of the solar wind (Chian *et al.*, 2006).

5. Acknowledgement

A. C.-L. Chian and R. A. Miranda wish to thank Professor Michio Yamada, Dr. Yoshitaka Saiki and RIMS at Kyoto University for their kind hospitality. This work is supported by CAPES and CNPq, Brazil.

References

Chian, A. C.-L., Kamide, Y., Rempel, E. L., & Santana, W. M. 2006, *J. Geophys. Res.* 111, A07S03, doi:10.1029/2005JA011396.

Chian, A. C.-L. & Miranda, R. A. 2009, *Ann. Geophys.* 27, 1789.

Dunlop, M. W., Balogh, A., Glassmeier, K. H., & Robert, P. 2002, *J. Geophys. Res-Space Phys.* 107, A11, 1384.

Foullon, C., Owen, C. J., Dasso, S., *et al.*. 2007, *Sol. Phys.* 244, 139.

Gosling, J. T., Skoug, R. M., McComas, D. J., & Smith, C. W. 2005, *J. Geophys. Res.-Space Phys.* 110, A01107.

Li, G. 2008, ApJ (Letters) 672, L65.

Li, G., Lee, E., & Parks, G. 2008, *Ann. Geophys.* 26, 1889.

Nishida, A. 2007, in: Y. Kamide, Y. & A. C.-L. Chian (eds.) *Handbook of the Solar-Terrestrial Environment* (Berlin, Springer), p. 280.

Priest, E. R. 2007, in: Y. Kamide & A. C.-L. Chian (eds.) *Handbook of the Solar-Terrestrial Environment* (Berlin, Springer), p. 50.

Retinò, A., Sundkvist, D., & Vaivads, A., *et al.* 2007, *Nature Phys.* 3, 236.

Sonnerup, B. U. O. & Cahill, L. J., 1967, *J. Geophys. Res.* 72(1), 171.

Sonnerup, B. U. O. *et al.* 1981, *J. Geophys. Res.* 86,10049.

Sundkvist, D. Retinò, A., Vaivads, A., & Bale, S. D. 2007, *Phys. Rev Lett.* 99, 025004.

Session VII

Effects of magnetic activity on planet formation and evolution

Chair: S. Hasan

Solar and Stellar Variability: Impact on Earth and Planets
Proceedings IAU Symposium No. 264, 2009
A. G. Kosovichev, A. H. Andrei & J.-P. Rozelot, eds.

doi:10.1017/S174392130999295X

Magnetic activity, high-energy radiation and variability: from young solar analogs to low-mass objects

Manuel Güdel

Institute of Astronomy, ETH Zurich
8093 Zurich, Switzerland
email: guedel@pastro.phys.ethz.ch

Abstract. Magnetic activity on cool stars expresses itself in a bewildering variety of radiative and particle output originating from magnetic regions between the photosphere and the corona. Given its origin in evolving magnetic fields, most of this output is variable in time. Radiation in the ultraviolet, the extreme ultraviolet, and the X-ray ranges are important for heating and ionizing upper planetary atmospheres and thus driving atmospheric evaporation. Additionally, stellar winds interact with the upper atmospheres and may lead to further erosion. The stellar high-energy output is therefore a prime factor in determining habitability of planets. We summarize our knowledge of magnetic activity in young solar analogs and lower-mass stars and show how the stellar output changes on evolutionary timescales.

Keywords. Sun: evolution, stars: activity, stars: evolution, stars: flare, stars: rotation, stars: magnetic fields, stars: winds, astrobiology

1. Introduction

The present-day Sun is a G2 V star with a surface effective temperature of approximately 5780 K. Stellar evolution theory indicates, however, that the Sun has shifted in spectral type by several subclasses, becoming hotter by a few hundred degrees and becoming more luminous; the bolometric luminosity of the Sun in its zero-age main-sequence [ZAMS] phase amounted to only about 70% of the present-day output (Sackmann & Boothroyd 2003). The lower luminosity poses serious problems with regard to habitability of the young Earth and Mars. While geological records indicate a warmer climate and the presence of liquid water on both planets at ages younger than 1 Gyr, both should have been in deep freeze with sub-zero temperatures. Although greenhouse gases could have made up for the deficit in the case of the young Earth, a higher-pressure CO_2 rich atmosphere on Mars would still not have retained enough energy (Kasting *et al.* 1993; Sackmann & Boothroyd 2003). One potential solution to this problem will be mentioned in Sect. 5 below.

However, the Sun's luminosity changed much more dramatically at shorter wavelengths in the course of its evolution. Direct and indirect evidence points to a much higher level of magnetic activity in particular in its pre-main sequence (PMS) phase, the subsequent epoch of its settling on the main sequence (MS), and the first few 100 Myr of its life on the MS. Direct evidence includes meteoritic traces and isotopic anomalies that require much higher proton fluxes at early epochs at least partly from *within* the solar system (e.g., Caffee *et al.* 1987); indirect evidence comes from systematic comparisons of the contemporary Sun with solar analogs of younger age. Atmospheres of solar system planets offer further clues to strongly elevated activity levels, among them the evidence for a warmer early climate on Mars but also the extremely arid atmosphere of Venus – a sister planet of the water-rich Earth.

We discuss three topics that are of some importance for our understanding of the young solar system and the evolution of planetary atmospheres. We first discuss the evolution of rotation and activity of solar-like stars and address their elusive ionized winds; we then summarize knowledge of the long-term evolution of the radiative output of the Sun and lower-mass stars; and third, we address the importance of flares in stellar activity. The last section discusses the importance of stellar high-energy output in shaping young planetary environments.

2. The rotation and mass-loss history of the Sun

Stellar magnetic winds are a crucial consequence of "stellar activity" but their detection in solar analogs is very difficult. The best – albeit indirect – proof of the presence of magnetized winds is the spin-down of convective stars on the main sequence as such a wind carries away angular momentum from the star.

Owing to its coupling to large-scale solar magnetic fields, the wind carries away angular momentum, leading to a steady decline of the solar rotation rate. The fractional change in the rotation rate, Ω, is,

$$\frac{\dot{\Omega}}{\Omega} \propto \frac{\dot{M}}{M}\left(\frac{R_{\rm A}}{R_\odot}\right)^m \tag{2.1}$$

where M is the stellar mass, $R_{\rm A}$ is the Alfvén radius, and m has a value between 0 and 1 depending on the magnetic field geometry (e.g., Mestel 1984). The surface magnetic field strength itself is coupled to the rotation rate via an internal dynamo and is expected to scale as $B_0 \propto \Omega^p$ with p being 1 or 2 (e.g., Mestel 1984). It has further been suggested that the wind mass loss rate itself scales with magnetic activity (see below); in other words, activity and rotation are coupled by a feedback loop (slower rotation $\rightarrow$ decreasing magnetic activity $\rightarrow$ smaller mass loss rate $\rightarrow$ less magnetic braking) that not only leads to a steadily declining rotation rate during the Sun's MS life but causes the rotation rate to converge toward a unique age-dependent value for a given stellar mass. This picture has found basic support from the observation of declining magnetic activity with increasing stellar age and rotation periods, as inferred from many open-cluster studies (e.g., Soderblom *et al.* 1993). This evolutionary scheme of rotation and magnetic activity overall holds for all late-type stars but the activity decay is considerably slower toward lower-mass stars (West *et al.* 2008). For solar analogs one finds,

$$\Omega \propto t_6^{-0.6\pm0.1} \text{ [d]} \tag{2.2}$$

(Ayres 1997). The rotation period thus decays by typically a factor of about 20 from ZAMS age to the end of the MS life.

Ionized winds have not yet been "directly" detected from any late-type main-sequence star other than the Sun. A potential direct detection method is provided by the measurement of thermal radio emission from the winds. Drake *et al.* (1993) obtained a radio upper limit to the wind mass loss of $\dot{M} < 1.7 \times 10^{-11}\ M_\odot\ {\rm yr}^{-1}$ for the *evolved* subgiant Procyon (albeit based on indirect arguments), while Lim *et al.* (1996) found an upper limit of $\dot{M} < 7\times10^{-12}\ M_\odot\ {\rm yr}^{-1}$ for the nearby, relatively old M dwarf Proxima Centauri. Although the rising spectra suggest high fluxes at millimeter wavelengths, the limited sensitivity available in that domain has not yielded definitive detections; tentative measurements (Doyle & Mathioudakis 1991; Mullan *et al.* 1992) suggested extremely high $\dot{M}$ revisited by Lim & White (1996), Lim *et al.* (1996) and van den Oord *et al.* (1997) who

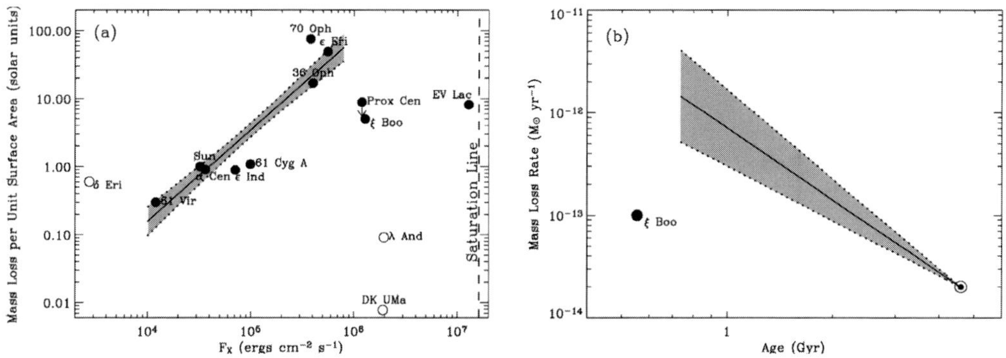

Figure 1. Left (a): Mass-loss rates per unit surface area vs. stellar X-ray surface fluxes inferred indirectly from Lyα analysis. MS stars are shown by filled circles. The trend for inactive stars (shaded area) is not followed by more active stars. – Right (b): Inferred mass-loss history of the Sun. The trend shown for inactive stars (shaded area) breaks down for the youngest, most active stars (from Wood *et al.* 2005), reproduced by permission of AAS.

argued against a wind interpretation. Interesting upper limits ($\dot{M} < 10^{-10}$–$10^{-9} M_{\odot}$ yr^{-1}) were also obtained for A and F-type stars, indicating that their mass loss is not critically altering their evolution (Brown *et al.* 1990). Gaidos *et al.* (2000) specifically focused on *young solar analogs.* The most stringent upper limits were set for the G stars π^1 UMa, κ^1 Cet and β Com, namely $\dot{M} < 4-5 \times 10^{-11}$ $M_{\odot}$ yr^{-1} (3σ), constraining the total mass loss in the solar history after arrival on the ZAMS to $\approx 6\%$ of $M_{\odot}$ (Gaidos *et al.* 2000).

Another method to detect ionized winds directly is the search for charge exchange signatures due to interactions between wind ions and neutrals from the interstellar medium within the "astrospheres" (see below) of nearby stars. This method has provided a tight upper limit of $\dot{M} < 3 \times 10^{-13}$ $M_{\odot}$ yr^{-1} for Proxima Centauri (Wargelin *et al.* 2002).

The most promising (indirect) approach to date makes use of Lyα absorption in so-called "astrospheres". When solar/stellar winds collide with the interstellar medium, they form, with increasing distance from the star, a termination shock (where the wind is shocked to subsonic speeds), a heliospause (separating the plasma flows from the star and the ISM), and the bow shock (where the ISM is shocked to subsonic speeds). The heliosphere is permeated by interstellar H I with $T \approx (2-4) \times 10^4$ K (Wood *et al.* 2002). Much of this gas is piled up between the heliospause and the bow shock, forming the so-called "hydrogen wall" that can be detected as an absorption signature in the Lyα line. The excess absorption from the Sun's own hydrogen wall is, due to the deceleration of the ISM relative to the star, redshifted, while that of other astrospheres is blueshifted.

The measurable absorption depths in Lyα are compared with results from hydrodynamic model calculations (Wood *et al.* 2002, Wood *et al.* 2005). The amount of astrospheric absorption should scale with the wind ram pressure, $P_{\mathrm{w}} \propto \dot{M}_{\mathrm{w}} V_{\mathrm{w}}$, where V_{w} is the (unknown) wind velocity (Wood & Linsky 1998). The latter is usually assumed to be the same as the solar wind speed, i.e., a few 100 km s^{-1}. Consequently, $\dot{M}_{\mathrm{w}}$ can be derived. A systematic study of of a sample of nearby stars at different activity levels shows that $\dot{\mathcal{M}}_{\mathrm{w}}$ per unit stellar surface correlates with the stellar X-ray surface flux,

$$\dot{\mathcal{M}}_{\mathrm{w}} \propto F_{\mathrm{X}}^{1.34\pm0.18} \tag{2.3}$$

(an equivalent relation therefore holds between $\dot{M}_{\mathrm{w}}$ and L_{X}); using the activity-age relation (see below), one finds

$$\dot{\mathcal{M}}_{\mathrm{w}} \propto t^{-2.33\pm0.55} \tag{2.4}$$

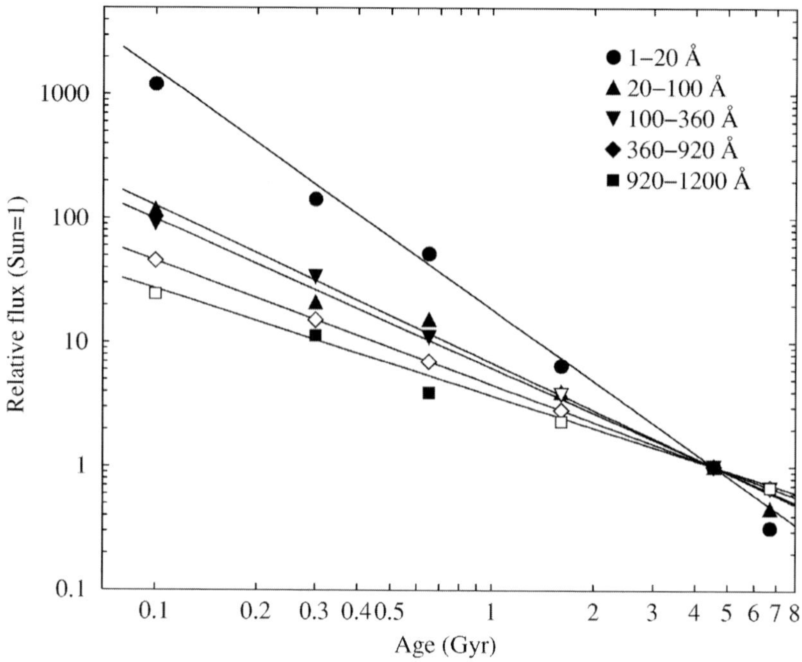

Figure 4. Power-law decays in time for various spectral ranges, normalized to the present-day solar flux. Note that the hardest emission decays fastest (from Ribas *et al.* 2005, reproduced by permission of AAS).

$$L_{\rm EUV} \propto P^{-2.0} \propto t^{-1.2} \tag{3.4}$$

$$L_{\rm X} \propto P^{-3.2} \propto t^{-1.9} \tag{3.5}$$

4. Flares and the highest energy photons

Coronal magnetic flares heat plasma explosively to $\gtrsim 10$ MK on time scales of minutes, and radiate at elevated levels for up to many hours. In a standard model of solar flares, a magnetic reconnection process taking place high in the corona accelerates electrons to (at least) hundreds of keV. Electron beams impact in the lower atmosphere where they deposit their energy in the chromospheric gas. Part of the energy is radiated in hard X-rays (HXR) and gamma rays. As a consequence of the heating, the high-pressure gas evaporates into the magnetic loops of the corona, where it loses its energy by radiation and conduction. This is the main phase of the soft X-ray flare. Some flares, in particular the most energetic examples, are accompanied by ejection of plasma clouds, the "Coronal Mass Ejections" (CMEs) that join the stellar wind at high velocities.

Most of these flare signatures can be observed on stars, the most crucial exception being the non-thermal HXR component for which sensitive detectors are not yet available. Osten *et al.* (2007) reported a case of a giant flare on an RS CVn binary detecting photons up to 100 keV with a spectrum compatible with a power-law although a thermal bremsstrahlung continuum from a superhot (> 100 MK) plasma component cannot be excluded. The HXR radiation was unusual in that it was recorded throughout the gradual phase of the flare. If interpreted as a power-law, the spectrum indicates an energy power-law index for the colliding electrons of $\delta \approx 3$, a value very similar to those inferred from solar HXR flares.

Flares may solve a long-standing riddle in solar physics, namely the problem of coronal heating. Many flares are too small to be recognized individually in disk-integrated light curves, but these are the most frequent events and may thus contribute significantly to coronal heating. This assumption is known as the "microflare" or "nanoflare" hypothesis in solar physics (Parker 1988). In the solar case, hard-X-ray studies have shown that flares are distributed in energy according to a power law,

$$\frac{dN}{dE} = kE^{-\alpha} \tag{4.1}$$

where dN is the number of flares per unit time with a total energy in the interval $[E, E+dE]$, and k is a constant. If $\alpha \geqslant 2$, then the energy integration (for a given time interval, $\int_{E_{\min}}^{E_{\max}} E[dN/dE]dE$) diverges for $E_{\min} \to 0$, that is, if the power law is extrapolated to small flare energies, a lower cut-off is required for the power-law distribution. On the other hand, any energy release power is possible depending on the value of $E_{\min}$.

Such distributions apply equally to stars. Studies have used X-rays and extreme ultraviolet radiation to find α mostly in the range of 2–2.5 for G–M dwarfs (Audard *et al.* 2000; Kashyap *et al.* 2002; Güdel *et al.* 2003) and for T Tauri stars (Stelzer *et al.* 2007), supporting the view that moderate flares are the dominant heating source for these active coronae. This observation may indeed be the basis for the correlation between radio and soft X-ray emission in magnetically active stars and flares, the high level of apparently constant emission being the result of a large number of comparatively small flares not individually detected in the light curves (Telleschi *et al.* 2005).

There is further evidence supporting this model. Specifically, the rate of EUV flares above a given threshold scales linearly with the average X-ray luminosity of a late-type star (Audard *et al.* 2000), a relation that was suggested to apply to T Tauri stars as well (Stelzer *et al.* 2007).

The statistical flare model has some implications for the production of HXR and gamma-ray emission as well as energetic particles and CMEs. As seen in the Sun, accelerated particles produce HXR bursts in the impulsive flare phase, with a total HXR energy scaling non-linearly with the soft X-ray energy (Battaglia *et al.* 2005). Combining this scaling with Eq. (4.1) predicts a quasi-steady, average HXR luminosity of a few times 10^{-5} L_X (Güdel 2009, see also Caramazza *et al.* 2009). This radiation could contribute to heating and ionization of upper planetary atmospheres, with a larger penetration depth than soft X-rays or EUV photons.

5. High-energy radiation and the stellar environment

The ionized and magnetized solar wind interacts strongly with planetary magnetospheres and atmospheres, inducing such phenomena as aurorae or "ionospheric disturbances", but at a more fundamental level, wind-planet interactions have played an important role in the formation and evolution of planetary atmospheres and consequently the formation of life on Earth. Interactions between the upper planetary atmosphere/ionosphere and the wind or related electric and magnetic fields lead to continuous atmospheric erosion as ions are carried away. The most crucial processes are related to ion pick-up, sputtering, ionospheric outflow, and also dissociative recombination; the ionizing source for the upper planetary atmosphere is the EUV and X-ray emission from the Sun (for a summary see, e.g., Chassefière & Leblanc 2004, Kulikov *et al.* 2007, and Lundin *et al.* 2007). These processes are thought to have fundamentally altered the atmospheres of Venus, Earth, and Mars, leading to strong loss of oxygen (Lammer *et al.* 2006) and therefore water (after photodissociation of water vapor in the upper atmosphere) from young Venus and Mars. A strong magnetosphere around the Earth, shielding the lower atmospheric layers, may have been instrumental in retaining much of its water and making it habitable. Nevertheless, the amount of shielding depends on the strength of the wind and its ability to compress the magnetosphere.

Planets such as "hot Jupiters", but also planets in the close-in "habitable zones" around low-mass M dwarfs, rapidly evolve toward slow, tidally locked rotation, i.e., a rotation period equal to the orbit period of several days. As a consequence, the dynamo-generated magnetic moment will be comparatively small, and the ram pressure of the solar/stellar wind will compresses the magnetosphere sufficiently to expose the upper atmosphere to the solar/stellar wind; this effect is particularly strong if the upper atmosphere is heated and expanded by strong EUV/X-ray flux from a magnetically active central star. Magnetic protection of the atmosphere is thus strongly reduced (Grießmeier *et al.* 2005; Lammer *et al.* 2007). At the same time, the cosmic ray flux above the planetary atmosphere will be enhanced given the weaker and smaller magnetosphere (Grießmeier *et al.* 2005).

Coronal mass ejections (CME), often accompanying large flares, add significantly to atmospheric erosion. An ejection rate of several CMEs per day will essentially act like an enhanced solar wind, which will further compress planetary magnetospheres and erode the upper atmospheres (Khodachenko *et al.* 2007). The combination of enhanced exospheric heating by EUV irradiation with consequent exospheric expansion and a CME "wind" could induce atmospheric losses of up to tens or hundreds of bars for close-in planets. This may again be particularly important for planets in the habitable zones around M dwarfs (Lammer *et al.* 2007).

Increased wind mass loss rates may solve the *Faint Young Sun Paradox* (FYSP) emerging from evidence that the young Earth (and Mars) maintained a mild climate and carried liquid surface water although standard evolutionary models indicate that the zero-age main-sequence (ZAMS) Sun was bolometrically 30% fainter than at present (leading to global freeze-out of water). A higher mass loss rate in the Sun's early evolution would suggest a higher initial mass, M_i, and therefore higher bolometric luminosity (Whitmire *et al.* 1995; Sackmann *et al.* 2003). M_i is bounded by $1.07 M_\odot$ to prevent the young Earth from losing its water reservoir through a moist greenhouse (Sackmann & Boothroyd 2003 and references therein, see Fig. 5). But to allow liquid water on the Martian surface *even in the presence of an extreme CO_2 greenhouse*, $M_i \gtrsim 1.03 - 1.04\ M_\odot$ is required, depending on the functional form of the mass-loss history (Sackmann *et al.* 2003; Whitmire *et al.* 1995). This would require $\dot{M}_i \approx 10^{-11} - 10^{-10}\ M_\odot\ \mathrm{yr}^{-1}$ (Sackmann *et al.* 2003), steadily declining to present-day levels according to Eq. (2.1).

The required mass loss rates are in a range that is becoming accessible to radio-astronomical observations. As mentioned in Sect. 2, Gaidos *et al.* (2000) constrained the total mass loss in the solar history to $\approx 6\%$ of $M_\odot$ using upper limits to radio fluxes from young solar analogs and a power-law decay law of the form given in Eq. (2.1), i.e., a wind-decay law following $\dot{M} \propto t^{-1}$.

Mass loss rates derived indirectly using the Lyα method (see Sect. 2) are too small for the above requirement. Following the results from the Wood *et al.* studies, the integrated mass loss during the past 4.5 Gyr history of the Sun would amount to only $0.003 M_\odot$. We note, however, that the apparent breakdown of the observed relation for very active (young) stars remains unexplained, and that the Lyα method relies on a rather indirect interpretation that should be confirmed using other approaches.

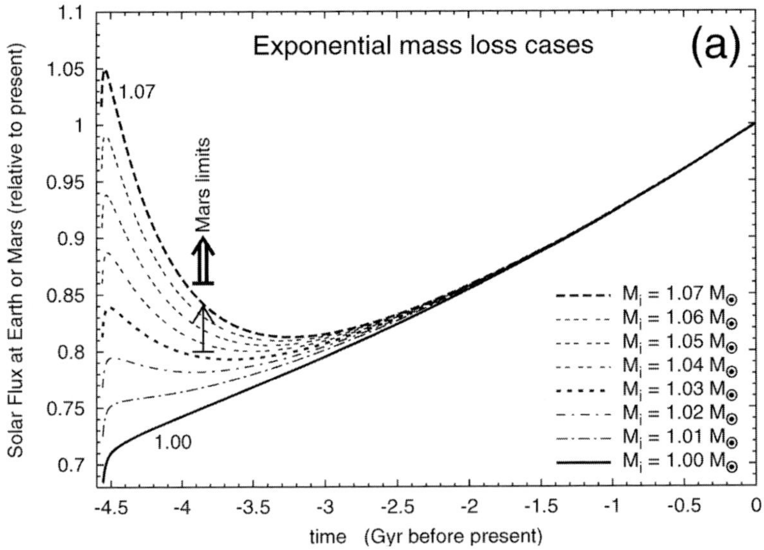

Figure 5. Solar flux in time with respect to present, assuming that the young Sun was subject to increased mass loss. Different curves show calculations for different initial masses, i.e., different mass loss rates that evolve to a mass of $1M_\odot$ after 4.6 Gyr. It has been assumed that the mass-loss rate declines exponentially in time. The flux increase at intermediate to high ages corresponds to the evolutionary increase of the Sun for nearly constant mass. The double arrow indicates the lower limit required for the presence of liquid water on early Mars, the thin arrow an (unrealistic) extreme lower limit. Preservation of water on Earth requires the solar flux to remain below 1.1 times the present-day flux (from Sackmann *et al.* 2003, reproduced by permission of AAS).

6. Summary

While the young Sun was bolometrically 30% fainter than the present-day Sun, its magnetic activity has declined continuously as it has been spinning down due to mass loss in a magnetized wind. As a consequence, magnetically induced short-wavelength radiation has decreased by orders of magnitude in particular at the shortest (X-ray) wavelengths. Similar trends apply to lower-mass stars although their magnetic evolution is slower. Young planets were subject to much higher XUV irradiation, inducing enhanced photochemistry, ionization and heating of their upper atmospheres, with serious consequences for the presence of water and habitability. Another key feature of active stars is their ionized wind that has also been suggested to steadily decrease during a star's MS evolution. It further contributes to erosion of planetary atmospheres. On the other hand, a more massive young Sun undergoing much stronger mass loss may have prevented the young Earth and Mars from a deep freeze.

References

Audard, M., Güdel, M., Drake, J. J., & Kashyap, V. L. 2000, *ApJ*, 541, 396
Ayres, T. R. 1997, *J. Geophys. Res.*, 102, 1641
Battaglia, M., Grigis, P. C., & Benz, A. O. 2005, *A&A* 439, 737
Brown, A., Veale, A., Judge, P., Bookbinder, J. A., & Hubeny, I. 1990, *ApJ*, 361, 220
Caffee, M. W., Hohenberg, C. M., Swindle, T. D., & Goswami, J. N. 1987, *ApJ*, 313, L31
Caramazza, M., Drake, J. J., Micela, G., & Flaccomio, E. 2009, *A&A*, 503, 505
Chassefière, E. & Leblanc, F. 2004, *Planet. Space Sci.*, 52, 1039
Doyle, J. G. & Mathioudakis, M. 1991, *A&A*, 241, L41

Drake, S. A., Simon, T., & Brown, A. 1993, *ApJ*, 406, 247
Gaidos, E. J., Güdel, M., & Blake, G. A. 2000, *Geophys. Res. Lett.*, 27, 501
Güdel, M. 2009, in Simbol-X: Focusing on the Hard X-Ray Universe, AIP Conf. Proc. 1126, 341
Güdel, M., Audard, M., Kashyap, V. L., Drake, J. J., & Guinan, E. F. 2003, *ApJ*, 582, 423
Guinan, E. F. & Ribas, I. 2002, in ASP Conf. Ser. 269 Eds. Montesinos, B., Giménez, A., & Guinan, E. F. (San Francisco: ASP), 85
Grießmeier, J.-M., Stadelmann, A., Motschmann, U., Belisheva, N. K., Lammer, H., & Biernat, H. K. 2005, *Astrobiology* 5, 587
Kashyap, V. L., Drake, J. J., Güdel, M., & Audard, M. 2002, *ApJ*, 580, 1118
Kasting, J. F., Whitmire, D. P., & Reynolds, R. T. 1993, *Icarus*, 101, 108
Khodachenko, M. L., Ribas, I., Lammer, H., Grießmeier, J.-M., Leitner, M., *et al.* 2007, *Astrobiology* 7, 167
Kulikov, Y. N., Lammer, H., Lichtenegger, H. I. M., Penz, T., Breuer, D., *et al.* 2007, *Space Sci. Rev.*, 129, 207
Lammer, H., Lichtenegger, H. I. M., Biernat, H. K., Erkaev, N. V., Arshukova, I. L., *et al.* 2006, *Planet. Space Sci.*, 54, 1445
Lammer, H., Lichtenegger, H. I. M., Kulikov, Y. N., Grießmeier, J.-M., Terada, N., *et al.* 2007, *Astrobiology* 7, 185
Lim, J. & White, S. M. 1996, *ApJ*, 462, L91
Lim, J., White, S. M., & Slee, O. B. 1996, *ApJ*, 460, 976
Lundin, R., Lammer, H., & Ribas, I. 2007, *Space Sci. Rev.*, 129, 245
Mestel, L. 1984, in *Cool Stars, Stellar Systems, and the Sun: Third Cambridge Workshop*, Eds. Baliunas, S. L., & Hartmann, L. (New York: Springer) 49
Mullan, D. J., Doyle, J. G., Redman, R. O., & Mathioudakis, M. 1992 *ApJ*, 397, 225
Osten, R. A., Drake, S., Tueller, J., Cummings, J., Perri, M., *et al.* 2007, *ApJ*, 654, 1052
Parker, E. N. 1988, *ApJ*, 330, 474
Pizzolato, N., Maggio, A., Micela, G., Sciortino, S., & Ventura, P. 2003, *A&A*, 397, 147
Randich, S. 2000, in Stellar Clusters and Associations: Convection, Rotation, and Dynamos, eds. R. Pallavicini, G. Micela, & S. Sciortino, (San Francisco: ASP), 401
Ribas, I., Guinan, E. F., Güdel, M., & Audard, M. 2005, *ApJ*, 622, 680
Scalo, J., Kaltenegger, L., Segura, A. G., Fridlund, M., Ribas, I., *et al.* 2007, *Astrobiology*, 7, 85
Sackmann, I.-J. & Boothroyd, A. I. 2003, *ApJ*, 583, 1024
Soderblom, D. R., Stauffer, J. R., MacGregor, K. B., & Jones, B. F. 1993, *ApJ*, 409, 624
Stelzer, B., Flaccomio, E., Briggs, K., Micela, G., Scelsi, L., *et al.* 2007, *A&A*, 468, 463
Telleschi, A., Güdel, M., Briggs, K., Audard, M., Ness, J.-U., & Skinner, S. L. 2005, *ApJ*, 622, 653
van den Oord, G. H. J. & Doyle, J. G. 1997, *A&A*, 319, 578
Wargelin, B. J. & Drake, J. J. 2002, *ApJ*, 578, 503
West, A. A., Hawley, S. L., Bochanski, J. J., Covey, K. R., Reid, I. N., *et al.* 2008, *AJ*, 135, 785
Whitmire, D. P., Doyle, L. R., Reynolds, R. T., & Matese, J. J. 1995, *J. Geophys. Res.*, 100, 5457
Wood, B. E. & Linsky, J. L. 1998, *ApJ* 492, 788
Wood, B. E., Müller, H.-R., Zank, G. P., & Linsky, J. L. 2002, *ApJ*, 574, 412
Wood, B. E., Müller, H.-R., Zank, G. P., Linsky, J. L., Redfield, S. 2005, *ApJ*, 628, L143

Solar and Stellar Variability: Impact on Earth and Planets
Proceedings IAU Symposium No. 264, 2009
A. G. Kosovichev, A. H. Andrei & J.-P. Rozelot, eds.

doi:10.1017/S1743921309992961

Stellar activity and magnetic shielding

J.-M. Grießmeier[1], M. Khodachenko[2], H. Lammer[2], J. L. Grenfell[3], A. Stadelmann[4] and U. Motschmann[4]

[1]ASTRON, Postbus 2, 7990 AA, Dwingeloo, The Netherlands
email: griessmeier@astron.nl

[2]Space Research Institute, Austrian Academy of Sciences, Schmiedlstr. 6, A-8042 Graz, Austria

[3]Zentrum für Astronomie und Astrophysik, Technische Universität Berlin (TUB), Hardenbergstr. 36, 10623 Berlin, Germany

[4]Technical University of Braunschweig, Mendelssohnstraße 3, 38106 Braunschweig, Germany

Abstract. Stellar activity has a particularly strong influence on planets at small orbital distances, such as close-in exoplanets. For such planets, we present two extreme cases of stellar variability, namely stellar coronal mass ejections and stellar wind, which both result in the planetary environment being variable on a timescale of billions of years. For both cases, direct interaction of the streaming plasma with the planetary atmosphere would entail servere consequences. In certain cases, however, the planetary atmosphere can be effectively shielded by a strong planetary magnetic field. The efficiency of this shielding is determined by the planetary magnetic dipole moment, which is difficult to constrain by either models or observations. We present different factors which influence the strength of the planetary magnetic dipole moment. Implications are discussed, including nonthermal atmospheric loss, atmospheric biomarkers, and planetary habitability.

Keywords. astrobiology, planetary systems, magnetic fields, stars: winds, outflows, Sun: coronal mass ejections (CMEs), cosmic rays

1. Introduction

One of the many fascinating questions in the field of exoplanet studies is the search for habitable worlds. Because of their relatively small mass, low luminosity, long lifetime and large abundance in the galaxy, M dwarfs are sometimes suggested as prime targets in searches for terrestrial habitable planets (Tarter *et al.* 2007; Scalo *et al.* 2007). Interestingly, M dwarfs also seem to have a larger number of Super-Earth planets (i.e. planets with a mass smaller than 10 terrestrial masses, i.e. $M \lesssim 10 M_{\mathrm{E}}$) than more massive stars (Forveille *et al.* 2009). Excluding planets orbiting around pulsars, the first super-Earth detected was GJ 876d, a planet with $\sim 7.5 M_{\mathrm{E}}$ orbiting an M star (Rivera *et al.* 2005). Thereafter, other super-Earth planets have been discovered (Beaulieu *et al.* 2006; Lovis *et al.* 2006; Udry *et al.* 2007; Ribas *et al.* 2008; Mayor *et al.* 2009; Forveille *et al.* 2009; Howard *et al.* 2009; Bouchy *et al.* 2009). In total, 12 planets with masses $\lesssim 10 M_{\mathrm{E}}$ are reported today, and more detections are expected for the near future (Howard *et al.* 2009; Mayor *et al.* 2009)†. The least massive planet has a mass of 4.2 M_{E}.

For M stars the habitable zone (HZ, defined by the range of orbital distances over which liquid water is possible on the planetary surface, see e.g. Kasting *et al.* 1993) is much closer to the star (depending on stellar mass, but typically $\leqslant$0.3 AU). Such close-in distances pose unique problems and constraints to habitability. For example, such planets are exposed to strong stellar winds in the early phases of the host star's evolution. Also, one

† Up-to-date numbers can be found at the Extrasolar Planets Encyclopaedia: http://www.exoplanet.eu

expects a large number of coronal mass ejections (CMEs) on active M stars (Houdebine *et al.* 1990; Scalo *et al.* 2007). The related interaction of the dense plasma flux (either from the stellar wind or a CME) with the atmosphere/magnetosphere environment of the exposed planets during the active stage of the stellar evolution could be strong enough to erode the atmosphere or the planets' water inventory via non-thermal atmospheric loss processes (Khodachenko *et al.* 2007; Lammer *et al.* 2007). Weak magnetic shielding of tidally locked planets will also lead to an increased influx of galactic cosmic rays and stellar energetic particles (Grießmeier *et al.* 2005, 2009), which can have important consequences for biological systems, both directly and indirectly via the modification of the planetary atmosphere (e.g. destruction of atmospheric ozone, see Grenfell *et al.* 2007, 2010).

This paper is organised as follows: The evolution of the conditions in the stellar vicinity over the timescale of the stellar evolution is presented in Section 2. Two cases are considered: stellar CMEs (Section 2.1), and the average stellar wind (Section 2.2). Planetary magnetic shielding against the influence of the star is discussed in Section 3. Consquences are discussed in Section 4, with special focus on the potential evaporation of the planetary atmosphere (Section 4.1) and the flux of cosmic rays to the planetary atmosphere (Section 4.2). Section 5 closes with a few concluding remarks.

2. Stellar variability

Stars are known to be variable in may different ways and on many different timescales. Effects such as stellar CMEs, flares, starspots and stellar rotation occur on timescales of hours or days, while stellar magnetic cycles are expected on the timescale of decades or even centuries. Stellar irradiation, the stellar CME activity and the stellar wind parameters evolve very slowly over Gyr. Many of these effects are discussed elsewhere in this volume. In this section we focus on the slow evolution of two of the stellar parameters over the stellar evolutionary timescale, namely the evolution of stellar CME parameters and of the average stellar wind.

2.1. *Stellar CMEs*

There are two reasons why in addition to the steady stellar wind CMEs have to be taken into account for the interaction with the planetary atmosphere: (a) Due to their limited propagation distance, CMEs are more frequent for close-in exoplanets than at larger distances, and (b) CMEs can be much denser and faster than the average stellar wind, thus exerting a considerable influence on the planet.

At 1 AU, CME collisions with the terrestrial magnetosphere are relatively rare events. However, this is due to the fact that most CMEs cannot be detected at orbital distances $\geqslant 0.1$ AU. Only $\leqslant$20% of CMEs can reach distances of ≈ 1 AU and more. Thus, for a planet at 0.1 AU, where all CMEs have to be taken into account, the number of CME-planet collisions is expected to be considerably higher (Khodachenko *et al.* 2007).

An estimation of CME parameters (combining in-situ measurements near the sun (e.g. by Helios) with remote solar observation by SoHO, Khodachenko *et al.* 2007) has shown that for small orbital distances, CMEs can be considerably denser and faster than the solar wind (roughly one order of magnitude denser and half an order of magnitude faster). Here, we assume that stellar CMEs behave similarly to solar ones ("solar-stellar analogy", see Khodachenko *et al.* 2007). If CME-planet collisions are frequent enough (such that the planet effectively moves through a constant background of CME plasma), the interaction of the planetary atmosphere with stellar CMEs can be estimated by replacing the stellar wind parameters $n(d)$ and $v(d)$ by the corresponding parameters

$n_{\rm CME}(d)$ and $v_{\rm CME}(d)$ for strong CMEs as given by Khodachenko *et al.* (2007). The consequences for atmospheric erosion will be discussed in Section 4.1.

2.2. *Stellar wind*

The stellar wind density n and velocity v encountered by a planet are key parameters defining the size of the magnetosphere. As these stellar wind parameters depend on the stellar age, the quality of magnetic shielding is strongly time dependent.

For stellar ages > 0.7 Gyr, the radial dependence of the stellar wind properties is well described by a Parker-like (Parker 1958) stellar wind model. In this model, the interplay between stellar gravitation and pressure gradients leads to a supersonic gas flow for sufficiently large substellar distances d. The free parameters are the coronal temperature and the stellar mass loss. They are indirectly chosen by setting the stellar wind conditions at 1 AU. More details on the model can be found elsewhere (e.g. Parker 1958; Mann *et al.* 1999; Prölss 2004; Grießmeier *et al.* 2007).

The dependence of the stellar wind density n and velocity and v on the age of the stellar system is measured by observing astrospheric absorption features of stars with different ages. The characteristic Lyα absorption feature (at 1216 Å) created by neutral hydrogen at the astropause was detected in high-resolution observations obtained by the Hubble Space Telescope (HST). Comparing the measured absorption to that calculated by hydrodynamic codes, these measurements allowed the first empirical estimation of the evolution of the stellar mass loss rate as a function of stellar age (Wood *et al.* 2002; Wood 2004; Wood *et al.* 2005). It should be noted, however, that the resulting estimates are only valid for stellar ages $t \geqslant 0.7$ Gyr (Wood *et al.* 2005). These results for the mass loss rate can be combined with the model for the age-dependence of the stellar wind velocity of Newkirk (1980) to obtain the age-dependent stellar wind density. One thus finds (Grießmeier *et al.* 2007):

$$v(1\text{AU}, t) = v_0 \left(1 + \frac{t}{\tau}\right)^{-0.43}, \tag{1}$$

and

$$n(1\text{AU}, t) = n_0 \left(1 + \frac{t}{\tau}\right)^{-1.86\pm0.6}, \tag{2}$$

with $v_0 = 3971$ km/s, $n_0 = 1.04 \cdot 10^{11}$ m^{-3} and $\tau = 2.56 \cdot 10^7$yr. The dependence of $n(1\text{AU}, t)$ and $v(1\text{AU}, t)$ as a function of stellar age is shown in Figs. 1(a) and 1(b) for the case of a sun-like star.

The procedure to obtain the stellar wind velocity $v(d, M_\star, R_\star)$ and density $n(d, M_\star, R_\star)$ at the location of an exoplanet (i.e. at distance d) for a host star of given mass $M_\star$, and radius $R_\star$ and age t from $v(1\text{AU}, t)$ and $n(1\text{AU}, t)$ is described in Grießmeier *et al.* (2005, 2009).

3. Magnetic shielding

For an efficient magnetic shield to exist, the planet requires a strong magnetic field. A necessary condition for a magnetic dynamo is a large, electrically conducting fluid region in non-uniform motion (i.e. the liquid outer core for terrestrial planets, or a layer of electrically conducting hydrogen for gas giants). According to planetary dynamo theory, this flow should be convective in nature (Stevenson 1983). In practice, all planets with core convection can be assumed to have a magnetic dynamo (Stevenson 1983). Conversely, convection can be regarded as a necessary requirement for a planetary magnetic field.

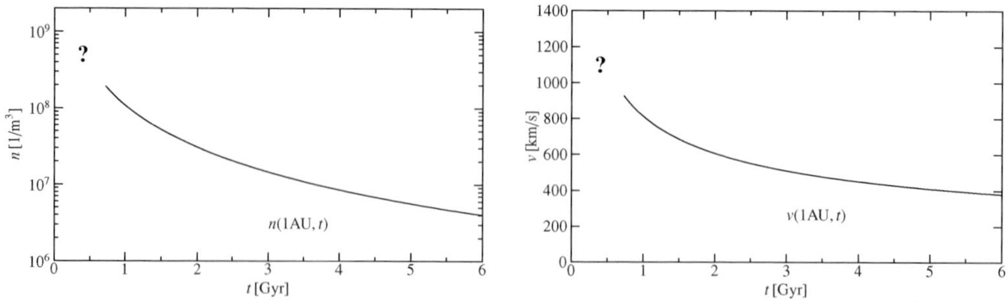

(a) Stellar wind density $n(1\mathrm{AU}, t)$. (b) Stellar wind velocity $v(1\mathrm{AU}, t)$.

Figure 1. Stellar wind density $n(t)$ and velocity $v(t)$ at 1 AU distance of a sun-like star ($M_\star = M_\odot$, $R_\star = R_\odot$), as a function of stellar age.

Convection is assumed to require plate tectonics in order to operate. The reason for this is that plate tectonics helps to cool the planetary interior more efficiently than a stagnant lid configuration (Stevenson 2003). Efficient cooling is required to maintain convection, and is thus essential to keep the action of a magnetic dynamo (Stevenson 1983). The lack of plate tectonics is also suggested as one of the possible reasons for the lack of an intrinsic magnetic field on Venus and Mars (Stevenson *et al.* 1983; Stevenson 2003). Thus, plate tectonics may be required to keep an Earth-like magnetic dynamo in operation over geological time spans (Lammer *et al.* 2009).

There are two reasons why the requirement for plate tectonics for convection might be important as far as exoplanets are concerned: (a) For close-in exoplanets, the small orbital distance usually means that the planet is tidally locked (e.g. Grießmeier *et al.* 2009). It is currently not clear how this influences plate tectonics, but it is sometimes suggested that tidal locking may inhibit plate tectonics partially or completely. (b) For Super-Earth planets, is it still being debated whether plate tectonics is likely to occur or not (Valencia *et al.* 2007; O'Neill *et al.* 2007).

Thus, we see that plate tectonics is essential for convection, which is a necessary ingredient for magnetic dynamo action. In the following, we will assume that a magnetic dynamo does exist. However, even if convection is present and a magnetic dynamo exists, the magnetic moment is not necessarily large enough to efficiently shield the planet. The tidally locked state of close-in exoplanets means that they have low rotation rates, which – according to scaling laws derived from theoretical considerations – leads to a reduction in the planetary magnetic moment (e.g. Grießmeier *et al.* 2005)†.

Following the approach of Grießmeier *et al.* (2005), one finds that for the slowly rotating tidally locked planets, the magnetic moment is much smaller than for freely rotating planets. For example, an Earth-like planet in an orbit of 0.2 AU around a star with 0.5 stellar masses, may have a magnetic moment in the range $0.02\mathcal{M}_\mathrm{E} < \mathcal{M} < 0.15\mathcal{M}_\mathrm{E}$ ($\mathcal{M}_\mathrm{E}$ is the value of Earth's current magnetic moment). In the following, the minimum value will be adopted to estimate an upper limit for atmospheric espace (Sect. 4.1), but for the cosmic ray flux (Sect. 4.2), $\mathcal{M}$ is assued to take the maximum allowed value.

† While the analytical models considered here predict a decrease of magnetic moment with decreasing rotation rate, numerical experiments indicate that the magnetic moment may be independent of the angular frequency. Christensen & Aubert (2006) and Olson & Christensen (2006) have studied numerically an extensive set of dynamo models, varying the relevant control parameters by at least two orders of magnitude. In these simulations, the magnetic field is mainly controlled by the buoyancy flux. This is in contradiction with the previously discussed concept. Future studies are required to clarify the relation between magnetic field strength, the planetary rotation rate, electrical conductivity, and buoyancy flux.

In the following, the magnetosphere is modelled as a cylinder topped by a half-sphere (Voigt 1981; Stadelmann 2005; Grießmeier *et al.* 2005). The size of the magnetosphere is characterized by the magnetopause standoff distance R_s, i.e. the extent of the magnetosphere along the line connecting the star and the planet. R_s can be obtained from the pressure equilibrium at the substellar point. This pressure balance includes the stellar wind ram pressure, the stellar wind thermal pressure of electrons and protons, and the planetary magnetic field pressure. For a given planetary orbital distance d, only the magnetospheric magnetic pressure is a function of the distance to the planet, while the other contributions are constant. Thus, the standoff distance R_s is found to be (Grießmeier *et al.* 2007):

$$R_s = \left[\frac{\mu_0 f_0^2 \mathcal{M}^2}{8\pi^2 \left(mnv^2 + 2nk_B T \right)} \right]^{1/6} . \tag{3}$$

Here, $f_0 = 1.16$ is the form factor of the magnetosphere and includes the magnetic field caused by the currents flowing on the magnetopause (Voigt 1995; Grießmeier *et al.* 2004). From eq. (3) it is evident that a small magnetic moment $\mathcal{M}$, a high stellar wind velocity v or a high stellar wind density n will lead to a small magnetosphere.

4. Consequences

The weak magnetic shielding of close-in planets against stellar wind of young stars and against stellar CMEs has important consequences for the planet. The implications that will be discussed in the following are strong atmospheric erosion via stellar wind and CMEs, and the enhanced flux of galactic cosmic rays onto the planetary atmosphere.

4.1. *Atmospheric escape*

Here, we explore the region in parameter space where the stellar wind and CMEs can compress the magnetosphere down to a certain altitude level. In the following we chose the critical altitude level to be equal to $R_{\text{crit}} = 1.15R_P = 1.15R_E$ (i.e. strong atmospheric loss when the magnetosphere is compressed to less than $0.15R_E$ = 1000 km above the planetary surface). The numerical study of Lammer *et al.* (2007, e.g. Figure 9) shows that under certain conditions, the critical altitude level may be as high as 2 R_E (i.e. 1 R_E above the planetary surface), so that this is a conservative assumption.

Fig. 2 shows the region in parameter space where the magnetosphere of an exoplanet with mass, radius, and internal parameters identical to those of the Earth can be strongly compressed (i.e. $R_s < R_{\text{crit}}$) under the action of strong (= dense) CMEs or under the action of a young stellar wind, respectively. In the tidally locked regime, the magnetic moment was taken to be equal to the minimum value, but tidal locking was assumed only when the timescale for synchronous rotation is < 100 Myr (denoted as "locked"). For timescales between 100 Myr and 10 Gyr ("potentially locked") and for timescales above 10 Gyr ("unlocked") the magnetic moment was estimated using the (larger) initial rotation rate of the planet.

In Fig. 2(a), it can be seen that almost all Earth-like exoplanets within habitable zones (HZs, grey shaded area) of M-stars will experience this extreme magnetospheric compression. In such cases, CMEs bring the magnetosphere down to the level of the ionopause, and the planetary atmospheres can be heavily eroded (Khodachenko *et al.* 2007; Lammer *et al.* 2007).

Fig. 2(b) is similar to Fig. 2(a), but it studies the effect of the age-dependent stellar wind rather than that of stellar CMEs. Three cases are considered: The stellar wind of a star of 0.7 Gyr age (all three dotted areas), the stellar wind of a star of 1.0 Gyr age

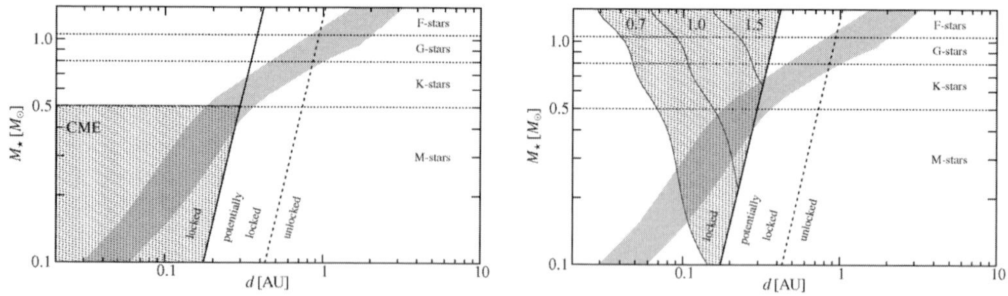

(a) Atmospheric erosion by CMEs. (b) Atmospheric erosion by stellar wind.

Figure 2. Comparison between the HZ (grey shaded area) and the areas where strong magnetospheric compression is possible by CMEs or stellar wind. The dotted area denotes the region where the magnetosphere can be compressed down to 1.15 Earth radii or less (i.e., 1000 km above the planetary surface). 4.1 Magnetospheric compression by strong CMEs. From Khodachenko *et al.* (2007). 4.1 Magnetospheric compression by stellar wind of a young star. Three cases are considered: Stellar wind of a star of 0.7 Gyr age (all three dotted areas), Stellar wind of a star of 1.0 Gyr age (upper two dotted areas), Stellar wind of a star of 1.5 Gyr age (top dotted area only).

(top two dotted areas), and the stellar wind of a star of 1.5 Gyr age (upper dotted area only). One can clearly see that for a young star (0.7 Gyr), all planets in a habitable zone between 0.1 and ~0.3 AU have magnetospheric altitudes lower than the critical altitude level, so that strong erosion is likely. As the star's age increases, the stellar wind becomes more tenuous, and the region of strong erosion in the d-$M_\star$-parameter space shrinks.

An additional effect of stellar age should be mentioned: Not only is the magnetosphere of a close-in planet around a young star strongly compressed, but also the planetary atmosphere is more extended due to the intense heating by the high stellar X-ray and EUV-flux (Lammer *et al.* 2003; Grießmeier *et al.* 2004; Ribas *et al.* 2005). Thus, the critical level the magnetosphere is allowed to reach, $R_{\rm crit}(t)$, in reality is age-dependent, and it is much more likely for a young planetary magnetosphere to reach the critical altitude than one would find using present day stellar irradiation.

From Figures 2(a) and 2(b) it becomes clear that the relatively weak intrinsic magnetic fields of tidally locked planets are not suffient to protect Earth-like exoplanets against significant atmospheric erosion caused by CMEs and the stellar wind. The sharp cutoff at the right edge of the dotted area demonstrates the importance of magnetic shielding: Within the dotted area, planets are tidally locked. They are assumed to have a small magnetic moment, and magnetic protection is weak. To the right of the dotted area however, the planets are assumed to be freely rotating, so that they can maintain a much higher magnetic moment and are well protected against strong atmospheric erosion.

4.2. *Cosmic ray flux*

In order to quantify the protection of extrasolar Earth-like planets against galactic cosmic rays, the motion of galactic cosmic protons through magnetospheres of extrasolar planets has been investigated numerically (Grießmeier *et al.* 2005, 2009). From these calculations, one finds that the cosmic ray flux is strongly enhanced in the case of a weakly magnetized planet. Fig. 3 shows the reference energy spectrum outside the magnetosphere (Seo *et al.* 1994) as a dash-dotted line. The energy spectrum at the of the Earth's atmosphere is shown as a solid line, and the cosmic ray energy spectrum for a weakly magnetized Earth-like exoplanet in an orbit of 0.2 AU around a 0.5 $M_\odot$ K/M star is shown as a dashed line. Due to tidal locking, the magnetic moment of that planet is reduced to (at most) 15 %

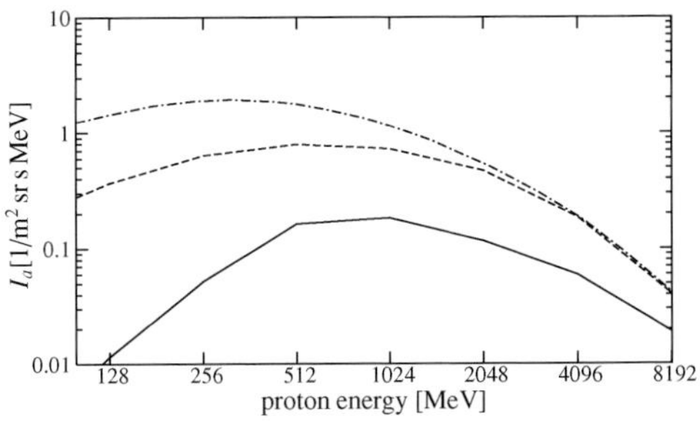

Figure 3. Dash-dotted line: energy spectrum outside the magnetosphere. Dashed line: energy spectrum of cosmic ray protons impacting the atmosphere (100 km above the surface) of an Earth-like exoplanet at 0.2 AU around a 0.5M K/M star. Solid line: energy spectrum of cosmic rays impacting the atmosphere of the Earth. From Grießmeier *et al.* (2005).

of the Earth's magnetic moment (Grießmeier *et al.* 2009, section 2.2), which translates to a strongly enhanced cosmic ray flux to the planetary atmosphere.

The increased flux of cosmic ray protons of galactic origin to the upper atmospheres of M-star planets has implications for the flux of (secondary) cosmic ray particles reaching the ground, which can be expected to have biological implications. It also influences the atmospheric composition and thus the remote detection of biomarkers as well as the surface UV flux.

4.2.1. *Direct biological impact*

While a quantitative treatment is not yet available, the increased flux of cosmic rays to the top of the planetary atmosphere has implications for the flux of secondary radiation to the surface. When galactic cosmic rays of sufficiently high energy reach the planetary atmosphere, they generate showers of secondary cosmic rays. Of the different components of such showers, slow neutrons have the strongest influence on biological systems. The propagation of high energy cosmic ray particles also depends on the composition and density of the planetary atmosphere. For an Earth-like atmosphere, the minimum energy which a proton must have to initiate a nuclear interaction that is detectable at sea level is approximately 450 MeV (Reeves *et al.* 1992; Shea & Smart 2000). Terrestrial planets with dense atmospheres like Venus (100 bar surface pressure) would be better shielded by the planetary atmosphere, so that no secondary radiation can reach the surface. On the other hand, for planets with thin atmospheres like Mars (6 mbar surface pressure), the surface would probably be totally sterilised, even for a relatively low cosmic ray flux.

Keeping the atmospheric pressure constant at $\sim$1 bar, the flux at energies $>$ 450 MeV is strongly increased for a weakly magnetized planet (see Fig. 3), and a large increase of secondary cosmic rays can be expected at the planetary surface.

Biological effects, namely an increase in cell fusion indices for different cell-lines, were found to be significantly correlated with the neutron count rate at the Earth's surface (Belisheva *et al.* 2005; Grießmeier *et al.* 2005). Similar, but much stronger and more diverse effects were observed during large solar particle events, where solar cosmic rays dominate over galactic cosmic rays. The effect of cosmic rays on cells and on DNA are potentially hazardous for life. On the other hand, changes at the genetic level are a necessary condition for biological evolution, so that cosmic rays may also play a role in radiation-induced evolutionary events.

As tidally locked Earth-like exoplanets inside the HZ of K/M stars are only weakly protected against high energetic cosmic rays, they can be expected to experience a higher surface neutron flux and stronger biological effects than Earth-like planets with a strong magnetic field. For this reason, it may be more difficult for life to develop on the surface of planets around such stars.

4.2.2. *Atmospheric modification*

The influence of a high flux of galactic cosmic rays (GCRs) and of stellar energetic particles on the atmospheres of planets in the HZ around M-stars is studied by Grenfell *et al.* (2007) and Grenfell *et al.* (2010). When cosmic ray particles travel through an Earth-like atmosphere, they have sufficient energy to break the strong N_2 molecule, and the chemical products react with oxygen to form nitrogen oxides (NO_x). If UV levels are intense, e.g. corresponding to the upper mesosphere and above on the Earth, then the NO_x cannot survive. NO_x affects biomarkers differently depending on altitude. In a stratospheric environment, NO_x depletes ozone by catalytic cycles (Crutzen 1970). Once ozone is affected, the other biomarkers can change too, because ozone affects the temperature profile (hence e.g. mixing processes, chemical sources and sinks) and UV levels (hence photolysis rates) for other chemical species.

This is important, because atmospheres of extrasolar planets will be studied from the point of view of possible biological activity in the near future. "Atmospheric biomarkers" are compounds present in an atmosphere which imply the presence of life and which cannot be explained by inorganic chemistry alone. The simultaneous presence of significant amounts of atmospheric reducing gases (e.g. methane) and oxidising gases (e.g. oxygen) as on the Earth is also an indicator of biological activity (Lovelock 1965; Sagan *et al.* 1993). Good biomarkers include oxygen (produced by photosynthesis), ozone (mainly produced from oxygen) and nitrous oxide (produced almost exclusively from bacteria). For the study of biomarkers in planetary atmospheres, it is important to know all inorganic effects which can reduce the abundances of these molecules. Modelling studies are required to rule out false conclusions in cases where inorganic chemistry, e.g. reactions triggered by the influx of cosmic ray particles, can mimic (or mask) the presence of life.

Using the cosmic ray spectrum of Fig. 3, Grenfell *et al.* (2007) found that for Earth-like planets in close orbits around M-stars, GCRs may only change the abundances of biomarker molecules (especially water and ozone) by a small factor. This can also be seen by comparing the cases "Earth GCR" and "Exoplanet GCR" in Fig. 4. The case is different for solar energetic particles (Grenfell *et al.* 2010), which can reduce the abundance of ozone in the planetary atmosphere by several orders of magnitude. This effect can be seen in Fig. 4, where "Exoplanet SPE" shows the effect of solar energetic particles (adapted from Grenfell *et al.* 2010). Such changes in atmospheric biomarker concentrations have to be taken into account when searching for biosignatures in the spectra of Earth-like exoplanets by future space missions like DARWIN (Fridlund 2004) or SEE-COAST (Schneider *et al.* 2006).

4.2.3. *Modification of UV shielding*

Both direct UV radiation (Buccino *et al.* 2007; Scalo *et al.* 2007) and indirect UV radiation generated by energetic particles (Smith *et al.* 2004) can have important consequences for biogenic processes on M-star planets. As was shown in Section 4.2.2, the planetary ozone layer of a weakly magnetized planet can be virtually destroyed during a solar energetic particle event. The corresponding enhancement of the UV flux has to be taken into account when biological effects of UV radiation are considered. The influence of UV radiation of different wavelenth on life is also discussed by Cuntz *et al.* (2010).

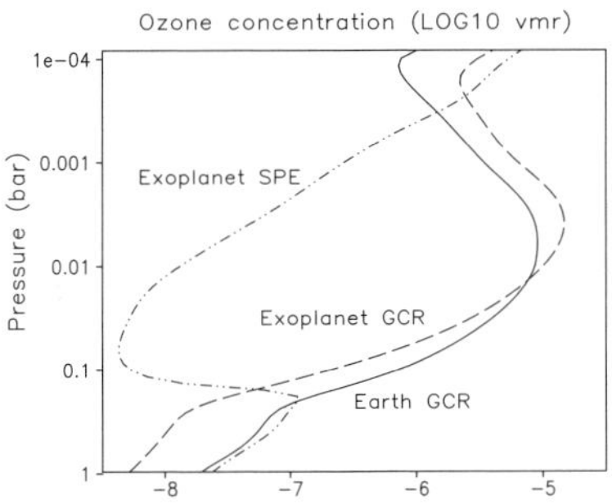

Figure 4. Atmospheric biomarker profile concentrations (log volume mixing ratio): ozone for three different cases (see text). From Grenfell *et al.* (2010).

5. Conclusions

Because of tidal locking, extrasolar planets with orbits in the habitable zone of K/M are likely to have much smaller magnetic moments than more distant planets. Combined with the strong ram pressure of dense and fast stellar winds and the high CME activity expected around young stars, this leads to the conclusion that such planets will have small magnetospheres which offer only limited magnetic protection.

The weak magnetic protection may lead to strong atmospheric erosion during the early stages of stellar evolution, which could be strong enough to erode the atmosphere or the planets' water inventory via non-thermal atmospheric loss processes. Another consequence of the weak magnetic protection is a high flux of galactic cosmic rays and stellar energetic particles to the planetary atmosphere. This has implications for potential habitability, but also for the atmospheric chemistry and composition (e.g. destruction of atmospheric ozone). These effects have also to be considered for missions studying biosignatures in the observed spectra of Earth-like exoplanets.

Overall, the magnetic protection of exoplanets in the habitable zone of K/M stars can be weak, so that these planets may only be weakly shielded against stellar activity.

Acknowledgements

This study was supported by the International Space Science Institute (ISSI) through the ISSI Team "Evolution of Exoplanet Atmospheres and their Characterisation" and by the Helmholtz association through the research alliance "Planetary Evolution and Life". MK acknowledges the Austrian Fond zur Förderung der wissenschaftlichen Forschung (project P21197-N16).

References

Beaulieu, J.-P., Bennett, D. P., Fouqué, P., *et al.* 2006, *Nature*, 439, 437

Belisheva, N. K., Kuzhevskii, B. M., Vashenyuk, E. V., & Zhirov, V. K. 2005, *Doklady Biochemistry and Biophysics*, 402, 254

Bouchy, F., Mayor, M., C., *et al.* 2009, *Astron. Astrophys.*, 496, 527

Buccino, A. P., Lemarchand, G. A., & Mauas, P. J. D. 2007, *Icarus*, 192, 582

Christensen, U. R. & Aubert, J. 2006, *Geophys. J. Int.*, 166, 97

Crutzen, P. J. 1970, *Q. J. R. Meteorol. Soc.*, 96, 320

Cuntz, M., Guinan, E. F., & Kurucz, R. L. 2010, this proceedings, p. 419

Forveille, T., Bonfils, X., Delfosse, X., *et al.* 2009, *Astron. Astrophys.*, 493, 645

Fridlund, C. V. M. 2004, *Adv. Space Res.*, 34, 613

Grenfell, J. L., Grießmeier, J.-M., Patzer, B., *et al.* 2007, *Astrobiology*, 7, 208

Grenfell, J. L., Grießmeier, J.-M., Patzer, B., *et al.* 2010, to be submitted

Grießmeier, J.-M., Preusse, S., Khodachenko, M., *et al.* 2007, *Planet. Space Sci.*, 55, 618
Grießmeier, J.-M., Stadelmann, A., Grenfell, J. L., Lammer, H., & Motschmann, U. 2009, *Icarus*, 199, 526
Grießmeier, J.-M., Stadelmann, A., Motschmann, U., *et al.* 2005, *Astrobiology*, 5, 587
Grießmeier, J.-M., Stadelmann, A., Penz, T., *et al.* 2004, *Astron. Astrophys.*, 425, 753
Houdebine, E. R., Foing, B. H., & Rodonò, M. 1990, *Astron. Astrophys.*, 238, 249
Howard, A. W., Johnson, J. A., Marcy, G. W., *et al.* 2009, *Astrophys. J.*, 696, 75
Kasting, J. F., Whitmire, D. P., & Reynolds, R. T. 1993, *Icarus*, 101, 108
Khodachenko, M. L., Ribas, I., Lammer, H., *et al.* 2007, *Astrobiology*, 7, 167
Lammer, H., Bredehöft, J. H., Coustenis, A., *et al.* 2009, *Astron. Astrophys. Rev.*, 17, 181
Lammer, H., Lichtenegger, H. I. M., Kulikov, Y. N., *et al.* 2007, *Astrobiology*, 7, 185
Lammer, H., Selsis, F., Ribas, I., *et al.* 2003, *Astrophys. J.*, 598, L121
Lovelock, J. E. 1965, *Nature*, 207, 568
Lovis, C., Mayor, M., Pepe, F., *et al.* 2006, *Nature*, 441, 305
Mann, G., Jansen, F., MacDowall, R. J., Kaiser, M. L., & Stone, R. G. 1999, *Astron. Astrophys.*, 348, 614
Mayor, M., Udry, S., Lovis, C., *et al.* 2009, *Astron. Astrophys.*, 493, 639
Newkirk, Jr., G. 1980, in The Ancient Sun: Fossil Record in the Earth, Moon and Meteorites, ed. R. O. Pepin, J. A. Eddy, & R. B. Merrill, 293–320
Olson, P. & Christensen, U. R. 2006, *Earth Planet. Sci. Lett.*, 250, 561
O'Neill, C., Jellinek, A. M., & Lenardic, A. 2007, *Earth Planet. Sci. Lett.*, 261, 20
Parker, E. N. 1958, *Astrophys. J.*, 128, 664
Prölss, G. W. 2004, Physics of the Earth's Space Environment (Berlin: Springer-Verlag)
Reeves, G. D., Cayton, T. E., Gary, S. P., & Belian, R. D. 1992, *J. Geophys. Res.*, 97, 6219
Ribas, I., Font-Ribera, A., & Beaulieu, J.-P. 2008, *Astrophys. J.*, 677, L59
Ribas, I., Guinan, E. F., Güdel, M., & Audard, M. 2005, *Astrophys. J.*, 622, 680
Rivera, E. J., Lissauer, J. J., Butler, R. P., *et al.* 2005, *Astrophys. J.*, 634, 625
Sagan, C., Thomson, W. R., Carlson, R., Gurnett, D., & Hord, C. 1993, *Nature*, 365, 715
Scalo, J., Kaltenegger, L., Segura, A., *et al.* 2007, *Astrobiology*, 7, 85
Schneider, J., Riaud, P., Tinetti, G., *et al.* 2006, in SF2A-2006: Semaine de l'Astrophysique Francaise, ed. D. Barret, F. Casoli, T. Contini, G. Lagache, A. Levacelier, & L. Pagani, 429–432
Seo, E. S., McDonald, F. B., Lal, N., & Webber, W. R. 1994, *Astrophys. J.*, 432, 656
Shea, M. A. & Smart, D. F. 2000, *Space Sci. Rev.*, 93, 187
Smith, D. S., Scalo, J., & Wheeler, J. C. 2004, *Icarus*, 171, 229
Stadelmann, A. 2005, PhD thesis, Technische Universität Braunschweig, ISBN 3-936586-42-X, Copernicus-GmbH Katlenburg-Lindau, URL: http://www.digibib.tu-bs.de/?docid=00000002
Stevenson, D. J. 1983, *Rep. Prog. Phys.*, 46, 555
Stevenson, D. J. 2003, *Earth Planet. Sci. Lett.*, 208, 1
Stevenson, D. J., Spohn, T., & Schubert, G. 1983, *Icarus*, 54, 466
Tarter, J. C., Backus, P. R., Mancinelli, R. L., *et al.* 2007, *Astrobiology*, 7, 30
Udry, S., Bonfils, X., Delfosse, X., *et al.* 2007, *Astron. Astrophys.*, 469, L43
Valencia, D., O'Connell, R. J., & Sasselov, D. D. 2007, *Astrophys. J.*, 670, L45
Voigt, G.-H. 1981, *Planet. Space Sci.*, 29, 1
Voigt, G.-H. 1995, in Handbook of atmospheric electrodynamics, ed. H. Volland, Vol. II (Boca Raton: CRC Press), 333–388
Wood, B. E. 2004, Living Rev. Solar Phys., 1, 2, URL: http://www.livingreviews.org/lrsp-2004-2, accessed on 23 February 2007
Wood, B. E., Müller, H.-R., Zank, G. P., & Linsky, J. L. 2002, *Astrophys. J.*, 574, 412
Wood, B. E., Müller, H.-R., Zank, G. P., Linsky, J. L., & Redfield, S. 2005, *Astrophys. J.*, 628, L143

Solar and Stellar Variability: Impact on Earth and Planets
Proceedings IAU Symposium No. 264, 2009
A. G. Kosovichev, A. H. Andrei & J.-P. Rozelot, eds.

doi:10.1017/S1743921309992973

Detailed characterization of stellar high energy (FUV/EUV/X-ray) radiation fields during protoplanetary system formation

Alexander Brown[1]

[1]Center for Astrophysics and Space Astronomy, University of Colorado, Boulder, CO 80309-0389, USA
email: Alexander.Brown@colorado.edu

Abstract. Young stars undergoing the conversion of pre-main-sequence circumstellar disks into protoplanetary systems are strong sources of high energy (FUV/EUV/X-ray) radiation that controls the physical and chemical processes in their circumstellar environment out to hundreds of AU from the star. The high energy radiation resulting from magnetic activity and accretion onto the central star controls the thermal structure of disks, the formation process of planetesimals, and the photoexcitation and photoionization of protoplanets and young planetary atmospheres. Modeling of the dust and gas evolution requires an accurate understanding of the local radiation field throughout the ultraviolet (UV) and X-ray spectral regions, even those parts of the spectrum that are impossible to observe from Earth.

Our current research efforts are directed towards developing a better understanding of UV (using *HST* and *FUSE*) and X-ray (using *Chandra*, *XMM-Newton*, and *Swift*) stellar activity and the resulting radiation fields during pre-main-sequence evolution from ages of a few to several hundred million years. These studies include extensive UV and X-ray spectral sampling of individual stars in nearby star formation regions and the various moving groups of the Local Association, including our *HST* Cycle 17 Large Project (GO-11616), which is using 111 HST orbits to observe 32 T Tauri stars with the COS UV spectrograph. Most young stars are well over 100 pc from the Sun and are consequently hard to observe in the UV and X-ray regions at even moderate spectral resolution. However, members of the Local Association, whose ages range from 7 Myr to a few hundred Myr, surround the Sun at distances of 50 pc or less and permit the detailed study of the later stages of the early evolution of stellar activity when gas giant and terrestrial protoplanets are forming. We illustrate our methodology using the 12 Myr old early-M dwarf AU Mic, which possesses a striking dust debris disk, as an example.

Keywords. stars: pre–main-sequence, stars: activity, planetary systems: formation, planetary systems: protoplanetary disks

1. Introduction

Protoplanetary systems are highly complex with a magnetically-active central star surrounded by an accretion disk and with the interaction between them driving strong directional outflows. The overall kinematics and energy of the system is controlled by conservation of angular momentum, but with accretion onto star and the outflow geometry constrained by magnetic fields. The physics and chemistry of the circumstellar environment for tens - hundreds of AU around the star are controlled by the stellar radiation field. Pre–main-sequence (PMS) circumstellar disks would be very cold (and uninteresting!) if only subject to stellar photospheric and interstellar radiation fields. However, stellar high energy (X-ray/UV) photons significantly alter the thermal structure of the disk with X-ray and far-UV (FUV — between the H Lyman limit at 912 Å and

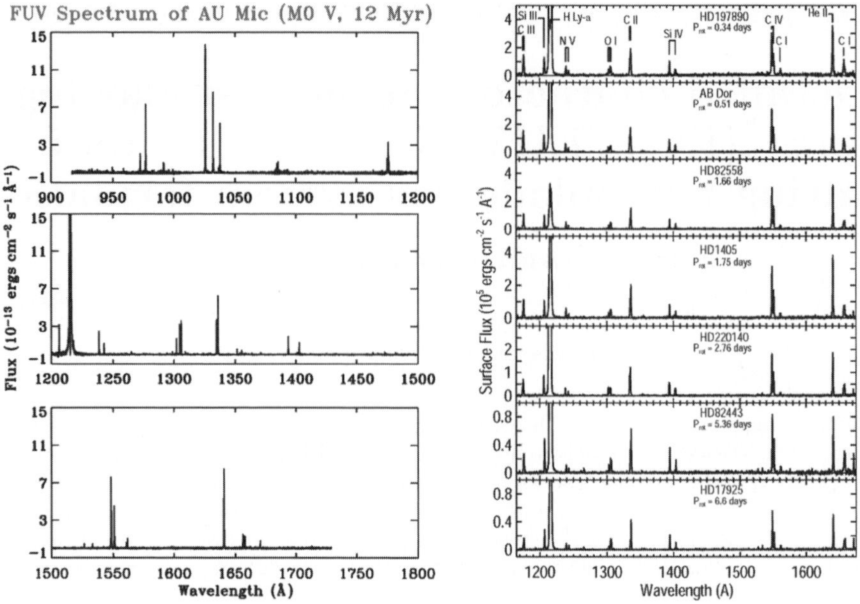

Figure 1. FUV spectra of young active PMS dwarf stars: **(left)** *HST-STIS* and *FUSE* spectra of the 12 Myr old M dwarf AU Mic; **(right)** *HST-GHRS* G140L FUV spectra of a sample of early-K Local Association dwarfs — the stars are ordered by decreasing rotation period rather than by age. Flux in these FUV energy distributions is concentrated in a few strong emission lines and is thus a highly selective photoexcitation source.

$\sim$ 2000 Å) photons penetrating to different depths in the disk and this produces a complex layered thermal structure and regions with different chemical reaction environments (for reviews see Dullemond *et al.* 2007 and Bergin 2009). Over time, circumstellar disks around young, low mass stars transform much of the dust and gas in their inner regions into planetesimals or larger solid bodies. The evolutionary trend is from gas-dominated circumstellar physics to a solid-dominated structure where the gas component exists as icy surfaces on planetesimals and larger bodies. In the latter case gravitational interactions control the evolution (see e.g. Wyatt 2008; Blum & Wurm 2008), but the radiation field controls the icy surface chemistry, both organic and inorganic.

2. Why does the Stellar Ultraviolet and X-ray Radiation Field Matter?

Young PMS stars are strong UV and X-ray emitters due to both strong stellar magnetic activity and the high temperature plasma generated within the accretion shock where magnetically-channeled matter impacts the star (Feigelson & Montmerle 1999). These high energy photons have a profound effect on the conditions around the star and play an important role in photoexciting atoms, molecules and dust grains in both the accretion disk (Glassgold, Najita, & Igea 2004) and lower density circumstellar gas. Stellar X-ray and extreme-UV (EUV) photons are the major ionization source for much of the disk (Alexander, Clarke, & Pringle 2006) with the very hardest X-rays perhaps penetrating right to the cold disk midplane. The high energy photons drive an acceleration of disk evolution through photoevaporation (Gorti & Hollenbach 2009; Owen *et al.* 2009).

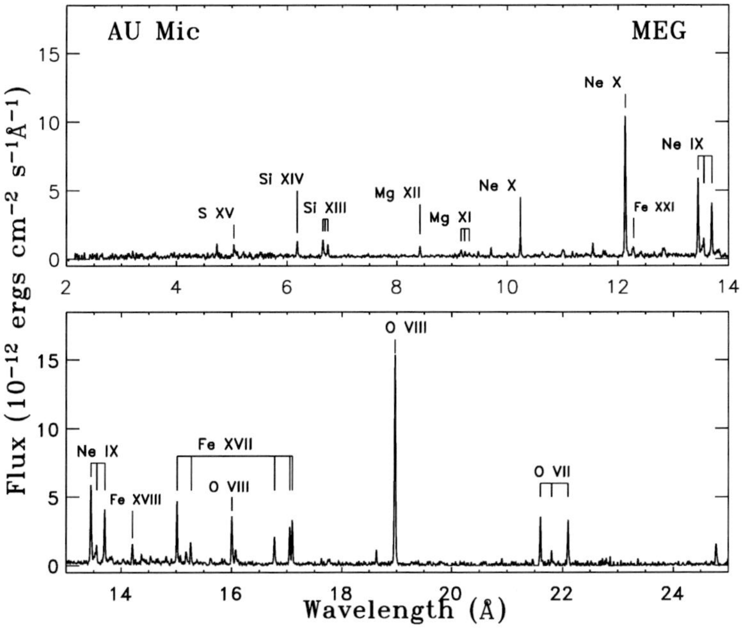

Figure 2. X-ray high resolution transmission grating spectrum of AU Mic (M0 V) as observed with the *Chandra HETGS* MEG spectrograph. The spectrum is dominated by strong coronal emission lines formed at temperatures between 10^6 and a few 10^7 K. A detectable X-ray continuum is present but comparatively weak.

The stellar radiation field is the primary driver of the physical conditions and chemical processes in the gas and dust surrounding the star. These effects are important for a wide variety of situations ranging from circumstellar disks, to accreting planetesimals, and to protoplanetary atmospheres. This UV/X-ray radiation controls:

- Photoexcitation of molecules and atoms.
- Photoionization of atoms and photodissociation of molecules.
- Drives more complicated chemistry in the gas phase and on dust grain surfaces, leading potentially all the way to complex organic molecules.
- Desorption of molecules from grain surfaces.
- Photoevaporation of material from disk surface layers.
- An increase in efficiency of magnetic coupling to disk material.

Information on the full spectral energy distribution of the central star is needed for accurate modeling of the influence of the radiation field, because photons with different energies have different effects on the circumstellar environment. Strong emission lines are present throughout the whole X-ray/EUV/FUV spectral range. For young active stars the flux in the FUV is very similar to that in the X-ray region.

FUV emission originates from upper chromospheric and transition region plasma (i.e. temperatures from just under 10^4 K to a few 10^5 K), while X-rays originate from corona plasma at a few to tens MK. The FUV region dominated by emission lines with only weak continuum. Fig. 1 shows the FUV spectrum of a young M dwarf AU Mic from the Lyman limit through much of the FUV — the emission line luminosities observed for this and all other young stars are ~1000 times stronger than those seen on the Sun. The other panel of Fig. 1 shows the spectra of a sample of Local Association (see below) dwarfs with similar masses but different ages and rotational periods — the strengths of

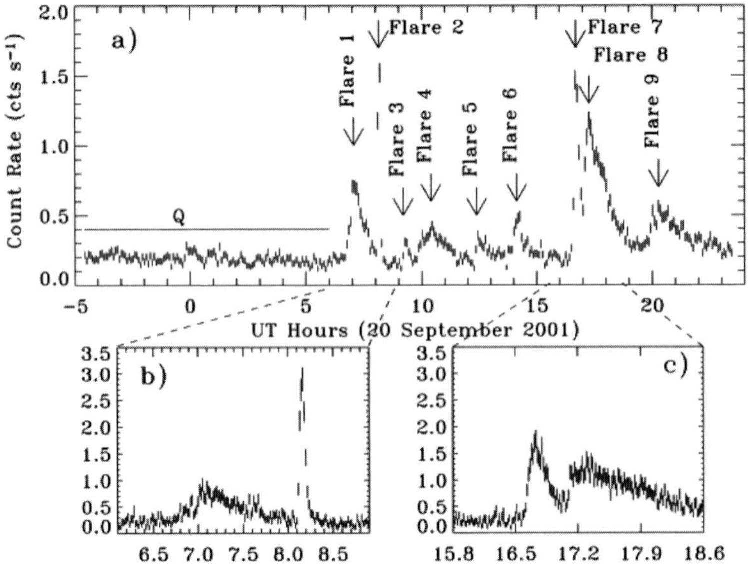

Figure 3. X-ray flaring on the 300 Myr old M dwarf EV Lac. Light curve of *Chandra* first order MEG count rate during a 100 kilosecond observation on 2001 Sep 19-20 from Osten *et al.* (2005). The ubiquitousness of frequent factor-of-a-few flare enhancements on timescales of a few hours is well illustrated.

the emission lines, and hence the level of stellar activity, depends on the rotational period (faster rotators are more active) rather than the age.

As is immediately obvious, this radiation field is a highly-specific excitation source with most of the flux restricted to narrow energy ranges (i.e. in the emission lines). In addition the excitation routes are equally specific — only when the two match is an excitation process strongly driven. The X-ray region is also mostly dominated by emission lines but a weak continuum is clearly present (see Fig. 2). Unlike the situation for the Sun where the full spectrum from X-rays through the EUV, FUV and NUV can be observed by Earth-orbiting satellites, for even the nearest stars much of the EUV spectrum is lost to interstellar absorption. The EUV is a vital component of the radiation field that must be known for accurate modeling as it plays a vital role in hydrogen ionization and photoevaporation. However, by observing a range of UV and X-ray emission lines that sample the transition region and coronal temperature distribution the full spectral energy distribution can be modeled, because the EUV is filled by emission lines that form at upper-transition-region and lower-coronal temperatures.

3. Variability Due to Stellar Flares

Low mass PMS stars are undoubtedly variable X-ray and UV sources with contributions from both flaring due to magnetic reconnection and to uneven activity distributions across the stellar surface, which then modulate the observed flux as the star rotates. From the XEST survey (Güdel *et al.* 2007) roughly half the stars showed measurable X-ray variability during observations that lasted typically 30 kiloseconds and with half of these showing obvious flares (Stelzer *et al.* 2007). Large flares occurred on individual stars every ten days or so. Flares continue to be common on young stars for hundreds of Myr after accretion ceases and factor of a few flare enhancements are typically seen on few hour time-scales. Flares can exhibit a wide range of energies and time-scales — this is well illustrated in Fig. 3 which shows the coronal emission from the 300 Myr old M dwarf EV Lac as observed over a single day (Osten *et al.* (2005)). Flare outbursts are complex with each individual flares disturbing the coronal magnetic topology and triggering further flares.

A hierarchy of flaring is seen with ever larger flares being seen less frequently. Vast long-duration multi-day flare outbursts have been observed, such as the 2 day (15x quiescent at peak) flare outburst seen by Cully *et al.* (1993) on AU Mic with the *EUVE* satellite. The very largest superflares have been very difficult to observe because they occur at frequencies of $\sim$ one per year per star at most. However the *Swift* BAT wide-field monitor has detected hard X-ray outburst for several active stars including EV Lac (Osten *et al.* 2009) and CC Eri, and the active binaries II Peg (Osten *et al.* 2007) and Algol. At the peak of superflares the flare X-ray luminosity can exceed the stellar photospheric bolometric luminosity. During superflares a significant hard X-ray continuum is detected out to at least 50 keV; such high-energy photons can significantly alter grain and planetesimals structure through annealing and melting. These large flare outbursts provide a vast increase in the X-ray radiation field with significant photon flux capable of penetrating deep into the dead zone of a protoplanetary disk.

4. Conclusions and Future Possibilities

The nearest examples of accreting PMS stars ("classical" T Tauri stars) are at distances of $\sim$ 150 pc in the nearby star formation regions, such as Taurus-Auriga, Chamaeleon, Upper Scorpius etc.; hence these are difficult targets for even moderate resolution FUV and X-ray spectroscopy. The situation is more favorable for older post-accretion stars because many nearby, young, single active dwarfs are members of the Local Association (Eggen 1992), which is a widespread aggregate of young stars related to star formation in the Sco-Cen association. By chance the Sun lies within the Local Association, thus its members cover the whole sky and many are very near ($\leqslant$ 50 pc). Many of these stars were identified because of their strong coronal activity (Montes *et al.* 2001; Zuckerman & Song 2004; Torres *et al.* 2008) The Local Association has been subdivided enthusiastically by many researchers, because, while Local Association ages span the range 7-300 Myr, the age spreads of the subgroups are much narrower. In our examples above we showed spectra of AU Mic, a particularly well-studied M1 dwarf member of the 12 Myr old β Pic Moving Group, which at a distance of only 9.9 pc from the Sun shows an easily resolvable debris disk.

So what are we doing to better understand PMS star high-energy radiation fields? Full datasets presently exist for only a few stars but much more can be done now using existing space-based observatories. Current instrumentation can now measure the UV and X-ray spectra of many critical example stars being studied extensively in other spectral regions. The COS spectrograph (Froning & Green 2009) on the *HST* is the ideal instrument to efficiently survey the FUV+NUV spectra of young stars. The COS medium resolution grating can measure both accretion diagnostics and the fluxes and profiles of more normal emission lines formed over a wide range of temperatures. Over the next year (2010) we shall be acquiring data for a carefully selected sample of young stars with accretion rates spanning 3-4 orders of magnitude and also probing stars with transitional disks and where dust disks have disappeared. [*HST* Large Program (GO11616; PI Herczeg) with 32 GO targets]. This sample is already well-studied in the infrared and X-ray spectral regions and detailed multi-wavelength modeling of these stars should provide significant insights into the evolution of PMS stars and their disks.

I thank Rachel Osten for extensive discussions relating to X-ray superflares and for permission to quote her results ahead of publication. I thank Edmund Hodges-Kluck for his assistance in the reduction and analysis of the *Chandra* HETGS spectrum of AU Mic. This work was supported by *Chandra* grants GO8-9005X, GO9-0015X, and GO9-0020B, *HST* GO grant 11336, *FUSE* grants NNX07AU48G and NNX07AU51G, and *NASA LTSA* grant NAG5-13058.

References

Alexander, R. D., Clarke, C. J., & Pringle, J. E. 2006, *MNRAS*, 369, 229

Bergin, E. A. 2009, in: P. Garcia (ed.), *Physical Processes in Circumstellar Disks Around Young Stars*, (Chicago: University of Chicago Press), in press

Blum, J. & Wurm, G. 2008, *ARAA*, 46,21

Cully, S. L., Siegmund, O. H. W., Vedder, P. W., & Vallerga, J. V. 1993, *ApJ*(Letters), 414, L49

Dullemond, C. P., Hollenbach, D., Kamp, I., & D'Alessio, P. 2007, in: B. Reipurth, D. Jewitt, and K. Keil (eds.), *Protostars and Planets V* (Tucson: University of Arizona Press), p. 555

Eggen, O. J. *AJ*, 104, 2141

Feigelson, E. & Montmerle, T. 1999, *ARAA*, 37, 363

Froning, C. S. & Green, J. C. 2009, *Ap&SS*, 320, 181

Glassgold, A. E., Najita, J., & J. Igea, J. 2004, *ApJ*, 615, 972

Gorti, U. & Hollenbach, D. 2009, *ApJ*, 690, 1539

Güdel, M., Briggs, K. R., Arzner, K., Audard, M., Bouvier, J., Feigelson, E. D., Franciosini, E., Glauser, A., Grosso, N., Micela, G., Monin, J.-L., Montmerle, T., Padgett, D. L., Palla, F., Pillitteri, I., Rebull, L., Scelsi, L., Silva, B., Skinner, S. L., Stelzer, B., & Telleschi, A. 2007, *A&A*, 468, 379

Montes, D, López-Santiago, J., Gálvez, M. C., Fernández-Figueroa, M. J., De Castro, E., & Cornide, M. 2001, *MNRAS*, 328, 45

Osten, R., Drake, S., Godet, O., Pye, J., Barthelmy, S. D., Cummings, J., Pal'shin, V., Golenetskii, S., Tueller, J., & Krimm, H. 2008, *ATel*, 1499

Osten, R. A., Drake, S., Tueller, J., Cummings, .J, Perri, M., Moretti, A., & Covino, S. 2007, *ApJ*, 654, 1052

Osten, R. A., Hawley, S. L., Allred, J. C., Johns-Krull, C. M., & Roark, C. 2005, *ApJ*, 621, 398

Owen, J. E., Ercolano, B., Clarke, C. J., & Alexander, R. D. 2009, *MNRAS*, in press

Stelzer, B., Flaccomio, E., Briggs, K., Micela, G., Scelsi, L., Audard, M., Pillitteri, I., & Güdel, M. 2007, *A&A*, 468, 463

Torres, C. A. O., Quast, G. R., Melo, C. H. F., & Sterzik, M. F. 2008, in: B. Reipurth (ed.) *Handbook of Star Forming Regions Vol. II* (San Francisco: Astronomical Society of the Pacific), p. 757

Wyatt, M. C. 2008, *ARAA*, 46, 339

Zuckerman, B. & Song, I. 2004, *ARAA*, 42, 685

Solar and Stellar Variability: Impact on Earth and Planets
Proceedings IAU Symposium No. 264, 2009
A. G. Kosovichev, A. H. Andrei & J.-P. Rozelot, eds.

doi:10.1017/S1743921309992985

Impact of the solar magnetic cycle on a protoplanetary disk

Andrey G. Tlatov

Kislovodsk Mountain Station, The Central Astronomical Observatory of RAS at Pulkovo, 357700, Box-145, Gagarina st., 100, Kislovodsk, Russia
email: tlatov@mail.ru

Abstract. We consider the influence of the periodic magnetic field of the Sun on the protoplanetary disk. Solar magnetic cycle may create a special orbit, which were formed main planets of the solar system. In orbits on which magnetic field accumulation occurs most effectively, there is a substance replacement. The Keplerian orbit with the period close to the period of solar magnetic cycle $T \sim T_M$ is most unstable for material accumulation. Two most close orbits where there is an accumulation of substance have periods $T = 1/2T_M$ and $T = 5/4T_M$. These orbits on are close to orbits of Jupiter and Saturn. Other planets were probably formed under influence of gravitation of Jupiter, Saturn and solar magnetic cycle. Perhaps, the effect of periodic magnetic field can explain the Titius-Bode rule.

Keywords. planetary systems: formation,protoplanetary disks; Sun: evolution

Magnetic fields likely play a key role in the dynamics and evolution of protoplanetary disks. They have the potential to efficiently transport angular momentum. Various rotation rates of central body and external disk resulted in increase an azimuthal component of magnetic field, increase angular momentum of protoplanetary disk and delay of central body rotation. Hardly tense magnetic spiral puts outward pressure upon gas. Thus gas goes on a spiral outside and planetosimales inside (Hoyle, 1960,Safronov & Ruzmaikina, 1985).

Let's consider a case if time of interaction of variable magnetic field protoplanet nebulae occurs to plasma for close to solar magnetic cycle period. It is possible, if the residual magnetic field perceived by plasma, will be kept for a commensurable time interval. The dissipation time of magnetic field in plasma with small magnetic numbers Reynolds is determined by diffusion equation $\partial B/\partial t = \eta \Delta B$, where η - magnetic diffusion coefficient which means, that variations of magnetic field with characteristic scale l_0 disappear for characteristic diffusion time: $\tau_d = {l_0}^2/\eta$. For in partially ionization plasmas $\tau_d = 3.7 \cdot 10^2 \cdot {l_0}^2 \cdot T^{-1/2} \cdot n_e/n_n$. For the characteristic size protoplanetary disk 10^{12} m, $\sim 10^3$ K^o and ionization level 10^{-12} dissipation time of magnetic field is about 100 years. Thus, the magnetic field can be kept and, under certain conditions, to grow on periods of same solar magnetic cycles. It can result in effective influence of magnetic field to nebulae plasma.

Let's assume that the solar magnetic cycle existed at the earliest stages of existence of solar system when planets have not been forming yet. The period of solar magnetic cycle, probably, differed from present time, but during evolution, period of magnetic cycle changed also. Let's consider interaction mechanisms of variable magnetic field with rotary protoplanets nebulae. Let's apply some simplifying assumptions. We shall take into account that during solar magnetic cycle there is change of magnetic polarity of each solar hemisphere. Also we shall notice that inclination angle D of the solar spin axis is not strictly perpendicular ecliptic planes and slopes for the majority planets makes $D \sim 5 - 7^o$.

In this case, in one half of ecliptic plane radial magnetic lines have one direction in and other half about an opposite direction. In these parts of ecliptic plane the Sun is visible under different angel D. During the magnetic cycle the direction of magnetic field half-planes will vary to opposite. Thus, the rotaries around the Sun plasma, hits in the magnetic field of various polarities (Tlatov, 2000). On major distances from the Sun force lines of magnetic field have preferentially radial direction B_r. At interaction with the gyred protoplanet disk from radial magnetic field can appear an azimuth component of magnetic field B. The magnitude B can be appreciated as $B_\varphi \sim Re_m \cdot B_r$ in case of constant magnetic fields B_r (Safronov & Ruzmaikina, 1985), where Re_m -magnetic Reynolds number. If change happen slowly it is possible to accept, that plasma and magnetic field are in magnetostatic equilibrium. Pressure of plasma p and components of a magnetic field and B_φ , B_r also are coupled by relation:

$$0 = \partial P/\partial R + \partial/\partial R(B_r^2 + B_\varphi^2)/2\mu + B_\varphi^2/\mu R - \rho\partial\Phi/\partial R, \qquad (1)$$

where P-pressure, Φ-gravitational potential, a ρ-densities. From this equation it is visible, that growth of magnetic intensity results in pushing out of substance. In case of constant magnetic field force of pushing out grows to the disk center. At a variable magnetic field the modification of a component of magnetic field B_φ, is described by the equation:

$$\partial B_\varphi/\partial t = \eta\nabla^2 B_\varphi + R \cdot B_r\partial\omega/\partial R, \qquad (2)$$

where η-coefficient of magnetic diffusion, R-distance from a spin axis, a ω-angular velocity, B_r-magnitude of radial magnetic field. The first item characterizes a temp of a dissipation of magnetic field, second - temp of transformation of radial component of magnetic field in poloidal magnetic field. In case of solar magnetic cycle component B_r is a variable. B_r depends from a phase of solar cycle and an angular position of plasma region. We can accept, that magnitude B_r is proportional $B_r \sim B_{r0} \cdot sin(\theta) \cdot cos(\omega_m \cdot t)$, where $\omega_m = 2/T_m$, T_m-period of magnetic cycle B_{r0}-intensity of dipole magnetic field near to the Sun, θ - angle between of part disk and direction on centre of the Sun. Being rotated on a Keplerian orbit regions of a disk change the relative position concerning solar equator, hence the angle θ will vary, i.e. $\theta = D \cdot sin(\omega t + \psi)$, where a ψ-phase describing an angular position of a region, ω- Keplerian angular velocity of a disk. Thus $B_r \sim B_{r0} \cdot cos(\omega_m \cdot t) \cdot D \cdot sin(\omega t + \psi)$. This relation reflects a change of magnetic field during magnetic cycle and rotation of protoplanet disk around of the Sun. For determination of component B_φ it is necessary to carry out integration the equation (2). Let's estimate the first term in a right part of the equation (2) time Parker's instabilities (Tout & Pringle, 1992). Diffusion transposition happens in basic through side areas protoplanet disk, we can suppose, that the equation (2) can be noted as:

$$\partial B_\varphi/\partial t = B_\varphi/\tau_p + B_{r0} \cdot cos(\omega_m \cdot t) \cdot D \cdot sin(\omega t + \psi), \qquad (3)$$

where τ_p- a temporal scale for Parkers instabilities $\tau_p = 2,2 \cdot h_d/v_A$, where h_d -a gravitational scale of height, v_A - aplfven velocity for B_φ components of a magnetic field, ψ a phase describing a position of a element, $f(r)$- function of distance from central body. The scale of height h_d can be estimated from an isothermal sound velocity and a velocity of rotation of a disk $h_d = c_s/\Omega$.

From the equation (3) it is visible, that the magnetic field differently influences on protoplanet substance depending on distance from the central body. The period of Keplerian circulation $T = T_m$ there is the most effective accumulation of magnetic field. In this orbit there is a linear growth of magnetic field and magnetic pressure. Concentration

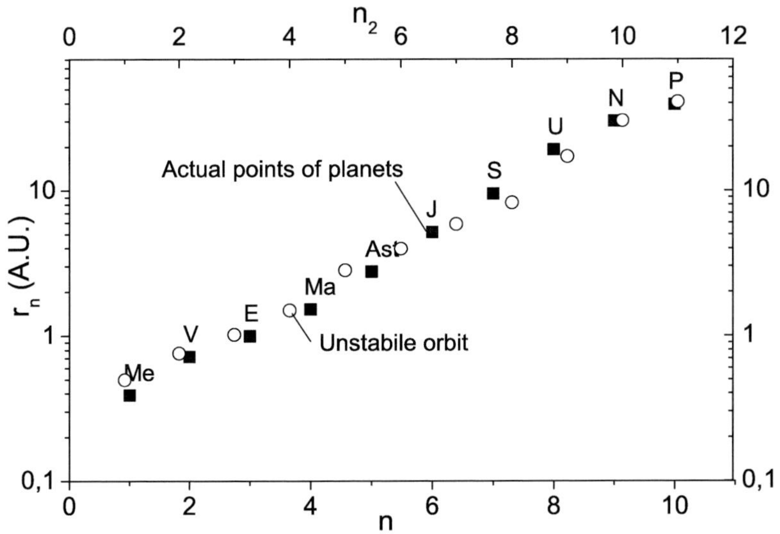

Figure 1. Actual points of planet of solar system expressed in Astronomical Units (squares) and distribution of orbits with unstable orbits in relation to accumulation of a magnetic field (circle).

of particles nebulae thus should decrease what to provide a magnetogasdynamic equilibrium (1). Thus, the substance will be aimed to abandon an orbit with period of rotation $T = T_m$. Also local maxima exist at periods $T = 1/3 \cdot T_m$ and $T = 3 \cdot T_m$ and others (Fig. 1). But it is possible to note, that the dissipative term restricts time of magnetic field accumulation. Thus, the substance will accumulation there where the demagnetization of plasma disk happens most fast. Let's consider an orbit with period $T = T_m/2$. For this time a complete reversal the solar magnetic dipole will replace the sign on opposite and the element of plasma disk will be acted with magnetic field of opposite polarity. It means, that the second member in a right member of the equation (3) will convert in a zero in time equal to a reversal period $T = T_m$. Residual magnetic fields of opposite polarity will cancel each other. The orbit with phase $T = T_m/2$ is preferable to substance accumulation. Now the period of magnetic cycle is about ~ 22 years. Hence, the orbit in which, forces of magnetic ejection on a plasma ring were minimal, was an orbit with the period about 11 years, i.e. an orbit close to Jupiter orbit. If, on the contrary, to accept, that a substance period which the Jupiter was formed corresponded to of half period of a magnetic cycle the full period was equal $T_m \sim 11.86 \cdot 2 = 23.72$ year.

References

Cameron, A. G. W. 1961, *Icarus*, 1, 13.

Hoyle, F. I. 1960, *QJRAS*, 1, 28.

Safronov, V. S. & Ruzmaikina, T. V. 1985, *in: Protostars and planets*, (Tucson, AZ, University of Arizona Press), 59.

Tout, C. & Pringle, J. E. 1992, *MNRAS*, 259, 604.

Tlatov, A. G. 2000, *in proc. conf. "Solar in maximum activity and solar-star anolg."*, (ed. A. V. Stepanov & S. Peterburg), 71.

Solar and Stellar Variability: Impact on Earth and Planets
Proceedings IAU Symposium No. 264, 2009
A. G. Kosovichev, A. H. Andrei & J.-P. Rozelot, eds.

doi:10.1017/S1743921309992997

Solar-terrestrial energy transfer during sunspot cycles and mechanism of Earth rotation excitation

Yavor Chapanov[1] **and Daniel Gambis**[2]

[1]Central Laboratory for Geodesy, Bulgarian Academy of Sciences
Akad. G. Bonchev Str. Bl.1, Sofia 1113, Bulgaria
email: chapanov@clg.bas.bg

[2]SYRTE, Observatoire de Paris, CNRS, UPMC
61, avenue de l'Observatoire, 75014 Paris, France
e-mail: daniel.gambis@obspm.fr

Abstract. The solar-terrestrial energy transfer, due to the total solar irradiance (TSI), solar wind and interplanetary magnetic field, has 11-year modulation during the sunspot cycles. Other oscillations of solar-terrestrial energy transfer are with periods of 22 and 45 year due to the magnetic reversal and equatorial solar asymmetry, which cause corresponding oscillations of all Earth systems, including climate and weather, atmosphere and ocean circulations, geomagnetic field and core processes. A part of this energy variation is transformed to oscillations of the Earth rotation. A model of indirect mechanism of Earth rotation excitation during sunspot cycles is proposed, which is based on global water circulation and periodical mass transfer between oceans and polar ice caps. The oscillations of the mean sea level (MSL) with periods 11, 22 and 45 years are determined by sea level data for the last two centuries from 13 maregraph stations. The necessary energy of water evaporation, corresponding to the observed MSL variations is provided by TSI oscillations with amplitudes between 0.2-0.5W/m^2, determined by means of reconstructed time series of the TSI since 1610. The determined mean Universal Time (UT1) amplitudes, corresponding to the 22-year and 45-year cycles of the solar activity are 185ms and 310ms.

Keywords. Sun: activity, solar-terrestrial relations, time

1. Introduction

The decadal variations of the Earth rotation are strongly affected by the cycles of the solar activity with a delay of about 1.5a. The explanation of the delayed UT1 response to the sunspot cycles is the appropriate change of the axial Earth moment of inertia, due to climatic variations, connected with the total solar irradiance change. The aim of this paper is to determine the mean amplitudes of the UT1 response to the solar activity cycles, according to the observed MSL variations and corresponding change of the Earth axial moment of inertia.

2. Total solar irradiance variations

Recently time series of Total Solar Irradiance (TSI) since 1610 was reconstructed by Lean *et al.*(1995)) and Lean (2000). These time series are shown in Fig. 1. The oscillations of the TSI with periods 11, 22 and 45 years (Fig.2, (a), (b), (c)) are determined by means of Fourier approximation and separation and combination of the oscillations with periods from the bands 10-12.5a for 11-year cycles; 19-25a for 22-year cycles and 34-56a for 45-year cycles. The time series of 11-year and 45-year cycles of the TSI are determined by means of the second solution of TSI (Lean, 2000). Unfortunately, in this solution the 22-year component is with very small amplitude due to applied new corrections, while the first version contain rather significant 22-year oscillations. So, the 22-year cycles of the TSI are derived from the first reconstruction (Lean *et al.*, 1995). The maximal amplitude of TSI oscillations with period 11a is 0.5W/m^2 and 0.2W/m^2 for 22-year and 45-year cycles.

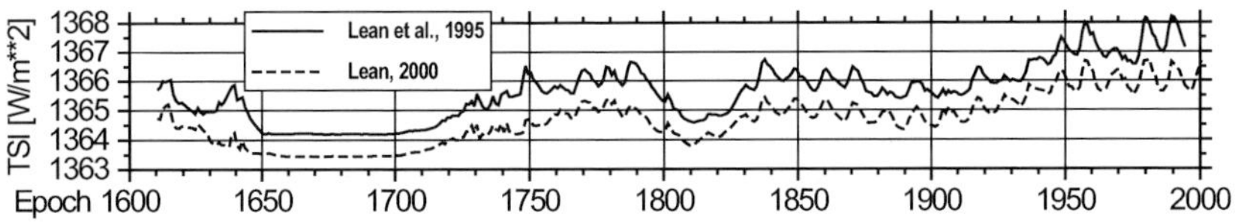

Figure 1. Time series of reconstructed Total Solar Irradiance (TSI) since 1610.

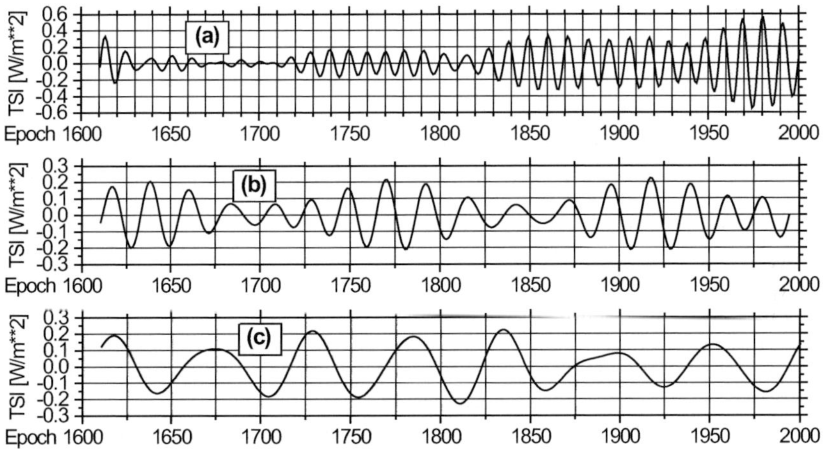

Figure 2. Time series of 11-year components of TSI (a), 22-year components of TSI (b) and 45-year components of TSI (c).

3. Mean sea level oscillations

The oscillations of the mean sea level with periods 11.25, 22.5 and 45 years are determined from maregraph data for the last two centuries by a model with fixed frequencies

$$\Delta L(t) = L_0 + L_1 t + \sum_{k=1}^{5} a_k \sin \frac{2\pi}{T_k} t + b_k \cos \frac{2\pi}{T_k} t, \tag{3.1}$$

$$A_k = \sqrt{a_k^2 + b_k^2} \tag{3.2}$$

where $\Delta L(t)$ are MSL variations for the epoch t; a_k, b_k are harmonically coefficients; A_k are amplitudes of the oscillations with periods $T_1 = 1\text{a}$, $T_2 = 18.6\text{a}$, $T_3 = 11.25\text{a}$, $T_4 = 22.5\text{a}$ and $T_5 = 45\text{a}$. The estimated amplitudes of MSL oscillations at 13 maregraph stations are given in Table 1. The corresponding mean amplitudes A_{11}, $A_{22.5}$ and A_{45} are $6.9 \pm 2\text{mm}$; $10.6 \pm 3\text{mm}$ and $8.9 \pm 2\text{mm}$.

4. Conclusions

• The 11-year, 22-year and 45-year solar activity cycles strongly affect earth climatic variations by the variations of the total solar irradiance with maximal amplitudes between 0.2–0.5W/m^2. These variations provide additional amount of energy (~1017MJ for the

Table 1. Amplitudes A_{11}, $A_{22.5}$, A_{45} of the 11-year, 22.5-year and 45-year oscillations of the mean sea level for some maregraph stations. Values less than 3mm are removed.

Station	Period	A_{11} [mm]	$A_{22.5}$ [mm]	A_{45} [mm]
Balboa	1908–1996	11.7	9.6	6.0
Barencburg	1948–2006	8.3	18.6	11.0
Brest	1807–2005	4.4	–	–
Fremantle	1897–1990	4.9	11.0	6.3
Genova	1884–1998	4.7	10.9	11.3
Ijmuiden	1871–2006	6.5	6.3	6.3
Harlingen	1865–2005	7.4	3.5	8.9
Honolulu	1905–2006	6.4	14.6	–
Maassluis	1848–2006	6.2	8.3	12.4
Mumbai	1878–1994	6.1	8.5	–
Prince Rupert	1909–2006	14.2	17.7	3.7
Stockholm	1774–2001	9.1	–	11.7
Sydney	1886–1994	4.5	8.7	9.8
Mean[1]	–	6.9 ± 2	10.6 ± 3	8.9 ± 2

Notes:
[1] The mean values are calculated after removing one minimal and one maximal amplitudes.

whole Earth and 130-260MJ per square meter on the equator) during these cycles. This energy is capable to evaporate 50-100l of water from 1 square meter equatorial ocean surface and 20-40l of water from 1 square meter ocean surface from the north latitudes.

- The periodical UT1 variations due to solar cycles are provided by additional water evaporation and global mean sea level oscillations, synchronized with the global water redistribution and corresponding change of the ice thickness at the polar caps. These events cause periodical change of the Earth axial moment of inertia and periodical oscillations of the Earth rotation, according the law of angular momentum conservation. The amplitude of 1mm of the mean sea level oscillation with periods 11, 22 and 45 years corresponds to UT1 oscillations with the same periods and amplitudes 12.5ms, 25ms and 50ms.
- The oscillations of the mean sea level with periods 11.25, 22.5 and 45 years at 13 maregraph stations are determined by a model with fixed frequencies. The corresponding mean amplitudes of these oscillations are 6.9 ± 2mm; 10.6 ± 3mm and 8.9 ± 2mm. The previously estimated UT1 response to 11-year sunspot cycles with amplitude 60ms need global water redistribution from oceans to the polar caps, corresponding to "pure" mean sea level oscillation with amplitude of 4.8mm (Chapanov & Gambis 2008). The rest of 2.1mm to the observed 11-year MSL amplitude is redistributed over the continents. So, according the above method of MSL oscillations determination with fixed frequencies, the proportion between the redistributed water over the polar caps and continents is 70% to 30%. The effectiveness of 70% means that the effective MSL amplitudes $A_{22.5}$ and A_{45} are 7.4mm and 6.2mm, so the expected mean UT1 amplitudes, corresponding to the 22-year and 45-year cycles of the solar activity, are 185ms and 310ms.

Acknowledgements

The work was supported by grant No. DO 02-275 with Bulgarian NSF.

References

Chapanov Ya. & Gambis, D. 2008, *Journées 2008 "Syst. de ref. spat.-temp.", Dresden*, 131–132
Lean, J., Beer, J., & Bradley, R. 1995, *Geophys. Res. Let.*, 22, 3195–3198
Lean, J. 2000, *Geophys. Res. Let.*, 27, 2425–2428

Solar and Stellar Variability: Impact on Earth and Planets
Proceedings IAU Symposium No. 264, 2009
A. G. Kosovichev, A. H. Andrei & J.-P. Rozelot, eds.

doi:10.1017/S1743921309993000

Common 22-year cycles of Earth rotation and solar activity

Yavor Chapanov[1], Jan Vondrák[2] and Cyril Ron[2]

[1]Central Laboratory for Geodesy, Bulgarian Academy of Sciences
Akad. G. Bonchev Str. Bl.1, Sofia 1113, Bulgaria
email:chapanov@clg.bas.bg

[2]Astronomical Institute, Academy of Sciences of Czech Republic
Boční II, 141 31 Prague, Czech Republic
email: vondrak@ig.cas.cz, ron@ig.cas.cz

Abstract. The 22-year oscillations of the Earth rotation due to several geophysical processes in the core-mantle system, oceans, atmosphere and geomagnetic field are excited mainly by 22-year cycles of the solar activity. These geophysical processes produce their own oscillations of the Earth rotation with different periods around 22 years. The direct and indirect influence of the solar activity on 22-year cycles of the Earth rotation are separated from the core effects and corresponding amplitudes are estimated by means of two approaches. The first, direct approach uses extended time series of Wolf's numbers with 22-year cycles, determined by sign alternation of even sunspot cycles. A linear regression between 22-year cycles of UT1 and solar activity is determined and this regression model is used to calculate the UT1 response to the 22-year cycles of the solar activity. The second, indirect approach uses 22-year oscillation of the mean sea level, caused by water evaporation due to variations of the total solar irradiance. The influence of the mean sea level variations on the Earth rotation is calculated by means of an empirical model of global water redistribution. The core-mantle effects on the 22-year UT1 variations are determined by excluding the UT1 response to the solar activity and core angular momentum due to the geomagnetic field variations, according to the solutions from the Special Bureau of the Core (SBC).

Keywords. Sun: activity, sunspots, solar-terrestrial relations, time

1. Introduction

The variations of the Earth rotation are mainly caused by the influence of the Moon, Sun and planets, displacements of matter in different parts of the planet and other excitation mechanisms. The most powerful source of these oscillations is Sun, which solar activity cycles can drive great number of geodynamical processes connected with the convections of the Earth fluids on the surface and inside the Earth. Many of climate and weather parameters are affected directly by the variations of the solar activity. The common 22-year cycles of Earth rotation, solar activity and mean sea level are investigated here by means of the existing long time series of the Universal Time UT1 (1623–2005), based on the solution of Stephenson & Morrison (1984); the Wolf's numbers since 1749 and the Mean Sea Level (MSL) in Stockholm since 1774 (Ekman 2003).

2. Common 22-year cycles of UT1 and mean sea level

A significant part of the UT1 oscillations with period 22 years are excited by the climatic variations, due to the solar activity cycles with the same period. The energy source of this process is the Total Solar Irradiance (TSI) variations with maximal amplitude $0.2\mathrm{W/m^2}$, which provide an additional water evaporation and global mean sea level oscillations, synchronized with the global water redistribution and corresponding change of the ice thickness at the polar caps (Chapanov & Gambis 2008). The common cycles of the UT1 and Mean Sea Level (MSL) variations are extracted from the Fourier approximation of the available data for the period 1800-2000 and combination of the oscillations with periods from band 20-25a (Fig. 1,a, Chapanov *et al.* 2008). UT1 data from star occultation and MSL data from Stockholm maregraph are used. The 22-year MSL and the delayed UT1 cycles are highly correlated with coefficient −0.81 and linear regression (Fig. 1,b) ΔUT1= -41ΔMSL + 2.4 [ms/mm]. This result points out to some dependence between the MSL variations and total UT1 oscillations, dominated mainly by the liquid Earth core effects. It is possible to determine UT1 response to 22-year solar cycles by applying a regression coefficient −17.5ms/mm to the MSL variations. This coefficient is determined in (Chapanov & Gambis 2009) with accounting the proportion of 70% of evaporated water, redistributed over the polar ice and 30% - over the continents. The resulting time series is shown in Fig. 1,c together with the 22-year core effects on UT1 oscillations, determined from the total UT1 variations after excluding the solar activity influences.

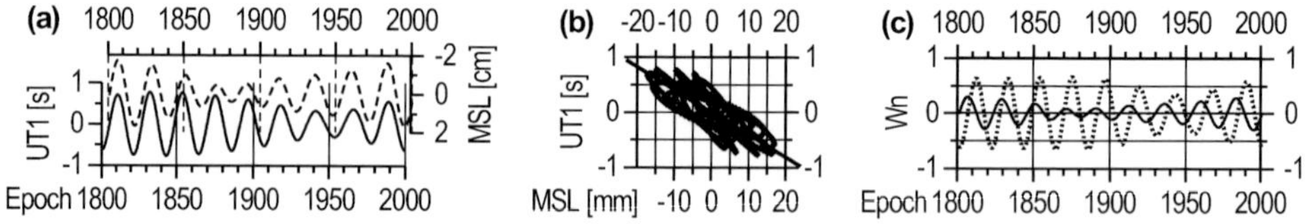

Figure 1. Common 22-year UT1 (solid line) and MSL (dashed line) cycles for the period 1800-2000 - (a); their linear regression - (b); separation of 22-year cycles of the Earth rotation, due to TSI variations (solid line) and liquid Earth core effects (dotted line) - (c).

3. Common 22-year cycles of UT1 and extended Wolf's numbers

The separation of 22-year cycles of the Earth rotation, due to solar activity and core effects by means of MSL data, shown in Fig. 1,c yields rather smoothed time series with no high-frequency details. It is possible to improve the resolution by means of extended 22-year Wolf's numbers cycles $W_{n/22a}$) determined by sign alternation of 11-year Wolf's numbers cycles (Fig. 2,a).

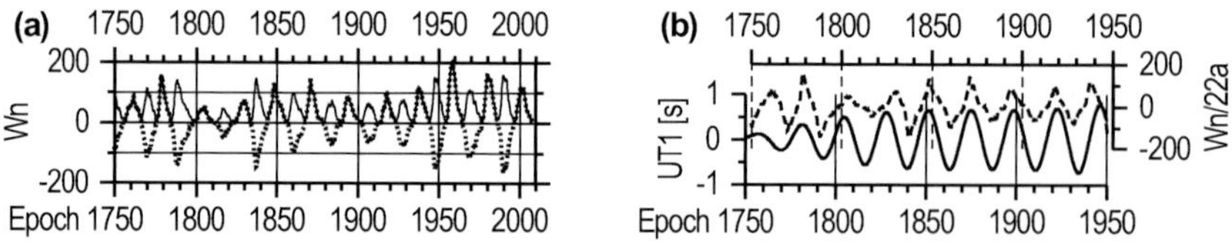

Figure 2. Extended time series of 22-year solar activity cycles (dotted line), determined by sign alternation of Wolf's numbers cycles (solid line) - (a); comparison between the 22-year cycles of UT1 (solid line) and extended Wolf's numbers (dashed line) for the periods 1750–2005 - (b).

Comparison between the 22-year cycles of UT1 and extended Wolf's numbers shows good agreement for the periods 1750-2005 (Fig. 2,b), where the linear regression is ΔUT1= $6.4W_{n/22a}$ + 1.1 [ms] (Fig. 3,c). This regression reflects the possible dependence between the solar activity and total UT1 oscillations, dominated mainly by the liquid Earth core effects. The UT1 response to 22-year solar cycles is determined by linear regression between the MSL and Wn/22a variations for the period 1800-2000 (Fig. 5)

ΔMSL$=-0.074W_{n/22a} + 0.26$ [mm], and according to the results in (Chapanov & Gambis 2009), the final regression coefficient between UT1 variations and $W_{n/22a}$ is 1.3ms. The separated solar and core effects of 22-year UT1 oscillations are presented in Fig. 3,d.

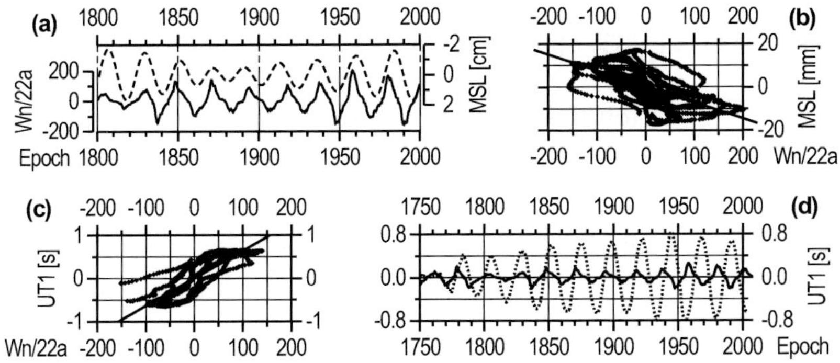

Figure 3. Comparison between $W_{n/22a}$ (solid line) and 22-year cycles of MSL (dashed line) - (a); their linear regression - (b); linear regression between UT1 and $W_{n/22a}$ - (c); separation of 22-year cycles of UT1, due to solar activity (solid line) and liquid Earth core effects (dotted line) - (d).

4. Conclusions

• The influence of 22-year cycles of the solar activity on the Earth rotation variations is determine by analysis of two time series - first with the mean sea level data from Stockholm maregraph station for the period 1800–2000 and second - with an artificial Index of 22-year solar cycles, based on the Wolf's numbers for the period 1750-2005. The mean sea level data analysis is more realistic, because the long-term MSL oscillations are physically connected with additional water evaporation due to the total solar irradiance variations and global water redistribution over the continents and polar ice, following by the changes of the Earth axial moment of inertia and Earth rotation velocity. The extended 22-year Wolf's numbers Index provide longer data, more short-term details and exact match of cycles duration, but it need future improvement in order to represent more realistic the amplitude of the TSI variations.

• The direct influence of 22-year solar cycles on the Earth rotation variations, determined by means of the mean sea level data for the period 1800–2000, is with slow variable amplitude of UT1 cycles and maximal value of about 0.3s. The core effects in 22-year UT1 oscillations are with constant amplitude of about 0.66s for the period 1800-1890 and non-linear changes after 1890; close frequency to the solar cycles and constant phase difference of about 65 degrees.

Acknowledgements

The work was supported by grants No.DO 02-275 with Bulgarian NSF and No.A300130702 with Grant Agency of Acad. of Sci. of the Czech Republic.

References

Chapanov, Ya. & Gambis, D. 2008, *Journées 2008 "Syst. de ref. spat.-temp.", Dresden*, 131–132

Chapanov, Ya. & Gambis, D. 2009, *Proc. IAU Symposium 264*, This Issue

Chapanov, Ya., Vondrák, J., & Ron, C. 2008, *Journées 2008 "Syst. de ref. spat.-temp.", Dresden*, 178–179

Ekman, M. 2003, *Small Publ. in Hist. Geophys.*, 12, 31

Stephenson, F. R. & Morrison, L. V. 1984, *Phil. Trans. R. Soc. of London, Ser. A.*, 313, 47–70

Solar and Stellar Variability: Impact on Earth and Planets
Proceedings IAU Symposium No. 264, 2009
A. G. Kosovichev, A. H. Andrei & J.-P. Rozelot, eds.

doi:10.1017/S1743921309993012

Manifestations of dark energy in the dynamics of the Solar system

Michal Křížek[1] and Jan Brandts[2]

[1] Academy of Sciences, Žitná 25, CZ-115 67 Prague 1, Czech Republic
email: `Krizek@math.cas.cz`

[2] Korteweg-de Vries Institute, Faculty of Science, University of Amsterdam, Science Park 904, 1098 XH Amsterdam, The Netherlands
email: `J.H.Brandts@uva.nl`

Abstract. The expansion speed of the Universe is increasing (Glanz 1998). This acceleration is attributed to dark energy which acts almost uniformly everywhere (including the Solar system) and thus essentially influences the Hubble constant. Its current value on a distance of 1 AU is $H_0 = 10$ m/(yr AU). This is quite a large number and thus, the impact of dark energy should be detectable in the Solar system. We will illustrate it by several examples. Dark energy may partially be caused by gravitational aberration of the Sun, planets and other bodies.

Keywords. dark matter, solar system: general, Earth, Moon

---

1. The Hubble constant. Recent observations suggest that the Hubble parameter $H(t)$ is the sum of a decreasing term $H_1(t) = C/t$, which models the Big Bang and subsequent gravitational interaction, and an increasing, but initially relatively small and later more influential term $H_2(t)$ of yet unknown origin and nature. The energy needed for the acceleration of the expansion represented by $H_2(t)$ is so-called dark energy, which, like dark matter, remains to be detected. The present value H_0 of the Hubble parameter is about 20 km $\mathrm{s}^{-1}\mathrm{Mly}^{-1}$. Since light travels 1 AU in about 500 seconds, we get

$$H_0 = \frac{2}{c}\,\mathrm{cm\,s^{-1}yr^{-1}} = 0.02 \times 500/(\mathrm{AU})\,\mathrm{m\,yr^{-1}} = 10\,\mathrm{m\,yr^{-1}AU^{-1}}. \tag{1}$$

This is quite a large number and therefore, the impact of dark energy should be observable not only on large scales, but also in the Solar system as we shall see.

2. The recession speed of the Moon is larger than accounted for. Laser retroreflectors installed on the Moon reveal that the mean distance $D = 384\,400$ km between Earth and the Moon increases about 3.84 cm per year. Note that only around 55 % of this value can be explained (Křížek 2009) by tidal forces caused by the Moon and the Sun. The remaining part is often attributed to less explicitly known effects, such as deglaciation, internal processes, and mass transfers in the atmosphere and hydrosphere. An alternative explanation is that the missing part is essentially due to dark energy, which, contrary to the term $H_1(t)$, we conjectured to act also locally within the system Earth-Moon. Since the sum $H(t) = H_1(t) + H_2(t)$ is known to be increasing for already seven billion years, the missing 1.73 cm per year should ideally be a non-trivial part $H_2(T)$ of the total present value $H_0 = H(T)$ of the Hubble parameter for $T =$ 13.7 Gyr. On the distance D this amounts due to (1) to a yearly expansion of $H_0 =$ 2.56 $\mathrm{cm\,yr^{-1}}D^{-1}$, i.e., $H_2(T) \approx 0.67H_0$. Thus, the unexplained part of the recession speed of the Moon is in magnitude indeed comparable to a non-trivial part of the total expansion speed of the Universe.

3. The ecosphere has been moving, but Earth remained within it. If the current radius of Earth's orbit around the Sun would be more than 5% larger or 2% smaller, photosynthesis and hence life as we know it would not be possible. The relatively small region in which life is feasible is called the ecosphere. Obviously, the position of the ecosphere depends on the luminosity of the Sun. Since the Sun is a star of the main sequence, its luminosity has increased approximately linearly during the last 4.5 billion years of its existence, starting at about 70% of its current value. Therefore, when life originated on Earth 3.5 billion years ago, the luminosity of the Sun was around 77 % of its current value. Since luminosity decreases quadratically with distance, the distance of the ecosphere to the Sun was about 130×10^6 km $\approx \sqrt{0.77}$ AU. Thus, Earth must have been much closer to the Sun than it is now, or it would have been frozen, in spite of a different composition of its atmosphere and other temperature increasing effects like a higher level of radioactivity, the impact of comets, and more vulcanism. It is, in fact, actually known that there were liquid oceans on Earth already 4.35 Gyr ago, when the luminosity of the Sun was even smaller. By investigation of fossils of thermophilic bacteria it was found that the temperature of oceans around 80° C when life emerged (Lineweaver & Schwartzman 2003). Taking the above into account, we conclude that Earth must have been around 20×10^6 km $\approx (1 - \sqrt{0.77})$ AU closer to the Sun then it is now. Using this value, we find that the average receding speed of Earth from the Sun to get an almost constant luminosity, and hence influx of solar energy over the past 3.5 Gyr, must have been about 5 m yr^{-1}, i.e., $H_2(T) \approx 0.5 H_0$ (cf. (1)). Note that Kepler's third law implies that Earth's orbital period increases less than 1 ms yr^{-1}. This, nor the yearly increase of 5 meters of Earth's distance to the Sun, can be verified by direct measurements. This is particular difficult because due to the influence of the larger planets, the barycentre of the Solar system describes an irregular orbit near the Sun, in which it travels more than 10^5 km per year (Bertotti *et al.* 2003, p. 542). Note that effects like tidal forces from the Sun, the solar wind, and the Sun's decreasing mass are able to explain a recession speed of no more than 0.01 m yr^{-1}.

The Hubble constant did not change too much during the last 3.5 Gyr. It seems that the magnitude of its repulsive part $H_2(t)$ is just right for a receding speed giving an almost constant influx of solar energy. This represents further support for the so-called *Anthropic Principle*, which states that basic physical constants are favorable to the emergence of life only if they are in very narrow intervals.

4. Mars must have been closer to the Sun to have had liquid water. At present, the temperature at the equator of Mars can become close to 0° C when the Sun is at its highest point and Mars is at its perihelion, but in average over the day over the whole planet it is with −63° C deeply below the freezing point. Nevertheless, there is convincing evidence for the presence of liquid water in the form of river valleys on Mars around three to four billion years ago. Thus, the average temperature and atmospheric pressure on Mars must have been significantly higher at that time. However, the luminosity of the Sun was then about 73 – 80 % of its current value, i.e., one third of the solar constant. Moreover, due to snow and ice also in regions away from the polar caps, and due to clouds, the albedo of Mars was higher than at present, implying that more solar energy was reflected back into space. Both facts support the claim that Mars also must have been much nearer to the Sun to have liquid water on its surface 3 – 4 Gyr ago. As a consequence we must conclude that Mars have been moving away from the Sun with an average speed of at least several meters per year (cf. (1)).

5. Fast inner satellites have not yet crashed on their mother planet. Each planet has a stationary orbit with radius R_{stat}, where the orbital and rotational periods are equal. Satellites that are below the stationary orbit are called fast. For instance, around Uranus there are at least eleven fast satellites and around Neptune five, all between 48 and 76 thousand kilometers away from the centre of their mother planet. All these satellites have nearly circular orbits with small inclinations and orbit their mother planet in the same direction, which indicates that most of them are genuine satellites and have not been captured. Although some may be parts of larger disintegrating satellites, there is no reason to assume that they have not been orbiting their mother planet, below their stationary orbit, for already 4.5 Gyr. However, during this period, tidal forces have acted on them. The resulting tidal bulges continually reduce their potential energy and orbital periods. Thus, all these fast satellites approach their planets along spiral trajectories. Tidal forces are proportional to M/r^3, where M is the mass of a planet and r is the radius of a given satellite orbit. For all the above-mentioned satellites the value M/r^3 is of the same order of magnitude as for Phobos, whose orbital period is decreasing. It was estimated that tidal forces are responsible for a decrease of 1.8 $\mathrm{cm\,yr^{-1}}$. If we would assume a similar speed 1–2 $\mathrm{cm\,yr^{-1}}$ for the other quick satellites, we find that during the last 4.5 Gyr, they must have gotten 45 000–90 000 km closer to their mother planet. However, all of them (except for Phobos) are in orbits of large radii 0.58–0.92 R_{stat}. This raises the intriguing question why these fast satellites have not fallen down onto their mother planet due to tidal forces. Again, it may be due to dark energy which causes a repulsive force.

6. Kuiper belt. By (Bertotti *et al.* 2003, p. 534) the Kuiper belt of comets must have been formed much closer to the Sun than its current position indicates. We can easily verify that one half of the value given in (1) is sufficient to have moved this belt over a distance of approximately 10 AU away from the Sun during the last 4.5 Gyr.

7. Delayed Neptune. One half of the value given in (1) would also have shifted Neptune about 5 AU further away from the Sun which explains more easily its formation than at the current region with very low velocities, 30 AU away from the Sun. Moreover, the observed delay of Neptune's position can be partly explained by dark energy.

8. Slow rotation of Mercury. There is also evidence that Mercury may have been closer to the Sun than its current distance of 46 – 70 million kilometers. Its sidereal rotation of 59 days is, even for its low mass, very slow. This may be the result of tidal forces that, ten million kilometers closer to the Sun, are twice as large as at its present location, because they decrease cubically with distance. In order to undergo these forces, Mercury must of course have been millions kilometers closer to the Sun than it is now.

9. Conclusions. All of the above strange phenomena may be explained by one source – dark energy. This also explains why the system Earth-Moon has at present such a large orbital momentum. From Sect. 2 we know that the distance Earth-Moon increased not only due to tidal forces. In (Křížek 2009) we illustrate that the dark energy needed for the accelerated expansion of the Universe may partially come from a finite speed of gravitational interaction that causes a positive value of gravitational aberration much smaller than the aberration of light.

Acknowledgements

The authors thank M. Brož, J. Grygar, and V. Pravda for the useful discussions. Supported by grant no. IAA 100190803 of the Czech Acad. Sci.

References

Bertotti, B., Farinella, P., & Vokrouhlický, D., *Physics of the Solar System*, *Kluwer*, 2003.
Glanz, J., *Science* 279 (1998), no. 5355, 1298–1299.
Křížek, M., *Comm. Comput. Phys.* 5 (2009), 1030–1044.
Lineweaver, C. H. & Schwartzman, D., *Cosmic thermobiology*, In Origins (ed. J. Seckbach), *Kluwer*, 2003, 233–248.

Solar and Stellar Variability: Impact on Earth and Planets
Proceedings IAU Symposium No. 264, 2009
A. G. Kosovichev, A. H. Andrei & J.-P. Rozelot, eds.

doi:10.1017/S1743921309993024

Polarimetic search for exoplanets with a tangential transit

L. V. Ksanfomality

Space Research Institute, Russian Academy of Sciences
email: ksanf@iki.rssi.ru

Abstract. This paper is devoted to the data obtained in measurements of the linear light polarization of few stars (that are definitely known to possess planets) and the results of the measurements processing. One may surmise with caution that the suggested method yields positive results and, if confirmed, can be of use both in searching for exoplanets in transiting or tangentially transiting orbits and in refining their masses.

Keywords. stars: planetary systems; techniques: polarimetric; methods: data analysis

1. Introduction

Optical polarimetry as a trial method of obtaining evidence of the presence of "hot Jupiter" planets around stars is suggested and tested experimentally by astronomical observations that have been performed. Polarimetry is convenient in its simplicity when compared with the radial velocity spectral method. Light polarization is retained in the stellar emission, irrespective of the degree of emission attenuation, unless there is a scattering medium on the line of sight.

2. Observations and data processing

Systematic observations of the linear light polarization of few stars (that are definitely known to possess planets) were performed in 2003 at the Steward Observatory of the University of Arizona, USA. The Kuiper telescope of Mount Bigelow Observatory (110°44.04 W, 32°24.59 N) has been used (a Ritchey-Chretien system, $D = 1.54$ m, $F = 20.8$ m). Based on the visibility conditions and the accessible duration of the observations, four short period exoplanetary systems, with an orbital period 3 – 4 days, were selected: HD 46375, 51 Peg, HD 46375, Ups And and HD 209458. Among them, only HD 209458 has transits. The measurement system consisted of a Wollaston prism, filters and a cooled CCD array camera. The measurements in both components were performed simultaneously, increasing significantly the accuracy of the result. The CCD polarimeter was mounted at the prime focus of the telescope. The polarization of light from the selected stars was systematically measured within a 5 days period, in a "blue" filter (SS5). Fig. 1 shows a sequence of six polarimetric pairs of measurements of υ And performed with an exposure time of 220 ms and with intervals of 100 ms. Since the orientation of the starlight polarization plane is not known in advance, the measurements were performed for three orientations of the Wollaston prism differed by $\alpha = 45°$. Therefore, if the light consists of a nonpolarized component B_n, and polarized component, B_p, and if the prism first have been installed at φ angle to the light polarization plane, then the measured values in each of the two beams when the prism is turned, should vary as:

$$b = B_n/2 + B_p cos^2(\alpha + \varphi).$$

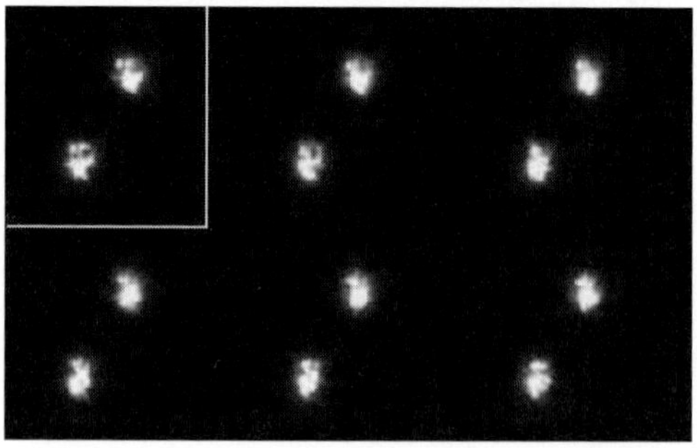

Figure 1. Example of one frame including 6 polarimetric pairs obtained during the observations of υ And. The position of the first pair is highlighted.

The observing time for each chosen star did not exceed 3 h. The following data volumes have been obtained: 15.2 Mbytes in the observations of 51 Peg, 12.1 Mbytes for υ And, and 15.2 Mbytes for HD 209458. The data for the faint star HD 46375 was not processed later.

The ultimate goal of the data processing was the search for periodic polarization variations correlated with the known exoplanet period. More special, using the intensity ratio of orthogonally polarized waves obtained for three different groups of measurements (for different prism orientations), one could calculate the degree of polarization for the star, averaged for a given night. Once the data processing has been performed, it yields in the data like image Fig. 2. Detailed information about the processing codes is given in Talyzin (2006).

3. Results

Fig. 3 shows the fitting function for the observations of υ And and the experimental data points. Polarization and time (in days) are along the vertical and horizontal axes, respectively. It is unlikely that the scatter of points can reveal any correlation. This is also suggested by their large rms deviation from the plotted harmonic function, 0.368. The data for the polarization variations in the star HD 209458 looks differently (Fig. 4). The rms deviation is low, 0.161, and the data points fall nicely on the curve. Thus, the periodic polarization variations by $\pm 1.7\%$ can be assumed to be actually attributed to the transit of the exoplanet HD 209458b and its comet-like tail across the stellar disk. The origin of the background or instrumental polarization (less than 2%) was not investigated specially. It could be introduced by both the telescope or the camera itself.

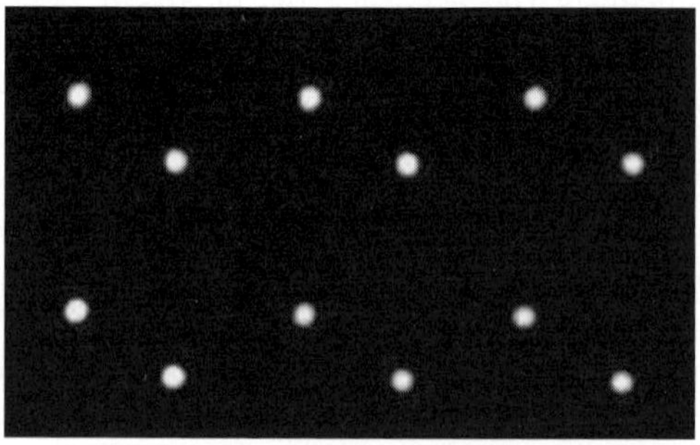

Figure 2. The image shown in Fig. 1 reconstructed from the digital array of calculated mean ratios of intensity pairs.

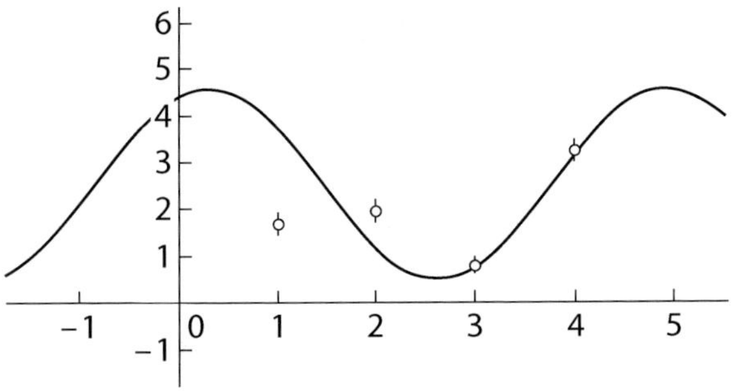

Figure 3. Search for a correlation in the data for υ And. Polarization (in percent) and time (in days) are along the vertical and horizontal axes, respectively. The rms deviation is 0.368.

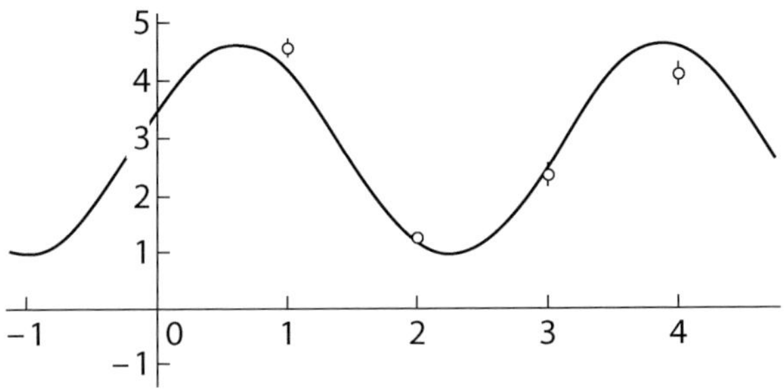

Figure 4. Polarization variations in the star HD 209458 (with exoplanet transits). The same quantities as those in Fig. 3 are along the axes. The rms deviation is 0.161.

The most interesting is the Fig. 5 which shows the polarization variations for the star 51 Peg that is known to be the first extrasolar planet discovered (Mayor & Queloz 1995). No transits are known for 51 Peg. Nevertheless, the rms deviation from the fitting harmonic function here is lowest, only 0.025, while the data points show virtually almost no deviation from the curve. If this is indeed the effect of light scattering in the cometlike tail of 51 Peg b, then one can assert that the angle i between the orbital plane of the planet and the observer's direction is small. Hence, the exoplanet's mass $M \sin i$, known to within $\sin i$, is actually M, i.e., 0.46 M_J (Jupiter mass). The ultraviolet range is much more sensitive to the transits of a gaseous tail. That's why it can be suggested that ultraviolet observations will allow transits of the tail of 51 Peg b to be detected photometrically.

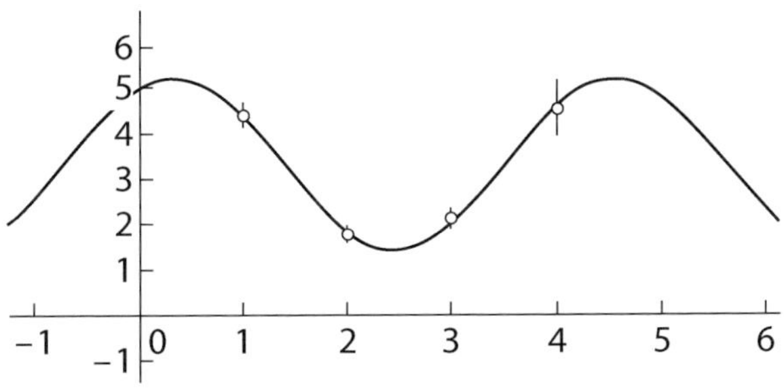

Figure 5. Polarization variations in the star 51 Peg (the same quantities as those in Fig. 3 are along the axes). The rms deviation is 0.025. There is no information on the transits for 51 Peg b in the literature. The obtained data suggest that during the revolution of the exoplanet 51 Peg b, tangential transit of its comet-like tail across the stellar disk takes place.

The depth of light attenuation in the UV during transits, for HD 209458 is almost an order of magnitude higher than that in the optical range.

4. Conclusion

The spectroscopic radial velocity method and the transit method remain the most productive for exoplanet search. Optical polarimetry is suggested as a trial method for detecting "hot Jupiter" planets around stars in transiting or tangentially transiting orbit. The method has been tested experimentally by the necessary astronomical observations and their processing (Ksanfomality 2007).

There are the following results of the first experiment (requiring confirmation by performing new astronomical observations):

- The suggested method allows to surmise with caution that it yields positive results and can be of use both in searching for exoplanets and in refining their masses.
- The possible errors cannot be ruled out completely. For example, nonidentical meteorological conditions during the observations can be the error sources.
- The high correlation between the polarization variations and the revolution period of the 51 Peg b exoplanet suggests that the angle between the orbital plane of the planet and the observer's direction is small ($\sin i \approx 1$) and the parameter $M\sin i$ turns into M. That means that the mass of first exoplanet 51 Peg b must be close to its estimated value 0.46 M_J (if the effect is actually related to a tangential transit of the exoplanet's comet-like tail).
- The above conclusions require confirmation by performing new astronomical observations to cover several complete periods of the planetary systems under study.

A possible mechanism(s) of influence of an exoplanet on the periodic variations in starlight polarization is not considered here. At least, one may suppose that light scattering in the atmosphere of an exoplanet and its comet-like tail during transit (or tangential transits) produces light polarization, same as well known effects in comet atmospheres. Apart from the light scattering in the planetary atmosphere and tail, other explanations are possible.

References

Ksanfomality, L. V. 2007, *Solar System Research*, 41 (4), 301.

Mayor D. & Queloz, D. A. 1995, *Nature*, 378, 355.

Talyzin, A. S. 2006, *Physical Peculiarities of Object HD 209458b. A New Method of Detection of Exoplanets. An Attempt to Find a Comet-like Tail of the Exoplanet in an Unusual Way.* Master's Thesis (Moscow: Mosk. Fiz.-Tekh. Inst., Kafedra Kosm. Fiz.).

Session VIII

Impact of solar and stellar variability on planetary atmospheres and climate

Chair: A.H. Andrei

Solar and Stellar Variability: Impact on Earth and Planets
Proceedings IAU Symposium No. 264, 2009
A. G. Kosovichev, A. H. Andrei & J.-P. Rozelot, eds.

doi:10.1017/S1743921309993036

Biological damage due to photospheric, chromospheric and flare radiation in the environments of main-sequence stars

Manfred Cuntz[1], Edward F. Guinan[2] and Robert L. Kurucz[3]

[1]Department of Physics, University of Texas at Arlington,
Arlington, TX 76019-0059, USA
email: cuntz@uta.edu

[2]Department of Astronomy and Astrophysics, Villanova University,
Villanova, PA 19085, USA
email: edward.guinan@villanova.edu

[3]Harvard-Smithsonian Center for Astrophysics,
Cambridge, MA 02138, USA
email: rkurucz@cfa.harvard.edu

Abstract. We explore the biological damage initiated in the environments of F, G, K, and M-type main-sequence stars due to photospheric, chromospheric and flare radiation. The amount of chromospheric radiation is, in a statistical sense, directly coupled to the stellar age as well as the presence of significant stellar magnetic fields and dynamo activity. With respect to photospheric radiation, we also consider detailed synthetic models, taking into account millions or hundred of millions of lines for atoms and molecules. Chromospheric UV radiation is increased in young stars in regard to all stellar spectral types. Flare activity is most pronounced in K and M-type stars, which also has the potential of stripping the planetary atmospheres of close-in planets, including planets located in the stellar habitable zone. For our studies, we take DNA as a proxy for carbon-based macromolecules, guided by the paradigm that carbon might constitute the biochemical centerpiece of extraterrestrial life forms. Planetary atmospheric attenuation is considered in an approximate manner.

Keywords. astrobiology, planets and satellites: general, stars: activity, stars: chromospheres, stars: evolution, stars: flare, stars: late-type

1. Theoretical Approach

The centerpiece of all life on Earth is carbon-based biochemistry. It has repeatedly been surmised that biochemistry based on carbon may also play a pivotal role in extraterrestrial life forms, if existent. This is due to the pronounced advantages of carbon, especially compared to its closest competitor (i.e., silicon), which include: its relatively high abundance, its bonding properties, and its ability to form very large molecules as it can combine with hydrogen and other molecules as, e.g., nitrogen and oxygen in a very large number of ways (Goldsmith & Owen 2002).

In the following, we explore the relative damage to carbon-based macromolecules in the environments of late-type main-sequence stars using DNA as a proxy. We focus on the effects of both photospheric and chromospheric radiation and comment on the significance of flare activity. Previous studies indicating the importance of UV radiation concerning the suitability of extrasolar planets for the existence of life and biological evolution has been given by, e.g., Cockell (1999), Guinan & Ribas (2002), Guinan *et al.* (2003), Ribas *et al.* (2005), Buccino *et al.* (2006), and Guinan & Engle (2009). With respect to the Sun-Earth system, a detailed investigation about the UV effects on biological systems has been

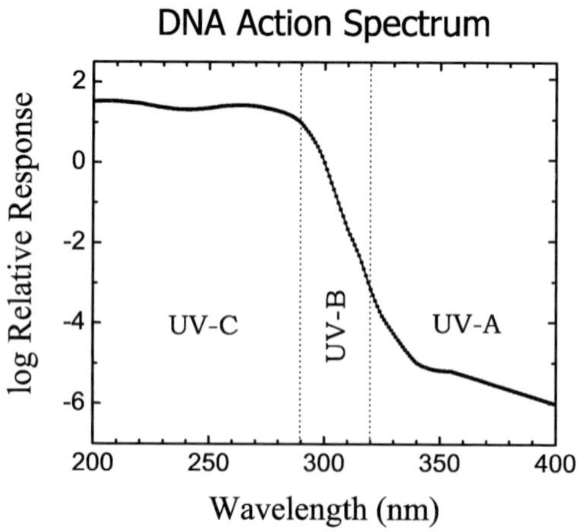

Figure 1. DNA action spectrum following Horneck (1995) and Cockell (1999).

Table 1. Stellar Data and Habitable Zones

Sp. Type	$T_{\rm eff}$	R_*	L_*	HZ-i (gen.)	HZ-i (con.)	HZ (E.)	HZ-o (con.)	HZ-i (gen.)
...	(K)	($R_\odot$)	($L_\odot$)	(AU)	(AU)	(AU)	(AU)	(AU)
F0 V	7200	1.62	6.33	1.826	2.251	2.516	3.222	3.710
F5 V	6440	1.40	3.03	1.370	1.614	1.740	2.316	2.741
G0 V	6030	1.12	1.49	1.002	1.152	1.220	1.657	1.991
G5 V	5770	0.95	0.90	0.799	0.904	0.948	1.302	1.580
K0 V	5250	0.83	0.47	0.604	0.665	0.685	0.961	1.188
K5 V	4350	0.64	0.13	0.342	0.359	0.363	0.525	0.670
M0 V	3850	0.48	0.05	0.207	0.213	0.213	0.313	0.407

Abbreviations: con. = conservative, gen. = general, E. = Earth-equivalent

pursued by Diffey (1991). This line of studies also includes the detailed modeling of the atmospheric attenuation of UV (especially due to an Earth-type ozone layer, e.g., Segura *et al.* 2003), photobiological effects and the effects of UV on biomolecules, particularly DNA, and microorganisms.

We consider the radiative effects on DNA by applying a DNA action spectrum (Horneck 1995) that shows that the damage is strongly wavelength-dependent, increasing by more than seven orders of magnitude between 400 and 200 nm (see Fig. 1). The different regimes are commonly referred to as UV-A (400–320 nm), UV-B (320–290 nm), and UV-C (290–200 nm). The test planets are assumed to be located in the stellar habitable zone (HZ). Following the analyses by Kasting *et al.* (1993) and Underwood *et al.* (2003), we distinguish between the conservative and generalized HZ (see Table 1). For stars of different spectral types, we also define planetary Earth-equivalent positions given as $R_{\rm E} = \sqrt{L/L_\odot}$. For the conservative HZ, the inner limit of habitability is given by the onset of water loss that occurs in an atmosphere warm enough to have a wet stratosphere, resulting in a gradual loss of water by photodissociation and subsequent hydrogen loss to space. Moreover, the outer limit of habitability is given by the first CO_2 condensation where for a surface temperature of 273 K, CO_2 begin to form. Concerning the general HZ, the inner limit of habitability is given by the runaway greenhouse effect.

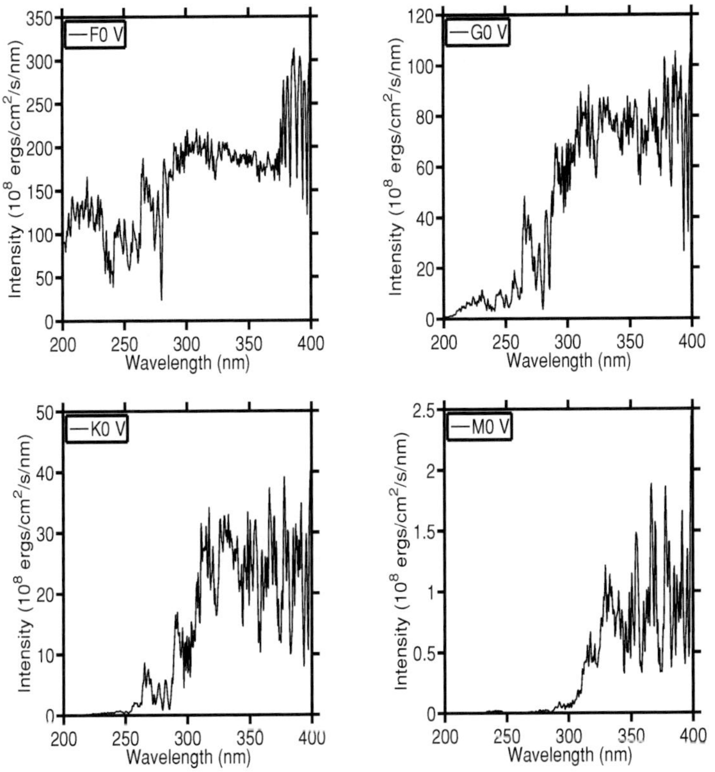

Figure 2. Kurucz models for the F0 V, G0 V, K0 V, and M0 V stars used in this study. For the sake of display, the spectral resolution has artificially been decreased to $\lambda/\Delta\lambda \simeq 300$, implemented by running means. Note the different scales of the y-axes.

In the following, we will present our results for different positions of Earth-type planets while considering both stellar photospheric and different levels of chromospheric radiation. We will also discuss the likely consequences of stellar flares.

2. Stellar Photospheric Radiation

For our study we employ a detailed consideration of stellar photospheric radiation for the UV-A, UV-B, and UV-C spectral regimes. The adopted target stars are: F0 V, F5 V, G0 V, G5 V, K0 V, K5 V, and M0 V with effective temperatures of 7200, 6440, 6030, 5770, 5250, 4350, and 3850 K, respectively (see Table 1). As stellar surface gravity, log g = 4.5 (cgs) is used, and the microturbulence broadening is considered. We make use of the spectral models by R. L. Kurucz which take into account millions or hundred of millions of lines for atoms and molecules; see Castelli & Kurucz (2004) and related publications for details.

Note that it is virtually impossible to display the complete photospheric spectra of our target stars owing to the richness in their spectra features, particularly the very large number of narrow lines due to the large number of elements and line levels taken into account. Thus, to provide a tutorial comparison between the different spectra in the 200 to 400 nm wavelengths regime, we created “fake spectra” obtained by installing a running mean. This has been done by smoothing the spectra for a length of 1 nm, corresponding to artificially decreasing the spectral resolution to $\lambda/\Delta\lambda \simeq 300$ (see Fig. 2).

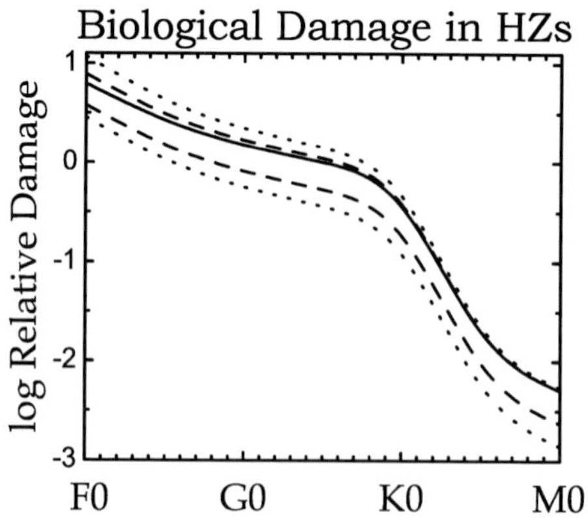

Figure 3. Biological damage to DNA for a planet (no atmosphere) at an Earth-equivalent position (solid line), and at the inner and outer limits of the conservative (dashed lines) and generalized HZ (dotted lines).

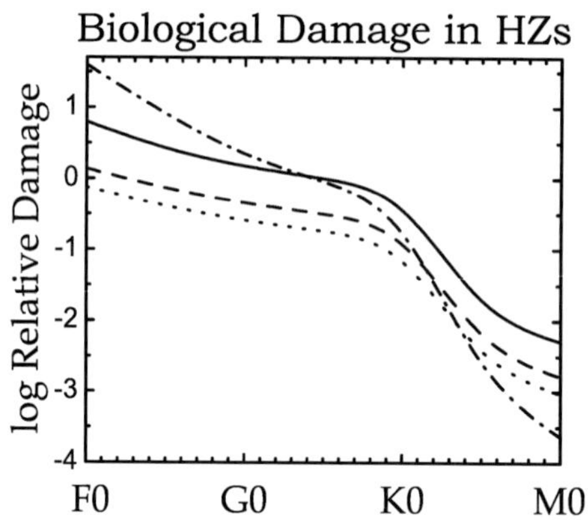

Figure 4. Biological damage to DNA for a planet at an Earth-equivalent position without an atmosphere (solid line), and an atmosphere akin to Earth 3.5 Gyr ago (dashed line) and today (dotted line). The dash-dotted line refers to a planet without an atmosphere at 1 AU from its host star, irrespectively of the position of the stellar HZ.

As part of our models, we also consider the effects of attenuation by an Earth-type planetary atmosphere, which allows us to estimate wavelength-dependent attenuation coefficients appropriate to the cases of Earth as today, Earth 3.5 Gyr ago, and no atmosphere at all (Cockell 2002). Improved simulations of the effects of planetary atmospheres must be based on the detailed treatment of atmospheric photochemistry, including the build-up and destruction of possible ozone layers.

Our results are presented in Figs. 3 and 4. They show the relative damage to DNA due to different types of main-sequence stars encompassing spectral spectral types between F0 and M0 normalized to today's Earth. We consider Earth-type planets at the inner and outer edge of either the conservative or generalized HZ as well as planets of different atmospheric attenuation. We obtained the following findings: (1) All main-sequence stars of spectral type F to M have the potential of damaging DNA due to UV radiation. The amount of damage strongly depends on the stellar spectral type, the type of the planetary atmosphere and the position of the planet in the HZ; see Cockell (1999) for previous work. (2) The damage to DNA for a planet in the HZ around an F-star (Earth-equivalent distance) due to photospheric radiation is significantly higher (factor 5) compared to

Table 2. Increase in Biological Damage due to Chromospheric Radiation

Sp. Type	log F_{basal}	Inact. Chr.	Act. Chr.
F0 V	5.890	...	...
F5 V	5.755	...	1.08
G0 V	5.590	...	1.12
G5 V	5.486	...	1.16
K0 V	5.316	...	1.4
K5 V	4.879	1.6	6.7
M0 V	4.498	2.6	17

Note: F_{basal} is given in ergs cm^{-2} s^{-1}.

planet Earth around the Sun, which in turn is significantly higher than for an Earth-equivalent planet around an M-star (factor 180). (3) We also found that the damage is most severe in the case of no atmosphere at all, somewhat less severe for an atmosphere corresponding to Earth 3.5 Gyr ago, and least severe for an atmosphere like Earth today, as expected. Moreover, any damage due to photospheric stellar radiation is mostly due to UV-C. The relative importance due to UV-B is relatively small, and damage due to UV-A is virtually nonexistent (see Cuntz *et al.* 2010 for details).

3. Effects by Stellar Chromospheres and Flares

The study of chromospheric radiation and its effects on circumstellar environments, including hosted planets, has been the focus of a large array of research projects. In our study, we added chromospheric emission as exemplified by Mg II h+k at 2803, 2796 Å to the photospheric emergent radiation. Previous studies including those associated with the "Sun in Time" project (see Sect. 4), have shown that for F and G-type stars, the photospheric UV continua are typically dominant relative to any chromospheric contributions implying that chromospherically induced biological damage for planets hosted by these stars will be insignificant.

The situation is, however, decisively different for K and M dwarfs. Any of these stars, including inactive stars, show noticeable chromospheric emission. Previous research (e.g., Cuntz *et al.* 1999, and references therein) point to the existence of two-component stellar chromospheres with the non-magnetic component (prevalent in case of old stars; see Sect. 4) heated by acoustic waves and the magnetic component heated by magnetic energy dissipation. The acoustically heated chromospheric component is usually also referred to as basal component F_{basal}, noting that the radiative energy flux ranges from 3×10^4 (M dwarfs) to 8×10^5 ergs cm^{-2} s^{-1} (F dwarfs); see Rutten *et al.* (1991).

Table 2 depicts the enhancement of biological damage due to chromospheric radiation relative to photospheric radiation for stars between spectral type F0 V and M0 V. This study has been pursued for stars with both basal (inactive) and significantly increased (active) chromospheric emission. For the latter, we considered an increase in chromospheric emission by a factor of 10, which for solar-type stars amounts to the maximal chromospheric emission observed in regard to fast-rotating (i.e., young) stars (Vilhu & Walter 1987). Our results show that for stars with basal chromospheric emission, chromosphere-induced biological damage only occurs for stars of spectral type mid-K and later, whereas for stars with relatively high chromospheric emission, chromosphere-induced biological damage can also be found in the environments of very late G-type and early K-type stars.

Other important agents for the delivery of highly energetic radiation, especially UV-C, are flares (e.g., Pettersen 1989, Redfield *et al.* 2002, Hawley *et al.* 2003, Robinson *et al.* 2005). Note that flare activity is most pronounced in K and M-type stars. It has also the potential of (partially) stripping the planetary atmospheres (e.g., Kulikov *et al.* 2007, Lammer *et al.* 2008), which is especially relevant for close-in planets, particularly planets located in the HZs of K and M-type stars (see Table 1). The maximal near-UV and far-UV radiative energies of flares are akin or larger than those of an active stellar chromosphere, although flare energy is typically provided episodically rather than continuously. Therefore, the results given in Table 2 may also be applicable to flaring as a first approximation, except that the actual biological damage due to flares might be enhanced further by a factor of 5 to 10 due to the possible damage (i.e., evaporation) of the planetary atmosphere.

Nonetheless a more detailed analysis of flare-related effects is still required, which should also incorporate a detailed treatment of planetary atmospheric photochemistry, including the build-up and destruction of ozone as pointed out by Segura *et al.* (2003, 2005) and others.

4. Summary: The 'Sun in Time' Project and Its Biology Expansion

We attempted to quantify the biological damage expected to occur in the environments of late-type main-sequence stars due to photospheric, chromospheric and flare radiation. Stellar photospheric radiation has been considered by utilizing spectral models given by R. L. Kurucz and collaborators, which take into account millions or hundred of millions of lines for atoms and molecules; see, e.g., Castelli & Kurucz (2004). Concerning chromospheric radiation, we considered models exemplifying basal chromospheric emission (e.g., Schrijver 1987, Rutten *et al.* 1991) and significantly increased chromospheric emission (e.g., Vilhu & Walter 1987). Note that basal emission is usually attributed to acoustic heating in the limiting case of magnetic energy dissipation to be small or negligible. Detailed models of one-component and two-component chromospheric heating for late-type stars have been given by, e.g., Buchholz *et al.* (1998), Cuntz *et al.* (1998, 1999), Fawzy *et al.* (2002), and Rammacher & Cuntz (2003).

It is highly noteworthy that there is a deeper underlying connection between the level of chromospheric heating and emission, on one hand, and more fundamental stellar physics involving processes concerning the stellar interior and phenomena associated with stellar winds, on the other hand. In case of the Sun, this connection has been explored in detail by, e.g., Guinan & Ribas (2002), Guinan *et al.* (2003), Ribas *et al.* (2005), and Güdel (2007). A highly crucial question, as gauged by its implied planetary, biological and societal percussions, is whether the Sun has always been a relatively inactive star or, in contrast, has experienced some periods of stronger magnetic activity.

Compelling observational evidence (Güdel *et al.* 1997) shows that zero-age main-sequence (ZAMS) solar-type stars rotate over 10 times faster than today's Sun. As a consequence of this, young solar-type stars, including the young Sun, have vigorous magnetic dynamos and correspondingly strong high energy emissions. From the study of solar-type stars of different ages, Skumanich (1972), Simon *et al.* (1985), Charbonneau & MacGregor (1993), MacGregor & Charbonneau (1994) and others showed that the Sun loses angular momentum with time via magnetized winds (magnetic breaking), thus leading to a secular increase in its rotational period (e.g., Durney 1972, Keppens *et al.* 1995). This rotation slowdown is well fitted by a Skumanich-type power law roughly proportional to $t^{-1/2}$ (e.g., Skumanich 1972, Soderblom 1982, Ayres 1997). In response to slower rotation, the solar dynamo strength diminishes with time, causing the Sun's high energy emissions and mass loss also to undergo significant decreases.

A direct consequence of this behavior concerns the generation of photospheric and chromospheric magnetic flux, which constitute the physical reason for the varying level of chromospheric heating and associated generation of UV and EUV radiation with stellar age or rotation period (e.g., Mathioudakis *et al.* 1995). Consequently, it is thus also possible to relate the photospheric magnetic flux to the stellar rotation period (Noyes *et al.* 1984; Marcy & Basri 1989; Montesinos & Jordan 1993; Saar 1996a) as well as to the emergent chromospheric emission (Saar & Schrijver 1987; Schrijver *et al.* 1989; Montesinos & Jordan 1993; Saar 1996b; Jordan 1997).

The acquisition, interpretation and modeling of data for any of these phenomena is the main focus of the "Sun in Time" project (originally solely focused on G0–G5 V stars) pursued under the leadership of E. F. Guinan, which relates, in a statistical sense, the strength of the emergent UV, EUV and X-ray flux and the stellar mass loss rate to the stellar age. More recently, this project has been extended to K-type and M-type dwarfs with the latter element of study now referred to "Living with a Red Dwarf". The critical ultimate connection to be targeted in the future, see Guinan & Engle (2009) for the dissemination of various intermediate results, is to relate the type and strength of stellar activity for different types of stars to planetary climatology, particularly for planets in stellar HZs, and to exosolar planetary biology, if existing.

Acknowledgements

M.C. would like to thank the American Astronomical Society and the IAU for travel support.

References

Ayres, T. R. 1997, *J. Geophys. Res.* 102 (E1), 1641

Buccino, A. P., Lemarchand, G. A., & Mauas, P. J. D. 2006, *Icarus*, 183, 491

Buchholz, B., Ulmschneider, P., & Cuntz, M. 1998, *ApJ*, 494, 700

Castelli, F. & Kurucz, R. L. 2004, in N. E. Piskunov, W. W. Weiss & D. F. Gray (eds.), *Modelling of Stellar Atmospheres*, IAU Symp. 210 (Astr. Soc. Pac., San Francisco), CD-ROM, Poster 20

Charbonneau, P. & MacGregor, K. B. 1993, *ApJ*, 417, 762

Cockell, C. S. 1999, *Icarus*, 141, 399

Cockell, C. S. 2002, in G. Horneck & C. Baumstark-Khan (eds.), *Astrobiology: The Quest for the Conditions of Life* (Springer, Berlin), p. 219

Cuntz, M., Ulmschneider, P., & Musielak, Z. E. 1998, *ApJ* (Letters), 493, L117

Cuntz, M., Rammacher, W., Ulmschneider, P., Musielak, Z. E., & Saar, S. H. 1999, *ApJ*, 522, 1053

Cuntz, M., Guinan, E. F., & Kurucz, R. L. 2010, *Icarus*, submitted

Diffey, B. L. 1991, *Physics in Medicine and Biology*, 36, 299

Durney, B. 1972, in Sonett, C. P., Coleman, P. J., & Wilcox, J. M. (eds.), *Solar Wind* (NASA, Washington), p. 282

Fawzy, D., Rammacher, W., Ulmschneider, P., Musielak, Z. E., & Stępień, K. 2002, *A&A*, 386, 971

Goldsmith, D. & Owen, T. 2002, *The Search for Life in the Universe*, 3rd ed. (University Science Books, Sausalito)

Güdel, M. 2007, *Liv. Rev. Sol. Phys.*, 4, 3

Güdel, M., Guinan, E. F., & Skinner, S. L. 1997, *ApJ*, 483, 947

Guinan, E. F. & Engle, S. G. 2009, *The Ages of Stars*, IAU Symp. 258 [arXiv:0903.4148]

Guinan, E. F. & Ribas, I. 2002, in Montesinos, B., Gimenez, A., & Guinan, E. F. (eds.), *The Evolving Sun and Its Influence on Planetary Environments* (Astr. Soc. Pac., San Francisco), Vol. 269, p. 85

Guinan, E. F., Ribas, I., & Harper, G. M. 2003, *ApJ*, 594, 561
Hawley, S. L., *et al.* 2003, *ApJ*, 597, 535
Horneck, G. 1995, *J. Photochem. Photobiol. B: Biology*, 31, 43
Jordan, C. 1997, *Astron. Geophys.*, 38, 10
Kasting, J. F., Whitmire, D. P., & Reynolds, R. T. 1993, *Icarus*, 101, 108
Keppens, R., MacGregor, K. B., & Charbonneau, P. 1995, *A&A*, 294, 469
Kulikov, Yu. N., Lammer, H., Lichtenegger, H. I. M., Penz, T., Breuer, D., Spohn, T., Lundin, R., & Biernat, H. K. 2007, *Space Sci. Rev.*, 129, 207
Lammer, H., Kasting, J. F., Chassefière, E., Johnson, R. E., Kulikov, Yu. N., & Tian, F. 2008, *Space Sci. Rev.*, 139, 399
MacGregor, K. B. & Charbonneau, P. 1994, in Caillault, J.-P. (ed.), *Cool Stars, Stellar System, and the Sun VIII* (Astr. Soc. Pac., San Francisco), Vol. 64, p. 174
Marcy, G. W. & Basri, G. 1989, *ApJ*, 345, 480
Mathioudakis, M., Fruscione, A., Drake, J. J., McDonald, K., Bowyer, S., & Malina, R. F. 1995, *A&A*, 300, 775
Montesinos, B. & Jordan, C. 1993, *MNRAS*, 264, 900
Noyes, R. W, Hartmann, L. W., Baliunas, S. L., Duncan, D. K., & Vaughan, A. H. 1984, *ApJ*, 279, 763
Pettersen, B. R. 1989, *Sol. Phys.*, 121, 299
Rammacher, W. & Cuntz, M. 2003, *ApJ* (Letters), 594, L51
Redfield, S., Linsky, J. L., Ake, T. B., Ayres, T. R., Dupree, A. K., Robinson, R. D., Wood, B. E., & Young, P. E. 2002, *ApJ*, 581, 626
Ribas, I., Guinan, E. F., Güdel, M., & Audard, M. 2005, *ApJ*, 622, 680
Robinson, R. D., *et al.* 2005, *ApJ*, 633, 447
Rutten, R. G. M., Schrijver, C. J., Lemmens, A. F. P., & Zwaan, C. 1991, *A&A*, 252, 203
Saar, S. H. 1996a, in Strassmeier, K. G. & Linsky, J. L. (eds.), *Stellar Surface Structure*, IAU Symp. 176 (Kluwer, Dordrecht), p. 237
Saar, S. H. 1996b, in Uchida, Y., *et al.* (eds.) *Magnetodynamic Phenomena in the Solar Atmosphere — Prototypes of Stellar Magnetic Activity*, IAU Coll. 153 (Kluwer, Dordrecht), p. 367
Saar, S. H. & Schrijver, C. J. 1987, in Linsky J. L. & Stencel R. E. (eds.), *Cool Stars, Stellar System, and the Sun V* (Springer, Berlin), p. 38
Schrijver, C. J. 1987, *A&A*, 172, 111
Schrijver, C. J., Coté, J., Zwaan, C., & Saar, S. H. 1989, *ApJ*, 337, 964
Segura, A., Krelove, K., Kasting, J. F., Sommerlatt, D., Meadows, V., Crisp, D., Cohen, M., & Mlawer, E. 2003, *Astrobiology*, 3, 689
Segura, A., Kasting, J. F., Meadows, V., Cohen, M., Scalo, J., Crisp, D., Butler, R. A. H., & Tinetti, G. 2005, *Astrobiology*, 5, 706
Simon, T., Herbig, G., & Boesgaard, A. M. 1985, *ApJ*, 293, 551
Skumanich, A. 1972, *ApJ*, 171, 565
Soderblom, D. R. 1982, *ApJ*, 263, 239
Underwood, D. R., Jones, B. W., & Sleep, P. N. 2003, *Int. J. Astrobiology*, 2, 289
Vilhu, O. & Walter, F. M. 1987, *ApJ*, 321, 958

Solar and Stellar Variability: Impact on Earth and Planets
Proceedings IAU Symposium No. 264, 2009
A. G. Kosovichev, A. H. Andrei & J.-P. Rozelot, eds.

doi:10.1017/S1743921309993048

Influence of the Schwabe/Hale solar cycles on climate change during the Maunder Minimum

Hiroko Miyahara[1], Yusuke Yokoyama[2,3,4], and Yasuhiko T. Yamaguchi[2,3,4]

[1]Institute for Cosmic Ray Research, The University of Tokyo,
5-1-5 Kashiwanoha, Kashiwa 277-8582, Japan
email: hmiya@icrr.u-tokyo.ac.jp

[2]Ocean Research Institute, The University of Tokyo,
1-15-1 Minamidai, Nakano-ku, Tokyo 164-8639, Japan

[3]Department of Earth & Planetary Science, The University of Tokyo,
7-3-1 Hongo, Tokyo 113-0033, Japan

[4]Institute of Biogeoscience, Japan Agency for Marine-Earth Science and Technology,
2-15, Natsushima-cho, Yokosuka 237-0061, Japan

Abstract. We have examined the variation of carbon-14 content in annual tree rings, and investigated the transitions of the characteristics of the Schwabe/Hale (11-year/22-year) solar and cosmic-ray cycles during the last 1200 years, focusing mainly on the Maunder and Spoerer minima and the early Medieval Maximum Period. It has been revealed that the mean length of the Schwabe/Hale cycles changes associated with the centennial-scale variation of solar activity level. The mean length of Schwabe cycle had been ~14 years during the Maunder Minimum, while it was ~9 years during the early Medieval Maximum Period. We have also found that climate proxy record shows cyclic variations similar to stretching/shortening Schwabe/Hale solar cycles in time, suggesting that both Schwabe and Hale solar cycles are playing important role in climate change. In this paper, we review the nature of Schwabe and Hale cycles of solar activity and cosmic-ray flux during the Maunder Minimum and their possible influence on climate change. We suggest that the Hale cycle of cosmic rays are amplified during the grand solar minima and thus the influence of cosmic rays on climate change is prominently recognizable during such periods.

Keywords. Sun: activity, solar-terrestrial relations, ISM: cosmic rays

1. Introduction

Both observational and proxy records of climate change often show quasi periodic variations similar to solar activity cycles over a wide range of time scales. However, the detailed mechanism and the extent of the influence of solar activity on climate change have not been clearly understood. Although the exact role of each of solar parameters on climate change has not been quantitatively clarified, several possible mechanisms are proposed; such as the forcing through total (e.g. Lean *et al.*, 1995) and spectral irradiance (e.g. Haigh 1996; Kodera and Kuroda, 2005), solar wind (e.g. Tinsley, 1996) and the galactic cosmic rays (Friis-Christensen and Svensmark, 1997; Svensmark, 2007).

Among these parameters related to solar activity, galactic cosmic rays possess characteristic variations depending on the polarity of solar dipole magnetic filed as shown in Figure 1. The polarity of solar dipole magnetic field reverses at every maxima of 11-year sunspot activity cycle, and so the polarity reversals of solar magnetic field possess 22-year

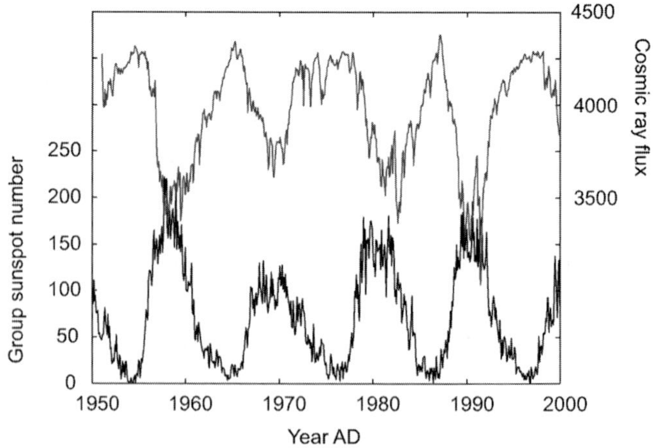

Figure 1. Monthly variation of the cosmic ray intensity observed by the Climax neutron monitor (upper gray line) and the group sunspot numbers (Hoyt & Schatten, 1998) (lower black line).

cycle. The cosmic rays are modulated by solar wind and the interplanetary magnetic field and hence the flux of cosmic rays at the earth varies with the 11-year solar activity cycle, while, the polarity of solar dipole magnetic field determines the trajectory of cosmic rays in the heliosphere and thus the flux of cosmic rays at the earth varies depending also on the polarity of solar dipole magnetic field (Kota and Jokipii 2001). As is shown in Figure 1, the patterns of cosmic ray flux over solar cycles slightly differ depending on the polarity of solar dipole magnetic field, resulting in the component of 22-year cycle in cosmic-ray variation. This feature is very helpful in distinguishing the effect of cosmic rays on climate change from the other effects caused by e.g. irradiative outputs of the Sun.

Extension of the record of cosmic rays back in time enable us to examine if the connection between cosmic rays and climate change suggested by Friis-Christensen and Svensmark (1997) and Svensmark (2007) for the recent two decades had also existed in the past. We had investigated the history of Schwabe and Hale solar and cosmic ray cycles based on the carbon-14 content in tree rings with annual time resolution, originally for understanding the mechanism of multi-decadal to multi-centennial variation of solar activity level. Such record is however also applicable to investigating the Sun-climate relationship at decadal time scale. Carbon-14 is produced by cosmic rays, and circulates in the form of carbon dioxide to be absorbed in trees by photo synthesis. Since the age determination of each annual data is assured in the case of using tree rings, it is possible to determine the history of solar cycles with accurate timing. The beryllium-10 in ice cores from polar region can be also used for the reconstruction of solar cycles in the past. In the case of using ice cores, it is often difficult to obtain the record with absolute age, while, it is possible to derive much clear signal than carbon-14 due to the difference in the circulation process. The combination of these two nuclides provides clear image of cosmic ray variation with reliable age.

2. Solar cycle during the Maunder Minimum

Figure 2 shows the carbon-14 content in tree rings with decadal resolution obtained by Stuiver *et al.* (1998) and the annually measured data by Miyahara *et al.* (2004; 2006; 2007; 2008). During the last millennium, the Sun has experienced several prolonged sunspot minima such as the Maunder and the Spoerer Minima, which appear as upward peaks in

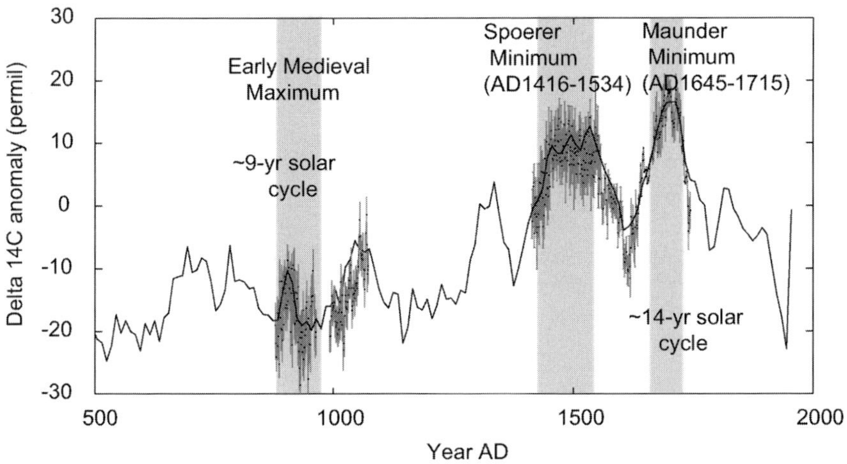

Figure 2. Variation of the carbon-14 content in tree rings with decadal resolution obtained by Stuiver *et al.*, 1998 (black line), and the annually measured data by Miyahara *et al.* (2004; 2006, 2007; 2008) (black dots). The error bars for the annual data have been shown together (gray bars). The frequency analyses of the annual data over the shaded areas for the early medieval maximum and the Maunder Minimum have revealed that the mean length of the 11-year solar cycle had been modulated to ∼9 years and ∼14 years, respectively.

carbon-14 content in Figure 2, in addition to the long-term relatively active era called the Medieval Maximum Period (the ∼10-12th century). Frequency analyses of the annually measured carbon-14 records for the Maunder Minimum (AD1645-1715) revealed that the mean length of the 11-year cycle over the period had been stretched to be sin14 years (Miyahara *et al.*, 2004; 2008). On the other hand, the mean length was sin9 years at the early Medieval Maximum Period (Miyahara *et al.*, 2008). sin28-year cycle, which is the double length of sin14 years, was also detected in the frequency spectrum of carbon-14 content during the Maunder Minimum, suggesting that the polarity reversals of solar dipole magnetic field had been carried on during the prolonged sunspot absence. Figure 3 shows the reconstructed Schwabe and the Hale cycles during the Maunder Minimum. The Schwabe cycle in Figure 3(a) was obtained by applying Fourier band-pass filter with the bandwidth of 10-16 years, while the Hale cycle in Figure 3(b) was obtained by applying the filter of 20-30 years. The reconstructed Schwabe cycle is consistent with the variation of group sunspot numbers by Hoyt and Schatten (1998), at the times they show distinct cyclic variation. Solar cycles can now be extended back to the early 17th century complemented by the carbon-14 record, revealing that the Maunder Minimum had started at Solar Cycle -8.

3. Solar and cosmic-ray Hale cycle during the Maunder Minimum

The Hale cycle of cosmic rays in the recent decades indicate that the mean flux of cosmic rays over solar cycle is relatively higher when the polarity of solar dipole magnetic filed is positive as shown in Figure 1. This profile is consistently reproduced by the numerically simulated cosmic ray flux as shown in Figure 4(b), where the profile is obtained based on the model by Kota and Jokipii (2001) shown in Figure 4(a) by assuming that the range of the tilt angle of solar neutral line over solar cycle is 10-70 degrees. For the Maunder Minimum, it had been suggested by Jokipii (1991) that the phase of the Hale cycle could have been 180 degrees reversed during the Maunder Minimum, which means that incoming cosmic ray flux is expected to be relatively higher when the polarity is

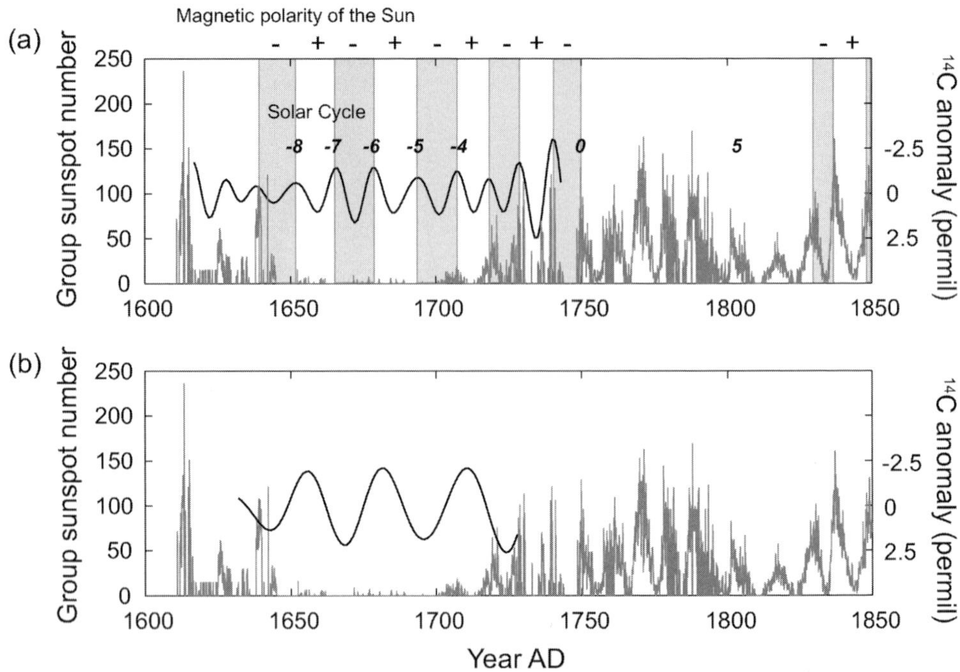

Figure 3. Reconstructed Schwabe and Hale cycles based on annually measured carbon-14 content in tree rings by Miyahara *et al.* (2004; 2007). (a) Band-pass filtered carbon-14 record with the bandwidth of 10-16 years, together with the group sunspot numbers by Hoyt & Schatten (1998) (gray line). The number of Solar Cycle is determined based on the band-pass filtered carbon-14 record. (b) Same with (a) but the bandwidth is 20-30 years. The magnetic polarity shown in (a) is determined by the phase of Hale cycle in (b) for the Maunder Minimum, while that for AD1830-1850 is determined by the number of solar cycles from present.

negative. Based on the variations of the Schwabe/Hale cycles detected by carbon-14 and also on the model in Figure 4(a), we obtained the time profile shown in Figure 4 (c) as the most probable variation of cosmic rays during the Maunder Minimum, where the range of the tilt angle was assumed to be 0-70 degrees. The profile is characterized by the sharp peak of cosmic ray flux at solar cycle minima when the magnetic polarity of the Sun is negative, resulting in the enhancement of the Hale cycle in cosmic ray variation. This profile has been recently supported by the beryllium-10 record from Greenland ice core (Berggren *et al.*, 2009). In the record of beryllium-10, three sharp peaks are recognized at the beginning of the Maunder Minimum, ~1700AD and in between. The periods of these three peaks are consistent with the timing that the content of carbon-14 in tree rings was also found to be relatively higher as is shown in Figure 3(a). It suggests that these three periods correspond to the minima of Schwabe cycle of the times solar magnetic polarity is negative. Both of our carbon-14 and the Greenland beryllium-10 records suggest for the Maunder Minimum that the tilt angle of solar neutral line had reached close to ~0 degree at every cycle minima, while the neutral line had certainly waved toward the suggested polarity reversals at solar cycle maxima. The transitions of the polarity of solar dipole magnetic field around the Maunder Minimum deduced from the carbon-14 record is summarized in Figure 3(a). The determined polarity is consistent with what we obtain from the counting of the number of solar cycles from present (as those for AD1830-1850 in Figure 3(a)). It had been pointed out that there is a possibility that Solar Cycle 4

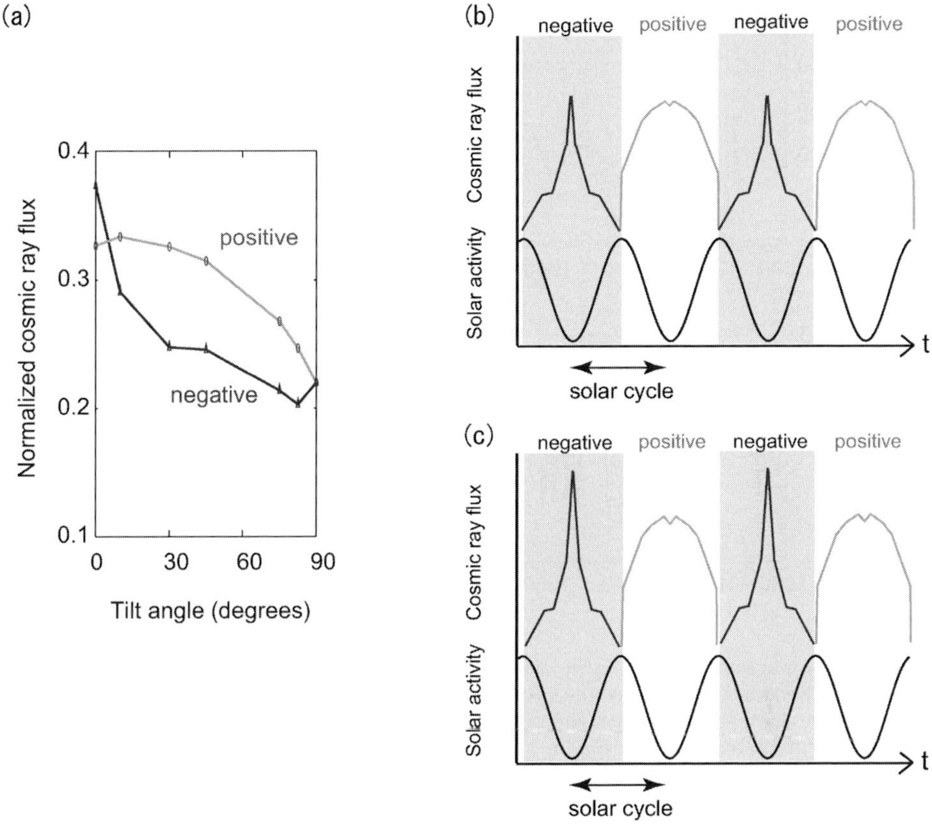

Figure 4. Numerically simulated variation of cosmic ray flux at the earth against solar activity and magnetic polarity by Kota and Jokipii (2001). (a) The flux of cosmic rays against the polarity of solar dipole magnetic field and the tilt angle of solar magnetic neutral line (based on Kota & Jokipii, 2001). (b) Model of the cosmic ray variation when the range of the tilt angle over solar cycle is 10-70 degrees, which is comparable to that of recent decades. (c) Model of the cosmic ray variation during the Maunder Minimum where the range of the tilt angle is assumed to be 0-70 degrees.

consists of two solar cycles (Usoskin *et al.*, 2001), however, the determined history of the polarity suggests that it consists of single solar cycle with long duration.

4. Influence of the Schwabe/Hale solar cycles on climate change

Figure 5 shows the comparison between the reconstructed solar cycle (same as in Figure 3(a)) and the reconstructed temperature based on oxygen-18/oxygen-16 ratio in Greenland Ice core (Vinther, 2003). The shaded areas correspond to cycle minima of negative solar polarity determined by carbon-14 record, where the flux of incoming galactic cosmic rays is suggested to be high as discussed in the previous section. It is recognized that the reconstructed temperature in Figure 5(b) shows rapid cooling around the shaded area, corresponding to the periods when the flux of cosmic rays is high. It is suggested that the climate change has some component dependent on the polarity of solar magnetic field, resulting in the appearance of the Hale cycle in its variation, and that solar magnetic cycle is playing important role in climate change probably through the mechanism involving cosmic rays. As the actual length of the Schwabe cycle is ~14 years,

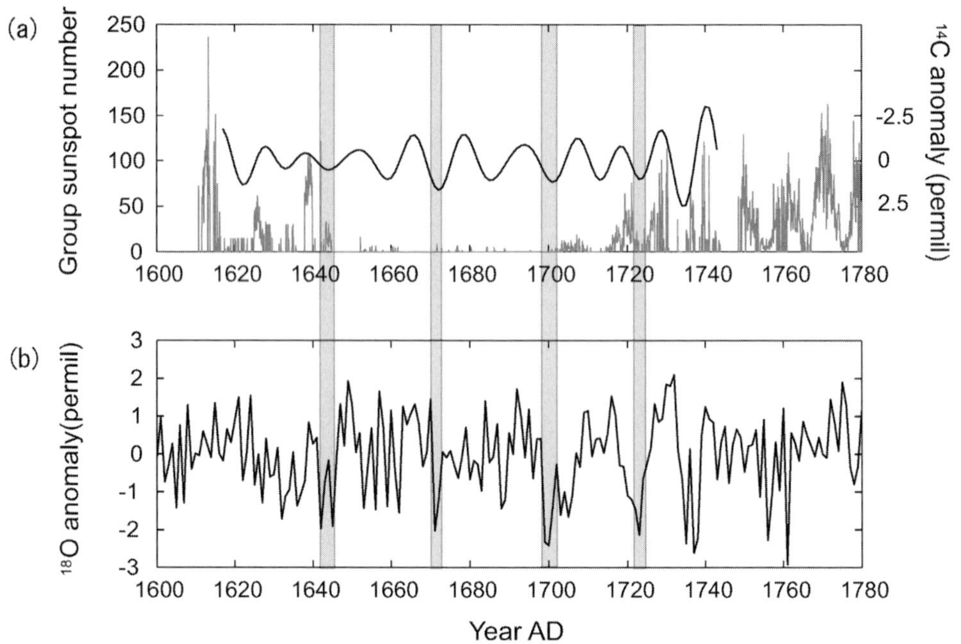

Figure 5. Comparison of solar cycle with reconstructed temperatures around the Maudner Minimum. (a) Reconstructed Schwabe cycle based on the band-pass filtered carbon-14 record as shown in Figure 3(a), together with the group sunspot numbers. (b) Reconstructed temperature based on oxygen-18/oxygen-16 ratio in the Greenland ice core by Vinther *et al.* (2003). The shaded areas correspond to the solar cycle minima of polarity negative determined by carbon-14 record as shown in Figure 3.

the length of the Hale cycle detected in the reconstructed temperature is also stretched to be ~28 years. As has been discussed in our previous paper (Miyahara *et al.*, 2008), the phase of the Hale cycle in climate change is reversed after the Maunder Minimum as well as that of cosmic rays, and the cooling is found to occur especially around the cycle minima of polarity positive. The length of climate Hale cycle also recovers to 22 years. It is noteworthy that the variability of climate change caused by cosmic ray variation can be more prominent during the grand solar minima compared to the other era (also see Miyahara *et al.*, 2008), since the variability of cosmic rays is likely enhanced during such periods, as is suggested in Figure 4(c), due to the change in the range of tile angle of solar neutral line. It is contradicting to the expected feature of the influence of irradiative outputs of the Sun, for which it is expected that the amplitude becomes less when solar activity is low. Such amplification of the "22-year" cycle could have brought additional occasional very severe climate during the Little Ice Age.

5. Summary

The mechanism of the influence of cosmic rays on the cloud formation is not fully understood, however, our proxy based analyses of cosmic rays and climate change during the Maunder Minimum exhibit the importance of cosmic rays as a medium of solar forcing of climate change at decadal to multi-decadal time scales. The complex features of solar magnetic and cosmic ray cycles, such as the variable length of the "11-year" cycle, the subsequent lengthening/shortening of the "22-year" Hale cycle, the amplification of the 22-year cycle in cosmic rays at grand solar minima, may be able to explain some of the complex features of climate change at this time scale.

Acknowledgements

We thank Dr. J. Kota (Univ. of Arizona) and Prof. K. Munakata (Shinshu Univ.) for fruitful discussions. This research was partly funded by the Japan Society for the Promotion of Science, the Grant-in-Aid for Scientific Research, and Global Environmental Research Fund.

References

Berggren, A.-M., Beer, J., Possnert, G., Aldahan, A., Kubik, P., Christl, M., Johnsen, S. J., Abreu, J., & Vinther, B. M. 2009 *Geophys. Res. Lett.*, 36, L11801

Friis-Christensen, E. F. & Svensmark, H. 1997 *Adv. Space Res.*, 20, 913

High, J. D. 1996 *Science*, 272, 981

Hoyt, D. V. & Schatten, K. H. 1998 *Sol. Phys.*, 181, 491

Jokipii, J. R. 1991, in: C.P. Sonnet, M.S. Giampapa & M.S. Matthers (eds.), *The Sun in Time*, (The University of Arizona Press, USA)

Kodera, K. & Kuroda, Y. 2005 *J. Geophys. Res.*, 110, D02111

Kota, J. & Jokipii, J. R. 2001 *Adv. Space Res.*, 27, 529

Lean, J. L., Beer, B., & Bradley, R. 1995 *Geophys. Res. Lett.*, 22, 3195

Miyahara, H., Masuda, K., Muraki, Y., Furuzawa, H., Menjo, H., & Nakamura, T. 2004 *Sol. Phys.*, 224, 317

Miyahara, H., Masuda, K., Muraki, Y., Kitagawa, H., & Nakamura, T. 2006 *J. Geophys. Res.*, 111, A03103

Miyahara, H., Masuda, K., Nagaya, K., Kuwana, K., Muraki, Y., & Nakamura, T. 2007 *Adv. Space Res.*, 40, 2060

Miyahara, H., Yokoyama, Y., & Masuda, K. 2008 *Earth Planet. Sci. Lett.*, 272, 290

Stuiver, M., Reimer, P., & Braziunas, T.F. 1998 *Radiocarbon*, 40, 1127

Svensmark, H. 2007 *Astron. Geophys.*, 48, 118

Tinsley, B. A. 1996 *J. Geophys. Res.*, 101, 29701

Usoskin, I. G., Mursula, K., & Kovaltsov, G.A. 2001 *A&A*, 101, 29701

Vinther, B. M., Johnsen, S. J., Andersen, K. K., Clausen, H. B., & Hansen, A. W. 2003 *Geophys. Res. Lett.*, 30, doi:10.1029/2002GL016193

Solar and Stellar Variability: Impact on Earth and Planets
Proceedings IAU Symposium No. 264, 2009
A. G. Kosovichev, A. H. Andrei & J.-P. Rozelot, eds.

doi:10.1017/S174392130999305X

STRESS – STEREO TRansiting Exoplanet and Stellar Survey

Vinothini Sangaralingam[1a], Ian R. Stevens[1], Steve Spreckley[1] and Jonas Debosscher[2]

[1]School of Physics and Astronomy, University of Birmingham, Birmingham, B15 2TT, UK. [1a]email: vs@star.sr.bham.ac.uk
[2]Instituut voor Sterrenkunde, Catholic University of Leuven, Belgium.

Abstract. The **H**eliospheric **I**mager (**HI**) instruments on board the two **STEREO** (**S**olar **TE**rrestrial **RE**lations **O**bservatory) spacecraft provides an excellent opportunity for space based stellar photometry. The HI instruments provide a wide area coverage ($20^\circ \times 20^\circ$ for the two HI-1 instruments and $70^\circ \times 70^\circ$ for the two HI-2 instruments) and long continuous periods of observations (20 days and 70 days respectively). Using HI-1A which has a pass band of 6500Å to 7500Å and a cadence of 40 minutes, we have gathered photometric information for more than a million stars brighter than 12th magnitude for a period of two years. Here we present some early results from this study on a range of variable stars and the future prospects for the data.

Keywords. methods: data analysis, techniques: image processing, stars: variables, survey, etc.

1. Introduction

At magnitude 12, it is estimated that at least 90% of the variables are unknown (Eyer & Blake, 2005). Even though we know the basic mechanism of pulsation in most of these objects, the models are still not adequate enough to explain the evolution of these objects across the HR diagram (Eyer & Mowlavi, 2008). In recent years, there has been an increased interest to study these variables due to the technical advancements which provides a high precision and long time-series data to the community. The **S**olar **TE**rrestrial **RE**lations **O**bservatory - STEREO, is a system of two identical spacecrafts in Heliocentric Earth orbit, one travelling ahead and another behind Earth. Each spacecraft houses a suite of identical instruments to study the origin and evolution of Coronal Mass Ejections (CME) from the Sun in 3D. There are two Heliospheric Imagers (HI), which are wide angle imaging systems designed to study the evolution of CMEs in the interplanetary medium and to enable a stereo-graphic reconstruction of CMEs propagating through the inner heliosphere up till 200 solar radii ($R_\odot$). Hence these instruments are also capable of continuously observing the background stars to a high precision, thus aiding us to do stellar photometry as well. The large cadence, high precision, wide magnitude range and broad sky coverage of this instrument makes it feasible to search for transiting extra solar planets. The 4 instruments together cover nearly 60% of the sky over a year, thus placing STEREO data as a compromise between other currently active surveys for transiting exoplanets and stellar variability studies. We have developed the basic data reduction and analysis pipeline to do aperture photometry on images from the HI-1A instrument so far. In this conference, we present the preliminary results obtained as well as the future prospects for the data.

2. Instrument overview

The HI-1 instrument is a multi-baffled imaging system sensitive to wavelengths ranging from 6500Å - 7500Å with a field of view of 20° centered at 50 $R_\odot$ and a resolution of 35.15″/pixel whereas the HI-2 instrument has a large passband varying from 4000Å to 10000Å , with a field of view of 70° centered at 200 $R_\odot$ and a resolution of 2.05′/pixel. A quick review of the main characteristics of the two HI instruments are given in Table 1.

Table 1. HI Instrument Characteristics

	HI-1 Characteristics	HI-2 Characteristics
Field of View	20° × 20°	70° × 70°
Plate scale	35.15″/pixel	2.05′/pixel
Passband	6500Å - 7500Å	4000Å - 10000Å
Nominal exposure time	40 sec	50 sec
Integrated cadence	40 min	2 hrs

The HI-1 and HI-2 cameras have individual exposure times (single shutter open/close cycle) of 40 and 50 seconds respectively separated only by the CCD read-out times of approximately 4.8 seconds. Thirty such exposures are summed up together, after correcting for cosmic ray effects (on-board), thus providing a cadence of 40 minutes for each HI-1 image. Similarly 99 exposures of HI-2 are summed to provide a digitally integrated image cadence of 120 minutes (Eyles *et al.*, 2009).

3. Data Analysis

The summed up images are publicly available for download from the UK Solar System Data Centre (UKSSDC) website of Rutherford Appleton Laboratory (RAL) in an interactive way (www.ukssdc.rl.ac.uk). All initial analysis of the data is done using IDL solar-soft packages and software basically developed by STEREO support team at RAL.

All the raw images have to be preprocessed with the pipeline developed by RAL to remove the various basic instrumental and spacecraft effects such as shutterless readout, saturated columns due to bright stars and planets, pre-flight flat fielding, bad files and also to update the spacecraft pointing and optical parameters in the header files (Brown *et al.*, 2009).

The data at this stage still has the solar F corona which dominates the image. A minimum background map for each day is then subtracted from the image. The photometric information is now extracted from these images by the processing pipeline developed by us (Fig. 1) with the input catalogue from NOMAD - Naval Observatory Merged Astrometric Dataset for objects brighter than 12 in the R magnitude.

The light curves thus obtained exhibits two characteristic modes of variability. The minimum background subtraction actually leaves a residual one day signal in the data due to the variations in F corona in time scales of a few hours. One solution of the problem is to make a more robust background map which is being currently investigated by us with the help of the team at RAL. The other is to model the F corona which is beyond the scope of this project. Hence at present it is not possible to study periodicities in the region of one day. Another trend exhibited by the data is due to the path of the object across the CCD. Since the contribution from the F corona varies as we move from left

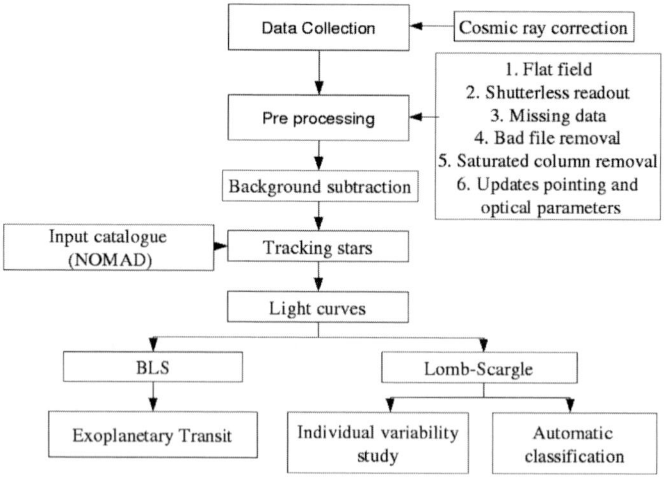

Figure 1. A schematic of the STEREO HI data analysis pipeline, showing the various stages of data preparation involved in producing light curves from raw images (see text for explanation).

to right of the CCD, a varying flux is subtracted during the minimum background step of the process. We are working on removing this trend using various filtering methods.

So far, we have processed data for nearly a million stars for two orbits of the spacecraft with 20 days of observation in each orbit for the HI-1A instrument. As a preliminary work, we analysed all the light curves using Lomb-Scargle periodogram (Scargle, 1982) for periods in the range of 0.1 days to 10 days (the Nyquist frequency is 18 cd^{-1}) excluding a region around one day. The lower cut-off of the point to point scatter of objects which are not flagged by Lomb-Scargle is less than a percent for stars of magnitude 6 to 8 and a few percentage for objects fainter than 8th magnitude. But these are just initial results, with the above mentioned systematic variabilities, hence more work needs to be done to get a very accurate estimate of these values. All these various parameters makes this data conducive to study variables like β Cepheids, δ Scutis, γ Doradus as well as exoplanetary transits. Another aspect of our work is to automatically classify these variables using neural networks (Debosscher *et al.*, 2007, Sarro *et al.*, 2009). A

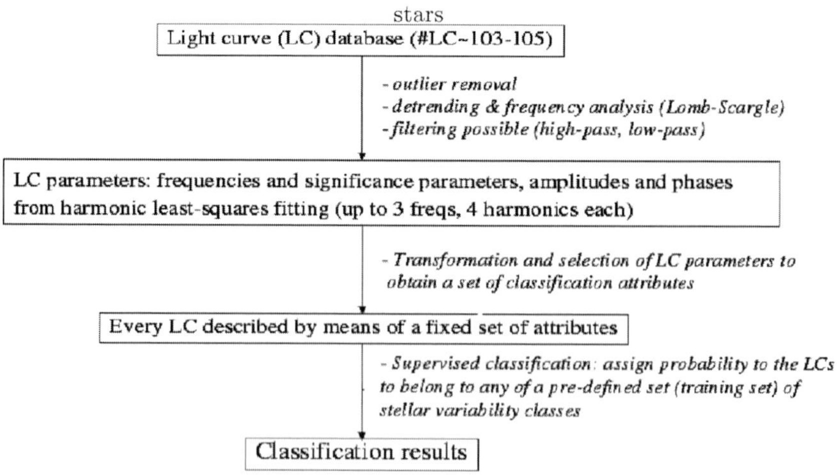

Figure 2. A Schematic of the steps involved in the automatic classification of these variable stars

brief schematic of the process involved is given in Fig. 2 (for more explanation please see the above references).

4. Results and Discussions

In this section, we present sample light curves of some well known variable types, obtained from our analysis pipeline. Also, to illustrate the potential of the data, we present light curve of a new variable object as well as a suspected micro-variable star. We advise caution when viewing these results since they are very preliminary and definitely more investigation is needed to confirm or dismiss these results. We also present an example from our automatic classification work.

These objects are selected at random from our database just to represent the various types of stars which can be studied as well as to describe the quality of the data at this early stage. We would like to draw your notice to the fact that these data contain all the various noise discussed in the previous section as well as some unknown factors.

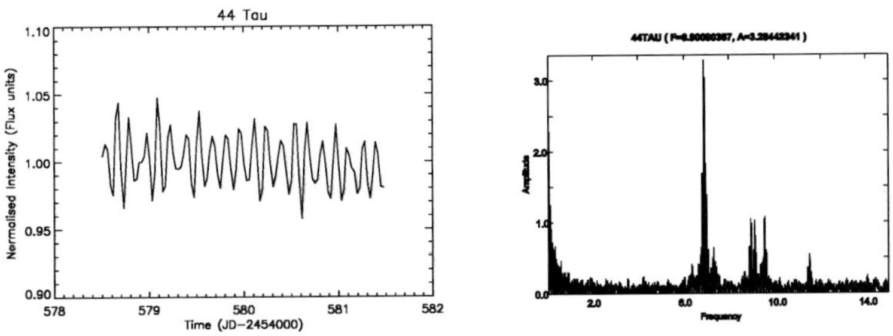

Figure 3. STEREO light curve of 44 Tau - a δ Scuti (Vmag = 5.39) and its frequency spectrum (produced by Period04).

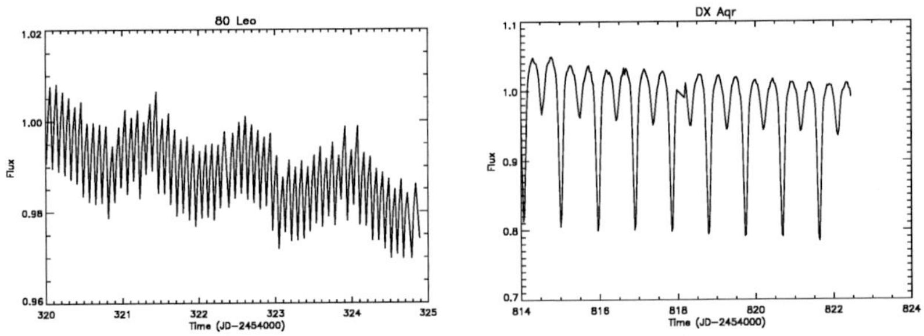

Figure 4. Left: STEREO light curve of 80 Leo - a γ Dor - Vmag = 6.35. Right: DX Aqr - Vmag = 6.37 - An Algol type eclipsing binary.

<u>**44 Tau**</u>: A δ Scuti variable with known 29 pulsational frequencies (Vmag = 5.39) in the region of 6 - 29 cycles per day (Antoci *et al.*, 2007). From STEREO data (Fig. 3), we can detect nine of those frequencies with an SNR of 26.9 at the prominent frequency of 6.9 cycles per day. Right panel of Fig. 3 is the power spectrum obtained using Period04 (Lenz & Breger, 2005), which clearly shows the different periods present. The amplitude of few of the frequencies detected were different from that in literature, indicating that more work needs to be done to understand this star.†

† The SNR of all objects are computed by the Period04 software.

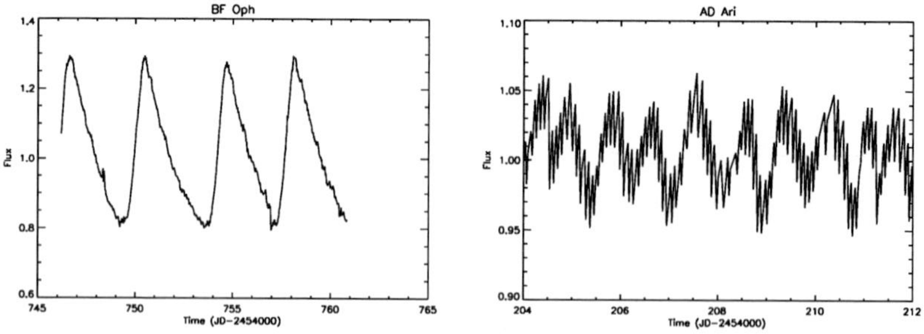

Figure 5. Left: BF Oph - Classical Cepheid - Vmag = 7.37. Right: AD Ari - δ Scuti - Vmag = 7.43

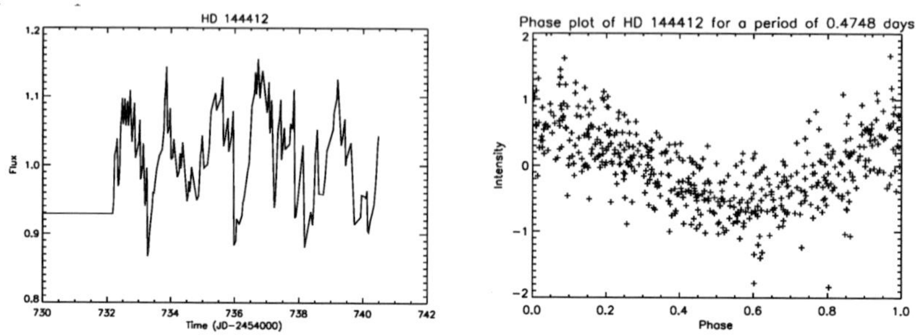

Figure 6. Left panel - STEREO light curve of HD 144412 which is found to execute a previously unreported variability. Right panel contains the phase plot at a period of 0.4748d.

80 Leo: A γ Doradus star (Vmag = 6.35) with a known period of 0.45286d (Henry & Fekel, 2002). We observe a period of 0.4529d with a SNR of 15.7. This light curve (Fig. 4 - Left) also shows the long term systematic effect due to the path of the object across the CCD as well as the one day periodicity.

DX Aqr: Known as a binary both visually and spectroscopically with an A0 and A1 companion and a period of 0.945d from GCVS. We detect a period of 0.9453d with an SNR of 46.65 (Fig. 4 - Right).

BF Oph(HD 154365): A Classical delta cepheid of Vmag = 7.37 and a period of 4.0677d (Hertzsprung, 1930). The STEREO light curve (Fig. 5 - Left) also reveals the quality of data in terms of continuous coverage.

AD Ari: A known but not very well studied δ Scuti F0 star with a Vmag of 7.4. GCVS has a period of 0.5397d and we also find the same period (Fig. 5 - Right).

HD 144412: A relatively not known A2 star with a Vmag of 8.70. From the STEREO light curves (Fig. 6 - Left panel). We found a period of 0.4748d with an SNR of 15.32. Right panel of Fig. 6 shows the phase plot at found period. But definitely more work needs to be done to confirm this variability.

HD 202369: A 5.34 mag M3 giant, suspected to be a variable by Rufener *et al.*, 1982. We find a prominent period of 0.4447d from STEREO data (Fig. 7 - left). The power spectrum (Fig. 7 - right) has a broad peak indicating the presence of more than one period as well as many modes.

Fig. 8 is an example result from our automatic classification method using neural networks. This is a well known δ Scuti star - VW Ari of Vmag = 6.71 and spectral class F0 (Liu *et al.*, 1996). Fig. 8 also shows the phase plot at the prominent frequency f1 and the corresponding amplitude spectrum.

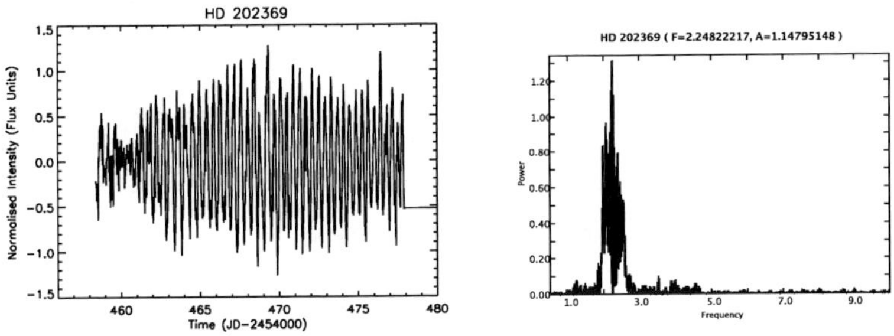

Figure 7. Left - STEREO light curve of HD 202369 which is previously suspected to be a variable. Right panel contains the power spectrum with the broad peak around the prominent period.

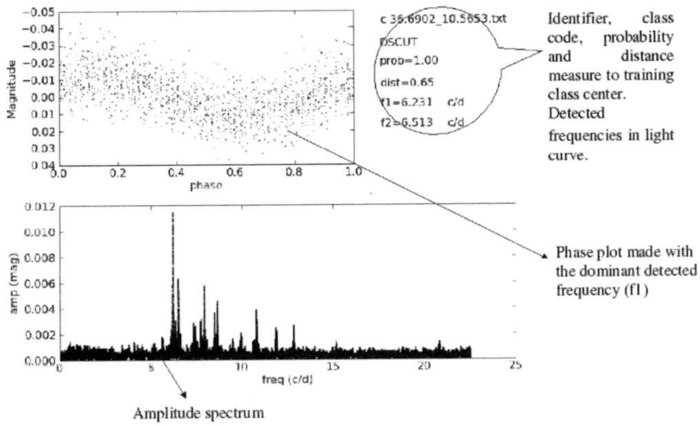

Figure 8. VW Ari - A F0 star of Vmag of 6.71 - δ Scuti

5. Summary

The HI instruments with their wide field, shallow magnitude coverage, continuous observation and a good PSF provides the compromise with other active transit search and stellar variability studies. Our current investigation involves studying in detail the new and suspected variables to confirm their periodicities apart from working on automated classification of these variable stars using neural networks. We are also working on correcting for the various instrumental and consistent variations which are present in the light curves. All these, we are sure will enable us to look for transiting exoplanets in our data in the near future.

References

Antoci, V. *et al.* 2007, *A&A*, 463, 225

Brown, D. S. *et al.* 2009, *Solar Phys.*, 254, 185.

Debosscher, J. *et al.* 2007, *A&A*, 475, 1159.

Eyer, L. & Blake, C. 2005, *MNRAS*, 358, 30.

Eyer, L. & Mowlavi, N. 2008, *Journal of Physics: Conference series*, 118, 2010.

Eyles, C. J. *et al.* 2009, *Solar Phys.*, 254, 387.

Henry, G. W. & Fekel, F.C. 2002, *PASP*, 114, 988.

Hertzsprung, E. 1930, *BAN*, 6, 9.

Lenz, P. & Breger, M. 2005, *CoAst*, 146, 53.

Liu, Y. Y. *et al.* 1996, *A&AS*, 120, 179.

Rufener, F. *et al.* 1982, *A&AS*, 48, 503.

Sarro, L. M. *et al.* 2009, *A&A*, 494, 739.

Scargle, J. D 1982, *APJ*, 263, 835.

Solar and Stellar Variability: Impact on Earth and Planets
Proceedings IAU Symposium No. 264, 2009
A. G. Kosovichev, A. H. Andrei & J.-P. Rozelot, eds.

doi:10.1017/S1743921309993061

The influence of starspots activity on the determination of planetary transit parameters

Adriana Silva-Válio[1]
[1]CRAAM, Mackenzie University, São Paulo, Brazil
email: avalio@craam.mackenzie.br

Abstract. As a planet eclipses its parent star, dark spots on the surface of the star may be occulted, causing a detectable variation in the transit light curve. There are basically two effects caused by the presence of spots on the surface of the star which can alter the shape of the light curve during transits and thus preclude the correct determination of the planet physical and orbital parameters. The first one is that the presence of many spots within the latitude band occulted by the planet will cause the depth of the transit in the light curve to be shallower. This will erroneously result in a smaller radius for the planet. The other effect is that generated by spots located close to the limb of the star. In this case, the spots will interfere in the light curve during the times of ingress or egress of the planet, causing a decrease in the transit duration. This in turn will provide a larger value for the semi-major axis of the planetary orbit. Qualitative estimates of both effects are discussed and an example provided for a very active star, such as CoRoTo-2.

Keywords. stars: activity, stars: spots, planetary systems

1. The spotted star model

During one of its transits, an extrasolar planet may pass in front of a stellar group of spots. Spots are regions of high concentration of magnetic fields, being cooler, and therefore darker, than the surrounding photosphere. Because a darker region of the stellar surface is being occulted when the planet crosses in front of the spot, a small bump may be observed in the bottom of the transit light curve.

I have used the model described in Silva (2003), that simulates the star as a 2-D image with quadratic limb darkening. The planet is assumed to be a dark disc of radius R_p, with no emission. Finally, the spots are modeled by three parameters: (i) Intensity, as a function of stellar intensity at disk center (I_c); (ii) Size, as a function of planet radius; and (iii) Position: Latitude (restricted to the transit path) and Longitude.

The transit light curve is calculated every two minutes, when the planet is centered at a given position in its orbit. The orbit, considered to be circular, is described by the semi-major axis (or orbital radius) and inclination angle. The integrated flux is calculated by summing the intensity in all the pixels, thus yielding the light curve. The three panels of Figure 1 show the effect of varying one of the model parameters:

• *Planet radius:* The observed transit depth is $\Delta I/I = (R_p/R_s)^2$, where R_s is the star radius. Planets with larger radius will produce deeper transits. Jupiter size planets produce a 1% decrease in the light curve intensity of a solar like star.

• *Orbital radius:* The orbit is assumed to be circular, that is zero eccentricity. The larger the orbital radius, the smaller the phase interval of the transit. This is equivalent to saying that larger orbital distances result in shorter transits (in time) for the same orbital period.

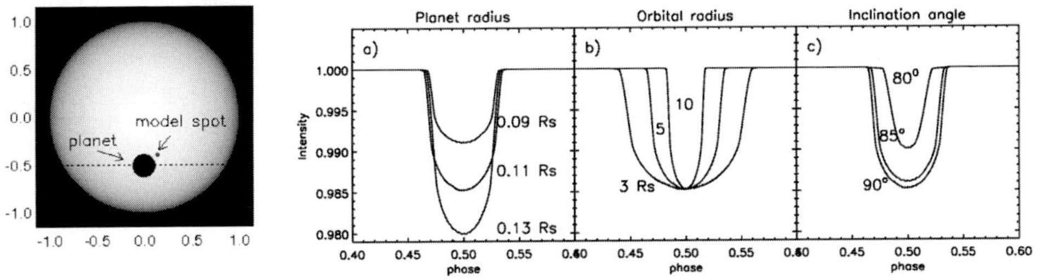

Figure 1. Left: An example of the simulate star with a small spot and the planet. The effect of varying the planet parameters: (a) planet radius; (b) semi-major axis; and (c) inclination angle.

- *Inclination angle:* For a transit to be detect, the orbital inclination angle has to be very close to 90°, otherwise no transit will be observed. The smaller the angle, the shorter the phase interval of the transit. Triangular shaped transits result from grazing eclipses, that is, inclination angles less than about 80°.

2. Transit depth and width

As mentioned above, the transit depth depends on the ratio of the planet to the star radii. As an illustration, the result from the simulation of a star with 9 spots of constant radius equal to 0.5 R_p and intensity of half the star central maximum intensity is shown as a dotted line in the left panel of Figure 2. A heavily spotted star will present a shallower transit. Therefore the fit to this light curve results in a smaller radius for the planet (gray solid line). In the present simulation, the estimated planet radius from the light curve changed from 0.1 to 0.093 R_s, a 7% decrease in the radius determination.

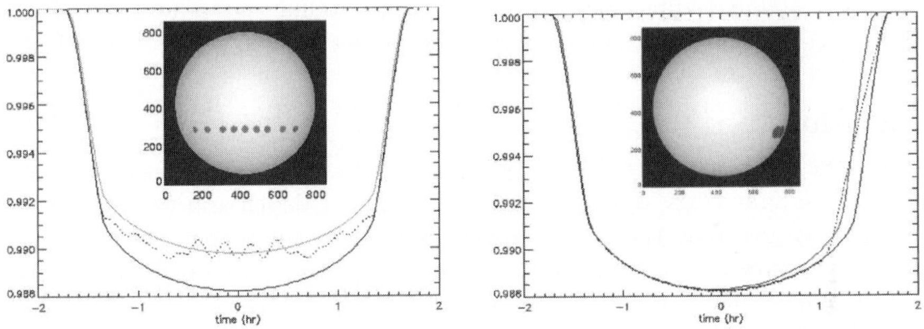

Figure 2. Left: Model light curves representing the simulated model of a spotless star (black line), and that of a star with nine spots (shown in the inset): data (dots) and fit (gray line). Right: Model light curves representing the simulated model of a spotless star (black line), and that of a star with two limb spots (shown in the inset): data (dots) and fit (gray line).

The next simulation includes two spots close to the limb of the star, as shown in the right panel of Figure 2. In this case, the width of the transit, or its duration will seem smaller (dotted line). The gray solid line in the figure represents a misleading fit to the transit light curve. Thus, when the transit fitting is performed the resulting semi-major axis of the orbit will be larger. In this simulation the inferred orbital radius increased from 7 to 7.3 R_s, a 4% increase in the orbital radius.

3. Application to CoRoT-2

CoRoT-2 is known to be a very young star and due to its intense spot activity, many spot on its surface are present at any given time. This star has an orbiting planet such that 77 transits of the planet were observed with 32 second resolution by the CoRoT † satellite (Baglin *et al.* 2006). According to Alonso *et al.* (2008) the planet has a radius of 1.47 R_{Jup} and orbits the star every 1.743 days on a circular orbit (Bouchy *et al.* 2008) of radius 6.7 R_s with an inclination angle of 87.84°. This inclination angle implies that the path of the transit overlays the star at a latitude of 14.6°.

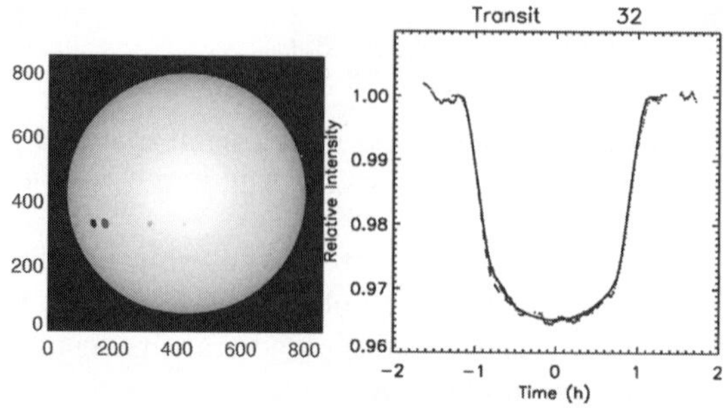

Figure 3. Left: Model of the star with a few spots. Right: The 32nd transit light curve for CoRoT-2 (dots) and the model for a spotless star (dashed line) and the fit to the data (solid line) for the model shown in the left panel.

To estimate the best radius for the planet, the deepest transit was sought. Of all the 77 transits detected, the 32nd transit displays the lower light variation (Figure 3). This transit is believed to have the least number of spots on its surface within the transit latitudes during the whole period of observation. The dashed curve on the figure represents the model without any spot whereas the solid line refers to the fit of the light curve with basically three small spots. To obtain this fit light curve it was necessary to use a planet radius of 0.172 stellar radius, instead of the 0.1667 stellar radius (Alonso *et al.* 2008). This represents an increase of about 3% in the planet radius, and is equivalent to 1.52 R_{Jup}.

4. Conclusions

Spots on the stellar surface are responsible for basically two effects that can alter the shape of the light curve during transits: (1) smaller depth and (2) shorter duration. These effects can preclude the correct determination of the planet physical and orbital parameters. The presence of many spots within the latitude band occulted by the planet will result in a light curve with shallower transit depth, thus erroneously yielding a smaller radius for the planet. The other effect is that caused by spots located close to the limb of the star that interferes in the light curve at times of ingress or egress of the planet, causing a decrease in the transit duration. This in turn will provide a larger value for the planetary orbital radius. In the simulations presented here, these effects are of the order of a few percent (3-7%) in both cases.

References

Alonso, R., Auvergne, A., Baglin, A. *et al.* 2008, *A&A*, 482, L21

Baglin, A. *et al.* 2006, 36th COSPAR Scientific Assembly, 36, 3749

Bouchy, F., Queloz, D., Deleuil, M., Loeillet, B., Hatzes, A. P. *et al.* 2008, *A&A*, 482, L25

Silva, A. V. R. 2003, *ApJ*, 585, L147

† CoRoT is a space project operated by the French Space Agency, CNES, with participation of the Science Programme of ESA, ESTEC/RSSD, Austria, Belgium, Brazil, Germany, and Spain.

Solar and Stellar Variability: Impact on Earth and Planets
Proceedings IAU Symposium No. 264, 2009
A. G. Kosovichev, A. H. Andrei & J.-P. Rozelot, eds.

doi:10.1017/S1743921309993073

UV habitability and dM stars: an approach for evaluation of biological survival

Ximena C. Abrevaya[1,2], Eduardo Cortón[2] and Pablo J. D. Mauas[1]

[1]Institute for Astronomy and Space Physics UBA-CONICET.
Buenos Aires, Argentina.
email: abrevaya@iafe.uba.ar

[2]Biochemistry Department, Faculty of Exact Sciences, UBA.
Buenos Aires, Argentina.

Abstract. Dwarf M stars comprise about 75 percent of all stars in the galaxy. For several years planets orbiting M stars have been discarded as suitable places for development of life. This paradigm now has changed and terrestrial-type planets within liquid-water habitable zones (LW-HZ) around M stars are reconsidered as possible hosts for life as we know it. Nevertheless, large amount of UV radiation is emitted during flares by this stars, and it is uncertain how these events could affect biological systems. In particular UV-C ($\lambda < 290nm$) exhibits the most damaging effects for living organisms. To analyze the hypothesis that UV could set a limit for the development of extraterrestrial life, we studied the effect of UV-C treatment on halophile archaea cultures. Halophile archaea are extremophile organisms, they are exposed to intense solar UV radiation in their natural environment so they are generally regarded as relatively UV tolerant. Halophiles inhabits in hipersaline environments as salt lakes but also have been found in ancient salt deposits as halites and evaporites on Earth. Since evaporites have been detected in Martian meteorites, these organisms are proposed as plausible inhabitants of Mars-like planets. Our preliminary results show that even after UV damage, the surviving cells were able to resume growth with nearly normal kinetics.

Keywords. astrobiology, stars: flare, ultraviolet: stars, methods: laboratory

1. Introduction

According to “Mediocrity principle”, which proposes that our planetary system, life on Earth, and technological civilizations are an average case in the Universe (von Hoerner, 1961), life as we know it could emerge in any place with conditions similar to Earth, and would develop following similar selection rules.

The Habitability Criteria define which places are appropriate for life according to physical and chemical environmental factors (Cockell, 2007). Usually, the presence of liquid water in the planet’s surface which is known as Liquid-Water Habitable Zone (LW-HZ)is the most important aspect for habitability (Huang, 1959; Dole, 1964; Hart, 1979). However, other important environmental factors as UV radiation can play a significant role in habitability. It is well known that UV have deleterious effects for cells, acting as a genotoxic agent and inducting DNA damage, or damaging a wide variety of proteins and lipids. In particular, high exposure to UV-C ($\lambda < 290nm$) is lethal to most terrestrial biological systems. Therefore, UV radiation should also be considered as an habitability criteria (Buccino *et al.*, 2006).

In particular, during the last decade there was a change in paradigm, which led to consider that terrestrial planets around dM stars can be suitable places for the emergence and evolution of life. However, many dMe stars emit large amount of UV radiation during flares, and it is uncertain how these UV events can affect life (Buccino *et al.*, 2007).

Extremophile organisms are subject of study in astrobiology due to their capacity to survive in extreme physicochemical environmental conditions. Between these group there is a kind of microorganisms which inhabits in environments with high salt concentrations (3.4 – 5.1 M NaCl range for optimally growth), commonly known as halophilic archaea. These halophiles inhabits in hipersaline environments as salt lakes but also have been found in ancient salt deposits as halites (salt sediments several million years old) and in other ancient deposits as evaporites or rock salt (Whitby *et al.* 2000; Fendrihan *et al.*, 2006). Since evaporites have been detected in Martian SNC meteorites (Gooding, 1992; Rieder *et al.*, 2004), these microorganisms are proposed as plausible inhabitants of Mars-like planets or other extrasolar planets (Stan-Lotter *et al.*, 2004). Moreover, because much halophiles are exposed to intense solar UV radiation in their natural environment, so they are generally regarded as relatively UV tolerant. In this work we examine the effect of UV-C on *Natrialba magadii* an halophile archaea initially isolated from Magadi Lake in Kenya, Africa.

2. Methods

Cultures of *N. magadii* were grown around optical density (OD) related to mid-exponential phase. Samples were divided in five groups: Control (non irradiated culture), and irradiated for 5, 10, 20 and 30 minutes (doses: 1110, 2220, 4440, 6660 J m^{-2}, respectively). Liquid culture was diluted to reach OD_{600nm}= 0.05, and irradiated being exposed to a UV-C source (Phillips 15W Hg lamp $\lambda = 254nm$, irradiance 3.7 W m^{-2}). Aliquots of the irradiated and control groups were withdrawn after different irradiation times and the effect of the UV treatment was assesed following the changes of the growth kinetics.

3. Results

To obtain growth curves, OD values for each sample were acquired at different times after irradiation and were plotted versus post-irradiation time. Then data points were fitted to a modified logistic equation (A. Mendez, personal communication), as is shown:

$$y = \frac{A}{1 + exp[\frac{4\mu_m}{A}(\lambda - t) + 2]} \quad (1)$$

where y is proportional to biomass (in our case OD), A is the maximum growth, μ_m is the maximum specific growth rate, λ is the lag time, and t is the post-irradiation time (see Fig. 1.a). Growth curves were obtained for each dose assayed as shown in Figure 1.b and parameters were calculated according to equation 1. Our results show that there is a dose dependent delay in the growth of the different samples (Table 1).

Experimental group	Irradiance Wm^{-2}	Irradiation time s	Dose Jm^{-2}	Lag time (λ) h
Control	0	0	0	20.13 ± 2.36
UV 5 min	3.7	300	1110	20.31 ± 2.41
UV 10 min	3.7	600	2220	27.95 ± 1.50
UV 20 min	3.7	1200	4440	34.35 ± 1.52
UV 30 min	3.7	1800	6660	48.24 ± 2.07

Table 1. Parameters obtained from the experimental data

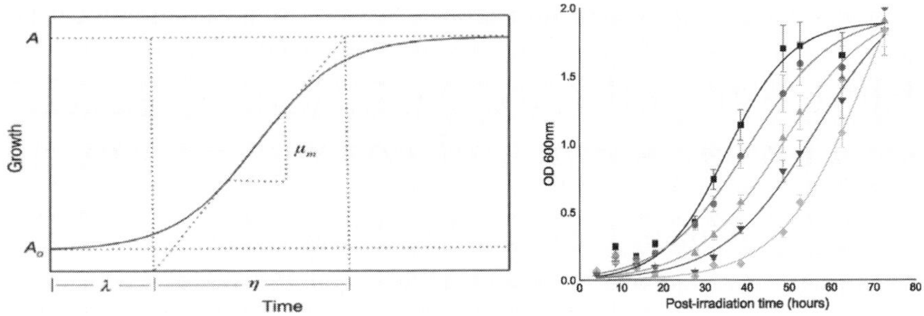

Figure 1. (a) : Growth curve showing the meaning of the parameters in equation 1, where "A_o" is the initial growth and "η" is the growth time. (b) : Growth curves for all samples. Each line shows different UV doses: control, 0 J/m^2(■) and irradiated 1110 J/m^2(•); 2220 J/m^2(▲); 4440 J/m^2(▼); 6660 J/m^2(♦).

4. Discussion

It is well known that UV induces diverse kind of damage in cells. In particular UV-C is a potent inductor of DNA chemical modifications like formation of cyclobutane pyrimidine dimers and 6-4 photoproducts (Mitchell *et al.*, 1989; Cadet *et al.*, 2005). If these alterations are not repaired they can cause blockage of diverse cellular processes, leading to mutations or cell death. Growth delay could be explained by some of these effects, and could be related to possible sublethal effects like cell cycle arrest (Rudolph *et al.*, 2007) induced by DNA damage or by other processes capable of blocking cell growth like tRNA photodamage (Opezzo & Pizarro, 2001). In such cases bacterial population remains in lag phase, which is visualized as a dose dependent displacement of the growth curve. In any case these preliminary results show that even after significant UV damage, as judged by the absence of detectable growth for more than 30 hours, the surviving cells were able to resume growth with nearly normal kinetics at all doses analyzed. UV irradiances used in the present work are of the same order to those received from a weak flare, such as those detected in EV-Lac (4.94 Wm^{-2}). In next experiments we will include as an important factor the flare frequency which usually is very high: e.g., one flare every 10 hours for Proxima Centauri and every 1-1.5 hours for AD Leo (Cincunegui *et al.*, 2007; Mauas & Falchi, 1994).

References

Buccino, A. P., Lemarchand, G. A., & Mauas, P. J. D. 2007, *Icarus*, 192, 582
Buccino, A. P., Lemarchand, G. A., & Mauas, P. J. D. 2006, *Icarus*, 183, 491
Cadet, J., Sage, E., & Douki, T. 2005, *Mutat.Res*, 571, 3
Cincunegui, C., Diaz, R. F., & Mauas, P. J. D. 2007, *aap*,461, 1107
Cockell, C. S. 2007, in: Horneck, G., & Rettberg, P., Eds. *C. C. in Astrob.* (Wiley-VCH)151, 177
Dole, S. H. 1964, *Habitable planets for man* (Blaisdell Pub. Co., 1st ed.), New York
Fendrihan, S., *et al.* 2006, *Rev.Environ.Sci.Biotechnol*, 5, 203
Gooding, J. L. 1992, *Icarus*, 99, 28
Hart, M. H. 1979, *Icarus*, 37, 351
Huang, S. S. 1959, *Am.Sci.*, 47, 397
Mauas, P. J. D. & Falchi, A. 1994, *aap*, 281, 129
Mitchell, D. L. & Nairn, R. S. 1989, *Photochem.photobiol.*, 49, 805
Opezzo, O. J. & Pizarro, R. A. 2001, *J.Photochem.Photobiol.B:Biology*, 62, 158
Rieder, R., *et al.* 2005, *Science*, 306, 1746
Rudolph, C. J., Upton, A. L., & Lloyd, R. 2007, *Genes &Dev.*, 21, 668
Stan-Lotter, H., *et al.* 2004, *Halophilic microorganisms*, (Springer Verlag, N.Y.), 89, 102
von Hoerner, S. 1961, *Science* 134, 1839
Whitby, J., *et al.* 2000, *Science*, 288, 1819

Solar and Stellar Variability: Impact on Earth and Planets
Proceedings IAU Symposium No. 264, 2009
A. G. Kosovichev, A. H. Andrei & J.-P. Rozelot, eds.

doi:10.1017/S1743921309993085

Evidence for climate variations induced by the 11-year solar and cosmic rays cycles

William Bruckman[1] **and Elio Ramos**[2]

[1]Department of Physics and Electronics, University of Puerto Rico at Humacao,
CUH Station, 100 Route 908, Humacao, PR 00791-4300
email: miguelwillia.bruckman@upr.edu

[2] Department of Mathematics, University of Puerto Rico at Humacao,
CUH Station, 100 Route 908, Humacao, PR 00791-4300
email: elio.ramos@upr.edu

Abstract. We analyzed data from PSMSL monthly mean sea level seeking correlations between sea level fluctuations and the solar and cosmic rays 11 year cycle. The data reveals decadal variability that could be causally connected to the solar and cosmic rays cycle, since these periodic changes are correlated. It is also found that the solar (cosmic rays) cycle correlates (anti-correlates) with the mean global surface temperature anomaly. A probable explanation of the above correlations is that the solar intensity and cosmic rays variations induce oscillations in the average temperature and precipitation, with corresponding changes in the continental water and snow accumulation. Thus, for instance, a higher than average snow and water over land, and lower temperatures produce oceans thermal contraction and lower mass, implicating lower mean sea level.

Keywords. Sun: activity, solar-terrestrial relations, cosmic rays

---

The relevance of the Solar Cycle on Earth climate became an important issue with the publication of an article by J.A. Eddy in the Journal Science in 1976. In this research strong evidence was presented about an historical period between 1645 and 1715, called The Maunder Minimum, where solar-spots virtually disappeared, coinciding this with much lower than average temperatures. The absence of solar-spots during the Maunder minimum was mentioned by Wolf (1856, 1868) and Sporer (1887). The investigations was continued by Maunder (1922), who strongly defended the observational evidence from the skeptical scientific community that questioned the reliability of the observations. Eddy was able to convince many that Maunders arguments were sound and robust, and he added new observational evidence, like the absence of solar corona and variations of the Carbon 14 isotopes in the atmosphere, during the Maunder minimum. Carbon 14 production in the atmosphere is sensitive to the intensity of the galactic cosmic rays flux (GCRF) which in turn is modulated by the solar activity cycle. Thus, the solar cycle have an associated GCRF cycle, that is almost perfectly anti-correlated to it.There is also evidence that the GCRF cycle and other solar irradiance indirect influence in the atmosphere affect climate, and we will briefly discussed this topic later on.

Modern measurements by artificial satellites demonstrate indeed, as suggested by the Maunder minimum and similar historical events, that during the solar cycle we have a variation of about 0.1% in the solar flux, which is very highly correlated with the solar-spot 11 year cycle. Although the variation is small (about 1 watt per square meter per cycle) we expect it to have a measurable effect on the Earth surface temperature, and we can see in Fig. 1 that the data of anomalies in sea surface temperature, since the beginning of the NOAA, NCDC record in the year 1880, show oscillations with periodicity seemingly in correlation with the solar cycle. To put that interpretation to a statistical test an analysis of the data was carried out, and we found that a cross-correlation function analysis between solar-spot number and the temperature global anomalies of Fig. 1 reveals a significant correlation of r=0.64, with a delay of about 1 year, using the data since 1950. There is a gradual deterioration of the correlation when data before 1950 is included. This reduction in correlation is to be expected if the number of quality measurements decrease with old data. Furthermore, during the interval 1880 to 1940 solar-spot activity was much lower than afterward, thus making perhaps solar variation effects less perceptible.

There is also indirect effects on climate coming from the ways in which the solar flux variations in the ultraviolet can affect temperature in the stratosphere (Haigh 1996, 1999, Schindell *et al.* 1999). Furthermore, another source for climate alteration could be the formation of ions by GCRF, that, as me mentioned earlier, is a flux modulated by the solar cycle. In fact there is observational evidence (Svensmark and Friis-Chritensen, 1997; Svensmark, 1998) for increments in cloud cover during GCRF maximum(solar-spot minimum), and also Kniveton and Todd (2001) correlated precipitation with GCRF. The above considerations indicate that temperatures can be affected by other mechanism besides direct solar radiation absorption at the surface of the Earth.

Another interesting consequence of solar flux and temperature oscillations is the variation of the mean sea level (MSL), resulting from the thermal contraction and expansion of the oceans and the variation in the accumulation of snow in the continents. We can estimate this sea perturbation by noticing that in about a 50 years interval we have an increase in MSL of approximately 100 mm (Jevrejeva *et al.* 2008), while we also have a corresponding change in mean sea surface temperature of near 0.4 K (NOAA, NCDC, global ocean temperature, in http://climexp.kmni.nl/selectindex.cgi). This information allows us to parametrized MSL change with temperature variation, and thus obtain dMSL/dT$\sim$100mm/0.4K =260mm/K. Therefore, for amplitude change in temperature according to Fig. 1 we should have discernible changes in MSL,since MSL variations of even few millimeters are within the observable capability of the permanent system for MSL(PSMSL), and we will describe below MSL variations of the expected magnitude and period correlated with the solar cycle.

A typical time series from PSMSL is shown in Fig. 2, from Magueyes station in Puerto Rico, where we can see a seasonal variation in MSL with a yearly minimum of variable amplitude. These minima are the result of the thermal contraction due to minima in winter sea surface temperature (that happened there around the beginning of march) plus the accumulation of more snow in the continents. We also note that this amplitude have a variation with a periodicity (triangles) similar to the 11 year sunspot cycle. In fact we found a high correlation of 0.71 between the curve define by the minima from Magueyes and the solar cycle, with a delay of 3 years, with similar results for San Juan P.R.(Bruckman and Ramos 2003, 2004, 2007), and individual stations in the Caribbean, and, collectively, in the North Atlantic,North Pacific, and in 1023 stations around the world (Fig. 3). The correlation of global MSL variations, in Fig. 3, with the solar cycle was r=0.64, with a delay of 4 years, for data after 1940. We also find a lowering in correlation when including data before 1940, similar to what we found for global surface temperature. The explanation of the observed delays require the study of complicated cause and effect inertial time lag. However, for thermal inertia of the oceans we have an analogous delay in the seasonal variation in temperature, where the minimum in temperature do not coincide with the beginning of the winter in the hemisphere. The thermal time lag value depends of thermodynamics parameters and the frequency of the oscillations.

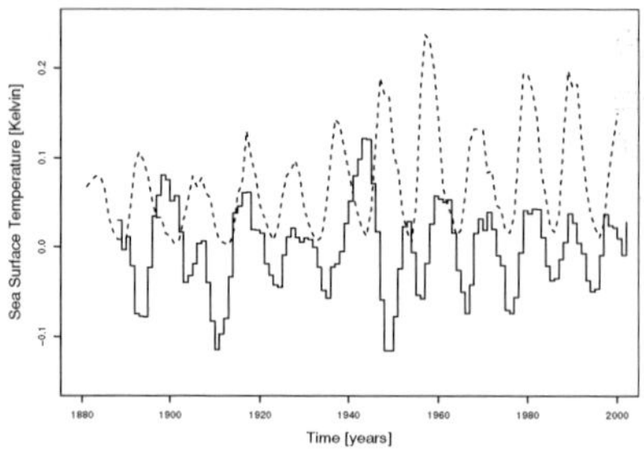

Figure 1. Detrended 3 years running mean (3RN) Sea Surface Temperature (solid) vs. Solar-spot Cycle (dash).

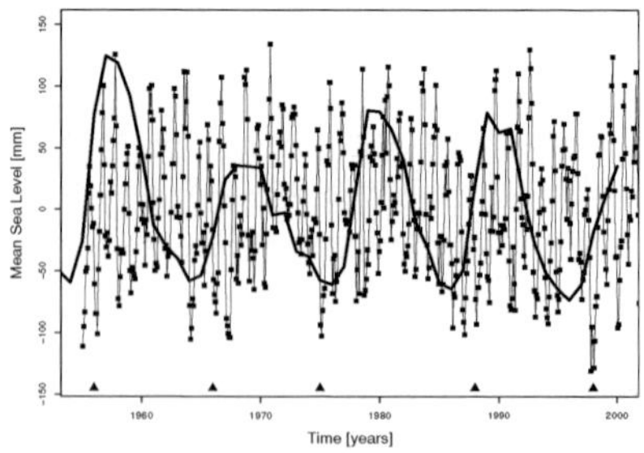

Figure 2. Detrended Mean Sea Level (MSL) for Magueyes, Puerto Rico vs. Solar Cycle (solid). The black triangles point to dipping in MSL.

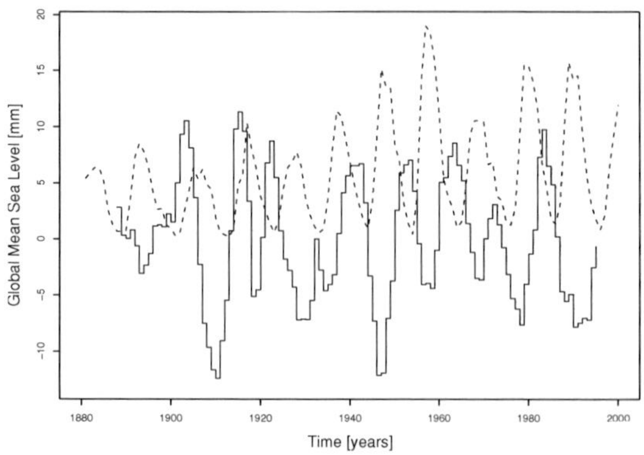

Figure 3. Detrended 3RN Global Mean Sea Level (solid) vs. Solar-spot Cycle (dash).

References

Bruckman, W. & Ramos, E. 2003, *Submitted to Geophys. Res. Lett. (unpublished).*

Bruckman, W. & Ramos, E. 2003, *Simposio: Cambio Climatico: Universidad de Puerto Rico(UPR) Bayamon. Proceedings UPR-Bayamon Library*

Bruckman, W. & Ramos, E. 2007, *Congreso Internacional Gaia y El Cambio Climatico 2007, UPR-Rio Piedras, PR. Revista Umbral,Primer tomo*

Eddy, J. A. 1976, *Science*, 192, 1189.

Haigh, J. D. 1999,*Quart. J. Roy. Meteor. Soc.*, 125, 871–892.

Haigh, J. D. 1999, *J. Atmos. Solar Terrest. Physics*, Gl, 63–72.

Haigh, J. D. 1996, *Science*, 272, 981–984.

Jevrejeva, S., Moore, J. C. Grinsted, A., & Woodworth, P. L. 2008, *Geophys. Res. Lett.*, 3, L08715.

Kniveton, D. R. & Todd, M. C. 2001, *Geophys. Res. Lett.*, 28, 1527–1530.

Maunder, E. W. 1890, *Mon. Not. R. Astron. Soc.*, 50, 251.

Shindell, D., Rind, D., Balabhandran, N., Lean, J., & Lonergan 1999, *Science*, 284, 305.

Sporer, F. W. 1887,*Vierteljahrsschr. Astron. Ges. Leipzig*, 22, 323.

Svensmark, H. & Friis-Christensen, E. 1997, *J. Atm. Sol. Terr. Phys*, 284, 59, 1225–1232.

Svensmark, H. & Friis-Christensen, E. 1998, *Phys. Rev. Lett.*, 81, 5027–5030.

Wolf, R. 1856, *Astron. Mitt. Zurich*, 1, VIII.

Wolf, R. 1868, *Astron. Mitt: Zurich*, 24, III.

Solar and Stellar Variability: Impact on Earth and Planets
Proceedings IAU Symposium No. 264, 2009
A. G. Kosovichev, A. H. Andrei & J.-P. Rozelot, eds.

doi:10.1017/S1743921309993097

Synchronous manifestation of 160-min pulsations of the ground pressure and Z-component of geomagnetic field at Moscow, Apatity, Oulu, Yakutsk and Tixie

V. Ye. Timofeev[1], D. G. Baishev[1], L. I. Miroshnichenko[2], S. N. Samsonov[1] and N. G. Skryabin[1]

[1]Yu.G. Shafer Institute of Cosmophysical Research and Aeronomy, SB RAS, 677980, Yakutsk, Russia
email: timofeev@ikfia.ysn.ru

[2]N.V. Pushkov Institute of Terrestrial Magnetism, Ionosphere and Radio Wave Propagation (IZMIRAN), RAS, Troitsk, Moscow Region, Russia

Abstract. Oscillations of the ground pressure and Z-component of geomagnetic field with a period of 160-min in December 2003 and March 2004 have been studied by regular measurements carried out over the territory of Europe and Siberia. As the analysis has shown, in December 2003 these specific oscillations are manifested synchronously from Europe up to Siberia, preferentially in the form of 'packages' ('quanta') consisting of two-five pulses. Their average amplitudes in December 2003 are of 0.0115±0.0023 mb in the ground pressure and of 0.323±0.070 nT in the Z-component. We suppose that the 160-min pulsations in those both terrestrial parameters represent the effect of a gravitational wave of the same period coming from the Galaxy center.

Keywords. Earth, Sun: activity, gravitational waves

1. Introduction

Gravitational waves in the Universe can be formed only by very dense bodies of major mass (so-called relativistic objects (see review by Rudenco 2006). Their nearest and strongest concentration is related to the central regions of our Galaxy. According to Kotov & Kotov (1996) the 160-min pulsations of the Universe can benefit the gravitational wave radiation when interacting with such objects.

If there exists a specific, preferred arrival direction of the gravitational waves from the central parts of the Galaxy (in particular, with a period of 160-min) then it should be manifested at the Earth by changes of the wave phase in a sidereal time when the Earth moves along its orbit around the Sun. In this case, one can expect some differences in the manifestations of pulsations between major areas in the meridian and latitudinal sections depending on the arrival direction of gravitational wave to the Earth. It may take place, for example, due to the gravitational wave changes its specific parameters (dimension, polarization etc.) that are still unknown.

To register gravitational waves it is suggested to use the planet Earth as a whole as a unique very massive device. We assume that this device is constructed by the analogy of heavy cylinder described by Rudenco (2006), with the only difference in that the central part of the cylinder is substituted by the mass of the whole Earth planet, and the edges of it - by the whole Earth's atmosphere. The gravitational wave covers the whole such a 'cylinder' instantly and completely. As a result, the pressure of terrestrial atmosphere will have the same phase on the major areas of the planet. One can register synchronous changes of the pressure using the system of pressure sensors distributed over the globe. Simple calculations show that the sensitivity of such 'device' will be greater by a factor of 10^{16} than it is specified by Rudenco (2006). By this analogy it is possible to construct a 'device' in which the central part of 'cylinder' will be a core of the Earth, and its edges - the rest terrestrial parts. The registration is carried out, in particular, by magnetometers that are operating synchronously and are distributed all over the terrestrial globe.

2. Processing and analysis of experimental data

As a preliminary, a raw observational data have been processed twice by a band probability filter with a band of oscillation periods from 40 to 180 minutes. Fig.1a shows the results of processing of 5-min values of the ground pressure using the superposed epoch technique in two positions of the Earth at its orbit, December 2003 and March 2004, at 5 stations (Tixie, Apatity, Oulu, Yakutsk, and Moscow). A zero meridian (December 1 at 00:00 UT) has been accepted as a reference point of superposition. In Fig.1b we preesent the results of processing of 1-minute values of Z-component for the same periods at Oulu, Moscow, Novosibirsk, Irkutsk, and Lvov. It may be seen from the Figures 1a,b that the manifestations of 160-min pulsations depend on sidereal time.

In Fig.1a (solid lines) one can see that in the ground pressure all phases obtained from the December data coincide at all stations all over the territory of Europe and Siberia. This figure shows also that a synchronism in March is worse manifested than in December (in March it is seen only at two stations - Tixie and Moscow). It means that there takes place a distinction in the phases of synchronism manifestation of 160-min pulsations in March and December, presumably because of the distinction of sidereal and solar time. In March the synchronism is kept only for the latitudes of Moscow and for the polar latitudes regions (for example, Tixie).

In the geomagnetic field, we deal with the response to some effects in the current systems existing at large depths and in the Earth's core, such effects being modulate a magnetic field at the terrestrial surface. Mechanisms of formation of such response are still unknown. But in December the synchronism takes place at all stations (see Fig.1b, solid lines). In March there is a change of phases but their character is manifested in another way than in the ground pressure. The effects calculated by a superposed epoch technique, on the average, are equal to 0.0115 ± 0.0023 mb in the ground pressure and to 0.323 ± 0.070 nT in the Z-component in December.

3. Some implications of the results

Since we suppose that pulsations are caused by the gravitational wave which instantly covers the whole terrestrial globe then, after taking the sum of the filtered data of ground pressure in December such pulsations at all stations will have the identical phases. Really, after such summation, the 160-min pulsations in December are manifested in the form of 'packages' of 2-5 pulses.

For any 'package', in the most cases, the distribution of amplitudes is characterized by a gradient (force) to the direction of the process development (i.e., to the direction of general course of time). If so, the 'packages' under the influence of such forces 'fly' to the future, apparently, with acceleration. It is not unlikely that this feature may be one of the experimental signatures of accelerated expansion of the Universe found out recently.

From the temporal change of average amplitude of 160-min pulsations in the ground pressure, another one effect seems to become apparent, this effect being caused by some

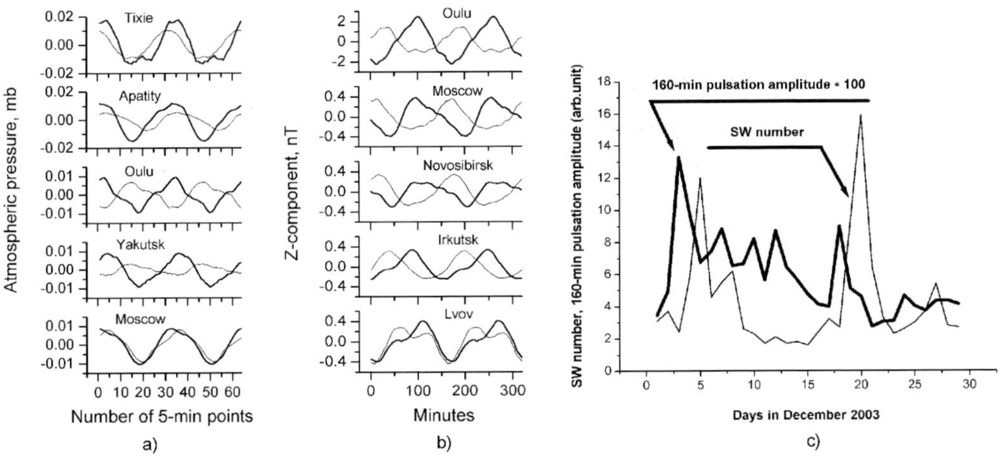

Figure 1. a) Average temporal changes of 160-min pulsations on the interval of two periods (in December 2003 - solid lines, and in March 2004 - thin lines) in the ground pressure; b) Z-component of geomagnetic field at the territory of Europe and Siberia; c) Comparison of variations of 160-min pulsation amplitude (solid line) and solar wind number density (thin line).

non-stationary processes in the Galaxy center that amplify the generation of gravitational waves. There are some grounds to expect that solar response to gravitational waves is amplified inside the Sun in the form of a sound wave. About 1 hour after, it manifests itself at the Sun 'surface' as an increase in the solar wind density and temperature. In the vicinities of the Earth those changes of density and temperature will manifest themselves after $\sim$2 days since the solar wind speed is relatively small (on the average, it is about 440 km/s, but can be more or less). By this time, the response to such processes in the center of Galaxy has already taken place on the Earth. Proposed effect of 2-4 days advance of solar activity in the ground pressure may be seen in Fig.1c. In fact, in the time profile of average amplitude of the 160-min pulsations in the ground pressure, one 'burst' in the pressure was observed on December 3 and in the solar wind - on December 5; the second burst was fixed on December 16 and 20, respectively. By their general view, the bursts in the ground pressure repeat, even in details, the bursts of solar wind density.

No telescope can observe what has happened at those moments in the center of our Galaxy because proposed effects are caused by changes of intensity of just gravitational waves, but not optical and any electromagnetic radiation. Such phenomena are very rare events (apparently, about one per year), but they have supposedly caused two bursts in solar wind density on December 5 and 20, 2003.

References

Kotov, V. A. & Kotov, S. V. 1996, *Radiophysics and Quantum Electronics*, 39, 807

Rudenco, V. N. 2006, *Earth and Universe* (in Russian), 6, 28

Solar and Stellar Variability: Impact on Earth and Planets
Proceedings IAU Symposium No. 264, 2009
A. G. Kosovichev, A. H. Andrei & J.-P. Rozelot, eds.

doi:10.1017/S1743921309993103

Manifestation of the Jupiter's synodic period in the solar wind, interplanetary magnetic field and geophysical parameters

S. N. Samsonov[1] **and N. G. Skryabin**[1]

[1]Yu. G. Shafer Institute of Cosmophysical Research and Aeronomy, SB RAS, 677980, Yakutsk, Russia
email: s_samsonov@ikfia.ysn.ru

Abstract. Studying by the authors of paper of solar wind parameters, namely: density, speed and temperature and also a module of interplanetary magnetic field (IMF) intensity has allowed to find out in them fluctuations with the period of 399 days. From references it is known that this period coincidence with the synodic period of Jupiter. So long as close by the given period another source of such fluctuations is not known we have assumed that fluctuations with the period of 399 days are fluctuations with the synodic period of Jupiter. The change of the solar wind plasma parameters and IMF intensity can lead to the change of the Earth's magnetic field parameters and, as a consequence, to the change of charged particle fluxes in the Earth's magnetosphere. On this assumption the IMF intensity in the Earth's vicinity, geomagnetic disturbance (Kp-index) and riometer absorption for the years of 1986-1996 have been analyzed. The analysis of the data has shown the presence of certain changes of these physical parameters with the period of 399 days. When the Earth and Jupiter were found to be on the same magnetic field line, the IMF intensity was decreasing up to 3.0±0.57%, the geomagnetic activity and riometer absorption were decreasing up to 5.2±1.46% and 9.4±2.63%, respectively.

Keywords. Sun: activity, solar wind, solar system: general

1. Introduction

It has been established in the early (for examples, Hill, Carbary & Dessler (1974) and modern (for examples, Ferreira *et al.* 2001) researches that Jupiter is a powerful source of charged particles. Hill, Carbary & Dessler (1974) using the observations of energetic electrons in the heliosphere from 1965 to 1974 showed that the increase of electron intensity is observed when the Earth appears to be close or crosses the IMF field line connecting the Sun and the Jupiter. But in works by Skryabin *et al.* (2005) and Timofeev *et al.* (2007) it was found that the variations with the Jupiter's synodic period of 399 days are observed in the plasma density of the solar wind, the intensity of cosmic rays and energetic electrons on the Earth's orbit.

The most favorable conditions for the propagation of the electrons from Jupiter towards the Earth are realized during their movement along the IMF field line. McDonald & Trainor (1976) showed that such conditions are practically realized every 13 months and last for 1-3 months.

So the authors of paper have decided to check up the following fact: whether there are such fluctuations in the solar wind parameters and interplanetary magnetic field.

The purpose of this work is to search for a variation with a synodic period in the IMF, geomagnetic disturbance index and riometer absorption.

2. Experimental data and results

Riometer measurements were carried on in Tixie from 1986 to 1996. The daily values of the absorption occurrences were used as riometer data. The event of absorption was considered to have taken place at the given moment if its intensity was >0.3 dB and the absorption lasted not less than 10 minutes. The cases caused by proton events and solar X-ray bursts were excluded from the experimental data.

The daily data of absorption occurrence frequency prepared in such a way are shown on the top panel (A) of Fig.1a. The ordinates axis shows the quantity of hours per day when the absorption was observed. The abscissa axis shows the ordinal number of the day beginning from January 1, 1986. The rare gaps in the observations "were restored" on the basis of observations made at neighbouring stations (Sokolov *et al.* 1988).

To exclude the short-period changes from the experimental data the method of smoothing was applied as a method of moving average with the period of averaging of 200 days. The smoothed experimental data of riometer absorption are given on the second panel (N) of Fig.1a. The verticals on this panel correspond to the moments of the opposition of the Earth and Jupiter. The axis of ordinates shows the daily frequency of occurrence of absorption, the abscissa axis - a through numbering of the days from January 1, 1986 to December 31, 1996 (4018 days). On the panels of Kp and IMF (OMNIWeb data) the data of daily Kp-index of the geomagnetic disturbance and IMF intensity module after smoothing are presented.

Then the experimental data given in Fig.1a were divided into 10 intervals equal to the average synodic period of Jupiter and after that they were processed using a superposed epoch technique. The results of such processing are presented in Fig.1b (for better clearness two periods are shown). In this Figure the solid verticals correspond to the moment of the opposition of the Earth and Jupiter, the dotted lines - to the expected location of Jupiter, the Earth and the Sun on the same IMF line. This moment comes, at an average, 156 days before the next opposition or 243 days after it. According to Skryabin *et al.* (2001) the average synodic period of Jupiter in the 21-st solar activity cycle is equal to 401±7 days changing from 372 to 437 days.

From Fig.1b it follows that the effect of Jupiter in the riometer absorption is 0,8 events a day or 9,4% from the average value equal to 8.5 events per day for the whole period of observation; in the geomagnetic disturbance (K-index) the effect is 1.0 or 5,2% from the average value equal to 19.3; in the IMF the effect is equal to 0.2 nT or 3,0% from the average value equal to 6.74 nT.

Fig.1c shows distributions of casual deviations from the results of application of superposed epoch technique. By these distributions the errors shown in Fig.1b have been determined using standard procedures. The amplitude distributions appear to be really similar to the law of normal distribution. That's why we consider that the errors in Fig.1b are caused by casual deviations and are equal to: σ=0.165 for the riometer absorption data and σ=0.20 for the Kp-index and σ=0.027 for the IMF.

3. Conclusion

The results presented above have allowed to find out the fluctuations with the synodic period of Jupiter in the IMF intensity, geomagnetic ctivity and riometer absorption in addition to the fluctuations with the synodic period of Jupiter in the solar wind parameters. The value of Jupiter's effect for the considered period accounts for 3.0±0.57% in the IMF, 5.2±1.46% in the geomagnetic activity and 9.4±2.63% in the riometer absorption.

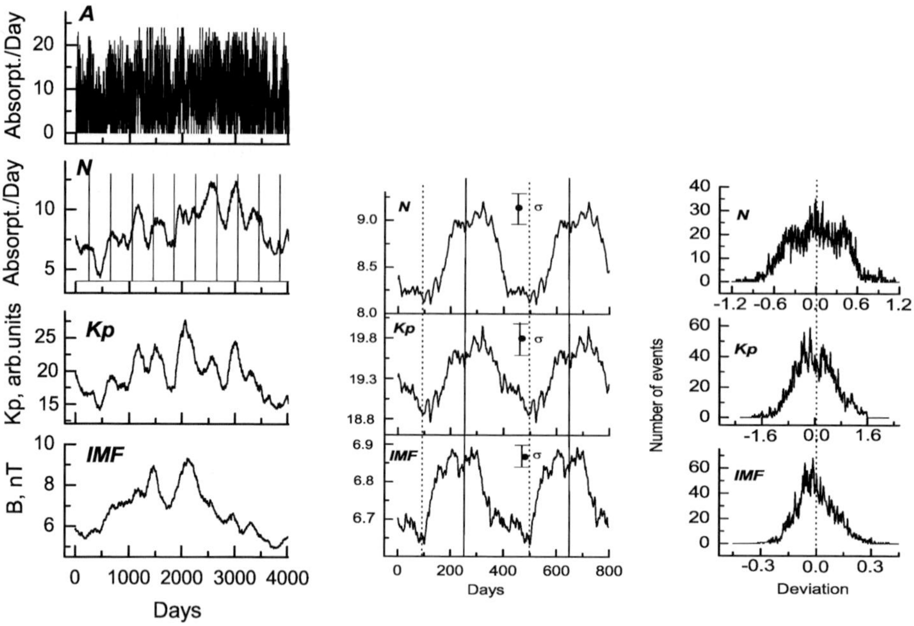

Figure 1. a) Raw riometer data (A) and smoothed data of the riometer absorption (N), Kp-index (Kp) and interplanetary magnetic field module (IMF); b) Variations of the riometer absorption (N), Kp-index (Kp) and IMF module (IMF) with the synodic period of Jupiter. σ - is the root-mean-square deviation.; c) Amplitude distribution of the following deviations for the riometer absorption (N), the Kp-index (Kp) and the IMF module (IMF).

The work was partly financed by RFBR (grants N 09-02-98505-p_vostok_a and 09-02-00284-a and the RAS Presidium Program for the basic research N16, part 3.

References

Ferreira, S. E. S., Potgieter, M. S., Burger, R. A., Heber, B., Fichtner, H., & Lopate, C. 2001, *Geophys. Res.*, 106, 29313

Hill, T. W., Carbary, J. F., & Dessler, A. J. 1974, *Geophys. Res. Lett.*, 1, 333

McDonald, F. B. & Trainor, J. H. 1976, in: T. Gerels (eds.), *Jupiter: Studies of the interior, atmosphere, magnetosphere and satellites* Tucson: University of Arizona Press, p. 961

Skryabin, N. G., Bezrodnykh, I. P., Plotnikov, I. Ya., & Ivanov, O. P. 2001, *Geomagnetism and Aeronomy*, 41, 432

Skryabin, N. G., Timofeev, V. E., Miroshnichenko, L. I., & Samsonov, S.N. 2005, *Astronomy Letters*, 31, 832

Sokolov, V. D., Samsonov, S. N., Ivanov, O. P., Yarygin, L. S., Vasil'ev, N. I., & Filippovich, V. N. 1988, *Observations of artificial celestial bodies* (in Russian), 84, 206

OMNIWeb data, *http://omniweb.gsfc.nasa.gov*

Timofeev, V. E., Miroshnichenko, L. I., Samsonov, S. N., & Skryabin, N. G. 2007, *Astronomy Letters*, 33, 63

Solar and Stellar Variability: Impact on Earth and Planets
Proceedings IAU Symposium No. 264, 2009
A. G. Kosovichev, A. H. Andrei & J.-P. Rozelot, eds.

doi:10.1017/S1743921309993115

Exoplanet environments to harbour extremophile life

Eduardo Janot-Pacheco[1], Claudia A. S. Lage and Ivan G. P. Lima[3]

[1]Departamento de Astronomia, IAG, Universidade de São Paulo,
Rua do Matão 1226, 05508-090, São Paulo, Brazil
email: janot@astro.iag.usp.br

[2]Instituto de Biofisica Carlos Chagas Filho, Universidade Federal do Rio de Janeiro,
A. Carlos Chagas Filho, 373, 21941-902 Rio de Janeiro, Brazil
email: lage@biof.ufrj.br

[3]Instituto de Biofisica Carlos Chagas Filho, Universidade Federal do Rio de Janeiro,
A. Carlos Chagas Filho, 373, 21941-902 Rio de Janeiro, Brazil
email: igplima@biof.ufrj.br

Abstract. In this contribution, we estimate the temperature at the surface of known exoplanets and of their putative satellites for two albedo extreme cases (Venus and Mars) and present a selection of extremophiles living on Earth that can live under those conditions. We examine also the possibility of survival of microorganisms in planetary systems of variable stars.

Keywords. planets and satellites: general, astrobiology

1. Introduction

Extremophiles are micro-organisms that dwell in extreme conditions of either high or low temperature, high or low pH, variable salt concentration, complete dessication, toxic metals, toxic organic chemical compounds and high levels of pressure or radiation. Extremophiles are thus excellent candidates to exist under extraterrestrial conditions, even outside habitable zones (HZ), as in the moons of the solar system Titan and Europa.

2. Exoplanets

There are more than 370 exoplanets discovered up to know, mainly through the velocimetry method (Schneider 2009). Many giant, probably gaseous planets close to their central stars - dubbed "Hot Jupiters" - have been found, showing that there are planetary systems quite different from the solar system. The CoRoT and KEPLER satellites using the method of transits will increase the population of known exoplanets by one order of magnitude and make the sample less biased. The method of transits allows in particular the detection of small, rocky planets that are more suitable to harbour life.

3. Temperature estimates

The surface temperature estimation depends not only on the stellar temperature but also on the planet's albedo and atmospheric chemical composition which will define the extent of the greenhouse effect and on how the heat is distributed around the planet.

The present sample of known exoplanets is strongly biased: long period planets and less massive ones are more difficult to detect. Surface temperatures of the known exoplanets are thus on the average higher than for planets in an Habitable Zone.

Surface temperatures of a number of Neptune-like exoplanets have been estimated (e.g., Rivera *et al.* 2005; Bonfils *et al.* 2007; Demory *et al.* 2007). They are supposedly mainly composed of icy/rocky material, being formed without the extended gaseous atmosphere or having lost it. Some of them have orbital periods between 2 and 6 days and surface temperature ranges from 400 to 700 K.

Even in these particular cases, extremophiles existing on Earth (the so-called "hyperthermophiles") could live in the coldests of them.

The most favourable case is the exoplanet Gliese 581c (Udry *et al.* 2007). It is a planet with about 5 Earth masses situated in the HZ of a MV star. In this case, for Earth-like or Venus-like albedos, the surface temperature of Gliese 581c is estimated to range between 270K and 313 K, respectively, a really conformable temperature for many extremophile existing on Earth.

One new, interesting alternative is to examine the temperature conditions in moons of gaseous giants in the HZ. We did this for seven gaseous giants in the HZ of their central stars. We give below the estimated surface temperatures at the surface of putative satellites of these exoplanets. HD 10697 (G5V; 6.35 M_J, 1072 d orbit) T$S \approx$ 264 K HD37124 (G4V; 1.04 M_J, 155.7 d orbit) T$S \approx$ 327 K HD134987 (G5V; 1.58 M_J, 260 d orbit) T$S \approx$ 315 K HD177830 (K2IV; 1.22 M_J, 392 d orbit) T$S \approx$ 362 K HD222582 (G3V; ?? M_J, 576 d orbit) T$S \approx$ 234 K Again, extremophiles existing on Earth could live confortably under these temperatures.

Finally, we consider two more extreme cases: a) Exoplanet temperatures for extreme albedos: considering that the planets have albedos as that of Venus ($A \approx 0.7$) and Mars ($A \approx 0.2$). In a planet like the Earth, the surface temperature would then be in the range 230 K ¡ T_{Earth} ¡ 283 K; b) An exoplanet orbiting a variable star whose radiation flux may change by up to $\approx$ 300%. The planet surface temperature may vary by $\approx$ 30% which means that extremophiles can still live under these conditions.

4. Conclusions

We have shown that extremophiles known on Earth could survive many of the different environments present in exoplanets, regarding extremes of temperature (but also, of radiation, pressure, pH, salinity, water content, etc...). Extremophiles can also live on the surface of planets orbiting variable stars, whose average energy may vary by up to $\approx$ 300%. Besides the "classical" Habitable Zones, one should also consider *extremophile zones* in exoplanetary systems.

References

Bonfils, X. *et al.* 2007, *A&A* 474, 293
Demory, B. O. *et al.* 2007, *A&A* 475, 1125
Rivera, E. J. *et al.* 2005, *ApJ* 634, 625
Schneider, J. 2009, *http://exoplanet.eu*
Udry, S. 2007, *A&A* 469, 43

Session IX

Current and future space missions and ground-based observing programs

Chair: C. Mandrini

Solar and Stellar Variability: Impact on Earth and Planets
Proceedings IAU Symposium No. 264, 2009
A. G. Kosovichev, A. H. Andrei & J.-P. Rozelot, eds.

doi:10.1017/S1743921309993127

Space solar missions

Jean-Claude Vial[1]

[1]Institut d'Astrophysique Spatiale, Université Paris-Sud,
Bâtiment 121, F-91405 Orsay Cedex, France
email: jean-claude.vial@ias.u-psud.fr

Abstract. In the frame of Symposium 264 which concerns Solar and Stellar Variability we address the space solar missions devoted to the various aspects of solar activity. We describe them in three time categories: missions ready for launch, missions which will operate in the 2012-2015 time frame and ambitious missions to be launched after 2015. We focus on the contributions of these missions according to the following criteria: Understanding mechanisms of activity, Improving detection and characterisation, Working out some prediction. Major activity contributors and manifestations are addressed: Coronal Mass Ejections, Flares, Solar winds, Magnetism (including dynamo), Irradiance.

Keywords. space vehicles: instruments, Sun: activity, Sun: coronal mass ejections (CMEs), Sun: flares, Sun: magnetic fields

1. Introduction

Outstanding space solar missions (from Yohkoh to more recent STEREO and Hinode) have been operating during the last 15-20 years and have brought a huge amount of information on the Sun, its atmosphere and its activity. Yohkoh obtained hard X-ray pictures of the Sun from the maximum of cycle 22 to the maximum of cycle 23. SoHO has provided a wealth of data (from the deep interior to the outer solar wind) over a time interval of more than 13 years quite uninterruptedly and can be considered as a powerful monitoring tool of solar activity over the whole cycle 23, e.g. a review by Fleck (2005). Since 2006, STEREO has provided "binocular" views of the solar outer atmosphere from corona to Earth and will continue to provide UV and white-light (WL) images of the Sun from very distinct viewpoints until the two spacecraft (S/C) will have a full view of the "other" side of the Sun in 2014 (for a review see: http://stereo-ssc.nascom.nasa.gov/publications.shtml). Since 2006 also, with its high resolution X-ray imager, EUV spectrometer and visible magnetometer, Hinode has provided a new view of the surface and coronal magnetism and of solar activity (see e.g. Tsuneta 2009). Others such as TRACE, CORONAS, CORONAS/PHOTON, RHESSI, have provided a local or global monitoring of the Sun in various wavelengths and energies.

These missions will continue their investigation of the solar atmosphere for still more years but new missions are being prepared which, through their lifetimes and objectives, will directly address the issue of solar activity. At a further horizon, missions are being prepared which will provide data before about 2015. They will partially overlap in time with present and immediate-future missions but will follow their own objectives. Finally, more ambitious missions for the post-2015 era are being discussed. They will be described because, as we shall see, their detailed contents is still a matter of compromise between scientific ambitions and actual capabilities.

What do we mean here by solar activity? Solar activity takes many forms in terms of electromagnetic radiation and particles and arises from various solar structures at all space and temporal scales. We limit our discussion to Coronal Mass Ejections, flares, solar winds, magnetism and irradiance in terms of events and physical quantities and we keep in mind what space can uniquely bring in terms of understanding the mechanisms of activity, of improving detection and characterisation, and finally on working out some prediction. Consequently, we exclude from this review, balloon and rocket experiments such as SunRise and Herschel. This does not mean in any way that they don't bring information on solar activity. Finally, the author apologizes in advance for space solar missions which would have been omitted in this limited overview.

2. Immediate space solar missions

Two of these missions (Solar Dynamics Observatory and PICARD) should be launched in February 2010 while PROBA2 was launched on November 2, 2009. This means that the solar community will have a fleet of three more fully operational S/C as soon as 2010.

Solar Dynamics Observatory. This NASA mission by itself offers splendid opportunities for doing excellent solar physics and for testing what could be done in the field of Space Weather (actually, it is launched in the frame of the NASA "Living With a Star" or LWS program). It is positioned on a geosynchronous orbit which allows for near uninterrupted observations all year long, except for a couple of weeks twice a year. This geosynchronous orbit will also provide an exceptionally high telemetry rate (about 160 Mbps), a first in space solar missions. The scientific goals of SDO are to improve our understanding of seven science questions:
1. What mechanisms drive the quasi-periodic 11-year cycle of solar activity?
2. How is active region magnetic flux synthesized, concentrated, and dispersed across the solar surface?
3. How does magnetic reconnection on small scales reorganize the large-scale field topology and current systems and how significant is it in heating the corona and accelerating the solar wind?
4. Where do the observed variations in the Sun's EUV spectral irradiance arise, and how do they relate to the magnetic activity cycles?
5. What magnetic field configurations lead to the CMEs, filament eruptions, and flares that produce energetic particles and radiation?
6. Can the structure and dynamics of the solar wind near Earth be determined from the magnetic field configuration and atmospheric structure near the solar surface?
7. When will activity occur, and is it possible to make accurate and reliable forecasts of space weather and climate?
As we shall see the three instruments (AIA, HMI and EVE) taken together address all of the above objectives.

1. AIA AIA stands for Atmospheric Image Assembly, which consists in four Science Telescopes offering eight Science Channels with four two-by-two common detectors (4096* 4096 CCD) providing a spatial resolution of 0.6 arcsec per pixel. The ten seconds cadence leads to about 1.1 Tbytes per day ! The Principal Investigator (PI) is A. Title (LMSAL). Main caracteristics of the instrument are presented in Table 1.

It should be noted that with its 13.1nm channel, AIA will observe "flaring" plasma at temperatures higher than $10^7 K$, which will allow to observe not only large flares but also flare-type activity at small scales as evidenced with RHESSI and Hinode/XRT (e.g. Reale *et al.* 2009). Such observations, will be critical for Parker-type heating models at small-scale, as proposed as early as 1994 by Cargill (Cargill 1994).

Table 1. AIA wavelength bands

Wavelength (nm)	Bandpass (nm)	Ion	Region of the atmosphere	Formation Temperature log(T)
Visible	–	Continuum	Photosphere	3.7
170	–	Continuum	Temperature minimum, photosphere	3.7
30.4	1.27	He II	Chromosphere, transition region	4.7
160	–	Continuum	Transition region, upper photosphere	5.0
17.1	0.47	Fe IX	Quiet corona, transition region	5.8
19.3	0.6	Fe XII, XXIV	Corona and Flare Plasma	6.1, 7.3
21.1	0.7	Fe XIV	Active Region Corona	6.3
33.5	1.65	Fe XVI	Active Region Corona	6.4
9.4	0.09	Fe XVIII	Flares	6.8
13.1	0.44	Fe XX, XXIII	Flares	7.0, 7.2

2. HMI HMI (Helioseismic and Magnetic Imager) will provide images, Dopplergrams and magnetograms in the Fe I line at 617.3 nm. It is a rather complex instrument (a follow-up of MDI on SoHO) with two 4096*4096 pixels CCD cameras providing a spatial resolution of 0.5 arcsec per pixel. The data acquisition cadence is 40 seconds. HMI will archive about 1000 Tbytes per year. It should be noted that because of its capacity to observe the full Sun, HMI will be very useful for local (and back-side) seismology and especially the emergence of magnetic flux from the convection zone into the outer atmosphere where it will be observed by AIA. More information can be found in T. Hoeksema's presentation in this Symposium. HMI P.I. is P. Scherrer (Stanford).

3. EVE EVE (Extreme u-v Variability Experiment) is a Full-Disk spectrometer with a complete EUV coverage from 0.1 to 122 nm with an excellent spectral resolution of 0.1 nm. Its high cadence (10s) will allow measuring the irradiance effect of flares. The low latency of data (3-5 min) will provide a high-value asset with respect to LWS and Space Weather activities. EVE PI is T. Woods (LASP/UC).

The launch of SDO is scheduled for February 2010.

Picard The payload of this French microsatellite mission consists in SOVAP (and BOS) which performs radiometric and bolometric measurements, PREMOS with a radiometer and 3 sunphotometers, and SODISM, a metrological imaging telescope dedicated to absolute and relative measurements of the solar radius. The interest of this instrumental space package has been demonstrated during the Symposium (see e.g. J.-P. Rozelot and S. Lefebvre contributions). The three instruments are built by IRM (B), PMOD (CH) and LATMOS (F) respectively. The mission PI is G. Thuillier (LATMOS). Launch is scheduled for February 2010.

PROBA2 This Project for On-Board Autonomy (PROBA) belongs to a series of technology demonstration missions of ESA. It includes a set of solar observation instruments:
1. a LYman-alpha RAdiometer (LYRA) built by ORB and CSL (B) which works in four bandpasses: 115-125 nm (including Lyman-alpha), 200-220 nm (Herzberg continuum range), 17-31 nm (including Helium II at 30.4 nm), and 1-20 nm. LYRA employs wide bandgap detectors based on radiation-hard diamond.
2. a Sun Watcher using AP-sensors and image Processing (SWAP) built by ORB and CSL (B): this EIT-type experiment provides a coronal image every minute after heavy data compression.

The PROBA2 payload also includes plasma instrumentation, a GPS receiver and a laser retroreflector array.

The above-mentioned missions (SDO, Picard and PROBA2) will be critical for the continuation of the International Living With a Star (IWLS) program as shown in Fig. 1. It is clear from Fig. 1 that these three missions will be essential contributions to irradiance variation measurements after 2010.

3. Space Solar missions in the 2012-2013 time frame

The missions we discuss now are supposed to be launched in 2012-2013 (or beyond). Their major feature is that they are approved.

PROBA3 The third PROBA mission of ESA is devoted to a technological test of Formation Flight (FF). A first S/C carries an external occulter about 1.2 m wide located 150 m in front of the main S/C which carries a coronagraph. Both S/C are precisely aligned along an axis centered on solar disk center at least during the FF phase (at apogee of the orbit). The resulting eclipse allows for a half field of view (FOV) which goes as close as 0.04 solar radius above the limb. The outer edge of the FOV is located at two solar radii above the limb. The pixel size is 3 arcsec for a 2K*2K detector, an exceptional spatial resolution in the corona. An A.O. for the payload was issued by ESA at mid-July 2009; answers were due for 15 September 2009, the selection should take place by the end of November 2009 and the flight is scheduled for Autumn 2013. Mission duration is two years.

IRIS IRIS (Interface Region Imaging Spectrograph) is a NASA SMEX mission led by A. Title (LMSAL). It addresses the issues of non-thermal energy and mass supply in the chromosphere and above, along with magnetic flux emergence in the chromosphere in relation with flares and coronal mass ejections (CMEs). Its 20 cm telescope allows for a pixel resolution of 1/6 arcsecond. Its channels include chromospheric lines such as Mg II h and k (excellent indicators of activity) with a resolution of 2.5 pm and TR lines such as C II, Si IV (with a 1.25 pm resolution). The Sun Synchronous orbit provides a 0.7 Mbits/s data rate. Launch is scheduled for December 2012.

4. Space Solar missions beyond 2015

The missions discussed here are being prepared by the relevant space agencies with inputs of the solar communities concerned. Most of them are very ambitious and should lead to major breakthroughs in solar physics. However, none of them has been formally approved at this date.

There are two major possible avenues for solar physics studies: either getting closer to our star in order to improve in-situ measurements (and combine them with remote sensing) or/and also to have polar views of the Sun, on one hand, or staying Earth-bound in order to increase the capabilities of remote-sensing instrumentation in terms of spatial, spectral and temporal resolution, of time coverage and of novel measurements (e.g. polarimetry), on the other hand.

With the first category, solar missions look like planetary exploratory missions which are constrained by low mass and volume, low telemetry and short observing time. As with planetary missions, they bring the ground truth for the local conditions which are being studied. The best example is the NASA Solar Probe Plus (SP+) mission which will diagnose the coronal plasma at less than ten solar radii from the Sun. Although it will not be discussed here in the frame of this Symposium devoted to solar and stellar activity (which can only be unravelled through continuous records) it is nevertheless a fascinating way of exploring our star.

With the second category, we have the possibility of bigger and more comprehensive missions which, step-by-step, and essentially by remote-sensing, improve the knowledge of the physics of the Sun. Increasing the apertures leads to increased spatial or/and temporal resolution. It allows for spectrography and possibly polarimetry with better Signal-to-Noise ratio. The Earth proximity allows for substantial telemetry (in the style of SDO) and a better control of the instrumentation in terms of flexibility, target choice, etc...

Solar-C This is a Japanese mission to be launched in 2016, whose PI is S. Tsuneta (NAOJ). Actually, there are two SOLAR-C mission concepts under study which epitomize the choice between exploratory and earth-bound missions.

In the first category ("Plan A"), Solar-C colleagues propose an out-of-ecliptic mission with an orbit staying at one astronomical unit (AU) but achieving an inclination (to the Solar equator) of 30 deg. in two years from launch and 45 deg. in five years from launch. Combined with other observations (e.g. Earth-bound), such a high-inclination orbit will allow to:

- Observe Doppler velocity at high latitudes without projection effects,
- Detect meridional flows at the pole,

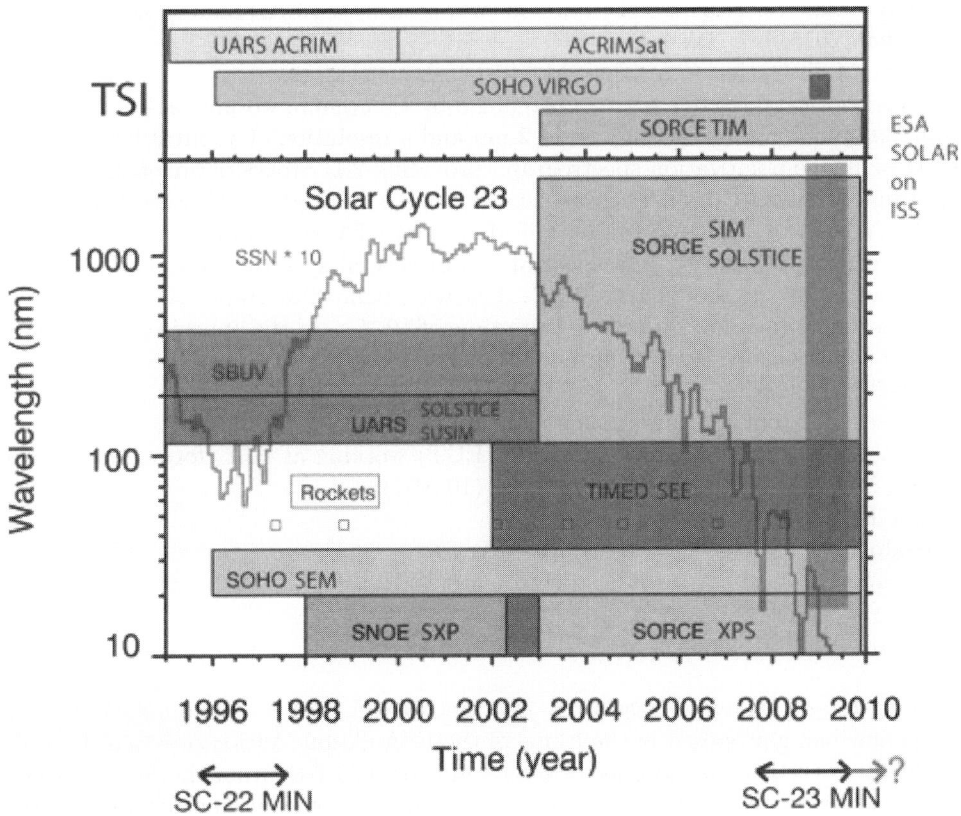

Figure 1. Irradiance missions in operations vs. time (abscissa) and wavelength (ordinate). The two horizontal sets of arrows indicate the time of activity minima for cycles 22 and 23. (Courtesy Tom Woods)

- Observe waves penetrating deep into the Sun: the magnetic field at the tachocline could be detected in this way (a 100kG module corresponds to a 25msec travel time difference).

The model payload would consist in: a Photospheric and chromospheric dopplergraph, a Stokes-polarimeter for photosphere and chromosphere and a X-ray/EUV imager. Optional is a total irradiance monitor.

In the second category ("Plan B"), the mission on Earth-bound orbit aims at high spatial resolution, high throughput, high cadence spectroscopic (polarimetric) and X-ray observations from photosphere to corona in order to investigate the magnetism of the Sun (and especially to solve the dynamo enigma) and its role in heating and dynamism of solar atmosphere. The model payload consists in:
a Near IR-Visible-UV telescope with an aperture of 50cm-1.5m ensuring that it is diffraction-limited, with a 0.1-0.4 arcsec resolution followed by a spectro-polarimeter,
an Ultra-high resolution EUV/X-ray telescope,
along with high resolution, high signal coronal spectroscopic capabilities.
Envisaged technology developments for SOLAR-C address Stokes polarimetry in UV (e.g. Ly-alpha) and EUV (e.g. 171 A) with Hanle effect, along with a Photon counting soft X-ray telescope with ultra-fast CMOS detector to obtain high-cadence high-quality global temperature maps. Working Groups have been established for evaluating the pros and cons of both "Plans".

Space Solar Telescope Continuing with projects from Eastern Asia countries, we have the Space Solar Telescope (SST) mission proposed by CNSA (China) for a launch no sooner than 2015. It consists in:

(*a*) The Main Optical Telescope (MOT) which consists in:
- a 1-meter optical telescope with a FOV of 2.8 by 1.5 arcmin, working at any wavelength within [393 nm-656 nm], a bandpass of 2 nm and a resolution of about 0.1 - 0.15 arcsec,
- a 2-D real-time polarization spectrograph providing 2-D Stokes parameter profiles and being tunable in wavelength. The wavelength range is 390-700 nm. Universal Birefringent Filters (UBF) have a Full Width at Half Maximum (FWHM) around 5 - 7.5 pm. Eight channels can be distributed in a spectral line or several spectral lines. The accuracy of polarization analyzer should be 2 10^{-4}. The wide-band filtergraph (380-550 nm) with bandpasses of 3 nm allow for an exposure time of 10^{-4} s. At the opposite, a set of vector magnetic field measurements requires up to 30s.
- A Correlation Tracker.

(*b*) A set of "auxiliary" instruments which include:
- an EUV Imager for the Solar Telescope (EUT) working at 17.1 (formed at 1.0 MK), 19.5 (1.3 MK), 21.1 (2.0 MK) and 12.9 nm (10 MK)
- a Wide Band Spectrometer (WBS)
- an H-alpha and White-light Telescope (HWT)
-a Solar and Interplanetary Radio-spectrometry (SIR).

The P.I. of SST is H. Zhang (NOAC).

Solar Orbiter This major mission by ESA and NASA, which was approved by ESA in 2000, is now being proposed in the frame of the ESA Cosmic Vision program. Its unique orbit features (getting as close as 48 solar radii -or 0.23 AU- from the Sun, viewpoints from about 30 deg. heliographic latitude) will allow for exploring unchartered innermost regions of the solar atmosphere, study the Sun from a co-rotation vantage point and provide images of the Sun's polar regions. Solar Orbiter (SO) will accommodate a set of remote-sensing and in-situ instruments which will have to overcome mission constraints such as increased thermal and particles exposures, limited telemetry and observing time (30 days at perihelion). Such constraints make the mission a challenge from both science and technology standpoints. Launch is scheduled for 2017 with operations starting about two years later. A suite of 10 instruments has been selected in March 2009 (see http://sci.esa.int/science-e/www/object/index.cfm?fobjectid=44469) as the scientific payload for the ESA Solar Orbiter mission, which are described below.

Remote sensing instruments The array of remote sensing instruments includes:

(*a*) EUI, an Extreme U-V Imager, whose PI is P. Rochus (Belgium). It consists in a package of three instruments: a dual Full Sun Imager (FSI) working sequentially in 30.4 and 17.4 nm, and two High Resolution Imagers (HRI), a dual one covering 17.4 and 33.5 nm and the other covering the Lyman alpha line (121.6 nm). Note that the HRI pixel size corresponds, at perihelion, to 80 km on the Sun, a resolution comparable to the best resolution obtained in the visible from the ground with adaptive optics (see e.g. the Swedish Solar Telescope).

(*b*) SPICE (SPectral Imaging of the Coronal Environment), funded by NASA, is a two-channel UV spectrograph working in the two bands: 70.2-79.2 nm and 97.2-105 nm (and 48.6-52.5 nm in the second order). The one arcsec resolution translates into a pixel resolution of 170 km on the Sun at perihelion. The spectral resolution should allow for a velocity determination of about +- 3 km/s through centroiding technique. The lines span the solar atmosphere from the chromosphere to the corona. SPICE spectroscopic data will be compared with in-situ composition measurements, they will allow to map the origins of solar wind streams, to diagnose the acceleration of the extended solar wind and to look after the source region of energetic particles. The P.I. of SPICE is D. Hassler (SwRI, USA).

(*c*) PHI (Polarimetric & Helioseismic Imager) is an HMI/SDO type of imaging magnetograph which will fully benefit from SO properties. Its PI is S. Solanki (MPS). As a linkage science, it will provide the magnetic field information to EUV imagers and spectrometer. It will provide the first direct view of magnetic and velocity fields at the poles. Through SO corotation, it will allow to follow surface and subsurface temporal evolution of solar features such as active regions. It will perform stereoscopic helioseismology to probe the deep interior, it will measure the complete field of the Sun (with a 360 deg view) and also provide the magnetic context for SP+. PHI consists in a High Resolution Telescope (HRT) (aperture 14 cm and 160 km resolution at perihelion) alternately operating with a Full Disk Telescope (FDT) (aperture 2.5 cm), a Vector magnetograph providing full Stokes polarimetry in Fe I 617.3 nm and solar LOS velocities with a spectral resolution better than 10 pm.

(*d*) STIX STIX (Spectrometer/Telescope for Imaging X-Rays) is led by A. Benz (ETH, CH). It provides location, structure, timing and imaging spectroscopy of hard X-ray sources and offers a strong link between remote-sensing and in-situ observations of high-energy events. As with the HXT of Yohkoh, it works on the basis of indirect imaging with 64 subcollimators (i.e. 32 Fourier components). Angular resolutions range from 7 arcsec to 8.8 arcmin. The full Sun is imaged from 4 to 150 keV, with an energy resolution between 1 and 15 keV. The time resolution is about 0.1 s (or larger). Of particular interest, will be the search of the link between X-ray sources and the particles recorded with in-situ instruments on SO.

(*e*) METIS METIS stands for Multi Element Telescope for Imaging and Spectroscopy (PI: E. Antonucci (OAT, I)) will simultaneously image the visible and ultraviolet emission of the solar corona and diagnose, with unprecedented temporal coverage and spatial resolution, the structure and dynamics of the full corona in the range 1.2 to 3.0 solar radii from Sun centre, at perihelion during the nominal mission. It relies upon broad-band imaging of the polarized visible K-corona (500-600 nm), narrow band imaging of the UV corona in the HI Ly-alpha, 121.6 nm, line and narrow band imaging of the EUV corona in the He II Ly-alpha, 30.4 nm. At perihelion, the cadence for a full measurement will be of one minute and the spatial resolution better than 3 arcsec.

(*f*) HI Heliospheric Imager (SoloHI) an instrument funded by NASA (PI: R. Howard, NRL, USA) is designed to pinpoint CMEs, shocks and other disturbances in the solar wind. The success of HI on the STEREO mission has shown the potential of such an instrument able to follow the propagation of CME plasma at a distance between 9 and 37 solar radii from the Sun with a 20 arcsec resolution. It will bridge the gap between remote-sensing (EUI, SPICE, XRT, METIS) and in-situ measurements (whether on SO itself or on other S/C in the interplanetary space, including SP+).

<u>In situ instruments</u>

(*a*) RPW The Radio and Plasma Waves (RPW) Instrument RPW will measure magnetic and electric fields at high time resolution, in order to determine the characteristics of electromagnetic and electrostatic waves in the solar wind from almost DC to 20 MHz. It will record waves from about 2-5 solar radii up to the SO orbit. It is, in a sense, both an in situ and remote sensing instrument. The PI is M. Maksimovic (LESIA/OP, F).

(*b*) EPD The Energetic Particle Detector will measure the properties of suprathermal and energetic particles: the sources, acceleration mechanisms, and transport processes of these particles. The PI is J. Rodriguez-Pacheco (SRG, Spain). It consists in SIS (Supra-thermal Ion Spectrograph) working in the range 8 keV/n to 10 MeV/n, EPT (Electron and Proton Telescope) (20 to 700-900 keV), LET (Low Energy Telescope) (1.5-12 MeV for protons), and HET (High Energy Telescope) whose range for electrons is 0.3 -20 MeV, for positrons 0.3-1 MeV, for protons 10-100 MeV and for ions heavier than He 50 -200 MeV.

(*c*) MAG This magnetometer (PI: T. Horbury (ICL, UK)) will measure the local magnetic field between 10^{-3} and 10 Hz with a signal much higher than the sensitivity of the instrument, a partial effect of the proximity to the Sun. It will be the first direct measurement of the coronal magnetic field at a distance of less than 50 solar radii.

(*d*) SWA Solar Wind Analysers (SWA) consist in a set of three instruments:

i. PAS This Proton -Alpha particle Sensor will measure the velocity distribution of the major ionic species.

ii. HIS This Heavy Ion Sensor will determine the charge states of O and Fe. It will also record pick-up ions of various origins.

iii. EAS This Electron Analyser System will determine the electron velocity distributions of the solar wind core, halo and strahl with high temporal resolution.

The PI is C. Owen (UCL/MSSL, UK).

<u>*InterHelio Probe*</u> The mission inherits from the Interhelios mission already proposed by Russia. It has a mission profile rather close to Solar Orbiter with the main following features: in-situ measurements close to the Sun (about 30 solar radii), corotation and out-of-ecliptic observations, observations of the invisible side of the Sun. A launch is scheduled in 2018; PIs are S. Kuzin and V. Kuznetsov (Izmiran), L. Zelenyi (IKI).
The P/L includes:

(*a*) a magnetograph with a spatial resolution of 1 arcsec and a sensitivity of about 2 Gauss (*b*) an EUV imager-spectrometer with a 1 arcsec resolution in three bands; 17.1, 29.-31 and 13.2 nm.

(*c*) an X-ray imager in the range 50-100 keV with a 10 arcsec resolution

(*d*) an optical coronagraph with a 1 arcsec resolution over a FOV of 8 deg.

(*e*) a multichannel optical photometer working between 280 and 1550 nm.

(*f*) a set of in-situ instruments (solar wind analysers, magnetometer, radio and gamma spectrometers, neutron detector).

Solar Orbiter, Solar Probe Plus, and other missions synergies It is difficult to describe all the possible synergies between SO, SP+, Solar C and other Earth-bound missions (including GBO). Let us take a simple example with the objectives achievable only with SO and SP+, when SP+ is at its perihelion. When SO and SP+ will have their perihelions aligned, they will record the same (low velocity) plasma along the Sun-S/C line. It will be possible to measure, e.g., the change of the distribution functions between 10 and 48 solar radii. SO will also provide some remote-sensing information about the surface sources candidate for providing the plasma measured by SP+. When SO will be behind SP+ perihelion by about 60 deg, SO and SP+ will record about the same Energetic Particles which follow the spiral magnetic field. SO remote sensing will observe the acceleration site of these Energetic Particles. On the contrary, when SO will be 90 deg ahead of (or behind) SP+ perihelion, SO remote sensing will observe the plasma analysed in-situ and remotely (HI) by SP+. Of course, this potential will be increased by Solar-C (especially with its "A" version) and InterHelio Probe capabilities of multi-points and multi-viewpoints measurements.

5. Conclusions

Many other projects in all countries are being studied and proposed within different time frames: ADITYA (India), KuaFu (China), four solar proposals for Cosmic Vision at ESA (COMPASS, DYNAMICS, POLARIS, HiRISE), LYOT/DESIR/HEBS. As mentioned earlier-on, they fall into two categories: either an encounter (close-up or out-of-ecliptic but P/L and TM are modest) or Earth-bound missions providing high resolution but with an in-ecliptic view point from one AU only. Actually, we need **both !** because the information from both are complementary: the respective measured scales of the observed structures differ by many orders of magnitude. They also differ in sizes and ambitions but it may be worth mentioning the value of "small" (SMEX-size or smaller) missions for the advancement of science, technology and training of young scientists and engineers.

Other missions are being operated (and proposed) in the frame of Space Weather such as the GOES of NOAA which will also provide very valuable information about the solar activity.

One can add that joint observations and science coordination between the above mentioned space missions and the Ground-Based Observatories (GBO) will provide a superior synergetic science. One can think, for instance, of the super spatial resolution brought by future telescopes either in the visible and the infra-red (ATST, EST) or in the radio domain (LOFAR, SKA, FASR, ALMA).

Finally, huge amounts of various data (think of SDO !) will be gathered and such a windfall, which also is a constraint, calls for user-friendly open software and data access, such as proposed by the Virtual Solar Observatory (VSO), involving both space and GBO missions.

Acknowledgements

I deeply thank colleagues (mostly PIs) who provided me much material about their instrument/mission: E. Antonucci, A. Benz, P. Boumier, S. Kuzin, M. Maksimovic, P. Rochus, P. Scherrer, S. Solanki, A. Title, S. Tsuneta, T. Woods, H. Zhang along with E. Buchlin and J. Leibacher who helped me in the preparation of the manuscript. I also thank the S264 organizers who invited me for this review.

References

Cargill, P. J. 1994, *ApJ*, 422, 381

Fleck, B. 2005, in: A. Hanslmeier, A. Veronig, & M. Messerotti (eds.), *Solar Magnetic Phenomena* (Astrophysics and Space Science Library), p. 139

Reale, F. and Mc Tiernan, J. M. and Testa, P. 2009, *ApJ*, 704, L58

Tsuneta, S. 2008, in: S. A. Matthews, J. M. Davis, & L. K. Harra (eds.), *Astronomical Society of the Pacific conference Series*, 397, p 3

Solar and Stellar Variability: Impact on Earth and Planets
Proceedings IAU Symposium No. 264, 2009
A. G. Kosovichev, A. H. Andrei & J.-P. Rozelot, eds.

doi:10.1017/S1743921309993139

First thoughts on stellar variability from Kepler commissioning data

Lucianne M. Walkowicz[1] **and Gibor Basri**[1]

[1]University of California Berkeley, 601 Campbell Hall,
Berkeley CA 94720 USA
email: lucianne@astro.berkeley.edu

Abstract. The Kepler mission will obtain high precision, continuous lightcurves for more than ~150,000 stars over the next four years. Prior to primary mission operations, ten days of commissioning data were obtained for the ~52,000 brightest targets in the Kepler field. While Kepler's main goal is the discovery of transiting low mass planets, it will also provide a rich dataset for studies of variable stars. These commissioning data give a first glimpse of the amazing diversity of stellar variability Kepler will observe. Here, we discuss the tools we are currently developing to quantify variability in the Kepler data, and show initial results on the distribution of target stars in these metrics. Ultimately these measures will be used both to characterize the data and to select active rotationally modulated stars for rotation period determination.

Keywords. stars: activity, stars: statistics

1. Introduction

The Kepler mission is a 1-meter photometer whose primary goal is to discover Earth-like planets in the habitable zones of Sun-like stars. Launched in March 2009, Kepler is now 6 months into operations, taking photometry of ~150,000 stars near Cygnus every 30 minutes. These high cadence, long duration lightcurves are not only capable of identifying small planets transiting their parent stars, they also provide opportunities for a wide variety of ancillary science. In particular, Kepler will observe stellar variability (in all its various flavors) with unprecedented coverage and precision. In this sense, Kepler will be as revolutionary for stellar astrophysics as it will be for planetary science.

Prior to the official start of Kepler science operations, the mission began with 10 days of commissioning observations. During this commissioning period, photometry was taken for the 52,496 stars in the field brighter than a Kepler magnitude† of 13.5, with the intent of characterizing the precision obtained through initial operations of the telescope and the first builds of the software pipeline. Just over half of the targets observed during this period were giants, which are typically excluded from the planetary search target list due to their large radii (making them poor candidates for observing transits of small planets). However, the giants are useful as bright targets by which to characterize the instrument, and some are kept on the observing list in subsequent quarters for use as astrometric calibration stars. The Kepler commissioning observations also allowed an extensive study of the photometric behavior of giants versus dwarfs (after the style of Gilliland 2008) – for cases where a particular target is not classified in the Kepler Input Catalogue, photometric behavior may be used to separate out desirable dwarf targets from the not-so-desirable giants. In addition to helping us understand the behavior of the instrument, the Kepler commissioning data also provides our first glimpse into the

† Kepler magnitudes are a broad bandpass optical magnitude between ~400-800 nm.

typical variability of a large sample of stars. The large sample of homogeneous space-based observations give us an unprecedented new view on stellar variability in its many varieties, from pulsations, to eclipsing systems, to magnetic activity.

As has long been observed on our Sun as sunspots and flares, magnetic activity is a source of both periodic and transient variability. There have been prior observations of similar phenomena on stars both like our Sun and considerably alien from it (e.g. Strassmeier 2002, Eyer & Grenon 1997), but these largely ground based observations suffer from such blights on observational astronomy as "daytime" and "weather", as well as the loss of precision that comes with observing through the Earth's atmosphere. As the magnetic field generation in stars is intimately linked with stellar rotation (Pizzolato *et al.* 2003), photometric modulation by magnetic features such as starspots provide a natural way to trace the rotation of the star. By measuring rotation periods for a large sample of stars, we can better constrain the relationship of rotation, activity, and age, as well as inform models of the magnetic field production.

However, prior to launching our full study of rotation for stars in the Kepler field, we must characterize the variability of the target stars as a whole. Because the sample size is so large, with ~50,000 lightcurves in commissioning alone and ~150,000 every quarter of regular operations, it is not feasible to characterize variability in each of the lightcurves by eye, nor is this a quantitative way to characterize the stars. We therefore developed a number of statistics and metrics that can be calculated automatically for the entire sample. These metrics may then be used to identify interesting subsamples (such as spotted stars whose rotation can be measured) that bear further investigation. We present these metrics, along with the results of our initial foray into characterizing photometric variability in the Kepler targets, below. This work is intended to demonstrate new methods of examining a large amount of qualitatively different stellar photometry in a quantitative way–truly physical results on stellar variability with Kepler await the main mission data and the further refinement of the methods described here.

The Kepler commissioning observations took place over ~10 days in May 2009. The target list was composed of 52,496 of the brightest targets in the field. Many of these targets had been previously classified in the Kepler Input Catalogue (KIC; Batalha *et al.* 2007), a compilation of available physical parameters for all targets in the field (for example, Kepler magnitude, T_{eff}, logg, etc.). Of the commissioning targets, the majority were giants (24,357 stars), with the rest classified either as dwarfs (20,535 stars) or unclassified (7,591 stars).

2. Characterizing variability in the Kepler commissioning data

The Kepler lightcurves reveal an amazing diversity of variable behavior. In Figure 1, we show Kepler lightcurves for four different examples of stellar variability, clockwise from top left: a magnetically active spotted star, a semi-periodic oscillating giant, a relatively quiet solar analogue, and a dwarf with periodic oscillations. Raw flux lightcurves are converted into differential lightcurves by normalizing the lightcurve by the median of the raw flux and subtracting 1. The lightcurves show both an overall trend and shorter timescale features as well. As the data reduction pipeline is still under revision (and has indeed already been updated since the computation of these results and publication of these proceedings), the commissioning data may contain systematic effects that are likely not physical in nature. The lightcurves shown in Figure 1 have therefore been de-trended by removing a fourth-order polynomial fit to the data, and we focus here on the bulk features of variability on timescales shorter than 10 days. While some of the long term

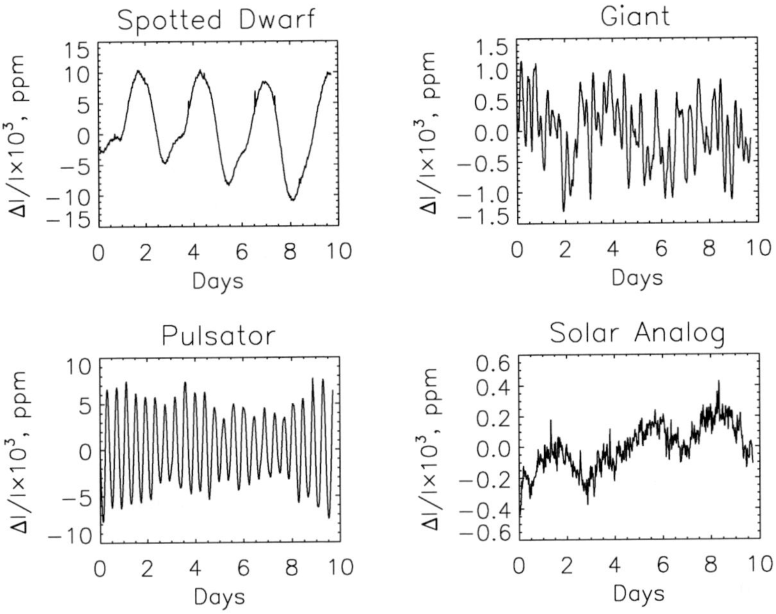

Figure 1. Example lightcurves for four different kinds of variable stars in the Kepler commissioning data, clockwise from top left: a spotted, rotationally modulated dwarf, a giant with small amplitude semi-periodic oscillations, a low activity solar analogue, and a hot pulsating dwarf.

trends our fit removes may be physical, we elected to err on the side of caution until the instrument is better understood.

From the rectified lightcurves, we calculate a set of statistics that can be used to quantify the variability. The maximum deviation above zero in the rectified curves is computed to capture the range of the largest feature in the lightcurve. To determine the point-to-point variability, we smooth the lightcurve on a 2 hour timescale and subtract this from the lightcurve itself, then calculate the standard deviation of the residuals. We use two metrics to determine the typical timescale of variability: the time separation between points where the differential lightcurve crosses zero, and the time separation between changes in the sign of the slope in the lightcurve. By searching the data using combinations of these statistics, it is possible to find populations of stars with similar variability characteristics.

3. Variability characteristics of the Kepler commissioning data

The statistics can be combined both to the characterize the sample and as filters to identify interesting populations for further study. The ability to isolate interesting categories of stars in this manner will be particularly important as the primary Kepler mission commences, when the number of targets will rise to ~150,000 stars per quarter of observations. In Figures 2 through 4, we show the distribution of the Kepler targets in several of our metrics– all statistics are shown in parts per million (ppm), corresponding to a 0.001 fractional change in intensity. Figure 2 plots the point-to-point variability in 0.5 hour time sampling for the giants and dwarfs in the sample (top and bottom panels, respectively) as a function of Kepler magnitude. The point-to-point variability rises towards fainter magnitudes as a result of the increasing noise floor for fainter stars. As is evident in this figure, giants are more variable than dwarfs at similar magnitudes– while some of them are quiet, the majority of the giants describe a locus above the main

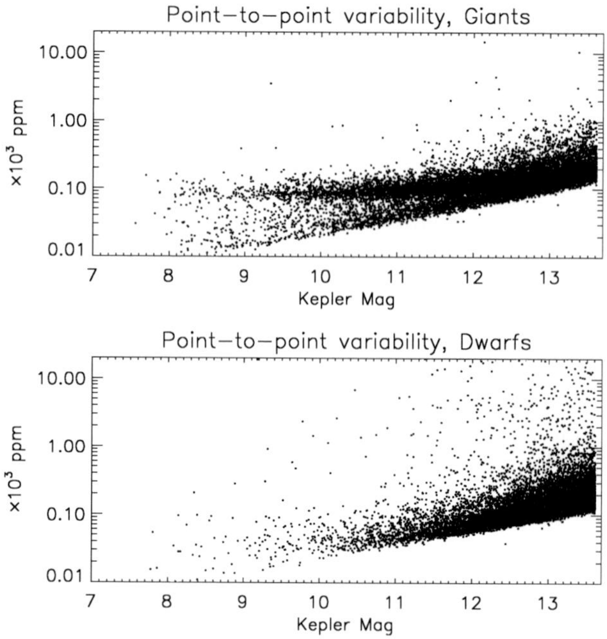

Figure 2. Point-to-point variability in parts per million (ppm) for giants (upper panel) and dwarfs (lower panel) in the Kepler commissioning data. The majority of giants are more variable on short timescales than the dwarfs, although some quite variable dwarfs (consisting of eclipsing binaries, pulsators and spotted stars) can be seen in the lower density cloud of points above the main locus of stars. The evident trend with magnitude is due to the rising noise floor for fainter stars.

density of dwarfs. The dwarfs are mostly quiet, with a lower density cloud of very variable stars lying above the main locus in the plot. These more variable stars consist of a mix of eclipsing/contact binaries, pulsators and spotted stars.

In Figure 3, we show the maximum deviation of the lightcurves from zero as a function of log (g). This metric quantifies the half amplitude of the largest features in the lightcurves– stars that are very variable have a high maximum peak height, while stars that are quiet have a low maximum peak height. Two main overdensities are seen here, the giants residing around a log (g) of ~ 2.5 and a typical maximum peak height of just under 1 part in a thousand, while the dwarfs are located around a log (g) of $\sim$4.3 and a typical amplitude of 0.2 parts in a thousand. These two main populations are bridged by a lower density region of subgiants in between them. The population of active, pulsating or eclipsing dwarfs can be seen around the same log (g) $\sim$ 4.3 as the rest of the dwarf sample, but at much higher maximum deviations.

Lastly, Figure 4 shows the typical timescale of variability as measured by the "slope separation", or the median timescale of changes in slope in the lightcurves. The utility of this metric is that it quantifies the timescale of the features in each lightcurve, regardless of whether they are periodic. Large slope separations correspond to larger bulk features in the lightcurves, whereas very small slope separations are typically the result of quiet lightcurves with few large features (if any). In the figure, giants and dwarfs both appear to be active on all timescales, but only the dwarfs have members with very small slope separations, quiet lightcurves with few features.

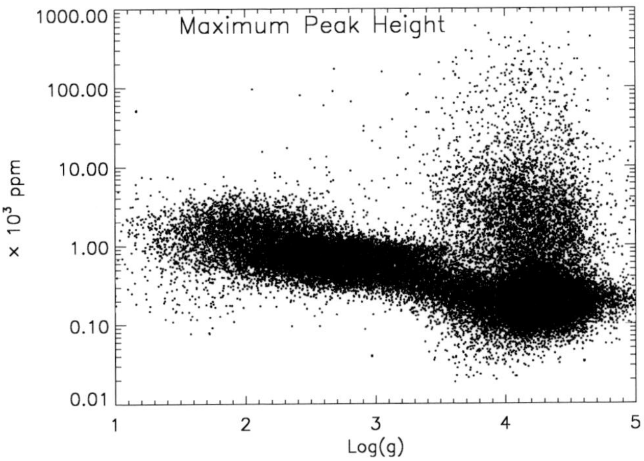

Figure 3. Maximum deviation from zero for all lightcurves in parts per million (ppm) as a function of log (g). There are two main overdensities of dwarfs and giants, bridged by a transitional population of subgiants. Active stars can be seen in the cloud of points at high gravity and high maximum deviation.

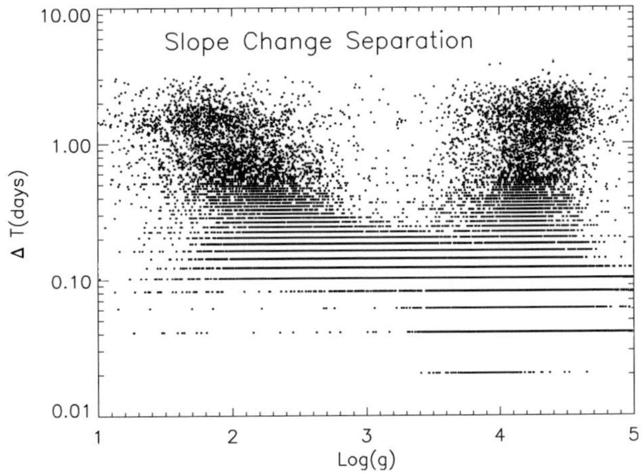

Figure 4. "Slope separation", or the median time between slope changes in the lightcurves, as a function of gravity. Slope separation shows the typical timescale of features in the lightcurve, whether they are periodic or not. Stars with large slope separation have features on longer timescales than those with small slope separation. Very small slope separation is characteristic of quiet stars without bulk features, where slope changes in the lightcurve are due to the point to point variability rather than large features.

4. Conclusions and Future Work

The Kepler mission holds great promise for investigations stellar variability in the years to come. Even in these relatively raw commissioning data, there are a multitude of interesting phenomena to whet the appetite for primary mission observations. The statistics and metrics we have developed and discussed here will be used to characterize variability for the full Kepler data set, and in particular to identify active spotted stars for rotation period determination. Ultimately, we intend to derive rotation periods, spot coverage, spot lifetimes and differential rotation parameters for as many of the Kepler targets as possible. In the coming months we expect to improve upon these metrics so that we may fully take advantage of the rich information content of the Kepler lightcurves.

References

Basri, G., Borucki, W. J., & Koch, D. G., 2005, *New Astron. Revs*, 49, 478
Batalha, N. M., *et al.* 2007, Bulletin of the American Astronomical Society, 38, 974
Eyer, L. & Grenon, M., 1997, ESA-SP 402 (Hipparcos Venice Workshop), p. 467
Gilliland, R. L. 2008, *AJ*, 136, 566
Horne, J. H. & Baliunas, S. L., 1986, *ApJ* 302, 757
Pizzolato, N., Maggio, A., Micela, G., Sciortino, S., & Ventura, P., 2003, A&A, 397, 147
Strassmeier, K. G., 2002, *AN* 323, 309

Solar and Stellar Variability: Impact on Earth and Planets
Proceedings IAU Symposium No. 264, 2009
A. G. Kosovichev, A. H. Andrei & J.-P. Rozelot, eds.

doi:10.1017/S1743921309993140

Does the lunar regolith contain secrets of the Solar System? Using the Moon as a cosmic witness plate

David S. McKay[1], Louise Riofrio[2] and Bonnie L. Cooper[2]

[1] NASA Johnson Space Center, Houston, Texas 77058, USA
[2] Oceaneering Space Systems, NASA Johnson Space Center, Houston, Texas 77058, USA
email: L.M.Riofrio@nasa.gov

Abstract. The lunar regolith (soil) has recorded a history of the early Moon, the Earth, and the entire solar system. A major goal of the developing lunar exploration program should be to find and play back existing fragments of that tape. By playing back the lunar tape, we can uncover a record of planetary bombardment, as well as solar and stellar variability. The Moon can tell us much about our place in the solar system and in the Universe. The lunar regolith has likely recorded the original meteoritic bombardment of Earth and Moon, a violent cataclysm that may have peaked around 4 GY, and the less intense bombardment occurring since that time. Decrease in bombardment allowed life to develop on Earth. This impact history is preserved as megaregolith layers, ejecta layers, impact melt rocks, and ancient impact breccias. The impact history for the Earth and Moon possibly had profound effects on the origin and development of life. Life may have arrived via meteorite transport from a more quiet body, such as Mars. The solar system may have experienced bursts of severe radiation from the Sun, other stars or from unknown sources. The lunar regolith has also recorded a radiation history in the form of implanted and trapped solar wind and solar flare materials and radiation damage. The Moon can be considered as a giant tape recorder containing the history of the solar system. Lunar soil generated by small impacts will be found sandwiched between layers of basalt or pyroclastic deposits. This filling constitutes a buried time capsule that is likely to contain well-preserved ancient regolith. Study of such samples will show us how the solar system has evolved and changed over time. The lunar recording can provide detailed snapshots of specific portions of solar and stellar variability.

Keywords. Moon; meteors, meteoroids; solar system: formation

The Moon is a witness plate for solar system history. Humans have an almost genetic need to understand how we got here, and where we may be headed. Answers to important questions may be preserved in the lunar regolith. Fossil regolith is affected by meteorite impacts, impact exotic clasts, implant galactic cosmic rays, solar flares radiation and particles, and solar wind. New studies of the lunar regolith should have much broader implications than just understanding regolith formation.

Lunar soil generated by small impacts will be found sandwiched between layers of basalt, layers of pyroclastic deposits, layers of impact ejecta, layers of impact melt sheets, and various combinations of these deposits. This sandwich filling has been trapped and removed from further processing at many places. It constitutes a buried time capsule likely to contain records of solar wind, giant solar flares, and impacting bodies such as comets or asteroids. This buried regolith may also a record of major, but rare events such as giant cosmic ray bursts or greatly enhanced meteorite bombardment derived from the fragmentation of large asteroids. Study of such samples will show us how the solar system has evolved and changed over time and provide us with detailed snapshots of specific portions of solar system history.

While we should not expect Moon-wide pristine layers to be preserved, isolated fragments of such layers may be found in many places protected from destruction or gardening by rapid subsequent burial. Regolith samples from the Apollo collection began with scoop samples collected from depths up to 10 cm. Trenches were dug to depths of 30 cm. Drive tubes were dug to depths ranging from 17-65 cm. The deepest samples were cores from depths ranging from 225-262 cm. The estimated time since burial of the trench samples ranged from 10 Myr to 100 Myr. Portions of the Apollo 15-17 drill cores had burial ages up to 1 Gyr.

However, much older regolith samples are within reach at appropriate lunar sites. Samples found beneath ejecta blankets of large craters may have burial or time capsule ages ranging up to 4 Gyr. Samples found beneath pyroclastic ash deposits have a lower limit exposure age defined by the underlying surface, and an upper limit given by the age of the overlying pyroclastic deposit. Based on Apollo data, typical ages range from 3.2-3.5 Gyr. Samples found between or beneath basalt flows have a lower and upper age ranges given by ages of the underlying and overlying basalt. These ages range from less than 3 to more than 4 Gyr. Similarly, samples located within breccia sheets in highland terrain have lower and upper age ranges corresponding to ages of the underlying and overlying breccia sheets. Ages of these breccia sheets range up to 4.4 Gyr. A key goal in our science of astrobiology is to determine whether life evolved independently on Earth. Some theories suggest that life arrived on Earth from elsewhere via meteorites. How long did planetary objects bombard the Moon and Earth during the early heavy bombardment? When did the rate of bombardment decrease, finally allowing life to catch hold on Earth? The impact history for the Earth, likely to be similar to that for the Moon, possibly had profound effects on the origin and development of life. The record of the first billion years has been destroyed on the Earth by geologic processing and resurfacing.

Did life come via meteorite transport from a quieter Mars? At the distance of Mars orbit, meteorite impacts were likely less severe and life could have started earlier. Mars life could become implanted on Earth via Martian meteorites, taking root in nutrient-rich oceans and ponds. The lunar regolith likely contains a record of the original bombardment earlier than 4 Gyr, the violent lunar cataclysm that may have peaked around 4 Gyr, as well as the less intense bombardment occurring since that time. Undamaged meteorite fragments have already been found on the Moon. Two examples were found at Station 9 in Hadley Rille by Apollo 15 astronauts and at Bench Crater by Apollo 12 astronauts. Many more fragments may be found in ancient regolith. The potential exists that fragments of Martian meteorites, Earth-derived meteorites, and even meteorites from moons of Jupiter and Saturn. Meteorites from Venus may also be buried in lunar regolith. During possible future mining operations to support lunar resource extraction and processing, a simple additional screening step might be added which could robotically detect non-lunar composition rocks and collect them for detailed examination. For example, based on the abundance of Na alone, it is possible to distinguish many Earth rocks from lunar rocks using a cutoff of about 0.5 Na_2O%. The potential exists that early meteorites from Earth found on the moon may preserve a record of the beginnings of life on Earth; these records are likely now destroyed on Earth by geologic processing, tectonics, and weathering.

Evolution of life would also be affected by radiation from space. In the course of its history, our solar system may have experienced strong bursts of severe radiation. Radiation might have come from the sun, other stars, supernovae, black holes, or other unknown sources. Severe radiation could blast away the ozone layer, causing major mutations in the development of life. Were these radiation bursts strong enough to kill early life on

Earth? Will we be able to recover a record of them trapped in pockets of ancient lunar regolith?

Solar wind particles have energies up to 1 keV/nucleon and penetrate the regolith to depths of 300 angstroms. More energetic solar electric protons (SEP) have energies from 1-100 KeV and can penetrate to depths of 1 cm. Galactic cosmic rays can have energies from 100 KeV to in excess of 3 GeV and can penetrate $>$ 1 meter into the regolith. A record of these strong radiation bursts could be preserved in ancient regolith samples on the Moon. It is likely that unusual fluxes could be detected in time capsule regoliths.

Burial of existing regolith by a hot basalt flow, a hot impact-generated melt flow, or even an impact-produced base surge of debris could preserve ancient regolith. Similar flows are found on Earth in locations like Craters of the Moon National Park and many locations in Hawaii, and these flows override and preserve underlying soil horizons. Large areas of presumed pyroclastic dark mantle are found on the front side of the moon, and small dark mantle regions have been identified on the back side. The contact between dark mantle and underlying normal regolith may be less disturbed than the contact between a basalt flow and underlying regolith and the dark mantle deposits may be less likely to heat and bake out the underlying regolith, thereby preserving the solar particulate profile and other details.

Plans are currently being made to return humans to the Moon by the early 2020's. Already a small armada of spacecraft from several nations has orbited the Moon. These orbiting spacecraft will soon be joined by automated landers. A major goal of the developing lunar exploration program should be to find and sample existing fragments of the Solar System record in the form of carefully located and sampled pockets of preserved ancient regolith. The lunar regolith is the only readily accessible place in the solar system for finding and revealing the detailed record of solar system history. The witness plate aspect of the moon may ultimately turn out to be its single most valuable scientific asset, and we need to focus a major exploration effort in this direction. This focus is readily understood by the public and can be the underlying scientific basis of much of our lunar program. The recovery, sample return and decoding of pristine ancient regolith may contribute to the most important scientific discoveries of the new lunar era.

Solar and Stellar Variability: Impact on Earth and Planets
Proceedings IAU Symposium No. 264, 2009
A. G. Kosovichev, A. H. Andrei & J.-P. Rozelot, eds.

doi:10.1017/S1743921309993152

X-exoplanets: an X-ray and EUV database for exoplanets

J. Sanz-Forcada[1], D. García-Álvarez[2,3], A. Velasco[1,4], E. Solano[1,4], I. Ribas[5], G. Micela[6] and A. Pollock[7]

[1]Centro de Astrobiología, CSIC-INTA, European Space Astronomy Center
Apartado 78, E-28691 Villanueva de la Canada (Madrid), Spain
email: jsanz@cab.inta-csic.es

[2]Instituto de Astrofísica de Canarias, Spain

[3]Grantecan CALP, Spain

[4]Spanish Virtual Observatory, Spain

[5]Institut de Ciences de l'Espai (CSIC-IEEC), Spain

[6]INAF- Osservatorio Astronomico di Palermo, Italy

[7]XMM-Newton SOC, European Space Agency, ESAC, Spain

Abstract. Extreme Ultraviolet (EUV) and X-ray emission is of great importance in several phenomena related to the formation of planetary systems and the atmospheres of planets. The atmospheric composition, and the mass of an exoplanet, are partly dependent on the X-ray and EUV radiation received during the first stages of formation and even during main sequence of the star. Biological life developing on exoplanets would depend severely on the high energy radiation arriving from its parent star.

Here we present a database of the X-ray and EUV emission of all the stars currently known to host exoplanets. The archive is public and accessible through the Spanish Virtual Observatory (SVO). The database gives the user the option to download observed X-rays and EUV spectra. Synthetic spectra covering the spectral range 1–912 Å are also available (present day telescopes do not give access to the EUV range at $\lambda > 180$ Å). These spectra are created using coronal models after fitting observed spectra.

Keywords. astrobiology, (stars:) planetary systems, stars: coronae, stars: abundances, X-rays: stars, atomic data

1. Introduction

Since the discovery of the first transiting exoplanet, it has been proposed that high energy stellar radiation should have some influence on the atmosphere of the planets (e.g. Seager & Sasselov 2000), in special the so called "hot Jupiters", planets with Jupiter-like mass at a short distance of the star. The effects of the high energy radiation (UV, EUV and X-rays) are important at different stages. First, during protoplanetary disc dissipation. Later, when the stellar rotation is fast and the planet keeps most of its original atmosphere, this radiation is believed to be a strong erosive agent for the atmosphere (e.g. Lammer *et al.* 2003). Finally, the developing of biological life in a planet will partly depend on the high energy (EUV and UV in special) radiation arriving to the surface.

During the evolution of a planet we have to consider also the evolution of its parent star. Once the protostellar disc dissipates, the rotation of the star will progressively slows down. The initial strong stellar rotation produces high coronal activity, with EUV and X-rays emission from the material heated at 1–30 MK. About ~100 Myr after the stellar birth, the X-rays and EUV emission flux from the corona decreases and cools down, as

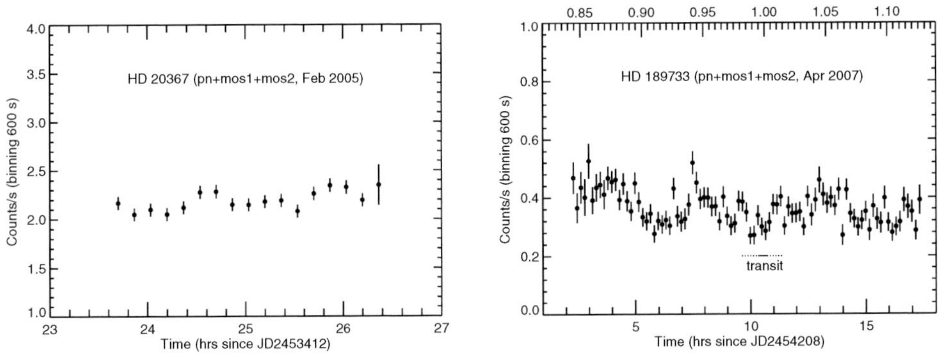

Figure 1. Two examples of X-ray light curves with some variability. *Left:* HD 20367, one of the most active stars in the sample. *Right:* HD 189733 hosts a transiting planet. The times for the total and partial (optical) transit are displayed, Upper axis show the orbital phase of the planet. Observations of evaporation signatures have been reported for this planet (Tinetti *et al.* 2007). High energy radiation is essential in this effect.

an effect of the slower rotation. The UV emission from the chromosphere and transition region follows the same pattern with age. It is possible to test the X-ray evolution with time by looking at the stellar clusters (Micela 2002). A direct application to the solar case was proposed by Ribas *et al.* (2005). A more detailed coronal model was employed by Cnossen *et al.* (2007) to evaluate the spectral energy distribution of κ Cet, a proxy of the Sun at the time in which life developed on Earth.

In the last years there has been an increasing interest in the use of the high energy emission from the parent star to know the possible effects on the planet, such as vaporization or atmospheric "erosion" (Lammer *et al.* 2003; Baraffe *et al.* 2004; Erkaev *et al.* 2007; Penz & Micela 2008; Penz *et al.* 2008; Cecchi-Pestellini *et al.* 2009; Lammer *et al.* 2009, Sanz-Forcada *et al.*, submitted). These models are limited to the use of X-ray or EUV fluxes calculated in the whole band. The use of a spectral energy distribution, with the highest resolution achievable, would be prefered to know better the effects in the planet atmosphere. The access to high resolution spectra in this spectral range is, however, very limited. For X-rays there are currently two large telescopes (XMM-Newton and Chandra) providing the ranges $\sim 4-38$ Å (XMM-Newton/RGS) and $\sim 1.5-180$ Å (Chandra/HETGS and LETGS). Such spectra can only be obtained for the brightest

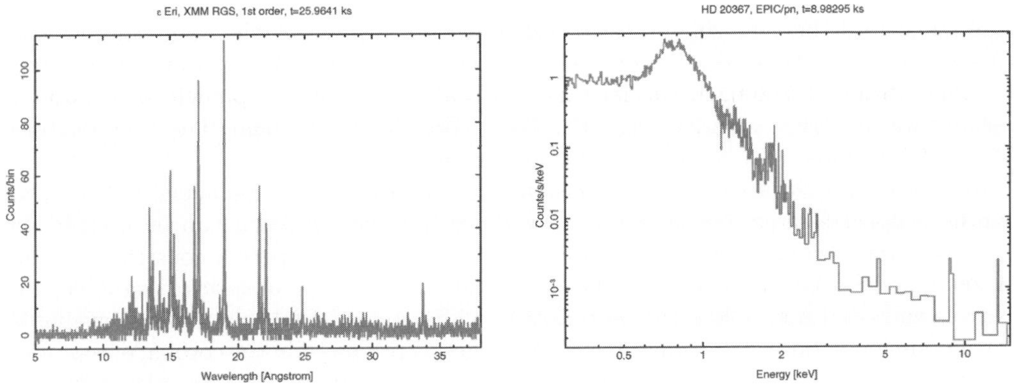

Figure 2. Real spectra observed with XMM-Newton. High resolution spectra (*left*) is only available for brightest stars. Most cases show, at most, a low-resolution spectrum (*right*).

Table 1. XMM-Newton flux of stars with exoplanets (0.12 – 2.48 keV)[1]

Planet name	Spectral type (star)	Stellar distance (pc)	log L_X (erg s^{-1})	S/N (L_X)	$m \sin i$ (m_J)	a_p (a.u.)	P_{orb} (d)
14 Her b	K0V	18.10 ± 0.19	26.92	4.9	4.64	2.77	1773.4
47 Uma b	G0V	13.97 ± 0.13	<25.45	...	2.60	2.11	1083.2
47 Uma c	"	"	"	"	0.46	3.39	2190.01
β Pic b	A6V	19.30 ± 0.19	25.63	5.75	8.00	8.00	6000
Cha Hα 8 b	M6.5	160	<27.87	...	18.00	1.00	1590.9
ϵ Eridani b	K2V	3.20 ± 0.01	28.20	297	1.55	3.39	2502
GQ Lup b	K7eV	140	29.45	32	21.50	103	...
HD 20367 b	G0	27.00 ± 0.79	29.30	140	1.07	1.25	500
HD 46375 b	K1IV	33.40 ± 1.19	27.16	7.3	0.25	0.04	3.02
HD 49674 b	G5V	40.70 ± 1.89	27.41	6.5	0.12	0.06	4.94
HD 50554 b	F8	31.03 ± 0.97	<26.59	...	4.90	2.38	1279
HD 62509 b	K0IIIb	10.34 ± 0.09	27.13	34	2.90	1.69	589.64
HD 70642 b	G5IV-V	29.00 ± 0.50	26.39	4.2	2.00	3.30	2231
HD 75289 b	G0V	28.94 ± 0.47	<26.16	...	0.42	0.05	3.51
HD 108147 b	F8/G0V	38.57 ± 1.03	27.39	4.2	0.26	0.10	10.90
HD 114762 b	F9V	39.46 ± 2.37	<26.61	...	11.02	0.30	83.89
HD 130322 b	K0V	30.00 ± 1.34	27.26	7.7	1.08	0.09	10.72
HD 187123 b	G5	50.00 ± 1.63	<26.80	...	0.52	0.04	3.10
HD 187123 c	"	"	"	"	1.99	4.89	3810
HD 189733 b	K1-K2	19.30 ± 0.32	28.18	93	1.13	0.03	2.22
HD 190360 b	G6IV	15.89 ± 0.16	<26.38	...	1.50	3.92	2891
HD 190360 c	"	"	"	"	0.06	0.13	17.10
HD 195019 b	G3IV-V	37.36 ± 1.24	<26.52	...	3.70	0.14	18.20
HD 209458 b	G0V	47.00 ± 2.22	<26.12	...	0.69	0.05	3.52
HD 216435 b	G0V	33.30 ± 0.81	27.74	12	1.26	2.56	1311
HD 216437 b	G4IV-V	26.50 ± 0.41	26.62	3.9	2.10	2.70	1294
HD 217107 b	G8IV	19.72 ± 0.29	<25.30	...	1.33	0.07	7.13
HD 217107 c	"	"	"	"	2.49	5.27	4210
HD 330075 b	G5	50.20 ± 3.75	26.51	3.1	0.76	0.04	3.37
NGC 2423 3 b		766	<29.47	...	10.60	2.10	714.3
τ Boo b	F7V	15.60 ± 0.17	28.94	317	3.90	0.05	3.31
VB 10 b	M8V	6.09 ± 0.13	25.83	20	6.40	0.36	271.56

Notes: [1] Planet data taken from The Extrasolar Planets Encyclopedia (http://exoplanet.eu)

X-ray sources. Among the planet-bearing stars only two have good spectra in this range, ϵ Eri and τ Boo (Sanz-Forcada *et al.* 2004; Maggio *et al.* 2009), with a few more cases with lower quality spectra, such as HD 20367 (Table 1). In most cases we will be limited to low-resolution spectra (range $\sim 1 - 125$ Å), and even upper limits of the flux for the lower activity stars. The situation is even worse in the EUV range ($\sim 100 - 900$ Å), where no current instrumentation can observe, except for the small range covered by X-rays telescopes. The Extreme Ultraviolet Explorer (EUVE) observed a few dozens of cool stars until the year 2000, most of them at close distances. The main limitation of that mission was the absorption by the hydrogen in the interstellar medium (ISM), that becomes stronger for longer wavelengths, up to 912 Å. Only one star with a known exoplanet was observed with EUVE, ϵ Eri (Sanz-Forcada *et al.* 2003a). The close distance of the star resulted in a good spectrum up to ~ 400 Å (Fig. 4).

In view of this situation we have constructed a database ("X-exoplanets") of real and synthetic spectra of stars with exoplanets. Since the real data that can be used in the models is very scant, we have created a set of high-resolution spectra constructed using coronal models. These models have been calculated using the real spectra available, and have been tested with ϵ Eri, the only case in which a good EUV spectrum is available. Most emission in both, the X-rays and EUV ranges, is produced in the corona, except for a few chromospheric and transition region lines. The database covers all the planet-bearing stars that have been observed with XMM-Newton or Chandra, and will be updated regularly as new data are incorporated. In the next lines we explain in detail how the

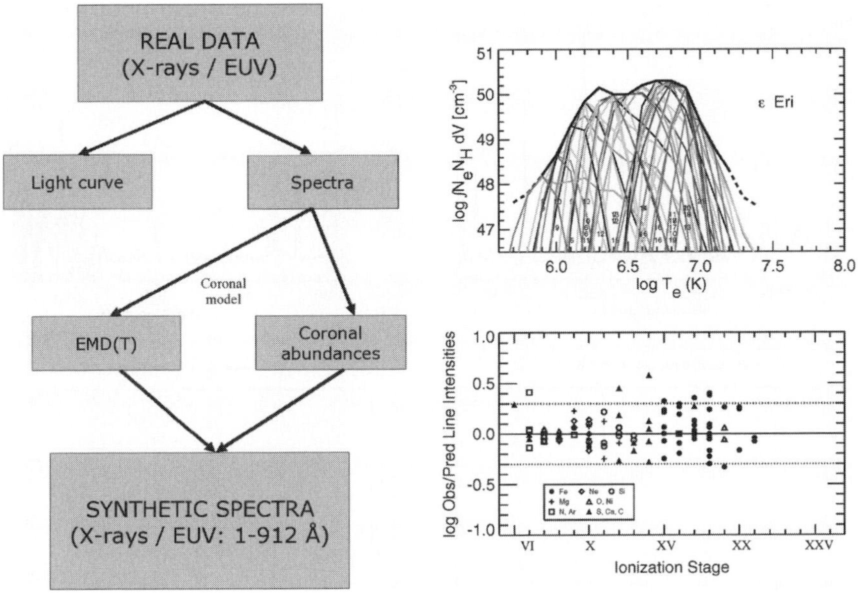

Figure 3. *Left:* schematic view of the technique employed to create the synthetic spectra from real data. *Right:* the high resolution data allow us to do a detailed reconstruction of the emission measure distribution. Synthetic spectra can be provided by use of this coronal model, in combination with the coronal abundances, also calculated during the process.

synthetic spectra are constructed, and what are the products that can be retrieved from X-exoplanets. The database can be accessed through the Spanish Virtual Observatory at the address http://sdc.cab.inta-csic.es/xexoplanets/

2. The database

The X-exoplanets database has the purpose to reduce all XMM-Newton and Chandra (eventually also EUVE) public data of stars with exoplanets reported, and to make these products available to the general public. We provide the light curves in different time bins (Fig. 1), and the low and high resolution spectra of the different instruments when available (Fig. 2). Some sources have not been detected, so circular regions have been selected at the expected position of the source, corrected from proper motion for the observing date; for those sources a light curve will be provided, and only an upper limit of the X-rays luminosity is included. For targets with enough statistics, the low resolution spectra have been fitted using rather simple coronal models, consisting on 1 to 3 temperature components, and a limited sample of element abundances, depending on the quality of the data (most sources can be well fitted with just 1 component). The spectra and necessary files are also provided, together with plotted spectra and residuals of the fit, and the coronal model used in the fit. For the few cases with high-resolution spectra, a detailed coronal model has been calculated using a line-based method, as described in Sanz-Forcada *et al.* (2003b), and references therein. A whole set of element abundances, and a coronal model with a $\Delta T = 0.1$ resolution are used in the fit.

The positions of the targets have been carefully calculated to improve former results. A comparison with past works show some inconsistencies with our X-ray fluxes (L_X). Since this quantity has little dependence on the exact parameters introduced in the fitting model, we attribute the discrepancy to wrong identifications of the sources. The case

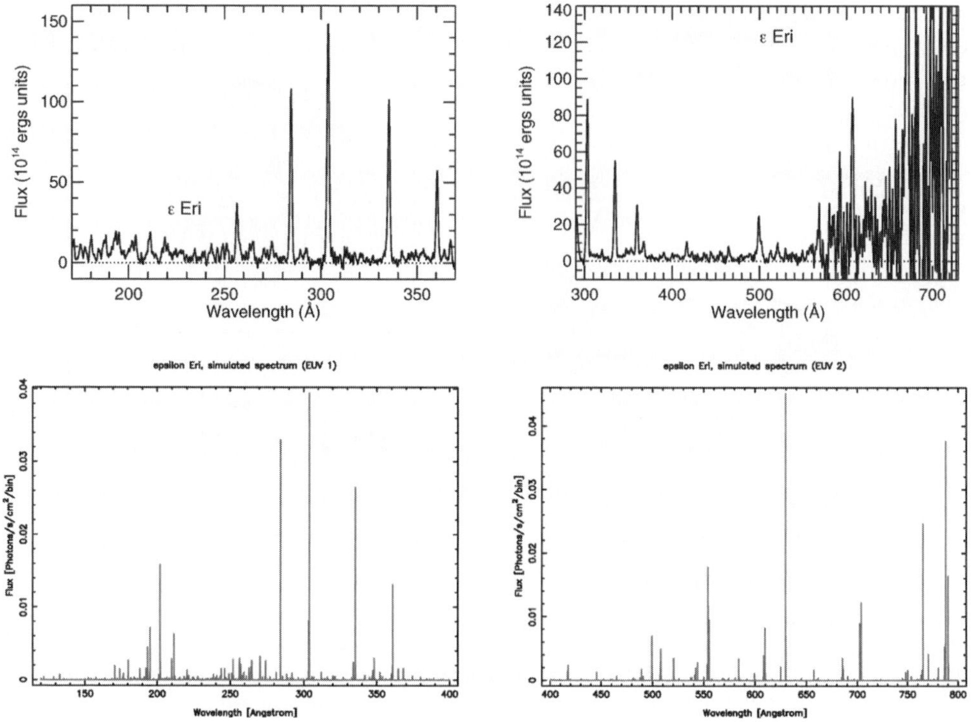

Figure 4. Real spectra (upper figures) of ϵ Eri observed with EUVE, in flux units. Note that the absorption in the interstellar medium allows very poor observations at longest EUV wavelengths. Synthetic spectra (lower figures) constructed with the coronal model successfully reproduces most spectral features of the real data.

of 47 UMa is particularly interesting. Kashyap *et al.* (2008) calculate in XMM-Newton $\log L_{\rm X} = 27.13$ erg s^{-1}. However the EPIC-pn image reveals two sources in the area of 47 UMa. The star itself (α = 10:59:28.0, δ = 40:25:46, $\log L_{\rm X} < 25.45$ erg s^{-1}) and a brighter source at α = 10:59:26.7, δ = 40:26:04, with a spectrum that resembles that of an AGN with z$\sim$ 0.2 and $\log L_{\rm X} \sim 43$ erg s^{-1} (G. Miniutti, private comm.). For the star HD 209458, that has been used in several studies because of its transiting planet, only an upper limit of $\log L_{\rm X} = 26.12$ erg s^{-1} could be calculated, while Kashyap *et al.* (2008) cite $\log L_{\rm X} = 27.02$ erg s^{-1}, probably confused with one of the nearby sources, well identified by us with 2MASS counterparts. This problem is even worse for ROSAT detections, given the poor resolution of ROSAT/PSPC.

3. Synthetic spectra

In order to have a better resolution spectra that can be used in the planetary atmospheric models, we have used our calculated coronal models to synthesize such spectra, using the atomic database APED (Smith *et al.* 2001). The method is described in Fig. 3. The central idea is that the emission in both, X-rays and EUV ranges, is originated in the corona of the star. The knowledge of the coronal model (basically the emission measure distribution, EMD, Fig. 3) and coronal abundances allows us to synthesize the whole spectral range, even if we have observations only in X-rays. The EMD is the distribution of coronal mass with temperature. The best is the model, the better will be the synthetic spectra; Low spectral resolution results only in a global fit to the spectrum and just 1–3 components of the EMD. But the high resolution data allow us to do a detailed reconstruction of the EMD, now calculated by comparing observed line fluxes with the fluxes predicted by the combination of the atomic data and different trial EMDs (see Sanz-Forcada *et al.* 2003b, for further details).

The atomic models are quite accurate for short wavelengths, but they have not been so well tested at longer wavelengths ($\gtrsim$ 500 Å). Besides, a few lines originated at lower temperatures (in the chromosphere or the transition region) are also present in the EUV range (most notably the He II 304 Å line). For a few cases there are UV observations and the coronal model can be extended to lower temperatures to reproduce these lines. For other targets we are using scaling laws from the coronal EMD, based on past observations available in Sanz-Forcada *et al.* (2003a). In order to test the validity of the method we have used ϵ Eri, the case with better spectral coverage. The coronal model of this star has been calculated using Chandra and EUVE high resolution spectra (see also XMM-Newton RGS spectrum in Fig. 2). International Ultraviolet Explorer (IUE) spectra extends the EMD to the transition region and the chromosphere, so a few "cooler" lines produced below 912 Å are properly modeled (Sanz-Forcada *et al.* 2003a, 2004). The results obtained are very promising (Fig. 4), with the models reproducing the main lines at $\lesssim$ 400 Å accurately. For longer wavelengths the synthetic spectra are still valid, but the quality of the real spectra falls at that range due to ISM absorption, and atomic models have not been tested so much as for shorter wavelengths.

This technique has been successfully applied to the star κ Cet (Cnossen *et al.* 2007) in order to see how the early Earth atmosphere would absorb the high energy emission from the star at the moment when life started on Earth. The use of this technique is particularly promising for transiting planets, in which some features, triggered as an effect of the incoming high-energy emission, might be observable in the transmitted spectrum of the planet.

References

Baraffe, I., Selsis, F., Chabrier, G., *et al.* 2004, *A&A*(Letters), 419, L13

Cecchi-Pestellini, C., Ciaravella, A., Micela, G., & Penz, T. 2009, *A&A*, 496, 863

Cnossen, I., Sanz-Forcada, J., Favata, F., *et al.* 2007, Journal of Geoph. Res. (Planets), 112, E02008

Erkaev, N. V., Kulikov, Y. N., Lammer, H., *et al.* 2007, *A&A*, 472, 329

Kashyap, V. L., Drake, J. J., & Saar, S. H. 2008, *ApJ*, 687, 1339

Lammer, H., Bredehöft, J. H., Coustenis, A., *et al.* 2009, *A&AR*, 17, 181

Lammer, H., Selsis, F., Ribas, I., *et al.* 2003, *ApJ*(Letters), 598, L121

Maggio, A., Sanz-Forcada, J., & Scelsi, L. 2009, in Proc. of the IAU Symposium No. 265, *Chemical abundances in the Universe: connecting first stars to planets*, ed. K. Cunha, M. Spite, & B. Barbury, in press

Micela, G. 2002, in ASP Conf. Series, Vol. 269, *The Evolving Sun and its Influence on Planetary Environments*, ed. B. Montesinos, A. Gimenez, & E. F. Guinan, p. 107

Penz, T. & Micela, G. 2008, *A&A*, 479, 579

Penz, T., Micela, G., & Lammer, H. 2008, *A&A*, 477, 309

Ribas, I., Guinan, E. F., Güdel, M., & Audard, M. 2005, *ApJ*, 622, 680

Sanz-Forcada, J., Brickhouse, N. S., & Dupree, A. K. 2003a, *ApJS*, 145, 147

Sanz-Forcada, J., Favata, F., & Micela, G. 2004, *A&A*, 416, 281

Sanz-Forcada, J., Maggio, A., & Micela, G. 2003b, *A&A*, 408, 1087

Seager, S. & Sasselov, D. D. 2000, *ApJ*, 537, 916

Smith, R. K., Brickhouse, N. S., Liedahl, D. A., & Raymond, J. C. 2001, *ApJ*(Letters), 556, L91

Tinetti, G., Vidal-Madjar, A., Liang, M.-C., *et al.* 2007, *Nature*, 448, 169

Solar and Stellar Variability: Impact on Earth and Planets
Proceedings IAU Symposium No. 264, 2009
A. G. Kosovichev, A. H. Andrei & J.-P. Rozelot, eds.

doi:10.1017/S1743921309993164

Surveying RFI for a new Brazilian Solar Spectroscope site

José R. Cecatto[1], Paulo C. G. Albuquerque[2], Ivan O. G. Vila[3], Alan B. Cassiano[1], César Strauss[1] and Francisco C. R. Fernandes[4]

[1]Astrophysics Division, National Institute for Space Research (INPE), 12227-010, São José dos Campos, Brazil
email: jrc@das.inpe.br

[2]Remote Sensing Division-INPE, 12227-010, São José dos Campos, Brazil
email: gurgel@ltid.inpe.br

[3]Aerospace Electronics Division-INPE, 12227-010, São José dos Campos, Brazil
email: geier@dea.inpe.br

[4]Institute of Research & Development-Univap, 12244-000, São José dos Campos, Brazil
email: guga@univap.br

Abstract. Brazilian Solar Spectroscope (BSS) carry out high time (0.01–0.1 s) resolution solar spectral investigations within the frequency range 1–2.5 GHz on a daily basis. At the moment, a new site is imposed to this facility. This site change became necessary due to both factors: the growing level of radio frequency interference (RFI) signals at actual site and requirements of rising up a new building at the BSS antenna location. We present results of RFI signals search at INPE-Cachoeira Paulista area as purpose to define a new BSS site.

Keywords. Sun: radio radiation; instrumentation: spectrographs; site testing

1. Introduction

Almost twenty years ago a radio spectrograph designed by Brazilian Solar Spectroscope (BSS) was installed at INPE (45° 51' 37" W, 23° 12' 29" S), São José dos Campos, Brazil. In its first phase, operated with narrow band (1600 ± 50 MHz) up to the starting of last solar cycle. By that time, the feeder was replaced. Then, BSS changed to operate in a broad band frequency range, 1–2.5 GHz (Sawant *et al.*, 2001). Since then, it has been carried out several thousand hours of solar observation. Also, over four hundred groups of solar radio bursts have been recorded and used for research. However, during recent years, radio frequency interference (RFI) signals became disturbing to the frequency band of BSS acquisition data. In parallel to this, a plan to rise up a new building in the place of BSS antenna imposed the instrument moving to a new site. In this paper we briefly describe RFI measurements and results as a purpose to define a new site.

2. Searching RFI signals and results

The process to choose a new radioastronomical site is complex and requires to be taken into account for several important aspects. The main is the RFI signal level at the site. An ideal site is isolated from all RFI. In practice, as an ideal site is unavailable a place with the lowest acceptable level of RFI must be chosen. As this we mean detectable signals whose level is below the nominal operational level or which do not mask or disturb acquired data. Also important for a new radioastronomical site are (NSF, 1956):

i) antenna should be out the view and farthest possible any close-by human sources of noise which might disturbed the daily acquisition data. Comercial radio, mobile phone, and wireless internet broadcast antennas are the most common noise sources; ii) antenna should be out the view of any high tension power lines; iii) antenna should be in a valley surrounded by hills as much high and in as many directions as possible; iv) site should be in a remote place, tens of kilometers distant from big concentrations of people and industries; v) site should be in a place not subject to violent winds.

Fortunately, at BSS frequency band the atmospheric water vapour has a negligible influence. Other factors are also important, as known, availability of basic infra-structure as well as technical support, cost of site installation, and relatively easy access.

Following, a brief description of the method used to measure RFI in five places at INPE-Cachoeira Paulista campus.

Radio Frequency Interference - RFI. RFI signals constitute disturbing either intermittent or constant radio signals identified within the band on which operates a radiotelescope. Among we can mention mobile phone, illegal radio broadcasts, wireless internet, and GPS satellite signals. By convenience, a search of RFI signals in five places within INPE-Cachoeira Paulista (INPE-CP) area has been done, during November 2008, as a purpose to define a suitable radioastronomical site for the BSS frequency range (1000-2500 MHz). Chosen places have the following coordinates: Place 1: 22° 41' 32" S, 45° 00' 56" W; Place 2: 22° 42' 07" S, 45° 00' 20" W; Place 3: 22° 40' 56" S, 44° 59' 12" W; Place 4: 22° 41' 43" S, 45° 01' 12" W; Place 5: 22° 41' 07" S, 44° 59' 06" W.

Figures 1 shows a satellite view of INPE-CP area with the five chosen places for RFI measurements numbered. Mounting used for the 5–6 days continuous measurements is composed of: a discone antenna, LNA with an approximately flat response in the above frequency range, portable spectrum analyzer (FSH6, 100 kHz - 6 GHz; R&S) and a portable computer as shown in Figure 2. Figure 3 shows the results of measurements at the better place, numbered 3. It has to be remarked that place presented the most clean spectrum with only one short duration (ms pulse) low level ($\leqslant -80$ dBm) signal recorded at 1875 MHz during one day of a 5–6 days period of measurements.

Figure 1. Satellite view (Google Earth, 2009) of INPE-Cachoeira Paulista (INPE-CP) area showing the five numbered places where RFI measurements have been done.

3. Discussions and conclusion

Defining a new radioastronomical site is a complex task which requires definite rules and care during the search process. Main rule refers to the existence of RFI signals

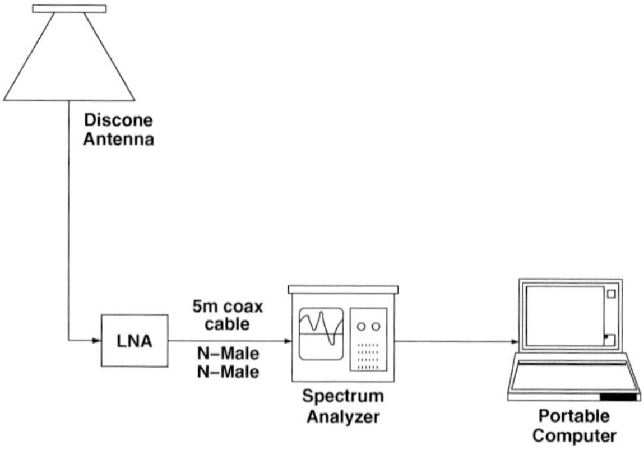

Figure 2. Mounting used for RFI measurements at INPE-Cachoeira Paulista.

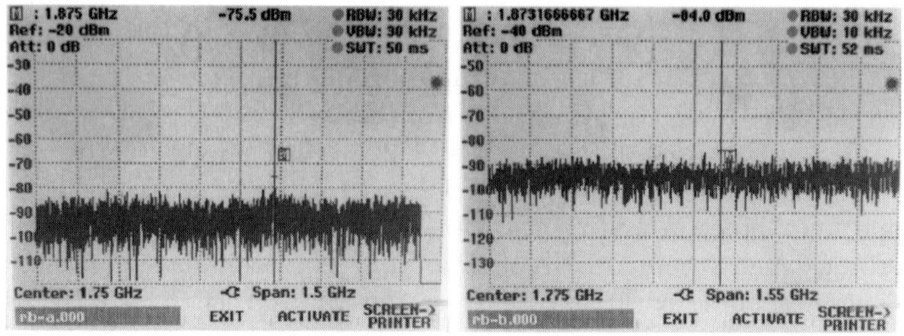

Figure 3. Measurements of RFI obtained on place 3 showing the lowest local level of RFI signals down to −80 dBm.

within the radiotelescope operational frequency range in a level for which the instrument operation becomes disturbed. Also, it can be detected by the radiotelescope harmonics of lower frequencies corresponding to some strong signal at a particular fundamental frequency. A classical one is mobile phone signal at 800–900 MHz. This rule has been taken into account. Then, by convenience as our first approach we carried out a continuous 5–6 days search for RFI signals in five places within INPE-CP area. Our measurements indicate three places with open horizon or higher altitude present RFI which can disturb BSS operation. Other two places are relatively quiet. However, one is subjected to strong wind, has hills just on the West side and implies high cost on infra-structure installation. Last place fulfilled the requirement of low-level RFI ($\leqslant -80$ dBm) associated to other requirements. Then, in principle even non-ideal that place within INPE-CP area seems to be adequate for a radioastronomical site. Details about all other aspects mentioned above relative to the definition of a new BSS site will be published somewhere else.

4. Acknowledgements

We acknowledge CNPq by Grant 475723/2004-0, INPE by financial-technical support.

References

NSF-Associated Universities Inc. 1956, *Planning document for Establishment and Operation of a Radio Astronomy Observatory*, NRAO Miscellaneous Unnumbered Internal Reports, 72p.

Sawant *et al.* 2001, *Solar Phys.*, 200, 167.

Solar and Stellar Variability: Impact on Earth and Planets
Proceedings IAU Symposium No. 264, 2009
A. G. Kosovichev, A. H. Andrei & J.-P. Rozelot, eds.

doi:10.1017/S1743921309993176

The development of the Heliometer of the Observatório Nacional

Victor d'Ávila[1,2], Eugênio Reis[1], Jucira Penna[1], Luiz Carlos Oliveira[1,3], Alissandro Coletti[4,5], Victor Matias,[1,3] Alexandre Andrei[1,6] and Sergio Boscardin[1]

[1]Observatório Nacional/MCT, R. Gal. Jose Cristino 77, Rio de Janeiro, Brasil
email: victor@on.br

[2]Uerj, [3]NGC-51, [4]Azeheb, [5]FACC, [6]OV/UFRJ

Abstract. The Observatório Nacional at Rio de Janeiro has developed a new heliometer to perform accurate measurements of the solar diameter. The equipment makes use of split parabolic mirrors, rather than split achromatic lenses as objective. In its first campaign this reflector heliometer generated a few thousands of images of the Sun through which we evaluate the precision of the solar diameter measurements as been 5 mas. The mechanical and thermal stability of the instrument is guaranteed by the use of a telescope tube manufactured in carbon fiber. The tube firmly holds the heliometric mirrors and the CCD camera that collects the images. The perfect lodging between the two half-mirrors over its supporting plate defines the angular instrumental separation between the two images of the sun. The heliometric mirrors and supporting plate have been made in ceramic material in order to ensure the stability of the optical configuration and to maintain the two images of the solar disk displaced by a fixed angle in relation to each other. To verify the stability of the instrument we designed the procedures of collimation and comparison. The collimation consists of using the heliometer for measuring the diameter of an artificial solar image specifically created for this purpose. The comparison is performed by measuring the residual unfold of a punctual image after been unfold by two opposed heliometers. The heliometric approach allows the measurement of the solar diameter at any heliolatitude and can be undertaken as often as desired. Computer programs have been developed for automated image acquisition and analysis. The optical and mechanical design has been performed by using 3D computer aided design software.

Keywords. Sun: fundamental parameters; instrumentation: heliometers; techniques: photometric

1. Introduction

In order to follow up and improve on the transit-time, Astrolabe CCD surveying of the solar at diameter at Observatório Nacional/MCT, a second approach was develloped, which consists of the direct astrometric measurement of the figure of the solar disk with a heliometer.

The fundamental difficulty of measuring a small angle by optical means comes from the fact that the measurement is directly dependent on the positioning of the focal plane of the instrument. In fact, the difficulty pointed out here for performing angular measurements is also present in reflecting telescopes. The solution presented in this paper consists of a using bisected mirror rather than achromatic doublet as has been adopted in the past (Geyer, 1984). Each half-mirror is tilted but an angle of 0.125 degrees in order to displace the images relatively to each other by one solar diameter approximately. In this way we will have opposite limbs of the sun almost in tangency in the focal plane

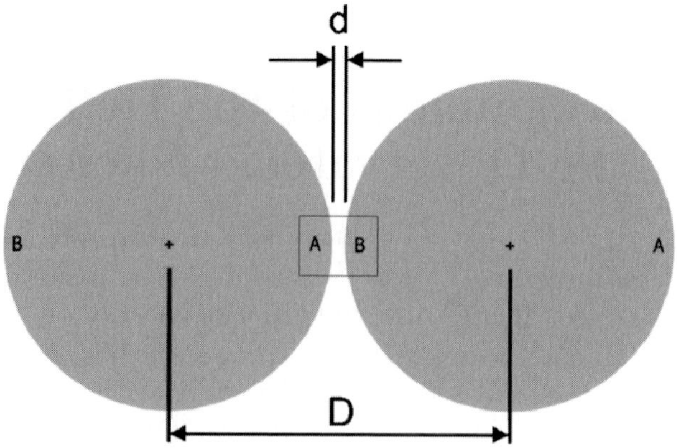

Figure 1. Schematic heliometric measurement: solar diameter = D - d.

(Fig. 1). With this setup we eliminate all chromatic defects, and also take advantage of the mechanical stability of mirrors.

2. Heliometric Concept

The basic idea behind the heliometric technique consists of doubling, in some way, the field of view of a telescope and applying a relative displacement between them, in such way to bring two nearby objects into close proximity. This approach is very useful as a way to reduce the measurement of finite angle, say of a few degrees, into something more convenient, and as small as 1 minute of arc, or so. Supposing that, in principle, the errors are proportional to the amount to be measured, we see that it is possible to get, a considerable increase in the measurements.

Another advantage of the method heliometric comes from the fact that we can perform measurements along any chosen heliolatitude and whenever desired. Of course we have to take into account the effects of the differential refraction between the limbs along different heliolatitudes. In order to have the observations made under similar conditions we have chosen the instrumental displacement of the images in such a way that the minimum distance between the limbs be approximately equal to the variation of the apparent solar diameter between aphelion and perihelion (Fig. 2).

In the first two years of the heliometer project we have built some prototypes to test different image duplication techniques:

(*a*) Two plane mirrors in angle positioned in front of the objective have been tried. That would offer the freedom to use a standard telescope to perform the task. It would imply, however, in more parts involved.

(*b*) The microscope prism approach also allows the duplication of the images. But that does not help in the dependency of the measurements with the focal distance.

(*c*) We also have made some experiments with the classical solution of bisecting refracting objectives. However, in general, objective lenses are chromatic doublet or even panchromatic triplet. It is very difficult to have all parts of one such objective maintained fixed in relation to each other. And we still have take into consideration chromatic effects of second order.

(*d*) Finally, we have worked with an entirely new approach, which consists of bisecting a primary mirror of a telescope. Now, two possibilities are presented and have been tried: The linear heliometer (in which we produce the displacement of the images by translating the two half-mirrors along the line of cut), and the angular heliometer (in

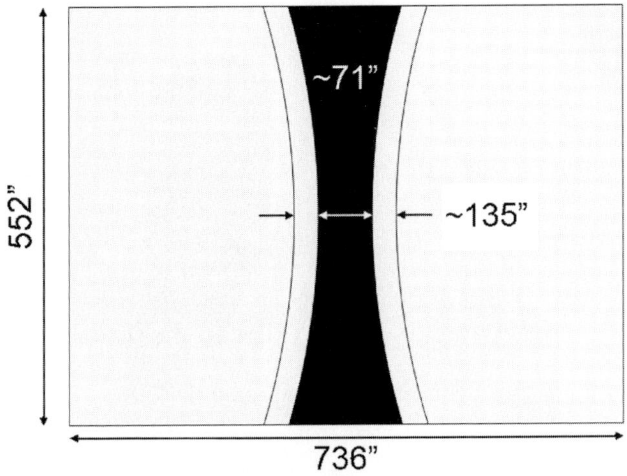

Figure 2. Aphelion and perihelion heliometric configuration.

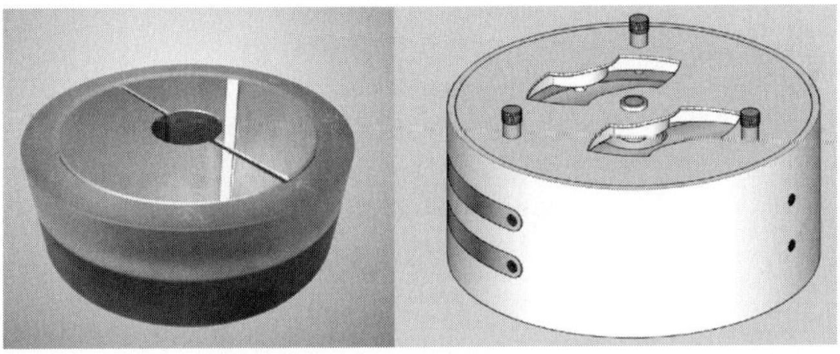

Figure 3. The actual heliometric mirror assembling and the project of its cell.

which we produce the displacement of the images by rotating the two half-mirrors along a line perpendicular to the line of cut).

The latter method was the choice for the construction of the Heliometer of the Observaório Nacional.

The heliometric mirror is all made of CCZ-HS, a ceramic material with very low thermal expansion coefficient ($0.0 \pm 0.2 \times 10^{-7}/°C$). The two half mirrors are immobilized, in relation to each other, by means of a external ring, all resting over an optical plate (Fig. 3a). Its cell guarantees the mechanical and geometrical stability for the entire set (Fig. 3b). The mask seen at the top of the cell has been designed to keep the two half-mirrors blocked in place and also assures that entrance pupil has a symmetric shape. The surface quality of the optical plate and the mirrors is better than $\lambda/12$ and $\lambda/20$, respectively.

The tube of the telescope is made of carbon fiber. This material, as well as extremely rigid, has very low coefficient of thermal expansion. It is mounted inside a stainless steel truss support and can rotate around its axis (Fig. 4).

In order to eliminate the secondary mirror the CCD chip was removed apart from the camera electronics and installed directly in the focal plane.

During the collimation and comparison procedures we observe the images of an artificial sun, which consists of a small ceramic disk placed in the optical axis of the instrument and immediately above the heliometric mirrors. A pair of thin nylon wires, in form of cross, have been installed over this artificial sun in order to provide a reference frame.

Figure 4. The Heliometer of the Observatório Nacional.

To perform the collimation procedure we install a parabolic mirror in front of the heliometer in such way to produce an image of the artificial sun at the infinite. During the collimation we observe the images of the two opposed limbs of the artificial sun in a way that is substantially similar to the real observation of the sun as can seen in Fig. 5a.

With this setup we are able to measure the displacement in two orthogonal direction. 1)- The angular displacement along the direction parallel to the cut is determined by the horizontal gap between the limbs. 2)- The angular displacement along the direction orthogonal to the cut can be evaluated by the vertical displacement of the images of the horizontal wire.

The comparison procedure, in its turn, is made by replacing parabolic mirror of the collimation by a secondary heliometer. This secondary heliometer produces an unfolded image of the artificial sun at the infinite. The primary heliometer then bring back these unfolded images to close proximity. In this configuration we observe the final images of nylon cross after been successively unfolded by the two opposing heliometers. We are then able to evaluate the remaining displacement along the directions of the two orthogonal wires (Fig. 5b).

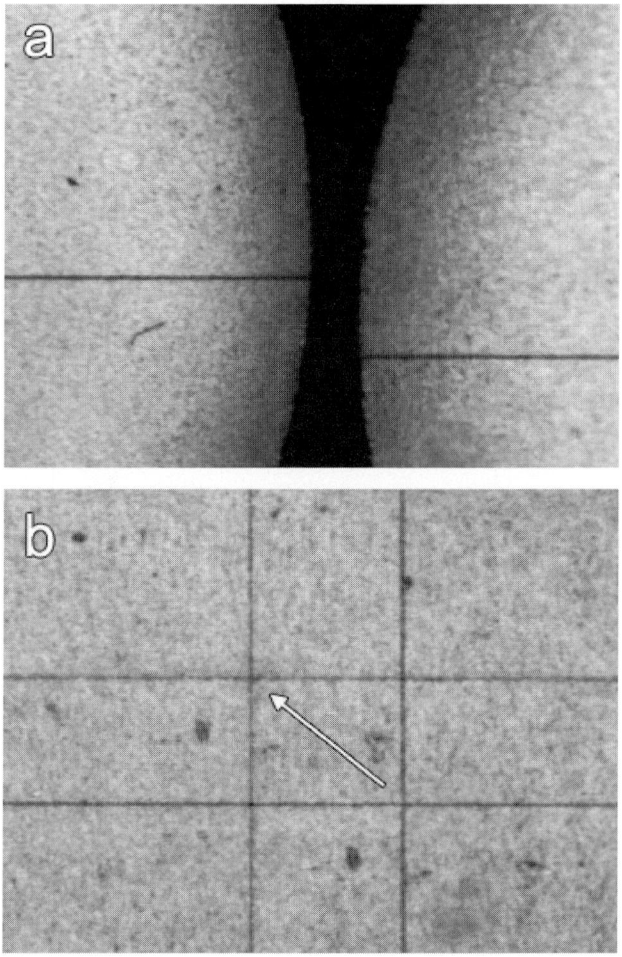

Figure 5. Collimation (a) and comparison (b) procedure images. The arrow indicates the displacement of the nylon cross center.

3. Results

In May 2009 there was an observational campaign that extended for 9 days, with over 70,000 heliometric images of the Sun. The heliometer exhibited a precision around 0.5 arc seconds for an isolated observation. And, we also see that in only one day of observations we can reach a precision about 5mas. The results are shown in Table 1.

These preliminary observations show a internal dispersion that is consistent with what should be expected to be due to the atmospheric seeing alone.

The results show that the values of the measured semi-diameters follow a normal distribution what indicates the absence of significant instrumental errors (Fig. 6). As the observational procedure is able to capture and process thousands of heliometric images per day, the measurements can achieve a significantly high accuracy in a short period of time.

Table 1. First heliometric measurements of the solar semi-diameter.

Day[1]	SD[2] (")	σ (")
13	959.084 ± 0.005	0.461
14	959.656 ± 0.006	0.559
18	959.440 ± 0.006	0.533
19	959.306 ± 0.005	0.459
20	959.367 ± 0.006	0.521
21	959.503 ± 0.007	0.639
25	959.646 ± 0.006	0.515
26	959.811 ± 0.006	0.501
27	959.823 ± 0.006	0.528
	959.519 ± 0.002	0.566

Notes:
[1] May 2009.
[2] Solar semi-diameter at 1 AU.

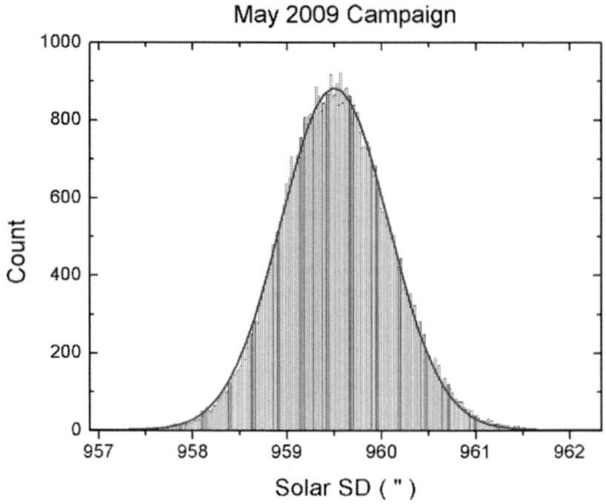

Figure 6. Solar semi-diameter distribution with a gaussian fit.

4. Conclusion

The materials that form the Heliometer have thermal and mechanical stability to dex-7. The number of optical parts is minimized and their quality is better than $\lambda/12$. Original software for the automated collection and analysis of the images was developed. All along the project was developed by actual implements on to a trial heliometer. By using the final version fully realized, we conducted an observational campaign through 9 days, deriving more than 70,000 heliometric images of the Sun These results indicate a precision of 0.5 arcseconds, with no instrumental bias, and limited only by the provisional atmospheric modeling. Therefore, as meant from start, an accuracy of 5 mas can be achieved. The first results indicate that we should pursue this approach and initiate a campaign of observations aiming to cover this solar cycle which is just in its beginning, at least.

Acknowledgements

The authors acknowledge partial financial support from the Brazilian Science Funding Agencies FAPERJ, CNPq, CAPES and FINEP. We also acknowledge the NGC-51 group.

Reference

Geyer, E. H. 1984, *Ap&SS*, 110, 183

Solar and Stellar Variability: Impact on Earth and Planets
Proceedings IAU Symposium No. 264, 2009
A. G. Kosovichev, A. H. Andrei & J.-P. Rozelot, eds.

doi:10.1017/S1743921309993188

Brazilian Decimetric Array (BDA) project - Phase II

C. Faria[1], S. Stephany[2], H. S. Sawant[2], J. R. Cecatto[2] and F. C. R. Fernandes[3]

[1]Department of Computer Science - DC/PUCMinas, Pocos de Caldas, Brazil
email: faria@pucpcaldas.br
[2]National Institute for Space Research - INPE, Sao Jose dos Campos, Brazil
[3]Institute of Research and Development - UNIVAP, Sao Jose dos Campos, Brazil

Abstract. The configuration of the second phase of the Brazilian Decimetric Array (BDA), installed at Cachoeira Paulista, Brazil (Longitude 45° 0' 20" W and Latitude 22° 41' 19" S), is a T-shaped array where 21 antennas are being added to existing 5 antennas of the first phase. In the third phase, in each arm of the T array, four more antennas will be added and baselines will be increased to 2.5×1.25 km in east-west and south directions, respectively. The antennas will be equally spaced at the distances of 250 meters from the central antenna of the T-array. Also, the frequency range will be increased to 1.2–1.7, 2.8 and 5.6 GHz. The Second phase of the BDA should be operational by the middle of 2010 and will operate in the frequency range of (1.2–1.7) GHz for solar and non solar observations. Here, we present the characteristics of the second phase of the BDA project, details of the array configuration, the u-v coverage, the synthesized beam obtained for the proposed configuration.

Keywords. instrumentation: interferometers; techniques: interferometric, image processing

1. Introduction

The Brazilian Decimetric Array (BDA) is a radio interferometeric array that is being developed by the National Institute for Space Research (INPE, Brazil) as an international collaborative program (Sawant *et al.* 2000). The images of active regions provided by BDA will be analyzed by using spectral tomography technique being developed for application to space weather forecasting (Rosa *et al.* 2000). Also analysis of the flare component will lead to better understanding of the fundamental problems in solar physics (Sawant *et al.* 2002; Sawant *et al.* 2003). BDA also will be useful for galactic and extra-galactic investigations of the southern sky that is not accessible to Very Large Array (VLA) (Napier *et al.* 1983). The BDA site is located at INPE campus in Cachoeira Paulista, SP. This site has a valley with dimensions 400m × 300m where will be constructed a central compact Tee array and a control room with facilities to have on line data processing. The BDA project is been developed in three phases:

(*a*) **BDA Phase I**: In the first phase of the BDA project an east-west five-element interferometer was developed and installed at INPE using a 4-meter diameter parabolic dish with alt-azimuth mount and complete tracking capability. BDA phase I operates in the frequency range of 1.2–1.7 GHz. The five antennas had been laid out over a distance of 216 meters in the west-east direction getting a spatial resolution about 3 minutes of arc at 1.5 GHz.

(*b*) **BDA Phase II**: in the second phase 21 antennas is being laid out over the distance of 252 meters in the east-west direction and 9 antennas will be laid out over a distance of 166 meters in the southern direction, forming a T-shaped array. The frequency range is being increased to 2.8 and 5.6 GHz.

(*c*) **BDA phase III**: finally, in the third phase, 8 more antennas will be added in the east-west direction and 4 antennas will be added in the south direction. The baselines will be increased in both directions to 2.5 km and 1.25 km, respectively, to increase the spatial resolution of the array up to approximately 4.5 arc seconds at 5.6 GHz.

2. Antennas Configuration for BDA - Phase II

Radio interferometric arrays measure the Fourier transform of the sky brightness distribution (visiblity) on a finite set of points in the Fourier plane. The sky brightness distribution can be reconstructed by the inverse Fourier transform of the sampled visibility. The quality of the reconstructed images strongly depends on the array configuration, since it determines the sampling function of the Fourier plane. Several approaches have been proposed to optimize the response of one radio interferometric array (Cornwell, 1988; Keto, 1997; Kogan, 2000). It is clear that the best array configuration is dependent on the scientific goals intended for the array. However, the optimization of the antenna positioning involves several others aspects, that may be conflicting, such as cost or geographic constraints. The BDA is being designed to obtain optimized images of the radio sources at the decimetric band with high temporal and spatial resolutions. This precludes the use of earth rotation synthesis, requiring snapshot image acquisition. The BDA "T" shape configuration is suitable since it yields a uniform sampling in a retangular region using a regular spacing between antennas. In addition, this shape is good considering site constraints and also implementation and maintenance costs. The proposed BDA array configuration presents a dense array near the intersection of the "T", with a fundamental spacing of 9 meters between adjacent antennas. In each arm of the "T" this spacing is increased to 18 and 36 meters, after the fourth and seventh antennas respectively. The Fig. 1 shows the BDA site and the antennas configuration for BDA - Phase II.

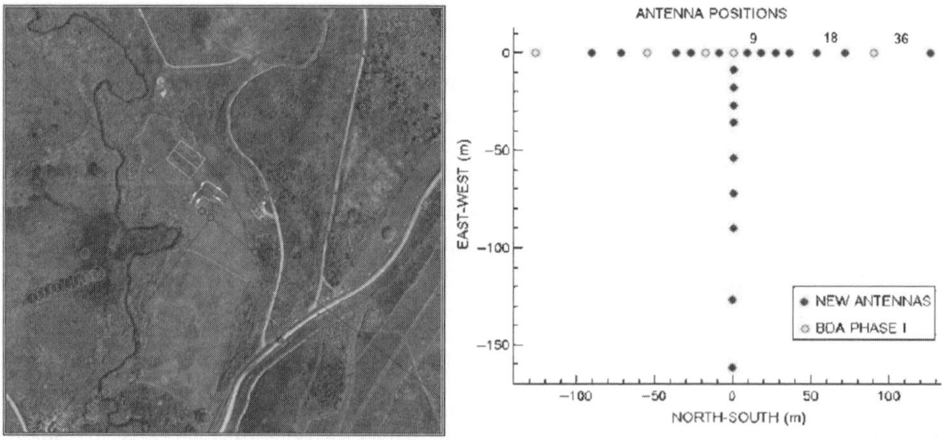

Figure 1. Aerial view of the BDA site (left) and proposed antennas configuration for BDA - Phase II (right). The red circles in the left image correspond to the antennas of BDA Phase I.

3. Configuration Results for BDA Phase II

The results obtained with proposed configuration is presented in Fig. 2, which shows the uv coverage (a), the synthesized beam (b) obtained with the configuration at frequency of 1.4 GHz and the one-dimensional profile of the beam (c) at 2.8 GHz.

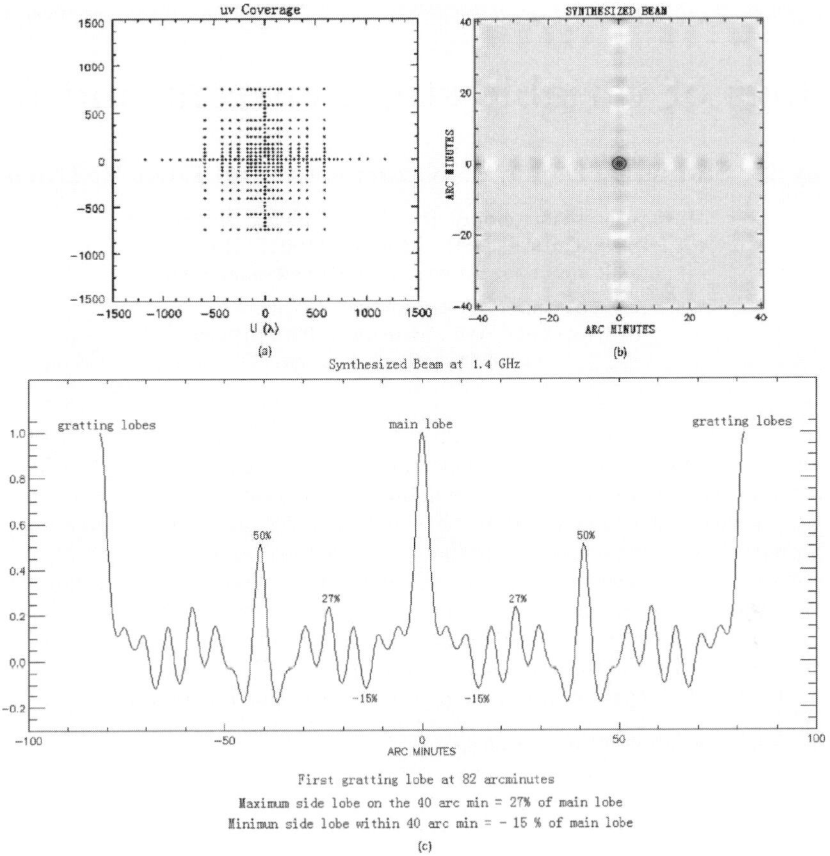

Figure 2. Resulting uv coverage (left) and synthesized beam (right) obtained with the proposed configuration at 1.4 GHz.

4. Final Considerations

This work presented the array configuration of the BDA phase II. The proposed configuration presents a regular spacing between adjacent antennas in the centre of the array. This compact array results in a good uniform uv coverage in the centre of the uv plane. Configuration results show that BDA phase II presents one synthesized beam with side lobes minimized below 20 percent of the main lobe inside of Field of View (FOV). The Central T will have spatial resolution 2x3 of arc minutes, field of view of 3 degrees at 1.4 GHz.

References

Cornwell, T. J. 1988, *IEEE Trans. Antennas Propagat*, 36

Keto, E. 1997, *ApJ*, 475

Kogan, L. 2000, *IEEE Trans. Antennas Propagat*, 48

Napier, P. J., Thompson, A. R., & Ekers, R. D. 1983, *Proceeedings of the IEEE*, 71, 11

Rosa, R. R. *et al.* 2000, *Adv. Sp. Res.*, 25, 9

Sawant, H. S. *et al.* 2000, *Adv. Sp. Res.*, 25, 9

Sawant, H. S. *et al.* 2002, *Proc.10th European Solar Physics Meeting*, 2

Sawant, H. S. *et al.* 2003, *Adv. Space Res.*, 32, 12

Solar and Stellar Variability: Impact on Earth and Planets
Proceedings IAU Symposium No. 264, 2009
A. G. Kosovichev, A. H. Andrei & J.-P. Rozelot, eds.

doi:10.1017/S174392130999319X

Catalogs of variable stars, current and future

Nikolay N. Samus[1,2], Elena V. Kazarovets[1] and Olga V. Durlevich[2]

[1]Institute of Astronomy, Russian Academy of Sciences,
48 Pyatnitskaya Str., Moscow 119017, Russia
email: samus@sai.msu.ru, helene@inasan.ru

[2]Sternberg Astronomical Institute,
13 University Ave., Moscow 119992, Russia
email: gcvs@sai.msu.ru

Abstract. To study effects of stellar variability, it is necessary to have complete lists of variable stars with good coordinates, correct variability ranges, reliable classification. The General Catalogue of Variable Stars (GCVS) is one of the main sources of information of this kind. Recently, the flow of information on stellar variability has increased strongly, requiring new approaches to GCVS compilation. New classification schemes are needed for variable stars, taking into account the development of our knowledge. We present information on the current state of the GCVS, its web data base, new Name-Lists containing thousands of new variables. We also discuss the problems of developing a new classification system for variable stars, of implementing catalogs of variable stars based on automatic surveys into the GCVS, of cooperation and interaction between the GCVS and other important sources of information on variable stars.

Keywords. stars: variables: other, catalogs

1. Introduction

The General Catalogue of Variable Stars (GCVS) is a project under way in USSR/Russia since 1946 on behalf of the International Astronomical Union. It replaced the series of variable-star catalogs that had been published before the World War II in Germany by the Astronomische Gesellschaft. The initiators and leaders of the Soviet GCVS project were Pavel Parenago (1906–1960), B. V. Kukarkin (1909–1977), P. N. Kholopov (1922–1988). Complete lists of variable stars with good coordinates, correct variability ranges, reliable classification are very important for any studies of impact of stellar variability on stars and planets.

The 4th GCVS edition, the last one in the paper form (Kholopov *et al.* 1985–1987) contains about 28 500 variable stars of our Galaxy plus about 14 800 suspected variables in the New Catalogue of Suspected Variable Stars (the NSV catalog, Kukarkin *et al.* 1982). A supplement to the NSV catalog (Kazarovets *et al.* 1998) added about 11 200 stars. With the recent Name-Lists of variable stars, the number of variables named in the GCVS system has become as large as about 41 500. Together with the catalogs of suspected variable stars, our files contain 64 324 stars (by mid-2009). At the same time, automatic surveys announce very many new discoveries. For instance, the ASAS-3 survey (Pojmanski 2002) has announced about 30000 new discoveries. Very soon, the number of newly discovered variable stars is expected to be millions. Old star-by-star approaches to variable-star catalogs have reached their limit and cannot be continued for very long. Meanwhile, there remain possibilities to improve the contents of the GCVS also for "old" variable stars.

2. Recent Improvements of the GCVS

After the fourth GCVS edition, we continue to support and improve the GCVS online version, also including catalogs of suspected variable stars.

The online GCVS version takes into account numerous corrections compared to the printed version. In particular, the coordinates have been greatly improved (see Samus *et al.* 2006 and references therein). Revised information has been provided for about 8000 suspected variables. Several search possibilities are available. Our online catalog can be found at the Sternberg Institute's web site (http://www.sai.msu.su/gcvs/gcvs/). This version also includes stars of variable-star Name-Lists published after the 4th GCVS edition. Recently we initiated publishing the Name-Lists, containing several thousand variables each, as complete catalogs of new variable stars (cf., for example, Kazarovets *et al.* 2009). The online information for stars from some of the older Name-Lists is incomplete; the situation is improved in the course of our systematic effort to revise astrophysical information of the GCVS. So far, this work has been completed for the constellations of Andromeda, Antlia, Aquarius, Ara, Auriga, and Telescopium.

The data from automatic surveys in open access in the Internet make it possible to transfer many stars from catalogs of suspected variables to the GCVS. Thus, about 1250 NSV stars were transferred to the GCVS in the recent 79th Name-List of Variable Stars (Kazarovets *et al.* 2008).

3. Problems for the Near Future

One of the important first steps leading to new approaches to variable-star catalogs is the improvement of the GCVS classification system. Most automatic surveys have simplified classification systems, not completely meeting the needs of the community; the classification system of the 4th GCVS edition is too complex. We have suggested a revised system to the IAU Commissions 27 and 42 during the IAU XXVI in Prague, it can be found at http://www.sai.msu.su/gcvs/future/future.html – the working group created in Prague (and continuing its activity after Rio de Janeiro) will be grateful for any additional suggestions. Note that the GCVS team receives conflicting signals from the community: while experts on particular variability types urge us to make the classification more detailed, most users would prefer a simpler system, easier to understand and remember. Prospects of automatic classification, badly needed for improving productivity, also require a simpler system with not too many different types of variable stars and no need of detailed studies to distinguish them.

One of the problems not permitting to immediately merge variable-star catalogs from automatic surveys with the GCVS is the quality of coordinates in the surveys based on observations with small cameras. Quite often, automatic surveys present good-quality light curves for variables discovered at positions between two stars of comparable brightness, $10-15''$ apart. To determine which of the two components really varies, additional observations or other kinds of supplementary information are needed. Note that, typically, the amplitudes quoted in automatic survey in such cases are also too low because of contamination from the non-varying neighbor. The GCVS team is not going to deteriorate its current rather high standard of accuracy of tabulated coordinates (see Section 2), so such stars will remain outside the GCVS until it becomes possible to find out which particular star varies.

One of possible ways to solve such identification problems is to use information contained in photographic plate stacks and still awaiting its evaluation. This information cannot be completely used unless the existing plate collections are digitized. The

Sternberg Institute plate collection (about 60000 plates) is being digitized since 2006 using two CREO EverSmart Supreme II scanners. So far, about 500 new variable stars have been discovered using these scans, almost 300 of them already announced in Kolesnikova *et al.* (2008).

Recently (Samus *et al.* 2009), we have determined sufficiently accurate equatorial coordinates for virtually all stars of the Catalogue of Variable Stars in Globular Clusters (Clement *et al.* 2001). This permits us to consider a possibility to change the obsolete rule to leave globular-cluster variables outside the GCVS, designed to be a catalog of all galactic variable stars.

The ever increasing flow of information on variable stars, including new discoveries, makes it very important to improve international cooperation in the field of variable-star catalogs. The International Variable Star Index (VSX; http://www.aavso.org/vsx/), supported by the American Association of Variable Star Observers (AAVSO), has the intention to become the main entry point to the GCVS system. It already provides information for variable stars from sources outside the GCVS, among them catalogs of automatic surveys. In April 2009, the GCVS and VSX teams have agreed to coordinate information provided by the two facilities and invited SIMBAD to join them in this coordination.

Our work on variable-star catalogs is supported by grants from the Russian Foundation for Basic Research (08-02-00375), from the Program "Origin and Evolution of Stars and Galaxies" of the Presidium of Russian Academy of Sciences, and from the Program of Support for Leading Scientific Schools of Russia (NSh 433.2008.2). We wish to thank Dr. C.M. Clement for fruitful cooperation. Many thanks are due to the AAVSO and personally to Dr. A. Henden for the invitation to discuss the problems of variable-star catalogs at the AAVSO headquarters.

References

Clement, C. M., Muzzin, A., & Dufton, Q., *et al.* 2001, *AJ*, 122, 2587

Kazarovets, E. V., Samus, N. N., & Durlevich, O. V. 1998, *Inform. Bull. Var. Stars*, No. 4655

Kazarovets, E. V., Samus, N. N., Durlevich, O. V. *et al.* 2008, *Inform. Bull. Var. Stars*, No. 5863

Kazarovets, E. V., Samus, N. N., Durlevich, O. V. *et al.* 2009, *Astronomy Reports*, 53, 1013

Kholopov, P. N., Samus, N. N., Frolov, M. S. *et al.* 1985–1987, *General Catalogue of Variable Stars*, Vols. I–III, (Moscow: Nauka)

Kolesnikova, D. M., Sat, L. A., Sokolovsky, K. V. *et al.* 2008, *Acta Astronomica*, 58, 279

Kukarkin, B. V., Kholopov, P. N., Artyukhina, N. M. *et al.* 1982, *New Catalogue of Suspected Variable Stars* (Moscow: Nauka)

Pojmanski, G. 2002, *Acta Astronomica*, 52, 397

Samus, N. N., Durlevich, O. V., Zharova, A. V., *et al.* 2006, *Astronomy Letters*, 32, 263

Samus, N. N., Kazarovets, E. V., & Pastukhova, E. N., *et al.* 2009, *PASP*, 121, 1378

Solar and Stellar Variability: Impact on Earth and Planets
Proceedings IAU Symposium No. 264, 2009
A. G. Kosovichev, A. H. Andrei & J.-P. Rozelot, eds.

doi:10.1017/S1743921309993206

The Indian National Large Solar Telescope (NLST)

S. S. Hasan
Indian Institute of Astrophysics, Bangalore 560034, India
email: hasan@iiap.res.in

Abstract. The Indian National Large Solar Telescope (NLST) will be a state-of-the-art 2-m class telescope for carrying out high resolution studies of the solar atmosphere. Sites in the Himalayan region at altitudes greater than 4000-m that have extremely low water vapor content and are unaffected by monsoons are under evaluation. This project is led by the Indian Institute of Astrophysics and has national and international partners.

NLST is an on-axis alt-azimuth Gregorian multi-purpose open telescope with the provision of carrying out night time stellar observations using a spectrograph. The telescope utilizes an innovative design with low number of reflections to achieve a high throughput and low instrumental polarization. High order adaptive optics is integrated into the design that works with a modest Fried's parameter of 7-cm to give diffraction limited performance. The telescope will be equipped with a suite of post-focus instruments including a high resolution spectrograph and a polarimeter. A detailed concept design of the telescope is presently being finalized and fabrication is expected to begin in 2010 with first light in 2014.

Keywords. instrumentation: high angular resolution, instrumentation: adaptive optics, site testing, Sun: magnetic fields

1. Introduction

Understanding the interaction of magnetic fields with plasma is central to interpreting various processes in the atmosphere of the Sun. These fields are organized in flux tubes with horizontal sizes ranging from thousands of kilometers down to a few kilometers. Recent numerical simulations suggest that crucial physical processes like vortex flows, dissipation of magnetic fields and the generation of MHD waves can occur efficiently on length scales even as small as 10 km. Such waves are likely candidates for transporting energy to the upper atmosphere of the Sun. Resolving these structures observationally is of utmost importance to study and improve our understanding of the different physical processes involved. Unfortunately, even the largest current solar telescopes are limited by their apertures to resolve solar features to this level at visible wavelengths. On the global scale, the energy stored in magnetic fields is eventually dissipated in the higher layers of the solar atmosphere, for instance in the form of flares and coronal mass ejections (CMEs) that release energetic solar plasma into the interplanetary medium.

Presently, the best spatial resolution that the existing generation of solar telescopes can attain during moments of good seeing and using adaptive optics is limited to about 0.13 arc sec. In addition to the requirement of good angular resolution, a high photon throughput is also necessary for spectropolarimetric observations to accurately measure vector magnetic fields in the solar atmosphere with a good signal to noise ratio. Consequently, in order to resolve structures with sub-arc sec resolution in the solar atmosphere as well as to carry out spectropolarimetry, a sufficiently large aperture telescope is required.

Taking a cue from recent simulations, one needs at least a 2-m class telescope, operating at its diffraction limit, to observe processes occurring on spatial scales of tens of kilometers (e.g. Schüssler 2009). Such a telescope would require about 2.5 s (at 630 nm) to carry out a single polarimetric observation (Keller 2003), which corresponds to an optimal exposure time of about 10 s to determine the 4 Stokes parameters needed for measuring the vector magnetic field.

Based on such considerations as well as practical reasons related to design and costs, we have proposed a 2-m class National Large Solar Telescope (NLST) for India. NLST will be larger than the telescopes which are now close to completion such as GREGOR (the 1.5-m German telescope on Tenerife). On the other hand NLST is small enough not to run into the design problems which are related to the 4-m class projects such as the ATST (Advanced Technology Solar Telescope) and EST (European Solar Telescope).

2. Science Objectives

NLST is envisaged as a multi-purpose instrument that will serve the needs of the national and international solar astronomers. Some of the main science goals are:

2.1. *Dynamics of the magnetic network*

Observations have revealed the presence of fine-scale flux tubes in the magnetic network on the Sun. In Ca H or K line images, the network shows up as a collection of "coarse mottles" or "network grains" that stand out against the darker background. These features are continuously bright with intensities that vary slowly in time, in contrast to the "fine mottles" or "cell grains" which are located in the cell interiors and are much more dynamic (e.g. Rutten & Uitenbroek 1991). The physical processes that produce the enhanced emission in the network are still not fully understood. Is the network heated by wave dissipation, and if so, what are the properties of these waves? Unambiguous observations of waves would be required to settle this and related questions.

2.2. *Internetwork magnetic fields*

A major finding has taken place recently regarding the nature of magnetic fields in the internetwork (IN). New observations from the Hinode Stokes Polarimeter (SP) (with a spatial resolution of 0.3) reveal the ubiquitous presence of horizontal fields with an average value of about 55 G (Lites 2008). These observations show that, whereas the vertical magnetic field mainly occurs in the intergranular lanes at the network boundaries, the field in the internetwork regions is dominantly horizontal and well separated from the vertical fields. However, the situation may be more complex as pointed out by Stenflo (2010) on the basis of an independent analysis of the same data. More observations with good spatial and high spectropolarimetric sensitivity are needed to settle this question.

2.3. *Nature of sunspots*

When viewed at high resolution, sunspots reveal a complex and intricate structure, such as umbral dots, light-bridges and the interlocking-comb structure in the penumbra. Despite noteworthy progress on the theoretical front particularly through sophisticated numerical simulations in 3-D, there is still no general agreement on many features including the overall picture of whether sunspots are monolithic flux tubes or consist of a cluster of several flux tubes as originally proposed by Parker (1977) (for a recent review see Thomas 2009). Sunspots exhibit a range of oscillatory motions, including umbral flashes, oscillations and running penumbral waves. Sunspot seismology can serve as a probe to study the internal structure of these features. Furthermore, even after a century of its discovery, there is no universally accepted model for the Evershed effect. The above topics will form a part of the NLST observational programmes that will attempt to accurately determine the magnetic field topology with high spatial and spectral resolution.

2.4. *Action region dynamics*

A study of active regions can provide useful clues to the solar dynamo believed to be located at the base of the convection zone. Preliminary observations show that newly emerging flux has nearly constant twist. The measurement of the twist is important, both to infer the dynamical evolution of the magnetic flux tubes while they rise through the solar interior to the surface as well as to understand the role of the twist leading to instabilities and eventual dissipation of magnetic energy in the solar atmosphere. Current vector magnetograms show a persistent pattern of electric currents and helicity associated with strong magnetic fields of active regions. The knowledge we have gained so far of active regions is limited. Systematic observations of magnetic helicity in these regions are lacking. Such observations require vector magnetic field measurements on spatial scales of a few tens of kilometers combined with a temporal resolution of few seconds. A large field of view of the order of 5 arc min is also essential in order to capture a full view of the entire active region.

2.5. *Coronal structures and transient phenomena*

The corona displays a myriad of phenomena that include loops, prominences, flares and CMEs, that are believed to be inherently magnetic in nature. Space observations from SoHO, TRACE and Hinode have provided considerable information on their properties. However, a detailed picture of the underlying physical mechanisms that are responsible for their occurrence is still lacking. A quantitative understanding of these processes requires an accurate determination of the magnetic topology through vector magnetograms at high spatial resolution. This would enable us to model the complex magnetic structure in the corona through a measurement of the field in the photosphere and corona, which provides the lower boundary for the field. Such investigations would also shed new information on mechanisms responsible for coronal heating.

2.6. *Night Time Astronomy*

We propose to use NLST for carrying out stellar observations during the night using a FEROS type high resolution spectrograph. The broad areas that will be investigated are:
- Activity monitoring in Ca, He and Balmer lines;
- Cycles on solar-like stars;
- Doppler imaging;
- Radial velocity monitoring;
- Extrasolar planets;
- Elemental abundances.

3. Technical Specifications

Keeping in mind the aforementioned science goals, the broad technical specifications of NLST are presented in Table 1.

3.1. *Design*

The guiding philosophy in the optical design of NLST is high optical efficiency which has been implemented by limiting the number of mirrors to only 6. NLST has a high throughput, 8 times more than GREGOR, which is highly desirable for polarimetry and

Table 1. Technical specifications of NLST.

Aperture (primary mirror M1)	:	2 metre
Focal length	:	3.5 metre
Optical configuration	:	3 mirror, Gregorian on-axis
Field of view(FOV)	:	300 arc sec
Final focal ratio of the system	:	f/40
Image scale	:	2.5 arc sec mm^{-1}
Optical quality	:	0.06 arc sec over limited FOV of 200 arc sec and < 0.3 arc sec within 300 arc sec
Wavelength of operation	:	380 nm to 2.5 microns
Polarization accuracy	:	10^{-4}
Active and Adaptive optics	:	to realize near diffraction limited performance
Strehl ratio within the isoplanatic patch	:	> 0.5
Spatial resolution	:	< 0.1 arc sec at 500 nm

speckle interferometry. The telescope has a high-order adaptive optics (AO) system to ensure diffraction limited performance. The optical design is shown in Figure 1.

A 2-m parabolic primary mirror M1 (f/1.75) forms an image of the solar disk with a diameter of 33 mm at the prime focus F1. Here a cooled heat stop rejects and dissipates all the energy which does not pass through the stop. The primary mirror is cooled from below to keep it close to the ambient temperature. A beam, providing a field of view (FOV) of 200 arc sec, passes through the center hole with 3.4 mm diameter. An elliptical mirror M2 creates a f/7.7 beam and forms a secondary focus F2 at a distance of 600 mm in front of M1 and about 200 mm above the elevation axis. Here a FOV of 200 arc sec corresponds to a scale of 15 mm. A weak negative field lens is also situated here (see below). The F2 image is picked up by another elliptical mirror M3 which changes the f-ratio to f/40 in the beam that produces a final image at F3. Here we have the desired image scale of 2.5 arc sec mm^{-1}.

M4 is a flat mirror with a central hole that reflects the beam into the elevation axis. The mirror group M5/M6 reflects the beam into the azimuth axis which in our design is besides the telescope. By means of the field lens in F2 the pupil is imaged on M6 which can serve as the tip tilt mirror of the AO. M5 is a deformable mirror. F3 is about

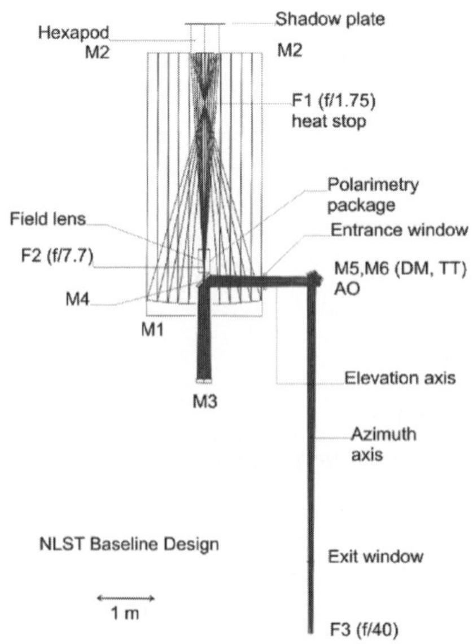

Figure 1. Optical layout of NLST

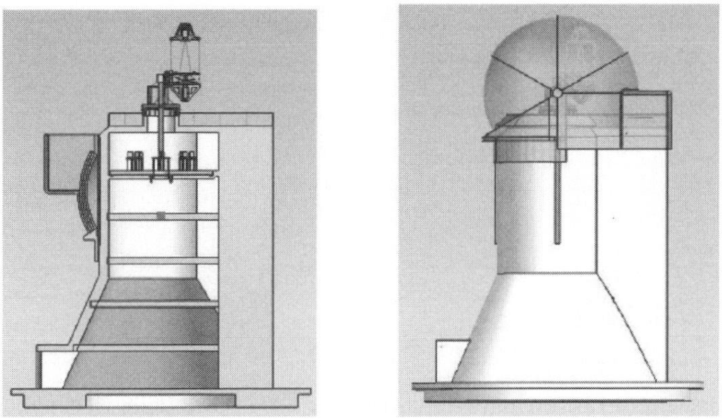

Figure 2. NLST with the dome retracted (left panel) and dome deployed (right panel)

6200 mm below the elevation axis which allows for convenient access of the focus stations in the building.

A mechanical turntable behind the telescope moves the whole post focus assembly and so compensates for the rotation of the image due to the alt-azimuth telescope system. Several ports for post focus instruments are provided. These instruments include a high resolution spectrograph and polarimeter, a tunable Fabry-Perot filter for narrow band imaging at multiple wavelengths, narrow pass band filters for H-alpha, Ca II K, CN band, G band and 1083.0 nm observations and a fibre-fed echelle spectrograph for night time astronomy.

The telescope has an open design with a simple retractable dome that will cover the telescope during the period when there are no observations. Figure 2 shows a vertical cross-section through the telescope, instrument platform and tower top with (a) the telescope open (left panel) and (b) the dome deployed (right panel).

3.2. *Polarimetric Package*

Polarimetric investigations form a major objective of NLST. We need to minimize instrumental polarization which will adversely affect the performance of the telescope. The F2 focus is unaffected by instrumental polarization because the layout is rotationally symmetric up to that point, which we find is the natural place for either a calibration unit or a modulation unit. In both cases such a device contains at least one polarizer and one retarder with variable retardance. The polarimetry package will be placed in a space which extends 400 mm in the vertical direction and has a width three times the beam diameter.

4. Site characterization

Site characterization for NLST is being carried out at the following three sites in north India: Hanle and Merak in Ladakh (Jammu & Kashmir) and Devesthal (Uttarakhand) using a Solar Differential Image Motion Monitor (SDIMM), Shadow Band Ranger (SHABAR), automatic weather station (AWS), all sky camera, an automatic sky radiometer and a micro-thermal data acquisition system, designed and built in-house.

Two years of results obtained so far suggest that the site at Merak has excellent seeing conditions, with an average of 2270 annual sunshine hours, a median wind speed of 4.9 m s^{-1} and about 540 hours annually of seeing with a Fried's parameter $r_0 > 7$ cm (at a height of 8-m). Figure 3 shows a histogram of the Fried's parameter using data during

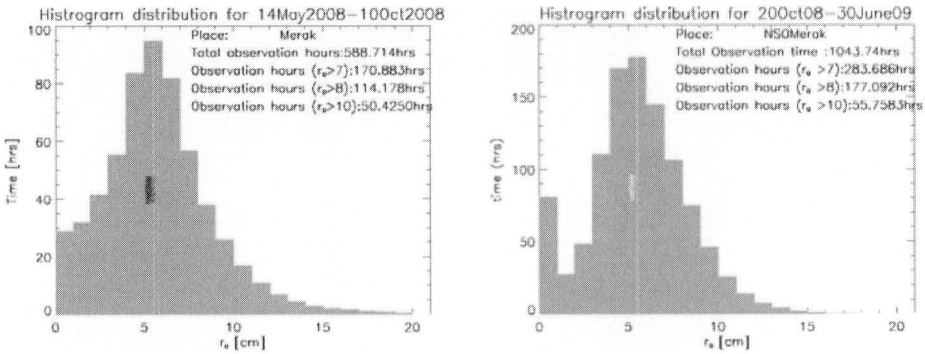

Figure 3. Histogram of r_0 for the periods (a) May 2008- Oct. 2008 (left panel), and (b) Oct. 2008 - June 2009 (right panel)

2008–09. It is worth pointing out that there are a significant number of hours (over 100 in a year) with $r_0 > 10$ cm. Consequently, Merak is comparable to the best sites in the world for solar observations such as Haleakala and Big Bear.

5. Current status

The detailed concept design of NLST has been carried out by MT-Mechatronics, Germany with technical support from the Kiepenheuer Institute, Freiburg. A detailed concept design report is now ready and a detailed proposal is under preparation for submission to our funding agency. The fabrication of NLST is expected to begin by late 2010 and be completed by early 2014. The backend instruments for day time observations will be made in house and work on a prototype spectropolarimeter has already commenced. The spectrograph for night time astronomy will be developed by the Hamburg Observatory, Germany.

6. Summary

NLST will be a state-of-the-art 2-m class telescope for carrying out high resolution studies of the solar atmosphere. An innovative design with minimal number of mirrors, high throughput and high order adaptive optics will provide close to diffraction limited performance. Its geographical location will fill the longitudinal gap between Japan and Europe. NLST will be the largest solar telescope with an aperture greater than 1.5 m, till ATST and EST come into operation.

References

Keller, C. U. 2003, in S. Fineschi (ed.), *Polarimetry in Astronomy*, SPIE, 4843, p. 100

Lites, B. W. 2008, *ApJ*, 672, 1237

Parker, E. 1977, *ApJ*, 230, 905

Rutten, R. J., Uitenbroek, H. 1991, *Sol. Phys.*, 134, 15

Schüssler, M. 2009, in: R. Schlichenmeier & O. von der Lühe (eds.), *2nd EAST-ATST Workshop on Large Telescopes*, Astron. Nachr., in press

Stenflo, J. 2010, in: A. H. Andrei, A. S. Kosovichev & J-P. Rozelot (eds.), *Solar and Stellar Variability - Impact on Earth and Planets*, IAU Symp. 264 (Cambridge), in press

Thomas, J. T. 2009, in: S. S. Hasan & R. J. Rutten (eds.), *Magnetic Coupling between the Interior and the Atmosphere of the Sun*, Astrophys. Sp. Sci. Proc. (Springer: Heidelberg, Berlin), in press

Session X

Summary, open discussion, and closing remarks

Chair: J.-P. Rozelot

Table 1. Evolution of the solar fundamental constants and their variability along the 11 yr solar cycle. The heavy elements fraction mass comes from Anders & Grevesse estimate, Grevesse & Noel one and today from Asplund *et al.* (2009). See text.

	Reference values Allen	Present values	Time Variability
Luminosity	1360.488 ($\pm 2.10^{-4}$)	1367.6 W/m^2 - 1361 W/m^2	1-4 W/m^2
Radius	695 990 km	693710 km (min)	10-160 km
Seismic radius (f modes)	-	695660 km	-
Radius shape	-	oblateness 6 to 10 km	6-14 km
Heavy element Z in fraction mass	0.02 then 0.0173	0. 0134	no evidence

- the solar internal variability, its emergence and other stellar variability
- the Stars and Planets interaction (s).

2. The solar (stellar) fundamental quantities and their variability

The solar fundamental quantities are the total irradiance, the radius and the composition. Table 1 summarizes their values and their variability with the 11 year solar cycle. It is quite amazing that the calculation of a solar standard model requires the adjustment of the initial helium, the initial metal composition and the mixing length to get the measured luminosity, radius and surface abundances at an accuracy level of 10^{-4}- 10^{-5} when the star of 1 $M_\odot$ star reaches the age of 4.6 Gyrs (including the contraction phase and the pre mainsequence). One notes that this accuracy is not yet reached today. In the standard framework, the solar constant does not vary by more than 10^{-8} during the last 100 years. Of course this is contradicted by the irradiance measurements.

2.1. *The solar irradiance*

Clearly the Sun does not agree with the standard model: its luminosity varies by less than 10^{-3} every 11 years, the Hale cycle, about 0.9 W/m^2, but in fact 3 or 4 times more near the maximum of the cycle due to the presence of sunspots and faculae (Figure 1). These last two years, the total irradiance has been measured at a real minimum, characterized by the absence of sunspot: a decrease of about 0.25-0.3 W/m^2 in comparison with previous cycles 21 and 22 minima has been obtained. But, if there is a general trend on the long term, it is not yet established. Of course the interpretation of the cyclic variability of the solar irradiance is attributed for a large fraction (at least 80%) to the radiative effects of the magnetic activity in the photosphere (Domingo *et al.*, 2009).

Does the total irradiance variability reflect only superficial magnetic effects and the Hale cycle ? If there are some other sources, what are they ?

2.2. *The detailed solar composition*

The solar abundances are measured through the photospheric absorption lines. Their determination has been improved several times and all suspicious problems (photospheric iron in disagreement with meteoric one, absence of comparison with meteoric elements for C, N, O, Ne) seem to be solved. Today, two teams are converging toward an abundance of heavy elements ($Z = 0.0134$) strongly reduced in comparison with what was used 20 years ago (Asplund *et al.*, 2009; Caffau *et al.*, 2008, 2009). This reduction is in fact a little bit smaller that announced 5 years ago and in reasonable agreement with Holweger estimate (Holweger *et al.*, 2001). This is an important result because of its consequences

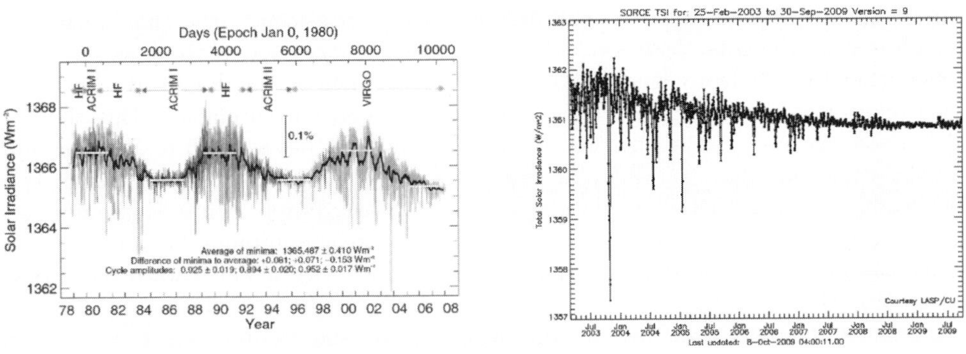

Figure 1. a) Reconstruction of the solar irradiance variability obtained by ACRIM and SoHO satellite observations. From Fröhlich (2006). b) Zoom on the TSI measured by SORCE from July 2003 to September 2009 (Kopp *et al.*, 2005). One notices a rapid strong variability due to spots and faculae on both datasets near the maxima. The absolute value difference between the two series has no explanation up to now.

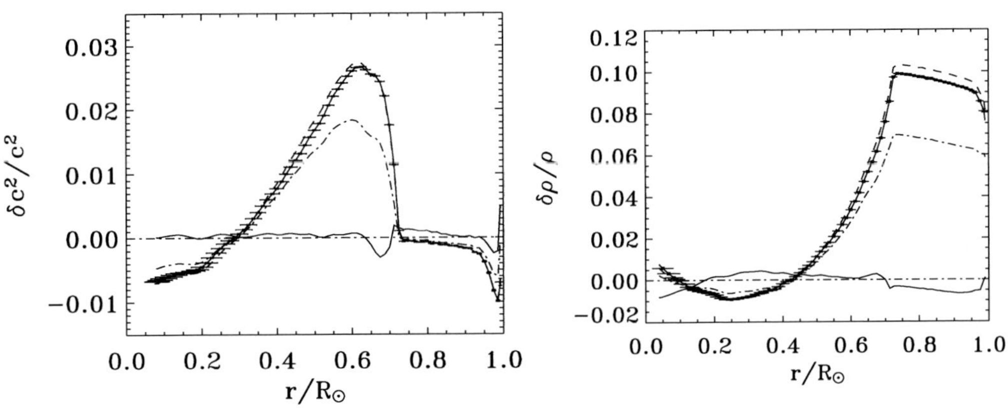

Figure 2. Squared sound speed and density differences extracted from the seismic inversions obtained with GOLF+MDI/SOHO acoustic modes and the standard model using the Asplund 2005 composition (solid line with seismic error bars) or using Holweger (2001) composition (dot dashed line) or Lodders (2003) one (dotted line). The difference with the seismic model is drawn in full line. The vertical error bars are too small to be visible, the horizontal ones are still large in the nuclear core (below 0.3 $R_\odot$ which contains more than half the mass of the Sun), they will be strongly reduced in adding several gravity mode frequencies. The present composition is compatible with such differences. From Turck-Chièze *et al.* (2004a).

on galactic evolution: the Sun is no more considered as enriched by a supernova (Turck-Chièze *et al.*, 2004a). This recent result confirms the inability of the standard model to predict properly the observed sound speed (see Figure 2). One now needs to look for the processes which justify this disagreement: microscopic diffusion, opacities, dynamical proocesses (Turck-Chièze *et al.*, 2004a; Turck-Chièze *et al.*, 2009a).

Up to now, there is no reason to seek an explanation based on microscopic processes rather than on macroscopic processes, both must be investigated (see below).

2.3. *The solar radius*

The time variability of the radius and the photospheric shape are rather difficult subjects because the effects are very small. Progress on this field is important these last years and several issues have been clarified: in using radiative transfer calculations, Habbereiter *et al.* (2008) estimate that the solar radius measured at the limb differs from the f-modes estimate by 0.3 Mm (+ 4.3 10^{-4}). It is not so easy to estimate the time variation of the radius with the solar Hale cycle but a detailed analysis seems to show an anticorrelation (Lefebvre & Kosovichev, 2005; Lefebvre, Nghiem & Turck-Chièze, 2009). This anticorrelation is also suggested by ground and balloon measurements. But the space and balloon variability differ by one order of magnitude (between 10 to 160 km) along the solar cycle but they use different techniques at different wavelengths (Egidi *et al.*, 2006; Emilio *et al.*, 2007). The MDI data point to nearly purely oblate shape near solar maximum, but with a significant hexadecapole shape near minimum. Of course it would be nice to follow the time radius variability on ground, with a well known technique as an astrolabe. The brazilian team has shown the difficulty due to atmospheric effects but the interest of continuity at low cost encourages to pursue ground measurements. They get a mean value of the radius of 959.163 ± 0.004 arc. (Andrei *et al.*, 2006), which can be compared to the solar minimum value obtained with the SDS balloon of 959.561 ± 0.111 (Djafer, Thuillier & Sofia, 2009). The aspheric shape of the Sun is not yet understood, the balloon shows values of oblateness of 4.3 to 10. 10^{-6} increasing with the cycle (Emilio *et al.*, 2007). This can be compared to the effect of the internal rotation alone which leads to values of the order of 2.-3. 10^{-7}. It is evident that the role of the deep and varying magnetic field must be added. See a complete review on the subject from Rozelot (2009).

The present situation seems today unclear. The methods used to extract the information must be compared, and very precise measurements from space might clean the present situation.

2.4. *The other stars*

The situation is largely different for the other stars. In most cases, the radius is not known, only its luminosity and effective temperature are determined. Consequently the mass is not precisely estimated, and depending on the detailed physical ingredients considered (composition, overshooting...) the mass can vary up to a factor 2. The radius is only measured for well deformed stars which rotate quickly. We have very good hope that the deployment of asteroseismology with COROT and KEPLER will largely improve the situation, the asteroseimic revolution arrives clearly with KEPLER which will deliver observations of 100000 stars during 3.5 years. The main interest related to this conference is to better establish which kind of activity is found in different stars.

3. Solar variability and other stellar types of variability

It is now relatively well established that the solar and stellar activity originates from the internal processes, but it is only recently that quantitative works have been devoted to the link between internal and external phenomena. We note presently very important specific results but not yet a general picture. Once again the Sun remains the best guide to study this connection in details.

3.1. *The Sun is highly observed globally*

We are lucky because the ground based instruments are now accompanied by space missions like Ulysses, SoHO, RHESSI and now STEREO and HINODE. Altogether, they offer a lot of complementary information on the solar external manifestation of activity. The variability in sunspot, γ, X and radio will be treated in the next section. I shall concentrate here on global effects. Figure 3 summarizes the evolution of ideas in the understanding of the solar magnetic field with the Ulysse spacecraft. Prior to Ulysses, the models of the solar magnetic field assumed that it was largely dipolar, field lines near the solar equator were thought to form closed loops whereas field lines from the poles

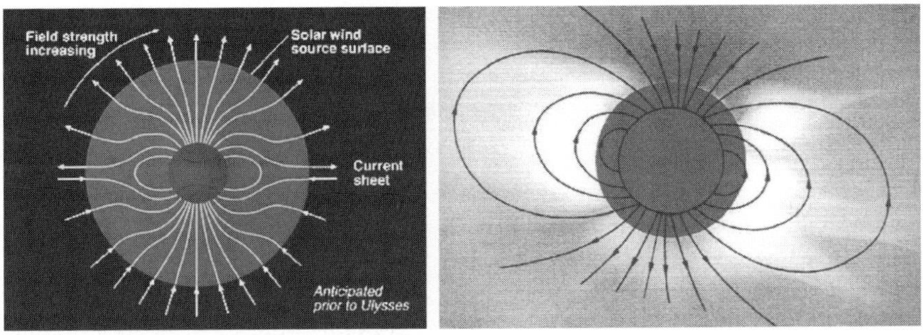

Figure 3. Evolution of the ideas on the configurations of the solar magnetic field thanks to the 18 years of ULYSSES spacecraft observations out of the ecliptic.

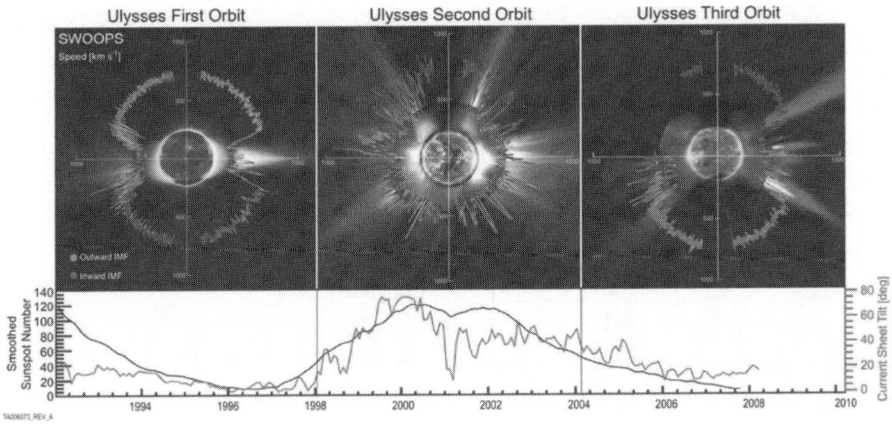

Figure 4. Polar plots of the solar wind seen by Ulysses during two successive minima a), c) and one maximum (b). Superimposed are images from EIT/SoHO. d) time variation of the smoothed sunspot number and of the heliospheric current sheet tilt. From McComas *et al.* (2008).

were dragged far into interplanetary space by the solar wind (Figure 3 left). Ulysses (Figure 3 right) found that the amount of outward magnetic flux in the solar wind did not vary greatly with latitude (Smith *et al.*, 2003). Nevertheless, the solar wind has been observed during three extrema and it is now measured daily aboard SoHO. The slow and high velocity components are not always present simultaneously, the high component depends on the importance of the coronal holes from where emerges the fast wind. Figure 4 illustrates how different are the two last minima. The present one is very long and is characterized by the absence of sunspots. It is accompanied by a low latitudinal wider wind and a less quadrupolar configuration of the wind. Its global pressure is decreased by 20% and the magnetic field at the level of the spacecraft reduced by 36% (McComas *et al.*, 2008). These results are of prime importance for the extension of the heliosphere which is presently reduced and consequently inhibits more the protection against the cosmic rays. This information is extremely useful for Space Weather (see below).

In parallel, W. Schmutz mentioned at the 3rd Climate Symposium in Lapland (see its oral presentation on website) that the interplanetar magnetic field B_{IMF} is a more justified indicator of activity than the solar sunspots which are proxies for active regions. The total irradiance TSI is well correlated with B_{IMF} which takes into account the quite Sun, network and active regions. He then deduces a value of 2 nT for B_{IMF} and

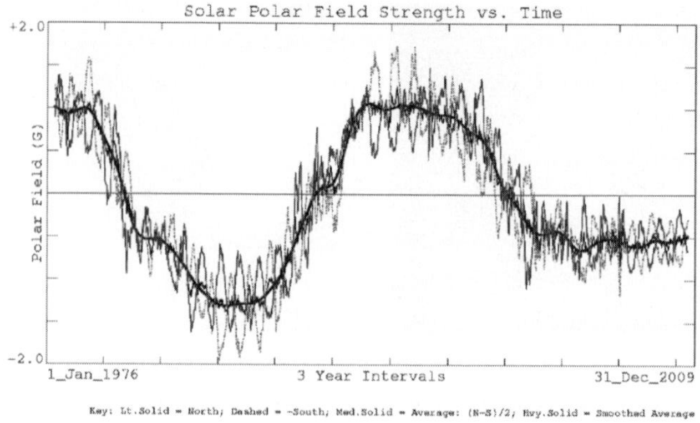

Figure 5. Solar polar field for latitude greater than 65 degrees measured at the Wilcox Observatory along the last 30 years. One observes a very different trend of this quantity between the two successive minima. See Hoeksema talk and Schatten (2005).

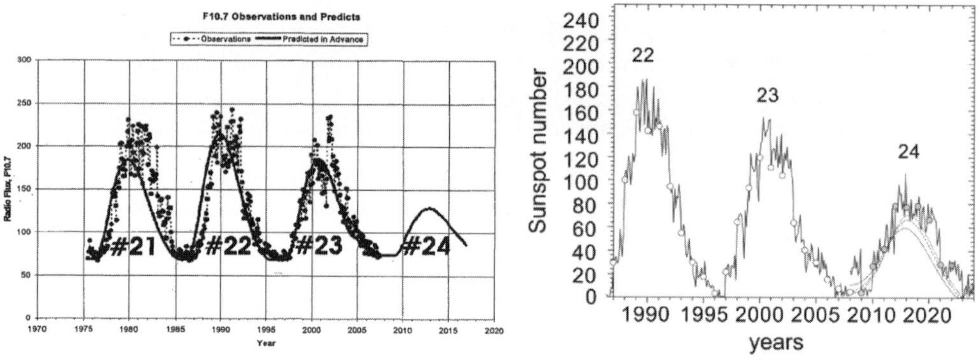

Figure 6. Observed F10.7 Radio Flux (circles) and solar flux predictions (solid lines) prior to each of the last three cycles. The radio flux is in $10^{-22}\mathrm{Js^{-1}m^{-2}Hz^{-1}}$. From Schatten (2009). Assimilation of data to predict cycle 24 done by Kitiashvili & Kosovichev, 2008.

a TSI of 1364.4 W/m^2 at the Maunder minimum, by extrapolation of the three decade observations of TSI obtained aboard the ACRIM and SoHO missions (Fröhlich, 2006).

Nevertheless, the prediction of the time evolution of the Hale cycle remains difficult. There has been very few predictions of the present very long low activity period. But a lot of indicators have shown*a posteriori* that 2004 has been an informative year. It is directly comparable to the previous minimum (cycle 22). At the Wilcox Solar observatory, 30 years of data have been accumulated on the antisymmetric and symmetric zonal flows. It is the same for the polar field (above 55 degrees) which varies between +1 and -1.5 G (see Figure 5) and the daily solar mean magnetic field which varies between 1 to -1 G. Now MDI and HINODE give also access to the polar field above 75 degrees and confirm a different behavior for the two cycles. Cycle 23 keeps a high value of about 7 G during at least 4 years in contrast with cycle 22. Zonal flows below the surface show also slower velocities than during the previous minimum (Howe *et al.*, 2009).

What is (are) the origin(s) of the polar field ? Fossil or Convective dynamo field ?

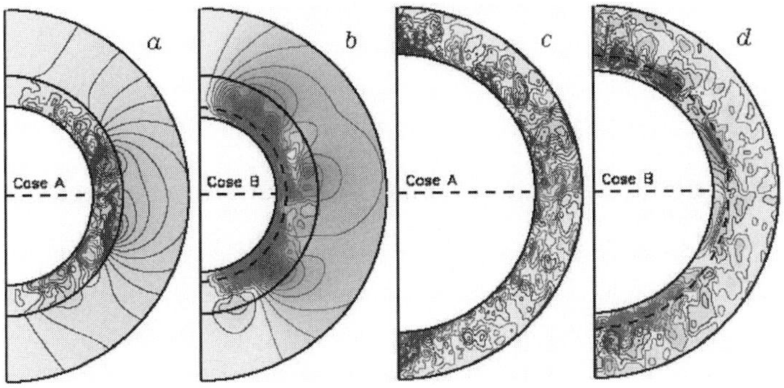

Figure 7. Mean solar convective poloidal magnetic fields (a, b) and toroidal magnetic fields(c, d) for two simulations differing by their magnetic Prandt numbrr (respectively 4 and 8), averaged over longitude and time (90 days). White (red in online version) tones denote a clockwise poloidal field orientation and eastward toroidal field whereas black (blue) tones de- note the opposite. Saturation levels for the poloidal and toroidal color ta- bles are 20G and 2000G respectively. Poloidal fields in (a, b) include a potential-field extrapolation out to 1.5R and the base of the convection zone is indicated by a dashed line in (b, d). From Miesch *et al.* (2008).

3.2. *The solar variability and the internal dynamical processes*

3.2.1. *The solar Hale cycle and the convective zone*

It is fundamental to understand the solar variability. The task is difficult because we have no direct access to the topology and strength of the internal fields. Important progress has been obtained during the last decade on the understanding of the Hale dynamo originally based on the works of Babcock (1961) and Parker (1984). It is interesting to notice that a shallow dynamo has a good predictive character for the next cycle. Effectively in the Babcock-Leighton Flux Dynamo theory, the polar field at the end of a cycle serves as a seed for the next cycle's growth. By examining 8 previous solar cycles and their impact on Earth, the method has successfully predicted cycles 21, 22 and 23 and 24 (Figure 6 left) using WSO and MWSO data. The unusually weak polar field around 2005 leads to forecast a small cycle 24 (Schatten, 2003; Schatten, 2005; Schatten, 2009). A more fundamental 2D approach explores a large scale dynamo that implies the whole convective zone and the tachocline. They solve the induction equation and consider not only the α and Ω effects but also the meridional circulation in the whole convective zone together with an estimate of the velocity on at least the last 3 solar cycles. They try to understand how the field is regenerated (Dikpati & Gilman, 2008). Such prediction is improving with time and the decelerating flow observed by MDI/SoHO and GONG (Howe *et al.*, 2009) leads to a delay in the onset of cycle 24. Moreover, these flux transport dynamo models begin to explore the asymmetry between hemispheres (see 4.3). One important ingredient among others, not yet accessible to observation, is the number of circulation cells that describe the meridional circulation (Jouve & Brun, 2007). A third method, the assimilation method, is also used to better understand what processes guide the predictions (Kitiashvili & Kosovichev, 2008). The three methods agree rather well but become predictable only when the minimum is reached (Figure 6 right).

The 3D simulations of the convective zone including rotation and magnetic field reproduce now reasonably well the latitudinal differential rotation and do not exclude that the circulation patterns could be complex (Brun, Miesch, Toomre, 2004; Miesch *et al.*, 2008). These 3D simulations are very promising (see Figure 7) to understand the interplay between the different processes. This is an opportunity for the community to have

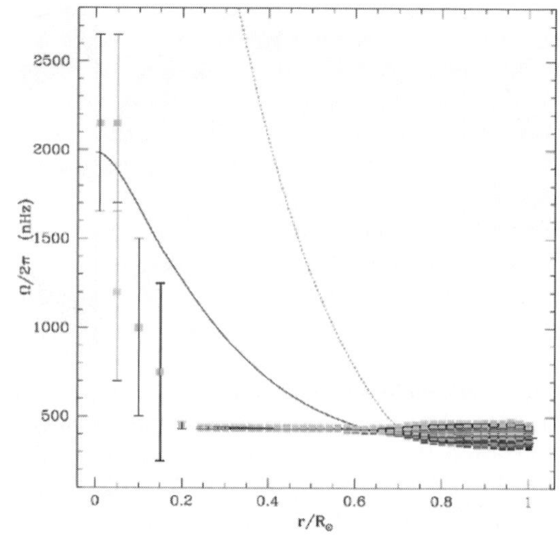

Figure 8. Solar internal rotation profile comparison between different models using different initial rotation rate, A: extremely slow (continuous line), B: 20km/s (dotted line) when it arrives on ZAMS, and the profile deduced from the acoustic mode splittings and potential observed gravity modes. Deduced from Turck-Chièze *et al.* (2009a).

two different numerical approaches, one using the Anelastic Spherical Harmonics (ASH) code, the other using the cartesian Pencil code (see Brun and Tavako papers, this issue). The tachocline is probably the seed of the large scale dynamo. The ASH simulations predict that the temperature anisotropy between pole and equator is of about 10 degrees and is due to the baroclinic effect. Such order of magnitude seems to be confirmed by HINODE observations. The multicells obtained in the CZ are favored to understand the cycle lengths of the solar analogs.

3.2.2. *The deep field and the radiative zones*

Understanding the 11 year solar dynamo is an important issue, but probably not the end of the story of stellar activity. Another fundamental question is the existence of fossil field in deep radiative zone of solar-like stars, knowing that the bulk of the stellar mass (95-98%) is contained in this part of the stars. The difficulty is due to the absence of any direct known signature of such a field, so rather few works have been dedicated to this fundamental subject during the last decade (Braithwaite & Spruit, 2004; Zahn, Brun & Mathis, 2007). Once again, the Sun is the best case where we may find some signature and helioseismology can help us to progress on such subject. An indirect evidence of deep dynamical motions is the solar rotation internal profile, which is extracted from acoustic modes for $R \geqslant 0.2R_\odot$ and from gravity modes below this limit. Figure 8 illustrates what the Sun seems to tell us, compared to solar models that take into account the transport of momentum and chemicals by rotation (Turck-Chièze *et al.*, 2009a). This paper shows also how different are the meridional circulation in comparison with the convective zone circulations.

Two regions are particularly interesting to comment: the tachocline and the core.

- The tachocline corresponds to the transition layers located between the deep region where the transport of energy is due to radiation and the external region where convection dominates. When photons reach these layers, the interactions with iron, oxygen...are so important that they inhibit the transport of energy, the convection plays the dominant role and partly due to this effect, this region is dominated by shear horizontal turbulence.

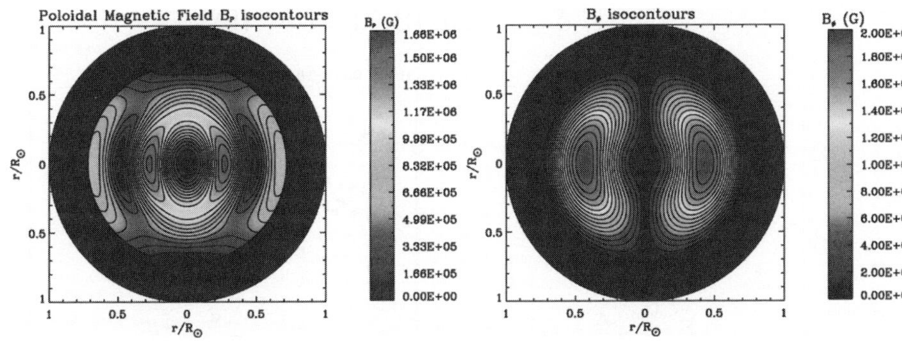

Figure 9. Isocontours of the poloidal magnetic field $B_{P(r,\theta)}$ (in G) in meridional cut. Isocontours of the azimuthal magnetic field $B_{\phi(r,\theta)}$ (in G) in meridional cut for a young Sun. From Duez, Mathis & Turck-Chièze (2009).

Moreover these specific layers probably maintain and restore the convective toroidal magnetic field. This results in a suppression of the differential rotation observed in the whole convective zone. As shown on Figure 8, it is now well established that the radiative zone rotates as a solid body at least down to the limit of the energy generation core (Eff-Darwich *et al.*, 2008).

The solar core has not yet received so much attention due to the absence of clear dynamical information up to recently. But the GOLF instrument aboard the SoHO satellite appears more and more promising to detect several individual gravity modes, the reason is given in Turck-Chièze *et al.* (2008). The present searches (Turck-Chièze *et al.*, 2004b; García *et al.*, 2007; Mathur *et al.*, 2007, 2008) favor a rotation in the core between 2 to 5 times the rotation of the rest of the radiative zone. This tendency might be confirmed before the end of the SoHO observations by detecting individual splitting with the signal of more than 13 years of data. In fact, individual components of dipolar modes become reachable at low frequency, these modes are strongly dependent on the central core rotation. Figure 8 shows that the induced rotation profile cannot be explained only by the effect of the transport of momentum and chemicals by rotation, so the natural first candidate to flatten the profile is the fossil magnetic field. Gough & McIntyre suggested that the uniformity of the rotation profile observed in the bulk of the radiative zone is due to the presence of a large-scale primordial magnetic field confined below the tachocline by flows originating from the convection zone (Gough, 2009). This idea has been recently numerically simulated by Garaud & Garaud (2008). They emphasize the key role of flows downwelling from the convection zone in confining the assumed internal field and the quenching of the large-scale differential rotation below the tachocline, including in the polar regions, as seen by helioseismology. A new approach illustrated by Figure 9 is to determine a stable magnetic configuration (a mixture of poloidal and toroidal fields) in order to follow its diffusivity and its transport of momentum along the long lifetime of the solar-like stars (Duez *et al.*, 2009). This work will lead to a complete and dynamical view of the solar interior.

The interaction between the fossil field and the dynamo field is not yet explored and certainly much work needs to be dedicated to the radiative zone.

3.3. *The young stars and the solar analogs*

Many stars show magnetic activities and more and more observational methods are available. Chandra and XMM allow observations of Alpha Centauri which are twins of the Sun and of a lot of other stars for which coronal observations are possible. Several other

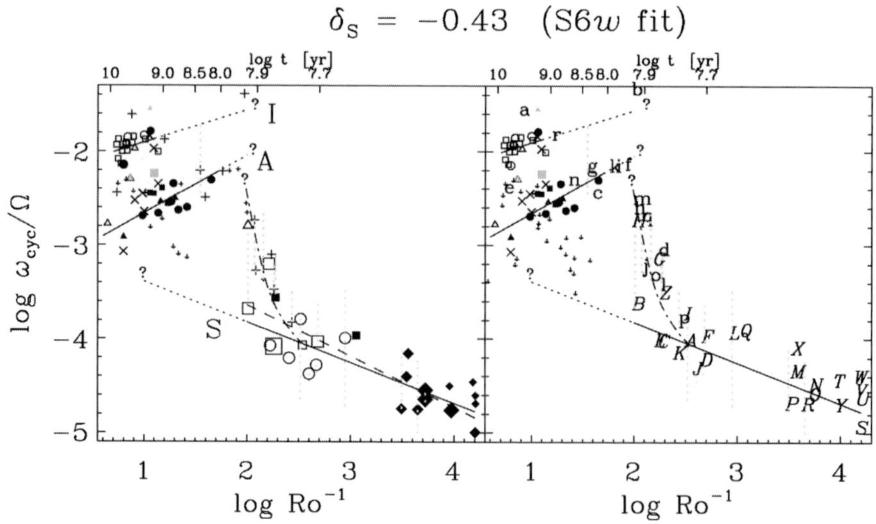

Figure 10. Relationship between cyclicity, rotation and the Rossby number ($R_o = P_{rot}/\tau_c$) for an extended "echantillon" of stars. From Saar and Brandenburg (1999).

indicators like CaII lines (variation of 2% in the solar case), Doppler imagers, spectropolarimetry with ESPADON, NARVAL and SEMPOL give complementary information like for example the presence of spots at the pole of young stars or some potential geometry of the external magnetic field. At Mount Wilson Observatory, on 100 stars followed along 25 years, 60 exhibit cyclicity, 25 are variables and 15 do not show any activity (Baliunas, 1995).

The study of Saar & Brandenburg (1999), summarized by Figure 10, shows that stars are not strictly segregated onto one or the other branch by activity level. The high-cycle branch is primarily composed of inactive stars. The extended dataset suggests that after 1 Gyr, stars can have cycles on one or both branches, though among older stars, those with higher (lower) mass tend to have their primary P_{cyc} on the lower (upper) cycle branch. The solar Gleissberg cycle agrees with this scenario, suggesting that long term activity trends, in many stars, may be segments of long cycles not yet resolved by the data. Most of the very active stars ($P_{rot} \geqslant 3$ days) appear to occupy a new third branch, with ω_{cyc}/Ω proportional to $R_o^{0.4}$. Many RS CVn variables lie in a transition region between the two most active branches. These authors predict P_{cyc} for three groups: stars with long-term Ca II HK trends, stars in young open clusters, and stars that may be in Maunder-like magnetic minima.

Observations of young stars have grown sharply, they are generally rapidly rotating stars and many possess substantial magnetic activity and strong axisymmetric magnetic fields. These stars are particularly interesting to follow in order to understand the young stage of the Sun, and also to better understand the impact of young stars on planets.

Figure 11 shows how the angular velocity profiles evolve with time for young solar-like star, potential young Sun. One notes that the radial contrast increases during the PMS phase even if the initial rotation is not so large due to the contraction phase. The advective term in the transport of momentum, connected to the meridional circulation, can significantly increase this contrast in the radiative zone, in opposite to the role of the magnetic braking which is generally invoked to explain the strong reduction of the superficial rotation during the main sequence (Turck-Chièze *et al.*, 2009a). But in the two cases, the radial differential rotation during the main sequence is increased by no more than 20% or 50% of the initial rotation rate achieved at the end of the PMS phase.

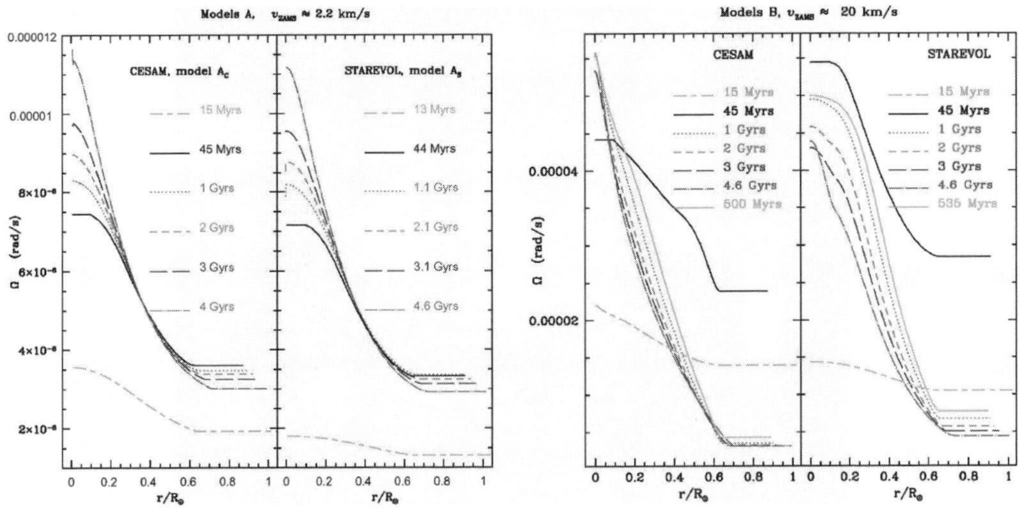

Figure 11. Evolution of the angular velocity profiles from the contraction phase of a star of 1 M$_\odot$ at six different ages obtained by two 1D evolution codes: CESAM and STAREVOL, for an extremely slow initial rotation rate and for a moderate case where the star arrives on the ZAMS with a surface rotation of 20 km/s and a sudden magnetic braking. The age of 45 Myr corresponds to the arrival on the main sequence. From Turck-Chièze *et al.* (2009a).

In parallel, simulations, with the 3D hydrodynamics ASH code, have shown that the young rapidly rotating stars may know intermittent convective phases (Ballot, Brun & Turck-Chièze, 2007). The team of *Sun in time* has studied about 13 stellar analogs of the Sun of about 1 M$_\odot$ but with rather different ages and rotation rates varying from 1 to 35 days. They exhibit very different types of dynamo, radial and often saturated dynamos (Güdel, 2007). Simulations of dynamo action in rapidly rotating suns with the ASH code explore the complex coupling between rotation, convection and magnetism. These simulations show that substantial organized global-scale magnetic fields are achieved by dynamo action in these systems. Striking wreathes of magnetism are built in the midst of the convection zone, coexisting with the turbulent convection and in the absence of tachocline. Some dynamos achieved in these rapidly rotating stars build persistent global-scale fields which maintain their amplitude and polarity for thousands of days. For five times the solar rate, the dynamo can undergo cycles of activity, with fields varying in strength and even changing polarity. As the magnetic fields wax and wane in strength, the primary response in the convective flows involves the axisymmetric differential rotation, which begins to vary on similar time scales. Bands of relatively fast and slow fluid propagate toward the poles on time scales of roughly 500 days. In the Sun, similar patterns are observed in the poleward branch of the torsional oscillations, and these may represent a response to poleward propagating magnetic field deep below the solar surface (Brown *et al.*, 2010).

The history of the magnetic field from the formation of the star to the solar present age is a very missing stone which must require more attention from both the observational side and the theoretical, modelling and simulation sides.

4. Stars and Planets interaction (s)

The interest to study Sun and solar-like star activity is amplified by the interest to describe the complex interaction between the Sun (Stars) and the Earth (Planets). Figure 12

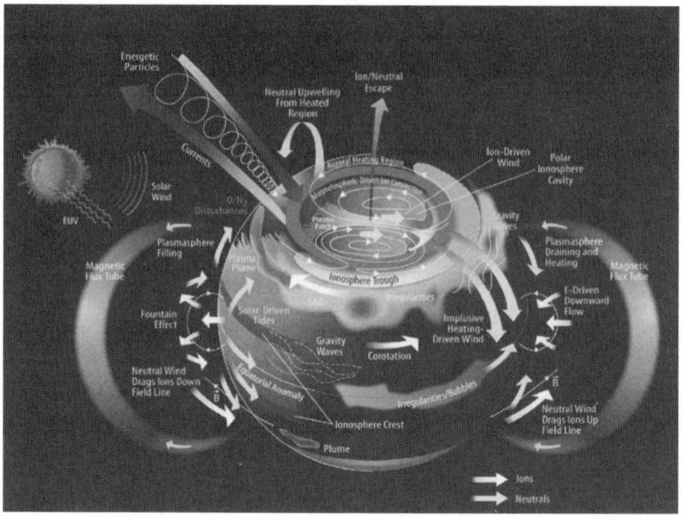

Figure 12. This figure illustrates the complexity of the phenomena which govern the Sun-Earth relationship. Courtesy of Joseph Grebowsky, NASA GSFC.

summarizes the complexity of these interactions. It also shows that the physical processes which describe the stellar and external stellar dynamics and the processes which need to be understood in the Earth atmosphere are inspired by the same physics. This field is rapidly growing thanks to the capability to detect planets by Doppler velocity techniques. Accordingly, we may hope that a common community will appear in the next decade to understand the real impact of the Sun (Stars) on the Earth (Planets) environment and climate. Precise space luminosity variations with MOST, COROT, KEPLER and in the future PLATO will contribute to the development of this field. So a new interest has been developed this last decade to understand the conditions of life on planets.

4.1. *Interaction between young stars and planets*

In this context, young stars are again under the spotlight. Indeed, it is important to observe analogous stars of the Sun to tentatively find the kind of stars which can favor life on planets. A sequence formed by EK Dra (100Myr), π^1 UMA (300 Myr), κ^{-1} Cet (650 Myr) and the Sun delivers many informations on the interaction between young stars and their planets (Kaltenegger *et al.*, 2009). The young stars are rotating faster than the Sun and a sequence of rotation can be considered (see Ribas talk). For the youngest, X rays and XUV are 100 to 1000 more intense than the solar case, the visible light is reduced by about 30%, FUV and UV are 5 to 60 times the present value, the solar wind 10 to 1000 the present solar one and the flares more frequent too (about 10 times per day). So clearly the young Sun or young solar-like stars interact with their planets rather differently and rather strongly.

As a natural consequence Lyman α, FUV and UV emissions produce photo dissociation reactions of CO_2, H_2O, CH_4, NH_3. X rays, EUV and Lyman α emissions heat, expand and photoionize the exosphere. The intense stellar wind carries away more atmospheric particles and erodes the atmosphere of the planet. If now one looks for the type of stars that can support habitable planets, it is noticed that due to biological damage by UV, it seems today that they are restrained to F and G stars. It is then interesting to look to twins of the Sun. Among others, κ Ceti appears particularly similar to the Sun (Porto de Mello *et al.*, 2006).

There is evidence that this science will largely progress with the KEPLER satellite these coming years. Building an history of the Sun-Earth connection along the evolution of the star will help to build an history of habitable planets.

4.2. *The Sun-Earth relationship and the Space Weather*

This discipline is rather young. It is composed of two parts. The first one, the quasi instantaneous relationship (at the level of seconds, hours or days) also called space weather, is now well organized since the launch of SoHO. The other one, at the level of months, years, decades or centuries, is still in its infancy and will gain to be developed. The frontier between these two aspects is not so clear and one can imagine that it will further reduce in the future (see the last section).

4.2.1. *Space weather, X, γ and radio emissions*

Space weather regroups the conditions of the Sun and in the solar wind, magnetosphere, ionosphere and thermosphere that can influence the performance and reliability of spaceborne and ground-based technological systems and can endanger human life or health. The most energetic phenomena of solar activity are flares and coronal mass ejections. Flares are large explosions that occur over the solar atmosphere and may last from a few seconds to hours. A solar flare is caused by a sudden, and yet unpredicted, energy release high above the magnetic loops. This magnetic energy is then transformed into particle acceleration (through their kinetic energy) and heat of the surrounding atmosphere. Both the energetic particles and the hot gas produce emission throughout the whole electromagnetic spectrum, from the very energetic gamma-rays all the way to long radio waves. From the observation of the emission produced during flares it is possible to infer the energetic particles spectrum and thus get a clue on the acceleration mechanism that produces these particles. The recent findings of flare observations, in the gamma-rays domain by the RHESSI satellite and at high radio frequencies by the Solar Submillimeter Telescope, are now analyzed. The observations of the geometry of the localized flux emergence need space and time resolution, the understanding of the underlying physics is now largely worked out (See Valio this issue; Wang *et al.*, 2008).

4.3. *The solar impact on the terrestrial climate*

This subject is extremely controversial and difficult due to the required accuracy, the use of proxies and the competition between the different influences.

- direct influence:

The solar irradiance contributes directly to the Earth temperature, consequently its variability is of prime importance. The Sun is not far from a black body so its total irradiance variability is dominated by the visible range variability. Such quantity is now followed since 3 decades, but with different intruments and with problems of ageing and calibration. The variability due to plages and rotation is well measured (up to 3-4 W/m^2) but the comparison between minima is still in debate but remains smaller than 0.5 W/m^2 (Scafetta & Willson, 2009; Krivova *et al.*, 2009) compared to 1361-1367 $W/^2$. Consequently the impact of the irradiance variability on the Earth climate is considered today as rather small, unless the variability was amplified during maxima by other phenomena.

The extrapolation towards the past is not easy and has been readjusted several times, leading each time to lower variation. From the ^{14}C measurements obtained in different sites (see Miyahara talk) showing clearly anti-correlation with the solar activity due to the solar dynamo, it has been shown that this dynamo persists even during long minima. Moreover the last millenium can be reasonably studied in using the open flux, deduced from the geomagnetic aa-index. After scaling on the last decades, it allows to extrapolate our knowledge of the total irradiance down to the Maunder minimum (at least) leading to a change of no more than 1.3 W/m^2 (Krivova *et al.*, 2007).

Could this irradiance change enough to justify the observed decrease of the mean Earth temperature during the Maunder minimum?

- indirect dynamical influence:

It is more and more believed that the Solar-Earth relationship on long timescale (decades or centuries) is largely more complex than previously thought (Haigh, 2005). It appears insufficient to consider only an energetic relationship. It is well known that the X, EUV, UV parts of the solar spectrum vary much more strongly than the total irradiance, but their impact cannot be seek in through the energetic changes, which are this time extremely small. So the magnetic interaction, the solar wind and the dynamics of the heliosphere are more and more studied to analyze the impact on the ionosphere and one the chemistry of the atmosphere (through solar wind or subsequent cosmic rays variation). Such variability may contribute to some indirect and dynamical effects. Today, this field of interaction begins to be explored and justifies to look to Earth latitudinal and seasonal effects. This supposes having robust and complete climatic models, like LMDz-Reprobus which includes full representations of dynamical, radiative, and chemical processes in the atmosphere and their interactions, especially feedbacks of the chemical tendencies on the dynamics : in particular, ozone is strongly affected by dynamics and transport (Jourdain *et al.*, 2008).

5. Open questions and Perspective

This review has tried to summarize the present situation of this thematic.

From this meeting, a lot of questions emerge naturally.

- Could we establish a proper relationship between luminosity variation and solar radius, can we measure solar surface deformation and deduce subsurface magnetic field variability and the order of magnitude of the deep field if it exists ?
- Is the solar dynamo sufficiently understood today ? Could many dynamos exist in the Sun ? one dynamo (global scale), two dynamos (shallow + CZ global scale) or three dynamos (with also a core dynamo) ?
- Is the magnetic diffusivity at the origin of the de-synchronization between the two hemispheres. Does it contribute to the lifetime of the cycle (Guerrero *et al.*, 2009) ?
- Is the radiative zone only a stable zone, or does it contain a fossil field which participates to the variability. Does it contribute to the greater cycles ($\geqslant$ 80 years) through convective plums exciting gravity waves ? Is the fossil field that initially creates the toroidal field of the tachocline ?
- Do we need to understand the whole dynamics of the Sun to build the bridge between active young stars and Gyr stars in exploring the different types of dynamos ?
- Is the Earth atmosphere more influenced by the solar dynamics and the particle fluxes than by the irradiance variability ? Is it also true for Star-Planet relationship just after the formation epoch ?
- Are the present studies for the existence of life on planets sufficiently mature to orientate the present search on exoplanets and are we ready for the next objectives ?

Some of these questions will find an answer soon. With the launch of PICARD and SDO in 2010, we hope to establish the absolute solar luminosity which is surprisingly not known

Figure 13. a) Inertial fusion study in the NIF hall https://lasers.llnl.gov/. b) The formation flying concept, an excellent perspective for the simultaneous development of Space Weather and Space climate (Turck-Chièze *et al.*, 2009b).

today better than 0.4%. The impact of this uncertainty on a solar model is important, such difference corresponds to 100 Myrs or a boron neutrino flux modified by 5%. We hope also to progress on the shape of the Sun which stays one of the best indicators with the solar wind that put interesting constraints on the strength and geometry of the internal field (s). The dynamics of the CZ will be largely improved.

In parallel to the observations, we need to cumulate numerical exercises and confront them to the observational facts. With the present capability of computers, we need to develop secular stellar evolution models with transport of momentum and chemicals by dynamical processes in 1 or 2D in parallel to 3D simulations which focus on the interplay between dynamical scales. For the solar-like stars, the 1D dynamical models are totally justified because the stellar structure is only slightly modified by the presence of internal dynamical phenomena and the dynamical processes evolve on secular scales.

Let me now express my personal thoughts:

- The standard solar model is no more the right framework for the comparison with observations. It remains nevertheless an excellent basis on which one must integrate step by step the effect of rotation, magnetic field and gravity waves. Until the arrival of the complete Dynamical Solar Model (DSM), the seismic model (SeSM), built to reproduce the sound speed profile (Couvidat *et al.*, 2003; Turck-Chièze *et al.*, 2004a; Mathur *et al.*, 2007), is the best solar model for structural and observable predictions.
- Secular dynamical models need to be developed (1D-2D) for a good understanding of all the sources of variability (in particular at larger scale than the 22 yrs). Solar (especially gravity mode detection), young stars, solar analogs observations will enrich the related physics (topology of the internal magnetic field). After COROT and ESPADON, KEPLER represents an impressive step forward. Cartesian and Spherical 3D simulations must be pursued, they will benefit from the increased power of the parallel computers to treat the very different scales of the solar (stellar) interiors.
- There was a lot of attention on solar sunspots (a long series) but dynamo exists without them, more focus on the radiative zone and probably on the corona flow is necessary to build the complete magnetic field story.
- There is a large need to build time series of other emissions: X rays, CME (in minima, not related to AR), big flares, UV, for their use with data assimilation methods for the development of Space Weather and Space Climate. One needs to establish clear relation between stars and planets: formation, atmosphere and possible life. For the present Sun-Earth relationship we begin to have good daily information. But the evolution of the

climatic relationship is not yet sufficiently established to give quantitative values of the influence of competitive processes: the Sun must certainly not be studied only as acting as a global energetic external input but one needs to build the seasonal and latitudinal influences of a lot of varying indicators. This requires the use of detailed 3D dynamical and chemistry terrestrial atmospheric models which are emerging nowadays.

- The facination for the Sun and Stars has not decreased due to fundamental and societal reasons (magnetic fusion, inertial fusion, space weather and space climate, life on planets). Big laser installations begin to produce stellar plasmas in laboratory to study micro and macro properties (Figure 13 left).
- We need to pursue continuous observation of the Sun. Rather than a lot of small satellites, it would be extremely exciting to organize ourselves to build a large world class mission with two Herschel-Planck platforms organized in formation flying (Figure 13 right) at L1 point. Such approach allows a permanent eclipse with an insight on the low corona from where the solar flux emerges. But the advantage of this concept is also to carry all the instruments we need in order to continuously look to the Sun from the core to the corona at different spatial and temporal scales. Such mission must be more ambitious than the project proposed in the framework of Cosmic Vision (Turck-Chièze *et al.*, 2009b). It is the only concept that contributes to the simultaneous development of Space Weather and Space climate. PROBA3 is the first step in this direction.
- The investigation of young stars is very important. It must be amplified together with properties of exoplanets. The whole stellar magnetic field story is an important objective for the next decade.

Acknowledgement

I would like to thank the organizers who have offered to all the participants, a program that shows the whole chain of processes. The result has been a very enjoying meeting. I would like to add personal warm thanks to all the participants who are the main actors of this wonderful week in Rio de Janeiro; they have inspired this present complementary review. I acknowledge the Wilcox Solar Observatory (WSO) of Stanford University directed by Prof. P. H. Scherrer for use of their polar field observations. A lot of results are due to space experiments from ESA, NASA and JAXA, we are grateful to these agencies and to national agencies like CNES as they not only contribute to the realization of instruments but they also accompany us in the scientific return.

References

Andrei, A., *et al.* 2006, IAU, JD 8, 36 *Space Science Review*, 145, 337
Asplund, M., *et al.* 2009, *Annu. Rev. Astro. Astrophys.*, 47, 481
Baliunas, S. L. *et al.* 1995, *ApJ.*, 438, 269
Babcock, H. W. 1961, *ApJ.*, 133, 572
Ballot, J, Brun, A. S. & Turck-Chièze, S. 2007, *ApJ.*, 669, 1190
Braitwaite, J & Spruit, H. C. 2004, *Nature*, 431, 819
Brown *et al.*, 2010, *ApJ.*, arXiv0906.2407
Brun, A. S., Miesch, M. & Toomre, J., 2004, *ApJ.*, 614, 1073
Caffau. E. *et al.*, 2008, *A & A*, 488, 1031; 2009, *A & A*, 498, 877
Couvidat, S., Turck-Chize, S, Kosovichev, A. 2003, *ApJ.*, 599.1434
Domingo, V., *et al.* 2009, *Space Science Review*, 145, 337
Dikpati, M. & Gilman, P. 2008, *J. Astrophys. Astr.*, 29, 29
Djafer, D., Thuillier, G. & Sofia, S. 2008, *Astrophys. J.*, 676, 651
Domingo, V., *et al.* 2009, *Space Science Review*, 145, 337

Duez, V., Mathis, S. & Turck-Chize, S. 2009, *MNRAS*, in press
Eff-Darwich, A., Korzennik, S. G., Jiménez-Reyes, S. J., & García, R. A. 2008, *ApJ.*, 679, 1636
Egidi, A., *et al.* 2006, *Sol. Phys.*, 235, 407
Emilio, M., *et al.* 2007, *Astrophys. J.*, 660L, 161
Fröhlich, C. 2006, *Space Science Reviews*, 125, 53-65
Garaud, P. & Garaud, J-D. 2008, *MNRAS*, 391, 1239
García, R. A. *et al.* 2007, *Science*, 316, 1591
Gough, D. 2009, in Magnetic Coupling between the Interior and the Atmosphere of the Sun, ed S. S. Hasan & R. J. Rutten, *Astrophysics and Space Science* Proceeding, p 67
Guerrero, G. *et al.*, 2009, *ApJ.*, 701, 725
Güdel, M. 2007, *Living Rev. Sol. Phys.*, 4, 3
Haigh, J. D. 2005, in *The Sun, Solar analogs and the climate*, ed I. Ruedi, M. Gudel and W. Schmutz, 1
Holweger, H. 2001, *Joint SOHO/ACE workshop "Solar and Galactic Composition*, 598, 23
Howe, R. *et al.* 2009, *ApJ.*, 701, L87
Jourdain, L., Bekki, S., Lott, F., & Lefèvre, F., 2008, *Ann. Geophys.*, 26, 1391-1413
Jouve, L. & Brun, A. S. 2007, *A & A.*, 474, 239
Kaltenegger, L. *et al.* 2009, *Astrobiology*, accepted
Kitiashvili, I. N. & Kosovichev, A. G. 2008, *ApJ.*, 688, L49
Kopp, G. 2005, *Sol. Phys.*, 230, 129
Krivova, N. A., Balmaceda, L., & Solanki, S. K., 2007, *A&A*, 467, 335
Krivova, N. A., Solanki, S. K. & Wenzle, T. 2009, *GRL*, submitted.
Lefebvre, S. & Kosovichev, A. G. 2005, *ApJ.*, 633, L149-152
Lefebvre, S., Nghiem, P. A. P., & Turck-Chièze, S. 2009, *ApJ*, 690, 1272-1279
Mathur, S. & Turck-Chieze, S. 2007, *ApJ.*, 668, 594
Mathur, S. *et al.* 2008, *A &A*, 484, 517
McComas, D. J. *et al.* 2008, *Geophys. Rev. Lett.*, 35, L18103
Miesch, M. S. *et al.* 2008, *ApJ.*, 673, 557; arXiv0811.3032 Proc GONG 2008/SOHO XXI
Parker, E. N. 1984, *ApJ.*, 276, 341; 281, 839; 286, 666 and 677
Porto de Mello, G. 2006, *Astrobiology*, 6, 308
Rozelot, J. P., in The Rotation of Sun and Stars, ed J. P Rozelot & C. Neiner, *Lecture Notes in Physics*, Vol. 765, Berlin: Springer, 2009., p. 15 and references therein
Saar, S. H. & Brandenburg, A 1999, *ApJ.*, 52, 295
Schatten, K. 2003, *Adv. Space Rev.*, 32, 451
Schatten, K. 2005, *GRL*, 32, L21106
Schatten, K. 2009, *Sol. Phys.*, 255, 3
Smith, E. J. *et al.* 2003, *Science*, 302, 5648, p 1165
Scafetta N. & Willson, N. C., 2009, *GRL*, 36, L05701
Turck-Chièze, S. *et al.* 2004a, *Phys. Rev. Lett.*, 93, 211102.
Turck-Chièze, S., *et al.* 2004b, *ApJ.*, 604, 455
Turck-Chièze, S. *et al.*, 2008, *Astron. Notes.*, 329, 521
Turck-Chièze, S. *et al.* 2009a, *ApJ.*, submitted
Turck-Chièze, S., *et al.* 2009b, *Exp Astron*, 23, 1017
Wang, H. N. *et al.* 2008, *Adv. Space. Res.*, 42, 1464
Zahn, J. P., Brun, A. S., Mathis, S. 2007, *A&A*, 474, 15

Author Index

Subject Index